Predictive Biomarkers in Oncology

Sunil Badve • George Louis Kumar
Editors

Predictive Biomarkers in Oncology

Applications in Precision Medicine

Springer

Editors
Sunil Badve
Department of Pathology
and Lab Medicine
Indiana University School of Medicine
Indianapolis, IN
USA

George Louis Kumar
Targos Inc.
Issaquah, WA
USA

ISBN 978-3-319-95227-7 ISBN 978-3-319-95228-4 (eBook)
https://doi.org/10.1007/978-3-319-95228-4

Library of Congress Control Number: 2018958983

This Springer imprint is published by the registered company Springer Nature Switzerland AG
The registered company address is: Gewerbestrasse 11, 6330 Cham, Switzerland

I would like to thank all the people who have guided, encouraged, and supported me throughout my career. Additionally, acknowledge the contributions of those who did not, but for them, I would not have learnt the value of success and the importance of character. A very humble thank you.

Sunil Badve, MD, FRCPath

"We are like dwarfs on the shoulders of giants, so that we can see more than they, and things at a greater distance, not by virtue of any sharpness on sight on our part, or any physical distinction, but because we are carried high and raised up by their giant size."

- Bernard of Chartres

To my dear father, Joseph, and my late mother, Miriam, for their unconditional love.
To my extraordinarily talented wife, Sujatha, for her continued support of my endeavors.
To my wonderful children, Vikram and Raj, for bringing so much joy to my life.

George Louis Kumar, PhD, MBA

Preface

"Precision/personalized or stratified medicine" refers to the tailoring of medical treatment or drug administration to the individual characteristics of each patient treatment. It does not literally mean that a pharmaceutical company makes a drug for an individual patient for consumption and treatment but rather means the ability to stratify (or classify) individuals into subpopulations that differ in their responsiveness to a specific drug. A marker that provides information on the likely response to therapy, i.e., either in terms of tumor shrinkage or survival of the patient, is termed "predictive biomarker." Examples include HER2 test to predict response to trastuzumab (Herceptin®) in breast cancer, the KRAS test to predict response to EGFR inhibitors like cetuximab (Erbitux®) and panitumumab (Vectibix®) in lung cancer, or the BCR-ABL oncogene detection to predict response to the tyrosine kinase inhibitor imatinib (Gleevec®) in chronic myelogenous leukemia.

Despite their promise in precision medicine and the explosion of knowledge in this area, there is not a single source on this subject that puts all this evidence together in a concise or richly illustrated and easy to understand manner. This book will provide a collection of ingeniously organized, well-illustrated, and up-to-date authoritative chapters divided into five parts that are clear and easy to understand.

Part I will provide an overview of biomarkers and introduce the basic terminologies, definitions, technologies, tools, and concepts associated with this subject in the form of illustrations/graphics, photographs, and concise texts.

Part II describes the signaling pathways controlling cell growth and differentiation altered in cancer. This part will analyze how predictive biomarkers are altered (expressed or amplified) across cancer types.

Part III will explore how predictive biomarkers play a role in patient stratification and tailored treatment in relationship to specific cancers (e.g., breast, gastric, lung, and other tumors).

Part IV will discuss how regulatory processes, quality and policy issues, companion diagnostics, and central laboratories help validate predictive biomarker assays.

Part V will wrap up with a description of precision medicine clinical trials around the world, and its successes and disappointments, challenges, and opportunities. This part will also summarize all FDA-approved drugs in oncology.

We hope that the proposed textbook will serve as a definitive guide for practicing pathologists, pathology residents, and personal in the pharmaceutical or diagnostic industry interested in learning on how "predictive biomarkers" are used in precision cancer therapy.

We wish to thank Sujatha Kumar, Yesim Gökmen-Polar, Bharat Jasani, Katherina Alexander, and Victoria Alexander for proofreading. Special thanks to Michael D. Sova, Developmental Editor at Deved, Inc., for superb editorial assistance during the production of this book.

Indianapolis
IN, USA
Sunil Badve

Issaquah
WA, USA
George Louis Kumar

Contents

Contributors

Esther Abels, MSc Digital Pathology Solutions, Pharma Solutions, Philips Digital Pathology Solutions, Best, The Netherlands

Mark Abramovitz, PhD Avera Cancer Institute, Sioux Falls, SD, USA

Kathrina A. Alexander, MD, FACP Targos Molecular Pathology GmbH, Kassel, Hessen, Germany

Katya Victoria Alexander Laboratory for Study Analytics 2, Targos Molecular Pathology GmbH,, Kassel, Germany

Eleni Andreopoulou, MD Department of Medicine, Division of Hematology & Medical Oncology, Weill Cornell Medicine/New York Presbyterian Hospital, New York, NY, USA

Sunil Badve, MD, FRCPath Department of Pathology and Lab Medicine, Indiana University School of Medicine, Indianapolis, IN, USA

Andrew T. Baker, PhD Department of Integrative Cell Biology, Loyola University Chicago, Maywood, IL, USA

Justin M. Balko, PharmD, PhD Department of Medicine and Cancer Biology, Vanderbilt University Medical Center, Nashville, TN, USA

Karla V. Ballman, PhD Department of Healthcare Policy and Research, Division of Biostatistics and Epidemiology, Weill Cornell Medicine, New York, NY, USA

Gudrun Bänfer, Dr. rer. nat. Department of Advance -Training and Consulting, Targos Molecular Pathology GmbH, Kassel, Germany

Peter Bankhead, BD, MSc, PhD Belfast Development Hub, Philips Digital Pathology Solutions, Belfast, UK

Sebastian Bauer, MD Department of Medical Oncology, Sarcoma Center, West German Cancer Center, University Duisburg-Essen, Medical School, Essen, Germany

German Cancer Consortium (DKTK), Heidelberg, Germany

Tim Beißbarth, Prof. Dr. Department of Medical Statistics, University Medical Center Göttingen, Göttingen, Germany

Jeffrey C. Bloodworth, MS Cardinal Bernardin Cancer Center, Loyola University Chicago, Maywood, IL, USA

Jose M. Bonnin, MD Department of Pathology and Laboratory Medicine, Indiana University School of Medicine, Indianapolis, IN, USA

Donald P. Bottaro, PhD Urologic Oncology Branch, National Cancer Institute, Bethesda, MD, USA

Catherine Bresson, MBA Win Consortium, Villejuif, Val de Marne, France

Alejandra Bruna, Bsc, PhD Cancer Research UK Cambridge Institute, University of Cambridge, Cambridge, Cambridgeshire, UK

Reinhard Buettner, Prof. Dr. med. Department of Pathology, University Hospital Cologne, Cologne, Germany

Zheng Cai, PhD Department of Pathology and Lab Medicine, University of Pennsylvania, Philadelphia, PA, USA

Felipe D'Almeida Costa, MD Department of Anatomic Pathology, A.C. Camargo Cancer Center, São Paulo, São Paulo, Brazil

Massimo Cristofanilli, MD Department of Medicine-Hematology and Oncology, Robert H Lurie Comprehensive Cancer Center, Feinberg School of Medicine, Chicago, IL, USA

Andrew A. Davis, MD Department of Medicine-Hematology and Oncology, Northwestern Memorial Hospital, Chicago, IL, USA

Pradip K. De, MS, PhD Department of Molecular and Experimental Medicine, Avera Cancer Institute, Sioux Falls, SD, USA

Louise De Brot, MD, PhD Department of Anatomic Pathology, A.C. Camargo Cancer Center, São Paulo, SP, Brazil

Dinuka M. De Silva, PhD Urologic Oncology Branch, National Cancer Institute, Bethesda, MD, USA

Nandini Dey, MS, PhD Department of Molecular and Experimental Medicine, Center for Precision Oncology, Avera Cancer Institute, Sioux Falls, SD, USA

Amar Gajjar, MD Department of Oncology, St. Jude Children's Research Hospital, Memphis, TN, USA

Yesim Gökmen-Polar, PhD Department of Pathology and Laboratory Medicine, Indiana University School of Medicine, Indianapolis, IN, USA

Mark I. Greene, MD, PhD, FRCP Department of Pathology and Laboratory Medicine, The Perelman School of Medicine at the University of Pennsylvania, Philadelphia, PA, USA

Wendy Greenwood, Bsc, MMedSci Cancer Research UK Cambridge Institute, University of Cambridge, Cambridge, Cambridgeshire, UK

Payal Grover, PhD Department of Pathology and Lab Medicine, University of Pennsylvania, Philadelphia, PA, USA

Marius Grzelinski, PhD Laboratory for Study Analytics 2, Targos Molecular Pathology GmbH, Kassel, Hesse, Germany

Sumeet Gujral, MD Department of Pathology, Tata Memorial Hospital, Tata Memorial Centre (TMC), Homi Bhabha National Institute (HBNI) University, Mumbai, Maharashtra, India

Peter Hamilton, BSc(hon), PhD Department of Digital Pathology, Philips UK, Belfast, Northern Ireland, UK

Amanda J. Harvey, PhD, BSc, PGCert Department of Life Sciences, Brunel University London, Uxbridge, Middlesex, UK

Petra Heinmöller, PhD Department of Quality Management, Targos Molecular Pathology GmbH, Kassel, Hesse, Germany

Thomas Henkel, PhD Targos Molecular Pathology GmbH, Kassel, Hessen, Germany

Juan C. Hernandez-Prera, MD Department of Anatomic Pathology, Moffitt Cancer Center, University of South Florida, Tampa, FL, USA

Bharat Jasani, BSc (Hons), PhD, MBChB, FRCPath Department of Pathology, Targos Molecular Pathology GmbH, Kassel, Hessen, Germany

Virginia Kaklamani, MD Department of Medicine, University of Texas Health Science Center San Antonio, San Antonio, TX, USA

Shyam Kalavar, MPH, CT(ASCP) Center for Devices and Radiological Health, Office of In Vitro Diagnostics and Radiological Health, Division of Molecular Genetics and Pathology, US Food and Drug Administration, Silver Spring, MD, USA

Takashi Kato, PhD Urologic Oncology Branch, National Cancer Institute, Bethesda, MD, USA

Li Yan Khor, MBBCh Department of Anatomical Pathology, Singapore General Hospital, Singapore, Singapore

Hartmut Koeppen, MD, PhD Research Pathology, Genentech, South San Francisco, CA, USA

Elie Kostantin, BSc Department of Biochemistry, Goodman Cancer Research Center, McGill University, Montreal, QC, Canada

Marcin Kowanetz, PhD Department of Oncology Biomarker Development, Genentech, South San Francisco, CA, USA

H. Krishnamurthy, MSc, PhD Central Imaging and Flow Cytometry Facility, National Centre for Biological Sciences, Tata Institute of Fundamental Research, Bengaluru, Karnataka, India

George Louis Kumar, PhD, MBA Targos Inc., Issaquah, WA, USA

Razelle Kurzrock, MD Moores Cancer Center, UC San Diego Moores Cancer Center, La Jolla, CA, USA

Tshering D. Lama-Sherpa, BS Department of Pathology, The University of Alabama at Birmingham, Birmingham, AL, USA

Kate Lathrop, MD Department of Medical Oncology and Hematology, University of Texas Health Science Center San Antonio, San Antonio, TX, USA

Vladimir Lazar, MD, PhD Win Consortium, Villejuif, Val de Marne, France

Andreas Leha, MD Department of Medical Statistics, University Medical Center Göttingen, Göttingen, Germany

Brian Leyland-Jones, MB BS, PhD Department of Molecular and Experimental Medicine, Center for Precision Oncology, Avera Cancer Institute, Sioux Falls, SD, USA

Victor T. G. Lin, MD, PhD Division of Hematology and Oncology, Department of Medicine, The University of Alabama at Birmingham, Birmingham, AL, USA

Na Luo, PhD Department of Medicine, Vanderbilt-Ingram Cancer Center, Vanderbilt University, Nashville, TN, USA

Department of Anatomy and Histology, School of Medicine, Nankai University, Tianjin, China

Steven Alexander Mann, MD Department of Pathology and Laboratory Medicine, Indiana University School of Medicine, Indianapolis, IN, USA

Perry Maxwell, PhD, FRCPath Precision Medicine Centre of Excellence, Queen's University Belfast, Belfast, UK

Mark L. McCleland, PhD Department of Oncology Biomarker Development, Genentech, South San Francisco, CA, USA

John Mendelsohn, MD Khalifa Institute for Personalized Cancer Therapy, The University of Texas MD Anderson Cancer Center, Houston, TX, USA

Funda Meric-Bernstam, MD Investigational Cancer Therapeutics, The University of Texas MD Anderson Cancer Center, Houston, TX, USA

Sheikh Khalifa Bin Zayed Al Nahyan Institute for Personalized Cancer Therapy, The University of Texas MD Anderson Cancer Center, Houston, TX, USA

Department of Breast Surgical Oncology, The University of Texas MD Anderson Cancer Center, Houston, TX, USA

Harikrishna Nakshatri, BVSc, PhD Departments of Surgery, Biochemistry and Molecular Biology, Indiana University, Indianapolis, IN, USA

Aejaz Nasir, MD, MPhil Diagnostic & Experimental Pathology, Tailored Therapeutics, Eli Lilly & Co., Indianapolis, IN, USA

BJ's Diagnostic & Precision Oncology, Tampa, FL, USA

Bharat N. Nathwani, MD Department of Pathology, City of Hope National Medical Center, Duarte, CA, USA

Paul O'Reilly, B.Eng., PhD Belfast Development Hub, Philips Digital Pathology Solutions, Belfast, UK

Clodia Osipo, PhD Department of Microbiology and Immunology, Stritch School of Medicine, Cardinal Bernardin Cancer Center of Loyola University Chicago, Maywood, IL, USA

Nicci Owusu-Brackett, MD Department of Surgical Oncology, The University of Texas MD Anderson Cancer Center, Houston, TX, USA

Nallasivam Palanisamy, MSc, MPhil, PhD Department of Urology, Henry Ford Health System, Detroit, MI, USA

Júlia Perera-Bel, MSc Department of Medical Statistics, University Medical Center Göttingen, Göttingen, Germany

Thao N. D. Pham, PhD Department of Biopharmaceutical Sciences, College of Pharmacy, University of Illinois at Chicago, Chicago, IL, USA

Aaron Phelan, MD Department of Pathology and Laboratory Medicine, Indiana University School of Medicine, Indianapolis, IN, USA

Reena Philip, PhD Center for Devices and Radiological Health, Office of In Vitro Diagnostics and Radiological Health, Division of Molecular Genetics and Pathology, US Food and Drug Administration, Silver Spring, MD, USA

Raju K. Pillai, MD Department of Pathology, City of Hope National Medical Center, Duarte, CA, USA

V. M. Pratt, PhD, FACMG Department of Medical and Molecular Genetics, Indiana University School of Medicine, Indianapolis, IN, USA

Jenifer R. Prosperi, PhD Department of Biological Sciences, University of Notre Dame, South Bend, IN, USA

Department of Biochemistry and Molecular Biology, Indiana University School of Medicine – South Bend, South Bend, IN, USA

Susan D. Richman, PhD, MSc, BSc Department of Pathology and Tumour Biology, Leeds Institute of Cancer and Pathology, Leeds, UK

Arpita Roy, PhD Urologic Oncology Branch, National Cancer Institute, Bethesda, MD, USA

Josef Rüschoff, Prof. Dr. med. Targos Molecular Pathology GmbH, Kassel, Germany

Manuel Salto-Tellez, LMS (MD), FRCPath, FRCPI Northern Ireland Molecular Pathology Laboratory, Centre for Cancer Research, Department of Cell Biology, Queens University Belfast, Belfast, Antrim, UK

Romil Saxena, MBBS, MD, FRCPath Department of Pathology and Laboratory Medicine, Indiana University School of Medicine, Indianapolis, IN, USA

Hans-Ulrich Schildhaus, MD Institute of Pathology, Universitätsmedizin Göttingen, Göttingen, Germany

Richard L. Schilsky, MD, FACP, FASCO American Society of Clinical Oncology, Alexandria, VA, USA

Maryam Shariati, MS Department of Investigational Cancer Therapeutics, Graduate School of Biomedical Sciences, The University of Texas MD Anderson Cancer Center, Houston, TX, USA

Lalita A. Shevde, PhD Department of Pathology, The University of Alabama at Birmingham, Birmingham, AL, USA

Jason K. Sicklick, MD, FACS Division of Surgical Oncology, General Surgery Residency, Biorepository and Tissue Technology Shared Resource, Moores Cancer Center, University of California San Diego (UCSD), School of Medicine, San Diego, CA, USA

Fernando Augusto Soares, MD, PhD Department of Pathology, Rede D'Or Hospital Network, São Paulo, SP, Brazil

Casey D. Stefanski, MS Department of Biological Sciences, University of Notre Dame, South Bend, IN, USA

Oliver Stoss, PhD Targos Molecular Pathology GmbH, Kassel, Hessen, Germany

W. Fraser Symmans, MD MD Anderson Cancer Center, Houston, TX, USA

Puay Hoon Tan, FRCPA Division of Pathology, Singapore General Hospital, Singapore, Singapore

Clive R. Taylor, MA, MD, DPhil Department of Pathology, University of Southern California, Los Angeles, CA, USA

Prashant Ramesh Tembhare, MD Hematopathology Laboratory, Advanced Centre for Treatment, Research and Education in Cancer (ACTREC), Tata Memorial Centre (TMC), Homi Bhabha National Institute (HBNI) University, Navi Mumbai, Maharashtra, India

Debra A. Tonetti, PhD Department of Biopharmaceutical Sciences, College of Pharmacy, University of Illinois at Chicago, Chicago, IL, USA

Michel L. Tremblay, PhD Department of Biochemistry, Goodman Cancer Research Center, McGill University, Montreal, QC, Canada

Simon J. P. Warren, MBBS Departments of Pathology and Dermatology, Indiana University, Indianapolis, IN, USA

Rosanne Welcher, BS, PhD, MBA, RAC Companion Diagnostics, Agilent Technologies, Carpinteria, CA, USA

Bruce M. Wenig, MD Department of Anatomic Pathology, Moffitt Cancer Center, University of South Florida, Tampa, FL, USA

Nicholas Shawn Whipple, MD, MPH Division of Hematology/Oncology, Department of Pediatrics, University of Utah and Primary Children's Hospital, Salt Lake City, UT, USA

Casey Williams, PharmD Center for Precision Oncology, Avera Cancer Institute, Sioux Falls, SD, USA

Scooter Willis, PhD Center for Precision Oncology, Avera Cancer Institute, Sioux Falls, SD, USA

Mahesh Yadav, PhD Department of Oncology Biomarker Development, Genentech, South San Francisco, CA, USA

Lixin Yang, PhD Department of Pathology, City of Hope National Medical Center, Duarte, CA, USA

Douglas Yee, MD Department of Medicine, Masonic Cancer Center, University of Minnesota, Minneapolis, MN, USA

Brandon Young, MS Center for Precision Oncology, Avera Cancer Institute, Sioux Falls, SD, USA

Hongtao Zhang, PhD Department of Pathology and Lab Medicine, University of Pennsylvania, Philadelphia, PA, USA

Zhiqiang Zhu, PhD Department of Pathology and Lab Medicine, University of Pennsylvania, Philadelphia, PA, USA

Andrei Zlobin, PhD Oncology Research Institute, Loyola University Medical Center, Maywood, IL, USA

Yevgen Zolotarov, BSc, MSc Department of Biochemistry, Goodman Cancer Research Center, McGill University, Montreal, QC, Canada

Part I

Basic Principles and Methods

Introduction to Predictive Biomarkers: Definitions and Characteristics

1

Clive R. Taylor

Biomarkers

The concept of "biomarkers" as indicators of health or disease is not new. Under the broadest interpretation, the use of biomarkers extends back to the "ancients," who elicited medical signs, measured the pulse, observed, and even tasted the urine and the like [1]. However, the use of the term biomarker is relatively recent in the field of medicine, where the definition continues to shift with context.

Certainly many clinical laboratory tests fall under a broad definition. Examples include hormone levels for endocrine disease, a succession of enzymes and proteins, up to present day troponin for myocardial infarction, and prostatic acid phosphatase, then PSA (prostate-specific antigen), for prostate cancer. Extending the definition to its limits, the structural changes observed in anatomic pathology, or in radiology, also meet the definitional criteria; a tissue diagnosis of prostate cancer, plus or minus grading (e.g., Gleason), is a biomarker in a very real sense. Other "biomarkers" of diverse variety also have long been applied in unrelated fields, such as archeology, geology, and the petrochemical industry.

This introductory chapter has a more restricted focus, namely, the utilization of "biomarkers" as identified by laboratory tests in relation to cancer; still more specifically, the focus is upon biomarkers detected directly in tissues from cancer patients (Table 1.1). Within this context of tissue and cancer, biomarkers include proteins and nucleic acids and derivatives and parts thereof. While the focus is narrow, the levels of complexity are manifold and growing day by day.

Biomarkers in Cancer

Tests for biological markers in malignant disease, for diagnosis, prognosis, and monitoring of progression, can be traced back at least a century and a half to the example of Bence-Jones protein in urine (Henry Bence-Jones 1813–1873) [1] for Kahler's disease (Otto Kahler 1849–1893), a surrogate for the detection and measurement of monoclonal (malignant-M) proteins that identify the condition that we now know as multiple myeloma. The modern era of biomarkers with respect to cancer in general may, on the one hand, be traced back to the discovery and use of CEA (carcinoembryonic antigen), a protein biomarker, and, on the other, to the Philadelphia chromosome, a genetic marker of chronic myeloid leukemia [1]. While CEA did not meet initial hopes of diagnostic utility in terms of sensitivity or specificity, measurement of CEA in the serum did find

C. R. Taylor
Department of Pathology, University of Southern California, Los Angeles, CA, USA

S. Badve, G. L. Kumar (eds.), *Predictive Biomarkers in Oncology*,
https://doi.org/10.1007/978-3-319-95228-4_1

Table 1.1 Biomarkers in the context of cancer

Biomarker: general definition	A characteristic that is objectively measured and evaluated as an indicator of normal biological processes, pathogenic processes, or pharmacological responses to therapeutic intervention
Diagnostic	Design and usage; primarily to assist diagnosis; commonly in IHC on tissue sections, but also sometimes indicative in serum
Prognostic	Design and usage; primarily as a guide to prognosis; the course and progress of disease –therapy unspecified
Predictive	Design and usage; specifically for classification of responders vs. nonresponders for a defined (usually targeted) therapy; assay and threshold developed in conjoint clinical trial with the specified drug
Companion	Predictive; co-developed with a specified therapy and "required" prior to use of said therapy
Complementary	Predictive; co-developed with a specified therapy; accepted as providing guidance for therapy but not required
Pharmacodynamic	Definitional within the pharmaceutical field, such as providing a surrogate marker for disease status, as in remission or progression
Monitoring	Design and usage; for evaluation of status, progression, and/or recurrence of established disease process

a place in monitoring of established disease and as a "biomarker" of recurrence, likewise for CA-125 and arguably PSA. Notably, in a different context that still is within the field of cancer, all three of these biomarkers maintain a (variable) role as diagnostic biomarkers when demonstrated in situ within tissue or cell by immunohistochemistry (IHC). Thus context matters.

The decade of the 1990s saw major developments in the measurement of estrogen (and progesterone) receptors (ER and PR) in breast cancer, with applications that were prognostic and, to a degree, predictive in terms of choice of therapy. Cytosol-based competitive assays, relying upon extracts of purported tumor tissue, gradually gave way to a different methodology based on the detection of ER (and or PR) in situ within tissue sections by labeled antibody methods, with IHC (immunohistochemistry) using FFPE (formalin-fixed paraffin-embedded) sections emerging as the standard.

This transition occurred in spite of the arguments levied against FFPE tissue, because of the unknown effects of protein "masking," and against IHC, because of subjectivity in interpretation and hence variability in scoring, and also because of the nonlinear relationship between signal intensity and target antigen (in this instance the estrogen receptor protein) [2]. The efforts of Craig Allred and others in the development of defined (but semi-quantitative) scoring methods were critical to acceptance of the IHC method for this purpose.

In the presence of proper controls of assay performance [2, 3], IHC brings exquisite specificity, by scoring only recognizable cancer cells, and extraordinary technical sensitivity, with the ability to detect one ER-positive cell among a 100 identifiable cancer cells (1%; the current threshold of a positive ER IHC test) or in fact 1 positive cell among 1000 or 10,000 or more cells. Expressed in these terms, namely, detection of positive cells, this level of sensitivity is far beyond anything that can be achieved by any method using an extract of tissue, which is necessarily an imperfectly known extract of an imperfectly known mixture of normal and cancer cells, themselves imperfectly identified.

In this mode of performance, the IHC ER "test" may be considered to represent the beginning of the current era of employment of biomarkers in cancer, for prognostic and predictive purposes.

The "First" Predictive Biomarker

However, the moment of critical impetus for the current explosion in interest and variety of cancer biomarkers was the day (September 25, 1998) upon which the FDA approved the HercepTest

(Dako, now Agilent, CA, USA) and simultaneously gave approval for the use of the companion drug Herceptin (Genentech, now Roche) for the treatment of patients with Her2-positive breast cancer (as measured by the HercepTest). A vitally important corollary message from the FDA was that drug and test should be developed in concert, during a combined clinical study, hence "companion diagnostic" (Table 1.1) (Fig. 1.1) [4–10].

From the beginning of the millennium to the present time, US and European regulatory and working groups [4–8] offered various definitions of a biomarker, including the following: "a characteristic that is objectively measured and evaluated as an indicator of normal biological processes, pathogenic processes, or pharmacological responses to therapeutic intervention." Subsequently the FDA went further with the definition of a "valid biomarker" – including that it should:

- Be measured in a test system with well-established performance characteristics
- Have a scientific background of evidence including clinical significance
- Be "fit to purpose"

A final consideration extended to a "clinically useful biomarker," which should in addition be reliable and clinically actionable in the specified setting.

The subsequent two decades have seen ongoing evolution of the term, with sub-definitions according to the design and use (Tables 1.1 and 1.2), accompanied by growing emphasis upon objectivity, reproducibility, and elements of true quantification, which reflect back upon methodology and ultimately performance of the "total test" from inception to interpretation, whichever the test modality employed (Table 1.3) [2, 3, 10, 11].

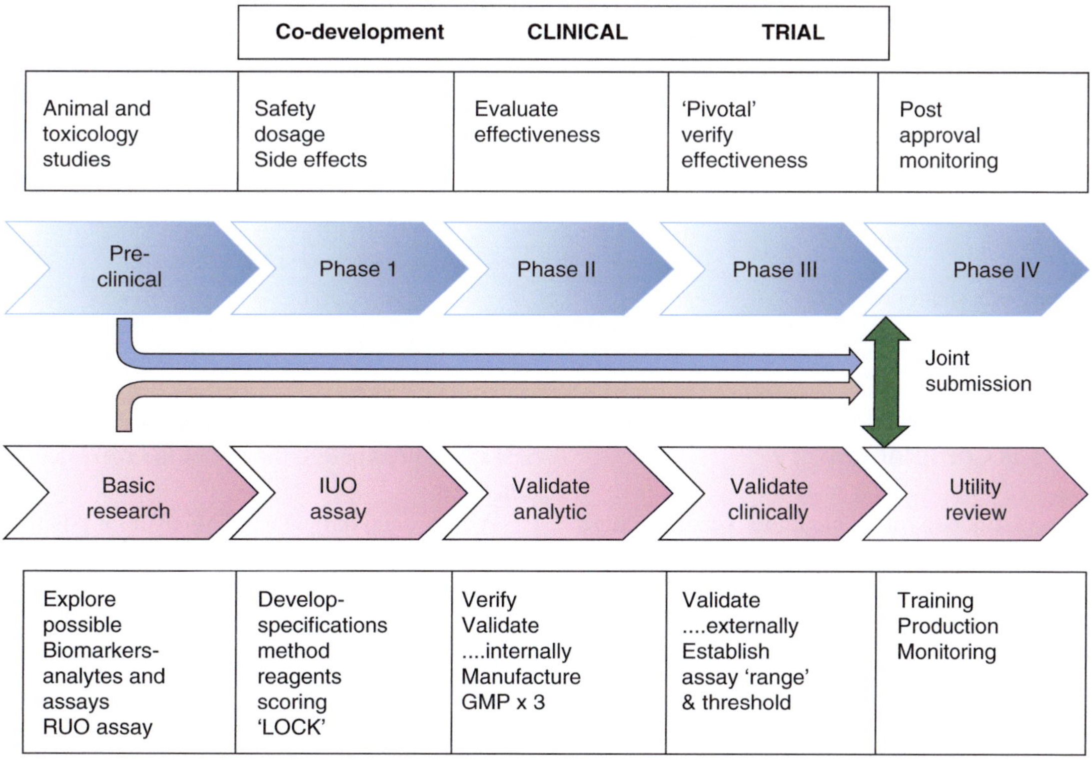

Fig. 1.1 Co-development process for "drug" and companion diagnostic. Time frame, up to 10 years; cost, up to 100 million dollars

Table 1.2 Laboratory reagents and tests; FDA categories

ASR	RUO	IUO	IVD	LDT
Analyte-specific reagent	Research use only	Investigational Use only	In vitro device	Lab developed test
No diagnostic claims	No diagnostic claims	No diagnostic claims	Specified claims FDA approved	Lab responsible for any claims[a]
FDA regulations	FDA regulations	FDA regulations	FDA regulations	CLIA[b] regulations FDA discretion
May be used as reagents for RUO, IUO, IVD, and LDT tests	Not for clinical use	Use restricted to specified study	Intended use define by trial Specified in labeling	For use only in the lab that developed the test

https://www.cms.gov/Clia/
[a]LDT may require FDA approval if used as a predictive marker; clinical utility must be validated
[b]*CLIA* Clinical Laboratory Improvement Amendments

Table 1.3 The "total test" approach

Pre-analytical (Sample preparation)	Test selection: indication for the test
	Specimen handling, from operating room to histology laboratory
	Fixation: total fixation time and type of fixative
	Paraffin embedding, storage, and sectioning
	Deparaffinization
Analytical (Reagents and protocol)	Antigen retrieval (exact method)
	Assay (staining) method and protocol
	Reagent validation
	Controls (reference standards)
	Technologist and laboratory certification
	Proficiency testing and quality assurance
Post-analytical (Interpretations and reporting)	Reading of result(s)/scoring/quantification
	Diagnostic, prognostic, or predictive significance
	Report
	Turnaround time
	Outcomes analysis/economics/reimbursement Pre-analytical

Based on data from Taylor [16]

Predictive Biomarkers: Companion Versus Complementary

The distinction of companion versus complementary biomarkers (Table 1.1) emerged from conjoint clinical studies, determined by the level of prediction of clinical response that the test rendered.

With a companion diagnostic, a positive result indicates treatment with the companion drug; a negative result indicates no treatment; and the test is required before the use of the corresponding drug.

With a complementary diagnostic, a positive result usually indicates treatment, but a patient having a negative result may or may not be treated according to an informed clinical decision.

For example, with PD-L1 tests, some "tests" emerged as companion diagnostics, and others as complementary, varying according to which anti-PD-L1 antibody was employed [8, 12, 13], by which method, and in which specified tumor type.

Intrinsic to the FDA definition of an approved IVD (in vitro diagnostic) companion diagnostic is that it "provides information that is essential for the safe and effective use of a corresponding therapeutic product" and that its use is "stipulated in the instructions for use in the labeling of both the diagnostic device and the corresponding therapeutic agent" (Table 1.2) [6–8]. The current EU definition is less rigorous, but similar in intent, and interestingly admits both "quantitative and qualitative determination of specific markers identifying subjects" [5, 8]. It specifically excludes monitoring.

The FDA definition carries with it an assignment of the IHC IVD to Class III (the highest level) requiring PMA (pre-market approval) in a co-development mode with the drug [4, 6–8, 12], whereas the EU regulations appear to leave companion diagnostics in the current general IVD category [5]; new regulations are afoot that likely will raise the level and may preclude the current self-certification route (for discussion of the subtleties of these definitions, see references 4 and 12 and later chapters in this book). The above statements apply specifically to companion diagnostics; there are as yet no corresponding written rules for complementary diagnostics; the definition of which is at present by precedent and usage, although proposals have been aired.

Method Development

These types of predictive biomarker tests have come to be of critical import in the context of targeted drug therapies, such that the majority of such agents now in clinical studies are following a co-development plan for "test" and "corresponding therapy." Detailed discussion of this co-development process is outside the scope of this chapter but is summarized in Fig. 1.1, examined in detail elsewhere in this book, and well-reviewed in a recent National Policy Workshop [4]. For drug development generally the process includes preclinical (animal) studies: phase 1, toxicity, in which potential biomarkers may also be assessed; phase 2, preliminary efficacy of drug, plus biomarker evaluation; phase 3, definitive efficacy and validation of biomarker; and phase 4, post market surveillance. Total patient accrual will be in the hundreds.

For the biomarker there is a preceding period of basic research and discovery that provides initial evidence of the potential utility of a molecule (biomarker) in the context of diagnosis or prognosis of cancer or a relationship to a potential therapeutic modality (drug – predictive) (Fig. 1.1). This discovery process is followed by evolution of a prototypic test using analyte-specific reagents (ASRs), through an investigational use only (IUO) test, on to an FDA-approved IVD (Table 1.2), which category includes all companion diagnostics. In some instances clinical laboratories may separately develop assays for clinical use, with internal validation under CLIA regulations (LDT, laboratory-developed test) (Table 1.2). The FDA has provided notice that it holds discretionary authority to regulate LDTs and has published guidelines, but not yet enforced them.

The total time span from bench discovery to approval and general clinical application is measured in years, and the total cost is counted in tens of millions of dollars, to be weighed by clinicians, and eventually by society at large, against the undoubted good sense of administering a targeted therapy only to those patients likely to benefit, and the avoidance of side effects and costs of inappropriate treatment of the remainder. This route to approval developed with reference to IHC tests, the most common method adopted for companion diagnostics to date; but other methods as they appear are constrained by similar rules.

As targeted therapies have proliferated, so of course have the corresponding biomarkers, and the methods applied for their detection

Table 1.4 FDA-approved biomarkers and LDTs

Test	Commonly applied tumor types
HER2	Breast, gastric
PD-L1	Melanoma, lung, kidney, head and neck, uterus
CTLA-4	Melanoma
CD 20	B lymphoma, CLL
CD 30	ALCL, Hodgkin L
ALK	Lung
TOPO1	Bladder, breast, colon, uterus, ovary
MMR (MLH1,MSH2, MSH6,PMS2)	Colon
EGFR	Colon, lung, pancreas, thyroid
VEGF	Lung, kidney, glioblastoma, colon,
TUBB3	Lung, bladder, uterus, kidney, prostate
PTEN	Breast, uterus, head and neck, lung, prostate
ER, PR	Breast, uterus, ovary
K-ras	Lung, colon
myc	Lymphoma
BCR-ABL 1	CML, (Ph chromosome)
BRCA 1	Breast, others
c-KIT protein	GIST
ERCC1	Bladder lung
BRAF	Melanoma, lung, colon, others
Immune cell profile[a]	Melanoma, lung, colon, breast, others
PSA CEA, p53, p21, Ki67	Various tumors, prognostic mainly
Multiple tissue biomarkers	Several hundred molecules demonstrated by IHC are use in diagnostic surgical pathology[b]

Multiple methods are applied [9–13]; to date the majority of FDA-approved biomarkers are demonstrated directly in tissues by IHC for diagnostic and or predictive use
[a]Immune cell profile, including CD3, CD4, CD8, CD20, CD68, FoxP3, and others (e.g., see Fig. 1.3)
[b]IHC tests (stains) used in surgical pathology as aids to diagnosis are considered Class 1 by the FDA. They require in lab validation

(Table 1.4). The practice of surgical pathology is being forced to change to meet these new demands (Fig. 1.2) [9–11]. Commensurately with these new assays, there has been a growing recognition of the need for higher standards of testing, in particular higher levels of control and reproducibility of test results from lab to lab (Tables 1.3 and 1.5). At long last the anatomic or surgical pathology laboratory that performs these tests, or at a minimum is involved in providing and preparing the tissues for these tests, is being held to the standards of the clinical laboratory.

Method Validation

For blood-based assays in the clinical laboratory, including serum biomarkers, a reference range usually is established that includes 95% of the "normal" population, with the "reference range" becoming the de facto definition of normalcy. Establishing a reference range is part of "routine" practice in the clinical laboratory and usually involves the testing of a defined population of "normal" subjects (may be a 100 or more), but not so in tissue-based anatomic pathology and not so with many of the newly developed companion diagnostics, where often only subcomponents of the "total test" (Tables 1.3 and 1.5) are validated, in spite of quite large case numbers incorporated into clinical trials.

In the validation of any new assay, and companion diagnostics are no exception, sample size is a matter of the clinical sensitivity and specificity of the test, variation in the population, confidence levels, and statisticians; it usually is accomplished during the transition from discovery (investigational use only (IUO)) to a validated assay (approved IVD) (Table 1.2) [4, 12, 13]. The matter is complex, beyond the compass of this introductory chapter, but is discussed in greater depth in succeeding chapters.

Suffice to say that for all assays that rely upon the use of tissue from cancer patients, the challenges in meeting these demands have been great, but not quite insurmountable. Effective sample (tissue) preparation has emerged as a neglected but key consideration for all assays, both IHC and those dependent upon extracts of FFPE tissues

(Table 1.3). In accommodating these demands, the practice of pathology has changed forever [9].

The Range of Methods

Viewed retrospectively, the first companion diagnostic of this present era was, as noted previously, an IHC–FFPE-based test for Her2 that incorporates cell line-based technical controls, a defined protocol and scoring guidelines derived from conjoint clinical studies. Subsequently, this prototypic IHC Her2 test has served as the model for a multitude of newly developed predictive biomarker tests, developed to match the burgeoning repertoire of targeted therapies [4, 6, 8, 9]. In addition, other technologies have been introduced to the companion diagnostic arena (Table 1.6), including ISH (in situ hybridization), PCR (polymerase chain reaction), and sequencing (Sanger or NGS – next-generation sequencing), with clear and imminent extension into RNA expression methods and proteomics (usually mass spectrometry or reverse-phase protein array) [9–12]. To date these methods have mostly been designed to detect molecular biomarkers, DNA (mutations), RNA (expression), or proteins (receptors, ligands, enzymes), either singly or in exploratory panels,

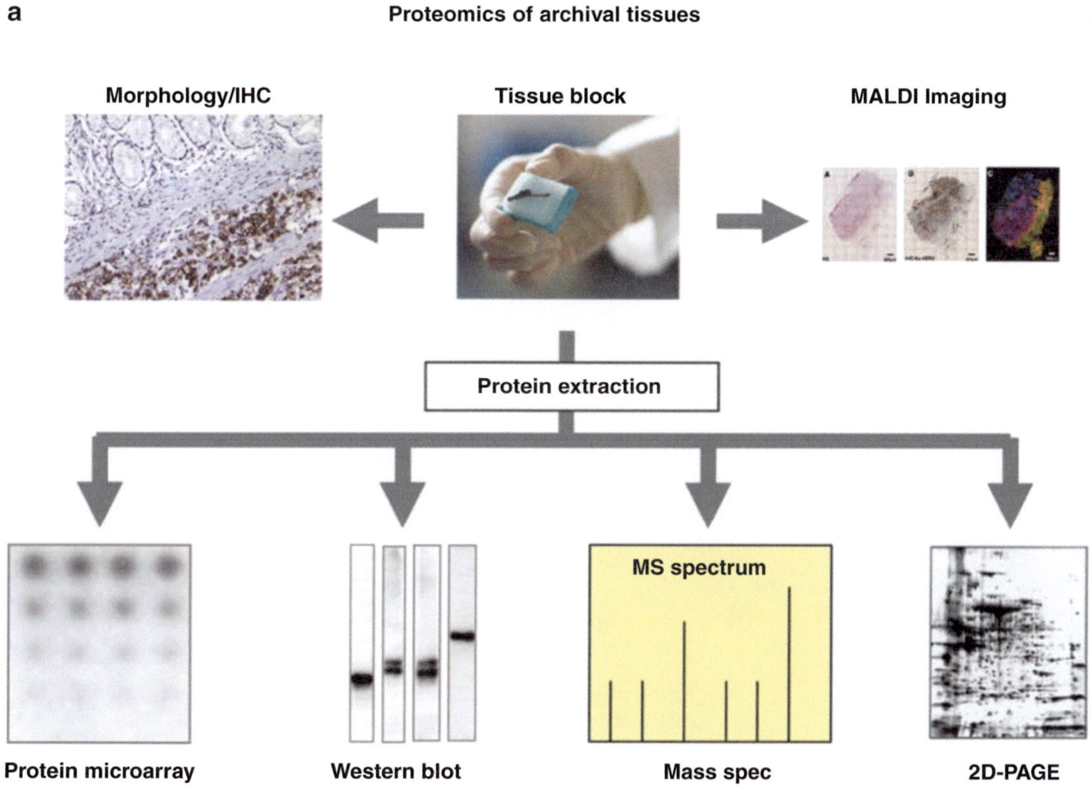

Fig. 1.2 Proteomics of archival tissue, and correlation with morphology, to capture cell origin of proteins of interest. (**a**) Many protein assay methods that are routinely used for frozen tissues can also be applied for FFPE tissues including immunohistochemistry (IHC), matrix-assisted laser desorption/ionization (MALDI) mass spectrometry (MS), Western blot, protein microarray, and two-dimensional (2D) gel electrophoresis. (**b**) Extraction-based protein analysis with parallel IHC studies to capture exact cell(s) of origin of protein(s) of interest. (Reprinted from Taylor and Becker [11]. With permission from Wolters Kluwer Health)

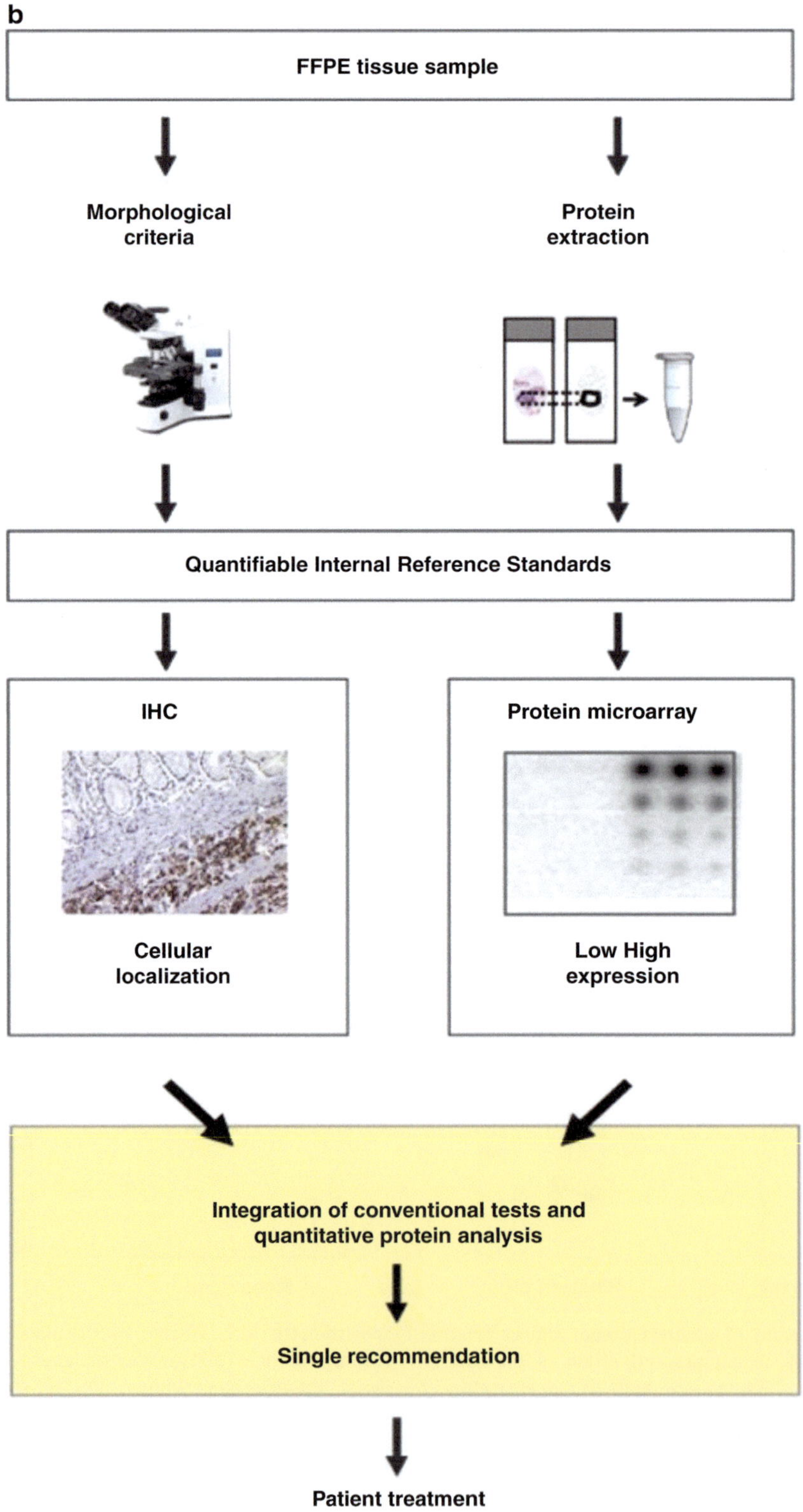

Fig. 1.2 (continued)

Table 1.5 Requirements for laboratory assays of cancer biomarkers

Total test approach – all aspects of test system should be encompassed, including sample preparation (Table 1.3)
Test method and analyte should have well-established performance characteristics
Test should be objective for read out/interpretation
Test ideally should produce a quantitative result (objective)
Threshold and reference range should be established
Test should be "fit to purpose", that is, designed and validated for the defined application
There should be well-developed control systems that are universally available
Test should be reproducible; run to run, day to day, lab to lab
Test should be readily performed and inexpensive

Based on data from Refs. [2, 3, 10, 11]

Table 1.6 Biomarker tests: commonly applied and developing methods

Sequencing: Sanger and NGS (next-generation sequencing)
Epigenetic differentiation
Laser capture microscopy
T and B cell receptor deep sequencing
Mass spectroscopy
Reverse phase protein arrays
RNA expression arrays
In situ hybridization (ISH)
Multiplex immunohistochemistry (IHC)

Based on data from Refs. [9–12]

exemplified by 40 plus mutation screens included in some NGS "tests" [4, 9, 12].

DNA and RNA sequencing methods can be traced back to the work of Frederick Sanger at the MRC Unit in Cambridge, England, in the 1970s [1]. Direct derivatives of his method provided the basis for the first sequencing of the human genome at the turn of the millennium. The achievement, time, and cost were extraordinary, but this success contributed to the development of multiple new approaches including the commercial availability of high-throughput sequencers, all of which together are known as next-generation sequencing (NGS). As a result the cost of sequencing a "cancer genome" has fallen dramatically and continues to fall, while availability, utility, and range of applications have enlarged so as to bring NGS from a discovery research mode into the realm of companion diagnostics. While the word genomics had been used half a century earlier, in practical terms this was the birth of the burgeoning field of "genomics" in medicine and in the public lexicon. Details of these various NGS approaches, instrumentation, reagents, methods, relative advantages, and disadvantages form the major topics of later chapters of this book.

The discovery of PCR, the polymerase chain reaction, is generally attributed to Kary Mullis in the 1980s [1]. It provided a means of almost infinite replication of defined DNA sequences that rapidly found an interface with Sanger DNA sequencing. Again numerous variants and derivative approaches have been described, and many have found major roles in the biomarker field, for highly sensitive detection of specific oncogenes, mutations, translocations, and the like in cancer, contributing to diagnosis, as well as much broader application in genetics as a whole.

DNA methods remain open to criticism in terms of clinical application, because not every change in DNA sequence is reflected in a change of cell function, a deficit that the biomarker field has attempted to repair through the use of RNA expression analysis, and studies of intermediate and end protein expression dubbed "proteomics." In the arena of cancer biomarkers, both transcriptional and posttranscriptional regulation have been studied extensively as described in later chapters. Proteomics as a concept, signifying both extensive and detailed analysis of tissue and cellular proteins, evolved also around the turn of the millennium as a companion of "genomics."

Detailed analysis of proteins has in many ways lagged behind related DNA and RNA analysis, for cogent reasons. Just as not every DNA sequence is translated to RNA, so not every RNA molecule is translated to protein, and RNA expression does not always correlate with protein expression. The whole process is increasingly recognized as being dynamic beyond earlier beliefs; in short, while the genome is relatively fixed and constant across time and across all of the cells of the organism (excepting a "few" mutations in cancer and in aging), the proteome dramatically is not, varying from tissue to tissue, cell to cell, and time to time.

Paradoxically, analysis of proteins by immunologic techniques has a long history, including, as noted at the beginning of this chapter, early biomarkers [1]. For example the ELISA (enzyme-linked immuno-sorbent assay) method devised by Stratis Avrameas has served as a gold standard for measuring individual proteins in fluids for well over half a century [1]. Detection of protein in a frozen section tissue environment by immunofluorescence was described by Albert Coons 80 years ago [1] and was adapted to FFPE sections for general routine use in the author's laboratory 40 years later and 40 years ago [1, 2]. However, these methods dependent as they are on the use of a specific antibody were directed to the protein of interest, typically detected only one protein at a time, until more recent developments as described subsequently.

Thus the advent of proteomics, in the context of "massive" analysis, awaited the use of techniques such as mass spectrometry, protein "chips," and reversed-phase protein arrays described in later chapters [11] (Fig. 1.2a, b). These methods initially proved difficult to standardize, for reasons of cell diversity and physiology as noted above and for technical reasons relating to extraction from FFPE tissue, principally unknown levels of degradation and loss, and in mass spectrometry, variable peptide recovery and detection (Fig. 1.3). Last but not least, interpretation of the huge data sets that were

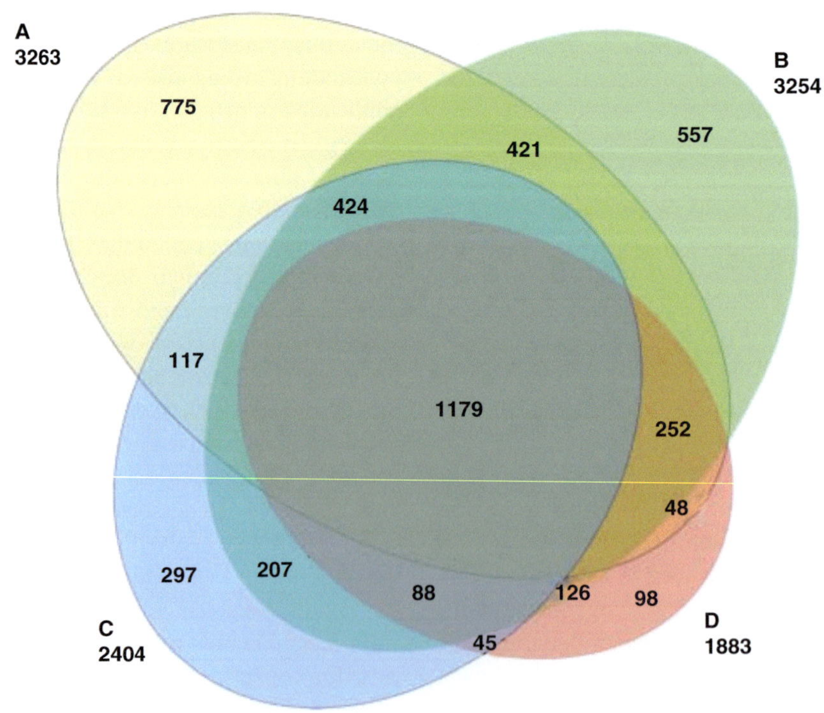

Fig. 1.3 Importance of validated sample preparation for mass spectrometry extraction-based proteomics. Four differently prepared extracts of the same renal carcinoma showing the number of distinct protein entries mapped by mass spectrometry using capillary isoelectric focusing (CIEF) with capillary reversed-phase liquid chromatography (RLPC). Samples A and B were extracted from FFPE tissue sections by using protocol of heat-induced retrieval with Tris-HCl buffer containing 2% SDS under different pH (pH 9 for A; pH 7 for B). Sample C was extracted from fresh tissue of the same case. Sample D was extracted from FFPE tissue by a protocol without heating treatment. (Reprinted from Shi et al. [17]. With permission from Sage Publications)

generated was a challenge. Much as with NGS, advancement of these methods was contingent upon the manifold increases in computer data analysis that occurred concurrently.

Each of these very different methods has inherent advantages and disadvantages. Most have been applied to extracts of FFPE tissues, or directly to FFPE tissue sections (IHC, ISH); all methods employed FFPE tissues – "because that is what we have" when the need for the test is recognized. Pathologists have long known that the process of formalin fixation and paraffin embedment compromises the integrity of all of the analytes tested by each of these methods, to differing degrees that are not yet completely understood. It is a significant problem that must be recognized and controlled whatever the method employed.

Extraction methods also require that the tissue that is subject to extraction contains a sufficient proportion of tumor cells versus normal cells (usually >20–30% for NGS), and mutated versus germ line DNA among the tumor cells (usually >10% depending upon method), in order to avoid a false-negative result [12]. Also for certain "biomarkers," such as "immune cell profiles," there are data that the use of tissue extracts necessarily sacrifices morphologic cellular and spatial information that may be critical to therapy choice and outcome. Selective extraction of tissue sections by microdissection or laser capture microscopy may also discard the very cell populations that subsequent tests seek to measure (e.g., immune cells). IHC has exquisite sensitivity on a cell to cell basis as already referenced but in the past has suffered from choice and quality of reagents, inefficient labeling methods, and subjective reading of the result. These shortfalls may be addressed by proper use of the method, coupled with computer-based analysis [2, 3, 10, 13].

With the current realization that the patient's immune response to their tumor, or lack thereof, affects the therapeutic efficiency of many drugs, it has become critically important to assess the patient's "immune cell profile." Determination of the immune profile is currently believed to be important for a broad range of new therapies, for which patient selection is critical to outcome (e.g., PD-1, PD-L1) (Fig. 1.4) (Table 1.4) [8, 9, 12]. While information on the nature and extent of any immune response to tumor may be derived from sequencing and proteomics studies, such information is inferential and may be compromised by extraction methods. The immune response and its constituent cells and molecular signals may be directly visualized in situ within the tissue by multiplex IHC, which accordingly has been added to the repertoire of methods now available (Figs. 1.4 and 1.5) [13].

Also notable are recent ventures into an area that has been by some termed "liquid biopsy," usually implying examination of blood components and or blood cells, although others have used liquid biopsy for various methods of examining tissue extracts [11]. Analysis of circulating DNA fragments and circulating tumor cells falls under the former definition. These methods hold great promise. Initial work is reviewed in later chapters but is yet to enter the mainstream of clinical care in a major way.

Multiple Biomarker Analysis

Until recently most of the approved companion diagnostics, as well as those in current ongoing trials, have been based upon detection of a single biomarker, although NGS and proteomics increasingly provide the potential for multiple parallel analysis. Now new demands have emerged, with an even higher order of complexity. The notion that clinical decisions may be based upon identification of the presence, or absence, of a single molecular target (exemplified by HER2, or PD-1) has extended to attempts at stratifying patients with respect to more than one biomarker. For example, with some targeted therapies the "drug labeling" states that it is necessary in arriving at a clinical decision to evaluate not only PD-L1 but also ALK and EGFR. The ultimate expression of this multi-marker trend has found immediate application in methods to assess the immune cell environment in and around the tumor. In real terms, this approach seeks to evaluate not simply the tumor itself but

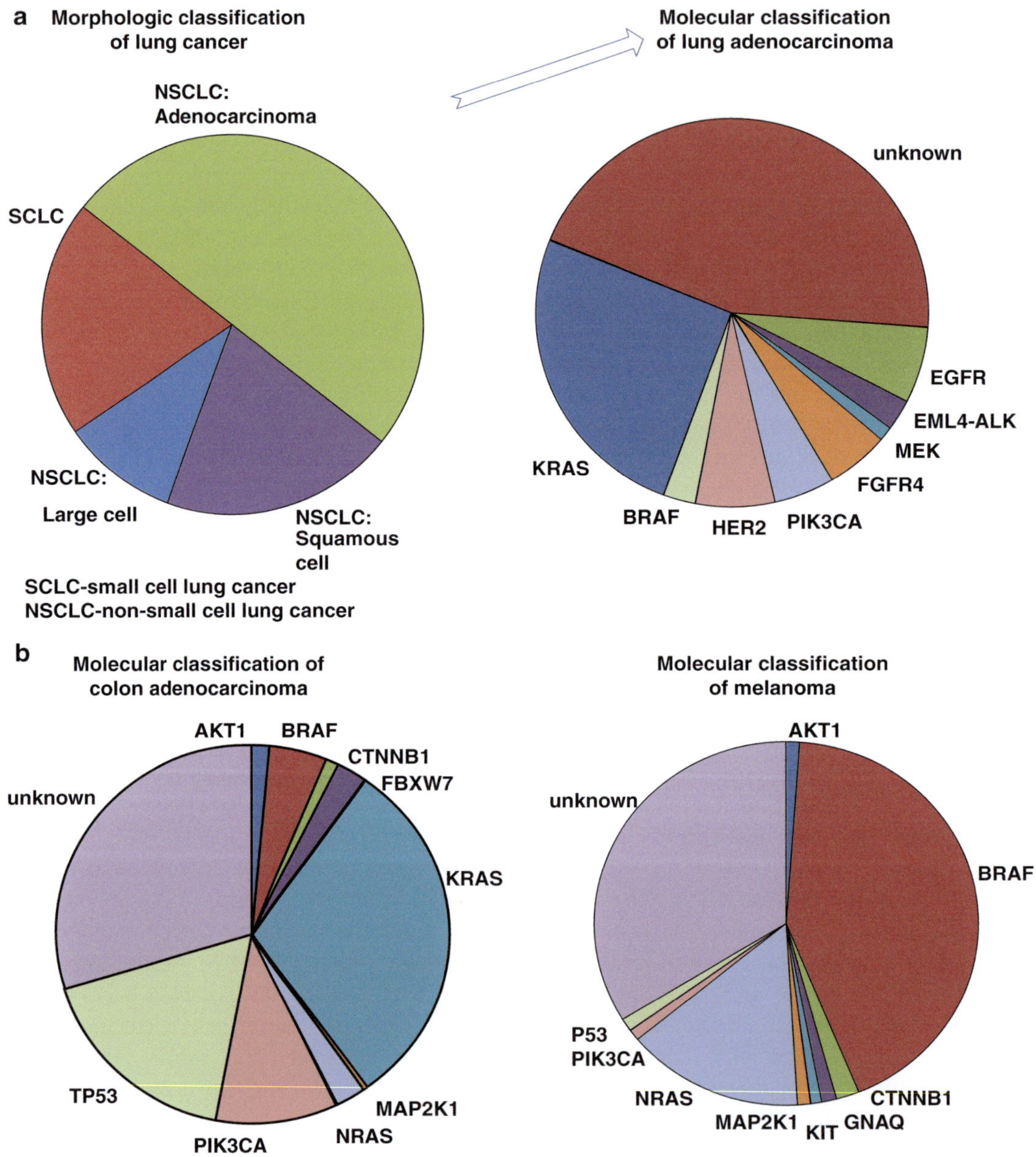

Fig. 1.4 Multiple "predictive biomarkers," exemplified by lung cancer, colon cancer, and melanoma. The "molecular" classification of these tumor types is superseding traditional morphologic classification as shown for lung cancer in (**a**); molecular profiles are shown for colon cancer in (**b**) and melanoma in (**c**). (Reprinted from Gu and Taylor [9]. With permission from Wolters Kluwer Health)

also the patient's immunologic response to the tumor, or lack thereof.

These studies have emerged primarily from evidence and resurgent enthusiasm for the "immunotherapy" of cancer, including the use of checkpoint inhibitors, exemplified by antibodies to CTLA-4 and PD-1, or its ligand PD-L1. Clinical trials, beginning with melanoma and extending rapidly to other solid tumors, indicated that patient responsiveness (or not) is dependent not only upon whether or not the tumor expresses the target (for the drug) but also whether there is an underlying immune response and whether such response is active or ineffective (suppressed).

Given the great complexity of the immune system in terms of both cellular and molecular

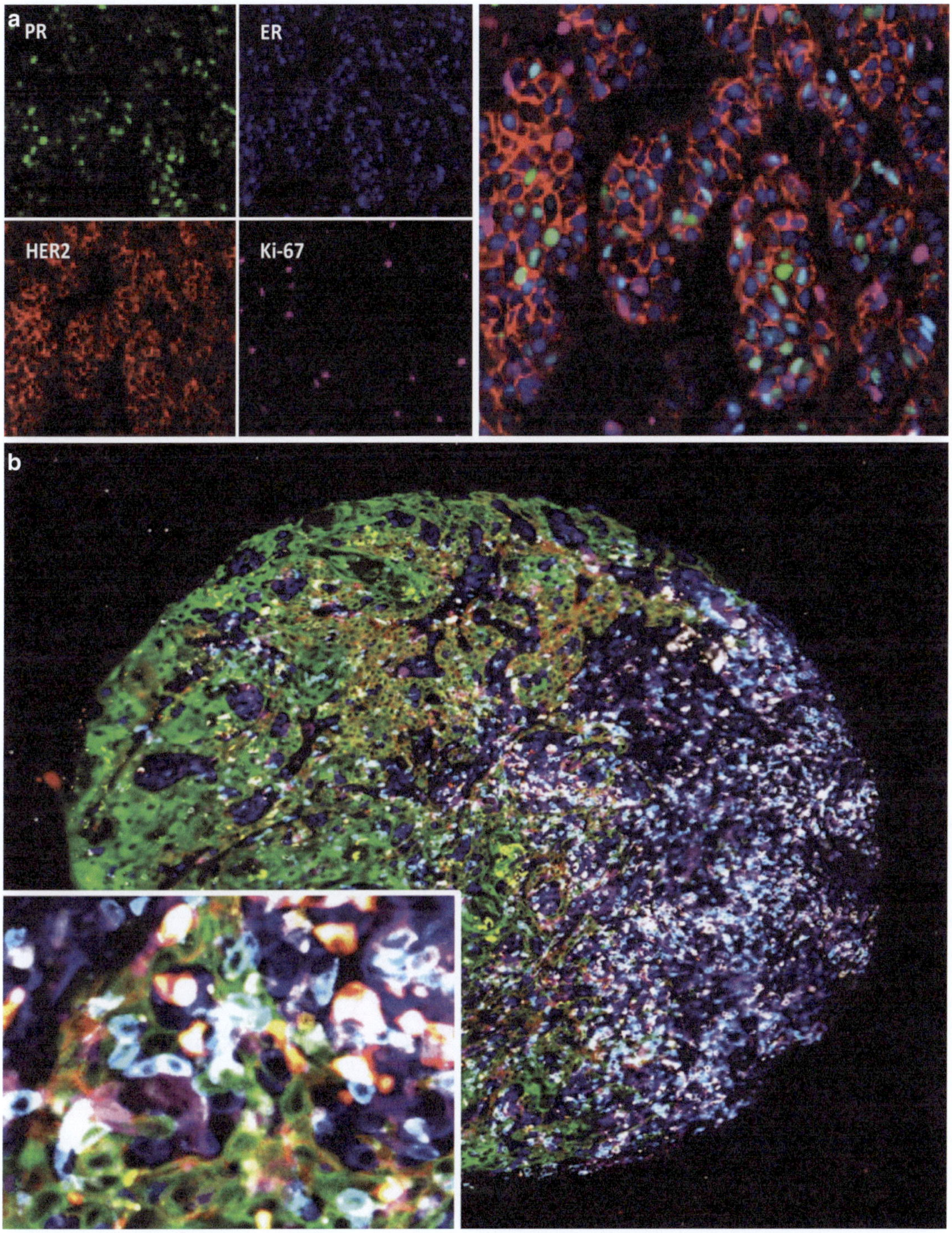

Fig. 1.5 Multiplex IHC ("Ultraplex"). (**a**) Quadraplex (four biomarkers) method. Triple positive breast cancer. On the left the four targets (colors) are displayed individually by the computer, allowing separate analysis. The composite image is on the right. PR, green; ER , blue; HER2, red; Ki67, magenta. (**b**) Decaplex (ten marker) method demonstrating cell identification, companion diagnostic and immune profile markers; squamous carcinoma, head and neck. Markers – cell identification: CK5, green; vimentin, blue. Companion diagnostic: EGFR, red. Immune cell profile: CD3, cyan; CD4, magenta; CD8, yellow; CD20, sepia; CD68, hot red; PD-1, gray; FoxP3, hot yellow. (Courtesy of David Schwartz, CEO, CSO (Cell IDx) with TMA samples provided by Mark Lingen, University of Chicago)

interactions, any "test" that evaluates only a single "biomarker" is unlikely to suffice. In addition, a means of evaluating the direct interface between the multiple types of immune cells and the tumor cells to which they are responding appear to be critically important. Lastly heterogeneity of biomarker expression in tumors has been recognized as a critical issue in terms of predictive value of testing, a concern that certainly includes evaluation of the immune cell infiltrate, not only variations in its intensity but also its character, focal or diffuse, and its location, intratumoral or at the invasive margin.

As noted, the presence of various immune cells and their state of activation may be inferred from proteomics or sequencing studies, including T cell receptor analysis, and information may be derived to class tumors as inflamed (hot) or non-inflamed [12]. However, numerical immune cell assessment, heterogeneity, and spatial relationships of multiple types of immune cells to each other and to tumor are necessarily compromised in any extraction-based assay and can only be fully assessed when considered in an undisturbed tissue-based context.

A Role for Multiplexed IHC Methods

"Multiplex" tissue-based IHC tests when performed in situ on FFPE sections of tumor tissue have the capability of displaying the "immune cell profile" (e.g., CD4, CD8, T regulatory lymphocytes, macrophages, myeloid-derived suppressor cells, etc.) and at the same time demonstrating the expression and distribution of regulatory molecules of interest, such as PD-1 and PD-L1, on tumor cells and associated immune cells (Fig. 1.5). On this basis tumors have been grouped into two broad categories, immunologically active (inflamed, hot) or immunologically silent (non-inflamed, ignorant, cold) (Table 1.7), which in turn have major implications for selection of classes of therapy, whether checkpoint inhibitors on the one hand or immune vaccines on the other.

Table 1.7 Two major classes of cancer as identified by immune profiling

	Immune silent/'ignorant'	Immunogenic/response suppressed
Class	"Non-inflamed"	'Inflamed'
Mechanisms	Lack of or tolerance to (self) tumor antigens (HLA)	"Tumor-induced" intrinsic suppression:
		Check point; PD-1; CTLA-4, Tim3, LAG3
		"Extrinsic" suppression: Tregs
		(CD25, FOXP3, Ki67), MDSC, blocking Abs
Tests		
Prognostic/predictive		
NGS/PCR	Low mutation load	High mutation load
NGS/RNA, protein, ISH/IHC	Targetable mutations – few	Targetable mutations – likely present
RNA, protein, IHC	Low check point expression	High check point expression; PD-1, PD-L1, CTLA-4, Tim3, LAG3
RNA, protein, IHC	Lack chemokines; immunomodulators	High immune modulators; suppressors dominate
Multiplex IHC	Lack – critical immune cells	High number critical immune cells; Tregs (CD25, FoxP3), MDSCs, macrophages (CD68)
Possible therapies	"Vaccines," immune activation modulators, BCG	Specific targeted therapy
		Checkpoint inhibitor blockade (PDL-1; PD-L1 block/deplete suppressor cells
	Recruit activated immune cells	Recruit and/or activate immune cells
	CAR T, CAR NK	CAR T CAR NK
Monitoring	Monitor immune profile change	Monitor immune profile change
	Monitor biodistribution	Monitor biodistribution
	CAR T, CAR NK, etc.	CAR T, CAR NK, etc.

These types of "immune profile" analyses clearly represent an entirely new class of assays for consideration, but equally clearly they are powerful "biomarkers" with both predictive and prognostic import.

Multiplex IHC is an extension of the basic IHC method, whereby several separate IHC protocols (four to eight or more) that are designed to detect different antigens (and cell types) are run on a tissue section in such a way that the results of all can be displayed and analyzed simultaneously. Several different approaches exist, either applying each separate antibody reaction sequentially, as in "Opal" (PerkinElmer), or "MultiOmyx" (Neogenomics) methods, the process taking 2 or more days to complete, or "UltraPlex" (Cell IDx) and "SigErMabs" (Calico Labs) that runs all reagents synchronously to complete a four- or ten-plex analysis in just 3 h (Fig. 1.5). Details of these methods are beyond the scope of this introductory chapter and are discussed elsewhere.

In brief, typically four or more differently colored fluorescent (or chromogenic) labels, each representing a different targeted molecule (protein, or nucleotide when combined with FISH), are developed on a single section. However, the human eye cannot distinguish the resultant kaleidoscope of colors (four to eight or more). Thus, this method has achieved practical utility only with the advent of high-resolution, high-speed tissue "scanners" that permit whole slide imaging and computer-based analysis of the complex multiple labels (Fig. 1.3a, b), coupled with sensitive, properly controlled, automated immune staining methods. Multiplex methods are evolving rapidly but are of course subject to similar standardization and total test requirements (Tables 1.3 and 1.5) as exist for other biomarker assays, including not only enhanced imaging and analysis methods but also high level controls for standardization [3].

The End of the Beginning

The challenges that this constellation of new test modalities presents to pathologists and clinicians should not be underestimated [2, 3, 10, 12, 13]. Neither should aspects of test availability and cost be neglected, for they may become the primary determining factors [14, 15]. There is ongoing debate with respect to choice of test, between "discovery-type tests" that assess multiple possible markers and generate huge data sets, but are very expensive, and tests that are specifically designed to answer a single question, to give the drug, or not, and are much less expensive. Some authors have explored the approach of using inexpensive, easy to perform tests, such as IHC, as screening tests, then following up with a more complex and expensive assay, only where clinically indicated [14].

Nonetheless, "precision" or "personalized medicine" appears to be an irresistible force, in turn requiring "precision pathology," which may be expected to result from further refinement and development of the methods, described briefly here, and discussed at greater length in the body of this book. Already the practice of pathology has been radically changed in the management of many malignant tumors (Fig. 1.4). Today we stand only at the end of the beginning of these changes; the ultimate end none of us as yet can foresee [9].

References

1. Van den Tweel J, Jiang J, Taylor CR. From magic to molecules: an illustrated history of disease: Beijing University Press; 2016.
2. Taylor CR. Quantitative in situ proteomics; a proposed pathway for quantification of immunohistochemistry at the light-microscopic level. Cell Tissue Res. 2015;360:109–20.
3. Cheung CC, D'Arrigo C, Dietel M, et al.; From the International Society for Immunohistochemistry and Molecular Morphology (ISIMM) and International Quality Network for Pathology (IQN Path). Evolution of quality assurance for clinical immunohistochemistry in the era of precision medicine: part 4: tissue tools for quality assurance in immunohistochemistry. Appl Immunohistochem Mol Morphol. 2017;25: 227–230.
4. Nass SJ, Phillips J, Patlak. Policy issues in the development and adoption of biomarkers for molecularly targeted cancer therapies. National Cancer Policy Forum. Workshop Summary. The National Academies Press. 2015. NAP.edu/10766.
5. European Parliament. Directive 98/79/EC of the European Parliament and of the Council of 27 October

1998 on in vitro diagnostic medical devices. 1998. http://eur-lex.europa.eu/LexUriServ/LexUriServ.do?uri=OJ:L:1998:331:0001:0037:EN:PDF.

6. U.S. Food and Drug Administration. List of cleared or approved companion diagnostic devices (in vitro and imaging tools). 2016. Other nucleic acid based tests are listed separately under an included link. Updated 6/09/16. http://www.fda.gov/MedicalDevices/ProductsandMedicalProcedures/InVitroDiagnostics/ucm301431.htm.
7. U.S. Food and Drug Administration. Guidance for industry and FDA staff. In vitro diagnostic (IVD) device studies – frequently asked questions. 2010. http://www.fda.gov/downloads/MedicalDevices/.../ucm071230.pdf.
8. Mahoney K, Atkins MB. Prognostic and predictive markers for the new immunotherapies. Oncology. 2014;28(suppl 3):39–48.
9. Gu J, Taylor CR. Practicing pathology in the era of big data and personalized medicine. Appl Immunohistochem Mol Morphol. 2014;22:1–9.
10. Taylor CR. Predictive biomarkers and companion diagnostics. The future of immunohistochemistry: "in situ proteomics," or just a "stain"? Appl Immunohistochem Mol Morphol. 2014;22:555–61.
11. Taylor CR, Becker KF. "Liquid morphology": immunochemical analysis of proteins extracted from formalin fixed paraffin embedded tissues: combining proteomics with immunohistochemistry. Appl Immunohistochem Mol Morphol. 2011;19:1–9.
12. Yuan J, Hegde PS, Clynes R, et al. Novel technologies and emerging biomarkers for personalized cancer immunotherapy. J Immunol Ther Cancer. 2016;4:3.
13. Gaule P, Smithy JW, Toki M, et al. A quantitative comparison of antibodies to programmed cell death 1 ligand 1. JAMA Oncol. 2016; https://doi.org/10.1001/jamaoncol.2016.3015. Published online August 18, 2016.
14. Murphy DA, Ely HA, Shoemaker R, et al. Detecting gene rearrangements in patient populations through a 2-step diagnostic test comprised of rapid IHC enrichment followed by sensitive next generation sequencing. Appl Immunohistochem Mol Morphol. 2017;25:513–523.
15. Yaziji H, Taylor CR. PD-L1 assessment for targeted therapy testing in Cancer: urgent need for realistic economic and practice expectations. Appl Immunohistochem Mol Morphol. 2017;25(1):1–3.
16. Taylor CR. Quality assurance and standardization in immunohistochemistry. A proposal for the annual meeting of the Biological Stain Commission. Biotech Histochem. 1992;67:110–7.
17. Shi S-R, Liu C, Balgley BM, Lee C, Taylor CR. Protein extraction from Formalin-fixed, paraffin-embedded tissue sections: quality evaluation by mass spectrometry. J Histochem Cytochem. 2006;54:739–43.

2 Introduction to Clinical Trials, Clinical Trial Designs, and Statistical Terminology Used for Predictive Biomarker Research and Validation

Karla V. Ballman

Introduction

Clinical research studies involving human patients or participants generally have two main variables of interest: *participant exposure* and *participant outcome*. In the context of biomarker studies in cancer research, the exposure would be the biomarker value for a patient, and the outcome might be survival. The distinguishing feature between a *retrospective study* and *prospective study* is what is known about the patient exposure and patient outcome at the time the study is designed. For a retrospective study, investigators look back into time to ascertain patient exposures (e.g., the biomarker value) and the patient outcome of interest (e.g., cancer survival). For a prospective study, the patient exposure of interest is known at the time the patient is included in the study (e.g., baseline biomarker value), and the patient is followed into the future to ascertain the outcome of interest (e.g., survival). As depicted in Fig. 2.1, in a retrospective study, the biomarker value and outcome for a patient are known by the start of the study.

In contrast, in a prospective study the outcome of interest has not yet occurred at the start of the study, and patients are followed into the future until the end of the study to determine their outcome.

Retrospective studies are limited by various confounding factors that introduce biases. In cancer biomarker studies, they are useful for the discovery of potential biomarkers to be explored in future studies but generally are not sufficient for biomarker validation. More definitive biomarker studies are based on data from prospective studies. For the purpose of establishing a treatment benefit of a predictive biomarker, the prospective study requires (1) a patient group that spans the biomarker outcomes (for a *dichotomous marker*, the study needs biomarker-positive and biomarker-negative patients; for a *continuous marker*, the study needs a group of patients that have biomarker values that represent the range of possible values), and across the biomarker values, it needs (2) patients treated with the treatment of interest and patient not treated with the treatment of interest (likely treated with a different treatment). The strongest design is one in which patients are *randomized* to the treatments as is done in a clinical trial. If patients are not randomized to treatment, the study will likely suffer from patient selection bias, similar to a retrospective study. The remainder of this chapter focuses on predictive biomarker studies in cancer that are based on clinical trial data. Sometimes, the biomarker study is conducted well after the clinical trial has been completed, but this still qualifies as a prospective study because at the time the patients were

K. V. Ballman
Department of Healthcare Policy and Research, Division of Biostatistics and Epidemiology, Weill Cornell Medicine, New York, NY, USA
e-mail: kab2053@med.cornell.edu

S. Badve, G. L. Kumar (eds.), *Predictive Biomarkers in Oncology*,
https://doi.org/10.1007/978-3-319-95228-4_2

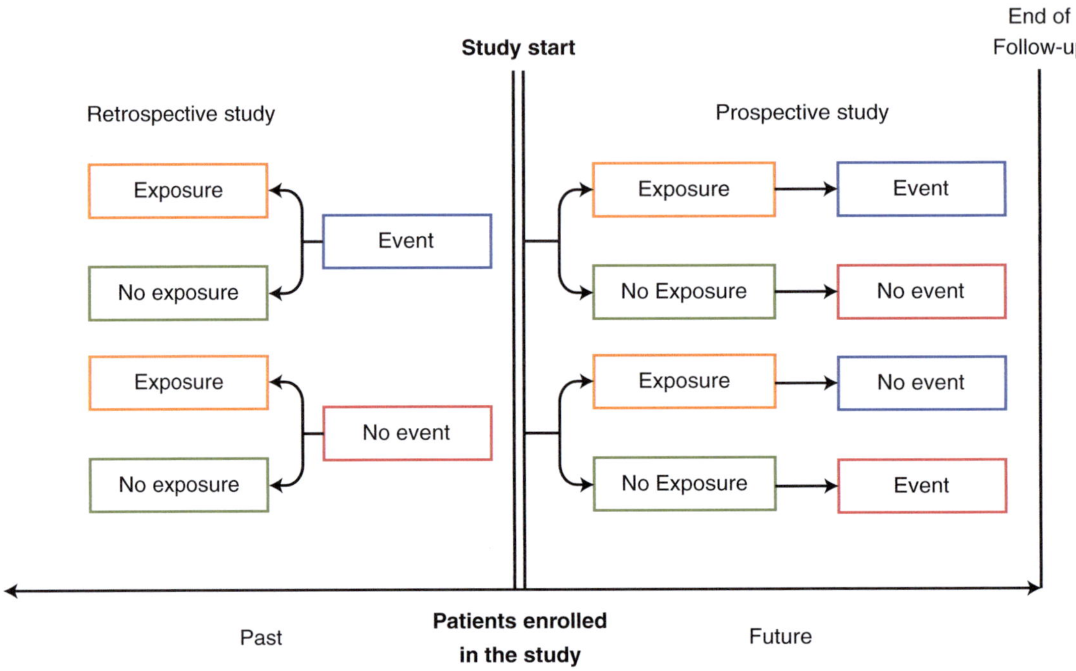

Fig. 2.1 Prospective studies identify patients, determine or assign their exposure, and then follow patients forward from that time until they have an event of interest or the end of the study. Retrospective studies enroll patients and then look backward in time from that point to ascertain their exposure status and whether they had the event of interest or not

enrolled on the trial, their baseline biomarker status was fixed (although it might not have been measured until much later), and patients were followed forward into the future for their outcomes.

A brief overview of the different phases of clinical trials is presented in section "An Overview of Oncology Clinical Trial Designs." Section "Analysis of Clinical Trial Data" provides a general description of clinical trial data analysis methods. The definition and characteristics of prognostic and predictive biomarkers are presented in section "Biomarkers in Clinical Trials." The interplay of biomarkers and clinical trial design is explored in section "Use of Forest Plots." Concluding remarks are made in section "Biomarker Clinical Trial Designs."

An Overview of Oncology Clinical Trial Designs

Oncology clinical trials are performed in different settings and by different groups. Some trials are initiated and led by an investigator that is a member of a cancer center within an academic medical center. These trials may be funded by a pharmaceutical company, the academic medical center, philanthropic funds, or a grant from the government (e.g., the National Cancer Institute, Department of Defense) or a nonprofit organization (e.g., Stand Up to Cancer). It is often the case that the funding comes from one or more of these sources. The principal investigator has control over the data, the data analyses, and the publication of results in investigator-initiated trials.

Pharmaceutical companies also conduct clinical trials. These trials are led and funded solely by the pharmaceutical company, and the company performs the data analysis and disseminates the trial results via publications. The National Cancer Institute (NCI) conducts the majority of government-funded trials, which includes internal trials as well as trials done by other institutions that are funded by NCI grants and contracts. Other government agencies that conduct or sponsor oncology clinical trials include the Department of Defense and the Department of Veteran's Affairs. Finally, the NCI also funds and supports the National Clinical Trial Network (NCTN) that includes four groups that conduct trials for adult cancer patients (Alliance for Clinical Trials in Oncology, ECOG-ACRIN Cancer Research Group, NRG Oncology, and SWOG) and one group that conducts trials for pediatric cancer patients (Children's Oncology Group). About half of all patients who participate in a cancer clinical trial in a given year do so in a NCTN-led trial. Trials conducted by the NCI NCTN often receive additional support from pharmaceutical companies and/or nonprofit organizations. However, the data analyses leading to publications are conducted independently of the other funding sponsors. Data from any trial funded by a government agency is required to be deposited in a public repository.

There are four general types of clinical trial phases used for drug development in oncology. A drug development plan usually starts with a phase I trial and proceeds through the other phases in a sequential manner if the previous phase is deemed to be a positive trial. A phase I trial is the first time the drug regimen (e.g., a single drug or a new combination of drugs) is being used in humans. These trials are generally small and are designed to find a safe dose to be used in a phase II trial. Typically, sample sizes for a phase I trial are between 10 and 80 patients. The number of patients depends on the number of dose levels to be tested. A positive phase I trial establishes a dose level that is tolerable (has limited adverse events) and thought likely to be active.

Phase II trials generally enroll on the order of 50–150 patients. The sample size is primarily driven by the number of treatment arms included in the trial. The purpose of a phase II trial is to further evaluate the safety of the drug regimen and to evaluate whether it has potential activity or efficacy. The decision rule is cast as a go/no-go decision. Specifically, if the clinical activity of the drug appears unpromising and/or the drug appears to be too toxic, the decision will be not to perform future trials with the regimen. On the other hand, if the activity level appears promising and the regimen appears to be relatively tolerable, the drug will likely be tested in a phase III trial. Measures of clinical activity depend on the patient population and the postulated mechanism of action of the drug regimen. Some examples include tumor shrinkage, often measured as the tumor response rate, or a decrease in an established biomarker such as PSA for prostate cancer. Phase II trials can be single-arm trials where all patients receive the drug regimen, or they can be multi-armed where patients are randomized to the arms. Examples of multi-armed trials are a comparison among several different new regiments to select the best one to test in a phase III trial, a comparison of the new regimen to a control arm or a comparison of several different dosing regimens in order to optimize the regimen delivery for a phase III trial.

The sample size for a phase III trial is generally in the range of a few hundred patients to a few thousand patients. The goal is to evaluate the efficacy of the drug regimen. In a phase III trial, patients are randomized to a new regimen or to a control group. Depending on the disease, the control group could be treated with a placebo, if the disease is not life threatening or if there are no approved treatments available for the patient population, or standard of care, in the case of life-threatening disease for which there is an established treatment available. A phase III trial could test several different interventions but always has a control arm. Phase III trials are generally considered to be definitive trials. A positive phase III trial shows that a new regimen has a beneficial effect compared to the current standard of care, i.e., the control arm. If a phase III trial is positive, it usually changes the standard of care and could be the basis for FDA approval of the drug for use in the patient population in which the trial was conducted.

Phase IV studies are conducted after a drug regimen has been marketed and typically involves several thousand patients. The focus of these studies is to monitor the effectiveness of the drug regimen in the general population. It also collects information regarding adverse effects. Phase IV studies have uncovered adverse events that where not observed in previous clinical trials that are due to patient comorbidities or drug-drug interactions.

Within the phase I–IV paradigm of drug development, biomarker discovery may start in

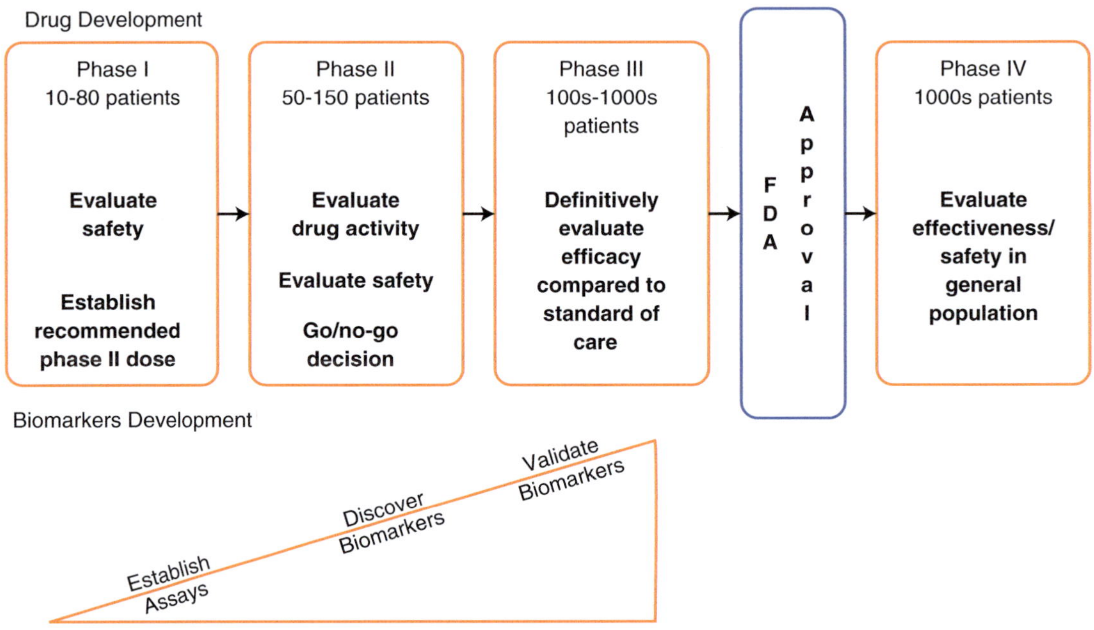

Fig. 2.2 Design phases for cancer drug development and an indication of the biomarker activities that parallel each of the drug development phases. The size of the triangle for the biomarker development represents the level of evidence for the utility of the biomarker as well as the number of samples typically involved

phase I trials but is often limited to preliminary exploration or proof-of-concept because of the small sample sizes. Phase II studies are generally the platform for initial biomarker discovery studies and identify markers to be evaluated further in phase III trials. The most informative biomarker studies are part of phase III trials because their larger sample sizes afford more power and because they randomize patients to the drug regimen of interest and a control arm. A phase III study could be used for biomarker discovery, it could be used to validate a proposed biomarker, or the biomarker could be used to determine patient treatment. Figure 2.2 summarizes the roles of the different stages of clinical trial design and biomarker development.

Analysis of Clinical Trial Data

The statistical method to be used in evaluating data from a clinical trial depends on the outcome of interest. For the sake of brevity, it is assumed the outcome of interest is a time-to-event measure such as overall survival (OS), disease-free survival (DFS), or progression-free survival (PFS). From this point the outcome will be described generically as survival but could be any measure that involves time from study start for a patient to an event where some patients are censored (i.e., they did not have the event by the end of the follow-up period). For a single-arm trial or the analysis of a single group, the survival time is summarized with a Kaplan-Meier (KM) curve. A KM curve estimates the proportion of patients who have survived as a function of time since treatment initiation (see Fig. 2.3). The median survival is often reported and represents the time point at which 50% of the patients have not survived (or had the event), implying that 50% have survived (or are event-free).

KM curves can be used to compare survival times of two or more groups when they are plotted on the graph. For example, Fig. 2.4 compares the survival times between patients randomized to a new experimental treatment (T) and patients randomized to a control group (C). It is clear that the T group has better survival in general than the C group. This is also demonstrated by comparing the estimated median survival times: 45.1 months

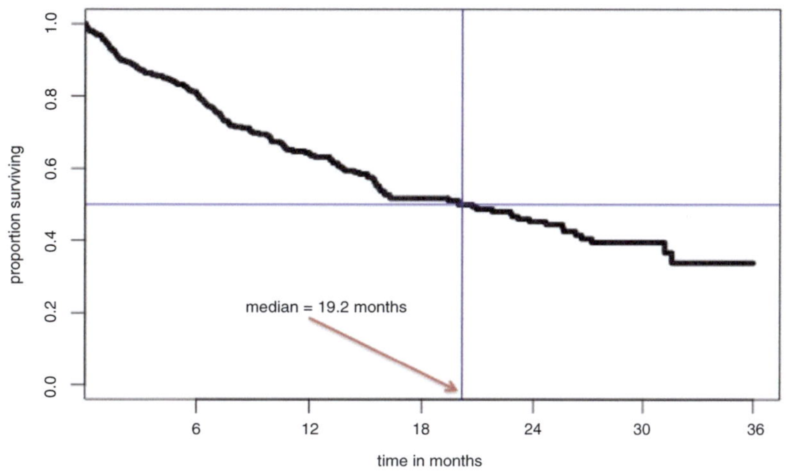

Fig. 2.3 An example of a Kaplan-Meier curve plot for a group of patients that have a median survival of 20.1 months. The median is the value for which 50% of the patients are still alive (equivalent to 50% have died)

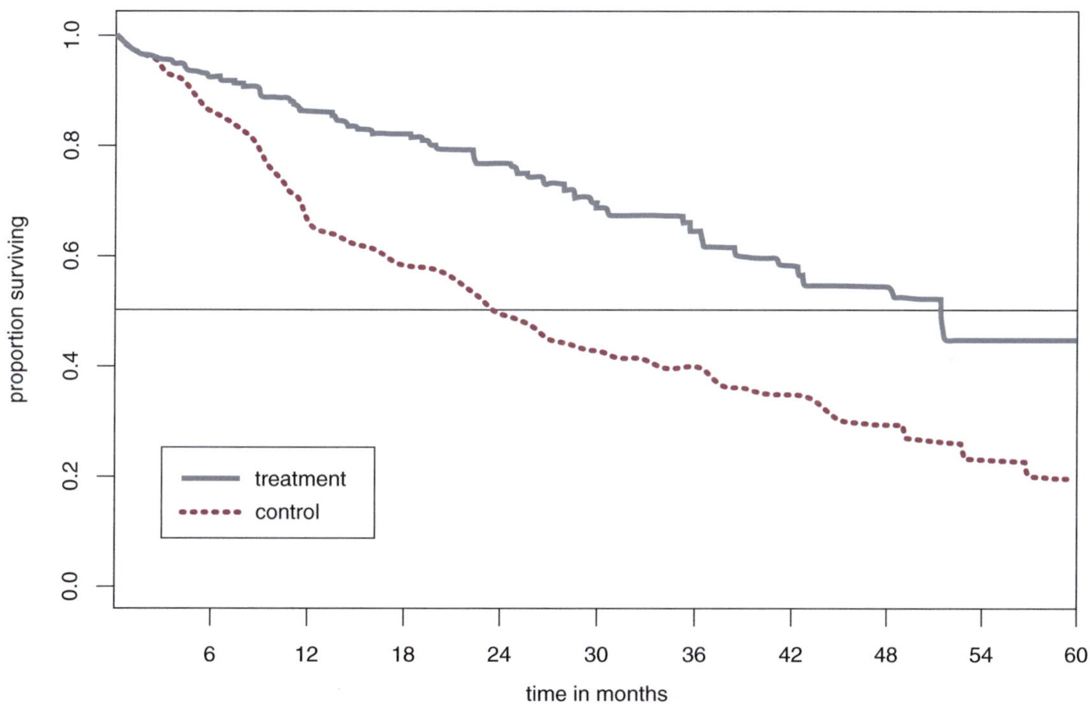

Fig. 2.4 A display of two Kaplan-Meier curves for survival with one corresponding to patients in the treatment group (solid gray line) and one corresponding to patients in the control group (dashed maroon line). The median survival for the treatment group is 49.2 months, and the median survival for the control group is 22.7 months

for group T compared to 26.3 months for group C. A *log-rank test* is used to determine whether the observed difference in the KM curves is likely due to chance alone (p-value ≥ 0.05) or is deemed statistical significant (p-value < 0.05), which implies there is a treatment effect. The log-rank p-value = 0.0035 for the curves in Fig. 2.4 shows that the patients in the treatment group appear to have a significantly better survival than patients in the control group. The log-rank test can also be used to evaluate whether there are differences in survival times among any number of groups.

Biomarker classification can also be used to define the patient groups to be compared. Suppose that a biomarker classifies patients into marker-positive (BM+) and marker-negative (BM−) groups. From Fig. 2.5 it appears as though the BM+ group has (very) slightly better survival compared to the BM- group; however, this difference is not statistically significant (*p*-value = 0.33). The conclusion in this case would be that the biomarker does not appear to be significantly associated with survival. An example of a biomarker that is not significantly associated with overall survival is PD-L1 protein expression in early-stage non-small cell lung cancer (NSCLC) [1] patients.

A question of interest might be whether there is an association of the biomarker and survival when adjusting for treatment group. Note that the biomarker analysis in Fig. 2.5 includes pooled patients across treatment groups meaning that the BM+ group contains patients in the treatment group as well patients in the control group and the BM− group contains patients in the treatment group as well as the control group. In the PD-L1 study referenced above, the BM+ group are all patients who are PD-L1 positive pooling across those who were and were not treated with adjuvant chemotherapy, and the BM- group are patients who are PD-L1 negative regardless of treatment. When the evaluation of the association with survival involves more than one variable, such as treatment group and biomarker status, statistical modeling is used, which in this case would be a *Cox proportional hazards model*. The relationship of each explanatory variable in the model and survival (the outcome variable) is summarized with a hazard ratio (HR), which is the ratio of the hazard of dying at a point in time for each group. The proportional hazard component of the model assumes that this ratio remains constant over all time points. A HR of 1.0 indicates there is no association between the variable and survival. Table 2.1 contains the *univariable HRs* for treatment group and biomarkers status.

The HR comparing the survival of the treatment group to the control group is HR = 0.62, which is less than one, and it is statistically significant (*p*-value = 0.0038). This means that patients in the treatment group are less likely to die than patients in the control group. (If the HR were greater than 1, this means that patients in

Table 2.1 Univariable estimates of the hazard ratio (HR) for treatment group and biomarker status group with 95% confidence intervals (CIs) and *p*-values

Variable	HR (95% CI)	*p*-value
Treatment group		0.0038
Control	1.00 (reference)	
Treatment	0.62 (0.45, 0.86)	
BM status		0.33
BM−	1.00 (reference)	
BM+	0.85 (0.61, 1.18)	

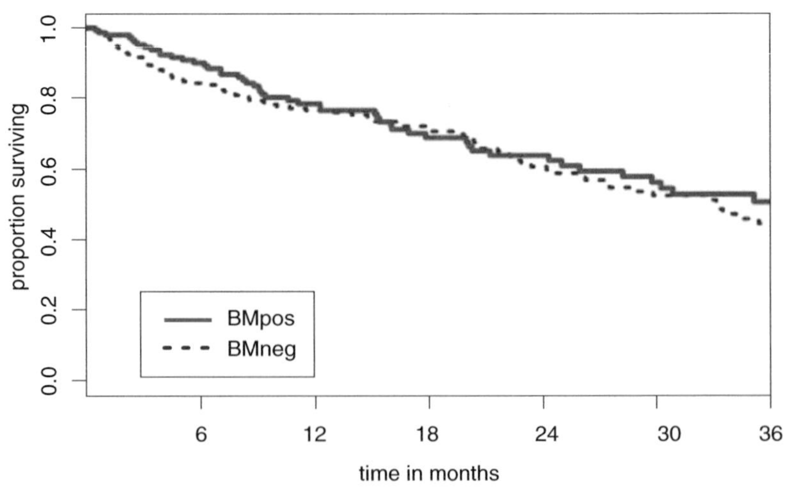

Fig. 2.5 The Kaplan-Meier curves for survival for the BM+ group (green solid line) and the BM− group (red dashed line)

the treatment group are more likely to die than patients in the control group.) The best estimate of the treatment HR is 0.62, but there is uncertainty associated with the estimate. Confidence intervals (CIs) are used to convey the precision of the estimate, and 95% CIs are the most commonly used. This is an interval for which there is a 95% probability that it contains the true HR. The 95% CI for the HR = 0.62 is 0.45–0.86. This interval does not contain one, which is consistent with the conclusion that the association of treatment with survival is statistically significant. The conclusion of the univariable analysis of the treatment variable is that it appears that the treatment is associated with longer survival compared to standard of care (control arm).

The univariable HR for the biomarker is HR = 0.85 (95% CI, 0.61–1.18) with a *p*-value of 0.33. The 95% confidence interval contains 1 and the *p*-value is not statistically significant. It appears as though the biomarker is not associated with survival. Note that the conclusions based on the univariable Cox models are consistent with those from the KM analysis with the log-rank test, which is almost always the case.

A multivariable Cox model is used to evaluate the association of the biomarker with survival while adjusting for the treatment to which the patient was randomized. The multivariable model has both the treatment group and biomarker group as explanatory variables. Table 2.2 contains the *adjusted HRs* for the variables in the multivariable Cox model.

The multivariable HR for the biomarker classification is HR = 0.85 (95% CI: 61–1.19), and its *p*-value is 0.35. The estimate of the association between the biomarker and survival did not change (only the upper value of the 95% CI changed slightly) when adjusting for treatment assignment, and the *p*-value did change slightly but is still not significant. The conclusion would be that the biomarker does not appear to be associated with survival when adjusting for the treatment to a patient received. The lack of change between the univariable and multivariable HR estimates indicates that the effects of treatment and biomarker are not related. Returning to the PD-L1 and NSCLC example, the univariable HR for the BM+ patients (PD-L1 positive) compared to BM− patients is HR = 0.91 (95% CI, 0.75–1.30; *p*-value = 0.91). When the model includes treatment, the adjusted HR for PD-L1-positive versus PD-L1-negative patients, adjusting for adjuvant treatment (chemotherapy versus none), is HR = 1.01 (95% CI, 0.76–1.35; *p*-value 0 0.93) [1]. The conclusion would be that PD-L1 status (positive versus negative) is not associated with overall survival in early-stage NSCLC patients because there is no significant association between PD-L1 status and overall survival, even after adjusting for treatment.

Biomarkers in Clinical Trials

A biomarker refers to a measurable indicator of a biological state. In cancer this includes indicators of cancer presence, of prognosis for patients with cancer, and of disease response to a specific treatment. A biomarker can be a single measurement

Table 2.2 Univariable and multivariable estimates of the HRs (with 95% CIs) and *p*-values for treatment group and biomarker status group. The univariable values are the same as in Table 2.1 and are the estimate of the HR for models that only have the indicated variable. The multivariable estimates come from a model that contains both variables at the same time

	Univariable models		Multivariable model	
Variable	HR (95% CI)	*p*-value	HR (95% CI)	*p*-value
Treatment group		0.0038		0.0039
Control	1.00 (reference)		1.00 (reference)	
Treatment	0.62 (0.45, 0.86)		0.62 (0.45, 0.86)	
BM status		0.33		0.35
BM−	1.00 (reference)		1.00 (reference)	
BM+	0.85 (0.61, 1.18)		0.85 (0.61, 1.19)	

(e.g., PSA level for men), or it can be computed form numerous measurements (e.g., Oncotype Dx for women with early-stage breast cancer which is based on 21 genes). The two types of biomarkers commonly used in cancer clinical trials are *prognostic* and *predictive* biomarkers.

A prognostic biomarker informs about a likely cancer outcome regardless of what treatment a patient receives (including no treatment); it is thought to reflect the natural history of the disease. In other words, a prognostic biomarker is significantly associated with survival when adjusting for treatment a patient received. In Fig. 2.6b it can be seen that the biomarker is associated with survival for patients in the treatment group and for patients in the control group (Table 2.3).

The magnitudes of the association of the biomarker and survival are the same for both groups. In Fig. 2.6d, it also can be seen that there is an association between the biomarker and survival for both groups. The difference between the scenarios depicted in Fig. 2.6d and that in 2.6b is that the magnitude of the association between the biomarker and survival depends on the treatment a patient received. For patients in the treatment arm, the magnitude of the biomarker association with survival is larger than for patients in the control group. In summary, if a biomarker is prognostic, there will be an association of the biomarker and survival regardless of treatment. If the magnitude of the association is the same in the groups, the biomarker is *purely prognostic*. If the magnitude differs between groups, the biomarker is both *prognostic and predictive*.

A biomarker is *predictive* when the treatment effect differs for BM+ patients and BM− patients. Figure 2.6c shows an association

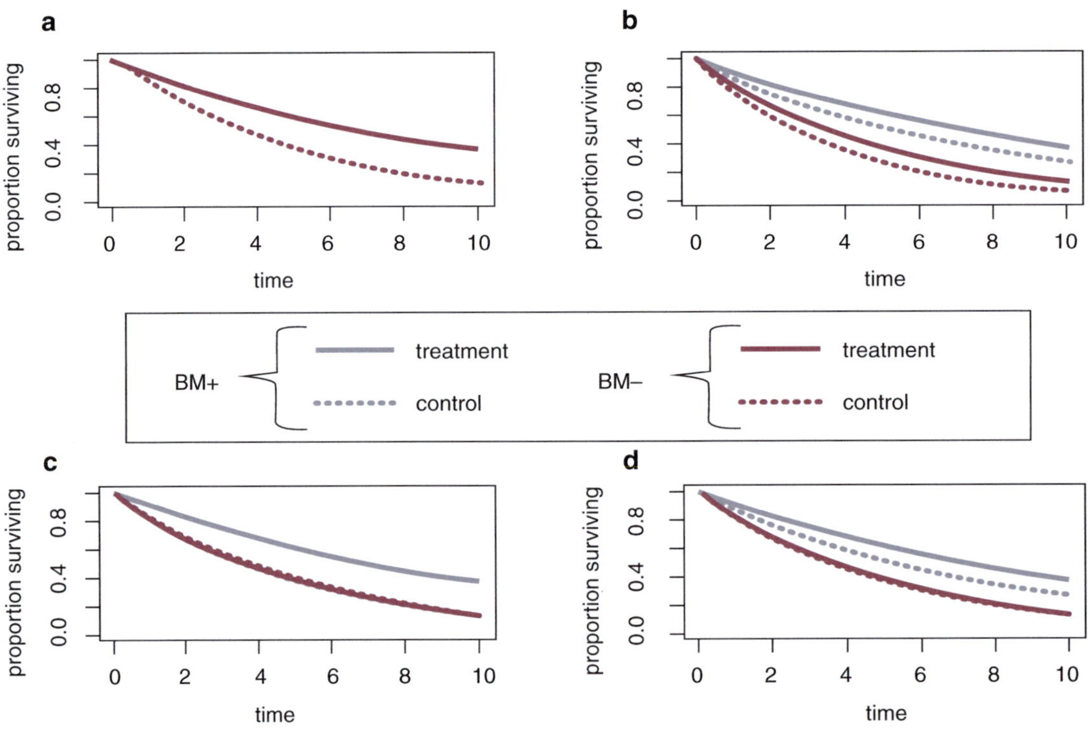

Fig. 2.6 Kaplan-Meier curves for different groups of patients where the color of the line denotes the biomarker group (BM+ is gray and BM− is maroon) and the line type denotes the treatment group (solid is the treatment group and dashed is the control group). (**a**) Illustrates the situation where the biomarker is neither prognostic nor predictive. (**b**) Illustrates the situation where the biomarker is prognostic but not predictive. (**c**) Illustrates the situation where the biomarker is predictive but not prognostic. (**d**) Illustrates the situation where the biomarker is both prognostic and predictive

Table 2.3 Definitions of different types of biomarkers with published examples of each

Biomarker	Definition	Example	Reference
Prognostic	A prognostic biomarker informs about a likely cancer outcome regardless of what treatment a patient receives (including no treatment)	Evaluation of PIK3CA mutation status for women with HER2-positive metastatic breast cancer	Baselga et al. [2]
Predictive	A biomarker is *predictive* when the treatment effect differs for biomarker-positive patients (BM+) and biomarker-negative (BM−) patients	RAS mutational status for treatment of metastatic colorectal cancer with an anti-EGFR antibody (cetuximab)	Van Cutsem et al. [3]
Prognostic and predictive	There is an association between the biomarker and survival for patients in the treatment and control groups	EGFR mutation status in NSCLC patients	Brugger et al. [4]
Neither prognostic nor predictive	Treatment is associated with survival, but within each treatment group, there is no association of the biomarker with survival	PD-L1 in early-stage NSCLC study	Tsao et al. [1]

between treatment and survival for BM+ patients; it appears as though patients in the treatment group have longer survival than patients in the control group. However, for BM− patients there is no association between treatment and survival. The same is true for Fig. 2.6d, where there appears to be a treatment benefit for BM+ patients but no treatment benefit for BM− patients. The difference between Fig. 2.6c, d is that the biomarker is *purely predictive* (and not prognostic) in Fig. 2.6c: there is no association between the biomarker and survival for patients in the control group. In Fig. 2.6d there is an association between the biomarker and survival for patients in the treatment and control groups indicating the biomarker is both predictive and prognostic. Figure 2.6a shows a case where the biomarker is neither predictive nor prognostic. Clearly, treatment is associated with survival, but within each treatment group, there is no association of the biomarker with survival.

In the era of precision medicine or individualized treatment, predictive biomarkers are more useful than prognostic biomarkers because they can be used to determine which patient will derive benefit from a treatment (say BM+ patients) and which will not (say BM− patients). In this case, a BM+ patient would receive the treatment because he/she would likely garner benefit, and a BM− patient would not be treated because he/she would potentially experience adverse events with no benefit. The goal is to discover and validate more predictive biomarkers so that patients are treated with regimens from which they benefit and spared those form which they will not benefit and may only be harmed.

KM curves such as those in Fig. 2.6 can be used to gain a preliminary indication of whether a biomarker is potentially predictive. To be able to evaluate if a biomarker is predictive, all four groups of patients are necessary: BM+ treated with drug of interest, BM- patients treated with drug of interest, BM+ patients treated with control, and BM− patients treated with control. A biomarker is potentially predictive if the treatment is associated with survival in one biomarker group (e.g., BM+) and not the other (e.g., BM−). However, this is not sufficient. There needs to be a formal test of whether the treatment effect differs between the different biomarker groups. Such a test is performed with a statistical model, such as the Cox model for a survival outcome. The model contains the explanatory variables of treatment group and biomarker status with the addition of a variable for the interaction between the treatment and biomarker, the treatment by

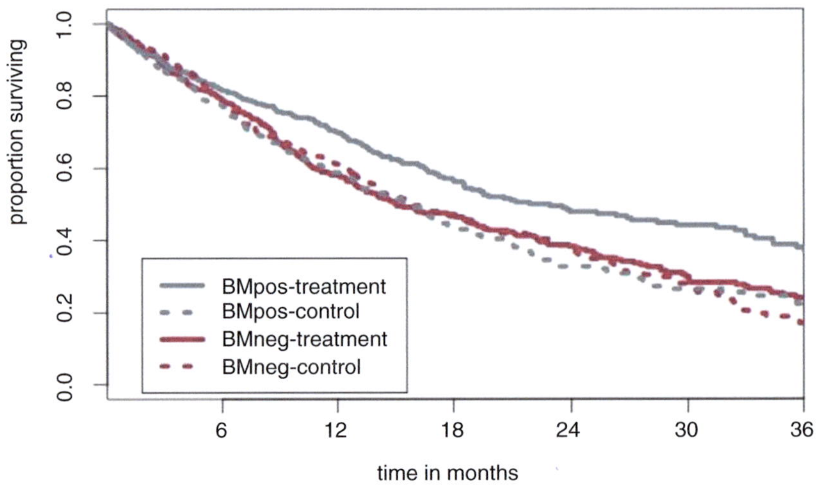

Fig. 2.7 A Kaplan-Meier plot summarizing the survival results for the four different biomarker and treatment group combinations. This plot suggests that the biomarker is predictive because BM+ patients derive benefit from treatment and BM− patients do not. The predictive nature of the biomarker is confirmed with a statistically significant biomarker-by-treatment interaction term in the Cox model (p-value for interaction = 0.0049)

biomarker interaction variable. To determine whether a biomarker is predictive, the treatment-by-biomarker interaction term in the Cox model needs to be statistically significant (e.g., p-value < 0.05). A significant treatment-by-biomarker interaction term indicates that the treatment effect differs by the biomarker group.

A Cox model that tests for an interaction between treatment groups by biomarker status will have three variables: treatment group, biomarker status, and the treatment-by-biomarker interaction. It is difficult to interpret and visualize the impact of the biomarker, treatment, and interaction based on the Cox model alone. In particular, the crude HRs that is produced by the software does not correspond to any of the four biomarker-by-treatment groups; the HRs for each of the four groups (one of which will be the reference group) are functions of the HRs of the model variables. KM curves can aid in understanding the relationship. Figure 2.7 contains the KM curves that correspond to a study of biomarkers and treatment. It appears as though BM+ patients drive benefit from treatment but BM− patients do not. The interaction term from the corresponding Cox model is statistically significant, p-value = 0.0049, indicating the biomarker is predictive.

If the treatment-by-biomarker interaction term is not statistically significant, then there is no evidence that the biomarker is predictive, even if it is the case that the log-rank test for treatment benefit is statistically significant in the BM+ group and not statistically significant in the BM− group. Often, investigators only analyze patients who were all treated with the drug of interest and conclude a biomarker is predictive if there is an association between the biomarker and survival. This is an inappropriate conclusion. Note that in Fig. 2.6b, for patients in the treatment group, there is an association between the biomarker and survival, BUT this is a purely prognostic biomarker because there is also an association between the biomarker and survival in the control group. Using only patients treated with the treatment of interest, it cannot be determined whether the situation is that in Fig. 2.6b (purely prognostic), Fig. 2.6c (purely predictive), or Fig. 2.6d (both prognostic and predictive).

The Use of Forest Plots

Often *meta-analysis* studies of predictive or prognostic biomarkers are conducted in order to garner more power, especially for testing for a biomarker status by treatment interaction that is required to establish a biomarker is predictive. A forest plot is a graphical display of estimated results from randomized trials that investigate the same question. A forest plot typically lists the names of the included trials on the left-hand side. The content of the plot is the measure of the effect, which for overall survival is the HR, for each of the studies. The confidence intervals for the effect estimate is represented by horizontal lines and is often the numerical values for the effect estimate and confidence interval boundaries are provided on the right-hand side of the graphic. The graph may be plotted on the logarithmic scale when using a HR so that the confidence intervals are symmetric around the estimated effect. Each square is centered on the effect size, and the area of the square is proportional to the size of the study, which dictates the study's weight or influence in the analysis. The overall meta-analysis estimate of effect is represented by a diamond, with the width of the diamond corresponding to the confidence interval. A vertical line corresponding to no effect (e.g., HR = 1) is often plotted.

Figure 2.8 is a forest plot taken from a study performed by Rowland et al. [5]. The authors performed a meta-analysis of randomized clinical trials that evaluated the effect of BRAF V600E mutation status, mutated (MT) versus wild type (WT), and benefit from anti-EGFR monoclonal antibody treatment (anti-EGFR mAB) in patients with metastatic colorectal cancer that was RAS wild type. From the figure, it can be seen that within these studies, patients with BRAF WT tumors obtained benefit from anti-EGFR mAB treatment, with a few studies yielding statistically significant results. On the other hand, it appears as though patient with BRAF MT tumors did not garner benefit from anti-EGFR mAb treatment with none of the studies having statistically significant results in this group. The meta-analysis estimate of anti-EGFR mAb benefit in patients with BRAF WT tumors is 0.81 (95% CI, 0.70–0.95; *p*-value = 0.009) and in patients with BRAF MT tumors is 0.97 (95% CI, 0.67–1.41;

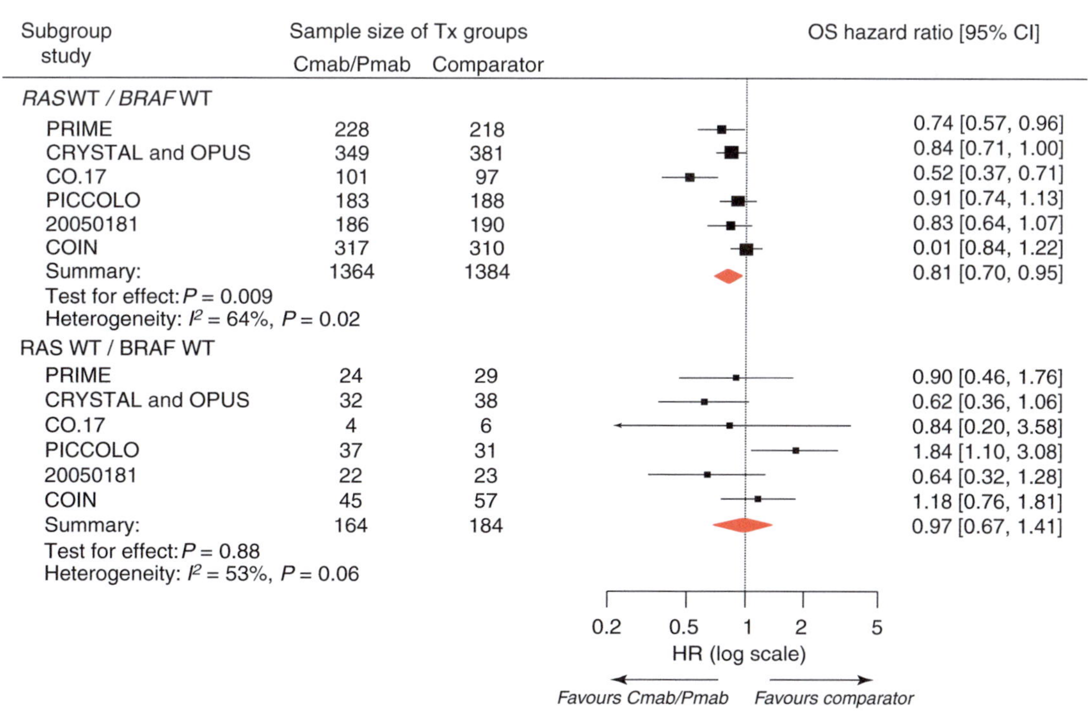

Fig. 2.8 A forest plot taken from Rowland et al. [5]. Note that there is an error in this plot in that the bottom portion is for RAS WT/BRAF MT patients. (Reprinted from Rowland et al. [5]. With permission from Nature Publishing Group)

p-value = 0.88). Although there appears to be differential treatment effects in the two biomarker groups, the test for interaction between BRAF status (WT versus MT) and treatment (anti-EGFR mAb treatment versus no anti-EGFR mAb treatment) was not statistically significant, p-value = 0.43. Hence, there is no evidence from this study that BRAF mutation status is a predictive biomarker for benefit from anti-EGFR mAb in patients with RAS WT metastatic colorectal cancer.

Biomarker Clinical Trial Designs

There are numerous clinical trial designs that incorporate biomarkers, validate biomarkers, and discover biomarkers. The *enrichment design* is used when there is compelling evidence that treatment benefit (if any) will be restricted to a subgroup of patients who do (or do not) have a particular biomarker. In this design, all patients are screened for the biomarker, and only those in the subgroup of interest (either have or do not have the biomarker) are enrolled on the trial (see Fig. 2.9).

This trial design cannot validate whether the biomarker is predictive for the treatment benefit since all patients are in the same biomarker subgroup. It can only provide evidence whether there is a treatment benefit in the selected biomarker subgroup. If there is benefit, it is unknown whether patients in the nonselected biomarker group may also have derived treatment benefit. Such a design should only be used in cases where there is persuasive evidence that the biomarker is predictive. A successful example of the use of this design was the trials for trastuzumab in patients with HER2+ breast cancer: the National Surgical Adjuvant Breast and Bowel Project (NSABP) B-31 and the North Central Cancer Treatment Group (NCCTG) N9831 trials [6]. These trials only included women with tumors that were found to be HER2 positive. There were strong preclinical data to indicate that only these patients would derive benefit from trastuzumab. The trials were successful and led to FDA approval for the use of trastuzumab to treat

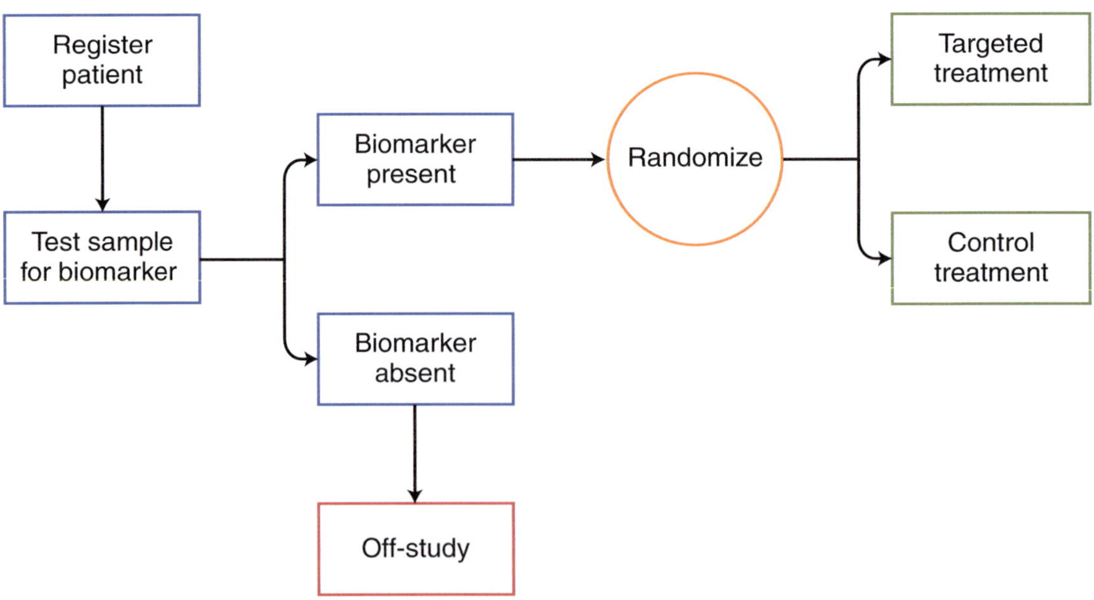

Fig. 2.9 A diagram of the schema for an *enrichment* trial design. Patients are registered (and consented) prior to their sample being tested for the biomarker. If the biomarker is "present" (either deemed positive or negative), the patient is then enrolled and randomized to the targeted treatment or the control treatment (usually standard of care). If the biomarker is "absent," the patient goes off-study and is no longer followed

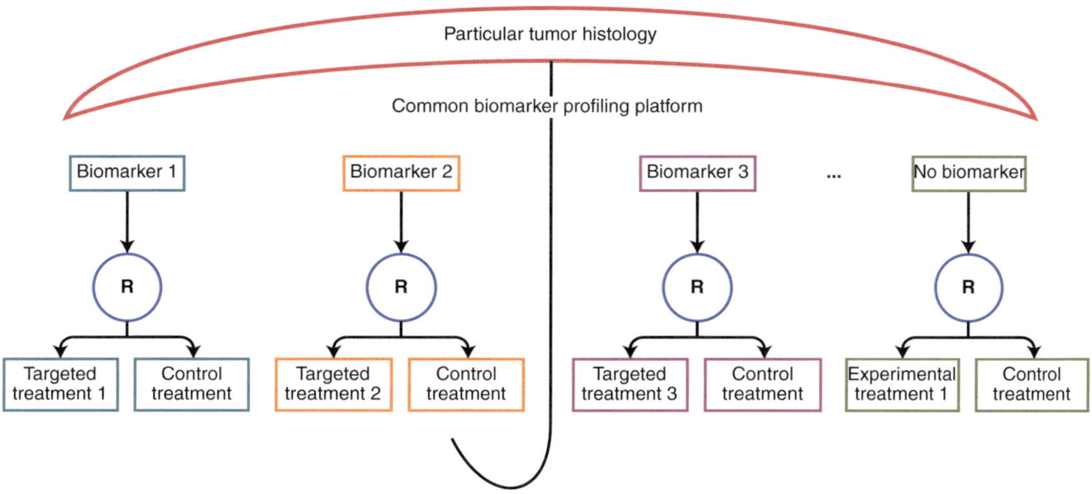

Fig. 2.10 A diagram for an umbrella trial. Tumor of a specific histologic type is tested for a panel of biomarkers on a common testing platform. Tumors that have biomarkers of interest are then randomized to a treatment that targets the biomarker or to a control treatment. If tumors have none of the biomarkers of interest, they either are off-study or are randomized between control and another (untargeted) experimental treatment

HER2-positive breast cancer in the adjuvant setting. The question of whether patients with HER2-negative tumors would benefit from trastuzumab is currently being investigated.

Two different enrichment designs have recently gained popularity: the umbrella trial and the basket (or bucket trial). The *umbrella design* tests the treatment benefit of multiple drugs on different mutations in a single tumor type or histology (see Fig. 2.10).

It provides a common infrastructure to facilitate patient screening and accrual. Patients are assigned or randomized to treatment arms based on their biomarker status. The intent of the trial is to evaluate the benefit of different drugs matched to their mutation in a single type of cancer. The biomarker testing is usually done at a central location prior to patient enrollment and randomization. Examples of recent umbrella trials include I-SPY2 [7, 8], BATTLE [9, 10], and Lung-MAP [11]. A *basket* or *bucket* trial includes cancers of different types that each has the same biomarker of interest (see Fig. 2.11).

This trial design tests the benefit of a treatment for which the biomarker is thought to be predictive. The design includes many different cancer types that belong to the same biomarker subgroup, and one targeted treatment (usually) is tested. Patients are tested for the biomarker prior to enrollment to the trial since the biomarker subgroup is an eligibility criterion. Examples of basket trials are MPACT [12], MATCH [13], and a vemurafenib trial for cancers with BRAF V600 mutations [14]. These are versions of enrichment trials and are designed to realize benefits of efficiency of using a single platform (umbrella trial) or to increase the number of patients eligible for treatment with a particular biomarker and to determine if the benefit is similar across tumor types (basket).

The *all-comer* (or *unselected*) design tests all patients for their biomarker status and enrolls all patients regardless of biomarker status. An eligibility criterion for this trial is adequate specimen availability and quality to perform the biomarker assay. The patients are randomized to the same set of treatment arms, for all the biomarker groups (see Fig. 2.12). The SATURN (sequential Tarceva in unresectable non-small lung cancer) trial [15] is an example of an all-comer trial. In this trial, all eligible NSCLC patients were randomly assigned to erlotinib or placebo plus standard of care, regardless of the EGFR status of their tumor. The trial was designed to evaluate the efficacy of erlotinib in all randomized patients as well as in the subgroup of patients that had EGFR-positive tumors.

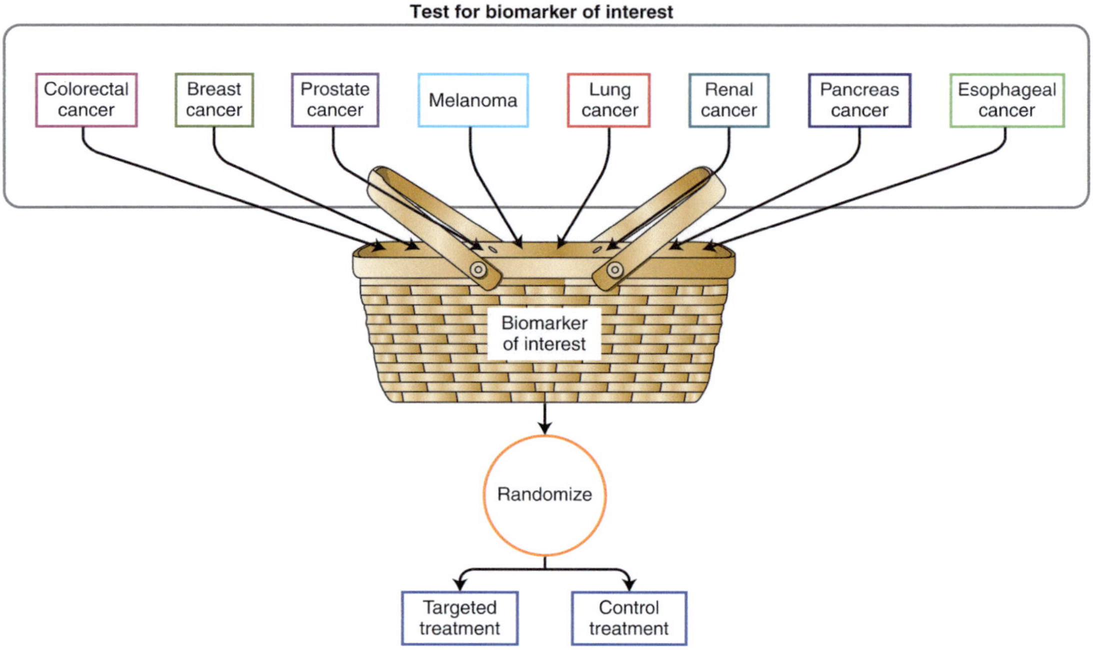

Fig. 2.11 A diagram for a basket trial. All types of tumors are tested for a specific biomarker. If they have the biomarker of interest, they are randomized to a targeted treatment or to a control treatment. If they do not have the biomarker of interest, they are not registered to the trial

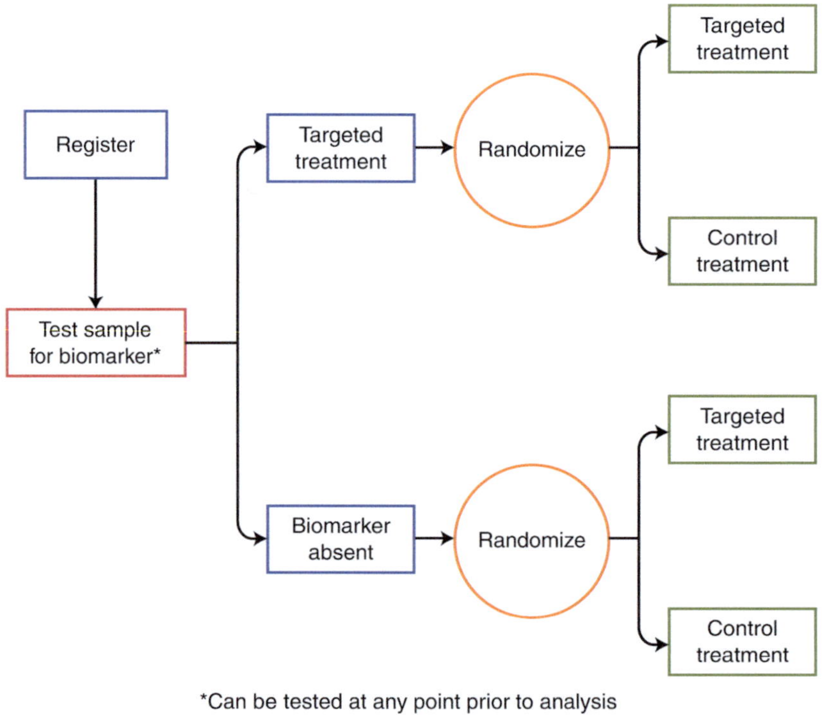

Fig. 2.12 A diagram of the schema for an *all-comers* trial design. Patients are registered and entered onto the trial regardless of their biomarker status. In the diagram, they are tested prior to randomization, but this does not have to be the case; the biomarker status only needs to be known prior to the analysis of the trial data. Both types of patients, those with the biomarker present or absent, are randomized to targeted treatment or to control treatment

The test for the biomarker can be performed before or after randomization. If the biomarker is a *stratification variable*, then to ensure the same distribution of biomarker subgroups among the treatment arms, it needs to be performed prior to patient randomization. If it is not used as a stratification factor, it can be performed at any time prior to the pre-planned trial analyses. There are several different ways the trial data could be analyzed, but the analysis method must be pre-specified at the time of trial design. If the primary interest is to validate that the biomarker is predictive, a biomarker by treatment interaction analysis will be the primary analysis. This formally tests for a biomarker by treatment interaction term in a Cox model as described above.

Another type of analyses determines which patient subgroups defined by the biomarker benefit from treatment, if any, by performing sequential analyses. One approach is to test for a treatment effect in the entire trial cohort (ignoring biomarker group). If this is not significant, then a test of treatment benefit will be done in a planned biomarker subset, which is the subset thought to be the most likely to derive benefit a priori. Another approach is to first test for treatment benefit in a biomarker subset (the one with the strongest a priori evidence it would benefit), and if this is statistically significant, perform a test of treatment benefit in the entire clinical trial cohort. The type of analysis plan that will be done is pre-specified during the trial planning stage, and the level of significance used for the planned sequential analyses are set to ensure the overall trial type I error is maintained at 0.05.

It is best to use the marker-by-treatment interaction analysis when there is uncertainty whether the biomarker is predictive or not. However, this analysis requires the largest sample size. The sequential testing approaches are also relevant for situations where there is uncertainty of whether the biomarker is predictive or not, but they are not powered to detect a biomarker by treatment interaction. The intent for the latter two approaches is to find subgroup(s) that benefit from treatment without formally establishing whether the biomarker is predictive. These trials are generally a bit smaller than what is needed for the maker-by-treatment interaction analysis.

Finally, there are refinements to the designs discussed above that incorporate a Bayesian aspect to perform exploratory analyses meant to discover biomarkers as the trial proceeds. These designs are sometimes called *exploratory platform designs* and usually are early phase (I or II) trials. Such designs are useful when there is uncertainty regarding the best biomarkers for the treatments under study. In this design, drug arms are pre-specified, and patients are initially randomized equally across the arms, regardless of the biomarker status of their tumor. Biomarker testing is performed on a tumor biopsy prior to randomized, and pre-specified biomarker cohorts are stratified evenly across treatment arms. After a sufficient number of patients have been assigned to each arm, the efficacy for each biomarker-treatment combination is evaluated, and the randomization is adapted so that future patients have a higher probability of being assigned to a treatment group that appears favorable for the biomarker in their tumor. Drugs that do not appear to be beneficial for any biomarker group are dropped. Biomarker-treatment combinations that surpass a pre-defined threshold of efficacy are brought forward in a larger enrichment trial (e.g., phase II or III). In these trials, only patients with tumors that have the identified biomarker are enrolled, and the patients are randomized to the experimental treatment or standard of care. Examples of exploratory platform trials with Bayesian adaptive randomization are BATTLE [16], for patients with previously treated lung cancer, and I-SPY2 [17], a neoadjuvant trial for breast cancer patients.

Concluding Remarks

For cancer treatments to be more individualized to patient and/or disease characteristics, it is necessary to develop predictive biomarkers. However, the success rate for finding predictive biomarkers has been disappointing. To increase the success rate, it is important to understand the evidence that is needed to determine whether a biomarker is predictive of treatment benefit. It is also important to understand the different roles of biomarkers in clinical trials and the implications of the different clinical trial designs for the evaluation of biomarkers.

Glossary

Adjusted (or multivariable hazards ratio (HR) A multivariable Cox model allows the evaluation of the association of multiple variables on the outcome (e.g., survival). This allows a more accurate assessment of the relationship of a variable of interest to overall survival by accounting for other variables that may be associated with survival. For example, when evaluating the association of a biomarker with survival, a treatment variable may be added to the model. This would allow the evaluation of the association of the biomarker with survival after accounting for the association of treatment with survival. The hazard ratio for a variable from a multivariable Cox model is referred to as a multivariable HR or an adjusted HR.

Continuous (bio)marker A continuous biomarker is one that has an infinite number of possibilities; in other words, it can take on any value between its minimum and maximum value if it could be measured to any desired degree of precision. An example of a continuous biomarker is PSA level for prostate cancer. The minimum value is 0 and there is no absolute maximum. If PSA could be measured to any desired degree of precision, all nonnegative values are possible.

Cox proportional hazards model A Cox proportional hazards model is a regression technique for time-to-event data (e.g., survival) where there is censoring (when some patients are alive at the time of analysis). It is a way to evaluate the association of a variable with the time-to-event outcome such as survival. The method is semi-parametric; that is, it does not assume a model for t survival but does assume that the effect of a variable on survival is constant over time. The association is measured by a hazard ratio (HR) where HR = 1 means no association, a HR <1 means increasing values of the variable reduces the chance of death, and HR >1 means that increasing value of the variable increases the chance of death.

Dichotomous (bio)marker A dichotomous biomarker is one that takes one of two possible values. It is used to split patient cohorts into two categories or groups. An example of a dichotomous biomarker is estrogen receptor (ER) status for women with breast cancer: ER positive versus ER negative.

Log-rank test A log-rank test is used to compare the survival distributions of two or more groups. The null hypothesis is that there is no difference among the groups. If the *p*-value is significant (e.g., less than 0.05), this is evidence that the groups have different survival experiences. Note this is only a test for a difference among the survival experiences and does not provide an estimate regarding the size of the differences between any two groups.

Meta-analysis A meta-analysis encompasses techniques for combining data from multiple studies. An underlying assumption is that the treatment effect is consistent across studies and combining results across studies yields increased power. Most meta-analysis approaches essentially compute a weighted average from the results of the individual studies, and larger studies tend to be given more weight.

Randomization or random assignment In randomized trials, the participants are assigned by chance to the treatment groups (arms) rather than by choice. Randomization serves to make the groups similar with respect to variables (e.g., patient characteristics, tumor traits) other than the treatment. This means if differences are observed for the outcome variable (primary endpoint), it can be attributable to the treatment since the groups balanced for the other variables. Randomization is accomplished with a chance procedure (e.g., flipping a coin) or a random number generator.

Stratification variable A stratification variable in a clinical trial is a variable that is used to group patients into strata corresponding to the values of the variable. Randomization is performed separately within each stratum. An example of a stratification variable is whether a patient has disease in his/her lymph nodes or not (e.g., lymph node status with values of lymph node positive and lymph node negative). Variables selected for stratification are those where it is important there is no imbalance between the treatment arms because they are highly prognostic of outcome.

Type I error Type I error is the error that occurs when the null hypothesis is rejected although it is true. It is a false-positive result. For example, suppose in reality there is no difference between the experimental treatment and standard of care with respect to overall survival. However, a clinical trial is performed, and it is found that the treatment arm had superior survival compared to the standard of care arm with a *p*-value of 0.03. The investigators conclude that the experimental treatment is better than the standard of care. In reality, this is an incorrect conclusion and an example of a type I error. (Note that the investigators would not know that their conclusion is incorrect.)

Univariable hazards ratio (HR) A univariable hazard ratio is the ratio of hazard rates for an event (e.g., death) corresponding to the different values of one variable of interest. For example, in a Cox model that contains only a treatment variable (experimental versus control), a HR = 0.50 for survival indicates that patients in the treatment group die at half the rate per unit of time as patients in the control group.

References

1. Tsao MS, Le Teuff G, Shepherd FA, Landais C, Hainaut P, Filipits M, Pirker R, Le Chevalier T, Graziano S, Kratze R, Soria JC, Pignon JP, Seymour L, Brambilla E. PD-L1 protein expression assessed by immunohistochemistry is neither prognostic nor predictive of benefit from adjuvant chemotherapy in resected non-small cell lung cancer. Ann Oncol. 2017;28(4):882–9. https://doi.org/10.1093/annonc/mdx003. PubMed PMID: 28137741.
2. Baselga J, Cortés J, Im SA, Clark E, Ross G, Kiermaier A, Swain SM. Biomarker analyses in CLEOPATRA: a phase III, placebo-controlled study of pertuzumab in human epidermal growth factor receptor 2-positive, first-line metastatic breast cancer. J Clin Oncol. 2014;32(33):3753–61. https://doi.org/10.1200/JCO.2013.54.5384. Epub 2014 Oct 20. PubMed PMID: 25332247.
3. Van Cutsem E, Lenz HJ, Köhne CH, Heinemann V, Tejpar S, Melezínek I, Beier F, Stroh C, Rougier P, van Krieken JH, Ciardiello F. Fluorouracil, leucovorin, and irinotecan plus cetuximab treatment and RAS mutations in colorectal cancer. J Clin Oncol. 2015;33(7):692–700. https://doi.org/10.1200/JCO.2014.59.4812. Epub 2015 Jan 20. PubMed PMID: 25605843.
4. Brugger W, Triller N, Blasinska-Morawiec M, Curescu S, Sakalauskas R, Manikhas GM, Mazieres J, Whittom R, Ward C, Mayne K, Trunzer K, Cappuzzo F. Prospective molecular marker analyses of EGFR and KRAS from a randomized, placebo-controlled study of erlotinib maintenance therapy in advanced non-small-cell lung cancer. J Clin Oncol. 2011;29(31):4113–20. https://doi.org/10.1200/JCO.2010.31.8162. Epub 2011 Oct 3. Erratum in: J Clin Oncol. 2011 Dec 10;29(35):4725. PubMed PMID: 21969500.
5. Rowland A, Dias MM, Wiese MD, Kichenadasse G, McKinnon RA, Karapetis CS, Sorich MJ. Meta-analysis of BRAF mutation as a predictive biomarker of benefit from anti-EGFR monoclonal antibody therapy for RAS wild-type metastatic colorectal cancer. Br J Cancer. 2015;112(12):1888–94. https://doi.org/10.1038/bjc.2015.173. Epub 2015 May 19. Review. PubMed PMID: 25989278; PubMed Central PMCID: PMC4580381.
6. Romond EH, Perez EA, Bryant J, Suman VJ, Geyer CE Jr, Davidson NE, Tan-Chiu E, Martino S, Paik S, Kaufman PA, Swain SM, Pisansky TM, Fehrenbacher L, Kutteh LA, Vogel VG, Visscher DW, Yothers G, Jenkins RB, Brown AM, Dakhil SR, Mamounas EP, Lingle WL, Klein PM, Ingle JN, Wolmark N. Trastuzumab plus adjuvant chemotherapy for operable HER2-positive breast cancer. N Engl J Med. 2005;353(16):1673–84. PubMed PMID: 16236738.
7. Rugo HS, Olopade OI, DeMichele A, Yau C, van 't Veer LJ, Buxton MB, Hogarth M, Hylton NM, Paoloni M, Perlmutter J, Symmans WF, Yee D, Chien AJ, Wallace AM, Kaplan HG, Boughey JC, Haddad TC, Albain KS, Liu MC, Isaacs C, Khan QJ, Lang JE, Viscusi RK, Pusztai L, Moulder SL, Chui SY, Kemmer KA, Elias AD, Edmiston KK, Euhus DM, Haley BB, Nanda R, Northfelt DW, Tripathy D, Wood WC, Ewing C, Schwab R, Lyandres J, Davis SE, Hirst GL, Sanil A, Berry DA, Esserman LJ, I-SPY 2 Investigators. Adaptive randomization of veliparib-carboplatin treatment in breast cancer. N Engl J Med. 2016;375(1):23–34. https://doi.org/10.1056/NEJMoa1513749.
8. Park JW, Liu MC, Yee D, Yau C, van 't Veer LJ, Symmans WF, Paoloni M, Perlmutter J, Hylton NM, Hogarth M, DeMichele A, Buxton MB, Chien AJ, Wallace AM, Boughey JC, Haddad TC, Chui SY, Kemmer KA, Kaplan HG, Isaacs C, Nanda R, Tripathy D, Albain KS, Edmiston KK, Elias AD, Northfelt DW, Pusztai L, Moulder SL, Lang JE, Viscusi RK, Euhus DM, Haley BB, Khan QJ, Wood WC, Melisko M, Schwab R, Helsten T, Lyandres J, Davis SE, Hirst GL, Sanil A, Esserman LJ, Berry DA, I-SPY 2 Investigators. Adaptive randomization of Neratinib in early breast cancer. N Engl J Med. 2016;375(1):11–22. https://doi.org/10.1056/NEJMoa1513750.
9. Kim ES, Herbst RS, Wistuba II, Jack Lee J, Blumenschein GR Jr, Tsao A, Stewart DJ, Hicks ME, Erasmus J Jr, Gupta S, Alden CM, Liu S, Tang X, Khuri FR, Tran HT, Johnson BE, Heymach JV, Li M, Fossella F, Kies MS, Papadimitrakopoulou V,

Davis SE, Lippman SM, Hong WK. The BATTLE trial: personalizing therapy for lung cancer. Cancer Discov. 2011; https://doi.org/10.1158/2159-8274.CD-10-0010.

10. Papadimitrakopoulou V, Jack Lee J, Wistuba II, Tsao AS, Fossella FV, Kalhor N, Gupta S, Byers LA, Izzo JG, Gettinger SN, Goldberg SB, Tang X, Miller VA, Skoulidis F, Gibbons DL, Li S, Wei C, Diao L, Andrew Peng S, Wang J, Tam AL, Coombes KR, Ja SK, Mauro DJ, Rubin EH, Heymach JV, Hong WK, Herbst RS. The BATTLE-2 study: a biomarker-integrated targeted therapy study in previously treated patients with advanced non–small-cell lung cancer. JCO. 2016;34(30):3638–47.
11. Steuer CE1, Papadimitrakopoulou V, Herbst RS, Redman MW, Hirsch FR, Mack PC, Ramalingam SS, Gandara DR. Innovative clinical trials: the LUNG-MAP study. Clin Pharmacol Ther. 2015;97(5):488–91. https://doi.org/10.1002/cpt.88.
12. Lih CJ, Sims DJ, Harrington RD, Polley EC, Zhao Y, Mehaffey MG, Forbes TD, Das B, Walsh WD, Datta V, Harper KN, Bouk CH, Rubinstein LV, Simon RM, Conley BA, Chen AP, Kummar S, Doroshow JH, Williams PM. Analytical validation and application of a targeted next-generation sequencing mutation-detection assay for use in treatment assignment in the NCI-MPACT trial. J Mol Diagn. 2016;18(1):51–67. https://doi.org/10.1016/j.jmoldx.2015.07.006.
13. Moore KN, Mannel RS. Is the NCI MATCH trial a match for gynecologic oncology? Gynecol Oncol. 2016;140(1):161–6. https://doi.org/10.1016/j.ygyno.2015.11.003. Review.
14. Cappuzzo F, Ciuleanu T, Stelmakh L, Cicenas S, Szczésna A, Juhász E, Esteban E, Molinier O, Brugger W, Melezínek I, Klingelschmitt G, Klughammer B, Giaccone G. SATURN investigators. Erlotinib as maintenance treatment in advanced non-small-cell lung cancer: a multicentre, randomised, placebo-controlled phase 3 study. Lancet Oncol. 2010;11(6):521–9. https://doi.org/10.1016/S1470-2045(10)70112-1. Epub 2010 May 20. PubMed PMID: 20493771.
15. Hyman DM, Pazanov I, Subbiah V, Faris JE, Chau I, Blay JY, Wolf J, Raje NS, Diamond EL, Hollebecque A, Gervais R, Elez-Fernandez ME, Italiano A, Hofheinz RD, Hidalgo M, Chan E, Schuler M, Lasserre SF, Makrutzki M, Sirzen F, Veronese ML, Tabernero J, Baselga J. Vemurafenib in multiple nonmelanoma cancers with BRAF V600 mutations. N Engl J Med. 2015;373:726–36. https://doi.org/10.1056/NEJMoa150230.
16. Kim ES, Herbst RS, Wistuba II, Lee JJ, Blumenschein GR Jr, Tsao A, Stewart DJ, Hicks ME, Erasmus J Jr, Gupta S, Alden CM, Liu S, Tang X, Khuri FR, Tran HT, Johnson BE, Heymach JV, Mao L, Fossella F, Kies MS, Papadimitrakopoulou V, Davis SE, Lippman SM, Hong WK. The BATTLE trial: personalizing therapy for lung cancer. Cancer Discov. 2011;1(1):44–53. https://doi.org/10.1158/2159-8274.CD-10-0010. Epub 2011 Jun 1. PubMed PMID: 22586319; PubMed Central PMCID: PMC4211116.
17. Barker AD, Sigman CC, Kelloff GJ, Hylton NM, Berry DA, Esserman LJ. I-SPY 2: an adaptive breast cancer trial design in the setting of neoadjuvant chemotherapy. Clin Pharmacol Ther. 2009;86(1):97–100. https://doi.org/10.1038/clpt.2009.68. Epub 2009 May 13. PubMed PMID: 19440188.

Overview of Methods Used in Predictive Biomarker Studies in a Molecular Anatomic Pathology Laboratory

3

Perry Maxwell and Manuel Salto-Tellez

Introduction

A major contribution to robust predictive biomarker development and utilization is the successful management of laboratory protocols, managing uncertainty and gaining the enhancement of knowledge regarding pre-analytical, analytical, and post-analytical factors. These include sample handling prior to testing, including the request workflow and chemical events during the process (pre-analytical), the workflow of specimen testing and limits of tests (analytical), and the workflow of reporting the results in a meaningful way (post-analytical). Aspects of pre-analytical management can be problematic for referral centers, but it is recognized that knowledge of individual laboratory practice is essential. While nucleic acid-based molecular technologies and tissue-based, *in situ* hybridization technologies (including *in situ* hybridization and so-called molecular immunohistochemistry) are conceptually different and are often delivered in different laboratories, the quality of the sample affects them similarly, and, as such, they are also considered in this chapter as a whole.

Tissue and Cell Fixation, Processing, and Handling: Pre-analytical Variable Management

The most common pre-analytical variables include fixation and tissue processing when dealing with formalin-fixed paraffin-embedded (FFPE) material and transport, handling, freezing conditions, and storage when dealing with fresh frozen material.

Timely fixation is probably the most essential preclinical element to biomarker use. As a working definition, formalin traditionally refers to the use of a 10% solution of formaldehyde, itself soluble to 40% in water; therefore, 10% formalin is a 4% formaldehyde solution. Variations include using formalin alone either buffered or unbuffered or in conjunction with saline. Formalin fixation is time-dependent, with the longer the period exposed, the harder the tissue becomes; thus, gross dissection of large specimens follows a protracted period of formalin fixation. A minimum period of 6 h and a maximum of 48 h have been cited as being optimal, at least for samples in block format (see below) [1]. The effect of formalin fixation protocols on respective biomarkers by individual laboratory practices needs to be determined by the molecular laboratory as part of

P. Maxwell
Precision Medicine Centre of Excellence, Queen's University Belfast, Belfast, UK

M. Salto-Tellez (✉)
Northern Ireland Molecular Pathology Laboratory, Centre for Cancer Research, Department of Cell Biology, Queens University Belfast, Belfast, Antrim, UK
e-mail: m.salto-tellez@qub.ac.uk

S. Badve, G. L. Kumar (eds.), *Predictive Biomarkers in Oncology*, https://doi.org/10.1007/978-3-319-95228-4_3

their validation process. The use of alternatives to formalin need to be tempered with the need for diagnostic morphological imperatives where limited tissue restricts the casual use of alternatives and where formalin is the fixative of choice.

Although most pathology laboratories refer to using a "standard" processing schedule, it has long been known that this so-called standard does not exist, with variations in solution content and exposure times being used by individual laboratories, each imparting not only different physical properties to tissue blocks but with the potential to have an effect on protein and nucleic acid integrity.

Fresh frozen samples require special handling and transport conditions. Snap freezing in liquid nitrogen is traditionally used, but problems associated with transport limit its use. Several solutions, e.g., isopentane or 30% sucrose are available, claiming to facilitate the maintenance of molecular integrity during the freezing process, but testing and knowledge of sample handling need to be identified by the molecular laboratory.

RNA and translocation targets will require transcription to cDNA which is more stable for storage than the original RNA sample, requiring monitored storage at −80 °C. Where RNA is the target in tissue sections, *in situ* RNAse activity needs to be minimized during all steps of the preanalytical and analytical processes.

In samples of bone content, either as whole bone samples or calcium content through normal or pathological processes, decalcification is a requirement prior to sectioning. Mineral acids are unsuitable prior to molecular testing. Solutions of EDTA or non-mineral acid-containing commercial solutions may be used and closely monitored for decalcification, minimizing exposure.

Whatever the system in place, it is incumbent upon the testing laboratory to gain knowledge regarding sample handling prior to nucleic acid extraction or slide staining. With such knowledge, the laboratory gains the ability to manage uncertainty and limit risks associated with these factors.

Types of Sample and Tumor Evaluation: Test Validation as Part of Analytical Variable Management

Irrespective of fixation and processing, the type of sample often determines the primary handling pathway for samples. For example, larger tissue, e.g., lobectomies, resection samples of the colon will need to be dissected to yield blocks of a size to facilitate optimum fixation and processing, typically blocks measuring approximately 2.5 × 2.5 × 1.0 cm. The optimal window for formalin fixation of between 6 and 48 h only applies to such blocks, and if larger resection cases are utilized, then gross dissection protocols need to account for biomarker downstream applications [2].

From the large to the small, any type of cell/tissue sampling may be utilized for biomarker stratification. For smaller biopsies and cytology cell blocks, optimal sectioning protocols need to account for the needs of diagnostic and therapeutic molecular studies. The sections taken between levels need to be retained and maintained under suitable conditions to reduce target degradation over time and the risk of contamination between stored samples. One of the most interesting paradoxes in modern diagnostics is that, in those referral centers with numerous feeding peripheral hospitals, and once the sample has been accepted as satisfactory for testing after morphological evaluation, the smaller samples (biopsy, cytology) typically have a smaller percentage of "unsatisfactory" tests because, despite of the smaller volume, controlling adequate fixation seems to be easier [3].

The use of a morphological assessment and tumor evaluation is essential in the identification of the material being tested. As a minimum, the tumor content and the approximate number of total cell population (both neoplastic and nonneoplastic) should be estimated. The tumor content will guide the validity of the test result within the lower limit of detection as determined for the test during the laboratory's validation protocol; the total cell/nuclei count will determine the

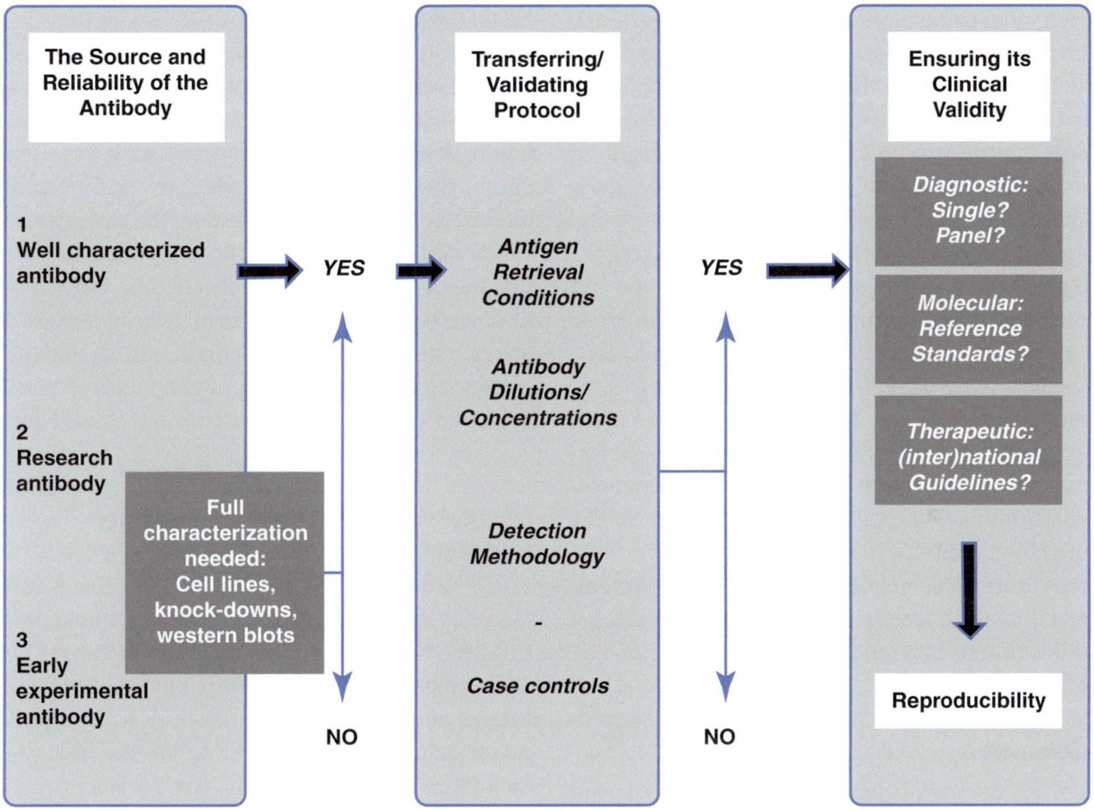

Fig. 3.1 Immunohistochemistry biomarker validation. (Reprinted from Maxwell and Salto-Tellez [5])

validity of the test result within the lower limit of nucleic acid content as determined by the manufacturer and/or the laboratory's validation protocol.

Analytical variables may be managed through a full validation protocol. Where a CE-IVD kit or test is provided, a thorough verification process may suffice. The determination of the level of verification and/or validation required needs to be determined by each laboratory in line with national or international accreditation standards [4]. For molecular test parameters such as accuracy, precision, and reproducibility, the lower limit of detection, the minimum amount of substrate required for a successful analyte, and the coverage of the test should be identified as part of the validation protocol.

For immunostaining in both tissues and cytology preparations, we have identified three aspects of the validation process: what is known about the primary antibody; is there a working protocol; and is the result adding to clinical benefit. This last point is particularly relevant when the antibody is not mainstream (as in the case of many clinical trials) or when it is early days in the delivery of a key new antibody (as is the case of PD-L1 immunohistochemistry at time of writing) (Fig. 3.1).

Biorepositories

Modern biorepositories are departing from the idea of existing as "collections of samples" to become "repositories of science" [6]. In any case, clinical studies in the testing of predictive biomarkers require robust clinical data associated with median survival time (MST) and/or other end-point parameters in order to develop robust predictive biomarker adoption and use. Such studies may use the biomarker to predict factors such as MST following the use of targeted therapy as a companion diagnostic test, as diagnosis, or as a means of independent prognostic significance. They also require sufficient numbers of the correct case selection with access to relevant clinical data. The use of well-maintained and well-governed biorepositories or tissue biobanks can provide such resources either initiated for a specific study or with biomarker discovery as its purpose. Of course, these need not be mutually exclusive, and the design of the banking facility may attempt to future-proof with a multipurpose constitution. Biorepositories and tissue banks, therefore, face challenges of consent; aims of the repository and conditions of storage; curation, governance of storage, and access; and managed utilization of facilities.

Consent

At the heart of all procedures is the concept of patient consent. It has been found that when questioned, patients overall are supportive of using tissue to benefit research activity although variation in willingness must be recognized to be extant within different socioeconomic and sociodemographic groups. Consent often comes with the price of guaranteed anonymity and de-identification of data and the nondisclosure of private healthcare data. These are manageable in prospective, single-purpose collections, but where secondary purposes arise as either foreseen or unforeseen opportunity, then secondary consent may be required or initial generic consent accounting for unspecified future research is desirable.

Aims of the Repository and Storage Conditions

The pre-analytical handling and subsequent processing and storage conditions of samples must match the aims and purpose of the repository or tissue bank. Moreover, the monitoring of the conditions including environmental humidity for FFPE blocks, continuous monitoring of temperatures of fridges and freezers with the aim of maintaining the integrity of samples in such a way as to allow them to be accessible to downstream molecular technologies need to be considered. The diverse nature of such technologies dictates that the quality metrics in use to assess suitability need to be identified and matched. As outlined above, the pre-analytical conditions should be recorded and available to researchers and recorded as part of procedure optimization and validation.

Curation, Governance of Storage, and Access

Once collected, the curation of samples and clinical data should record and audit the processes by which these are used by individual researchers. Ethical and scientific committees sit with respective organizations to determine access and it is incumbent upon the bank to provide material in format suitable for downstream application but it is the responsibility of individual researchers to ascertain that these conditions and material are suitable for the purpose intended. The collection and curation of clinical data should be robust with access restricted to researchers with appropriate permissions.

Managed Utilization of Facilities

Management of the utilization of the repository should include the recording of use by the respective researchers and how this utilization matches with the aims of the repository and are in keeping with original ethics governing the use for the respective samples (Table 3.1).

Technologies

Predictive biomarker medicine requires, therefore, the identification if not total control of pre-analytical factors and may resolve half of the

Table 3.1 Examples of organizations hosting biorepository collections

Repository/biobank/organization	Country	Website(s)
NIH NeuroBioBank	USA	https://neurobiobank.nih.gov/
		https://neurobiobank.nih.gov/contact/
Cooperative Human Tissue Network (NIH/NCI-supported)	USA	https://www.chtn.org/
NIH/NCI, Biorepositories and Biospecimen Research Branch (BBRB)	USA	https://biospecimens.cancer.gov/default.asp
Centre for Cancer Research and Cell Biology, Queen's University Belfast	UK	http://www.nibiobank.org
British Association for Tissue Banking (BATB)	UK and Ireland	http://www.batb.org.uk/
European Association of Tissue Banks (EATB)	Europe	http://www.eatb.org/
NCT, Nationales Centrum für Tumorerkrankungen	Germany	https://www.nct-heidelberg.de/forschung/nct-core-services/nct-tissue-bank.html
National University Hospital (NUH) Tissue Repository	Singapore	http://medicine.nus.edu.sg/tissue/links.htm
RIKEN Center for Genomic Medicine	Japan	http://www.src.riken.jp/english/outline/index.html
ABN, Australasian Biospecimen Network	Australia	http://abna.org.au

equation in the delivery of analysis; the provision and application of the technology(ies) appropriate to the analysis are essential. Detailed technology descriptions can be found elsewhere in this book. Where gel-based technologies such as Southern and Western blotting have largely been replaced by PCR- and digital-based technologies, we briefly describe respective technology definition and successful application at time of writing. In a fast-evolving area such as precision medicine, the introduction of new single and rational combinative therapies will make use of matching combinative technologies for cost-effective therapeutic prediction strategies.

Immunohistochemistry

First introduced in the 1940s, protein analysis *via* antibody-antigen reactions *in situ* on the slide became a staple in routine pathology services since the mid-1970s.

Immunohistochemistry (IHC) has been a very cost-effective means of predictive biomarker use in diagnostic, prognostic, and therapeutic scenarios. It is traditionally applied to formalin-fixed paraffin wax-embedded (FFPE) tissues from resection, biopsy, and cytology samples although formalin alternatives are sought such as generic fixatives (e.g., alcohol) or off-the-shelf, commercial fixatives. Modern pathology tends to differentiate between the descriptive (diagnostic) IHC and the semiquantitative (therapeutic) IHC, the latter being a key component of the molecular diagnostic armamentarium [7]. It has been successfully used for the stratification of patients likely to benefit from trastuzumab therapies in breast and gastric cancers. Where there are equivocal results, combining with *in situ* hybridization can identify patients for therapy and perhaps more importantly, patients where alternative standards of care may be beneficial and cost-effective. More recently, the identification of immune checkpoint proteins by IHC may be used in the predicting response to therapies where immune surveillance may be switched off.

In Situ Hybridization

Genomic *in situ* hybridization technology either in tissue sections (interphase nuclei) or as part of a cytogenetic analysis (metaphase nuclei) for gene amplification, traditionally utilizes fluorescent labels; alongside the identification of the carrier chromosome, it provides a ratio greater than 2.2 indicating amplification or, in the case of translocation, the identification of the affected gene locus in isolation as a break-apart label or in a new combination as a combinative label. Such technologies have the advantage of semiquantitation and have been successfully used for Herceptin therapy and ALK and MET inhibitors for targeted therapies or alongside quantitative PCR (QPCR) technologies in sarcoma or lymphoma.

Sanger Sequencing, Quantitative PCR, and Pyrosequencing

Biomarker identification using PCR reactions depends upon the amplification of gene sequences, the length and coverage of which determine the type of gene testing ranging in target from "hotspot" short sequences as single gene tests or longer whole gene, exome, or genome tests. From quantitative PCR (QPCR) or real-time PCR (RT-PCR) and pyrosequencing through Sanger sequencing to next-generation sequencing (NGS), the range of DNA-based technologies has been successfully used for precision medicine targets such as the stratification of patients for tyrosine kinase inhibitors (TKIs), selective monoclonal antibody therapy, and other small molecule inhibitor therapies. RNA-based technologies from fusion transcripts in sarcoma, lymphoma, and leukemia can be used for diagnostic purposes using reverse-transcriptase PCR (also confusingly known as RT-PCR) along with the stratification of patients in cancers such as adenocarcinoma of the lung for the prediction of likely benefit from ALK inhibitors.

Next-Generation Sequencing (NGS): Targeted, Exome, and Whole Genome

The establishment of NGS in routine diagnostic settings is still an area of interesting debate [8], as is the discussion of the minimum requirements for NGS diagnostic validation [9]. NGS technologies offer the means by which a number of genes may be identified. NGS has the advantage over single gene tests in that there is fixed cost in their preparation and sequencing, whether it be for 10 genes or 100. Only in the panel does the cost alter, whereas the cost of ten single gene tests is proportionately more expensive. The gene panels which go into an NGS design for predictive biomarker use have been classified into those genes where the identification of mutations may be actionable, potentially actionable, and awaiting appropriate therapy development or as a means of discovery. Targeted known "hotspot" regions of between 150 and 250 bases have the benefit of use in FFPE material due to the fragmentation of DNA brought about through formalin fixation. Typically, FFPE is not suitable for targets longer than 250 bases and alternative fixatives or fresh tissue need to be sourced [10]. These of course have the disadvantage of requiring bespoke pre-analytical protocols. The potential of PARP inhibitors where large genes such a BRCA1 and 2 require such technologies may also necessitate robust post-analytical protocols in the handling of patients and relatives in the light of results. Findings incidental to the purpose of the test may also be a consequence of handling larger NGS panels, and these, too, should have robust policies in place in describing the bioinformatics pathway protocols and clinical reporting environments.

Gene Expression Arrays and Sequencing

Predictive biomarker discovery from RNA utilizes important potential technologies in gene expression arrays and sequencing. Array technologies and associated bioinformatics pathways and programs offer insights into potential transcription targets which can be matched with proteomic studies. Moreover, RNA sequencing elucidates transcription variants which when combined with robust clinical data management and curation leads to robust biomarker discovery.

Microarray-based comparative genomic hybridization (aCGH) is a technology which can be used to analyze the complex karyotypes comprising multiple genomic abnormalities. This technology is based on the same principles as metaphase CGH where both techniques allow for the genome-wide study of DNA copy-number alteration. Interrogation of the genome can identify molecular drivers in cancer and hence potential therapeutic targets at an initial stage of biomarker discovery [11].

The evolution of a cancer can lead to tumor heterogeneity [12], and discovering new biomarkers which have the potential as therapeutic targets requires sensitive and reproducible technologies which may be in the interface between large targeted analyses and microarray technology, such as NanoString and digital droplet PCR [13]. Digital droplet technology, for example, separates individual molecules in oil droplets in increasing dilutions and the absolute quantification of relative amplified analytes, a key advantage over conventional PCR technologies with increased reproducibility although of similar sensitivity [14]. Such technologies are being used at the time of writing to explore the range of neoantigens produced during the evolution of cancer, especially in the field of immune therapy [15].

Circulating Tumor, Cell-Free DNA

Cells have the potential to shed DNA fragments either through induced or natural processes into circulating peripheral blood. In the presence of malignancy therefore, tumor DNA fragments enter the bloodstream but at a small fraction of the total DNA in circulation. The potential, however, to use the plasma as biomarker either in the discovery or clinical setting remains to be fully evaluated, but what must be considered are all the pre-analytical factors such as sample integrity and false-negative and false-positive rates, analytical factors such as technology limitations, and post-analytical factors such as reporting context. Single gene QPCR techniques can be used for gene hotspot detection such as those introduced to detect mutations resistance in the EGFR gene in response to TKI therapy. Multiple gene technologies such as digital droplet PCR can be used for larger biomarker profiling or biomarker discovery. To emphasize, the efficacy of using cell-free (cfDNA) or circulating tumor DNA (ctDNA) as a biomarker will depend on the integrity of the sample, especially in eliminating non-tumor DNA from the sample and the relative sensitivities of the methods used.

Conclusion

The demise of morphological pathology in favor of a molecular taxonomy of disease has been announced several times in the last two decades, and, despite these learned opinions, the morphological analysis of the disease remains the cornerstone of tissue-based diagnostics. Indeed, there is no good-quality molecular diagnostics without good-quality morphological assessment of the disease. This morphomolecular approach [16], which is at the heart of much of the content of this book, may dictate modern pathology and modern diagnostics for many years to come.

References

1. Yildiz-Aktas IZ, Dabbs DJ, Cooper KL, Chivukula M, McManus K, Bhargava R. The effect of 96-hour formalin fixation on the immunohistochemical evaluation of estrogen receptor, progesterone receptor, and Her2 expression in invasive breast carcinoma. Am J Clin Pathol. 2012;137:691–8.
2. Babic A, Loftin IR, Stanislaw S, Wang M, Miller R, Warren SM, et al. Methods. 2010;52:287–300.
3. Pang B, Dettmer M, Ong CW, Dhewar AN, Gupta S, Lim GL, et al. The positive impact of cytological specimens for EGFR mutation testing in non-small cell lung cancer: a single South East Asian Laboratory's analysis of 670 cases. Cytopathology. 2012;23:229–36.
4. Mattocks CJ, Morris MA, Mathijs G, Swinnen E, Corveleyn A, Dequeker E, et al. A standardized framework for the validation and verification of clinical molecular genetic tests. Eur J Hum Genet. 2010;18:1276–88.
5. Maxwell P, Salto-Tellez M. Validation of immunocytochemistry as a morphological technique. Cancer Cytopathol. 2016;124:540–5.
6. Lewis C, McQuaid S, Hamilton PW, Salto-Tellez M, McArt D, James JA. Building a "repository of science": the importance of integrating biobanks within molecular pathology programmes. Eur J Cancer. 2016;67:191–9.
7. McCourt CM, Boyle D, James J, Salto-Tellez M. Immunohistochemistry in the era of personalised medicine. J Clin Pathol. 2013;66:58–61.

8. Hynes S, Pang B, james J, Maxwell P, Salto-Tellez M. Tissue-based next generation sequencing: application in a universal healthcare system. Br J Cancer. 2017;116:553–60.
9. Salto Tellez M, Gonzalez de Castro D. Next-generation sequencing: a change in paradigm in molecular diagnostic validation. J Pathol. 2014;234:5–10.
10. Wong SQ, Li J, AY-C T. Sequence artefacts in a prospective series of formalin-fixed tumours tested for mutations in hotspot regions by massively parallel sequencing. BMC Med Genet. 2014;7:23–33.
11. Tan SP, Natrajan R, Reis-Filho JS. Microarray-based comparative genomic hybridization. In: Hannon-Fletcher M, Maxwell P, editors. Advanced techniques in diagnostic cellular pathology. Oxford: Wiley-Blackwell; 2009.
12. Turajlic S, Swanton C. Metastasis as an evolutionary process. Science. 2016;352:169–75.
13. Geiss GK, Bumgarner RE, Birditt B, et al. Direct multiplexed measurement of gene expression with color-coded probe pairs. Nat Biotechnol. 2008;26:317–25.
14. Hindson CM, Chevillet JR, Briggs HA, et al. Absolute quantification by droplet digital PCR versus analog real-time PCR. Nat Methods. 2013;10:1003–5.
15. Brown SD, Warren RL, Gibb EA, et al. Neo-antigens predicted by tumor genome meta-analysis correlate with increased patient survival. Genome Res. 2014;24:743–50.
16. Salto-Tellez M. A case for integrated morpho-molecular diagnostic pathologists. Clin Chem. 2007;53:1188–90.

Significance of Immunohistochemistry and *In Situ* Hybridization Techniques for Predictive Biomarker Studies

4

Hans-Ulrich Schildhaus

Introduction

This chapter outlines the basic principles of clinical applications of immunohistochemistry and various *in situ* hybridization techniques in the context of predictive biomarkers. The text is confined to general and technical aspects of the individual methodologies and their applications on patients' biomaterials. Common malignancies have been used to serve as examples for the clinical use of these techniques and in-depth information on specific tumor entities are detailed in the respective chapters of this book.

Immunohistochemistry (IHC) and *in situ* hybridization (ISH) both share the basic principles of keeping the morphologic context (tissues or cells) to investigate the disease-associated mechanisms for diagnostic or predictive use. For example, in IHC, protein expression can be directly evaluated to include the identification and localization of differentially expressed proteins and biomarkers within normal and diseased tissues, whereas, in ISH, changes to nucleic acids, i.e., detection of numerical or structural chromosomal or gene level aberrations, can be detected either using fluorescent or chromogenic markers. Both these methods continue to be widely used, and their applications have been rapidly expanding. IHC and ISH have a fundamental advantage over the grind and bind techniques such as sequencing. In sequencing, the genomic findings are averaged among all cell types (tumor and microenvironment) in a sample, whereas in IHC and ISH, aberrations within individual cells can be accounted. However, modalities and complexity of changes which are detectable by IHC and ISH are limited. Technological progress in the next few years will lead to fundamental improvements in both methods and, thus, widen clinical applications. Multiplexing strategies will allow detection of multiple proteins and/or multiple genes simultaneously and directly on the same slide in the same tumor cell.

Currently, immunohistochemistry and *in situ* hybridization represent the most used technologies for predictive biomarker assays, and the majority of companion diagnostics are based on these methods. Despite the fact that most pathologists are quite familiar with these types of assays, specific knowledge and training is mandatory to achieve the highest level of reliability. Technical aspects, as well as assessment and interpretation of findings, require continual improvement – always keeping in mind that the clinical management of patients is directly dependent on these assays. For a cartoon of IHC and ISH, refer to Figs. 4.1 and 4.2.

H.-U. Schildhaus
Institute of Pathology, Universitätsmedizin Göttingen, Göttingen, Germany
e-mail: hans-ulrich.schildhaus@med.uni-goettingen.de

S. Badve, G. L. Kumar (eds.), *Predictive Biomarkers in Oncology*,
https://doi.org/10.1007/978-3-319-95228-4_4

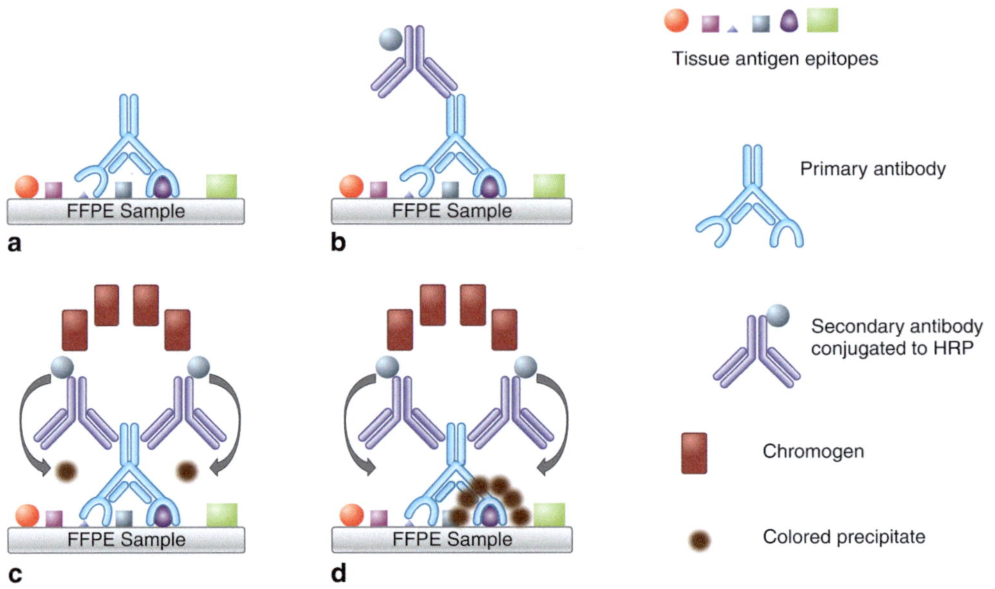

Fig. 4.1 Basic principle of immunohistochemistry. (**a**) Primary antibody recognizes specifically epitopes in the tissue (purple). (**b**) Secondary antibody binds to primary antibody. Application of large polymers can magnify and enhance staining signals. Secondary antibodies carry an enzyme, e.g., horse radish peroxidase (HRP). Another frequently used enzyme is alkaline phosphatase. (**c**) Enzymatic activity changes the color of the chromogen. A commonly used chromogen for HRP is 3,3′-diaminobenzidine (DAB). (**d**) Colored substrate precipitates around the epitope. There is a magnitude of chromogens in various colors available which enable double and multiplex stainings

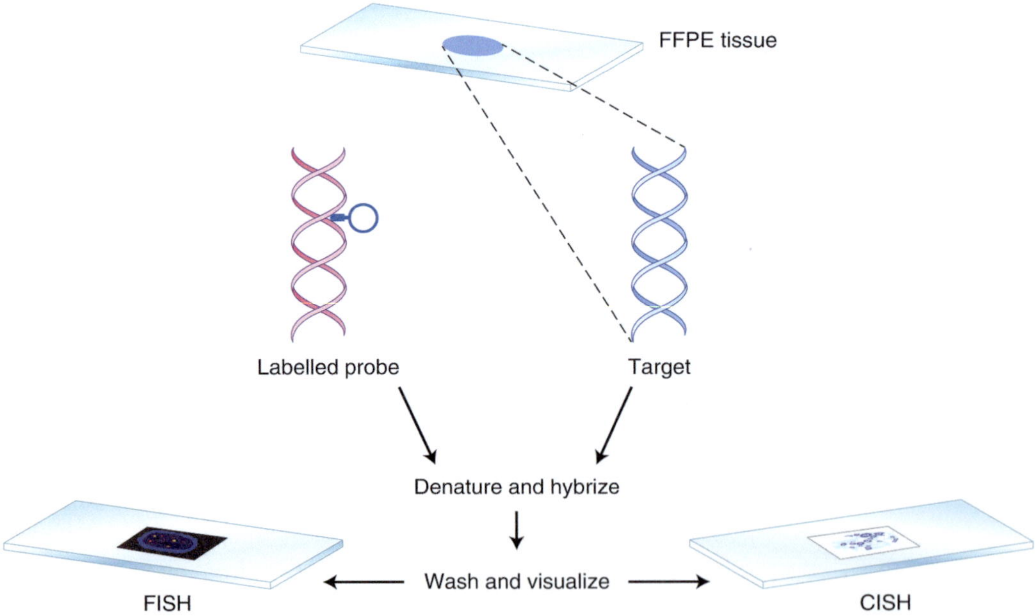

Fig. 4.2 Principle of *in situ* hybridization (ISH) techniques. Probes are DNA fragments (single or double stranded) which can bind specifically to nucleic acids (DNA or RNA) in the FFPE tissue. Probes are labeled with chromogens or fluorescent dyes to enable microscopic visualization. After simultaneous denaturation of both probe and tissue nucleic acids directly on the slide, probe molecules bind (hybridize) to complementary sequences in the tissue. Stringent washing steps remove nonspecifically bound probe molecules. Specific hybridization signals can be detected either as a fluorescent signal under a fluorescence microscope (FISH) or as a chromogenic signal (CISH) in a bright-field microscope. See text for details

General Considerations on Immunohistochemistry in the Context of Predictive Biomarker Assay: Staining Methods and Evaluation

Since decades, immunohistochemistry (IHC) has been widely used in most pathology departments. Many institutes stain several tens of thousands of slides per year, and there is a voluminous and still growing body of scientific literature on IHC applications. This high level of knowledge and personal experience together with the availability of highly specific primary antibodies and improved staining protocols makes immunohistochemistry a highly accessible method. IHC represents a technique to detect small regions (epitopes) of proteins or peptides in target tissues, mostly cancer cells or structures of the surrounding stroma. Epitopes can be localized in all cell compartments, i.e., the nucleus including nuclear membranes, in the cytoplasm or in the cell membrane. Knowledge on the intracellular distribution and trafficking of proteins helps to recognize specific staining. For example, transcription factors or steroid hormone receptors are usually expressed in the nucleus, and only this type of staining is regarded specific. However, certain disease-associated alterations may lead to a change in the subcellular localization of proteins. So, rearrangements of the anaplastic lymphoma kinase (*ALK*) gene in lung cancer usually lead to a more cytoplasmic expression of this receptor tyrosine kinase since the loss of the membrane binding domain is part of this gene fusion.

In routine clinical application, IHC has two different aims:

(i) *Diagnostic antibodies* (sometimes referred to as class I antibodies) are used as an adjunct in general diagnostic procedures. These antibodies provide general qualitative information on a yes-or-no basis. Staining results are used as an argument for or against a suspected diagnosis, e.g., of a specific entity, of tumor spread, etc. These findings are usually interpreted in the light of clinical information, morphologic appearance of a tumor, and other IHC results. Therefore, high specificity and an appropriate sensitivity should characterize these antibody-based stains.

(ii) *Antibodies for predictive biomarkers* (or class II antibodies), however, are used to select patients for certain cancer treatments in the context of personalized therapies or precision oncology (Table 4.1). Many of these markers may also have prognostic connotation(s). Class II antibodies frequently provide quantitative or semiquantitative measurements which help to categorize cancer patients (with an already established diagnosis of a certain entity) for tailored treatments. However, these biomarkers usually serve as single assay without additional supporting tests. Therefore, the quality requirements for these immunostains are especially high. Pathologists should demonstrate not only a high specificity and sensitivity but also a broad dynamic range of the assay to enable reliable (semi)quantitative evaluation. Furthermore, repeatability and robustness of the assays are mandatory. Many antibody assays in this context are used as part of companion diagnostics or complementary biomarkers for certain drugs.

Given the utmost importance of a high level of standardization and quality assurance of predictive IHC biomarkers, it is noteworthy that there are several factors that may influence and hamper the correct use of these assays [1].

Pre-analytical Factors

Pre-analytical factors have been proven to have a dramatic impact on the staining results. These factors include time of cold ischemia (i.e., the time between tissue removal from the patient's body and fixation), time and conditions of fixation, and tissue dehydration protocols. In general, it is recommended to fix tissues immediately in neutrally buffered formalin. Other fixatives including alcohol-based formulations have been shown to hamper integrity of epitopes in the tissue. Since a 4% formalin solution fixes human

Table 4.1 Commonly used immunohistochemistry assays for predictive biomarker studies

Disease	Protein detection	Aberration	Result
Gliomas	IDH1	IDH-R132H (mutation-specific antibody)	Mutated protein (similar to other IDH1/2 mutations which are not detectable by IHC) is an indicator of better outcome. Predictive value for treatment with small molecules is under debate
Breast cancer	Estrogen receptor (ER)	Semiquantitative measurement of expression level	High ER/PgR levels and absence of HER2 expression indicate a better outcome. ER/PgR expression provides the basis for endocrine treatment. Her2-positive cancers (IHC 3+, or IHC2+ with ISH assay demonstrating gene amplification) are subject to antibody treatment
	Progesterone receptor (PgR)		
	HER2		
Malignant melanoma	BRAF	BRAF-V600E (mutation-specific antibody)	BRAF-V600 mutated melanomas respond to BRAF inhibitors, e.g., vemurafenib. There are more genomic variants at codon V600 which may also be predictive
	PD-L1	PD-L1 (tumor proportion score, TPS; other methods to quantify PD-L1 expression, e.g., MEL score)	The role of PD-L1 testing in melanomas is still debatable. Data from a clinical trial indicate that low or absent PD-L1 expression (TPS < 5%) is associated with better response to combination therapy of nivolumab plus ipilimumab compared with nivolumab alone
Head and neck cancer	p16	Strong and diffuse p16 expression serves as a surrogate marker for HPV infection in oropharyngeal squamous cell carcinomas	HPV infection is basically a prognosticator for a better outcome. Based on this observation, de-escalating treatment regimens are applied (less modalities, de-intensified doses, local/robotic surgery)
	PD-L1	PD-L1 expression in tumor and inflammatory cells (tumor proportion score [TPS] and combined positivity score [CPS])	I/O treatments have been introduced to head and neck cancer. Early data from clinical trials point toward a response to pembrolizumab in PD-L1-positive tumors. As in many other indications, PD-L1 can also serve as a complementary biomarker for alternative I/O drugs
Gastric cancer and adenocarcinomas of esophagogastric junction	HER2	HER2 overexpression (IHC 3+) or amplification (IHC2+ plus gene amplification measured by ISH)	HER2-positive tumors respond to trastuzumab (caveat: also, focal overexpression/amplification is predictive)
	PD-L1	PD-L1 expression in tumor and inflammatory cells (measured by CPS; cutoff ≥1)	PD-L1-positive tumors may respond to I/O treatment. Pembrolizumab has been demonstrated to be effective in PD-L1-positive tumors
	Loss of DNA mismatch repair proteins (dMMR)	Loss of MLH1, PMS2, MSH2, or MSH6 (dMMR) is indicative of MSI	MSI carcinomas may respond to I/O treatment. Pembrolizumab is approved for this indication (entity agnostic)
Bladder cancer (urothelial carcinoma)	PD-L1	PD-L1 expression in tumor and/or immune cells	PD-L1 can serve as a complementary biomarker for I/O drugs
Endometrial carcinomas	Loss of DNA mismatch repair proteins (dMMR)	Loss of MLH1, PMS2, MSH2, or MSH6 (dMMR) is indicative of MSI	MSI carcinomas may respond to I/O treatment. Pembrolizumab is approved for this indication (entity agnostic)

Table 4.1 (continued)

Disease	Protein detection	Aberration	Result
Lung cancer	ALK	ALK protein overexpression is indicative of an underlying ALK gene rearrangement (NSCLC, adenocarcinomas)	Lung cancers with ALK rearrangement respond to TKI treatment, e.g., alectinib, crizotinib, or ceritinib. D5F3 antibody staining is approved as a predictive biomarker. Other stainings are used as prescreening test and require subsequent confirmation by ISH or sequencing techniques
	ROS1	ROS1 expression is indicative of a ROS1 gene rearrangement (NSCLC, adenocarcinomas)	Patients with ROS1 gene rearrangement-positive tumors respond to TKI treatment, e.g., crizotinib. Additional TKI is under clinical development. ROS1 IHC is currently used as a prescreening test. Confirmation by ISH or sequencing is recommended
	EGFR	EGFR protein expression in squamous cell carcinomas	EGFR expression is independent of EGFR mutation. Effect of necitumumab, a monoclonal anti-EGFR antibody, in combination with chemotherapy has been shown in pulmonary squamous cell carcinomas. Approval by the European Medicines Agency requires proof of EGFR expression by IHC
	PD-L1	PD-L1 expression on tumor cells, NSCLC (tumor proportion score, TPS)	Antitumor effect of several I/O drugs in NSCLC has been demonstrated to be associated with but not restricted to PD-L1 positivity. Therefore, PD-L1 can serve as a complementary biomarker for I/O therapies. Pembrolizumab is approved for the first-line treatment of PD-L1 highly expressing NSCLC (TPS ≥ 50%) and for second or higher therapy lines in PD-L1-positive NSCLC (TPS ≥ 1%)
	DLL3	DLL3 expression in small-cell and large-cell neuroendocrine carcinomas	Emerging biomarker, currently explored in the context of treatment with rovalpituzumab tesirine, an anti-DLL3 antibody-drug conjugate
	NTRK	NTRK1-3 expression in NSCLC, measured with a pan-NTRK antibody (surrogate marker for rearrangements of NTRK1 or NTRK3)	Emerging biomarker, currently explored in the context of treatment with entrectinib, a TKI. IHC is used as prescreening with subsequent confirmation by sequencing
Lymphomas	CD20	CD20 expression in B-cell lymphomas	Treatment of B-cell lymphomas with the monoclonal antibody rituximab is based on proof of CD20 expression
	CD30	CD30 expression in Hodgkin's lymphoma (HL), anaplastic large-cell lymphomas (ALCL), and cutaneous large-cell lymphoma or mycosis fungoides	Brentuximab vedotin, an anti-CD30-directed antibody-drug conjugate, is approved for the treatment of relapsed CD30-positive lymphomas

ALK anaplastic lymphoma kinase, *CPS* combined positivity score, *DLL3* delta-like 3 protein, delta-like canonical notch ligand 3, *dMMR* mismatch repair deficiency, *EGFR* epithelial growth factor receptor, *HPV* human papillomavirus, *IHC* immunohistochemistry, *I/O* immuno-oncology, *ISH* in situ hybridization, *MSI* microsatellite instability, *NSCLC* non-small cell lung cancer, *NTRK* neurotropic tropomyosin receptor kinase, *PD-L1* programmed death-ligand 1, *TKI* tyrosine kinase inhibitor tissues at 1 mm per hour, a minimum fixation time of 6 h is generally recommended. On the other hand, overfixation (longer than 48 h) should be avoided since formalin causes structural changes of proteins by cross-linking of amino groups. Whenever possible, decalcifications should be avoided, as harsh protocols (at low pH) can result in loss of epitopes.

It may be of interest for the practicing pathologists that tissue-specific changes of IHC assays due to pre-analytical factors cannot be monitored or recognized by on-slide controls or run controls.

Factors that Influence Staining

These factors include mainly the choice of an appropriate antibody and a matching and suitable staining protocol. Monoclonal antibodies provide a higher level of specificity compared with polyclonal antibodies and are therefore generally preferred. For many applications, there are sometimes dozens of monoclonal antibodies available, which may generate highly discordant staining results. Therefore, suitable antibodies for class II assays should be carefully selected based on generally accepted recommendations for specific clinical applications of interest. Moreover, staining protocols, e.g., selection of staining platform, duration and pH of heat-induced epitope retrieval, time and concentration of primary antibody incubation, and selection and application of detection systems (such as two- or three-step polymers, use of enhancers and amplifying reagents), and counterstaining may severely influence the staining result. In the context of predictive biomarkers, ready-to use staining kits which include all components for an immunostain should generally be preferred. Recommended protocols are also available from ring trial organizations, such as NordiQC (www.nordiqc.org). It is now widely accepted that fully automated immunostainers are superior to half-automated staining platforms or manual stainings in terms of standardization, reproducibility, and overall quality of assays. Use of stored slides should be avoided since aging may cause degradation of epitopes.

Reading and Scoring of a Staining

Immunohistochemistry for predictive biomarkers should be evaluated strictly following generally accepted rules and guidelines. This may include different scoring approaches for the same antibody staining depending on tumor entity and clinical question (see examples of HER2 and PD-L1 in Figs. 4.3 and 4.4) [2]. Evaluating pathologists should undergo specific training to ensure acceptable intra- and interobserver variability.

Reporting of Staining Results

Also, reporting of IHC findings in the setting of precision oncology should be standardized to facilitate their translation into clinical actions. Knowledge about therapeutic consequences of specific results helps to precisely describe IHC findings. Several guidelines encourage pathologists to include also characteristics of the staining methodology in the report (e.g., designation of antibody and staining platform, applied scoring rules, etc.).

Specifically Modified Antibodies

In the past few years, mutation-specific antibodies have been introduced. One commonly used example is the BRAF-V600E specific staining which is used to detect malignant melanomas, colorectal carcinomas, gastrointestinal stromal tumors, and other malignancies with this therapeutically actionable mutation. The specificity of this staining is usually high. However, the sensitivity is highly dependent on the staining protocol. Since patients with a BRAF-mutated metastatic melanoma can benefit from an anti-BRAF compound (such as vemurafenib), this staining needs to be vigorously validated if the final clinical decision is solely based on it. Many

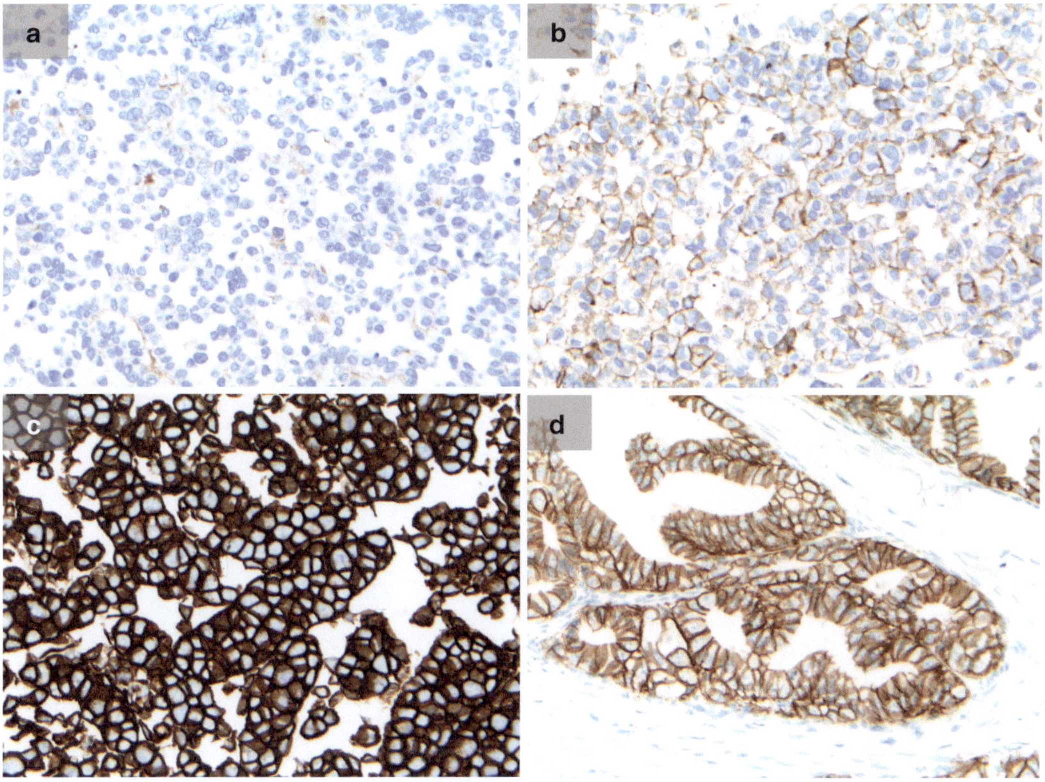

Fig. 4.3 Predictive IHC assays frequently require adherence to specific rules for assessment and evaluation. (**a–d**) Various expression levels of her2. In breast cancer, only complete circumferential membranous staining is regarded as positive. (**a**) Incomplete and weak staining, however, is interpreted as negative (score 1+). Moderate (**b**, score 2+) and strong (**c**, score 3+) complete staining in breast cancer. In the setting of predictive biomarkers, cases with 2+ staining are tested by ISH for Her2 gene amplification. (**d**) Contrarily, incomplete basolateral staining ("U-type") is regarded to be predictive in gastric cancer

of these mutation-specific antibodies have been developed by neuropathologists. Among these, IDH1 antibody, which detects specifically the R132H-mutant IDH1 variant in gliomas, is widely used. This example nicely illustrates the advantage as well as a major limitation of mutation-specific IHC. It simplifies genetic testing by detecting mutated protein by a fast and easy surrogate marker. In contrast, the staining is "false negative" if another mutation with the same or a comparable biologic effect occurs due to replacement by another amino acid. At least seven additional genomic variants of IDH1 and IDH2 exist in gliomas, which are not detectable by an IDH1-R132H specific antibody (these variants are, however, by far less frequent).

Another modification of IHC is the use of phospho-antibodies which specifically detect phosphorylated proteins. This allows recognition of activated enzymes such as phosphorylated tyrosine kinases or downstream elements in the signaling cascade. These phospho-antibodies are well recognized in research applications (e.g., in Western blot experiments) with cell lines or fresh unfixed tumor material. Although the purity of these antibodies has increased over time, their use in clinical routine is still limited. Phosphorylated proteins in tissues are rather unstable, and even short times of cold ischemia or suboptimal fixation and tissue processing can cause a dramatic decrease of phospho-proteins. Moreover, many proteins have multiple phosphorylation sites. Thus, the selection of the right antibody may be challenging, and attention should be paid to the phosphorylation site.

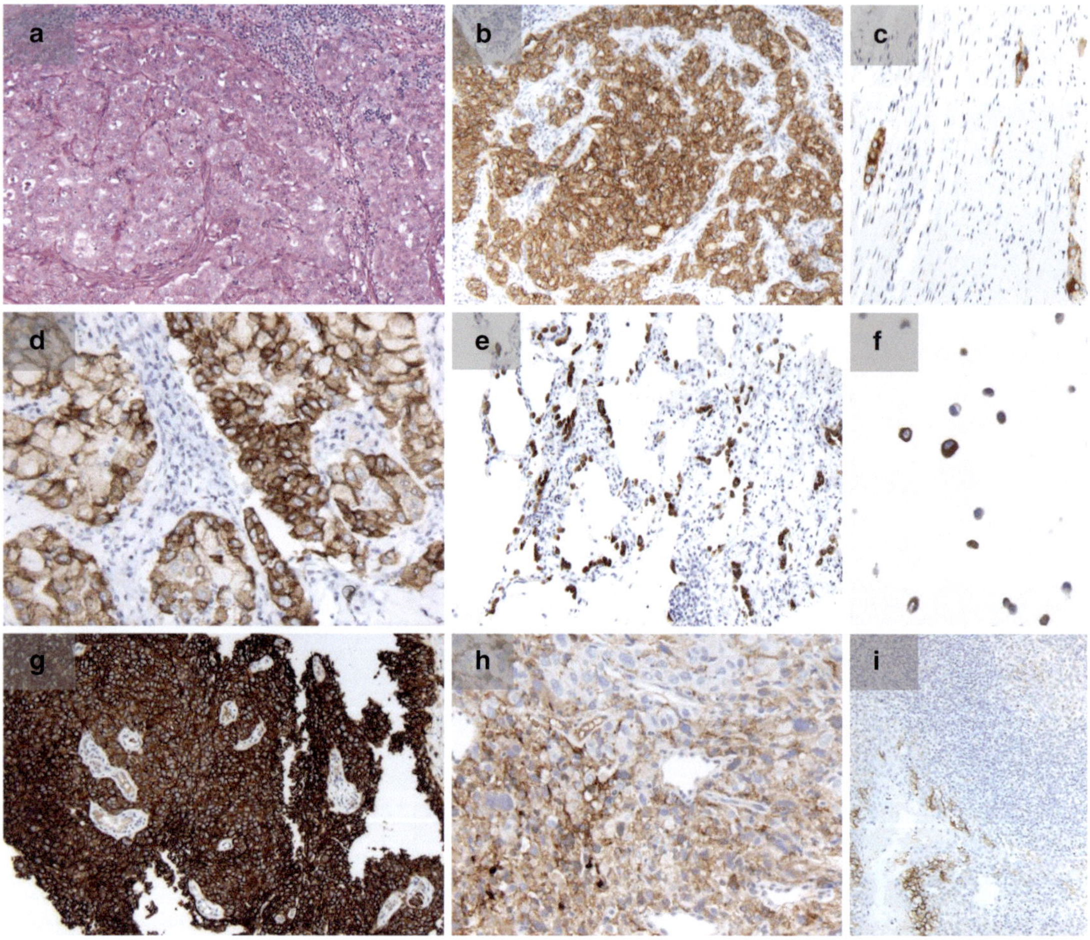

Fig. 4.4 IHC-based predictive biomarkers – clinical examples for assessment and controls (**a–c**) ALK immunohistochemistry. Pulmonary adenocarcinoma (**a**, H&E) showing diffuse moderate ALK staining (**b**, clon 1A4). The clinical relevance of predictive IHC assays depends on the specific antibody and staining intensity. Strong staining has been proven to be indicative of an underlying ALK gene rearrangement and, thus, to be predictive for tyrosine kinase inhibitor treatment. Moderate or weak stainings frequently require confirmation by ISH or sequencing. (**c**), Ganglion cells in the outer wall of an appendix serve as controls. (**d–f**), ROS1 staining (clone D4D6). (**d**) diffuse ROS1 staining in a ROS1 rearranged adenocarcinoma of the lung. (**e**) Readers should be aware that non-neoplastic cells such as activated pneumocytes may mimic positive carcinoma. (**f**) ROS1rearranged cell line can be used as on-slide control (cell block). There is no constant physiologic ROS1 expression in normal tissues. (**g–i**) PD-L1 staining (clone 28–8). (**g**, **h**) Two lung cancer cases showing PD-L1 staining. Note that – in contrast to other biomarkers such as her2 – staining intensity is not a relevant diagnostic category for PD-L1. Therefore, both cases fall in the same clinical category since more than 50% of tumor cells show a (complete or partial) membranous staining (tumor proportion score, TPS, >50%). (**i**) Tonsils are used as controls for PD-L1 staining. Crypt epithelia should display a moderate to high expression, cells in the germinal centers weak staining

Clinical Applications of IHC-Based Predictive Biomarker Assays

Immunohistochemistry-based biomarkers are inherent parts of cancer diagnostics. Treatment of a large and steadily growing number of cancer entities is dependent on these assays (Table 4.1). Many of them are applied to different tumor entities, a fact which requires knowledge on specific assessment rules. One well-recognized example is HER2 with different approaches of evaluation in breast cancer and gastric carcinomas (Fig. 4.3) [2].

In the evolving era of immuno-oncology, additional biomarkers have been introduced, among them is PD-L1 immunostaining. Also, for this marker, different scoring approaches exist which are directly related to clinical questions. PD-L1 is a mandatory predictive biomarker (i.e., a companion diagnostic) for treatment of non-small cell lung cancer with pembrolizumab. In contrast, treatment of the same cancer type with other immuno-oncology drugs such as nivolumab or atezolizumab is basically independent of PD-L1 expression. However, this biomarker may provide adjunct information for oncologists. This type of assays in this context is, therefore, designated as complementary biomarkers.

PD-L1 is, moreover, another good example of how complex the evaluation of IHC-based biomarkers has become. One approach is to determine the relative number of stained tumor cells which is measured by the tumor proportion score (TPS). This assessment is solely based on membranous staining (either circumferential or partial) in cancer cells and neglects staining intensity as well as staining in other cell types such as immune cells. There are different cutoff definitions of TPS for different cancer subtypes and even for different lines of treatments (Table 4.1). Recently, alternative ways to measure PD-L1 expression have been approved, e.g., assessments which also cover PD-L1 expression in certain inflammatory cells additionally to tumor cells (combined positivity score, CPS, in gastric or bladder cancer). Another approach for urothelial carcinomas is to measure only PD-L1 positive inflammatory cells in relation to the tumor area which is covered by them (IC-scoring).

Both HER2 and PD-L1 represent IHC-based assays which are evaluated in a semiquantitative manner. There are, however, also predictive tests which allow a rather qualitative measurement. Among the latter group are immunostains which detect DNA mismatch repair deficiency (dMMR). Lack of IHC expression of MLH1, PMS2, MSH2, or MSH6 in tumor cells is indicative of microsatellite instability, a feature which makes cancer vulnerable to immuno-oncology treatments, irrespective of the tumor entity (Table 4.1). Furthermore, IHC-based biomarkers represent surrogate markers for underlying genomic changes, e.g., ALK and ROS1 staining in lung cancer (Fig. 4.4).

In terms of correct application of predictive biomarker assay, it is also important to select appropriate biomaterials. Cytology specimens such as smears and cytospins may or may not be used for many IHC-based assays. Formalin fixed and paraffin embedded cell blocks, however, can overcome this issue. Continuous quality controls are mandatory to maintain high reliability of assays. One important measure in this context is the use of staining-specific on-slide controls (Fig. 4.4).

In Situ Hybridization (ISH): Technical Considerations

The term hybridization describes the detection of nucleic acids (DNA or RNA) by specific probes, i.e., by small DNA molecules. Since this hybridization takes place on a slide directly in a piece of tissue, the method is designated as *in situ* hybridization. Comparable probes are also used *in vitro*, in methods such as Southern or northern blotting, polymerase chain reaction (PCR) assays and in certain sequencing techniques. Thus, *in situ* hybridization helps in the detection of specific genetic information within a morphologic context. This makes the ISH method a perfect molecular application for anatomic pathologists in addition to increasing the significance of certain molecular findings. For example, in a breast cancer case with *HER2* amplification, it is important to morphologically clarify whether this amplification is confined to the DCIS component or whether it also occurs in the invasive cancer cells. Only the latter constellation is predictively meaningful; this discrimination cannot be achieved by a sequencing approach.

Probes are single- or double-stranded DNA pieces which are complementary to a specific target sequence in the genome. A single probe molecule usually has a size of several hundred nucleotides. Commercially available ISH probes consist, however, of a library of these DNA pieces which span a much larger genomic region – usually in the range of several hundred kilobases or even megabases. This increases the size of the visible signals and, thus, facilitates detection of hybridization under the microscope. In this context, it is noteworthy that so-called gene-specific ISH probes do not only detect the sequence of the gene of interest but also 3′ and 5′ flanking genomic regions which may sometimes also contain additional genes. For microscopic visualization, these probes are labeled – either with fluorescent dyes (fluorescent *in situ* hybridization, FISH) or with chromogens or silver pigments (chromogenic *in situ* hybridization, CISH; dual-color dual-haptene bright-field *in situ* hybridization, DDISH; or bright-field *in situ* hybridization, BrISH; Fig. 4.5). Most assays use at least two different dyes to simultaneously visualize different genomic regions of interest (see below).

Fluorescent *In Situ* Hybridization (FISH)

Fluorescent probes are mostly directly labeled with dyes which are covalently bound to nucleotides. This is achieved by the method of nick translation where induced DNA gaps, "nicks," are repaired by incorporating fluorescent nucleotides. FISH signals can be observed through the microscope by the use of high-transmission and numerical aperture objectives when the dyes are excited by light of a shorter wave length ("excitation spectrum"). The excitation light is generated by filtering white light and reflecting it at a dichroic filter. This finally leads to an emission of light at a higher wave length ("emission spectrum") by the fluorescent dyes which can pass the dichroic mirror in the microscope and can be seen in the eyepieces or with a camera. Excitation and emission spectra are specific for every fluorescent dye and, therefore, pathologists need to make sure that FISH probes and microscope equipment, i.e., the light source, microscope objectives, filter sets, and dichroic mirror, are all matched and optimized. FISH technique (in contrast to chromogenic *in situ* hybridization (CISH)) allows the pathologist to capture signals through the better spatial resolution of signal(s) and superior color discrimination using specific filter sets and aberration-free objectives. Under ideal conditions, it is often possible to detect the fluorescence emission from a single dye molecule, provided that the optical background and detector noise are sufficiently low. A disadvantage of FISH is that stained slides fade over time, from several months up to a few years depending on storage conditions. They are also subject to quenching and photobleaching during imaging.

Chromogenic *In Situ* Hybridization (CISH)/Bright-Field *In Situ* Hybridization

CISH probes are always indirectly labeled; they frequently contain incorporated biotin or streptavidin molecules. The staining itself is enzymatically produced by alkaline phosphatase or horseradish peroxidase in the same way as in immunohistochemistry. CISH has the added advantage of using regular light microscopes as opposed to the FISH technique which requires the use of an expensive fluorescence microscope (however, a 100× oil immersion objective may be useful for some CISH assays as well). It is also thought that morphology is better retained in CISH assays compared with FISH. However, in the authors' personal experience, FISH stainings usually provide sufficient or even better morphologic control if appropriate DAPI counterstaining and suitable objectives (40× or 63× oil immersion lenses) are used.

Method and Factors that Influence Assay Quality

As already mentioned, ISH assays are performed *in situ* by using cut tissue slides. The thickness of these slides should be taken into consideration because thicker sections (>3–4 μm) will result in a limited number of evaluable non-overlapping nuclei. Additionally, the quality of the staining itself is dependent on slide thickness since the effectiveness of digestions and pre-hybridization steps are reduced if sections are too thick. Generally, sections should be in the range of 2–3 μm for optimal imaging (similar to IHC slide

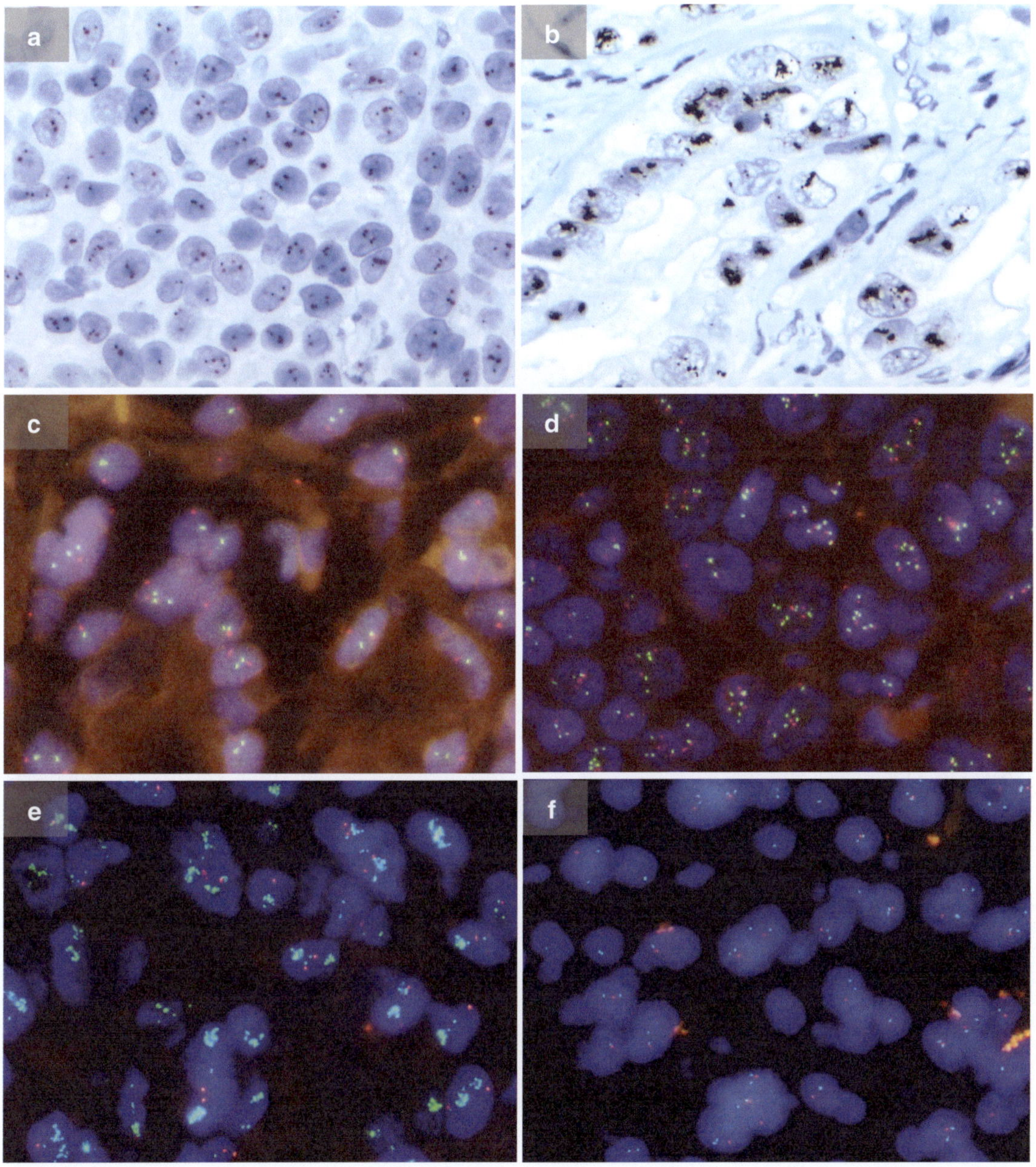

Fig. 4.5 Detection of amplifications and deletions by *in situ* hybridization. (**a**, **b**) Dual-color dual-haptene in situ hybridization (DDISH) for *Her2* amplification. Probe set consists of a red centromeric signal and a silver-black signal for the *Her2* locus. (**a**) Normal and unamplified tumor cells display one to two signals of each sort. (**b**) Breast cancer case with high level amplification showing 10–30 *Her2* gene copies per tumor cell. (**c–e**) Fluorescence in situ hybridization (FISH) for *Her2* showing green *Her2* and orange centromere 17 signals. (**c**) Nuclei which are negative for amplification. (**b**) Breast cancer case which falls in the borderline category defined by a slight increase in average gene copy number (4–6 *Her2* signals per tumor cell nucleus) with a *Her2*/CEN 17 ratio <2.0. (**e**) Highly amplified cancer with cluster amplification of *Her2*. (**f**) Detection of deletions by FISH. Sample shows exemplarily 1p deletion in an oligodendroglioma. Tumor shows only 0–1 orange 1p signals against the background of two green signals of the reference probe (at chromosome 1q) in most tumor cells

thickness). During pre-hybridization, the tissue is digested with pepsin or proteinases and incubated with buffers which increase the tissue permeability. After applying the probe mix at the slide, both tissue and probe are denatured at high temperature (DNA becomes single stranded) and then incubated at a lower hybridization temperature (mostly 37 °C) in a hybridization chamber for a couple of hours to allow renaturation and binding of probes to genomic DNA. This is followed by a stringent wash to remove nonspecifically bound probe molecules and to keep the specificity of signals. The stringency of this step is highly dependent on the salt concentration in the buffer and the temperature (the latter should be measured directly in the buffer). The higher the temperature and the lower the salt concentration, the more specific is the hybridization. Too high stringency, however, will result in the complete loss of bound probe. For FISH, an antifade solution which prevents the hybridized slides from rapid photobleaching together with a DAPI counterstain (4′,6-diamidin-2-phenylindol; a blueish fluorescent dye which nonspecifically labels DNA) is finally applied.

The quality and overall evaluability of ISH assays are mainly determined by the signal-to-noise ratio of the hybridization signals, i.e., the balance between specific signals of each color and unspecific background. The use of commercially available kits increases the staining quality by providing matching buffers and high-quality probes. However, critical factors remain – especially slide thickness, duration of enzymatic digestion, denaturation temperature as well as maintaining the exact temperature of the stringent wash. Probe suppliers provide useful recommendations; however, sometimes it is worth establishing own protocols and settings such as optimum denaturation temperature. Semi-automated or automated systems help to standardize protocols and to maintain a high hybridization quality of both FISH and non-fluorescent ISH assays. Some immunostainers can be also used for FISH or dual-color bright-field *in situ* hybridization (DDISH).

Nonneoplastic cells such as fibroblasts, immune cells, or endothelia can be utilized as internal controls. These cells should display one or two signals of each sort per cell. If readers of ISH assays evaluate and document the hybridization quality in these cell types properly, on-slide controls or even run controls can be omitted.

Types of Detectable Genomic Aberrations and Probe Settings

ISH is basically capable of detecting three types of gene aberrations, namely, (i) amplifications, (ii) large deletions, and (iii) rearrangements (synonymous with gene fusions or translocations). Point and indel mutations, small deletions, and many other genomic changes cannot be detected by these techniques.

The exact biologic definition of *amplification* is controversial. For practical reasons, and in the context of predictive biomarkers, gene amplification can be simply regarded as an increase in the copy number of a certain gene of interest which is meaningful in terms of tumor biology and which has been proven to be associated with response to a certain drug. Dual-color probe sets for detecting amplifications usually consist of a locus-specific (gene-specific) probe and a second probe which detects a reference locus (mostly the centromere of the respective chromosome; Fig. 4.5). For most genes, signals of gene and reference locus are counted individually in a defined number of tumor cells, and average gene copy number per tumor cell and gene/reference ratio is calculated. It is of special interest that the cutoffs vary considerably between genes and tumor entities. The definition of *HER2* amplification in breast cancer is not a universally applicable description of "amplification." Furthermore, also reference sequences may be included in large amplicons (or, in contrast, be part of a focal deletion) which can influence the calculated ratio. Therefore, both ratio and gene copy number are considered [3].

The detection of gene *deletions* is quite similar to the above-described approach. Locus-specific and reference (centromere) probes are co-hybridized and counted. Homozygous (defined by a complete loss of any target signal in a predefined percentage of tumor cells) or hemizygous deletions (mostly defined by a target/reference ratio below a predefined cutoff) are considered (Fig. 4.5f). Again, the definition of gene deletion in the context of predictive biomarkers is a matter of a gene- and entity-specific definition based on clinical trials or clinical experience.

Gene *rearrangements* include reciprocal translocations between different chromosomes as well as inversions and interstitial deletions within the same chromosome. The common effect of these changes is the creation of a novel chimeric fusion gene which consists of parts of two different genes. These chimeric genes and their encoded proteins can be oncogenic and can represent a target of anticancer drugs.

There are basically two ways to detect gene rearrangements by ISH (Figs. 4.6 and 4.7): break-apart probes and so-called fusion probes. Break-apart (or split) probes consist of two differently

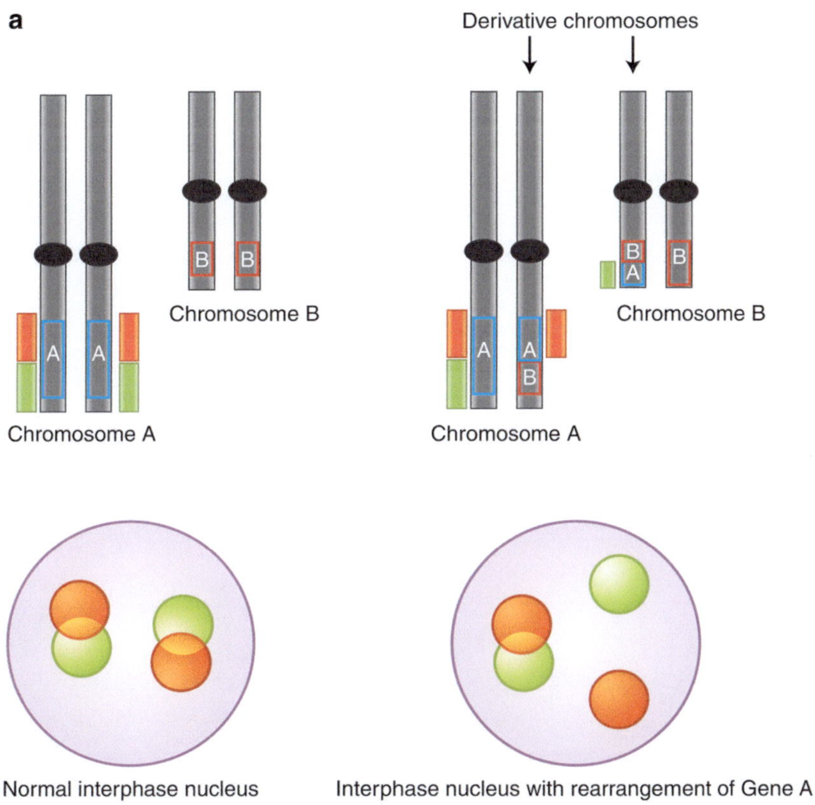

Fig. 4.6 Different ways to detect gene rearrangements by *in situ* hybridization. (**a**) Break-apart probe sets consist of two probes which flank 3′ and 5′ regions around the break point cluster region in gene A. In normal cells (left), both signals are fused or very close to each other. In case of a reciprocal translocation (right panel), orange and green signals are split. Some gene rearrangements are accompanied by loss of chromosomal material. In these cases, an isolated signal (3′ probe) may indicate rearrangement. (**b**) Fusion probe. In this approach, orange and green probes span the breakpoint cluster regions on both genes which are involved in that rearrangement. Normal cell(s) (left) display two signals of each sort. In case of a reciprocal rearrangement (right), parts of orange and green probes will fuse. (**c**) Triple-color combined fusion/break-apart probe set. This approach consists of a break-apart component with orange and green probes flanking the break point cluster region in gene A. The blue probe spans gene B. In normal cells (left), two sets of fused orange/green and two blue signals are seen. If an intrachromosomal inversion occurs (top right), parts of gene A fuse with parts of gene B, this is displayed by fusion of orange and blue signals. Not infrequently, this process is accompanied by an interstitial deletion of chromosomal material (causing loss of green signals). Triple-color combined fusion/break-apart probe sets can also be utilized to detect reciprocal translocations (with gene C, bottom right) where simply a split of orange and green signals occurs – without involvement of blue signals

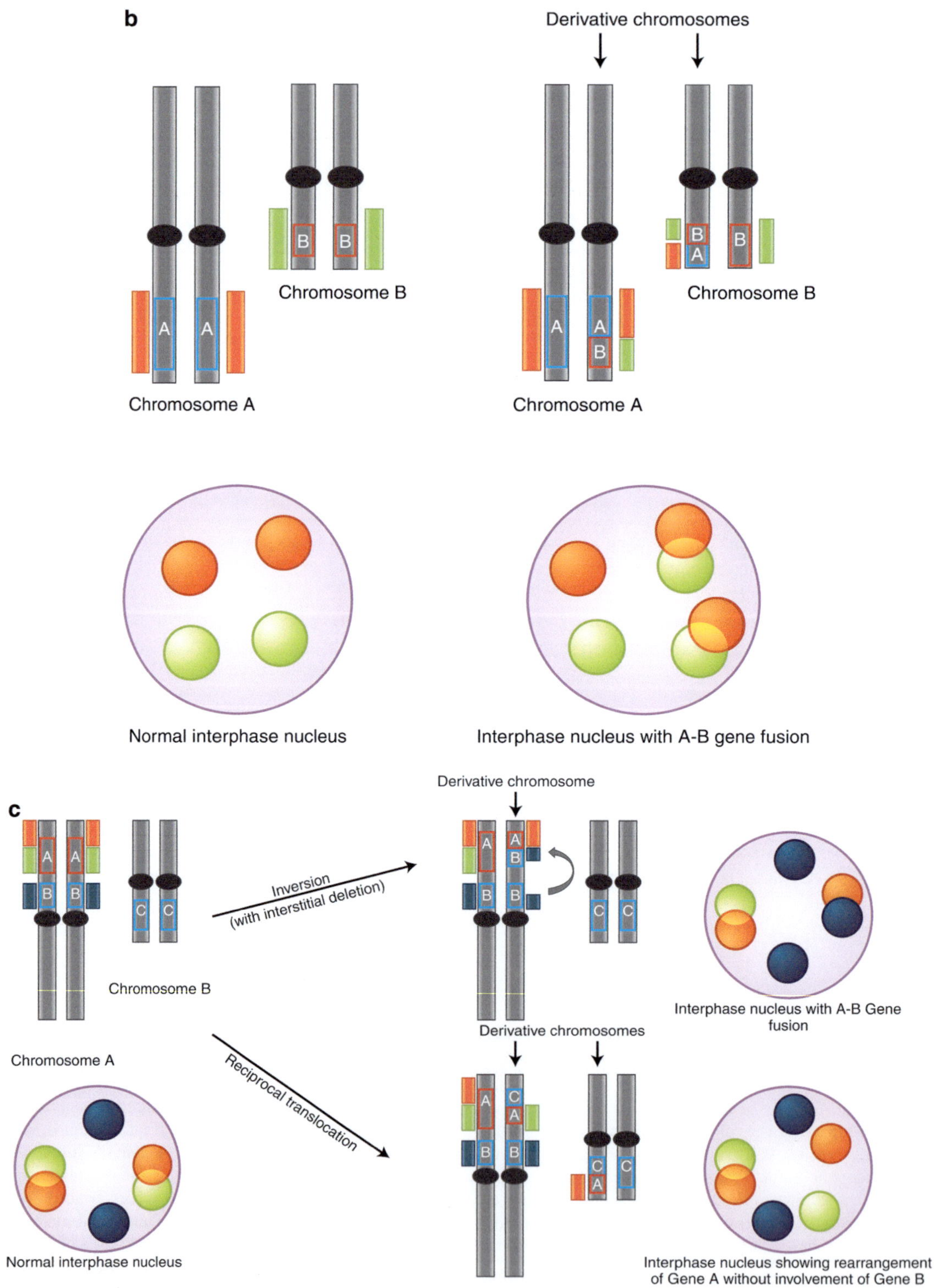

Fig. 4.6 (continued)

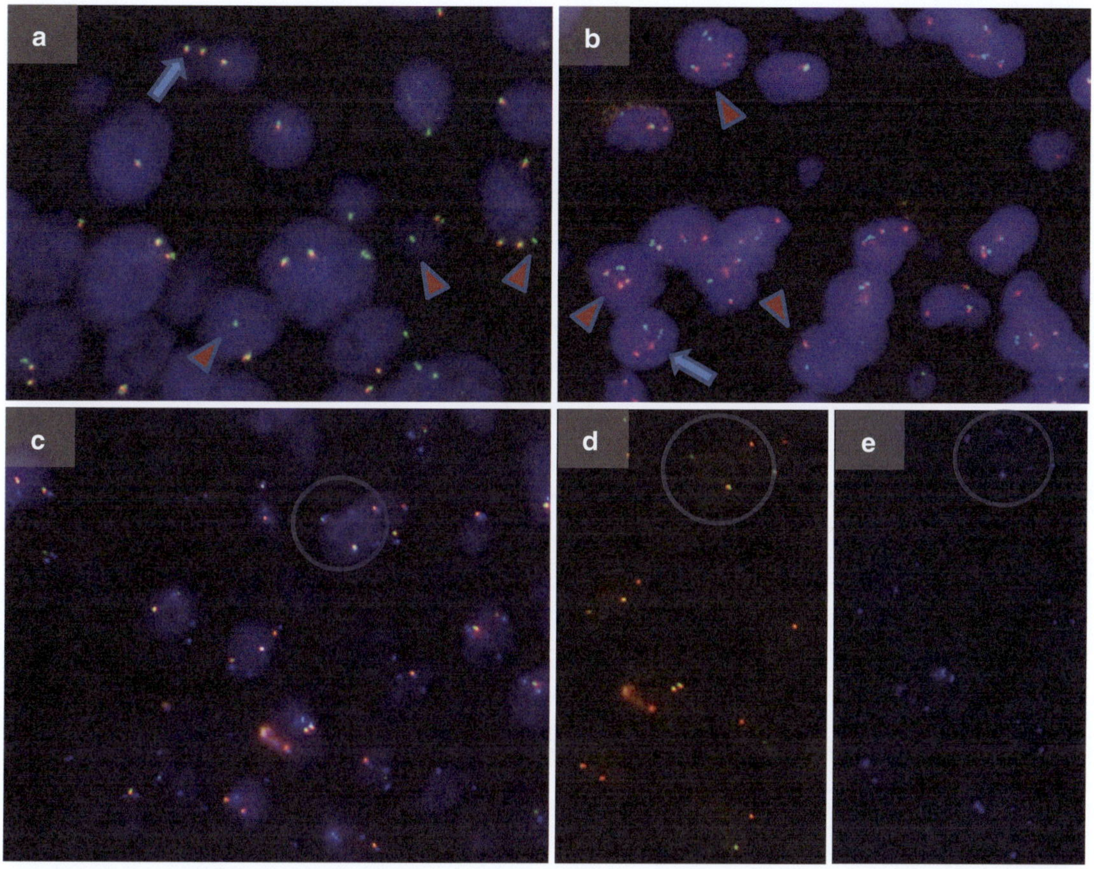

Fig. 4.7 Different ways to detect gene rearrangements by FISH. (**a**) Break-apart probe (*ROS1*). Normal nonneoplastic cells show fused signals consisting of orange and green probe signals which flank the break point cluster region within the *ROS1* gene (blue arrow). Tumor cells display rearrangement ("chromosomal break") indicated by isolated 3′ signals (green; arrowheads). (**b**) Dual-color dual-fusion probe (*bcr-abl*). Normal cells (blue arrow) showing to two copies of bcr (green) and abl genes (orange). Cells with a reciprocal translocation resulting in a bcr-abl gene fusion gain two fused signals and retain one copy of bcr and abl from the unaffected alleles. (**c–e**) Combined break-apart/fusion probe set (*ALK-EML4*). (**d**) Orange and green probes flank the break point in the *ALK* gene and form the break-apart component. Fused signals represent normal alleles, split signal indicate *ALK* rearrangement (circle). (**e**) Blue signals span the EML4 gene. (**c**) In case of an *EML4-ALK* inversion, one blue *EML4* signal fuses with the *ALK* 3′ signal

labeled probes (mostly orange and green) which span 3′ and 5′ flanking regions around the break point cluster region (an area in a gene – mostly certain introns – where the strand breaks occur which lead to that rearrangement).

In normal nuclei, these two signals appear indistinguishable close to each other which sometimes results in a superimposed yellowish signal (Figs. 4.6a and 4.7a). In case of a rearrangement, the signals of the affected allele are disrupted ("split" or "break-apart" signal). Since these split signals can also occur randomly, readers have to determine the percentage of tumor cell nuclei which are affected by this phenomenon. There are predefined cutoffs for specific genes and probe sets. For many genes, cases are regarded as being rearranged if at least 15% (or 20%) of nuclei display a break-apart signal. It is important to realize that break-apart probes recognize only one partner gene in a rearrangement. Frequently genes have multiple potential translocation partners with dozens of variant breakpoints which result in the same biologic and oncogenic effects. Therefore, break-apart FISH probes which cover the major component of a rearrangement (e.g., a gene encoding a receptor tyrosine kinase) are good candidates to recognize all relevant changes.

Another way to detect gene rearrangements is the use of so-called fusion probes. There are dual-color dual-fusion or dual-color single-fusion variants differing in terms of the number of resulting fusion signals which characterize the aberrant allele. Usually, two probe sets labeled in two different colors span the break point cluster regions of both genes of interest which are potentially involved in that rearrangement. Therefore, normal cells display two signals of each color. In case of a reciprocal translocation, one or two (depending on the probe architecture) fused signals will appear (Figs. 4.6b and 4.7b). Fusion probes provide full information on both genes which are involved. However, similar rearrangements with another partner won't be detectable. Fusion probes have only limited extension in the setting of predictive biomarker assays.

A more recent innovation was the introduction of three- or multicolor probe set which combines a classical break apart with a fusion probe component. This approach benefits from the full flexibility of break-apart probes and integrates additional information provided by the fusion probe component (Figs. 4.6c and 4.7c–e). Moreover, novel complex multiplex probes have been developed which detect multiple gene rearrangements (e.g., *ALK* and *ROS1* fusions in lung cancer) simultaneously in the same hybridization.

Another emerging trend is the detection of RNA by *in situ* hybridization techniques. This technology allows precise analysis of gene expression (also multiple genes in a multicolor assay) as well as the expression on noncoding RNA molecules in a tissue of interest. RNA *in situ* hybridization has also been shown to detect gene mutations, rearrangements, and splice variants.

Clinical Applications of *In Situ* Hybridizations

Currently, *HER2* amplification is still the most frequently used ISH assay. However, with improvements of HER2 IHC, the numbers are probably decreasing. In contrast, novel assays such as *MET* amplification are evolving (Table 4.2). Detection of gene rearrangements by ISH has been restricted to diagnostic tests for a long time. They were mainly related to entity-defining translocations in lymphomas, hematologic neoplasms, and sarcomas (e.g., *BCR-ABL* translocation in chronic myeloid leukemia, *BCL2-IgH* fusion in follicular lymphoma, or *EWS* rearrangement in Ewing sarcoma; these changes also occur in other entities). In the past few years, more and more recurrent rearrangements in solid tumors have been discovered using ISH which defines therapeutically relevant subgroups of larger entities. One example includes *ALK*- or *ROS1*-rearranged pulmonary adenocarcinomas [4, 5]. Many of these changes can be detected basically in various tumor entities but only in a very small proportion of patients. Given the low frequency of *ROS1* fusions in lung cancer (which is in the range of 1.5% in Western populations), a pathologist needs statistically to test 100 adenocarcinomas to detect one single positive patient. However, the dramatic and durable response of these tumors to tyrosine kinase inhibitor treatment justifies any effort. Many more comparable assays are currently evolving; further examples are summarized in Table 4.2.

Concluding Remarks and Future Directions

Immunohistochemistry and *in situ* hybridization techniques represent widely used methods to detect predictive biomarkers in cancer samples. However, both techniques have some limitations in the context of modern and timely cancer diagnostics where a magnitude of analytes needs to be analyzed simultaneously in one single tissue sample. Current classical IHC/ISH methods recognize only one parameter/analyte per assay. This severe limitation may be overcome by multiplex-IHC (and in parts by oligo-plex ISH) methods in the future. Furthermore, sequencing approaches will be increasingly utilized to capture information which has been historically provided by ISH.

On the other hand, both assays – IHC and ISH – keep full morphologic control, integrate

Table 4.2 Frequently applied examples of predictive ISH assays in solid tumors

Disease	Gene	Probe type	Aberration	Clinical impact
Breast cancer	*HER2*	Amplification	High *HER2*/CEN17 ratio (≥2.0) and/or high average gene copy number (≥6.0) define amplification	*HER2* amplification is a prognosticator of the worse outcome but provides the basis for antibody treatment
NSCLC	*ALK, ROS1*	Translocation (break-apart or break-apart/ fusion or multiplex probes)	The rearrangement is defined by the presence of (i) break-apart signals, or (ii) isolated 3′ signals in ≥15% of evaluated tumor cells	See Table 4.1
	RET	Translocation (break apart)	Translocation defined by break-apart or isolated 3′ signals (≥15% or 20% of nuclei depending on institutional cutoffs)	Emerging biomarker. *RET* rearranged NSCLC may respond to multi kinase-TKIs, e.g., cabozantinib or vandetanib. Efficacy may be related to translocation partners of *RET* (*RET-KIF5B* rearrangement may indicate less durable response)
	MET	Amplification	Definition of *MET* amplification is still debatable. High average gene copy number (≥6.0 or ≥10.0) is currently discussed	Emerging biomarker. Efficacy of TKI treatment (e.g., capmatinib, crizotinib) is currently explored in clinical trials. Note that also activating exon 14 skipping mutations of *MET* are predictive.
Gastric carcinoma/ adenocarcinoma of gastroesophageal junction	*HER2*	Amplification	Amplification defined by high average gene copy number and/or high *HER2*/CEN17 ration (comparable to breast cancer; minor differences exist)	Amplified tumors may respond to antibody treatment; trastuzumab is approved. Note that amplification is only predictive in the context of at least a 2+ immunostain
Cholangiocarcinoma	*ROS1, NTRK*	See above	See above	Emerging biomarker. Cholangiocarcinomas with *ROS1* or *NTRK* rearrangements may respond to TKI treatment (off label)
Soft tissue tumors	*ALK, ROS1*	See above	See above	*ALK* or *ROS1* rearrangements in IMFT provide a rationale for TKI treatment
	PDGFb	Translocation (break-apart or fusion probes)	Rearrangement is detected either by *PDGFb-COL1A1* fusion probes or by using a break-apart probe detecting one of both partner genes according to local institutional guidelines	Imatinib treatment is approved for relapsed DFSP showing PDGFb rearrangement. *For more detailed information on mesenchymal tumors, see chapter on predictive biomarkers in sarcomas*

ALK anaplastic lymphoma kinase, *DFSP* dermatofibrosarcoma protuberans, *IMFT* inflammatory fibrous tumor, *NSCLC* non-small cell lung cancer, *PDGFb* platelet-derived growth factor beta, *TKI* tyrosine kinase inhibitor findings directly in the histologic appearance of tumors, and are robust and relatively easily applied. Furthermore, IHC and ISH are very fast methods which provide clinically relevant information within the shortest turnaround time. There are many more predictive biomarkers to come in the near future, and a significant number of them will be related to either protein expression or chromosomal aberrations and will, therefore, represent potential subjects to IHC or ISH assays.

References

1. Wolff AC, Hammond EH, Hicks DG, Dowsett M, McShane LM, Allison KH, Allred DC, Bartlett JMS, Bilous M, Fitzgibbons P, Hanna W, Jenkins RB, Mangu PB, Paik S, Perez EA, Press MF, Spears PA, Vance GH, Viale G, Hayes DF. Recommendations for human epidermal growth factor receptor 2 testing in breast cancer: American Society of Clinical Oncology/College of American Pathologists Clinical Practice Guideline Update. J Clin Oncol. 2013;31:3997–4013.
2. Rüschoff J, Hanna W, Bilous M, Hofmann M, Osamura RY, Penault-Llorca F, van de Vijver M, Viale G. HER2 testing in gastric cancer: a practical approach. Mod Pathol. 2012;25:637–50.
3. Stoss OC, Scheel A, Nagelmeier I, Schildhaus HU, Henkel T, Viale G, Jasani B, Untch M, Rüschoff J. Impact of updated HER2 testing guidelines in breast cancer – re-evaluation of HERA trial fluorescence in situ hybridization data. Mod Pathol. 2015;28:1528–34.
4. Tsao MS, Hirsch FR, Yatabe Y, editors. IASLC Atlas of ALK and ROS1 testing in lung cancer. 2nd ed. North Fort Myers: Editorial Rx press; 2016.
5. Lindeman NI, Cagle PT, Aisner DL, Arcila ME, Beasley MB, Bernicker E, Colasacco C, Dacic S, Hirsch FR, Kerr K, Kwiatkowski DJ, Marc Ladanyi, Nowak JA, Sholl L, Temple-Smolkin R, Solomon B, Souter LH, Thunnissen E, Tsao MS, Ventura CB, Wynes MW, Yatabe Y. Updated molecular testing guideline for the selection of lung cancer patients for treatment with targeted tyrosine kinase inhibitors: guideline from the College of American Pathologists, the International Association for the Study of Lung Cancer, and the Association for Molecular Pathology. J Mol Diagn. 2018. pii: S1525-1578(17)30590-1. [Epub ahead of print].

Overview of PCR-Based Technologies and Multiplexed Gene Analysis for Biomarker Studies

5

Yesim Gökmen-Polar

Abbreviations

DNA	Deoxyribonucleic acid
dNTPs	Deoxyribonucleotide triphosphates
PCR	Polymerase chain reaction
qPCR	Quantitative real-time PCR
RT-qPCR	Reverse transcriptase quantitative real-time PCR
RT-PCR	Reverse transcriptase polymerase chain reaction

Introduction

The polymerase chain reaction (PCR) is a technique that amplifies a specific deoxyribonucleic acid (DNA) sequence from one or few copies to billions of copies by separating the DNA into two strands and incubating it with oligonucleotide primers and DNA polymerase in vitro. Since its discovery in the 1980s [1, 2], the conventional PCR has benefited many molecular biology techniques including DNA cloning, gene expression studies in recombinant systems, and molecular genetic analyses such as the detection of mutations (point or missense mutations) as well as in forensic sciences and detection of infectious diseases. In addition to the conventional PCR, real-time PCR [3], also called quantitative real-time PCR (qPCR), has further revolutionized the field of biological and clinical sciences due to its sensitivity, and accuracy in a real-time quantification of gene expression. Throughout the years, the basic principles of PCR have remained the same, but various methods have evolved to improve its efficiency, yield, sensitivity, specificity, cost, and applications within the medical field. PCR has widespread applications in detecting genetic alterations in tumor tissue.

This chapter aims to provide the basic principles and some applications of PCR-based technologies in measuring the changes in biomarker studies in clinical oncology.

Basic Principles of PCR

The PCR technique involves the primer mediated enzymatic amplification of DNA consisting of three major steps: (1) denaturation of double-stranded (ds) DNA template, (2) annealing of primers (forward and reverse primers), and (3) extension/elongation of ds DNA molecules (Fig. 5.1).

Step 1 consists of heating the reaction mixture to 94–95 °C for 15–30 s. During this step, the double-stranded DNA is separated to single strands due to breakage in weak hydrogen bonds. In step 2,

Y. Gökmen-Polar
Department of Pathology and Laboratory Medicine, Indiana University School of Medicine, Indianapolis, IN, USA
e-mail: ypolar@iu.edu

S. Badve, G. L. Kumar (eds.), *Predictive Biomarkers in Oncology*,
https://doi.org/10.1007/978-3-319-95228-4_5

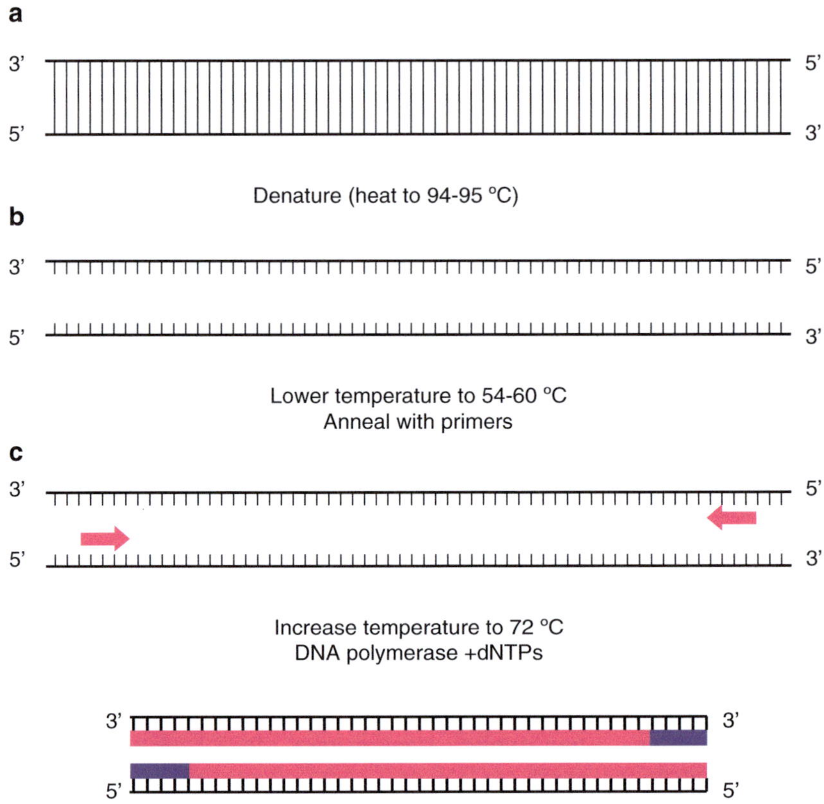

Fig. 5.1 Schematic diagram of basic steps of polymerase chain reaction (PCR) technique. (**a**) Denaturation, (**b**) annealing, and (**c**) extension. Pink arrows represent primers (left-forward primer and right-reverse primer)

incubation of DNA is required at 54–60 °C for 20–40 s to allow the primers, the short pieces of single-stranded DNA that are complementary to the target sequence, to bind (anneal) to their complementary sequence in the template DNA. In step 3, extension/elongation occurs at the end of the annealed primers to create a complementary copy strand of DNA. The DNA polymerase, a thermostable polymerase enzyme (mostly Taq polymerase), synthesizes new strands of DNA complementary to the target sequence by sequentially adding nucleotides to the 3′ end of each primer, extending the DNA sequence in the 5–3′ direction. This cycle is repeated 30 or 40 times exponentially amplifying the sequence of interest.

Components of PCR

The major components of a PCR reaction consist of the template DNA, DNA polymerase, primers, deoxyribonucleotide triphosphates (dNTPs), magnesium ion (Mg^{2+}), and buffer (15 mM $MgCl_2$, 500 mM KCl, 100 mM Tris-HCl, pH 8.3 at 25 °C). Nowadays, most of these components are sold by commercial vendors as "PCR master mix," which is a premixed concentrated solution with the components of a PCR reaction that is not sample-specific. A master mix usually contains a thermostable DNA polymerase, dNTPs, $MgCl_2$, and proprietary additives in a buffer optimized for PCR.

PCR Template

The source of a PCR template can be genomic DNA (gDNA), complementary DNA (cDNA), or plasmid DNA. Based on the complexity of DNA, the starting amount for a PCR reaction may differ such as gDNA (5–50 ng) or plasmid DNA (0.1–1 ng), respectively. Copy number can be calculated using the following formula, $L \times$ number of moles = $L \times$ (total mass/molar mass), where L represents Avogadro's constant. The molar mass of a particular DNA strand is computed from its size or total number of bases (i.e., a combination of DNA length and single-stranded or double-stranded nature).

DNA Polymerase

The choice of a DNA polymerase is also critical to determine the DNA template input for the better sensitivity and specificity of PCR reaction. *Taq* DNA polymerase is suitable for standard PCR. However, new generations of DNA polymerases have been introduced that improve the fidelity of the PCR reaction. For example, Platinum® *Taq* DNA polymerase High Fidelity (Thermo Fisher Scientific) is designed for amplification of DNA fragments when high yields, fidelity, and robust amplification are required. PCR specificity is also improved by preventing the polymerase from binding nonspecifically at room temperature using blocking agents such as antibodies (e.g., Platinum® automatic "hot-start" technology that uses Platinum *Taq* DNA Polymerase from Thermo Fisher Scientific).

Primers

PCR primers are synthetic DNA oligonucleotides of 15–30 bases. Certain criteria have to be met to design the PCR primers for an efficient PCR reaction. Besides the nucleotide length, the melting temperature (T_m) of the primers is important, which is defined as the temperature at which 50% of that same DNA molecule species form a stable double helix and the other 50% have been separated to single strand molecules [T_m 55–70 °C (within 5 °C, for two primers)]. The GC content should be between 40% and 60%, one C or G should be at 3′ end, and sequences that cause primer-dimers or secondary structures should be avoided. The primer concentrations for a typical PCR reaction range between 0.1 and 1 μM. There are several vendor-based software programs that can help to design the primer and probe sequences for PCR. RealTime Design (Biosearch Technologies), GenScript Real-time PCR (TaqMan) Primer design, and Primer Express (Thermo Fisher Scientific) are the major software programs for real-time PCR.

Deoxyribonucleotide Triphosphates (dNTPs)

dNTPs consist of four basic nucleotides—dATP, dCTP, dGTP, and dTTP—which are the building blocks of new DNA strands. Usually, equimolar amounts of these four nucleotides are added to the PCR reaction. The recommended final concentration is usually 0.2 mM. In specific situations such a random mutagenesis, unbalanced dNTP concentrations are added to promote a higher degree of mis-incorporation by a non-proofreading DNA polymerase. Other exceptions include the substitution of dTTP with deoxyuridine triphosphate (dUTP) using a uracil DNA glycosylase to prevent carryover PCR contamination.

Magnesium Ion (Mg^{2+})

Magnesium ion (Mg^{2+}) works as a cofactor for the activity of DNA polymerases and help the incorporation of dNTPs during polymerization. The magnesium ions at the enzyme's active site catalyze phosphodiester bond formation between the 3′-OH of a primer and the phosphate group of a dNTP. Mg^{2+} also facilitates formation of the complex between the primers and DNA templates and stabilizes negative charges on their phosphate backbones [4]. It is important to optimize the magnesium concentration together with dNTPs, primers, and DNA templates to increase the yield and maintain the specificity of the PCR reaction. A range of 1–4 mM Mg^{2+} is recommended. Low Mg^{2+} concentrations result in little or no PCR product, due to the polymerase's reduced activity, whereas high Mg^{2+} concentrations often end up with nonspecific PCR products increasing in replication errors from mis-incorporation of dNTPs.

Buffer

PCR reaction is carried out in a buffer, which provides the most optimal chemical environment for the activity of DNA polymerase. The buffer pH is usually between 8.0 and 9.5 and is often stabilized by Tris-HCl.

Types of PCR

Conventional or Endpoint PCR

Conventional PCR uses agarose gels for detection of PCR amplification at the final phase or endpoint of the PCR reaction. This method assesses changes in gene expression levels commonly using a fluorescent dye such as ethidium bromide. Conventional PCR is used mostly for research purposes rather than diagnostic purposes.

Reverse Transcriptase Polymerase Chain Reaction (RT-PCR)

PCR is a technique for DNA amplification. To amplify ribonucleic acid (RNA), reverse transcriptase PCR was developed. Briefly, RNA is first transcribed into complementary DNA (cDNA) by reverse transcriptase enzyme from total RNA or messenger RNA (mRNA). The cDNA then serves as the template for exponential amplification using PCR. RT-PCR can be performed in one-step or two-step (Table 5.1). The one-step approach combines the first-strand cDNA synthesis (reverse transcription) and subsequent PCR in a single reaction tube. This method requires gene-specific primers for amplification. Two-step reaction may be preferred for detecting multiple genes from a single sample, as the prepared cDNA from the same sample can be used subsequently with different gene-specific primers. The separation of RT and PCR reactions allows using the reverse transcription priming oligo(dT) primers, random hexamers, or gene-specific primers. It also provides a choice of *Taq* Polymerase with high fidelity that can improve the specificity of PCR reactions.

Table 5.1 Comparison of one-step and two-step PCR

PCR type	One-step RT-PCR	Two-step RT-PCR
Assay setup	Combined assay	Separate reverse transcription and PCR assays
Primers	Gene-specific primers	Oligo(dT), random hexamers (step 1) and gene-specific primers (step 2)
Purpose	Analysis of one or two genes	Analysis of multiple genes
Pros	Convenient and rapid	Flexible, better optimization for each step
Con	Difficult to troubleshoot RT step	More pipetting steps

Nested PCR

Two sets of primers are used in two successive reactions instead of one set of primers to increase the specificity of DNA amplification and decrease nonspecific amplification. In the first PCR, one pair of primers is used to generate DNA products, which will be the target for the second reaction using the inner or nested primers.

Quantitation of PCR

Quantitation of PCR or RT-PCR can be measured in three different ways: relative, competitive, and comparative. *Relative* quantifications involve the co-amplification of an internal control simultaneously with the gene of interest. The signal of the samples is normalized using the internal control. After normalization, relative DNA or transcript abundances will be compared across multiple samples for DNA or RNA, respectively. The expression of the internal control needs to be stable in assay conditions for each sample so that the normalization is not biased. The results are expressed as ratios of the gene-specific signal to the internal control signal. This provides a corrected relative value for the gene-specific product in each sample.

Competitive PCR is used for absolute quantification measuring the absolute amount of a specific DNA or RNA in a sample. A synthetic "competitor" DNA or RNA with various dilutions is included in the sample and is co-amplified with the endogenous target. To calculate the amount of the target DNA or RNA, a concentration curve of the competitor DNA or RNA is generated to compare the PCR or RT-PCR signals produced from the endogenous DNA or transcripts.

Comparative PCR is similar to *competitive RT-PCR* in that target DNA or RNA competes for amplification reagents within a single reaction with an internal standard of unrelated sequence. The results are compared to an external standard curve to determine the target RNA concentration.

Quantitative Real-Time Polymerase Chain Reaction (qPCR)

Quantitative polymerase chain reaction (qPCR), also known as real-time PCR [3], is based on the same principle of amplification, but it measures the amplification of a targeted DNA in real time, as the reaction progresses for each cycle unlike conventional PCR, where the amplified DNA is detected in an endpoint analysis. For detection, specialized thermal cyclers (real-time PCR instruments) are required to measure the fluorescence signal of amplification. The change in fluorescence over time is used to calculate the amount of amplified target sequence (amplicon) generated in each cycle. By plotting fluorescence against the cycle number, the real-time PCR instrument generates an amplification plot that represents the accumulation of product over the duration of the entire PCR (Fig. 5.2) [5]. The *baseline* of the real-time PCR reaction refers to the signal level during the initial cycles of PCR (usually cycles 3–15), in which there is little change in fluorescent signal. The low-level signal of the baseline can be determined as the background or the "noise" of the reaction. The baseline is set to allow accurate determination of the threshold cycle (Ct). The *threshold* of the real-time PCR reaction distinguishes relevant amplification signal from the background. The threshold cycle (Ct) is the cycle number at which the fluorescent signal of the reaction crosses the threshold. This assumes that the PCR is operating at 100% efficiency (i.e., the amount of product doubles perfectly during each cycle) in both reactions. As the template amount decreases, the cycle number at which significant amplification is seen increases.

Real-Time PCR Fluorescent Detection Systems

Real-time PCR chemistries have been categorized into two major groups. The first group consists of double-stranded DNA intercalating agents, such as SYBR Green I and EvaGreen, whereas the second group comprises fluorophore-labeled oligonucleotides [6, 7]. Among the DNA-binding agents, SYBR® Green I (Applied Biosystems/Thermo Fisher Scientific) is the most commonly used for real-time PCR. The fluorescent-based oligonucleotides have been further divided into three subgroups (1) primer-probes (Scorpions, Amplifluor®, LUX™, Cyclicons, Angler®); (2) probes; hydrolysis (TaqMan, MGB-TaqMan, Snake assay) and hybridization (Hybprobe or FRET, Molecular Beacons, HyBeacon™, MGB-Pleiades, MGB-Eclipse, ResonSense®, Yin-Yang or displacing); and (3) analogues of nucleic acids (PNA, LNA®, ZNA™, non-natural bases: Plexor™ primer, Tiny-Molecular Beacon). Figure 5.3 shows the principles of the detection systems representative for SYBR® Green I (A), hydrolysis probe (B), and hybridization probe (C) [7]. TaqMan probes are the most widely used hydrolysis probes in clinical practice, while other types of probes have been described (e.g., Beacons and Scorpions); the latter probes do not have significant market penetration and are not discussed further (see Navarro et al. [6] for detailed information about real-time PCR detection chemistry).

SYBR Green I

SYBR® Green I (Applied Biosystems/Thermo Fisher Scientific) is a commonly used fluorescent dye that binds double-stranded DNA mol-

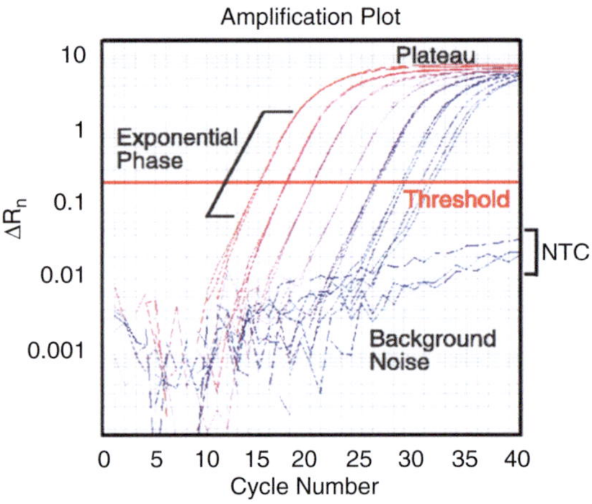

Fig. 5.2 Graphical representation of real-time PCR amplification plot. Rn is the fluorescence of the reporter dye divided by the fluorescence of a passive reference dye; i.e., Rn is the reporter signal normalized to the fluorescence signal of Applied Biosystems™ ROX™ Dye. An amplification plot shows the variation of log (ΔRn—Rn minus the baseline) with PCR cycle number. The CT (cycle threshold) value is the cycle number at which the fluorescence signal (ΔRn) of a given sample crosses the threshold value given by the software or decided by the user. The CT value is inversely proportional to the starting concentration of DNA material. DNA amplicon is doubled after every CT value until the reaction reaches plateau. Therefore, CT values are used for absolute and relative quantification of DNA and RNA. *NTC* No Template Control. (Reprinted from Ref. [5]. With permission from Applied Biological Materials, Inc. ©ABM)

ecules by intercalating between the DNA bases (Fig. 5.3a). The resulting DNA-dye-complex absorbs blue light at λ_{max} = 497 nm and emits green light at λ_{max} = 520 nm. Utilization of this technique for real-time PCR fluorescent detection is cost-effective, as it requires only the design of primers. However, one needs for caution when using this dye, since the dye does not discriminate the double-stranded DNA from the PCR products and those from the primer-dimers. In other words, target specificity may present a problem. Additional analysis such as dissociation curve analysis (or melting curve analysis) needs to be performed to determine the specificity of the amplified target, the amplicon. This provides a measurement of the melting temperature or Tm, taken as the point at which 50% of the double-stranded DNA (dsDNA) (dsDNA) molecules are dissociated. All real-time PCR software programs include this analysis.

Hydrolysis Probes (TaqMan Probes)

TaqMan probe is a dual-labeled oligonucleotide conjugated with a reporter fluorochrome (e.g., FAM, VIC, or JOE) and a quencher fluorochrome (e.g., TAMRA) and positioned within the target sequence (Fig. 5.3b). The fluorescent reporter probe is attached to the 5′ end and the quencher to the 3′ end of the oligonucleotide. During PCR, the primers and TaqMan probe anneal to the target DNA. DNA polymerase extends the primer upstream of the probe. If the probe hybridizes to the correct target sequence, the polymerase's 5′ nuclease activity cleaves the probe, releasing a fragment containing the reporter dye. Once cleavage takes place, the released reporter molecule is no longer quenched. Well-designed TaqMan probes are preferred for biomarker analysis due to the accuracy to the target sequence. TaqMan Gene Expression assays are available as single gene assays as well as custom-designed multigene formats (Applied Biosystems/Thermo Fisher Scientific).

Hybridization Probes

Hybridization probes technology uses two juxtaposed sequence-specific probes. One of these probes is labeled with a donor fluorochrome at the 3′ end, and the other probe is labeled with an acceptor fluorochrome at its 5′ end (Fig. 5.3c). When the two fluorochromes are in close vicinity (i.e., within 1–5 nucleotides), the emitted light of the donor fluorochrome will excite the acceptor fluorochrome. This results in the emission of fluorescence, which subsequently can be detected during the annealing phase and first part of the extension phase of the PCR reaction. After each subsequent PCR cycle, more hybridization probes can anneal, resulting in higher fluorescence signals.

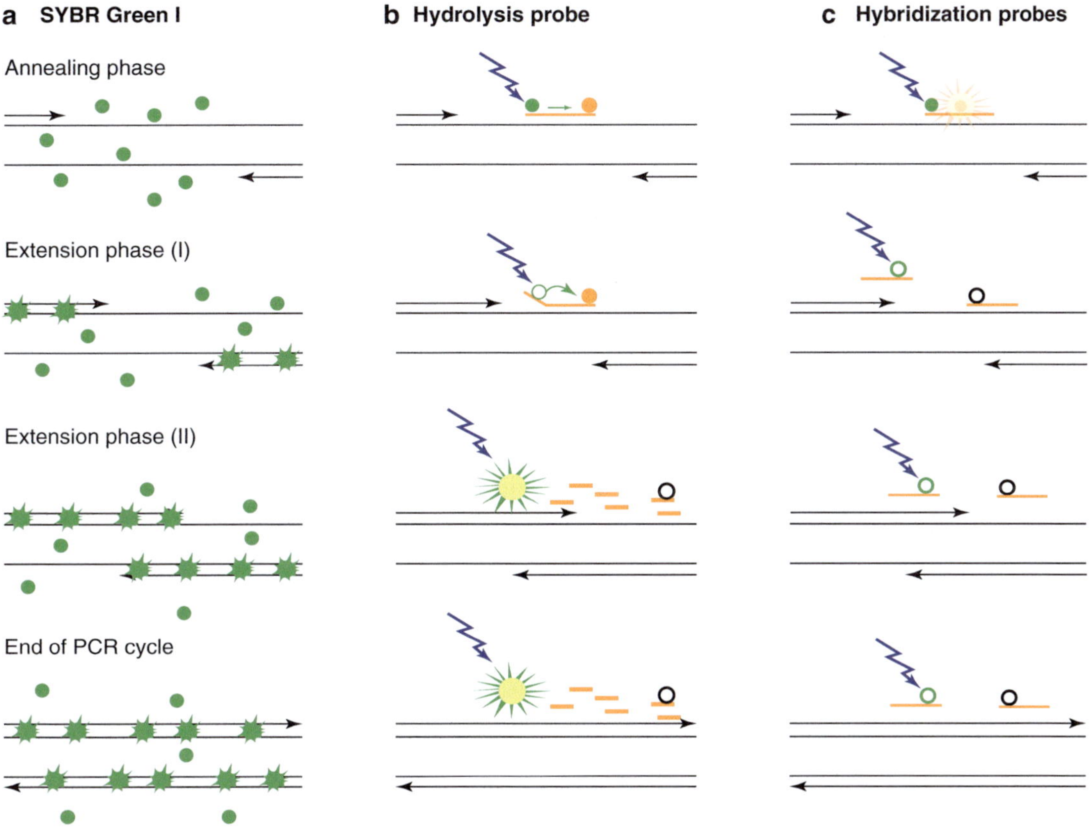

Fig. 5.3 Types of real-time PCR fluorescent detection systems that are commonly used in medical research and/or clinical practice. (**a**) SYBR Green I technique. SYBR Green I fluorescence is enormously increased upon binding to double-stranded DNA. During the extension phase, more and more SYBR Green I will bind to the PCR product resulting in an increased fluorescence. Consequently, during each subsequent PCR cycle, more fluorescence signal will be detected. (**b**) Hydrolysis probe technique. The hydrolysis probe is conjugated with a quencher fluorochrome, which absorbs the fluorescence of the reporter fluorochrome as long as the probe is intact. However, upon amplification of the target sequence, the hydrolysis probe is displaced and subsequently hydrolyzed by the Taq polymerase. This results in the separation of the reporter and quencher fluorochrome and consequently the fluorescence of the reporter fluorochrome becomes detectable. During each consecutive PCR cycle, this fluorescence will further increase because of the progressive and exponential accumulation of free reporter fluorochromes. (**c**) Hybridization probes technique. In this technique one probe is labeled with a donor fluorochrome at the 3′ end, and a second probe is labeled with an acceptor fluorochrome. When the two fluorochromes are in close vicinity (i.e., within 1–5 nucleotides), the emitted light of the donor fluorochrome will excite the acceptor fluorochrome. This results in the emission of fluorescence, which subsequently can be detected during the annealing phase and first part of the extension phase of the PCR reaction. After each subsequent PCR cycle, more hybridization probes can anneal resulting in higher fluorescence signals. (Reprinted from van der Velden et al. [7]. With permission from Springer Nature)

Digital PCR

Digital PCR (dPCR) is a newer method over traditional PCR method that is used to quantify and amplify nucleic acids including DNA, cDNA, and RNA [8]. It is used for absolute quantification and rare allele detection relative to conventional qPCR. It directly counts the number of target molecules rather than depending on reference standards or endogenous controls. A sample is diluted and partitioned into separate reaction chambers so that each contains not more than one

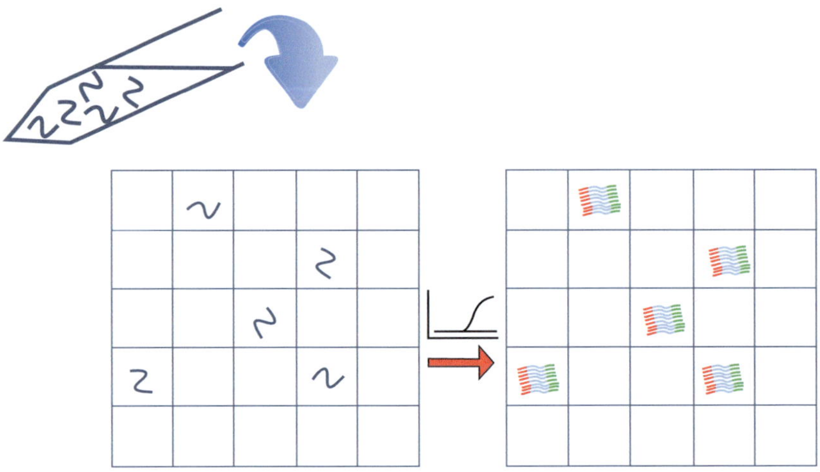

Fig. 5.4 Basic concept of digital PCR. Digital PCR is performed by diluting a sample into many partitions and counting up the number of partitions in which a reaction occurs. (Reprinted from Baker [8]. With permission from Springer Nature)

copy of the sequence of interest (Fig. 5.4). Copies of a DNA molecule can be counted by the number of "positive" partitions (in which the sequence is detected) versus "negative" partitions (in which it is not). A sample can be fractionated into approximately 20,000 oil droplets, where PCR amplification occurs (digital droplet PCR-ddPCR). Although its reagents and workflows are similar to TaqMan probe-based assays, the sample partitioning is the key aspect of the dPCR or ddPCR technology.

Comparison of qPCR Versus Digital PCR

Quantitative real-time PCR (qPCR) is the "gold-standard" method for quantification of gene expression (Table 5.2). qPCR can be used for both qualitative and quantitative analysis. Quantitative PCR is also the application of choice for relative gene expression if the differences are >2-fold. Current automation and high-throughput capabilities of qPCR are superior to dPCR.

Although qPCR can be used for a wide variety of applications, certain aspects of dPCR may present a better choice. In particular, sample partitioning and absolute quantification (instead of standard curve relative quantification) in dPCR increase the precision, high sensitivity, and reproducibility. Examples of these applications include rare allele detection, copy number variation, gene expression for <2-fold differences, quantification of NGS libraries, detecting low-abundance RNA, pathogen detection, and viral load detection. The first commercially available dPCR system was introduced in 2006 by Fluidigm followed by other companies such as Bio-Rad and RainDance. Fluidigm offers two systems that mix samples with reagents, partition the reaction mixture, perform thermocycling, and read results within each partition.

Table 5.2 Comparison of qPCR and dPCR

Parameter	qPCR	dPCR
Standard curve	Yes	No
Sample partitioning	No	Yes
Real-time vs endpoint	Real-time	Endpoint
Quantitation	Real-time	Absolute

Diagnostic Applications of PCR in Clinical Oncology

Quantitative real-time PCR is a valuable technology used in the diagnostics of cancer. In oncology, this technique is widely used in the detection and quantification of gene fusions (e.g., BCR-ABL gene fusions). In spite of the real-time quantitative nature of the technique compared to regular PCR, quantification of reference genes and their reliability pose some challenges in clinical samples. Despite these challenges, some of the most successful applications in oncology include the following.

Diagnostic DNA Mutation Analysis by PCR

Single Gene Analysis

Treatment of colorectal cancer with anti-epidermal growth factor receptor (EGFR) therapy requires demonstration of *RAS* mutation status (both *KRAS* and *NRAS*). In non-small cell lung cancer (NSCLC) and melanoma, assessment of *RAS* mutation status can be helpful for therapeutic decision strategies. Currently, some PCR assays that are FDA-approved are available for FFPE tissue determination of *KRAS* mutation status [9] (Table 5.3). KRAS mutation assays are important companion diagnostic tests to guide anti-EGFR antibody treatment for metastatic colorectal cancer. Similarly, qPCR assays can be used for detection of minimal residual disease in leukemia/lymphoma. A classic example is the quantification of BCR-ABL-positive cells post-induction chemotherapy/transplantation in acute lymphoblastic leukemia (ALL) [10]. Table 5.3 further lists other FDA-approved PCR or real-time PCR-based tests in cancer biomarker research.

Next-Generation Mutational Panels in Lung and Colon Cancer

Mutation profiling of non-small cell lung cancer (NSCLC) and metastatic colorectal cancer (mCRC) can be challenging. It requires frequent, real-time monitoring to detect secondary mutations that confer acquired resistance to therapy. To overcome these challenges, lung and colon cancer gene panels have been established that detect mutations with minimal input DNA. These panels enable mutation profiling of circulating tumor DNA (ctDNA) and circulating tumor cells (CTC) from plasma, in addition to solid tumor tissue. The highly multiplexed panels target mutations of known significance in BRAF, EGFR, ERBB2, KRAS, NRAS, and PIK3CA genes from the blood. These panels cover the initial stratification and monitoring disease progression.

Lastly, PCR has also been used to identify single-nucleotide polymorphisms in DNA as exemplified in patients with breast cancer harboring *ESR1* and *PGR* polymorphisms [11].

Real-Time Reverse Transcription-qPCR (RT-qPCR) as Multiple Gene Signature Assays in Cancer

Mostly, RT-qPCR assays are used in prognostic-multiple gene signatures such as Onco*type* DX® test for breast, colon, and prostate cancers. In RT-qPCR, RNA is first converted to cDNA and qPCR is performed subsequently. These assays are optimized for quantification of RNA extracted from formalin-fixed paraffin-embedded (FFPE) tumor tissue. The Oncotype DX® is a CLIA-approved RT-qPCR assay that measures the expression of 21 genes consisting of 16 cancer-related genes belonging to proliferation, invasion, and HER2 and ER pathways in addition to 5 reference genes [12].

Summary

PCR technologies have revolutionized laboratory-based analysis of nucleic acids. In particular, qPCR-based assays have become the gold standard for biomarker analysis (prognostic and predictive) due to their quantitative nature in clinical practice. Multigene mutational and expression panels are being incorporated in "Standard of Care" guidelines for the treatment of cancer. The scope of these assays will expand in the clinical decision-making, as they are being evaluated in prospective clinical trials.

Table 5.3 FDA-cleared or FDA-approved PCR-based tests

Test[a]	PCR method	Application
cobas® EGFR Mutation Test v2 Roche Molecular Systems, Inc.	Real-time PCR	Qualitative detection of defined mutations (Exon 19 deletions and L858R, T790M) of the epidermal growth factor receptor (EGFR) gene in non-small cell lung cancer (NSCLC) patients
ARUP Laboratories, Inc.	Qualitative PCR	*KIT* D816V Mutation Detection by PCR for Gleevec Eligibility in Aggressive Systemic Mastocytosis (ASM)
Abbott RealTime IDH2	Qualitative PCR	(PCR) assay for the qualitative detection of single-nucleotide variants (SNVs) coding nine isocitrate dehydrogenase-2 (IDH2) mutations (R140Q, R140L, R140G, R140W, R172K, R172M, R172G, R172S, and R172W) in DNA extracted from human blood (EDTA) or bone marrow (EDTA) in acute myeloid leukemia (AML) patients for treatment with enasidenib (IDHIFA®)
LeukoStrat® CDx FLT3 Mutation Assay. Invivoscribe Technologies, Inc.	PCR	In vitro diagnostic test designed to detect internal tandem duplication (ITD) mutations and the tyrosine kinase domain mutations D835 and I836 in the FLT3 gene in genomic DNA extracted from mononuclear cells obtained from peripheral blood or bone marrow aspirates of patients diagnosed with acute myelogenous leukemia (AML)
therascreen® EGFR RGQ PCR Kit Qiagen Manchester, Ltd.	Real-time PCR	Qualitative detection of exon 19 deletions and exon 21 (L858R) substitution mutations of the epidermal growth factor receptor (EGFR) gene in DNA derived from formalin-fixed paraffin-embedded (FFPE) non-small cell lung cancer (NSCLC) tumor tissue
The cobas® KRAS Mutation Test Roche Molecular Systems, Inc.	Real-time PCR	Detection of seven somatic mutations in codons 12 and 13 of the KRAS gene in DNA derived from formalin-fixed paraffin-embedded human colorectal cancer (CRC) tumor tissue. The test is intended to be used as an aid in the identification of CRC patients for whom treatment with Erbitux® (cetuximab) or with Vectibix® (panitumumab) may be indicated based on a no mutation detected result
BRACAnalysis CDx™ Myriad Genetic Laboratories, Inc.	PCR/ multiplex PCR	In vitro diagnostic device intended for the qualitative detection and classification of variants in the protein coding regions and intron/exon boundaries of the BRCA1 and BRCA2 genes using genomic DNA obtained from whole blood specimens collected in EDTA
		Large deletions and duplications in BRCA1 and BRCA2 are detected using multiplex PCR. Results of the test are used as an aid in identifying ovarian cancer patients with deleterious or suspected deleterious germline BRCA variants eligible for treatment with Lynparza™ (olaparib)
therascreen KRAS RGQ PCR Kit	Real-time PCR	Detection of seven somatic mutations in the human KRAS oncogene, using DNA extracted from formalin-fixed paraffin-embedded (FFPE), colorectal cancer (CRC) tissue
THxID™ BRAF Kit bioMérieux Inc.	Real-time PCR	Qualitative detection of the BRAF V600E and V600K mutations in DNA samples extracted from (FFPE) human melanoma tissue used as an aid in selecting melanoma patients for treatment with dabrafenib (Tafinlar) and trametinib (Mekinist)
cobas EGFR Mutation Test Roche Molecular Systems, Inc.	Real-time PCR	Qualitative detection of exon 19 deletions and exon 21 (L858R) substitution mutations of the epidermal growth factor receptor (EGFR) gene in DNA derived from formalin-fixed paraffin-embedded (FFPET) human non-small cell lung cancer (NSCLC) tumor tissue. The test is intended to be used as an aid in selecting patients with NSCLC for whom Tarceva® (erlotinib)
COBAS 4800 BRAF V600 Mutation Test	Real-time PCR	Qualitative detection of the BRAF V600E mutation in DNA extracted from FFPE melanoma tissue used as an aid in selecting melanoma patients for treatment with vemurafenib

[a]Modified from the FDA site: https://www.fda.gov/MedicalDevices/ProductsandMedicalProcedures/InVitroDiagnostics/ucm330711.htm

References

1. Mullis KB, Faloona FA. Specific synthesis of DNA in vitro via a polymerase-catalyzed chain reaction. Methods Enzymol. 1987;155:335–50.
2. Saiki RK, Scharf S, Faloona F, Mullis KB, Horn GT, Erlich HA, et al. Enzymatic amplification of beta-globin genomic sequences and restriction site analysis for diagnosis of sickle cell anemia. Science. 1985;230(4732):1350–4.
3. Bustin SA, Benes V, Garson JA, Hellemans J, Huggett J, Kubista M, et al. The MIQE guidelines: minimum information for publication of quantitative real-time PCR experiments. Clin Chem. 2009;55(4):611–22.
4. Steitz TA. A mechanism for all polymerases. Nature. 1998;391(6664):231–2.
5. Real-time PCR/quantitative PCR (qPCR) – an introduction. Available from: https://www.abmgood.com/marketing/knowledge_base/polymerase_chain_reaction_qpcr.php.
6. Navarro E, Serrano-Heras G, Castano MJ, Solera J. Real-time PCR detection chemistry. Clin Chim Acta. 2015;439:231–50.
7. van der Velden VH, Hochhaus A, Cazzaniga G, Szczepanski T, Gabert J, van Dongen JJ. Detection of minimal residual disease in hematologic malignancies by real-time quantitative PCR: principles, approaches, and laboratory aspects. Leukemia. 2003;17(6):1013–34.
8. Baker M. Digital PCR hits its stride. Nat Methods. 2012;9(6):541–4.
9. Cree IA. Diagnostic RAS mutation analysis by polymerase chain reaction (PCR). Biomol Detect Quantif. 2016;8:29–32.
10. Cazzaniga G, Lanciotti M, Rossi V, Di Martino D, Arico M, Valsecchi MG, et al. Prospective molecular monitoring of BCR/ABL transcript in children with Ph+ acute lymphoblastic leukaemia unravels differences in treatment response. Br J Haematol. 2002;119(2):445–53.
11. Hertz DL, Henry NL, Kidwell KM, Thomas D, Goddard A, Azzouz F, et al. ESR1 and PGR polymorphisms are associated with estrogen and progesterone receptor expression in breast tumors. Physiol Genomics. 2016;48(9):688–98.
12. Paik S, Shak S, Tang G, Kim C, Baker J, Cronin M, et al. A multigene assay to predict recurrence of tamoxifen-treated, node-negative breast cancer. N Engl J Med. 2004;351(27):2817–26.

Introduction to Microarray Technology

6

Nallasivam Palanisamy

Introduction

A DNA microarray is a small analytic device (plastic chip or glass slide) with thousands of short, single-stranded pieces of DNA from known genes that have been "printed" at precise locations. Microarray technology allows scientists to look through thousands of known genes at a glance enabling the rapid and quantitative analysis of gene expression patterns, genotypes, and disease onset and progression on a genome-wide level. From the clinical perspective, microarrays are key tools in genetic and cancer diagnosis allowing clinicians to identify specific subtypes within an overall disease category based on differences in gene expression, thus aiding in therapeutic interventions. Specifically, the field of oncology has greatly benefited with the advent of gene expression profile, copy number, and SNP microarrays. Genetic modifications such as insertions and deletions and signal transduction pathway analysis are providing researchers precious clues into the cause and effect of many different types of cancers, leading to appropriate diagnoses and treatment. Microarray technology has already shown its promising application in discerning heterogeneity of large B cell lymphoma in patients with respect to their gene expression and survival rates [1, 2]. Additionally, in heterogeneous diseases like breast cancer, partitioning of cancers into distinct molecular subtypes based on gene expression patterns leads to the development of new therapeutic successes [3]. Gene expression microarray technology has also been offered as a multi-panel gene test, which is approved by insurance companies and Medicare. For example, the Afirma® gene expression classifier (from Veracyte, Inc., South San Francisco, CA) measures the expression of 142 genes to classify thyroid nodules as benign or suspicious for malignancy. The 23-gene "Percepta Bronchial Genomic Classifier" (also from Veracyte, Inc.) covered by Medicare MolDx program is used to improve the accuracy and safety of lung cancer screening and diagnosis. Another gene expression assay covered by Medicare includes MammaPrint®, a 70-gene breast cancer recurrence assay for early-stage breast cancer patients that assesses the risk for metastatic disease. Other traditional microarray-based tests include ColoPrint®, to assess recurrence risk in stage II colon cancer patients; TargetPrint® for quantitative assessment of ER, PR, and HER2; and BluePrint® (Agendia®, Amsterdam, The Netherlands) for recurrence risk and molecular subtyping of breast cancer tumors.

N. Palanisamy
Department of Urology, Henry Ford Health System, Detroit, MI, USA
e-mail: npalani1@hfhs.org

S. Badve, G. L. Kumar (eds.), *Predictive Biomarkers in Oncology*,
https://doi.org/10.1007/978-3-319-95228-4_6

Principles of Microarray

Microarray technology rests on the ability to deposit many tens of thousands of different DNA sequences as short oligonucleotides, on a small surface, usually a glass slide referred to as a chip. Thousands of spotted sequences from known genes or genomic locations known as probes are immobilized on a chip. Each spot represents a genomic location within a gene or intergenic regions (Figs. 6.1 and 6.2).

Microarray Database

A microarray database is a repository containing microarray gene expression data from different experiments. For easing the accessibility to this data, the National Center for Biotechnology Information (NCBI) has formulated the Gene Expression Omnibus (GEO). It is a data repository facility, which includes data on gene expression from varied sources. This facility was developed to permit researchers to share data so that the same experiments are not repeated. Most of the data, deposited here, is freely available, although some might need additional permissions. The data permits basic scientists to identify the clinical significance of their gene of interest in various human conditions, without the need to find samples and perform gene analyses. In addition to GEO, ArrayTrack, National Cancer Institute mAdb, ImmGen database, Genevestigator, ArrayExpress, The Cancer Genome Atlas (TCGA), GeneNetwork system, and the Lineberger Comprehensive Cancer Center Genomics Core, are the additional resources containing datasets from thousands of experiment types (see Table 6.1).

In order to study the expression pattern of genes in a given sample, RNA is extracted from the cells of interest and either labeled directly, converted to a labeled cDNA, or converted to a T7 RNA promoter tailed cDNA which is further converted to cRNA. A wide variety of methods have been developed for labeling of cDNA or cRNA including incorporation of fluorescently labeled nucleotides during the synthesis, incorporation of biotin-labeled nucleotide which is subsequently detected with fluorescently labeled streptavidin, incorporation of a modified reactive nucleotide to which a fluorescent tag is added later, and a variety of signal amplification methods. The two most commonly used methods are the incorporation of fluorescently labeled nucleotides in the cRNA or cDNA synthesis step and the incorporation of a biotin-labeled nucleotide in the cRNA synthesis step. The labeled cRNA or cDNA is then hybridized to the microarray, the array is washed, and the signal is detected by measuring fluorescence at each spot (Fig. 6.2). In the case of biotin-labeled samples, the array is stained post-hybridization with fluorescently labeled streptavidin. Laser-induced fluorescence is typically measured with a scanning confocal microscope. The intensity of the signal(s) on each spot is taken as a measure of the expression level of the corresponding gene [5].

Applications of Microarrays

Gene Expression Analysis

Given the ready availability of a vast amount of gene expression data in the public domain, extensive analysis of multiple datasets for each type of cancer or disease has led to the development of diagnostic tests based on the expression pattern of gene or gene sets. Clinical tests based on gene expression are developed for the classification of disease into distinct molecular subtypes, to select patients to appropriate treatments and to assess treatment response. Selected examples of such tests are listed in Table 6.2.

Comparative Genomic Hybridization (CGH) Array

Comparative genomic hybridization (CGH) array allows for a locus-by-locus measure of copy number variation with increased resolution as low as 1 kb depending on the number of probes in array design. In this technique, genomic DNA (not RNA) is fluorescently labeled and used to determine the presence of gene loss or amplification (Fig. 6.3). Using CGH only unbalanced copy number changes (gains/losses) can be detected.

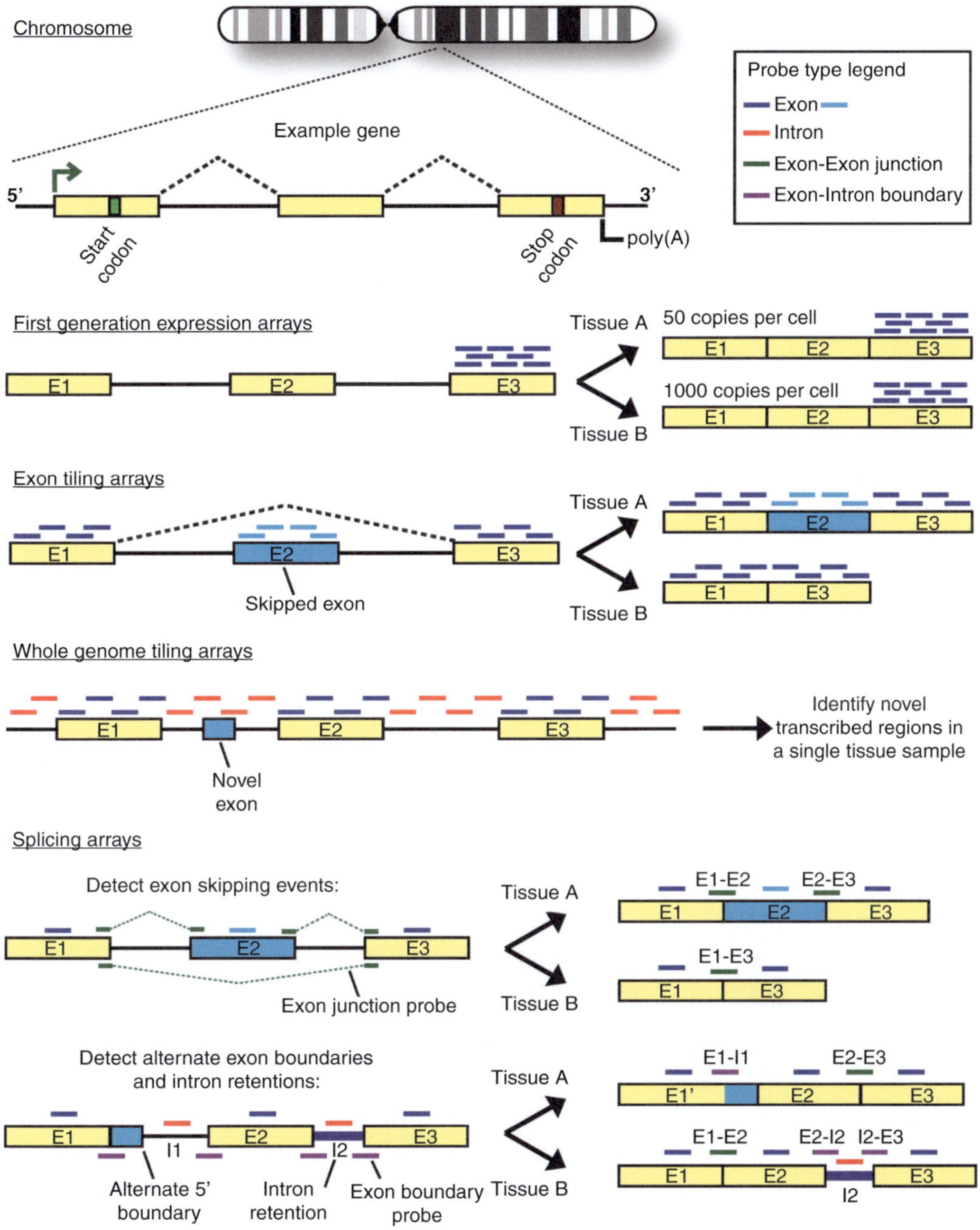

Fig. 6.1 Gene models are depicted as exons (colored rectangles) connected by introns (black lines). Hypothetical differences in mRNA products which can be detected by each array method are depicted to the right of each gene model. In each model, yellow exons are constitutive (i.e., exons that are present in all mature mRNAs), and blue exons are alternative (that may or may not be present in any particular mature mRNA). Differences in array design strategy, particularly the position and types of oligonucleotide probes used, are shown above each gene model as colored horizontal lines. Microarray chips can be used to measure mRNA expression levels for tens of thousands of genes from a sample by hybridizing fluorescently labeled cDNA to the chip, thus allowing parallel analysis for gene expression and gene discovery. An orderly arrangement of the probes on the array is important as the location of each spot on the array is used for the identification of a gene [4]. A similar approach is used to generate copy number and SNP arrays in which array designs contain probes from the specific genomic locations and SNPs, respectively. The fluorescent intensity of each probe will be converted to a numerical value to generate raw data that can be analyzed using bioinformatics tools (Fig. 6.2). (Courtesy of Malachi Griffith, PhD and Marco Marra, PhD, FRSC)

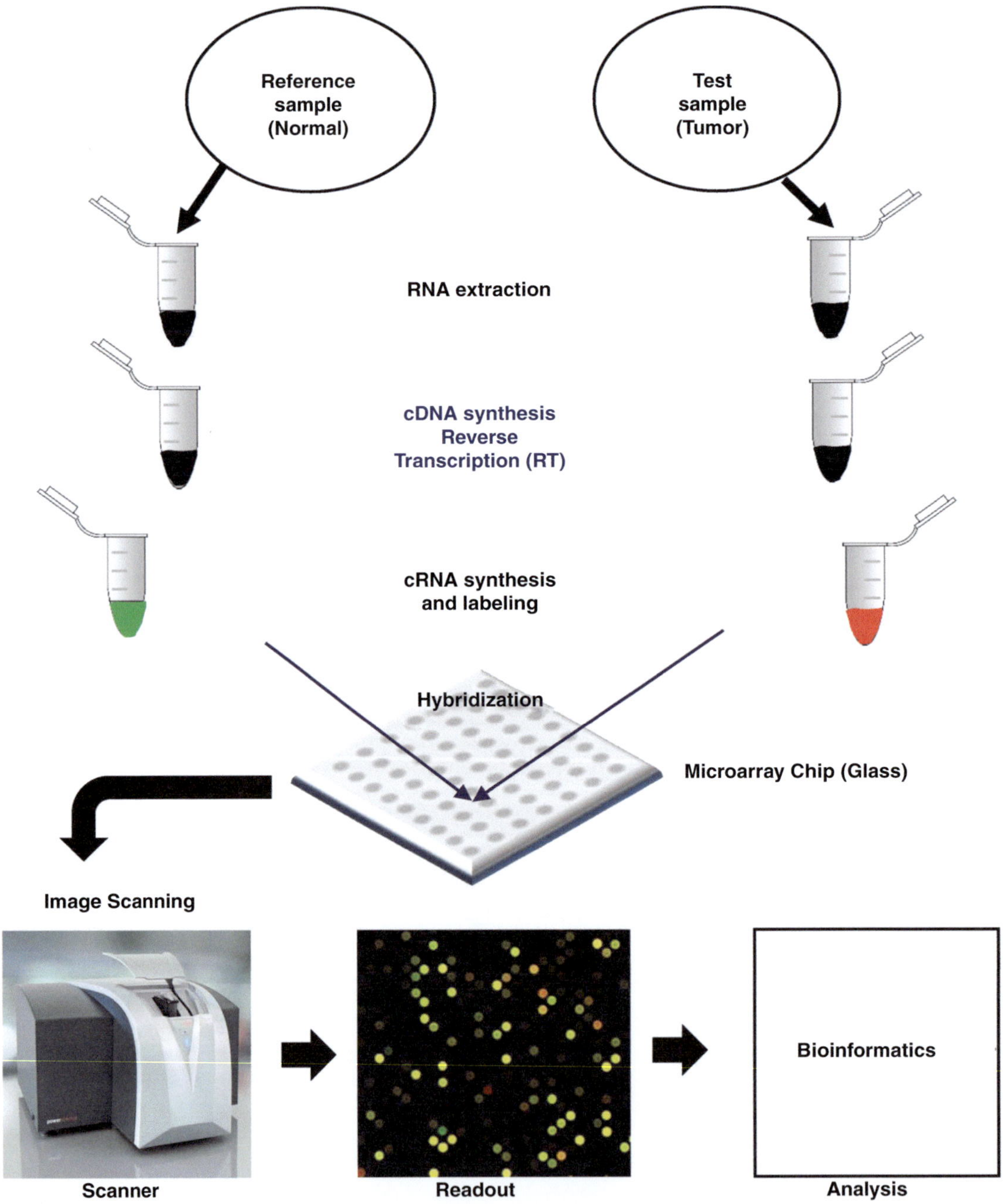

Fig. 6.2 Schematic diagram illustrates the processes involved in gene expression microarray. In order to perform a gene expression array experiment, total RNAs from test (tumor) and reference (normal) samples are extracted and converted into cDNA by reverse transcription, and reporter dyes (test RNA, red; reference RNA, green) are incorporated and hybridized to microarray chip. Following post-hybridization washes, each microarray chip will be scanned to capture the fluorescent intensity for each probe and converted into numerical values to evaluate the expression pattern of genes. Spots with excess green indicate loss of expression in tumor, spots with excess red indicate increased expression in tumor, and yellow spots indicate equal or normal expression. Dark spots indicate no or poor hybridization signal which are eliminated in the analysis. Data from multiple samples will be collected to perform in-depth bioinformatics analysis

Table 6.1 Microarray database sites

Microarray database	Uniform resource locator (URL)
National Center for Biotechnology Information (NCBI)	https://www.ncbi.nlm.nih.gov/guide/genes-expression/
Gene Expression Omnibus (GEO)	https://www.ncbi.nlm.nih.gov/geo/
ArrayTrack	https://www.fda.gov/ScienceResearch/BioinformaticsTools/ArrayTrack8482HCA-PCAStandalonePackage/default.htm
National Cancer Institute mAdb	https://madb.nci.nih.gov/
ImmGen database	http://www.immgen.org/
Genevestigator	https://genevestigator.com/gv/
ArrayExpress	https://www.ebi.ac.uk/arrayexpress/
The Cancer Genome Atlas (TCGA)	https://cancergenome.nih.gov/
GeneNetwork system	http://gn2.genenetwork.org/
The Lineberger Comprehensive Cancer Center Genomics Core	https://www.med.unc.edu/genomicscore
National Cancer Institute mAdb	https://madb.nci.nih.gov/
ImmGen database	http://www.immgen.org/
Genevestigator	https://genevestigator.com/gv/
ArrayExpress	https://www.ebi.ac.uk/arrayexpress/

Table 6.2 List of gene expression array-based tests used in clinical diagnosis

Gene expression test	Disease indication	Readout
MammaPrint®: Agendia®, Amsterdam, the Netherlands	Breast cancer	It is a prognostic test providing information about the likelihood of tumor recurrence in breast cancer. The test also provides information on how the patient would respond to systemic chemotherapy
Afirma® gene expression classifier: Veracyte, Inc., South San Francisco, CA	Thyroid cancer	Preoperative identification of thyroid nodules that are clearly benign based on fine-needle aspiration biopsy in order to potentially avoid surgery on these nodules
ThyraMIR™ microRNA Classifier: Interpace Diagnostics Group, Inc., Parsippany, NJ, USA	Thyroid cancer	miRNA gene expression classifier for testing thyroid cancer patients for assessing the risk of thyroid nodules being either benign or malignant—in order to potentially avoid and reduce unnecessary surgeries
Decipher® Prostate Cancer Classifier: GenomeDx, San Diego, CA	Prostate cancer	Genomic test that informs postoperative decision-making for men who have undergone a radical prostatectomy. Predicts risk of recurrence and metastasis
Percepta™ Bronchial Genomic Classifier: Veracyte, Inc., South San Francisco, CA	Lung cancer	Identify patients with a low risk of lung cancer, significantly reducing the need for invasive, risky, and expensive procedures to ensure a correct diagnosis and management strategy
MyPRS™ (Myeloma Prognostic Risk Signature): Miragen Therapeutics, Boulder, CO	Multiple myeloma	Initial evaluation of newly diagnosed patients with multiple myeloma in order to accurately determine the patient's risk of relapse and, thereby, ascertain which patients are truly "high risk" and unlikely to benefit from standard induction chemotherapy
Tissue of Origin™ Cancer Genetics, Inc., Rutherford, NJ, USA	Cancer of unknown primary	Provides a definitive diagnosis for poorly differentiated or metastatic cancers without a clear primary origin
RosettaGX™ Cancer Origin Test, Rosetta Genomics. Philadelphia, PA	Cancer of unknown primary	Allows for distinguishing between cancers of different origins

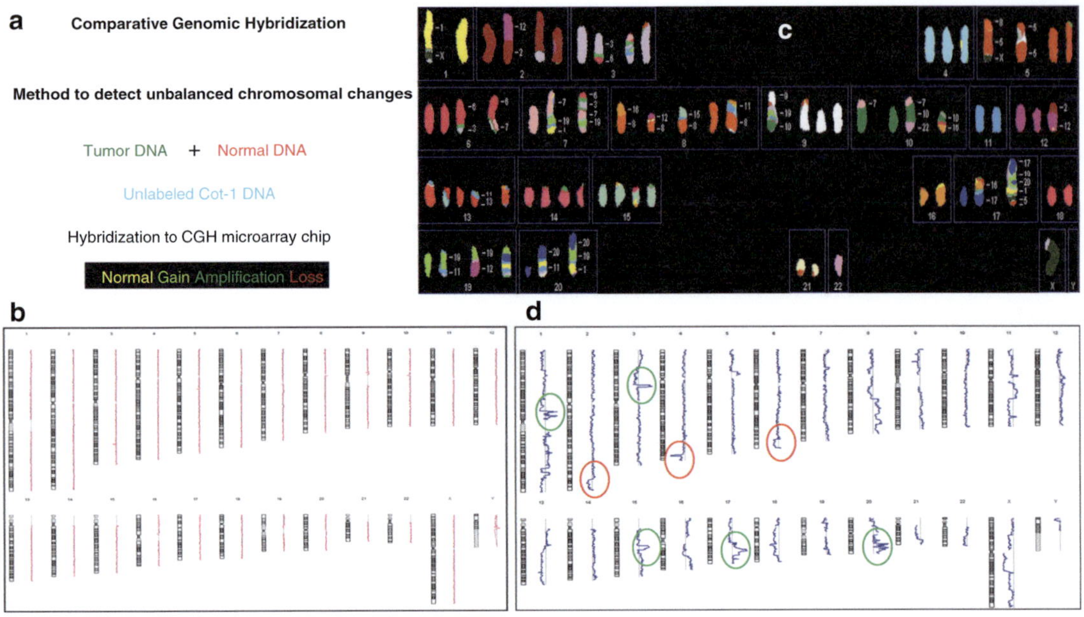

Fig. 6.3 Schematic representation of comparative genomic hybridization array method. (**a**) Genomic DNA is isolated from cancer and normal tissues and labeled with Cy3 (red) and Cy5 (green) fluorochromes. Equal quantities of normal and test DNA samples are mixed and hybridized to microarray chip spotted with a specific DNA probe set followed by the collection of images by a laser scanner. Both cancer and normal samples competitively bind to the spots and result in different fluorescent intensities which are read and captured by specialized software. Fluorescent ratios are plotted against each chromosome to assess the copy number changes. (**b**) Ratio profile analysis of aCGH in normal human chromosomes. Fluorescent intensities plotted against each chromosome show no copy number changes. (**c**) Spectral karyotype image of breast cancer cell line (MCF7) showing multiple chromosome rearrangements and unequal chromosome complement. (**d**) DNA isolated from MCF7 cell line analyzed by aCGH shows copy number gains (green circles) and losses (red circles) in different chromosome regions

Balanced rearrangements (e.g., translocations, inversions) are not detected by this method. Array-based comparative genomic hybridization has been used to map genetic abnormalities in a wide range of cancers as well as constitutional chromosome changes in various genetic diseases. CGH array can be performed using the DNA from fresh frozen tissue, cell lines, and formalin-fixed paraffin-embedded (FFPE) samples. Thus, a large collection of clinical samples stored in the archives of histopathology laboratories are suitable for use for clinical follow-up and clinical trial study. The latter samples are particularly valuable, for testing the influence of specific gene amplification events in response to therapy [4].

Array CGH (aCGH) application is mainly directed at detecting genomic imbalances relating to chromosome abnormalities in cancer, particularly deletions, amplifications, breakpoints, and ploidy abnormalities in human genetic disorders. Prior to microarray-based application, comparative genomic hybridization (CGH) method was developed based on the hybridization of probes to normal metaphase chromosomes, and the fluorescent intensity was measured along the entire length of the chromosome to assess the changes in copy number changes [6, 7]. The major limitation of this approach was the lower limit of the resolution of about 20 Mb which leads to the development of microarray-based method [8]. Identification of recurrent copy number changes at a higher resolution leads to the identification of different molecular subsets of cancer based on focal amplifications and deletions [9–11] and gene fusions [12–16]. Some of the advantages and limitations of array CGH are dependent on the array design with varying densities of probes. Higher probe numbers yield a higher resolution of copy number changes within small genomic interval as small as 1 kb [17–19]. Arrays that use bacterial artificial chromosome (BAC) clones provide sufficient intense signals to detect single-copy number changes and to locate aberration boundaries accurately. However, initial DNA yields of isolated BAC clones are low, and DNA amplification techniques are necessary. These techniques include ligation-mediated polymerase chain reaction (PCR), degenerate primer PCR using one or several sets of primers, and rolling circle amplification [20]. Arrays can also be constructed using cDNA. These arrays yield a high spatial resolution, but the number of cDNAs is limited by the genes that are encoded on the chromosomes, and their sensitivity is low due to cross-hybridization. This results in the inability to detect single-copy changes on a genome-wide scale. The latest approach is spotting the arrays with about one million short oligonucleotides (oligos). The amount of oligos is almost infinite, and the processing is rapid, cost-effective, and easy. Although oligos do not have the sensitivity to detect single-copy changes, averaging of ratios from oligos that map next to each other on the chromosome can compensate for the reduced sensitivity. Mutations in single gene at exon level can be detected using custom-designed targeted aCGH [21]. Distinct molecular subtypes of cancer can be identified using microarrays [1, 22–25].

Limitations of aCGH

aCGH only detects numerical copy number changes rather than structural chromosome aberrations. Even with high-density oligo arrays, the major limitation of the aCGH method is its inability to detect balanced chromosome aberrations such as translocations or inversions, certain forms of polyploidy (addition of full haploid sets of chromosomes), or imbalances not covered by the clones on the array. In addition, chromosomal regions with short repetitive DNA sequences are highly variable between individuals and can interfere with CGH analysis. Therefore, repetitive DNA regions like centromeres and telomeres need to be blocked with unlabeled repetitive DNA (e.g., Cot1 DNA) and/or can be omitted from screening.

Single Nucleotide Polymorphism (SNP) Array

Single nucleotide polymorphisms (SNPs) are the most commonly identified genetic variations in the human genome. Each SNP is recognized as a difference in a nucleotide with the replacement of a nucleotide with another nucleotide, for example, guanine (G) to adenine (A). In the human genome, SNP can occur at every 300 bases on average accounting for about 10 million SNPs. SNPs are used as biological markers to identify the potential disease locus. Most of the SNPs occur at intergenic regions, and the SNPs within a known gene may play a role in disease due to altered function of the gene. In general SNPs in the intergenic regions do not have a direct impact on health or normal development; they are used as markers to identify disease susceptibility locus. Therefore, given the known location of the SNPs in the genome, microarrays can be developed using only oligos from the SNP locus. SNP arrays are a unique type of DNAmicroarrays used to detect SNPs associated with cancer and other genetic diseases. An SNP variation at a single site in DNA is the most frequent (1%) variation in the genome [26]. SNP microarrays are constructed using known SNPs described in the human genome and provide the global analysis of these variations, adding insight into both disease-associated variations through genome-wide association studies (GWAS) and the delineation of copy number variations (gains/losses) that are also common throughout the genome. Therefore, SNP array allows greatly in the facilitation of predictive biomarker research. Once an association between SNP and a cancer is found, SNP microarrays are used as an ideal biomarker to predict the response of a patient to that particular treatment. For example, SNP in CCR5 gene promoter confers resistance/susceptibility to human immunodeficiency virus (HIV-1) infection [27]. The rising interest in the role of SNPs in prostate cancer development and progression is illustrated by a number of extensive genome-wide association studies wherein SNPs are involved in the association of prostate cancer risk. Two susceptibility loci based on SNP evaluation have been identified to be associated with aggressive prostate cancer [28]. SNPs in microRNAs and microRNA targets were also known to predict clinical outcome in prostate cancer [29, 30].

Comparison of CGH Array and SNP Array

Array CGH can only detect nonbalanced chromosomal DNA alterations such as gains and losses. They cannot detect balanced translocations. On the other hand, SNP arrays offer many advantages over CGH platforms including the ability to detect copy number variations and neutral loss of heterozygosity (LOH) events, which comprise 50–70% of the LOH detected in human tumors [31]. The advantages of SNP arrays over other techniques include the ability to assess copy number and genotype in one assay, in addition to high-resolution analysis of changes in the chromosome, scalability and automation, ease of scoring, single-primer assay methodology, minimal total genomic DNA needed, stringent quality control manufacturing, and relatively low cost. SNP arrays can also readily detect characteristic chromosomal lesions in paraffin-embedded renal tumors and can be used to correctly categorize the common subtypes with performance characteristics that are amenable for routine clinical use [32] (Fig. 6.4).

Future Directions

Ever since the development of microarray and comparative genomic hybridization methods in the early 1990s, improvements in the method and the devices have enabled the measurement of relatively small amounts of changes in DNA and RNA in a morphologically similar tissue in addition to identifying gene expression profiles that link disease characteristics to patient outcome. Although these methods have been proved to be useful, they do have certain limitations. Unlike the next-generation sequencing technology which allows a direct measure of DNA and RNA changes in the genome and the detection of novel genes and mutations, microarray technologies are limited with the availability of known information about a given genome at a high-resolution level. Moreover, due to the variation in the hybridization of the probe to targets and saturation limits, the information gathered at a given locus may not be proportional to the actual concentration of the molecules in the sample. Furthermore, due to the complexity of the human genome and significant homology in gene sequences, nonspecific signals due to cross-hybridization are unavoidable, particularly within gene family and splice variants. Next-generation sequencing technology alleviates many of these disadvantages associated with microarray technology by offering the unique advantage of unbiased assessment of DNA or RNA. With the identification of several noncoding RNA in the human genome, current microarrays are not suitable for the evaluation of noncoding RNA expression. Given the radical decline in sequencing costs and the enormous potential of next-generation sequencing (NGS) for the unbiased assessment of DNA, RNA, noncoding RNA, microRNA, and pseudogenes and its accepted use in clinical practice, it is a possibility that microarray technology will be expected to be phased out in the near future and replaced with NGS. That said, microarrays are still utilized in clinical Dx and are considered to be easier to use with less complicated and labor-intensive sample preparation and data analysis than NGS.

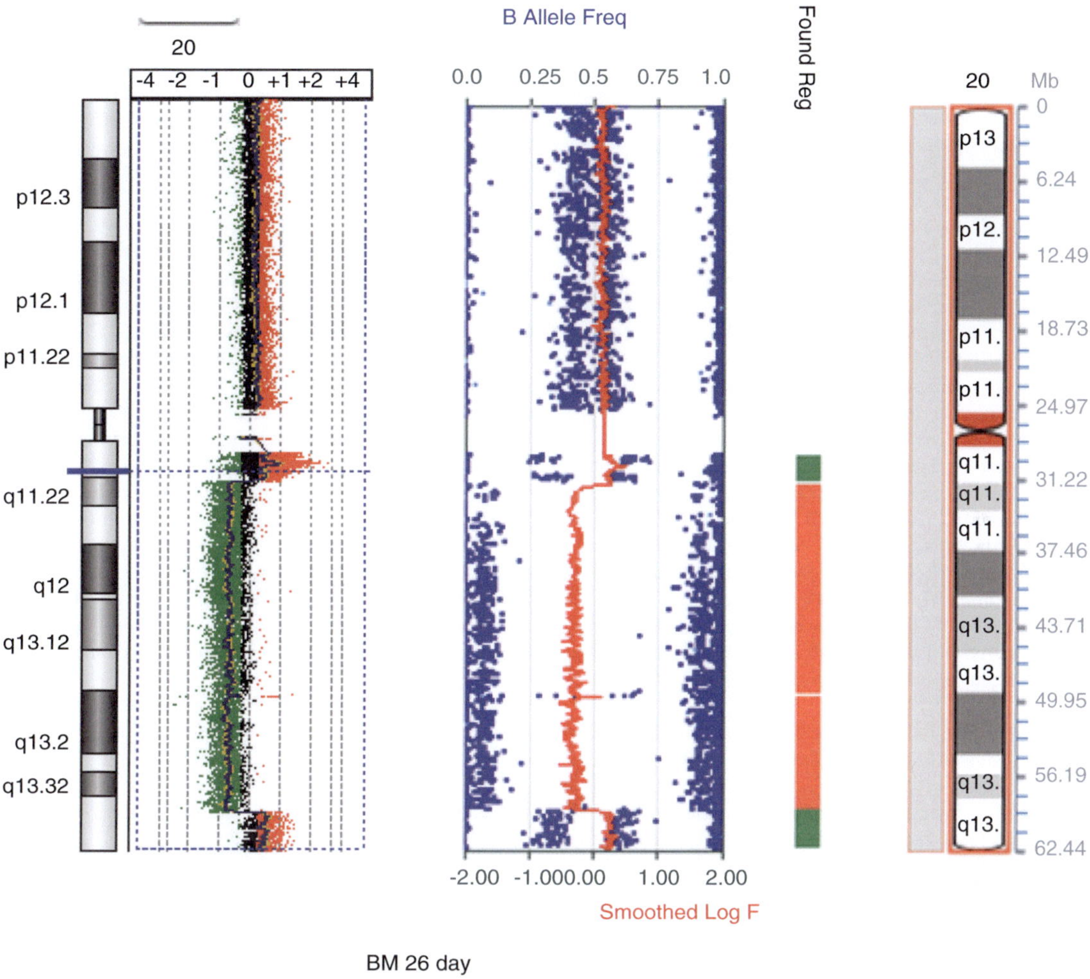

Fig. 6.4 Comparison of CGH vs. SNP array: a comparison of array CGH (left) and SNP array (right) results for chromosome 20 from 26-day-old bone marrow. Deletion, gain, or amplification of different regions of 20q and low-level gain of 20p are shown in green and red colors, respectively. The probes in the custom Agilent array are spaced between 200 bp and 9 kb apart, and the Illumina CytoSNP 12 array probes have a median spacing of 6.1 kb.0. (Reprinted from MacKinnon et al. [33]. With permission from Creative Commons Attribution: http://creativecommons.org/licenses/by/2.0)

References

1. DeRisi J, Penland L, Brown PO, Bittner ML, Meltzer PS, Ray M, Chen Y, Su YA, Trent JM. Use of a cDNA microarray to analyse gene expression patterns in human cancer. Nat Genet. 1996;14:457–60.
2. Alizadeh AA, Eisen MB, Davis RE, Ma C, Lossos IS, Rosenwald A, Boldrick JC, Sabet H, Tran T, Yu X, Powell JI, Yang L, Marti GE, Moore T, Hudson J Jr, Lu L, Lewis DB, Tibshirani R, Sherlock G, Chan WC, Greiner TC, Weisenburger DD, Armitage JO, Warnke R, Levy R, Wilson W, Grever MR, Byrd JC, Botstein D, Brown PO, Staudt LM. Distinct types of diffuse large B-cell lymphoma identified by gene expression profiling. Nature. 2000;403:503–11.
3. Yam C, Mani SA, Moulder SL. Targeting the molecular subtypes of triple negative breast cancer: understanding the diversity to progress the field. Oncologist. 2017;22:1086–93.

4. van de Rijn M, Gilks CB. Applications of microarrays to histopathology. Histopathology. 2004;44:97–108.
5. Bumgarner R. Overview of DNA microarrays: types, applications, and their future. Curr Protoc Mol Biol. 2013. Chapter 22:Unit 22 21.
6. Kallioniemi A, Kallioniemi OP, Sudar D, Rutovitz D, Gray JW, Waldman F, Pinkel D. Comparative genomic hybridization for molecular cytogenetic analysis of solid tumors. Science. 1992;258:818–21.
7. Palanisamy N, Imanishi Y, Rao PH, Tahara H, Chaganti RS, Arnold A. Novel chromosomal abnormalities identified by comparative genomic hybridization in parathyroid adenomas. J Clin Endocrinol Metab. 1998;83:1766–70.
8. Pinkel D, Albertson DG. Array comparative genomic hybridization and its applications in cancer. Nat Genet. 2005;37(Suppl):S11–7.
9. Jonsson G, Staaf J, Vallon-Christersson J, Ringner M, Holm K, Hegardt C, Gunnarsson H, Fagerholm R, Strand C, Agnarsson BA, Kilpivaara O, Luts L, Heikkila P, Aittomaki K, Blomqvist C, Loman N, Malmstrom P, Olsson H, Johannsson OT, Arason A, Nevanlinna H, Barkardottir RB, Borg A. Genomic subtypes of breast cancer identified by array-comparative genomic hybridization display distinct molecular and clinical characteristics. Breast Cancer Res. 2010;12:R42.
10. Lenz G, Wright GW, Emre NC, Kohlhammer H, Dave SS, Davis RE, Carty S, Lam LT, Shaffer AL, Xiao W, Powell J, Rosenwald A, Ott G, Muller-Hermelink HK, Gascoyne RD, Connors JM, Campo E, Jaffe ES, Delabie J, Smeland EB, Rimsza LM, Fisher RI, Weisenburger DD, Chan WC, Staudt LM. Molecular subtypes of diffuse large B-cell lymphoma arise by distinct genetic pathways. Proc Natl Acad Sci U S A. 2008;105:13520–5.
11. O'Hagan RC, Brennan CW, Strahs A, Zhang X, Kannan K, Donovan M, Cauwels C, Sharpless NE, Wong WH, Chin L. Array comparative genome hybridization for tumor classification and gene discovery in mouse models of malignant melanoma. Cancer Res. 2003;63:5352–6.
12. Chinnaiyan AM, Palanisamy N. Chromosomal aberrations in solid tumors. Prog Mol Biol Transl Sci. 2010;95:55–94.
13. Heselmeyer-Haddad K, Sommerfeld K, White NM, Chaudhri N, Morrison LE, Palanisamy N, Wang ZY, Auer G, Steinberg W, Ried T. Genomic amplification of the human telomerase gene (TERC) in pap smears predicts the development of cervical cancer. Am J Pathol. 2005;166:1229–38.
14. Przybytkowski E, Ferrario C, Basik M. The use of ultra-dense array CGH analysis for the discovery of micro-copy number alterations and gene fusions in the cancer genome. BMC Med Genet. 2011;4:16.
15. Tao J, Deng NT, Ramnarayanan K, Huang B, Oh HK, Leong SH, Lim SS, Tan IB, Ooi CH, Wu J, Lee M, Zhang S, Rha SY, Chung HC, Smoot DT, Ashktorab H, Kon OL, Cacheux V, Yap C, Palanisamy N, Tan P. CD44-SLC1A2 gene fusions in gastric cancer. Sci Transl Med. 2011;3:77ra30.
16. Williamson SR, Eble JN, Palanisamy N. Sclerosing TFEB-rearrangement renal cell carcinoma: a recurring histologic pattern. Hum Pathol. 2017;62:175–9.
17. Coe BP, Ylstra B, Carvalho B, Meijer GA, Macaulay C, Lam WL. Resolving the resolution of array CGH. Genomics. 2007;89:647–53.
18. Garnis C, Coe BP, Lam SL, MacAulay C, Lam WL. High-resolution array CGH increases heterogeneity tolerance in the analysis of clinical samples. Genomics. 2005;85:790–3.
19. Tucker T, Montpetit A, Chai D, Chan S, Chenier S, Coe BP, Delaney A, Eydoux P, Lam WL, Langlois S, Lemyre E, Marra M, Qian H, Rouleau GA, Vincent D, Michaud JL, Friedman JM. Comparison of genome-wide array genomic hybridization platforms for the detection of copy number variants in idiopathic mental retardation. BMC Med Genet. 2011;4:25.
20. Coe BP, Lockwood WW, Chari R, Lam WL. Comparative genomic hybridization on BAC arrays. Methods Mol Biol. 2009;556:7–19.
21. Tanner AK, Chin EL, Duffner PK, Hegde M. Array CGH improves detection of mutations in the GALC gene associated with Krabbe disease. Orphanet J Rare Dis. 2012;7:38.
22. Garber ME, Troyanskaya OG, Schluens K, Petersen S, Thaesler Z, Pacyna-Gengelbach M, van de Rijn M, Rosen GD, Perou CM, Whyte RI, Altman RB, Brown PO, Botstein D, Petersen I. Diversity of gene expression in adenocarcinoma of the lung. Proc Natl Acad Sci U S A. 2001;98:13784–9.
23. Perou CM, Sorlie T, Eisen MB, van de Rijn M, Jeffrey SS, Rees CA, Pollack JR, Ross DT, Johnsen H, Akslen LA, Fluge O, Pergamenschikov A, Williams C, Zhu SX, Lonning PE, Borresen-Dale AL, Brown PO, Botstein D. Molecular portraits of human breast tumours. Nature. 2000;406:747–52.
24. Pollack JR, Sorlie T, Perou CM, Rees CA, Jeffrey SS, Lonning PE, Tibshirani R, Botstein D, Borresen-Dale AL, Brown PO. Microarray analysis reveals a major direct role of DNA copy number alteration in the transcriptional program of human breast tumors. Proc Natl Acad Sci U S A. 2002;99:12963–8.
25. Sorlie T, Perou CM, Tibshirani R, Aas T, Geisler S, Johnsen H, Hastie T, Eisen MB, van de Rijn M, Jeffrey SS, Thorsen T, Quist H, Matese JC, Brown PO, Botstein D, Lonning PE, Borresen-Dale AL. Gene expression patterns of breast carcinomas distinguish tumor subclasses with clinical implications. Proc Natl Acad Sci U S A. 2001;98:10869–74.
26. Engle LJ, Simpson CL, Landers JE. Using high-throughput SNP technologies to study cancer. Oncogene. 2006;25:1594–601.
27. Zapata W, Aguilar-Jimenez W, Pineda-Trujillo N, Rojas W, Estrada H, Rugeles MT. Influence of CCR5 and CCR2 genetic variants in the resistance/susceptibility to HIV in serodiscordant couples from Colombia. AIDS Res Hum Retrovir. 2013;29:1594–603.

28. Berndt SI, Wang Z, Yeager M, Alavanja MC, Albanes D, Amundadottir L, Andriole G, Beane Freeman L, Campa D, Cancel-Tassin G, Canzian F, Cornu JN, Cussenot O, Diver WR, Gapstur SM, Gronberg H, Haiman CA, Henderson B, Hutchinson A, Hunter DJ, Key TJ, Kolb S, Koutros S, Kraft P, Le Marchand L, Lindstrom S, Machiela MJ, Ostrander EA, Riboli E, Schumacher F, Siddiq A, Stanford JL, Stevens VL, Travis RC, Tsilidis KK, Virtamo J, Weinstein S, Wilkund F, Xu J, Lilly Zheng S, Yu K, Wheeler W, Zhang H, African Ancestry GC, Prostate Cancer J, Sampson A, Black K, Jacobs RN, Hoover MT, Chanock SJ. Two susceptibility loci identified for prostate cancer aggressiveness. Nat Commun. 2015;6:6889.
29. Bao BY, Pao JB, Huang CN, Pu YS, Chang TY, Lan YH, Lu TL, Lee HZ, Juang SH, Chen LM, Hsieh CJ, Huang SP. Polymorphisms inside microRNAs and microRNA target sites predict clinical outcomes in prostate cancer patients receiving androgen-deprivation therapy. Clin Cancer Res. 2011;17:928–36.
30. Preskill C, Weidhaas JB. SNPs in microRNA binding sites as prognostic and predictive cancer biomarkers. Crit Rev Oncog. 2013;18:327–40.
31. Monzon FA, Hagenkord JM, Lyons-Weiler MA, Balani JP, Parwani AV, Sciulli CM, Li J, Chandran UR, Bastacky SI, Dhir R. Whole genome SNP arrays as a potential diagnostic tool for the detection of characteristic chromosomal aberrations in renal epithelial tumors. Mod Pathol. 2008;21:599–608.
32. Hagenkord JM, Parwani AV, Lyons-Weiler MA, Alvarez K, Amato R, Gatalica Z, Gonzalez-Berjon JM, Peterson L, Dhir R, Monzon FA. Virtual karyotyping with SNP microarrays reduces uncertainty in the diagnosis of renal epithelial tumors. Diagn Pathol. 2008;3:44.
33. MacKinnon RN, Selan C, Zordan A, et al. CGH and SNP array using DNA extracted from fixed cytogenetic preparations and long-term refrigerated bone marrow specimens. Mol Cytogenet. 2012;5:10.

7 Digital and Computational Pathology for Biomarker Discovery

Peter Hamilton, Paul O'Reilly, Peter Bankhead, Esther Abels, and Manuel Salto-Tellez

Digital and Computational Pathology: Transformative Technologies

One of the major turning points in pathology in recent years has been whole slide imaging (WSI) and the development of digital pathology: the rapid digital scanning of entire glass slides and tissue samples at diagnostic resolution, converting tissue sections into large digital images that contain all the necessary morphological, contextual and stain information to interpret, analyse and make diagnostic decisions. This transformation has been driving fundamental changes in digital education, research and primary diagnostics in pathology. Of importance is the pivotal role digital pathology is now playing in tissue biomarker development which includes the ability to digitally archive tissue biomarker WSI, share these online for remote scoring by experts and analyse them using an emerging powerful array of imaging tools.

Known now as *computational pathology*, the development of image analytics to provide objective, quantitative evaluation of tissue patterns and associated molecular biomarkers has progressed enormously in recent years. This includes the development of advanced image analysis toolboxes, tumour recognition algorithms, FDA/CE-IVD-cleared breast IHC algorithms and more recently deep learning neural networks which can make reliable automated computer vision in tissue pathology and biomarker measurement a reality.

These rapid advances in both digital and computational pathology in the last 10 years have been underpinned by the convergence of many disparate technologies into an integrated technology platform that could be truly transformative in pathology and biomarker discovery (Fig. 7.1).

Such is the power of digital and computational pathology technologies that it cuts across the entire spectrum of the drug/companion diagnostic co-development pipeline, driving improvements in biomarker development from early discovery to validation to clinical translation (Fig. 7.2). Many applications of digital and computational pathology are being used today in the field for a wide range of purposes. These are summarized in Table 7.1.

P. Hamilton (✉)
Department of Digital Pathology, Philips UK, Belfast, Northern Ireland, UK
e-mail: peter.hamilton@philips.com

P. O'Reilly · P. Bankhead
Belfast Development Hub, Philips Digital Pathology Solutions, Belfast, UK

E. Abels
Digital Pathology Solutions, Pharma Solutions, Philips Digital Pathology Solutions, Best, The Netherlands

M. Salto-Tellez
Northern Ireland Molecular Pathology Laboratory, Centre for Cancer Research, Department of Cell Biology, Queens University Belfast, Belfast, Antrim, UK

S. Badve, G. L. Kumar (eds.), *Predictive Biomarkers in Oncology*,
https://doi.org/10.1007/978-3-319-95228-4_7

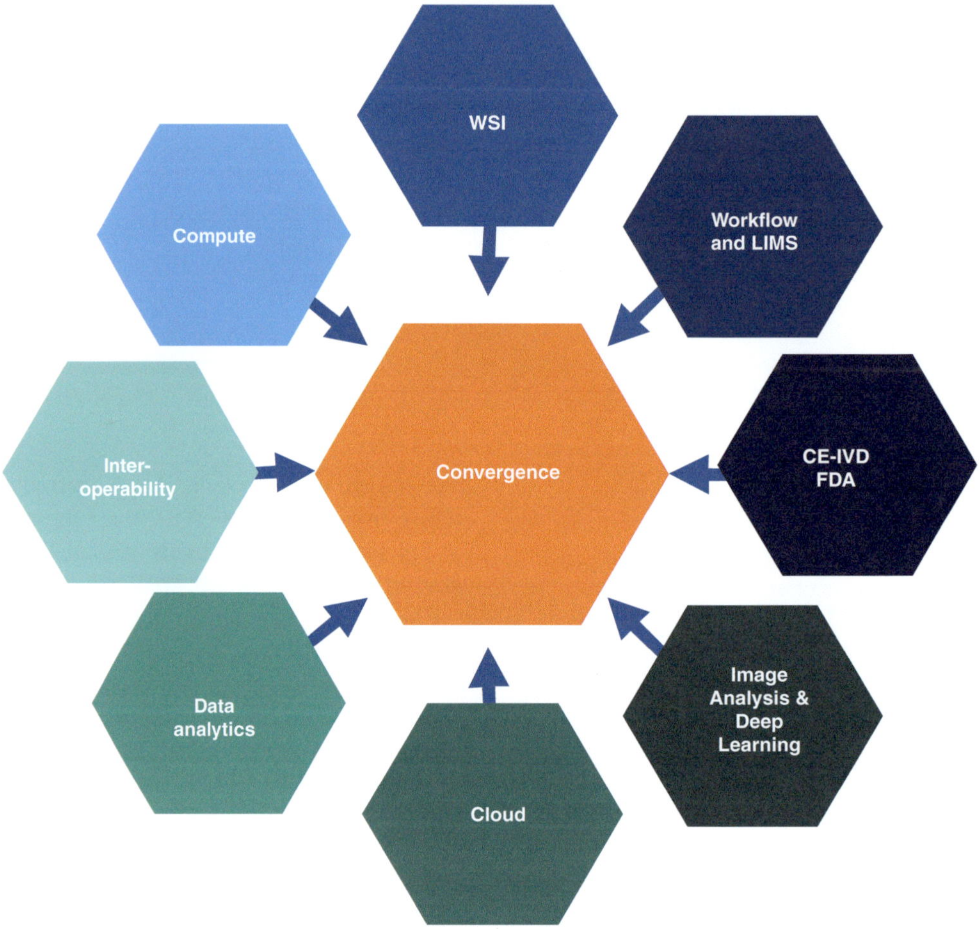

Fig. 7.1 The convergence of technologies has underpinned the growth of digital pathology: whole slide imaging (WSI) providing the tool to image entire tissue samples at high resolution; workflow and LIMS integration allows technologies to be embedded in discovery and diagnostics; regulatory clearance allowing new image technologies to be used for clinical applications; image analysis and machine learning for quantitative tissue biomarker analysis; cloud-enabled platforms for remote access and sharing of image and data; data analytics and informatics for the purposes of data mining and biomarker discovery; interoperability between digital pathology and third-party systems to support communications; and compute power to manage, stream and analyse image content

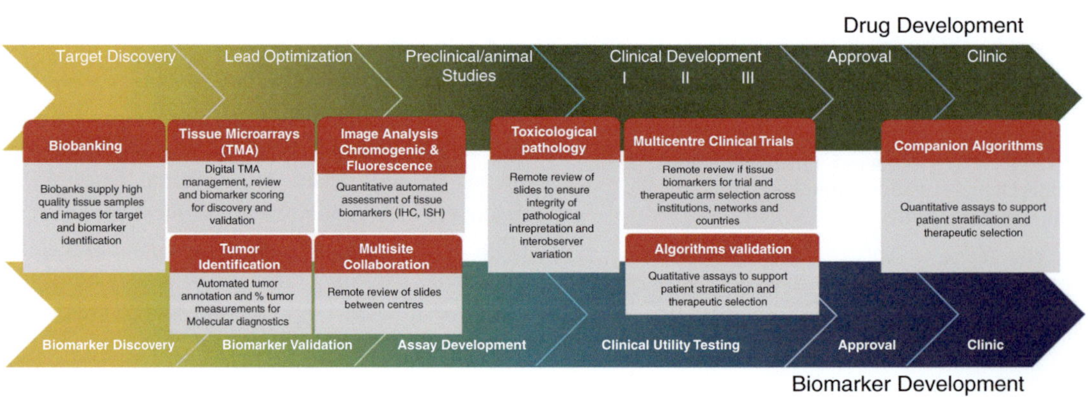

Fig. 7.2 Digital pathology provides a range of valuable tools to support the development of predictive and prognostic biomarkers across the drug/biomarker development pipelines

Table 7.1 Range of emerging applications for digital pathology and image analysis spanning the biomarker discovery process

Digital pathology applications	Value
Biobanking	Archiving of digital images for research
	Digital biobanking
	Automated analysis of archived digital slides
Digital image archives	Archives for training and education in biomarker discovery
	Digital data lakes for discovery
Multisite collaboration	Digital collaboration on images
	Multicentre/multinational clinical trials
	Outsourcing of pathology services
	Centralization of pathological review
Biomarker image analytics	Quantitative biomarker discovery
	Automation
	High-throughput Tissue Microarray (TMA) analysis
	Validation and clinical translation
Tissue microarray scoring	Digital management of TMA experiments
	Remote scoring of TMAs
	Remote access to pathologists
	Quantitative analysis of TMAs using computational algorithms
Immuno-oncology	Quantitative counts of inflammatory cells
	Spatial measurement of inflammatory cells
	Quantitative IHC of immune checkpoint markers, e.g. PD-L1 IHC (fluorescence)
Molecular pathology	Automated tumour annotation
	Quantitation analysis of tumour purity
	Improved reliability of molecular profiling in solid tumours
Companion diagnostics	Validation and verification of new tissue markers
	Regulatory approval of digital pathology
	Companion biomarker algorithms in clinical practice

Precision Diagnostics Using Computational Pathology

Computational Pathology and Image Measurement in Biomarker Research

High levels of subjectivity and poor reproducibility are associated with many aspects of manual visual interpretation of disease classification and associated tissue biomarkers [1]. Computational pathology is a term increasingly being used to describe the application of complex image analysis algorithms to WSI as a means of driving the objectivity and reliability of morphological and biomarker interpretation, in both research and diagnostics. Digital images inherently contain numerical data on each pixel within the image. Computer software in the form of algorithms can be designed to interrogate those image pixels, to extract quantitative data and to classify and segment pixels into well-defined tissue structures, components, cells and nuclei, and provide a rich array of measurements that describe tissues, tissue components, morphological patterns and cancer variants [2]. The advantages of computational pathology and image analytics are clear: together they can provide numerical data that are more objective, reproducible, consistent and reliable than visual interpretation.

Image analytics can also be used to quantitatively measure tissue marker expression, such as IHC detected proteins, within and across different cellular compartments (Fig. 7.3). This can then be expressed quantitatively as a labelling index for the nuclei/cells detected, as an intensity score or as a combined quantitative measure of immunopositivity. Analytical algorithms can also

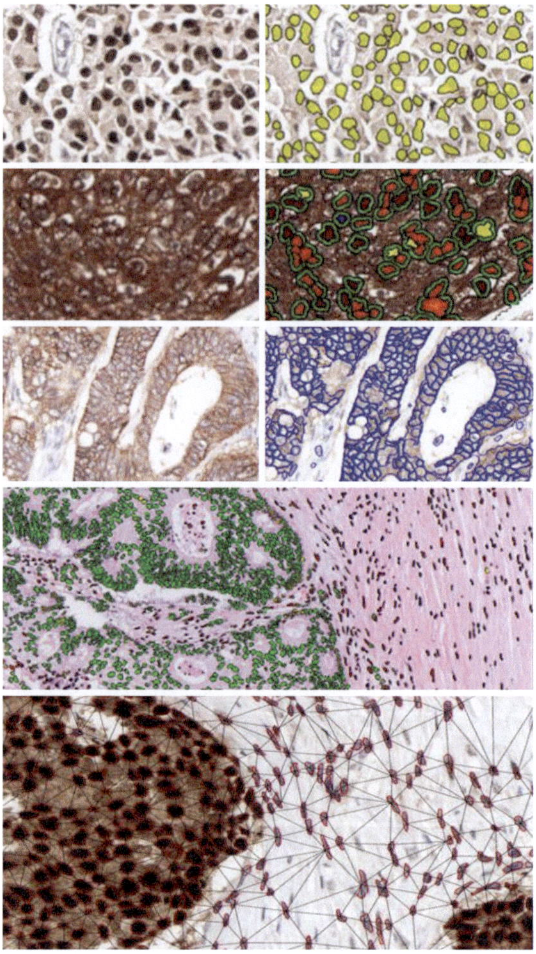

Fig. 7.3 Algorithms can be used to identify and measure a range of biomarkers in tissue components including (i) nuclei, (ii) cytoplasm and (iii) cell membrane. (Images generated using a range of pathology image analysis software platforms)

be used to measure ISH labelling of nucleic acid sequences in both chromogenic and fluorescence images as well as a range of new DNA and RNA labelling techniques.

Numerous studies have shown the improvement that computational pathology and digital image analysis can bring to the quantitative interpretation of tissue biomarkers [2, 3]. These studies have shown improved consistency for assessment of ER, PR, HER2 and Ki67 in breast cancer. This can be further improved by more novel segmentation algorithms that estimate cell membrane boundaries based on membrane boundary modelling following nuclear identification [4, 5] and deep learning [6]. Numerous other established and new biomarkers have been extensively measured in different tissue types using the quantitative image analytics including p53, BCL2, MED1, CD39 and PD-L1 [7, 8], and image analytics is now becoming the benchmark required to ensure consistency, reproducibility and reliability in tissue biomarker experiments.

Despite all the clear advantages of image measurement by computational pathology, extreme care needs to be taken in the use of image analysis for biomarker discovery. Validation studies need to be carefully designed and overseen by experienced pathologists to ensure that the tissue compartments and cellular objects being detected using image analysis are correct and that the expression status reflects the underlying biology of the biomarker. The impact of biomarker heterogeneity needs to be considered to reduce tissue sampling errors and avoid bias. Pre-analytical variables need to be minimized and kept consistent across experiments to avoid the introduction of noise that may impact on the ability to detect biological changes in expression that have clinical correlations. However, if sufficient attention is given to experimental conditions and analytical variables are carefully controlled, computational pathology can drive enormous improvements in tissue biomarker discovery and validation and as a means to benchmark biomarker assessment and standardize large-scale biomarker trials.

Computational Pathology, Standardization and Tissue Quality

As with any form of tissue biomarker evaluation, a variety of *pre-analytical* factors need to be considered (Table 7.2).

Table 7.2 Examples of key pre-analytical variables in tissue analytics including digital pathology that could impact on the quantitative imaging of tissue morphology and molecular markers

Pre-analytical factors in digital and computational pathology
Tissue processing:
Warm ischemia time
Time to fixation
Fixation concentrations
Fixation pH
Dehydration and cleaning times
Section thickness
Temperature and duration of slide drying
Temperature of FFPE block storage
Duration of FFPE block storage
Antibody and hybridization variants
Antibody and hybridization protocols
Counterstain concentrations
Length of glass slide storage
Internal and external controls
Digital pathology:
Scanner type
Scan magnification
Scanner configuration
Image format
Image preprocessing
Image compression

These pre-analytical variables are important considerations when undertaking retrospective digital and computational pathology studies on biomarkers from multiple centres or from single laboratories. Unless strict sample preparation protocols are adhered to within a lab, the use of archival tissue samples in biomarker studies can potentially introduce background variation in expression which is not due to the underlying biology of the tissue, but which is due to how it was handled prior to analysis. This noise will also impact on the quality of the image analytic data. Similarly, in multicentre biomarker trials, lab-to-lab variation in sample preparation will inevitably contribute to variation in the data. Computational pathology can provide reliable data and associated tools to disentangle the noise introduced by lab-to-lab variation from the clinical correlates that are being studied – however, attempts to ensure consistency across laboratories must still be prioritized.

Digital pathology hardware and optics also need to be controlled and processes standardized prior to use in biomarker studies. A few important considerations are listed in Table 7.2. For example, different scanner manufacturers can use different optics, scanning technology and digital preprocessing algorithms to generate WSI. These can look very different to the eye in terms of their background, contrast and colour densities and could impact on visual scoring of biomarkers. Different scanners can also radically impact on computational algorithm data generated in measuring biomarkers, and it is essential that the same scanning platform with the same configuration is used throughout. Some scanners can be configured to use different magnifications, preprocessing methods and compression ratios. This can have an impact on consistency of visual and computational analysis, even when slides are scanned on the same scanner. To avoid this situation, instrumentation settings must be kept consistent throughout the process and control slides used to calibrate the instrument optics where possible.

Computational pathology and biomarker imaging can actually be used to assess tissue quality. It has been shown that image analysis of fluorescently labelled breast cancer panels can measure biomarker antigenicity as a function of time to fixation [9]. A *tissue quality index (TQI)* has been proposed using image analysis to precisely measure the expression of baseline biomarkers in breast cancer which showed correlations with fixation delay [10]. By measuring the ratio of these image measurements, an objective metric on tissue quality can be assessed. This represents a potential step forward in the use of image analysis for quality assessment in large-scale biomarker trials, biobanking and other studies which rely on tissue quality for discovery.

Defining Biomarker Thresholds

One consistent challenge in IHC biomarker imaging is the definition of the appropriate threshold for IHC positivity. Since the data is now numerically defined and continuous, a range of possible positivity thresholds could potentially be used. In most experiments, an effective threshold is defined by eye and "locked" across the analysis of multiple samples. However, some software platforms now allow real-time "mining" and selection of thresholds based on survival impact. Here, thresholds can be scrolled through in real time to identify that point where outcome measures (e.g. survival) show strongest statistical significance and separation. This can provide a more robust and reliable means of defining thresholds which have clinical significance – something which cannot be undertaken using visual scoring methods. However, as with any classification assay, this should be tested on an independent set of test cases to ensure repeatability and avoid overtraining on a specific cohort of cases.

Digital Tissue Microarray (D-TMA) Analysis for Biomarker Discovery

Tissue microarrays (TMAs) provide a widely used platform for the evaluation of new biomarkers allowing the rapid evaluation of expression across multiple samples in a single assay. Digital whole slide scanning of the TMAs and segmentation of the tissue cores into digital maps to create digital TMAs (D-TMA) allow the use of a range of new digital tools for visualization and analysis to further enhance and accelerate biomarker discovery.

Using D-TMA software, pathologists and scientists can review and score TMAs from anywhere in the world using the remote web-based viewing, supporting multisite collaboration across multiple laboratories and large biomarker trials. Viewing high-resolution D-TMAs on screen (as opposed to a microscope) allows the rapid digital review of tissue cores, the movement from one tissue core to the next at the click of a button and the ability to select a location on the TMA map and move to the digital image of that core, and digitally track which tissue cores have been reviewed and which have yet to be scored. If multiple TMA sections from the same block have been labelled with different biomarkers, digital images from these can be co-registered to allow the simultaneous side-by-side review of multiple different markers simultaneously on the same core.

D-TMA software also usually allows the configuration of scoring criteria and forms which can be displayed alongside the digital image of the core. Electronic capture of scores on screen supports rapid remote digital scoring of TMAs with centralized storage and management of biomarker expression data. This is becoming increasingly important in large-scale biomarker trials where multiple pathologists need to review and score potential biomarkers across multiple TMAs. Remote web-based scoring of D-TMAs is also important in capturing interobserver reproducibility data from multiple and geographically dispersed pathologists in multicentre biomarker trials (Fig. 7.4).

Capturing structured TMA images and associated metadata in a digital map provides access to other unique tools that cannot be achieved with glass. One example is the creation of virtual TMAs. Here tissue cores can be selected from across multiple TMAs to create a new "virtual"

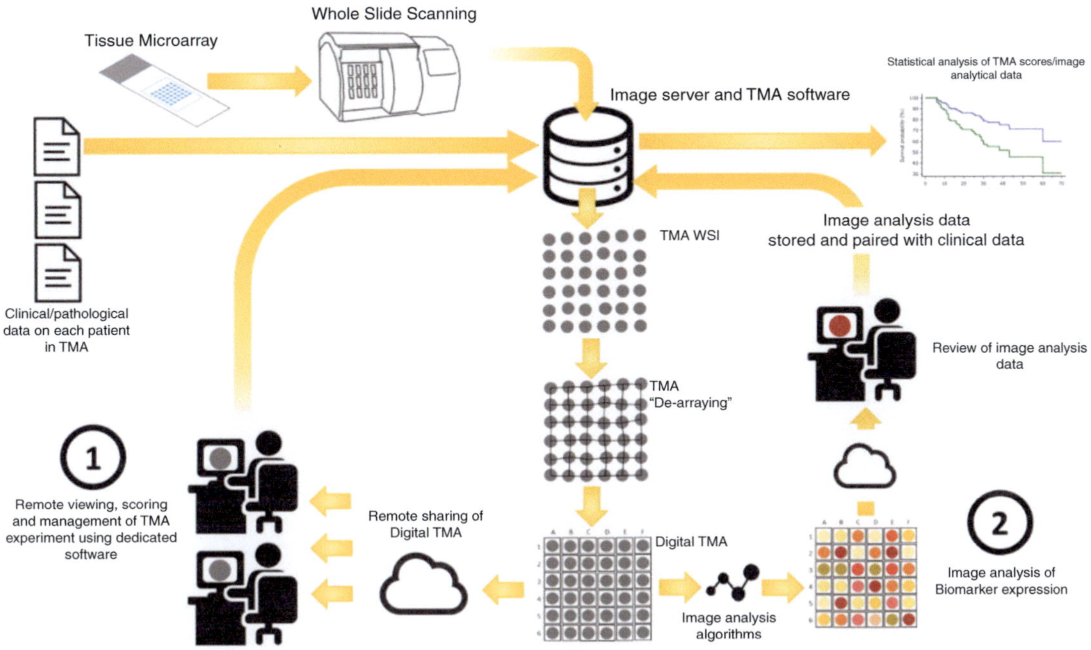

Fig. 7.4 Illustrates the role of digital TMA analysis and TMA management software to facilitate remote scoring by pathologists (1) and automated computational analytics of TMAs (2) to accelerate biomarker discovery and improve objectivity of biomarker interpretation

D-TMA for subsequent evaluation, avoiding distribution bias arising from TMA construction and which can be reviewed and scored remotely as if they were regular TMAs. Alternatively, tissue cores from multiple experiments, selected on the basis of their clinical, pathological or biomarker characteristics, can be retrieved and used to construct a new virtual D-TMA. This is enormously useful in reviewing staining patterns, grouping different cohorts of patients with similar staining patterns, ordering cases based on the biomarker expression, mining biomarker data to identify specific outliers and identifying patterns, expression profiles and relationship with clinical outcomes.

D-TMAs are particularly amenable to automated analysis of biomarkers using image analytics and algorithms – as outlined in section: "Precision Diagnostics Using Computational Pathology". Indeed, the discrete nature of TMAs lends itself to rapid high-throughput screening and quantitative scoring of tissue cores using image analysis (Figs. 7.4 and 7.5) as a means of screening candidate biomarkers and rapidly determining clinical correlations. Most commercial image analysis platforms have developed TMA modules for the purposes of automated TMA analytics, and open-source platforms such as QuPath (section: "Open Platforms and Software in Digital Pathology") also provide this capability.

Standard nuclear, cytoplasmic and cell membrane algorithms can also be applied to D-TMAs to quantitatively measure the expression of biomarkers (see Fig. 7.3) but as before need to be validated by experienced pathologists to ensure precise segmentation and quantitation.

One additional approach, often required in the rapid analysis of D-TMAs, is the automated identification of tumour regions to sub-select cells for biomarker expression measurement. Some organizations have approached this by double staining TMAs with a pan-cytokeratin marker to identify epithelial components, segment these with image analysis and then superimpose these regions on digital images of sequential IHC-stained

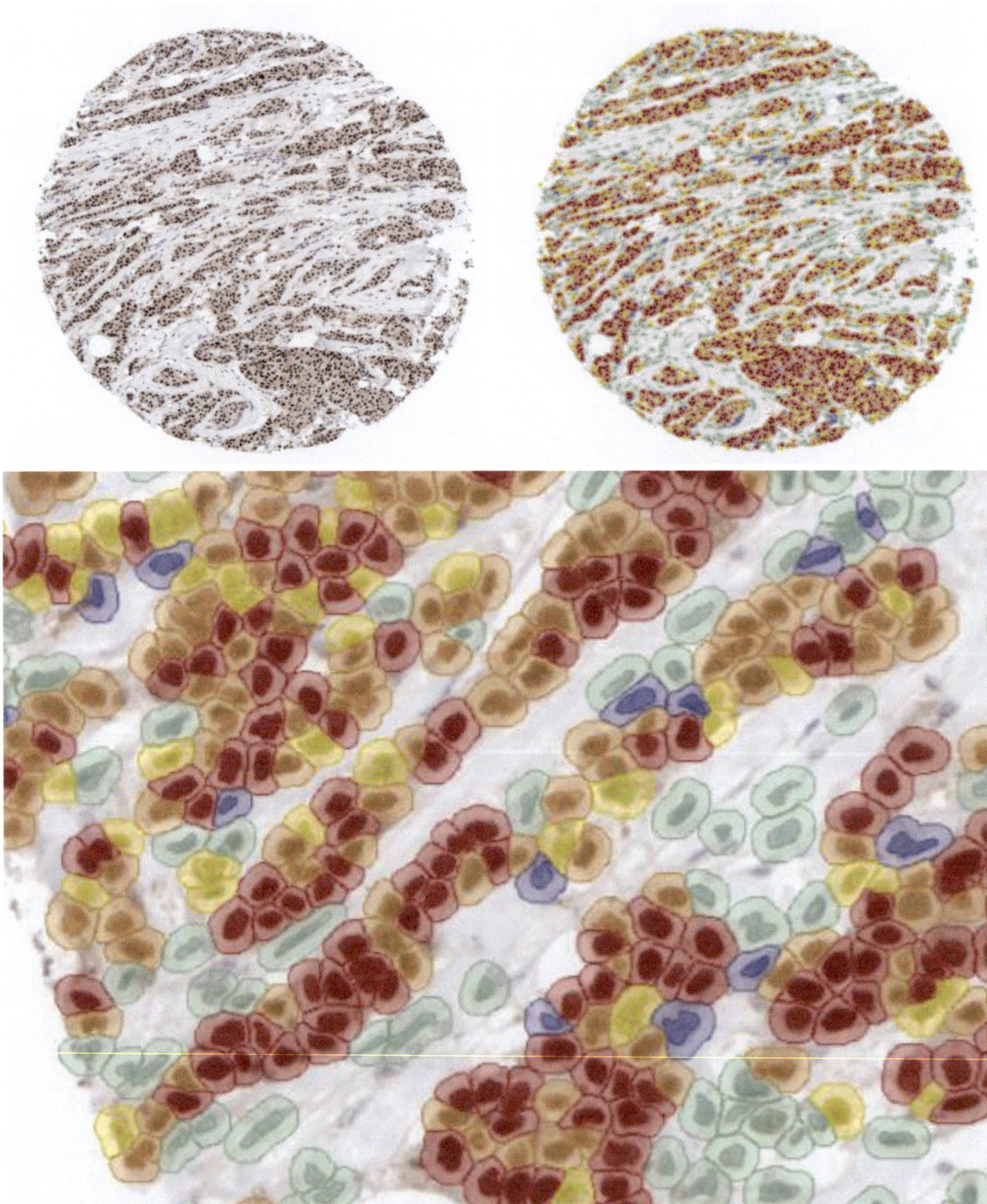

Fig. 7.5 (Top) Automated tumour identification on TMAs for the purposes of tumour cell-specific biomarker analysis. (Bottom) Illustration showing the segmentation of a tissue microarray core boundary, nuclei within a tissue core and automated identification of IHC-positive nuclei. This can be used to calculate a positive cell scores. (These images are generated using QuPath software [5])

sections, as masks for biomarker analytics [11]. Increasingly however, tumour identification algorithms are being applied to H&E images or directly to IHC images to preselect key regions on TMAs from which IHC and other biomarkers can be measured (Fig. 7.5).

Finally, TMA image analysis also lends itself to parallelization and high-performance computing [12]. By analysing individual TMA tissue cores across multiple computer processors, the speed at which tissue analysis can be achieved can be significantly accelerated. This can allow the quantitative evaluation of potentially thousands of tissue cores and hundreds of candidate biomarkers to be analysed very rapidly and mined for clinical significance.

Digital Pathology in Immuno-oncology

The field of immuno-oncology (IO) in cancer discovery and therapeutics has grown rapidly in the last number of years, and computational pathology is now finding important applications to support IO biomarker evaluation.

The expression of checkpoint markers such as PD-1/PD-L1 and inflammatory response in tumours are emerging as important tissue biomarkers in prognosis and the selection of patients for IO therapies with the burden falling primarily on pathologists to interpret these markers. For example, PD-L1 expression has become a key factor in determining the treatment for patients with non-small cell lung cancer (NSCLC). There is increasing interest in the use of PD-L1 immunohistochemical staining to guide the use of PD-1-targeted treatments in a variety of cancers, and inevitably pathologists must play a key role in interpreting these new biomarkers [13].

However, PD-L1 stain interpretation is particularly challenging given (i) its heterogeneous expression, (ii) the need to assigning clinically meaningful cut-off thresholds for positivity and (iii) the importance of ensuring consistency and reproducibility between pathologists and between laboratories [14]. A variety of thresholds for positivity are recommended, but assessing PD-1 and PD-L1 staining proves to be particularly challenging with variation and poor reproducibility reported to be high. Depending on the antibody and the method used, the kappa statistic for the scoring of PD-L1 can be low [14].

Improvements can be possible with better definitions, standards and pathologist training. However, an increasing number of studies are applying image analytics to support the measurement of PD-L1 and other immune checkpoint markers, across a range of tissues [8, 15]. Image analysis is also increasingly being used to measure immunophenotype in cancer. This approach demands a contextual evaluation of the tissue structures and the location of inflammatory cells relative to the tumour boundary (Fig. 7.6), with a variety of algorithms have been developed, including deep learning approaches, to automatically identify and digitally map immune cells in cancer (Fig. 7.7) [16].

In addition to simple single or dual staining approaches in research, multispectral IHC and fluorescence is being widely used to measure multiple inflammatory phenotypes simultaneously on a single section. Multiplex approaches to measuring the immune landscape require a unique combination of complex sample preparation, cellular immunohistochemistry, digital image capture and cellular analytics. Some approaches use multicolour kits that allow simultaneous detection of up to six different IHC targets simultaneously on the same FFPE tissue section, including co-localization following spectral unmixing methods. In these more complex experimental settings, with multiple overlapping markers, image analytics and the ability to map the intricate play of different cell type simultaneously is important. Here, digital pathology and computational algorithms are driving discovery and development.

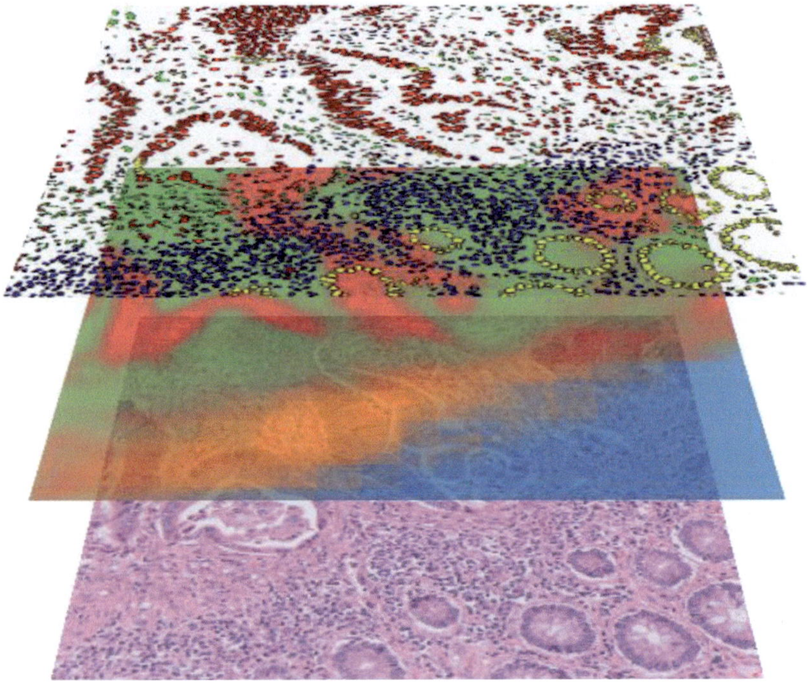

Fig. 7.6 Analysis generally needs to happen at multiple resolutions to identify morphological components and tissue patterns. This illustration shows how visualization and a multilayered approach to analysis might be used to identify biomarker expression in specific tissue compartments

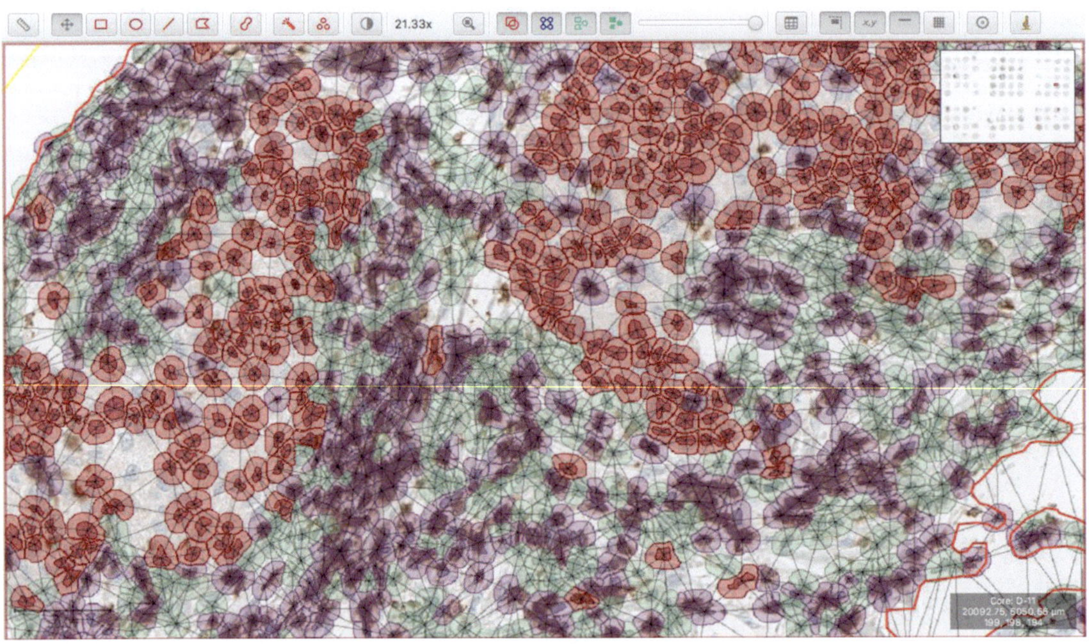

Fig. 7.7 Showing how cellular locational information can be used to measure relationships between cellular populations, co-localization and tumour boundary infiltration of inflammatory cells. This is becoming a crucial measurement in immuno-oncology. (Generated using QuPath [5])

Open Platforms and Software in Digital Pathology

In the past, digital pathology scanners and systems have been relatively closed systems where proprietary image formats could only be read, viewed and shared by the vendor's specific software applications. As the software industry has matured, there has been an increasing recognition that digital pathology hardware and software must interoperate to support the broad range of requirements in both research and diagnostic labs. This has resulted in a move towards more open platforms in digital pathology, an absolute requirement in order to provide the wider community with the tools and third-party applications which can be embedded within other platforms.

The use of open-source software has become firmly established for bioimage analysis, particularly in the case of fluorescence microscopy. These are summarized in Table 7.3. Popular tools include *ImageJ, Fiji, CellProfiler, icy* and *ilastik*. All of these packages offer a selection of general-purpose image processing methods and algorithms that might be meaningfully applied to digital pathology images, ranging from basic operations (e.g. filtering and thresholding) to more sophisticated techniques (e.g. pixel classification using machine learning). However, the widespread adoption of these for digital pathology has been limited by their lack of support for ultralarge 2D image analysis, frequently necessitating a variety of different open-source components to be used in combination to achieve the desired results. For example, *OpenSlide* and *Bio-Formats* are open-source software libraries that each provide the ability to extract pixels from a range of proprietary whole slide image formats, making it possible to pull out cropped or down-sampled image tiles of a manageable size for processing elsewhere – albeit at a loss of the wider contextual information present within the slide.

Two ImageJ plugins of particular relevance to pathology are *ImmunoRatio* [18] and *ImmunoMembrane* [19], which may be used for the quantification of nuclear and membranous IHC biomarkers, respectively. For the application of CellProfiler workflows to whole slide data, a collection of batch scripts (*slideToolkit*) has been made available that also incorporates the generation of image tiles suitable for analysis. Elsewhere, *CellProfiler* has been used in conjunction with R for predicting non-small cell lung cancer prognosis based on image features [20], whilst *CRImage* offers an R solution for segmenting and classifying cell types in H&E images. A combination of scripts, including Fiji, R and custom code, has also been applied to

Table 7.3 Summary of open-source image analysis software tools and libraries that can support quantitative tissue biomarker imaging

Open-source software	URL
ImageJ	https://imagej.nih.gov/ij/
Fiji	http://fiji.sc/
CellProfiler	http://cellprofiler.org
Icy	http://icy.bioimageanalysis.org/
Ilastik	http://ilastik.org/
OpenSlide	http://openslide.org/
Bio-Formats	http://www.openmicroscopy.org/bio-formats
Cytomine	http://www.cytomine.be/
ASAP	https://github.com/GeertLitjens/ASAP
Orbit	http://www.orbit.bio/
CRImage	https://www.bioconductor.org/packages/release/bioc/html/CRImage.html
QuPath	https://qupath.github.io/

discriminate benign from malignant intra-ductal proliferations of the breast [21].

Given the considerable effort involved in managing whole slide images and their analysis results, several open-source solutions also exist to enable slides to be viewed – and in some cases analysed – from within a standard web browser. *OMERO* has long been used for the management of microscopy images, offering support for large data sets including whole slide images. *Cytomine* represents a new platform for the distributed collaboration of multidisciplinary teams working on large-scale studies based on gigapixel images.

All the open-source image analysis approaches described above depend on a relatively high level of technical expertise for application to digital pathology, either in terms of setup or use. Recently, several new applications have made it possible to work directly with whole slide images on a standard desktop computer, with minimal setup required. *ASAP* primarily offers visualization and annotation functionality and has been used as part of the CAMELYON Grand Challenge on lymph node section analysis (https://camelyon17.grand-challenge.org). *Orbit* is a Java application offering a wider range of algorithms for image segmentation and classification, with optional links to OMERO and CellProfiler.

As a more comprehensive image analysis platform, *QuPath* [5] combines a greater variety of image analysis algorithms with a particular emphasis on novel biomarker discovery, user-friendliness, interactivity, annotation and visualization.

Artificial Intelligence, Deep Learning and Image Analytics

In recent years, computational image analysis has been revolutionized with the advent of deep learning technologies, particularly those based on convolutional neural networks (CNN). CNN architectures use a combination of various operations (convolutional kernels, nonlinear activation functions, subsampling) stacked in several layers with the objective of segmenting or classifying images, based on example ground truth images (Fig. 7.8). The strength of the connections between these layers (or weights) is adjusted (or learned) in order to optimize the segmentation or classification accuracy of the network across the full training set [22].

Since the operators used in CNN architectures can be applied in parallel, the training of CNNs is amenable to being carried out using "General Purpose Graphic Processing Units" (GPGPUs). It is this amenability to implementation on GPGPUs which has driven adoption throughout a range of problem domains. There are a number of deep learning frameworks such as Theano, Tensorflow, Keras and Caffe, which allow the use of the GPU to be abstracted away, allowing models to be trained, tested, shared and deployed [24], and the availability of these has also driven adoption.

In the domain of computational pathology, deep learning approaches have started to prove their worth in a range of competitions [25, 26] and are now under study in a range of use cases within pathology.

In essence, deep learning approaches use the data to learn the image features to be used for classification, as well as the classifier itself. This contrasts with traditional image processing techniques where the features are specified and then the classifier trained for those features.

One of the challenges with deep learning is the need for large numbers of annotations for training and validation. Many early image analysis examples have been applied to extremely large-scale image sets such as ImageNet [27], which are not possible to replicate in the pathology domain. However, there are a number of mitigation strategies which may be applied through deep learning to such problems.

1. Data Augmentation

The amount of training data may be artificially boosted by augmenting the available data (in terms of colour transformation, rotations and other image transformations). This creates multiple input images for training, based on, but different from the smaller available data set.

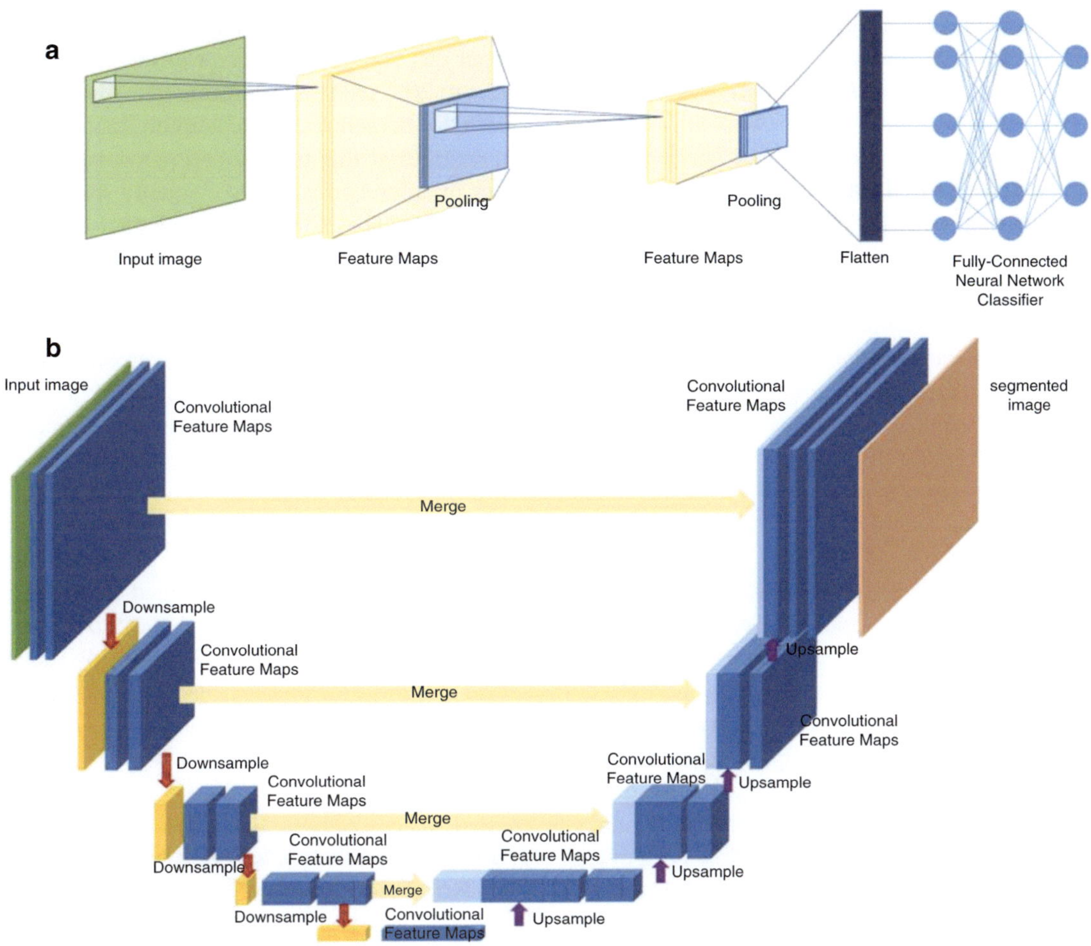

Fig. 7.8 Deep learning architectures for (**a**) classification and (**b**) segmentation. (**a**) An illustration of a convolutional neural network with two convolutional and pooling layers, using a fully connected neural network to perform the classification. In this example, the outputs of the convolutional stages are the learned features. (**b**) The architecture of a U-Net network [23] developed for biological segmentation problems. Since it is made up of convolutional (plus up-/downsampling layers), it can be very efficiently implemented on a GPU processor

2. Fine-Tuning Existing Networks

There are often freely available networks which have been pre-optimized for tasks such as classification of the ImageNet database. These can be used as the basis for more computational pathology-specific networks and be fine-tuned for the domain using a comparatively smaller amount of data from that used to train them originally. In such a way, it is possible to obtain much better performance using less data.

One aspect that can be seen in the application of deep learning to computational pathology segmentation is the need for high-quality annotations of ground truth. For segmentation problems the use of high-quality annotations of ground truth is necessary for training and validation of networks. These annotations may be at the level of tissue components or may be at a lower level, such as annotations of nucleus boundaries for training nucleus segmentation algorithms [28]. The acquisition of such annotations, alongside the strategies mentioned above should be used to train accurate algorithms, which are sufficiently robust to differences in image characteristics (e.g. staining,) to be of broad use.

Computational Pathology and Molecular Pathology

An increasing number of tissue-based biomarkers are based on nuclei-acid-based assays in the form of FFPE preparations of solid tumours, including next-generation sequencing (NGS). Whole genome sequencing in FFPE samples is also being developed rapidly in both retrospective and prospective specimens. The challenge here is that solid tumour samples are inherently heterogeneous, consisting of mixtures of different cell types including lymphocytes, stromal cells, endothelial cells, normal epithelial cells and tumour cells. When the target mutation is present in tumour cells, contamination from background non-tumour cells can dilute the ability to detect the molecular biomarker when present. In addition, with increasingly smaller sample sizes, the overall cellularity of the sample (and DNA sufficiency) also determines the success of the test. For this reason, measuring overall cellularity and proportion of tumour cells is vital in ensuring tumour purity and the quality of the subsequent molecular test. To achieve this, labs often need to apply a tumour enrichment step by macrodissecting the tumour component of the tissue section, thereby isolating the cells from the tumour region for downstream molecular analysis. This is commonly done by annotating the tumour region on an H&E using a pen and estimating the tumour cell percentage within that annotated region to determine if there is sufficient tumour DNA. Different molecular methods demand different percentages of tumour cells depending on their sensitivity.

Importantly, however, leading research is providing new evidence that the estimation of tumour cell percentage in tissue sections for molecular tests is highly subjective and prone to error. This can have a negative impact on the validity of established tumour biomarkers – such as EGFR, RAS and BRAF – but also on the discovery and validation of new tissue-based biomarkers in cancer. In lung cancer tissue samples, the percentage tumour can show interobserver differences ranging from 20% to 80% [29]. Similarly, Viray et al. [30] have demonstrated in a multi-institutional diagnostic trial that tumour cell percentage estimates in colorectal cancer are equally subjective and variable. These errors are particularly worrying as they could result in false-negative molecular test results for a mutation and could have a serious negative impact on patient care.

Computational pathology can overcome many of these shortfalls. Hamilton et al. [29] described a method called *TissueMark*[†] specifically designed for molecular pathology, aimed at annotating lung cancer H&E samples for macrodissection and precisely measuring the tumour cell percentage using image analysis. This work has subsequently been further developed by *PathXL* (a Philips-owned company) as workflow software to automate H&E analytics using deep learning algorithms and accelerate and improve accuracy of tumour sample annotation and measurement in molecular pathology (Fig. 7.9).

Translating Biomarker Algorithms into Routine Practice

Despite the wide application of image analytics in biomarker discovery programmes, there are relatively few biomarker algorithms that have made it into mainstream diagnostic applications.

Algorithms for Her2, ER, PR and Ki67 have FDA 510k/CE-IVD clearance, by some manufacturers, and are tied to specific scanning instruments and protocols under defined conditions as specified by the manufacturer to be compliant with the intended use. However, despite availability of this restricted subset of algorithms, adoption in routine practice has been slow. One of the key reasons for this is the lack of suitable platforms for the deployment of image analytics in routine pathological practice where algorithms can be fully embedded within the workflow and integrated with conventional digital slide review for primary diagnostics [17].

However, over the last number of years we have seen a distinctive (albeit slow) shift towards the

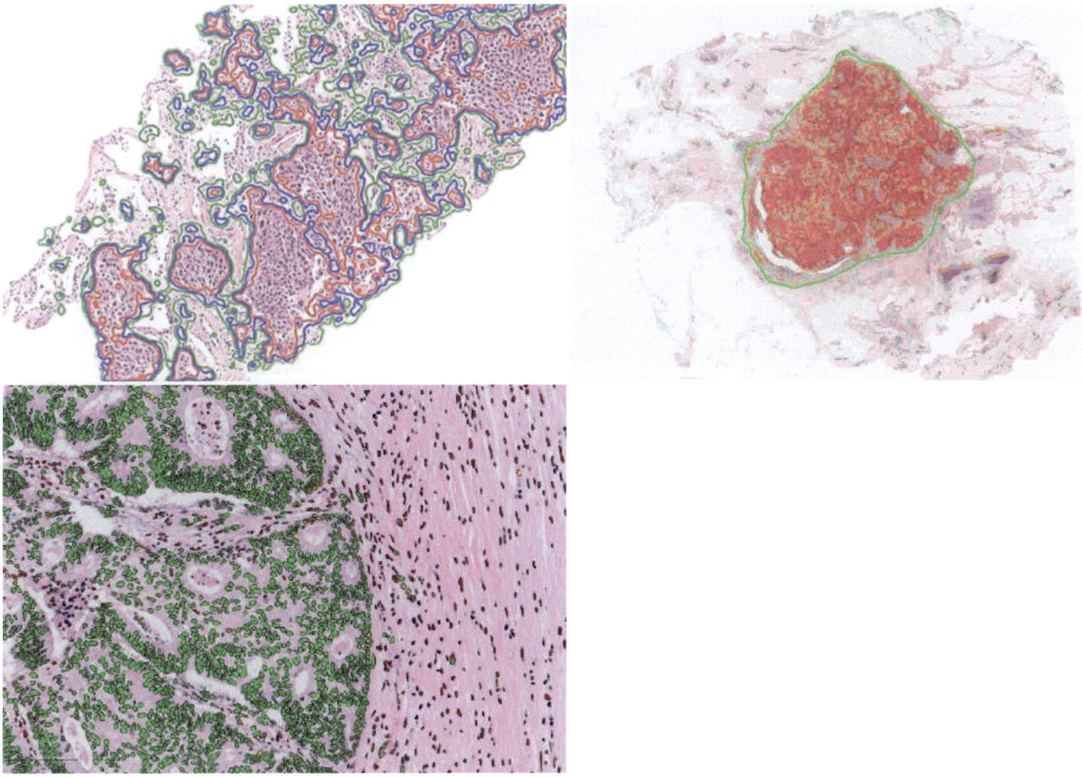

Fig. 7.9 Automated identification of tumour from H&E images using TissueMark* (top and middle). Using pattern recognition and machine learning, tumour texture can be distinguished reliably from background non-tumour regions. This can be applied routinely in molecular pathology to annotate the tissue sample for macrodissection. Nuclear segmentation and cell counts is applied for the measurement of total cellularity and the percentage of tumour cells. **Xplore and TissueMark are Research Applications; PathXL is the legal manufacturer*

adoption of digital pathology for routine primary diagnostics. At the time of writing, this is now growing at an impressive rate. This has been driven by (i) the integration of optical, digital, communication and software technologies (illustrated in Fig. 7.1), (ii) the development of high-throughput scanners with scan rates that can cope with large high throughput routine workloads, (iii) storage achitectures that facilitate routine high-volume image management, (iv) associated workflow software to facilitate workload management and routine digital review and (v) evidence that digital pathology can be used reliably for primary diagnostic interpretation. A number of studies from small- to medium-sized academic groups have shown a strong concordance between conventional diagnosis from glass slides and diagnosis from digital WSI (reviewed in [31]). However, prior to 2016, the FDA had given digital pathology a class III rating (highest risk), requiring the digital pathology industry to undertake large studies to demonstrate safety. Recently, it was recommended that manufacturers of WSI devices for primary diagnosis in surgical pathology submit their applications to the FDA through their *de novo* proces and that these devices should classified into class II instead of class II as previously proposed. A major milestone was accomplished when the FDA allowed the first vendor (Philips Medical Systems Nederland B.V) to market their device for primary diagnostic use in the USA and substantially equivalent devices of this generic type. As part of this clearance, Philips (Philips Medical Systems Nederland B.V) participated in a large pivotal trial in digital pathology and submitted this as a *de novo* FDA submission in 2017 on their Philips

IntelliSite Pathology Solutions (PIPS) product [32]. As one of the largest non-inferiority trials of its type, PIPS has demonstrated that digital pathology can be used in primary diagnostic with a small range of exceptions (https://www.accessdata.fda.gov/cdrh_docs/reviews/DEN160056.pdf). Clearance is currently restricted to the complete PIPS platform from slide capture to image storage and WSI viewing using the Philips product. This now represents the first authorized de novo predicate devise in WSI and provides a platform for other manufacturers to show substantive equivalence [32]. This represents an enormous step forward in digital pathology and the opportunity to establish digital workflows where computational biomarker algorithms can be easily translated into clinical practice and seamlessly embedded in routine digital diagnostics.

Historically, there is a major gap between early biomarker imaging R&D and clinical practice. The shift from glass to digital slides in primary diagnostic laboratories will underpin the future *translation* of powerful companion diagnostic algorithms into practice, putting them in the hands of pathologists for routine use. Industry has a strong part to play in providing the technology pipeline to close the translation gap, provide a conduit for clinical algorithms and support the regulatory frameworks to drive clinical studies and clearance by regulatory organizations (Fig. 7.10). This will help streamline the delivery of clinical companion and complementary algorithms in the future.

The wider deployment of WSI devices in diagnostic pathology laboratories across the world will build the foundation for advances in AI and deep learning in pathology and the routine adoption of image analytics for biomarkers that has not been possible to date. This will open doors for these digital pathology platforms to be deployed within larger enterprise image ecosystems that facilitate next-generation applications

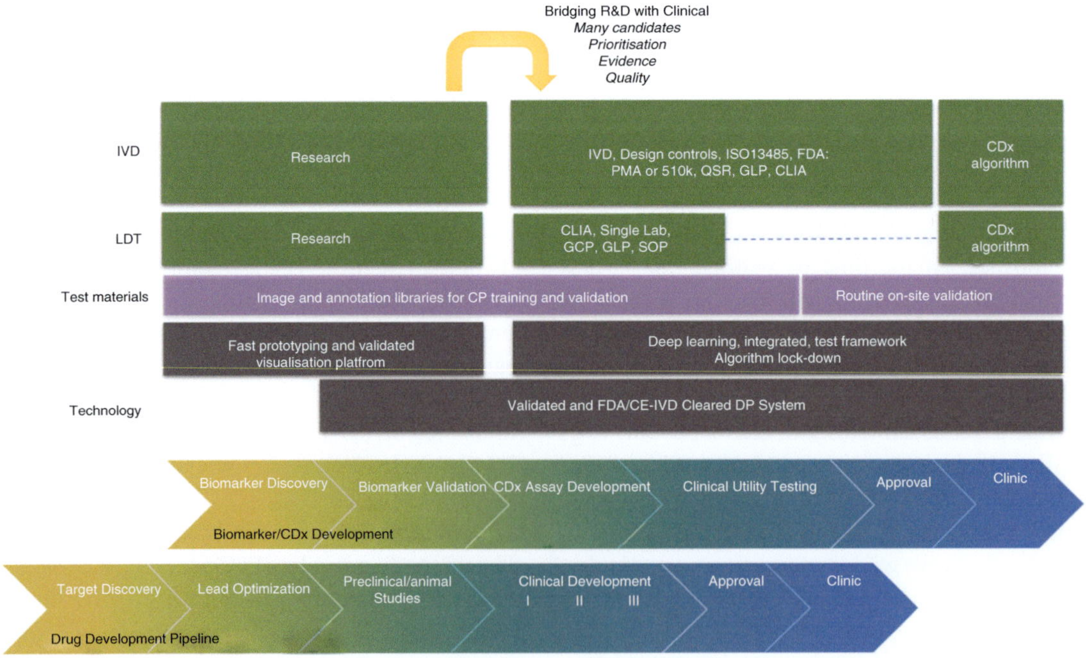

Fig. 7.10 An illustration of the workflows necessary to translate early-stage biomarkers into companion/complementary diagnostics. Bridging the R&D/clinical divide is essential to translate new companion diagnostic algorithms into digital pathology (CDx companion diagnostic, DP digital pathology, CP computational pathology, IVD in vitro diagnostic, LDT lab developed test, PMA premarket approval, GLP good laboratory practice, GCP good clinical practice, QSR quality systems regulation)

such as image analysis, streaming analytics and computational pathology. The next challenge awaiting the digital pathology community will be to resolve regulatory issues surrounding the use image algorithms for tissues and biomarkers that employ deep learning.

Conclusions

The power of digital and computational pathology is growing rapidly and bringing enormous advantages to tissue analytics and biomarker discovery in modern translational research laboratories. Digital pathology can help streamline many of the processes and workflows needed to support collaborative multicentre biomarker discovery programmes and manage complex tissue-centric research needs. Computational pathology through the use of automated image analytics can accelerate the discovery cycle, interrogating many thousands of cells and tissue samples very rapidly, screening for candidate biomarkers and mining large data sets to understand tumour biology to identify new therapeutic targets and predictive/prognostic biomarkers. Technologies such as deep learning will continue to drive improvements in tissue recognition giving the research community even more powerful tools for tissue discovery. Together with clinically approved digital pathology scanning and viewing platforms the integration of FDA/EU Market authorized algorithms is likely to drive the translation of companion diagnostics into companion algorithms for routine assessment of tissue signatures. Understanding this complex landscape of pheno-genotypic signatures, through digital integration, precise measurement and tissue mapping, will underpin the future of tissue biomarker discovery and drive improvements in patient stratification and precision medicine.

Disclosure The opinions expressed in this presentation are solely those of the author or presenters and do not necessarily reflect those of Philips. The information presented herein is not specific to any product of Philips or their intended uses. The information contained herein does not constitute, and should not be construed as, any promotion of Philips products or company policies.

References

1. Hamilton PW, van Diest PJ, Williams R, Gallagher AG. Do we see what we think we see? The complexities of morphological assessment. J Pathol [Internet]. 2009 [cited 2017 Jun 4];218(3):285–91. Available from: http://www.ncbi.nlm.nih.gov/pubmed/19291709.
2. Hamilton PW, Bankhead P, Wang Y, Hutchinson R, Kieran D, McArt DG, et al. Digital pathology and image analysis in tissue biomarker research. Methods [Internet]. 2014 [cited 2017 Jun 5];70(1):59–73. Available from: http://www.sciencedirect.com/science/article/pii/S1046202314002370.
3. Prescott JW. Quantitative imaging biomarkers: the application of advanced image processing and analysis to clinical and preclinical decision making. J Digit Imaging [Internet]. 2013 [cited 2017 Jun 5];26(1):97–108. Available from: http://www.ncbi.nlm.nih.gov/pubmed/22415112.
4. Masmoudi H, Hewitt SM, Petrick N, Myers KJ, Gavrielides MA. Automated quantitative assessment of HER-2/neu immunohistochemical expression in breast cancer. IEEE Trans Med Imaging [Internet]. 2009 [cited 2017 Jun 11];28(6):916–25. Available from: http://ieeexplore.ieee.org/document/4752752/.
5. Bankhead P, Loughrey MB, Fernández JA, Dombrowski Y, McArt DG, Dunne PD, McQuaid S, Gray RT, Murray LJ, Coleman HG, James JA, Manuel Salto-Tellez PWH. QuPath: open source software for digital pathology image analysis. bioRxiv. 2017; https://doi.org/10.1101/099796.
6. Vandenberghe ME, Scott MLJ, Scorer PW, Soderberg M, Balcerzak D, Barker C. Relevance of deep learning to facilitate the diagnosis of HER2 status in breast cancer. Sci Rep [Internet]. 2017 [cited 2017 Jun 12];7:45938. Available from: http://www.ncbi.nlm.nih.gov/pubmed/28378829.
7. Klümper N, Syring I, Vogel W, Schmidt D, Müller SC, Ellinger J, et al. Mediator complex subunit MED1 protein expression is decreased during bladder cancer progression. Front Med [Internet]. 2017 [cited 2017 Jun 11];4:30. Available from: http://www.ncbi.nlm.nih.gov/pubmed/28367434.
8. Parra ER, Behrens C, Rodriguez-Canales J, Lin H, Mino B, Blando J, et al. Image analysis–based assessment of PD-L1 and tumor-associated immune cells density supports distinct intratumoral microenvironment groups in non–small cell lung carcinoma patients. Clin Cancer Res [Internet]. 2016 [cited 2017 May 16];22(24). Available from: http://clincancerres.aacrjournals.org/content/22/24/6278.
9. Neumeister VM, Anagnostou V, Siddiqui S, England AM, Zarrella ER, Vassilakopoulou M, et al. Quantitative assessment of effect of preanalytic cold ischemic time on protein expression in breast cancer tissues. JNCI J Natl Cancer Inst [Internet]. 2012 [cited 2017 Jun 4];104(23):1815–24. Available from: http://www.ncbi.nlm.nih.gov/pubmed/23090068.

10. Neumeister VM, Parisi F, England AM, Siddiqui S, Anagnostou V, Zarrella E, et al. A tissue quality index: an intrinsic control for measurement of effects of preanalytical variables on FFPE tissue. Lab Invest [Internet]. 2014 [cited 2017 Jun 4];94(4):467–74. Available from: http://www.nature.com/doifinder/10.1038/labinvest.2014.7.
11. Lykkegaard Andersen N, Brugmann A, Lelkaitis G, Nielsen S, Friis Lippert M, Vyberg M. Virtual double staining. Appl Immunohistochem Mol Morphol [Internet]. 2017 [cited 2017 Jun 12];1. Available from: http://www.ncbi.nlm.nih.gov/pubmed/28248729.
12. Wang Y, McCleary D, Wang C-W, Kelly P, James J, Fennell DA, et al. Ultra-fast processing of gigapixel tissue MicroArray images using high performance computing. Cell Oncol [Internet]. 2011 [cited 2017 Jun 6];34(5):495–507. Available from: http://www.ncbi.nlm.nih.gov/pubmed/21559926.
13. Kerr KM, Nicolson MC. Non–small cell lung cancer, PD-L1, and the pathologist. Arch Pathol Lab Med [Internet]. 2016 [cited 2017 Jun 6];140 5858. Available from: http://www.archivesofpathology.org/doi/pdf/10.5858/arpa.2015-0303-SA?code=coap-site.
14. McLaughlin J, Han G, Schalper KA, Carvajal-Hausdorf D, Pelekanou V, Rehman J, et al. Quantitative assessment of the heterogeneity of PD-L1 expression in non-small-cell lung cancer. JAMA Oncol [Internet]. 2016 [cited 2017 May 16];2(1):46–54. Available from: http://www.ncbi.nlm.nih.gov/pubmed/26562159.
15. Xing X, Li Z, Wang J, Ji J. Analysis of PDL1 expression and T cells infiltration in 1014 gastric cancer patients. J Clin Oncol [Internet]. 2017 [cited 2017 May 16];35(4_suppl):50–50. Available from: http://ascopubs.org/doi/10.1200/JCO.2017.35.4_suppl.50.
16. Turkki R, Linder N, Kovanen P, Pellinen T, Lundin J. Antibody-supervised deep learning for quantification of tumor-infiltrating immune cells in hematoxylin and eosin stained breast cancer samples. J Pathol Inform [Internet]. 2016 [cited 2017 Jun 12];7(1):38. Available from: http://www.ncbi.nlm.nih.gov/pubmed/27688929.
17. Hulsken B. Philips IntelliSite open pathology platform. https://thepathologist.com/app-notes/0016/philips-intellisite-open-pathology-platform/.
18. Tuominen VJ, Ruotoistenmaki S, Viitanen A, Jumppanen M, Isola J. ImmunoRatio: a publicly available web application for quantitative image analysis of estrogen receptor (ER), progesterone receptor (PR), and Ki-67. Breast Cancer Res [Internet]. 2010 [cited 2017 Jun 11];12(4):R56. Available from: http://www.ncbi.nlm.nih.gov/pubmed/20663194.
19. Tuominen VJ, Tolonen TT, Isola J. ImmunoMembrane: a publicly available web application for digital image analysis of HER2 immunohistochemistry. Histopathology [Internet]. 2012 [cited 2017 Jun 11];60(5):758–67. Available from: http://doi.wiley.com/10.1111/j.1365-2559.2011.04142.x.
20. Yu K-H, Zhang C, Berry GJ, Altman RB, Re C, Rubin DL, et al. Predicting non-small cell lung cancer prognosis by fully automated microscopic pathology image features. Nat Commun [Internet]. 2016 [cited 2017 Jun 11];7:12474. Available from: http://www.ncbi.nlm.nih.gov/pubmed/27527408.
21. Dong F, Irshad H, Oh E-Y, Lerwill MF, Brachtel EF, Jones NC, et al. Computational pathology to discriminate benign from malignant intraductal proliferations of the breast. In: Sapino A, editor. PLoS One [Internet]. 2014 [cited 2017 Jun 11];9(12):e114885. Available from: http://dx.plos.org/10.1371/journal.pone.0114885.
22. LeCun Y, Bengio Y, Hinton G. Deep learning. Nature [Internet]. 2015 [cited 2017 Sep 1];521(7553):436–44. Available from: http://www.nature.com/doifinder/10.1038/nature14539.
23. Ronneberger O, Fischer P, Brox T. U-net: convolutional networks for biomedical image segmentation. Int Conf Med Image Comput Comput Interv 2015 (p. 234–41) Springer Int Publ. Available from https://arxiv.org/abs/150504597.
24. Erickson BJ, Korfiatis P, Akkus Z, Kline T, Philbrick K. Toolkits and libraries for deep learning. J Digit Imaging [Internet]. 2017 [cited 2017 Sep 1];30(4):400–5. Available from: http://link.springer.com/10.1007/s10278-017-9965-6.
25. Paeng K, Hwang S, Park S, Kim M, Kim S. A unified framework for tumor proliferation score prediction in breast histopathology. arXiv Prepr arXiv161207180 2016. Available from https://arxiv.org/abs/161207180.
26. Wang D, Khosla A, Gargeya R, Irshad H, Beck AH. Deep learning for identifying metastatic breast cancer. 2016 [cited 2017 Sep 1]; Available from: http://arxiv.org/abs/1606.05718.
27. Deng J, Dong W, Socher R, Li L-J, Li K, Li F-F. ImageNet: a large-scale hierarchical image database. In: 2009 IEEE Conference on Computer Vision and Pattern Recognition [Internet]. IEEE; 2009 [cited 2017 Sep 1]. p. 248–55. Available from: http://ieeexplore.ieee.org/document/5206848/.
28. Xing F, Yang L. Robust nucleus/cell detection and segmentation in digital pathology and microscopy images: a comprehensive review. IEEE Rev Biomed Eng [Internet]. 2016 [cited 2017 Jun 11];9:234–63. Available from: http://www.ncbi.nlm.nih.gov/pubmed/26742143.
29. Hamilton PW, Wang Y, Boyd C, James JA, Loughrey MB, Hougton JP, et al. Automated tumor analysis for molecular profiling in lung cancer. Oncotarget [Internet]. 2015 [cited 2017 Jun 12];6(29):27938–52. Available from: http://www.ncbi.nlm.nih.gov/pubmed/26317646.
30. Viray H, Li K, Long TA, Vasalos P, Bridge JA, Jennings LJ, et al. A prospective, multi-institutional diagnostic trial to determine pathologist accuracy in estimation of percentage of malignant cells. Arch Pathol Lab Med. 2013;137(11):1545–9.

31. Williams BJ, DaCosta P, Goacher E, Treanor D. A systematic analysis of discordant diagnoses in digital pathology compared with light microscopy. Arch Pathol Lab Med [Internet]. 2017; [cited 2017 Sep 1];arpa.2016-0494-OA. Available from: http://www.ncbi.nlm.nih.gov/pubmed/28467215.

32. Abels E, Pantanowitz L. Current state of the regulatory trajectory for whole slide imaging devices in the USA. J Pathol Inform [Internet]. 2017 [cited 2017 Jun 11];8(1):23. Available from: http://www.ncbi.nlm.nih.gov/pubmed/28584684.

Detection of Predictive Biomarkers Using Liquid Biopsies

8

Andrew A. Davis and Massimo Cristofanilli

Abbreviations

BEAMing	Beads, emulsions, amplification, and magnetics
CEA	Carcinoembryonic antigen
CTC	Circulating tumor cell
ctDNA	Circulating tumor DNA
ddPCR	Digital droplet PCR
EGFR	Epidermal growth factor receptor
EMT	Epithelial-mesenchymal transition
EpCAM	Epithelial cell adhesion molecule
ESR1	Estrogen receptor 1
FDA	Food and Drug Administration
FISH	Fluorescent in situ hybridization
HR	Hormone receptor
IBC	Inflammatory breast cancer
MAF	Mutant allele frequency
NGS	Next-generation sequencing
NSCLC	Non-small cell lung cancer
OS	Overall survival
PD-1	Programmed cell death-1 protein receptor
PD-L1	Programmed cell death-1 protein ligand
PFS	Progression-free survival
PSA	Prostate-specific antigen
SCLC	Small cell lung cancer

A. A. Davis
Department of Medicine-Hematology and Oncology, Northwestern Memorial Hospital, Chicago, IL, USA

M. Cristofanilli (✉)
Department of Medicine-Hematology and Oncology, Robert H Lurie Comprehensive Cancer Center, Feinberg School of Medicine, Chicago, IL, USA
e-mail: Massimo.Cristofanilli@nm.org

Introduction

In previous decades, a tumor biopsy of tissue from either a primary or metastatic site has served as the basis of making treatment decisions for patients. These treatment decisions were formulated based on the architecture, histology, and molecular features associated with a single tissue biopsy. However, cancer is a heterogeneous, dynamic, and evolving disease. As we have learned from countless studies, the genetic landscape of cancer is highly complex, characterized by many genomic alterations associated with cell growth, proliferation, survival, and resistance [1]. While a small number of these genomic alterations exist as early clones, many subclones compete and emerge in response to the selective pressures of tumor growth and resistance making the use of blood-based diagnostics an appealing approach for longitudinal monitoring.

The concept of circulating tumor cells (CTCs) first emerged in 1869 by Ashworth based on analysis of the blood of a patient with metastatic cancer [2]. However, it was not until 1948 that Mandel and Métais first reported the detection of

S. Badve, G. L. Kumar (eds.), *Predictive Biomarkers in Oncology*,
https://doi.org/10.1007/978-3-319-95228-4_8

extracellular nucleic acids in the blood of healthy individuals [3]. Nearly 30 years later, scientists reported that cancer patients had elevated concentrations of DNA in the blood, particularly in patients with metastatic disease. At the time, it was not known whether this DNA was the same as tumor DNA. Eventually, specific oncogenic point mutations were identified in serum that aligned with well-known oncogenic mutations in tissue. Later studies confirmed that the exact same nucleotide sequencing mutations in blood were also present in the tissue. These nucleic acids eventually became known as circulating tumor DNA (ctDNA). Collectively, based on these findings over a century, the potential existed to noninvasively detect and monitor cancer in the peripheral blood (Fig. 8.1).

Difference Between cfDNA and ctDNA

DNA is released into the bloodstream as fragments from inflamed, lytic, apoptotic, or necrotic cells or from macrophages through active secretion mechanisms. Although the majority of the released (extracellular) DNA is adsorbed to the surface of leukocytes or erythrocytes or other cells, a portion of the DNA remains unbound and can be identified in the plasma, known as cell-free DNA (cfDNA). The portion of cfDNA that is derived from tumor cells is called ctDNA. ctDNA has a short half-life in the circulation, ranging from 15 min to several hours.

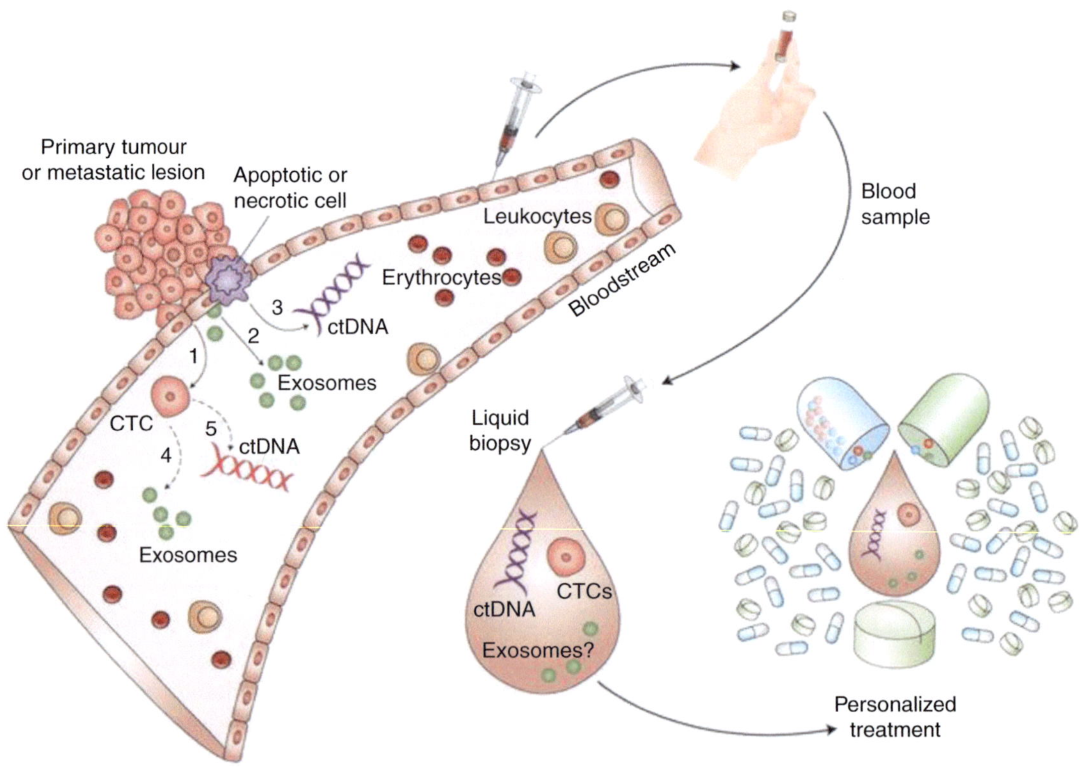

Fig. 8.1 The schematic depicts a primary tumor and/or metastatic lesion releasing the components of a liquid biopsy into the bloodstream including ctDNA, CTCs, and exosomes. These components are collected via a peripheral blood draw, analyzed, interpreted, and utilized to inform personalized treatment. (Reprinted from Alix-Panabières and Pantel [15]. With permission from SpringerNature)

These seminal observations led to the emergence of "liquid biopsies" as tools to monitor tumor evolution in *real time*. Liquid biopsies, which consist of cells or nucleic acids, are obtained primarily from blood but can also be accessed via other compartments such as sputum, stool, urine, pleural, peritoneal, or cerebrospinal fluid. These tools have allowed researchers and clinicians to explore solid tumors without accessing a single tissue site, an opportunity that was only previously available in leukemia, lymphoma, and multiple myeloma. Importantly, the contents in blood may arise from multiple disease locations, indicating a potential to capture tumor spatial heterogeneity that cannot be achieved with tissue biopsies at the primary tumor site or a single metastatic site. In this regard, a liquid biopsy could serve as a surrogate for the cellular and genetic contents of the entire tumor. In addition, while tissue biopsies have functional limitations with respect to potential complications and inability to sample multiple sites for serial biopsies, liquid biopsies can noninvasively monitor aspects of the tumor as a mechanism to understand the dynamic progression of cancer temporally in response to treatment. Therefore, solid tumors, in fact, possess a liquid phase that can serve as an important tool to capture tumor heterogeneity and metastasis over time. Clinical applications of liquid biopsies include potential roles as diagnostic, prognostic, and predictive biomarkers. The remainder of this chapter will focus on the two most studied aspects of liquid biopsies, including CTCs and ctDNA.

Circulating Tumor Cells (CTCs)

CTCs are present in low concentrations in the peripheral blood—on the order of one to ten cells per 10 mL or one CTC per 10^6–10^7 leukocytes [4]. While relatively rare in blood, studies have indicated that aggressive tumors can release thousands of these cells into the blood each day. Therefore, the number of CTCs is much greater when compared to the number of cells that enter and seed metastatic sites. Estimates of CTC half-life in the blood have been on the order of 1–2.4 h. The cells are detected from multiple primary and/or metastatic sites in the blood compartment. Various cell surface markers cytokeratin and EpCam distinguish CTCs from other components of serum and plasma.

The mechanism of CTC release into the peripheral blood appears to be both passive via tumor shedding and active through dynamic intravasation into the blood from either a primary or metastatic site. After entering the blood stream, these cells can seed metastatic sites through active trans-endothelial migration and either lay dormant or proliferate. Many studies examining the biology of these processes and how cells transition from dormancy to active growth at distant sites are ongoing (Fig. 8.2).

Prognostication

In 2004, the immunomagnetic sorting and detection of CTCs emerged as a promising biomarker in metastatic breast cancer with respect to progression-free survival (PFS) and overall survival (OS) based on a prospective study conducted at 20 centers around the United States [5]. This landmark study was based on measurements of CTC counts at baseline and at the first follow-up visit after initiation of therapy. The potential for this technique as both an independent prognostic and predictive tool was established by its early capability as a dynamic biomarker. Specifically, the number of CTCs changed in response to effective treatments (often with tumor regression) with serial measurements and was not modified (below five CTCs) in patients who would progress or experience a short-term stability. Since that time, CTCs have shown to be prognostic biomarker in a number of other solid tumors including lung, colorectal, prostate, bladder, kidney cancer and many others. In addition to the metastatic setting, detection of CTCs using

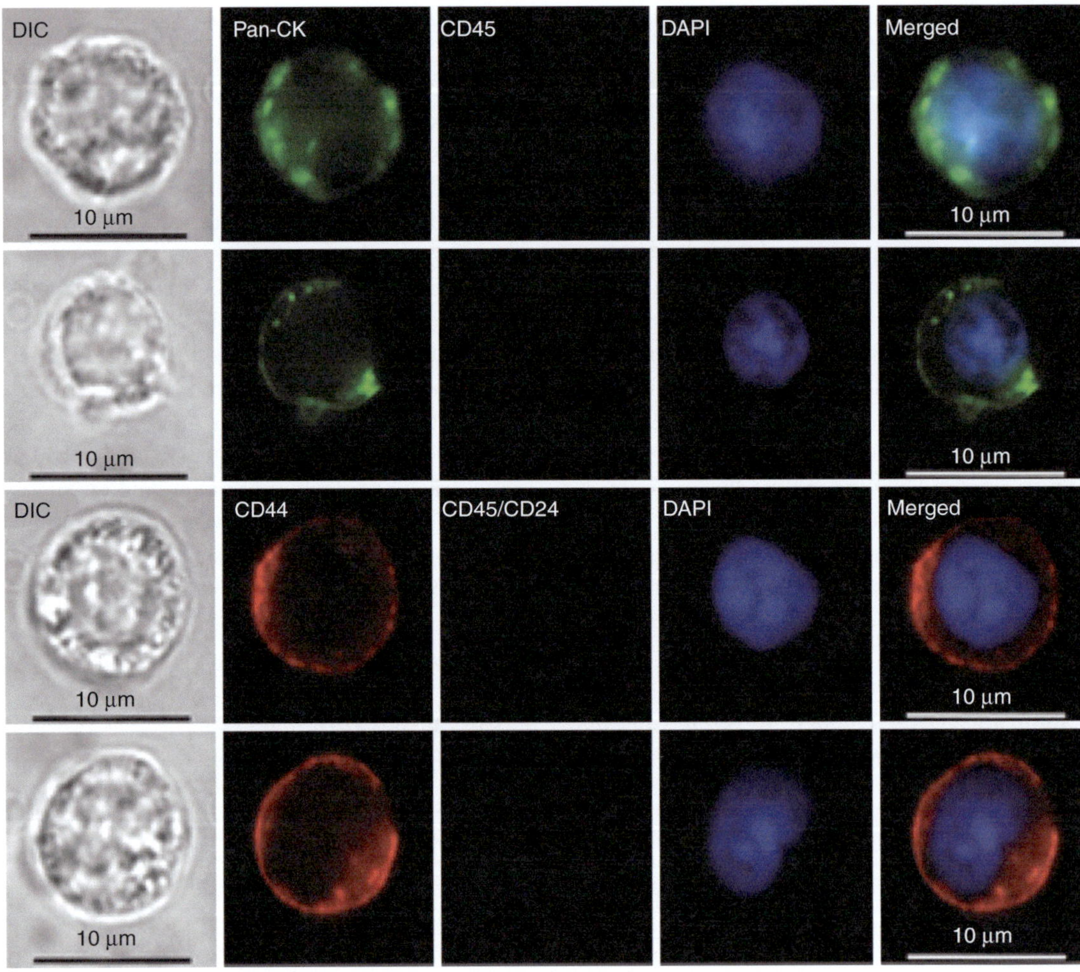

Fig. 8.2 CTCs isolated from breast cancer patients with Pan-CK+, CD44+, CD24/CD45 cells. DAPI indicates nuclear staining. Scale bar indicates 10 μm. (Reprinted from Boral et al. [16]. With permission from Creative Commons CC-BY)

the CellSearch™ system has been associated with worse prognosis in early breast cancer. While the number of CTCs has significant impact on prognostication, these studies opened up further work exploring how CTCs could be used to further refine treatment decisions as predictive biomarkers.

Response to Therapy and Resistance

Early work has utilized CTCs in combination with other laboratory measures to assess response to treatment, such as an androgen inhibitor, abiraterone, in metastatic prostate cancer. In breast cancer, the ongoing international Treat CTC Trial is exploring whether trastuzumab can eliminate chemotherapy-resistant CTCs in HER2-negative patients. One study (SWOG S0500) found that while CTCs were prognostic in patients with metastatic breast cancer, early switching to an alternative cytotoxic therapy based on elevated CTCs did not prolong OS [6].

Studies investigating concordance between tissue analysis and CTCs have often demonstrated differences in expression with respect to hormone receptor (HR) and HER2 expression. For example, early experimental data have shown that some patients with negative HER2 amplification by traditional tissue methods may, in fact, have HER2-positive CTCs in the blood and respond to adjuvant trastuzumab. This phenotypic heterogeneity may indicate spatial and temporal heterogeneity not easily captured via evaluation of a single-site tissue biopsy. Other mechanisms include HER2 arising as a resistance mechanism or providing therapeutic benefit via a different mechanism, such as acting on cancer stem cells or via immune-related mechanisms. Similar resistance mutations in other tumor types using CTCs have been discovered, including *EGFR* mutations in non-small cell lung cancer (NSCLC) receiving tyrosine kinase inhibitors and androgen sensitivity in prostate cancer.

Early preclinical work indicates the potential to utilize CTCs ex vivo in mice to mirror responsiveness of matched patients to chemotherapy in small cell lung cancer (SCLC) and other tumor types. Therefore, the potential exists to deliver precision medicine by using CTCs as tools for personalized drug testing, prior to choosing therapies in the clinic. These studies are also important in supporting that a proportion of CTCs do, in fact, have tumor-forming capabilities.

Monitoring for Early Disease

Some types of tumors, such as luminal breast cancer, may remain dormant for years before detectable recurrence on imaging. Furthermore, certain high-risk subtypes of cancer including inflammatory breast cancer (IBC) may spread via tumor microemboli that are not easily eliminated with surgery and remain as persistent micrometastatic disease. The current standard of care, which includes interval monitoring with imaging, has significant limitations in these high-risk populations. Serial monitoring of blood CTCs has the potential to identify residual disease early on, prior to radiographic changes associated with disease progression. Therefore, treatment could be initiated earlier before the accumulation of additional genomic alterations to dramatically improve patient outcomes. Studies have demonstrated that CTCs change not only in response to treatment, but they may also indicate minimal residual disease if found after surgery with curative intent.

Recent work has demonstrated the potential to use CTCs to monitor early cancer development. A proof-of-principle study monitored the cytopathologic and immunocytochemical characteristics of CTCs in high-risk individuals with chronic obstructive pulmonary disease [7]. The study demonstrated that CTCs could be detected in blood in a small proportion of patients prior to radiographic detection on CT scans. Interestingly, all of these patients went on to develop lung nodules with eventual biopsy confirmed lung cancer 1–4 years later. Furthermore, at resection, all were found to have stage I disease, indicating the potential to use CTCs as a screening strategy in high-risk individuals for early cancer detection. Further data are necessary to study CTCs in control populations and patients at high risk for detection of particular cancers. While CTCs are strongly correlated with metastatic disease, in rare cases, a limited number of CTCs can be detected in patients with normal or benign breast conditions in at least 1% of patients. Monitoring of these rare cases is necessary to further explore CTC biology and to determine whether these individuals may develop cancerous lesions at long-term follow-up or whether these CTCs were false positives.

Circulating Tumor DNA (ctDNA)

Cell-free DNA consists of non-cancerous nucleic acids and a smaller proportion of ctDNA. ctDNA represents a low quantity of cfDNA (often less than 1%), which varies based on tumor burden [8]. The amount of ctDNA depends on a number of factors including the tumor cell of origin and stage of malignancy. Similar to CTCs, the mechanisms of ctDNA release into the blood occurs via apoptosis and necrosis, while there is also likely an active mechanism of nucleic acid secretion to facilitate metastasis and gene expression at distant sites [9]. In either case, ctDNA must be shed or actively released in combination with a suitable method to detect small quantities of ctDNA in the blood. ctDNA has a short half-life, which is estimated at approximately 16 min to 2.5 h, and ctDNA is rapidly cleared by hepatic, renal, and nuclease mechanisms, indicating that ctDNA is a highly dynamic biomarker [10].

Prognostication

Detection of ctDNA varies by stage with a greater proportion of patients with detectable ctDNA in a stepwise fashion with a higher stage of the disease. In addition, certain types of cancer are more (such as breast or lung) or less (such as glioblastoma multiforme) likely to release ctDNA into the peripheral blood. Interestingly, many patients with detectable ctDNA do not have immunomagnetic detected CTCs, suggesting in some instances that these biomarkers are independent. Within particular histologies, several studies have demonstrated that higher mutant allele frequency (MAF) of ctDNA correlates with worse PFS and OS, independent of intervening treatment. The mechanism of this is likely the quantitative aspect of ctDNA, which correlates with both tumor stage and tumor burden. Higher quantity of ctDNA, therefore, may reflect more advanced disease with poor prognosis. Furthermore, data have specifically linked ctDNA MAF with volumetric imaging measurements of tumor burden in particular tumor types. Therefore, both the presence and quantity of ctDNA appear to be important prognostic biomarkers.

Response to Therapy and Resistance

A tremendous advantage of ctDNA biopsies is the ability to detect and monitor genomic alterations in the peripheral blood. In contrast to protein biomarkers such as the cancer antigen 125, carcinoembryonic antigen (CEA), or prostate-specific antigen (PSA), which lack specificity and remain in circulation for weeks, ctDNA has greater potential as a *real-time*, dynamic biomarker given the shorter half-life. Currently, ctDNA biopsies can detect a wide variety of genomic alterations with high specificity including single-nucleotide variants, indels, copy number alterations, rearrangements, and fusions. Serial monitoring of known driver mutations and subclones has demonstrated the potential to track quantitative changes in ctDNA and MAF over time and to detect emergence of new resistance mutations in comparison to the baseline mutational profile.

Serial ctDNA biopsies have demonstrated the capability to target known resistance mutations as they develop in the blood. While concordance between tissue and blood has been variable in different studies due to the differences in sequencing and sampling methods with respect to tumor heterogeneity, ctDNA biopsies invariably have high specificity. Therefore, in populations of cancer patients, early evidence indicates that detecting a known driver mutation using ctDNA assays can be used to change therapy with clinical benefit. Examples include initiation of osimertinib in *EGFR* T790M NSCLC and monitoring of acquired endocrine resistance based on *ESR1* mutations in breast cancer (Fig. 8.3).

However, a negative blood biopsy does not necessarily eliminate the possibility of the mutation existing given the lower sensitivity. Therefore, a subsequent tissue biopsy could be clinically indicated if these mutations were not detected in blood.

A variety of sequencing methods have been utilized including digital droplet PCR (ddPCR) and next-generation sequencing (NGS), which enable initiating therapy changes on the order of days to weeks, often without the need for repeat tissue biopsy. Importantly, these tools enable changing therapy in *real time* as these resistance mechanisms develop in order to optimally evolve treatment in response to tumor resistance. In addition to known resistance mutations, early data have suggested the potential for this technique to be used more broadly to monitor response to chemotherapy and immune checkpoint blockade. Ongoing work is searching for molecular signatures in the blood to noninvasively predict response to these therapies.

Monitoring for Early Disease

Serial measurements of ctDNA in blood have demonstrated the ability to detect both minimal

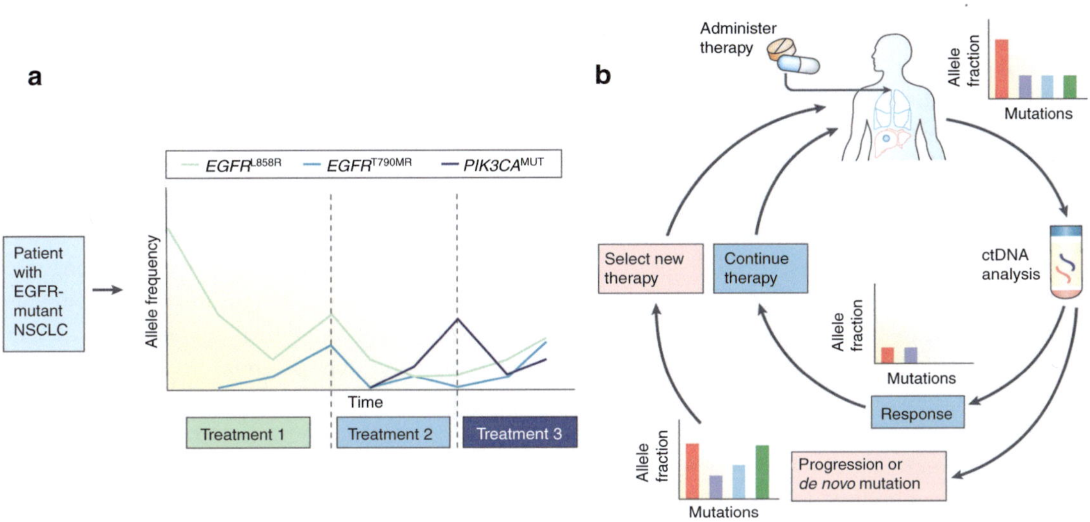

Fig. 8.3 (**a**) Serial monitoring of ctDNA enables detection of resistance mutations such as *EGFR* T790M and emergence of new genomic alterations such as *PIK3CA*. ctDNA allele frequency varies in response to treatment. (**b**) ctDNA analysis enables monitoring of allele fraction of ctDNA as a surrogate for response to therapy and emergence of de novo mutations to enable treatment decisions. This cycle may repeat using serial ctDNA biopsies. (Reprinted from Wan et al. [10]. With permission from Nature Publishing Group)

residual disease and disease recurrence after surgery with curative intent. Landmark studies have shown that blood biopsies collected prior to surgery and at intervals after surgery identify two populations of patients [11]. The first group of patients completely eliminates detectable ctDNA in the blood. In contrast, the second group has residual ctDNA after surgery, indicating a greater need to target these patients with early, aggressive therapy to eliminate minimal residual disease. Examples of these studies have been shown in multiple tumor types including lymphoma, stage II colorectal cancer, early-stage breast cancer, and NSCLC. These studies clearly link the quantity of ctDNA as a surrogate for tumor burden and micrometastasis. As the cost of blood biopsies decreases, pre- and postsurgical ctDNA measurements could eventually become routine in clinical practice. While the potential exists to use ctDNA to stratify patients who will benefit most from adjuvant therapy, further prospective studies are necessary to validate the clinical utility of this to demonstrate long-term survival benefit using this technique.

Detection of CTCs and ctDNA

Clinical detection of CTCs has been available for over a decade. The most commonly utilized clinical detection method is the CellSearch™ system. This is the only approved system by the Food and Drug Administration (FDA) for selection and enumeration of CTCs. The technique involves enriched serum for epithelial cell adhesion molecule (EpCAM) antigens to positively select CTCs that are fluorescently labeled and can be detected via basic microscopic methods [12]. The various techniques involve depletion of erythrocytes by lysis and leukocyte depletion, tumor cell isolation, cell staining, and detection via cytometry, microscopy, PCR, or fluorescent in situ hybridization (FISH), among others. However, the current FDA-approved CellSearch™ system is limited in the ability to detect cells that have undergone epithelial-mesenchymal transition (EMT), and therefore several mechanisms exist to identify this subpopulation of cells. For example, microfluidic devices using captured antibodies can use antibodies directed against EpCAM. However, some limitations exist with respect to downregulation of EMT cells that may occur during tumor progression. Additional studies have utilized size-dependent selection of CTCs using filtration techniques that enable detection of cells undergoing EMT. These methods are also promising to isolate a single CTC for genomic and functional analyses. Current laboratory-based techniques will likely be available for broader clinical and commercial use in the future, expanding the types and characteristics of CTCs detectable for analysis.

Improvement in DNA sequencing techniques has transformed the ability to detect ctDNA based on either rapidity of sample processing or the breadth and depth of sequencing. Current techniques to detect ctDNA involve candidate gene analysis or deep sequencing [13]. The former technique requires a priori knowledge of known hotspot mutations. Examples of this technology include digital droplet PCR (ddPCR) and beads, emulsions, amplification, and magnetics (BEAMing), among others. The advantage of these methods is the lower cost, speed (on the order of hours to days), minimal need for bioinformatics, and very high sensitivity to detect known driver mutations.

Deep sequencing through next-generation sequencing (NGS) can involve either a targeted or nontargeted approach, in both cases enabling the investigation of multiple genes in a single sample. Targeted deep sequencing by amplicon-based and hybrid capture methods extends the loci interrogated up to 50 Mb pairs. Sensitivity varies to some extent based on the number of exons assessed. In contrast, the nontargeted approach involves sequencing an even longer DNA length through whole-exome or whole-genome sequencing. Given the longer sequencing length, sensitivity tends to be lower as compared to targeted sequencing. NGS techniques involve aligning and comparing millions of ctDNA sequences compared to either a known reference genome or concurrently collected germline DNA. The considerable advantage of this approach is the ability to sequence longer lengths of the genome to monitor multiple genomic alterations, concurrently. Disadvantages associated with deep sequencing include that the method is more expensive and requires longer processing (on the order of days to a few weeks) due to intensive bioinformatics.

A number of commercial applications currently exist for clinical ctDNA analyses. These range from ddPCR of known resistance mutations to broader sequencing of over 70 oncogenic genomic alterations. Importantly, these assays report very high analytical specificity to detect mutations and the potential to monitor these alterations over time using mutant allele frequency as a surrogate for tumor burden. As existing companies scale up the technology and new companies enter the industry, we anticipate costs for comprehensive ctDNA sequencing panels will decrease to enable serial monitoring for a greater number of patients (Table 8.1).

Future Directions

Even with the considerable development and understanding of CTCs and ctDNA, liquid biopsies are likely in their infancy and significant potential will be achieved in the coming decade to utilize these assays to improve patient outcomes. A number of important future applications will be discussed.

CTCs and ctDNA each possess unique properties that will enable complementary studies to potentially change the way that we understand tumor biology. First, because CTCs are captured as intact cells, future studies will enable using this method as a biomarker for response to immune checkpoint blockade targeting programmed cell death-1 protein receptor (PD-1)/programmed cell death-1 protein ligand (PD-L1). For example, detection and analysis of cell surface markers on CTCs, such as PD-L1, could help select patients more likely to respond to immune checkpoint blockade without the need for tissue biopsy. In addition to cell count and enumeration of CTCs, CTCs can be used for "omic" studies. These cells could be explored more in depth by combining whole-exome sequencing with functional analyses including transcriptomic and proteomic analyses [14].

Table 8.1 Comparing characteristics of CTC, ctDNA, and tissue biopsies

	CTC	ctDNA	Tissue
Biology	Reflect tumor cells from multiple sites	May better reflect spatial tumor heterogeneity	Site specific
Logistics and practicality	Minimally invasive	Minimally invasive	Invasive
Pre-analytical, analytical, and post-analytical variables	Commercial or laboratory-based methods required	Often requires bioinformatics tools	Pathology review necessary
Sensitivity and specificity	Relatively low	Moderate sensitivity, high specificity	Relatively high
Method	CellSearch™ system, microfluidic devices, filtration techniques	Candidate gene analysis via ddPCR or BEAMing	Standard tissue biopsy techniques
		Deep sequencing	
		Next-generation sequencing	
Applications	Prognosis	Prognosis	Tissue architecture and histology for initial diagnosis
	Minimal residual disease	Minimal residual disease	
	Response to therapy	Detection of genetic alterations	Staging
	Single-cell molecular and genetic studies	Tumor evolution	Re-biopsy enables detection of emerging genetic alterations and response to therapy
		Monitoring disease burden and response to therapy	
Challenges	Determining utility as a predictive biomarker	Cost associated with serial biopsies	Capturing spatial tumor heterogeneity
			Feasibility of serial biopsies
Future	Single-cell transcriptomic and metabolomic studies	Understanding how targeting particular mutations in the blood may improve patient outcomes	Optimizing techniques to minimize complications
	Ex vivo testing of therapeutics	Early cancer detection	

These studies could improve understanding of gene expression and further enable drug sensitivity testing. For example, *ESR1* mutations can be detected on CTCs and sensitivity to various aromatase inhibitors could be tested in combination with other therapies ex vivo. Further work will also explore the potential to subclassify CTCs after detection into intact CTCs, circulating tumor microemboli, circulating tumor cell clusters, and circulating tumor materials. Through subtyping of CTCs, a better understanding of why a small proportion of these cells achieve metastasis could be explored. This is important because metastasis appears to be a relatively inefficient process in terms of the number of cells that survive at distant sites. In addition, with improvements in our understanding of ctDNA tumor burden and tumor mutational burden, preliminary evidence suggests that ctDNA serves as a surrogate for quantifying disease burden. If validated in further studies, serial blood biopsies could be utilized as a new mechanism for disease monitoring, enabling less frequent CT and PET imaging. Further research should also explore how dynamic changes in MAF in the blood in response to treatment affect PFS and OS to validate that targeting specific genomic alterations improves patient outcomes. In addition, optimal timeframe of measuring ctDNA in response to therapy must be further explored.

Finally, in order to make dramatic improvements in patient survival, a paradigm shift is necessary toward novel methods for early cancer detection. The issue with standard, guideline-based screening techniques is that detection is significantly delayed until radiographic (e.g., mammogram or CT scan) or visualized tissue (e.g., colonoscopy) may demonstrate cancerous lesions. In the future, liquid biopsies may possess the potential to revolutionize early detection of cancer. We envision this to occur through serial monitoring and detection of signatures in the blood via CTCs, ctDNA, and other blood-soluble signatures to identify patients who have developed cancer prior to radiographic detection of disease. In this regard, patients could receive surgery earlier or neoadjuvant therapy to enable surgery with curative intent for a significantly larger proportion of patients.

A fundamental challenge of using ctDNA for early cancer detection in asymptomatic individuals is considerations of the sensitivity characteristics of the test. While current ctDNA detection limits are on the order of 0.1%, detection of much smaller quantities of ctDNA would be necessary with greater depth of sequencing. Interestingly, it has been reported that individuals without known cancer have increased cell-free DNA concentrations in certain non-cancerous disease states (e.g., trauma, infection, or exercise) or detectable mutations in a small proportion of healthy controls. Many of these control patients likely possess genomic alterations that may never become oncogenic. Therefore, to improve theoretical sensitivity, more accurate early detection of cancer may require the combination of multiple assays from blood (ctDNA and CTCs), as well as other fluid compartments (urine, stool, sputum, etc.). Currently, several initiatives from companies, including GRAIL (Menlo Park, CA), Project LUNAR from Guardant Health (Redwood City, CA), and others, have invested millions to billions of dollars and plan to enroll hundreds of thousands of patients to begin to make these ambitious dreams a reality.

Conclusions

Liquid biopsies in the form of CTCs and ctDNA provide complementary information that will likely be used together to optimize sensitivity and improve the breadth of prognostic, predictive, genetic, and functional data to significantly impact patient care. These techniques will enable researchers and clinicians to redefine minimal residual disease based on the liquid phase of solid tumors, as opposed to current radiographic and tissue methods. The fundamental advantage of CTCs is capturing intact tumor cells to enable whole-exome and whole-genome sequencing, as well as functional studies. In contrast, ctDNA detects genomic alterations with high specificity and early evidence suggests using this technique can detect residual disease after surgery and targetable resistance mutations in blood and other fluid compartments.

Certainly, the potential for these techniques is considerable. However, careful, prospective studies are necessary to validate that these liquid biopsies do, in fact, improve clinical outcomes for patients. Furthermore, cost of these assays, particularly when used at multiple intervals for serial biopsies remains an important consideration. Theoretically, cost-effectiveness could be optimized by matching patients with treatments with a higher likelihood of benefit and then modifying therapy as the tumor evolves to save costs associated with ineffective treatment and unnecessary imaging. Additional resources would need to be invested into molecular tumor boards for multidisciplinary discussions and bioinformatics to optimally match patients with treatments based on complex genetic, functional, and clinical data.

As our understanding and detection of CTCs and ctDNA continue to improve, liquid biopsies will transform the way oncology is practiced. These biopsies have multiple potential clinical applications including screening at-risk populations prior to cancer development and monitoring disease response and resistance after diagnosis. These minimally invasive biopsies have the potential to allow *real-time* monitoring as predictive biomarkers and even potentially as a tool to test therapeutics ex vivo. Therefore, these assays may serve to represent the precise genetic, functional, and treatment landscape of the entire tumor.

References

1. Vogelstein B, Papadopoulos N, Velculescu VE, Zhou S, Diaz LA Jr, Kinzler KW. Cancer genome landscapes. Science. 2013;339(6127):1546–58.
2. Ashworth TR. A case of cancer in which cells similar to those in the tumours were seen in the blood after death. Aust Med J. 1869;14:146.
3. Mandel P, Metais P. Les acides nucléiques du plasma sanguin chez l'homme. C R Seances Soc Biol Fil. 1948;142(3–4):241–3.
4. Hong B, Zu Y. Detecting circulating tumor cells: current challenges and new trends. Theranostics. 2013;3(6):377–94.
5. Cristofanilli M, Budd GT, Ellis MJ, Stopeck A, Matera J, Miller MC, et al. Circulating tumor cells, disease progression, and survival in metastatic breast cancer. N Engl J Med. 2004;351(8):781–91.
6. Smerage JB, Barlow WE, Hortobagyi GN, Winer EP, Leyland-Jones B, Srkalovic G, et al. Circulating tumor cells and response to chemotherapy in metastatic breast cancer: SWOG S0500. J Clin Oncol. 2014;32(31):3483–9.
7. Ilie M, Hofman V, Long-Mira E, Selva E, Vignaud JM, Padovani B, et al. "Sentinel" circulating tumor cells allow early diagnosis of lung cancer in patients with chronic obstructive pulmonary disease. PLoS One. 2014;9(10):e111597.
8. Diaz LA Jr, Bardelli A. Liquid biopsies: genotyping circulating tumor DNA. J Clin Oncol. 2014;32(6):579–86.
9. Heitzer E, Ulz P, Geigl JB. Circulating tumor DNA as a liquid biopsy for cancer. Clin Chem. 2015;61(1):112–23.
10. Wan JC, Massie C, Garcia-Corbacho J, Mouliere F, Brenton JD, Caldas C, et al. Liquid biopsies come of age: towards implementation of circulating tumour DNA. Nat Rev Cancer. 2017;17(4):223–38.
11. Tie J, Kinde I, Wang Y, Wong HL, Roebert J, Christie M, et al. Circulating tumor DNA as an early marker of therapeutic response in patients with metastatic colorectal cancer. Ann Oncol. 2015;26(8):1715–22.
12. Toss A, Mu Z, Fernandez S, Cristofanilli M. CTC enumeration and characterization: moving toward personalized medicine. Ann Transl Med. 2014;2(11):108.
13. Siravegna G, Marsoni S, Siena S, Bardelli A. Integrating liquid biopsies into the management of cancer. Nat Rev Clin Oncol. 2017;14(9):531–48.
14. Ignatiadis M, Dawson SJ. Circulating tumor cells and circulating tumor DNA for precision medicine: dream or reality? Ann Oncol. 2014;25(12):2304–13.
15. Alix-Panabières C, Pantel K. Clinical prospects of liquid biopsies. Nat Biomed Eng. 2017;1(4). https://doi.org/10.1038/s41551-017-0065.
16. Boral, et al. Molecular characterization of breast cancer CTCs associated with brain metastases. Nat Commun. 2017;8:196. https://doi.org/10.1038/s41467-017-00196-1.

Measurement of Predictive Cancer Biomarkers by Flow Cytometry

9

Prashant Ramesh Tembhare, Sumeet Gujral, and H. Krishnamurthy

Technical Considerations

The basic principle of flow cytometry is the measurement (metry) of cellular (cyto) properties as they are moving in a fluid stream (flow). A flow cytometry is constructed of the four main components: (i) *fluidics* that generates a coaxial stream of inner core of cell suspension and outer layer of sheath fluid using the optimized differential pressures termed as "hydrodynamic focusing," (ii) *excitation optics* which include multiple lasers and lenses, (iii) *collection optics* that includes dichroic mirrors and filters which help in transmitting the emitted wavelengths to the respective photomultiplier tube (PMT)/detector, and (iv) *an analyzer* that converts the electrical signals to digital signals (Fig. 9.1a, b). The target cells coated with the fluorescent dye-conjugated antibodies are hydrodynamically focused as a row of single cells through one or more laser beams using the layer of isotonic sheath fluid. The tiny flashes of scattered light and fluorescence signals produced by the cells and fluorochrome-conjugated antibodies that adhered to the cell surface or inside the cells while passing through the laser beam are collected by respective optics and detectors (Fig. 9.1a, b). These analogue signals are converted into digital signals and displayed as frequency distribution histogram in the case of single parameter analysis or bivariate dot plot if it is a dual parameter analysis (Fig. 9.1c, d). The analogue to digital conversion is commonly based on pulse height as shown in Fig. 9.1c but can be based on pulse area or width. The data collected can be in linear or log scale depending on the difference in the signal intensities. Most of the immunocytometry data is collected on a log scale as the difference in the intensities of unstained and stained populations is substantial.

Since 1990, the technique of flow cytometry has undergone tremendous advancements with increasing complexity of the optics. In addition to the blue and red lasers, the newer instruments are equipped with a yellow, violet, and UV lasers. Simultaneously, the flow cytometry industry has provided a wide variety of fluorochromes with a broad range of emission spectra that are commercially available that enables the use of more than 12 fluorochromes in currently existing high-end

P. R. Tembhare (✉)
Hematopathology Laboratory, Advanced Centre for Treatment, Research and Education in Cancer (ACTREC), Tata Memorial Centre (TMC), Homi Bhabha National Institute (HBNI) University, Navi Mumbai, Maharashtra, India
e-mail: ptembhare@actrec.gov.in

S. Gujral
Department of Pathology, Tata Memorial hospital, Tata Memorial Centre (TMC), Homi Bhabha National Institute (HBNI) University, Mumbai, Maharashtra, India

H. Krishnamurthy
Central Imaging and Flow Cytometry Facility, National Centre for Biological Sciences, Tata Institute of Fundamental Research, Bengaluru, Karnataka, India

S. Badve, G. L. Kumar (eds.), *Predictive Biomarkers in Oncology*, https://doi.org/10.1007/978-3-319-95228-4_9

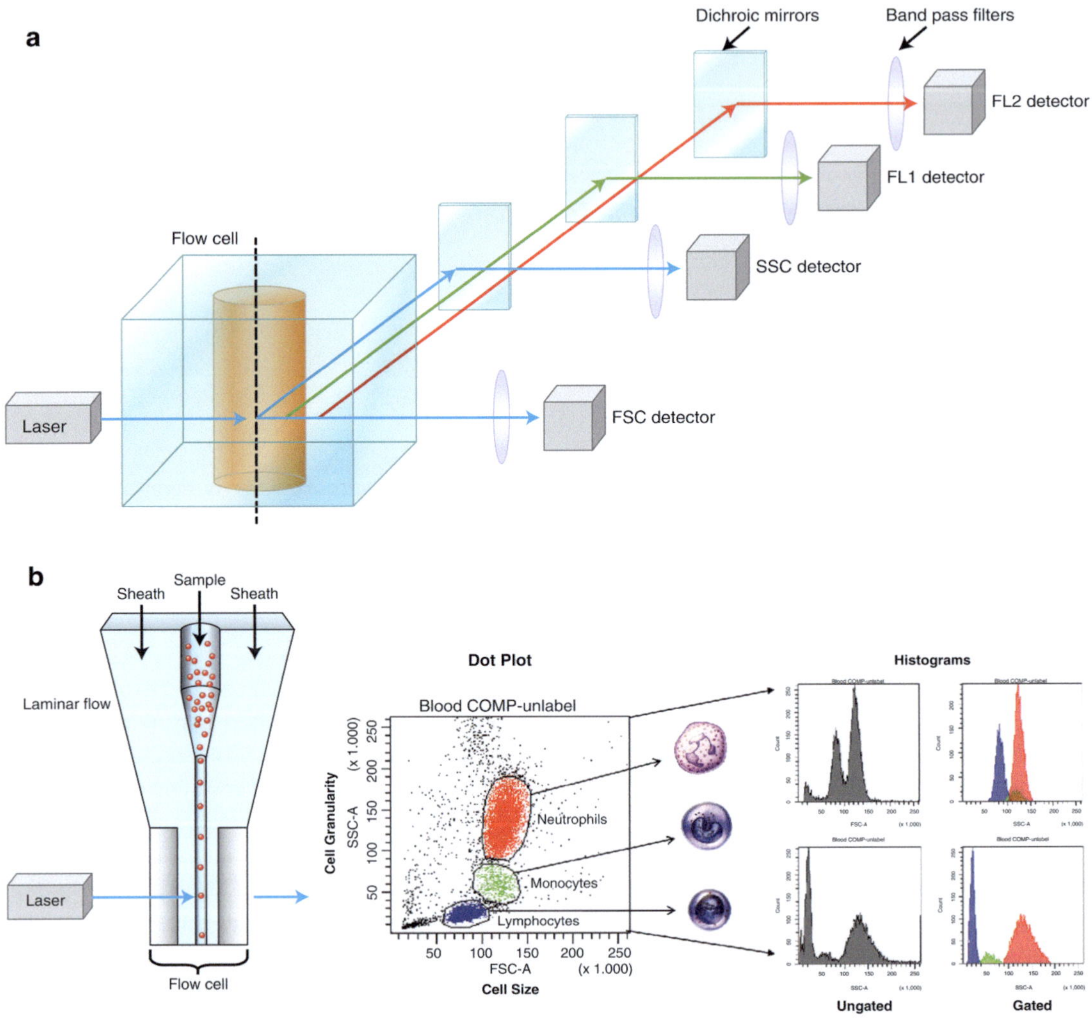

Fig. 9.1 (**a**) Shows LASER, flow cell, optics, detectors, and the light path for collecting scatter and fluorescence signals. (**b**) RBC-lysed blood cells showing lymphocytes, monocytes, and neutrophils on a dual parameter forward scatter (FSC) vs. side scatter (SSC) dot plot as single parameter histograms. The gray histograms are "ungated," while the colored histograms are "gated." (**c**) Analogue to digital conversion using pulse height of signal intensity. (**d**) Color compensation: The excitation and emission spectra of FITC and PE show spillover of FITC in PE channel. The spillover FITC in PE channel is compensated using PE-% FITC to match the PE median of unstained and FITC stained populations. (**e**) Bivariate density-cum-contour plots "i], & ii]" are showing the extent of spillover between the parent dye and its tandem dyes. Plot "i]-iii]" shows marked spillover between electron coupled dyes (ECD) PE-Texas Red and PE and between APC and APC-Alexa Fluor 750 and spillover between brilliant violet (BV)-421 and BV510, respectively. The later dyes are both excited with the same laser, i.e., "violet laser." Bivariate density-cum-contour plots "iv], v] & vi]" showed post compensation corrections. Notably, after compensation, the tandem dyes show the bidirectional spread of the events (indicated by arrows). Hence, use of mutually exclusive markers like CD4 vs. CD8 or CD36 vs. CD20 and adequate knowledge of markers with their expression on target cells like CD7 vs. CD3 can avoid interpretational errors

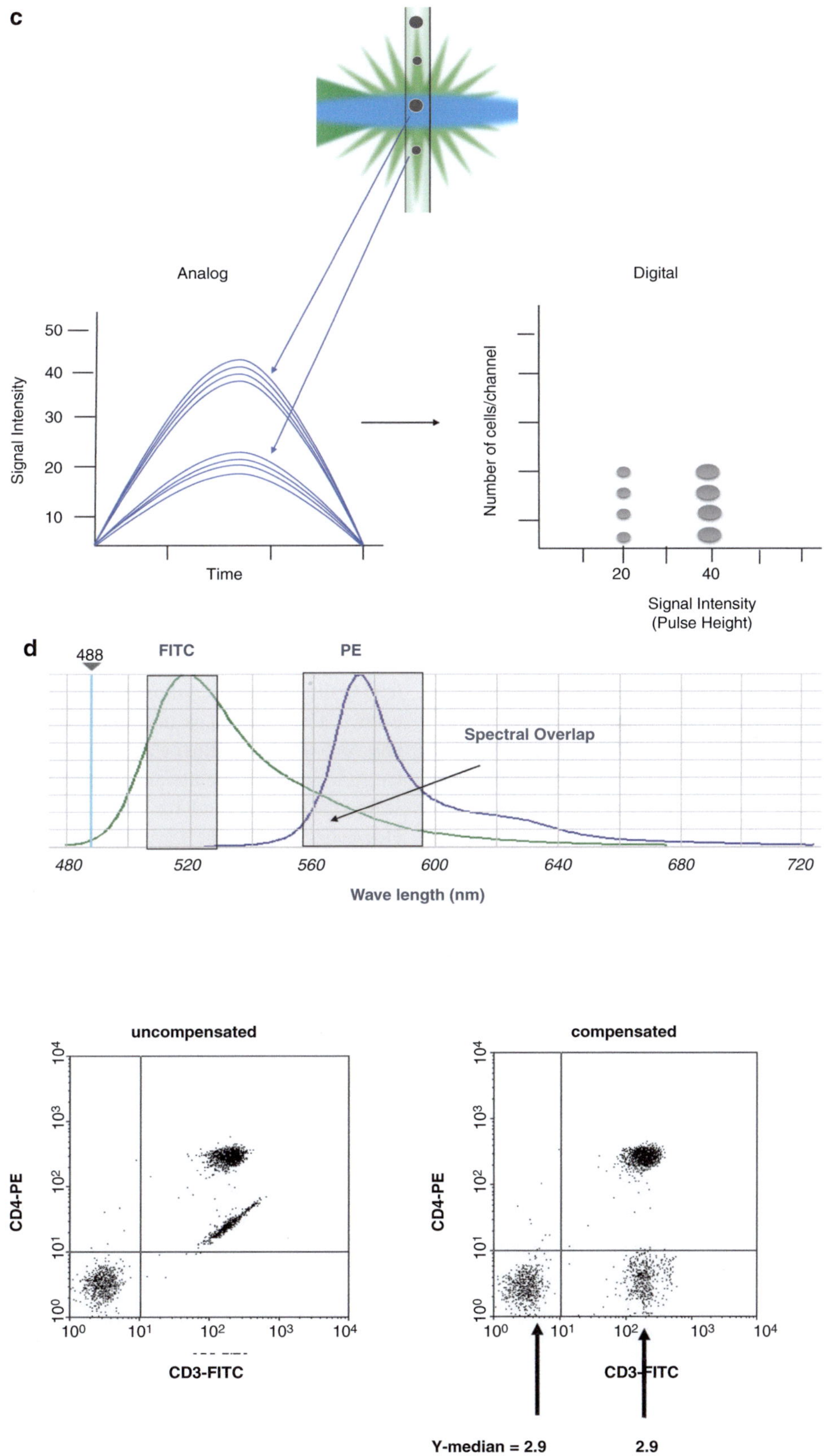

Fig. 9.1 (continued)

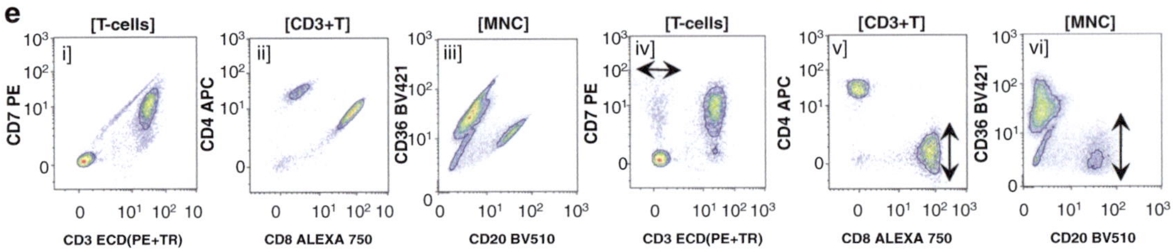

Fig. 9.1 (continued)

flow cytometers. Recently available advanced flow cytometers use up to 5 lasers and allow detection as many as 18 colors.

Critical Aspects of Multicolor Flow Cytometry

Panel Designing

Antibody panel designing for flow cytometric immunophenotyping involves selection of a combination of antibodies. Equally important is to select the right fluorochrome for each antibody. Although 18-color immunophenotyping has been successfully performed in a few research laboratories, it nevertheless poses some challenges. One of such challenges is choosing the right combination of antibodies and fluorochromes due to the significant overlap between the emission spectra of the fluorochromes.

A general rule is that fluorochromes excitable with the same laser exhibit significant overlap in their emission spectra. For example, when fluorochromes such as FITC and PE are excited by a 488 nm argon ion laser, they will emit fluorescence at 520 and 576 nm, respectively. Therefore, some of the FITC emission will be collected by the PE detector and vice versa. See Fig. 9.1d. This spillover of fluorochromes is called as spectral overlap. Such spectral overlap can be significantly high between an original dye and its tandem dyes such as between PE and its tandem dyes like ECD (PE and Texas Red), PE-CY5, and PE-Cy7 and between fluorochromes excitable with red laser, i.e., APC, and APC-tandem dyes like APC-Alexa Fluor 700, APC-Alexa Fluor 750, APC-H7, etc. (refer to Fig. 9.1e). Spectral overlap between emission spectra of the fluorochromes can cause serious interpretational mistakes. Hence, it needs to be corrected digitally by a function of the flow cytometry software that subtracts the amount of one fluorescence light spilled into the spectrum of other. This mathematical correction done by using flow cytometry software is known as "compensation." Figure 9.1d demonstrates the emission spectral overlap between CD3-FITC and CD4-PE. If this overlap is not corrected, it can lead to the misinterpretation of all CD3+ as CD4+ events. Therefore, it is important to compensate and remove the spillover as shown in Fig. 9.1d. Current flow cytometers are supported with software that is designed for digital compensation. However, due to the variable fluorochrome-to-antibody (F/P) ratio in a variety of commercially available reagents, sometimes the generic compensation may not correct the spillover and, hence, need antibody-specific compensation. After compensation, some of the dyes can show bidirectional spread (refer to Fig. 9.1e). To minimize interpretational mistakes in such situations, it is useful to either include mutually exclusive antibodies like CD4 and CD8 or CD36 and CD20 or markers with a known expression on target cells like CD3 and CD7 on T lymphocytes.

In multicolor immunophenotyping, the amount of overlap between the fluorochromes can be determined with a "fluorescence-minus-one (FMO)" experiment in which all fluorochrome-conjugated antibodies from a panel are added to the cells except one. This provides a fair idea of the amount of spillover of adjacent fluorochromes in the emission spectrum of the one that is not included and thus guides in the proper panel designing. Another basic rule that helps to reduce this issue is to use strong fluorochrome with an antibody against weakly expressed antigen and vice versa. For example, CD8 is strongly expressed on the T cells, and it can be studied using FITC-conjugated anti-CD8 antibody.

Another aspect of multicolor panel designing is selecting the right combination of antibodies. Selection of antibodies solely depends on the results one expects from the experiment or the aim of the assay. The panel for a given assay or experiment must include the combination of markers that would allow (a) adequate gating (isolation) of target cells from the rest of the events; (b) separation of their subsets; (c) differentiation between normal and abnormal (tumor) cells, or resting and functionally active cells, or mature and immature cells; and (d), ultimately, the correct quantitation. The inclusion of unnecessary reagents should be avoided as it increases the cost of the assay. While selecting the antibody panel for the diagnosis or monitoring in the clinical laboratory, the detailed clinical information and morphological features of the cells to be studied are always helpful. For example, history of young age, acute presentation, and high WBC count with the blastic morphology of the tumor cells indicates acute leukemia and helps to choose acute leukemia panel for immunophenotyping.

Designing the advanced multicolor panel (>10-color panel) is more challenging since the commercial availability of antibodies conjugated with the newer fluorochromes such as brilliant violet or quantum dots is limited. Moreover, not all the reagents which are available in these spectra show good signal to noise ratio (SN ratio) to get satisfactory signals. If at all, one managed to put together more than 12-color panel, it gives relatively higher noise due to the addition of signals from so many fluorochromes which can affect the capacity of the instrument to separate the weak signals from negative one reducing the sensitivity of the assay. Thus, one has to be careful and perform adequate standardization exercise while designing more than ten-color flow cytometric assay.

Sample Collection, Transportation, Storage, and Processing

Hematological samples which are already included in a single cell suspension must be collected in proper anticoagulants like EDTA or heparin. EDTA sample can be processed up to 48 h, and heparinized samples can be processed up to 72 h of collection, and beyond this duration, the integrity of the sample may get compromised and may not provide reliable results. The samples should be stored or transported at 18–20 °C.

The processing of the samples for immunophenotyping includes two main steps, i.e., enrichment of the WBCs by red cell lysis and density gradient-based separation followed by incubation of the cells with antibody cocktails. For the immunophenotyping in the solid tissue, cells need to be isolated as single cell suspension by either mechanical methods (slicing, mincing, and teasing) or enzymatic methods like trypsinization. Mechanical methods are favored as it avoids destruction of epitopes and alterations of the cell surface. Before staining of the cells with antibodies, it is highly desirable to evaluate the viability of the cells since nonviable cells are prone for non-specific binding antibodies due to exposure of Fc-receptors on the surface and compromised cell membrane. Hence, it is also advised to incorporate viability dye like 7-AAD in the panel to exclude the dead or nonviable cells from the analysis, especially in the functional or quantitative assay requiring higher sensitivity.

The antibodies selected for the experiment should be adequately titered to optimize the volume of antibody per test as the excess of antibody can result in non-specific binding, as well as increased noise levels (reduced S/N ratio); conversely low quantity of antibody can result in weak signals.

Acquisition and Data Analysis

Before the acquisition of the sample, it is necessary to perform all daily quality control measures as per the manufacturer's guidelines. Modern flow cytometers allow high-speed acquisition (up to 30,000 events/second); however, it is important to optimize the speed of acquisition for any given experiment. For example, in cell cycle or DNA ploidy analysis, the cells need to be acquired at a lower speed, i.e., 200–500 cells/second, as higher acquisition speed results in high CV and lower sensitivity of the assay. High-speed acquisition can also increase the number of doublets which can cause erroneous results. The exception to this rule may be the recently developed acoustic-hydrodynamic focusing which claims of improved high-speed acquisition without increasing CV or number of doublets.

One of the most critical aspects of flow cytometric immunophenotyping is data analysis and interpretation. There are many commercially available third-party softwares (e.g., FlowJo, Kaluza, FCS Express, Infinicyt) for data analysis. These softwares are relatively simpler to use and allow post-acquisition compensation adjustments if required. The essential step of the data analysis is a proper gating of the target cells. *Gating* is a function of the flow cytometry software that allows creating a digital window which makes only selected events visible and hides the remaining events and thus allows analysis of the cells of interest in reducing the interference due to other background events. It is very important to gate the right cells for data analysis which otherwise can lead to erroneous results. If the gating of the target cells is based on the light scatter characteristics like side scatter (SSC) and/or forward scatter (FSC), then it is important to know the light scatter characteristics of the target cells. For example, lymphoid cells usually have low SSC and FSC, but tumor cells in hairy cell leukemia possess relatively higher FSC and SSC equivalent to that of monocytes. In this sample, gating the events with low SSC and FSC can lead to false-negative results. In samples with high event acquisition, it is common to have doublets which need to be excluded initially using a bivariate dot plot of FSC height versus FSC width. Similarly, based on the SSC versus FSC plot, it is possible to exclude the majority of the nonviable cells such as dying or apoptotic cells which show SSC and FSC lower than that of lymphocytes. In clinical laboratories, it is recommended to use predesigned and verified controls for data analysis.

Comparison of Flow Cytometry and Other Equivalent Techniques

Flow cytometry (FC) works on a basic principle of antibody-antigen binding that results in the chemical attraction of an antibody to its antigen. Other common immunological methods that also work along the same principle are immunohistochemistry (IHC) and enzyme-linked immunosorbent assays (ELISA). While FC is applied on cells in suspensions like blood, bone marrow, body fluids, or fine needle aspirate, IHC and ELISA are applied on the cells fixed in paraffinized tissue sections or in the examination of soluble antigens present in body fluids such as serum, respectively. The major advantage of FC is that it allows analysis up to 30,000 cells per second and simultaneous determination of more than 18 antigens in that cell population which makes it a fast and very sensitive technique to identify multiple subsets of the cells simultaneously as well as a rare cell population in the background of millions of other cells [1]. On the contrary, cell-based ELISA and IHC typically cannot stain more than three proteins or antigens at a time. In fact, double and triple staining is also limited to a few markers due to cross-reactivity of primary and secondary antibodies with other antigens and endogenous immunoglobulins [2]. Hence, these techniques need multiple tissue sections or cell preparation to determine more than two markers. Immunofluorescence methods are being developed to address this need, but these are still in the research arena [3, 4]. Additionally, like ELISA, 96-well plate system can also be used on flow cytometers with the high-throughput system. FC can also be adapted for evaluation of extracellular proteins or antigens present in the serum, plasma, and body fluids using bead-based assays [5].

Predictive Cancer Biomarkers by Flow Cytometry: An Introduction

In the 1960s and 1970s, Lloyd Old and Ted Boyse introduced the concept of cell surface differentiation antigens that revolutionized the understanding of the immune system [6]. The knowledge of cell surface antigens enabled distinction of lineages and different subsets of leukocytes. These advances ultimately led to the precise and systematic classification of cell surface antigens, known as cluster of differentiation (CD) classification. Development of hybridoma technology further provided an easy way to recognize the cell surface antigens using fluorescent-conjugated monoclonal antibodies against these antigens with the analytical tools such as FC. Since the 1960s, FC technology has gone through continuous advances in instrumentation, which have led to the development of multicolor flow cytometers. These instruments have a high-resolution capacity enabling detection of 18 or more markers simultaneously (immunophenotyping) [1]. It has been widely used in the preclinical and clinical setting for generating biomarker data that is decisively used in the field of oncology. FC is thus a rapid and effective technique to identify a protein or a molecule present on the cell surface and intracellularly (e.g., myeloperoxidase (MPO) in myeloid blasts or terminal deoxynucleotidyl transferase (TdT) in lymphoid precursor cells) with a further advantage of providing quantitative data [7]. Hence, FC is a valuable tool in the discovery of several prognostic and predictive biomarkers. This chapter is focused on the application of multiparametric FC (MFC) to determine predictive biomarkers in the management of hematolymphoid and solid cancers.

Predictive Biomarkers

FC is the primary tool used in the diagnosis (acute vs. chronic leukemia) and lineage (B-cell, T-cell, or myeloid lineage) identification and classification of leukemia and lymphoma. Using a variety of monoclonal antibodies against the antigens expressed on the cell surface or intracellularly or both, FC identifies different cells types (B and T lymphocytes, NK cells) and cells at different stages of maturation (stem cells, immature or mature myeloid cells) and differentiates between normal and neoplastic tumor cells. Currently, more than 400 "CD"s are identified, and the antibodies against these antigens are available commercially. A list of the lineage associated/specific markers (CDs) for different cell types is displayed in Table 9.1.

In hematolymphoid malignancies, FC is used in the determination of predictive biomarkers for targeted therapy. In the last two to three decades, oncology field has seen many newer revolutionary therapeutic options that include targeted monoclonal antibody-based therapy and cellular immunotherapy. Monoclonal antibody-based (mAb) treatment is based on a principle that mAb binds to the cell surface antigen present on the

Table 9.1 List of commonly used markers for the immunophenotyping of hematopoietic cells by flow cytometry

Type of cells	Markers (i.e., antibodies against antigens mentioned below)
Common leukocyte antigen (LCA)	CD45
Blasts or stem cells	CD34, CD133, HLADR
Myeloid lineage cells	CD11b, CD13, CD15, CD33, CD117, myeloperoxidase[a] (MPO)
Monocytic lineage cells	CD11c, CD14[a], CD33, CD36, CD64, HLADR
Erythroid lineage cells	CD36, CD71, CD105, CD235a[a] (glycophorin-A)
Megakaryocytic cells	CD36, CD41[a], CD61[a]
Dendritic cells	CD123, CD303[a], HLADR,
B lineage cells	CD10, CD19, CD20, CD22, CD79a, CD79b, HLADR, kappa, lambda
Plasma cells	Strong CD38, CD138[a], CD229, CD319
T lineage cell	CD1a, CD2, CD3[a], CD4, CD5, CD7, CD8
NK cells	CD16, CD38, CD56, CD57, CD94[a], NKp46[a], CD161

[a]Indicates lineage-specific markers

tumor cells and provides more specific targeting of the tumor cell. Monoclonal antibody therapy works through a variety of mechanisms that induce tumor cell killing. These mechanisms include (1) activation or inhibition of cell signaling pathways or (2) Fc-dependent mechanisms such as antibody-dependent cell-mediated cytotoxicity (ADCC), complement-dependent cytotoxicity (CDC), and antibody-dependent cellular phagocytosis (ADCP) [6, 8]. The primary requirement of mAb therapy is the presence of enough antigen density on the tumor cells against which the mAb is directed. FC provides detailed information regarding the expression of a specific antigen on tumor cells and its normal counterpart cells along with its precise quantitation. Furthermore, this data may also enable monitoring the impact of treatment on residual normal counterparts. The most successful mAb therapy used in the treatment of hematolymphoid malignancies is "rituximab therapy" based on anti-CD20 mAb.

The flow cytometric assessment of antigen expression, its level of expression in the tumor cells in a specific malignancy, and antibody binding capacity (ABC) provides valuable information for selection of a specific mAb therapy. For example, rituximab is only effective in B-cell malignancies with CD20 antigen expression on its tumor cells with adequate antigen levels. FC can easily confirm the expression of CD20 and measure the levels of CD20 expression on the tumor cells and its ABC (Fig. 9.2a, d).

Similarly, the knowledge of expression of CD52 on the chronic lymphocytic leukemia or T-cell non-Hodgkin's lymphoma cells using FC can determine the response to "alemtuzumab," a humanized anti-CD52 monoclonal antibody. Thus, FC provides a good technological tool for the assessment of some predictive biomarkers in hematolymphoid malignancies utilizing mAb therapies.

FC is a good tool for studying "chimeric antigen receptor-modified T cells (CAR T cells)" for cancer therapy. For the successful application of the CAR T-cell therapy, the knowledge of targeted-antigen expression of CD19 on the tumor cells is very important. Thus, FC plays an important role as a tool for studying cellular immunotherapy. Anti-CD19 CAR T-cell therapy is being evaluated and has shown some promising results in the CD19-expressing B-cell malignancies like CLL and B-cell acute lymphoblastic leukemia. The CAR T-cell therapies against other antigens like BCMA, CD22, CD33, CD123, ROR-1, and NKG2D ligands are also in development [9]. Clinical trials on CAR T-cell therapy for few solid tumors like neuroblastoma (anti-GD2 CAR T-cells) and breast carcinoma (anti-HER2 CAR T cells) are under investigation.

In addition, a bead-based flow cytometric assay has been shown to be a rapid and reliable technique for the detection of abnormal oncoproteins resulting from the fusion of BCR-ABL gene in chronic myeloid leukemia (CML) and acute lymphoblastic leukemia [10]. In this assay, the leukemic cells are lysed releasing BCR-ABL fusion proteins in the sample. The BCR part of the protein is recognized by an anti-BCR antibody bound to a bead and ABL part by a phycoerythrin (PE)-labeled anti-ABL antibody. So if BCR-ABL fusion protein is present after the acquisition, these beads provide the PE fluorescence signals, and if the fusion protein is absent, then beads do not show PE fluorescence signals. Targeted inhibitors directed specifically against the BCR-ABL tyrosine kinase (TKI) are proven to be highly successful in these malignancies. A similar assay is also being evaluated for PML-RARA fusion proteins for the diagnosis and treatment of acute promyelocytic leukemia. Thus, the innovative advances in the field of flow cytometric technology provide reliable and widely useful predictive biomarkers in the rapidly advancing field of targeted therapies for the management of cancers.

Flow Cytometric Minimal Residual Disease (MRD) Is the Strongest Predictor of Therapeutic Response in Hematological Malignancies

Posttreatment relapse of leukemia is one reason for the poor clinical outcome and death. The major cause of this problem is the non-detection

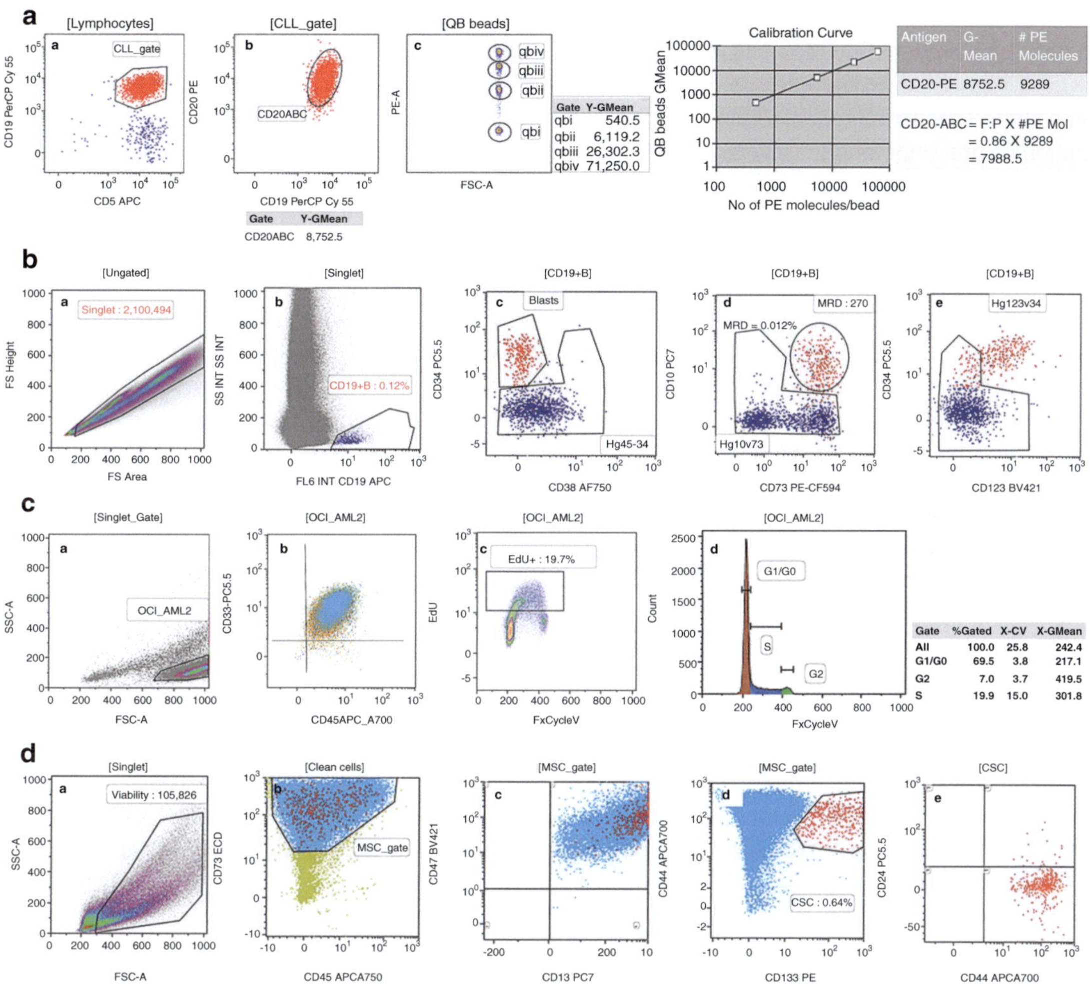

Fig. 9.2 (**a**) The figure demonstrates the measurement of CD20 antigen density and antibody binding capacity (ABC) on the cells of B-cell malignancy (chronic lymphocytic leukemia, CLL) using QuantiBRITE (QB) beads and QuantiCALC software (BD Biosciences). CLL cells with typical co-expression of CD5 and CD19 are gated (isolated) (bivariate plot "A"), and geometric mean fluorescence (GMF) of CD20 expressed by CLL cells is determined (plot "B"). Plot "C" shows GMF of QuantiBRITE beads. Beads of different PE molecule levels are shown as qbi, qbii, qbiii, and qbiv. The number of CD20 PE molecules per CLL cell was determined using "GMF," and known "PE molecules per bead" values are shown using QuantiCALC software (BD). ABC of CD20 is then calculated by multiplying PE molecule per cell values and fluorescence/protein (F/P) ratio. (**b**) The bivariate plots "A" to "E" demonstrate an example of minimal residual disease (MRD) in a B-cell precursor acute lymphoblastic leukemia (B-ALL). Bivariate plot "A" shows more than two million cell events acquired in the sample of which 0.12% are CD19+ B cells (blue dots Fig. 2II b) that were gated out using side scatter versus CD19 plot. Plots "C, D, and E" show the detection of MRD events (red dots) which are falling outside the gates defined for normal B-cell precursors (hematogones) using different markers. In this sample, MRD levels were 0.012%. (**c**) Plots "A" to "D" demonstrate an example of cell cycle analysis in the acute myeloid leukemia cell line "OCI_AML2." Density plot "C" demonstrates a subset of cells (19.2%) in the proliferation phase of cell cycle stained with EdU against FxCycle Violet dye, and histogram "D" shows cell cycle of these cells with G0/G1, S, and G2 phases. (**d**) The bivariate plots "A" to "E" demonstrate an example of cancer stem cell (CSC) evaluation in cells in culture from osteosarcoma. In plot "B," CD73-positive and CD45-negative mesenchymal cells were gated with "MSC_gate" (light blue dots) which were positive for strong expression of CD13 and CD47 (plot "C"). Of these cells, CSC were gated (red dots) using the positive expression of CD44 and CD133 in plot "D" and negative expression of CD24 in the subsequent plot "E"

of residual malignant cells by morphology. These can only be identified using techniques with a higher sensitivity such as multicolor FC and allele-specific oligonucleotide-PCR amplification of immunoglobulin (IG) and T-cell receptor genes (TCR), i.e., by Ig/TCR gene rearrangement analysis [11]. Multiparametric FC (MFC) is a fast, affordable, well-established, and widely available technique for the detection of MRD. Recent improvements in flow cytometers and availability of a broad range of fluorochromes have allowed more than eight-color immunophenotyping and analysis of millions of cells in each sample, thereby allowing detection of a rare population of leukemia blasts with the limit of detection of 1 in 10^5 cells [12]. The absence of MRD using either of the above mentioned sensitive techniques confirms the maximum clearance of leukemic cells. On the other hand, the presence of MRD in the bone marrow or peripheral blood sample indicates the existence of chemoresistant leukemic cells. Several studies have shown that MRD is the most powerful indicator of relapse in the many hematolymphoid malignancies like acute leukemia, i.e., BCPALL, T-ALL, and AML, as well as non-Hodgkin's lymphoma (NHL) like chronic lymphocytic leukemia (CLL) and multiple myeloma (MM). Recent studies have shown the better clinical outcome with MRD-based reduction of the treatment intensity (i.e., reduction of treatment intensity in early MRD-negative patients) to reduced drug toxicities [12]. Thus, MRD monitoring has become a standard practice in the management of ALL and is being used as an indicator to adjust the toxicity/benefit ratio of the therapy. Moreover, MRD monitoring is also guiding the treatment decisions in the relapsed cases of ALL and patients undergoing allogenic stem cell transplant [12]. An example of FC-MRD in B-ALL is shown in Fig. 9.2a.

MRD is also being used as a part of clinical trial studies for the monitoring of drug efficacy and therapeutic response at different stages of the trial. In fact, it is being considered as an end point for the future clinical trials of promisingly effective therapeutic agents to reduce the need for long-term follow-up.

Use of FC in Solid Tumors

FC has been used for measuring the DNA content and cell cycle analysis in the prognostication of a variety of solid malignancies; however, its role in the monitoring of predictive biomarker is limited and applied to a few targeted therapies.

Flow cytometric enumeration of circulating endothelial cells (CECs) is demonstrated in the selection of patients for the addition of neoadjuvant therapy of antiangiogenic agent bevacizumab in the treatment of epithelial cancers like breast carcinoma [13]. CECs are defined as DNA stain (+), CD45(−), CD34(+), CD31(+), and CD146(+) using FC. CEC count has been shown to be a useful surrogate marker of angiogenesis and antiangiogenic drug activity and to adjust the dosage of antiangiogenic drugs like bevacizumab [13]. More recently, flow cytometric measurement of γH2AX (an indicator of DNA double-strand breaks) and MRE11 (an indicator of DNA double-strand repair) in peripheral blood mononuclear cells has been used in the poly(ADP-ribose) polymerase inhibitors (PARPi) therapy for germline BRCA1 and BRCA2 mutation-associated ovarian cancers [14]. FC also has utility in the determination of surface antigens like GD2 for the planning of antiGD2-CAR T-cell therapy or antiGD2-mAb therapy in neuroblastoma. Thus, FC is increasingly used for developing newer predictive biomarkers in the effective application of novel targeted therapies in solid cancers.

Apart from the DNA and cell cycle analysis [15, 16], FC plays a major role in the evaluation of immune composition of tumor microenvironment, which includes the T lymphocyte subset quantitation like CD8+ cytotoxic T cells, natural killer (NK) cells, T regulatory cells (CD4+, CD25+ FoxP3+ T cells), and myeloid-derived suppressor cells (MDSC). T-cell subsets and MDSC have been proven as the predictors of therapeutic resistance and poor clinical outcome of solid tumors [17]. Besides that, researchers are now focused on the study of cancer-initiating cells, i.e., "cancer stem cells (CSCs)," which can be isolated using the positive expression of the markers like CD44 and CD133 and negative expression of markers like CD24 and lineage-specific markers (−) by FC (Fig. 9.2d). It has been hypothesized that CSC is a minute population of self-renewing cancer cells that cause tumor cell proliferation and are resistant to conventional chemotherapy [18]. This concept is being investigated to develop a CSC-targeted therapy to improve the clinical outcome. Lastly, as discussed in Chap. 8, circulating tumor cells (CTCs) are being used to make therapeutic decisions.

In summary, the current state of development in the field of FC has made it a powerful multiparametric technology, which has been extensively applied for the measurement of predictive biomarkers in the personalization of therapy and prognostic markers for the better risk stratification of hematolymphoid malignancies and many solid tumors. It has been proved to be a widely useful method in the identification of various cell types, subtypes, their origin, and developmental stages and monitoring of a variety of physiological and pathological functions of tumor cells and its microenvironment.

Glossary

Parameter This is either an antibody marker (e.g., CD45) or a physical parameter (e.g., side scatter or forward scatter or time).

Gating It is an application of a "gate" on the events of interest. "Gate" is a digital window in flow cytometry software that allows visualization of selected events.

References

1. Mair F, Hartmann FJ, Mrdjen D, Tosevski V, Krieg C, Becher B. The end of gating? An introduction to automated analysis of high dimensional cytometry data. Eur J Immunol. 2016;46(1):34–43.
2. Friis T, Pedersen KB, Hougaard D, Houen G. Immunocytochemical and immunohistochemical staining with peptide antibodies. Methods Mol Biol. 2015;1348:311–25.
3. Gerdes MJ, Sevinsky CJ, Sood A, et al. Highly multiplexed single-cell analysis of formalin-fixed, paraffin-embedded cancer tissue. Proc Natl Acad Sci U S A. 2013;110(29):11982–7.
4. Gerdes MJ, Gökmen-Polar Y, Sui Y, et al. Single-cell heterogeneity in ductal carcinoma in situ of breast. Mod Pathol. 2018;31(3):406–17.
5. Moncunill G, Campo JJ, Dobaño C. Quantification of multiple cytokines and chemokines using cytometric bead arrays. Methods Mol Biol. 2014;1172:65–86.
6. Scott AM, Allison JP, Wolchok JD. Monoclonal antibodies in cancer therapy. Cancer Immun. 2012;12:14.
7. Gujral S, Tembhare P, Badrinath Y, Subramanian PG, Kumar A, Sehgal K. Intracytoplasmic antigen study by flow cytometry in hematolymphoid neoplasm. Indian J Pathol Microbiol. 2009;52(2):135–44.
8. Leslie LA, Younes A. Antibody-drug conjugates in hematologic malignancies. Am Soc Clin Oncol Educ Book. 2013; https://doi.org/10.1200/EdBook_AM.2013.33.e108.
9. Firor AE, Jares A, Ma Y. From humble beginnings to success in the clinic: chimeric antigen receptor-modified T-cells and implications for immunotherapy. Exp Biol Med (Maywood). 2015;240(8):1087–98.
10. Raponi S, De Propris MS, Wai H, et al. An accurate and rapid flow cytometric diagnosis of BCR-ABL positive acute lymphoblastic leukemia. Haematologica. 2009;94(12):1767–70.
11. Campana D. Minimal residual disease monitoring in childhood acute lymphoblastic leukemia. Curr Opin Hematol. 2012;19(4):313–8.
12. van Dongen JJ, van der Velden VH, Brüggemann M, Orfao A. Minimal residual disease diagnostics in acute lymphoblastic leukemia: need for sensitive, fast, and standardized technologies. Blood. 2015;125(26):3996–4009.
13. Starlinger P, Brugger P, Reiter C, et al. Discrimination between circulating endothelial cells and blood cell populations with overlapping phenotype reveals distinct regulation and predictive potential in cancer therapy. Neoplasia (New York, NY). 2011;13(10):980–90.
14. Lee J-M, Gordon N, Trepel JB, Lee M-J, Yu M, Kohn EC. Development of a multiparameter flow cytometric assay as a potential biomarker for homologous recombination deficiency in women with high-grade serous ovarian cancer. J Transl Med. 2015;13:239.
15. Chang Q, Hedley D. Emerging applications of flow cytometry in solid tumor biology. Methods. 2012;57(3):359–67.
16. Tembhare P, Badrinath Y, Ghogale S, Patkar N, Dhole N, Dalavi P, Kunder N, Kumar A, Gujral S, Subramanian PG. A novel and easy FxCycle™ violet based flow cytometric method for simultaneous assessment of DNA ploidy and six-color immunophenotyping. Cytometry A. 2016;89(3):281–91.
17. Becht E, Giraldo NA, Dieu-Nosjean MC, Sautès-Fridman C, Fridman WH. Cancer immune contexture and immunotherapy. Curr Opin Immunol. 2016;39:7–13.
18. Rycaj K, Tang DG. Cell-of-origin of cancer versus cancer stem cells: assays and interpretations. Cancer Res. 2015;75(19):4003–11.

Personalized Cancer Treatment and Patient Stratification Using Massive Parallel Sequencing (MPS) and Other OMICs Data

10

Mark Abramovitz, Casey Williams, Pradip K. De, Nandini Dey, Scooter Willis, Brandon Young, Eleni Andreopoulou, W. Fraser Symmans, Jason K. Sicklick, Razelle Kurzrock, and Brian Leyland-Jones

Introduction

The transition from "evidence-based" (empiric) to "information-based" (precision) medicine (PM) has ushered in the era of personalized medicine in which breakthroughs in biological knowledge and technological capacity are used to stratify cancer patients so that they receive the most appropriate treatment. Biomarkers play a key role in this process and can be prognostic and/or predictive, with the difference being that prognostic biomarkers help in predicting the progress of the disease, while predictive biomarkers are connected with the response to a treatment.

Several challenges in achieving PM are well known. First, targeted therapeutic agents are increasingly available for clinical applications, but many of these drugs fail when used in a mono-therapeutic context and with inadequate patient selection in clinical trials. Second, proper patient stratification increasingly depends on the

M. Abramovitz
Avera Cancer Institute, Sioux Falls, SD, USA

C. Williams · S. Willis · B. Young
Center for Precision Oncology, Avera Cancer Institute, Sioux Falls, SD, USA

P. K. De
Department of Molecular and Experimental Medicine, Avera Cancer Institute, Sioux Falls, SD, USA

N. Dey · B. Leyland-Jones (✉)
Department of Molecular and Experimental Medicine, Center for Precision Oncology, Avera Cancer Institute, Sioux Falls, SD, USA
e-mail: brian.leylandjones@avera.org

E. Andreopoulou
Department of Medicine, Division of Hematology & Medical Oncology, Weill Cornell Medicine/New York Presbyterian Hospital, New York, NY, USA

W. F. Symmans
MD Anderson Cancer Center, Houston, TX, USA

J. K. Sicklick
Division of Surgical Oncology, General Surgery Residency, Biorepository and Tissue Technology Shared Resource, Moores Cancer Center, University of California San Diego (UCSD), School of Medicine, San Diego, CA, USA

R. Kurzrock
Moores Cancer Center, UC San Diego Moores Cancer Center, La Jolla, CA, USA

S. Badve, G. L. Kumar (eds.), *Predictive Biomarkers in Oncology*,
https://doi.org/10.1007/978-3-319-95228-4_10

development of innovative clinical trials in which biopsies are mandated to obtain clinically relevant and timely tumor specimens, enabling the discovery of new biomarkers predictive of response to treatments. This concept is also true in metastatic patients, where biopsies of the metastases are needed to identify context-relevant biomarkers. Third, the implementation of these biopsy-driven trials will require the establishment of a new generation of multidisciplinary translational and clinical research teams, which integrates molecular biologists, clinicians, bioinformaticians, and pathologists who are already involved at the study design stage. Forth, more predictive ex vivo and in vivo preclinical models will also be essential in helping define "drivers" and prioritize selecting from among the >700 drugs currently in clinical development for cancer. Last, a system for careful evaluation of both the patient and tumor biomarkers is essential to design optimal therapeutic strategies that can overcome potential acquired resistance and will best treat the patient with the least toxicity while reducing healthcare costs.

A comprehensive cancer omics and "Massive Parallel Sequencing"/next-generation sequencing (NGS) approach is paramount to be at the leading edge of the revolution in personalized cancer care. This great challenge will require putting in place the necessary molecular pathology and computational infrastructure and creating specialized basic, translational, and clinical multidisciplinary research teams that will transform the omics revolution and place it squarely at the forefront of personalized healthcare. Through the use of biopsy-driven novel clinical trial designs, novel statistical and computational analysis of tumor and host-derived molecular omics datasets, and the application of in vivo preclinical models underpinned by solid basic research, new tailored individualized therapies can be developed more rapidly and with much greater efficacy. Thus, the payoff of this omics revolution has the potential to be enormous, which will undoubtedly have a tremendous impact on patient care globally. Currently, however, less than 5% of cancer patients are enrolled in biopsy-driven clinical trials, and thus, far greater access for patients to biopsy-driven trials with new targeted therapies is imperative if we are to fuel the bilateral flow of information between bench and bedside.

This chapter will focus on how PM in cancer is being driven by genomics and prognostic and predictive biomarkers, as well as the role that NGS and omics play in all aspects of cancer treatment.

Predictive Biomarkers and Genomics in Cancer

A biomarker generally refers to a measurable indicator of some biological state or condition. Biomarkers can include genes, proteins, genetic variations, and differences in metabolic expression from different sources such as body fluids and tissues. Early biomarkers include the colon carcinoma tumor-specific antigen, the carcinoembryonic antigen (CEA), and the prostate-specific antigen (PSA), the latter two still used today in the clinical setting. Subsequently, a number of additional important tumor biomarkers have come to the forefront, many of which have been targeted by specific drugs including estrogen receptor/progesterone receptor (ER/PR) and human epidermal growth factor receptor 2 (HER2) in breast cancer; EGFR, KRAS, and UGT1A1 in colorectal cancer; HER2, GIST, and c-KIT in gastric cancer; p53 and LOH/microsatellite instability in head and neck cancer; CD20 antigen, CD30, FIP1L1-PDGRFalpha, PDGFR, Philadelphia chromosome (BCR/ABL), PML/RAR alpha, TPMT, and UGT1A1 in leukemia/lymphoma; AFP, AFLP, and DCP in liver cancer; ALK, EGFR, and KRAS in lung cancer; BRAF in melanoma; and HPV infection and oncogene E6 and E7 expression in uterine and cervical cancers.

Genome instability is at the heart of the hallmarks of cancer, as described so articulately in two seminal papers by Hanahan and Weinberg [1]. They initially proposed that human tumors are governed by a common set of six acquired capabilities: (1) self-sufficiency in growth signals, (2) insensitivity to anti-growth signals, (3) evasion of apoptosis, (4) limitless replicative potential, (5) sustained angiogenesis, and (6) tissue invasion and metastasis [1] (Fig. 10.1).

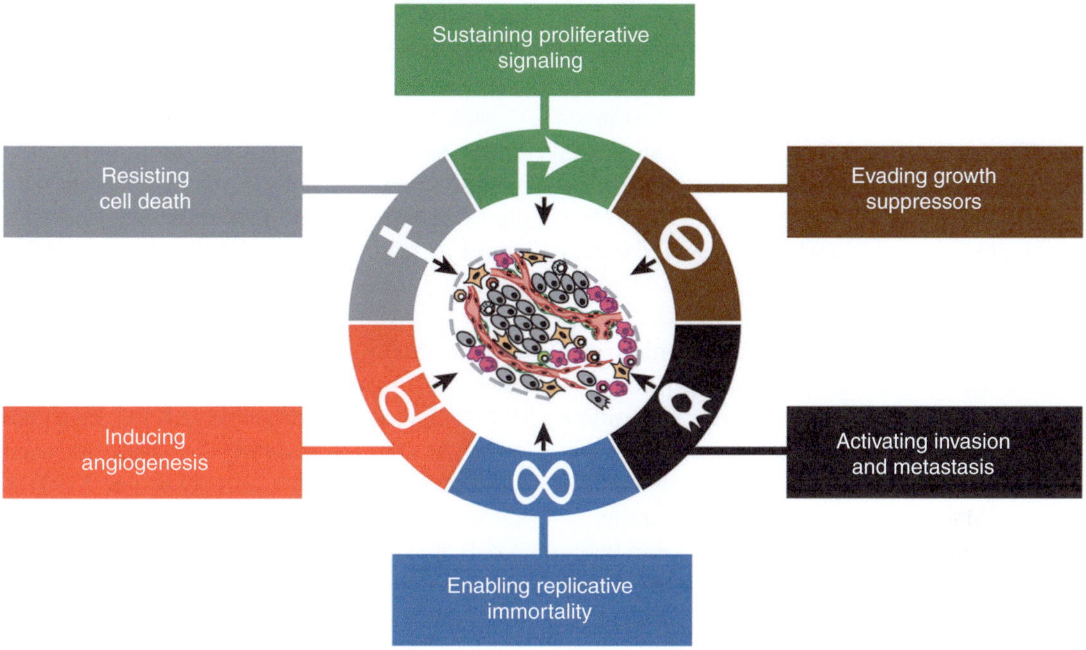

Fig. 10.1 Depicted are the six hallmarks of cancer, as originally proposed by Hanahan and Weinberg, who have subsequently updated them [1]. Over the past decades, great strides have been made in understanding how each hallmark contributes to cancer progression. (Reprinted from Hanahan and Weinberg [1]. With permission from Elsevier)

Subsequently, the authors added two additional emerging general hallmarks to the list, namely, reprogramming of energy metabolism and evading immune destruction [1].

It is this genetic diversity that accelerates the acquisition of these hallmarks in every cancer, making each one unique. Applying NGS to cancer has provided an improved understanding of the molecular mechanisms that are involved in tumorigenesis. Genomic alterations usually result in the generation of oncogenic drivers that are involved in the initial steps of oncogenesis [2].

Drugs that target these oncogenic drivers or biomarkers, including imatinib in chronic myeloid leukemias carrying the BCR-ABL fusion, trastuzumab in HER2-amplified breast cancer, and vemurafenib in BRAF-mutated melanoma, can be effective treatments and have been associated with several successes over the past decades, but not without their limits and unresolved issues.

One of the main problems in cancer treatment is that most tumors will eventually develop resistance, possibly due to intratumor heterogeneity and additional genomic and/or molecular events. Because most tumors comprise varying numbers of rare genomic events, administration of only a single medication, in most cases, is not sufficient. PM will take advantage of being able to identify tumor biomarkers that enable safe and effective therapy for every individual patient. Having the genetic profile of a patient's tumor will help oncologists select the proper medication(s) or therapy at the optimal dose(s) or regimen. The use of NGS is starting to make this a reality, and recent studies have shown that biomarkers that encompass driver oncogenes can be readily detected across numerous tumor types with a majority of those that are amenable to targeted therapy [3].

Use of NGS in Personalized Cancer Treatment

The Human Genome Project was a monumental achievement that ushered in the genomic age and the subsequent avalanche of downstream effects.

Sequencing cancer genomes was the next logical step [4], and with sequencing becoming increasingly affordable and reliable, this has led to the integration of genome science into clinical practice. The implementation of NGS, also known as massively parallel sequencing, has enabled the capture of large amounts of genomic data from a tumor; allowed for the comprehensive identification of alterations, genes, and pathways involved in the tumorigenic process; and allowed it to be integrated into a clinical workflow. Tumor samples can be used to derive increasingly complex genomic data along with a patient's germline DNA data determined using peripheral blood. The ability to use NGS to generate such data has pushed PM to the forefront of cancer therapy.

Importantly, NGS has raised the hope of being able to identify all cancer driver events in a tumor that are potential targets of existing and novel future drugs. Developing and stockpiling a vast arsenal of anticancer targeted drugs will provide oncologists with the ability to precisely assign the most efficacious targeted therapy to the individual patient based on the genomic events that are driving the tumor. The feasibility of this approach has been recently explored by Rubio-Perez et al. [3], whereby they developed a three-step in silico drug prescription strategy: (1) identify the driver events that include mutation, CNAs, and gene fusions; (2) find drugs, which include FDA-approved drugs and those being tested in clinical trials, targeting the driver gene protein products; and (3) assign the appropriate drug(s) to the patient based on his or her genomic driver events (Fig. 10.2). For this purpose they developed a Cancer Drivers Database (this database can be downloaded from the following website: https://www.intogen.org/downloads) that contains a list of genomically altered genes driving tumorigenesis in different tumor types and a Cancer Drivers Actionability Database (this database can be downloaded from the following website: https://www.intogen.org/downloads) that contains a comprehensive list of current and prospective anticancer targeted agents. They also describe a set of rules that is used to select the appropriate drug(s) to prescribe to patients. Both databases will be continuously updated and improved as knowledge of driver genes and anticancer therapies advances. The goal is to be able to establish a toolbox of tailored drugs that can deliver the promise of PM.

With the recent advances in NGS, the use of comprehensive whole-genome profiling has led to considerable changes in our understanding of the extensive genomic landscape that underlies cancer pathogenesis and has shifted the treatment paradigm from standard to personalized treatment in oncology. In breast cancer, several somatic driver mutations and alterations have been confirmed including ERRB2, PIK3CA, PTEN, alpha serine/threonine (AKT1), P53, cadherin 1 (CDH1), transacting T-cell-specific transcription factor GATA3, retinoblastoma 1 (RB1), mitogen-activated protein kinase 3 kinase 1 (MAP3K1), mixed lineage leukemia 3 (MLL3), and cyclin-dependent kinase (CDKN1B), along with many additional driver genes. Another amplified gene that has recently been detected in breast cancer is FGFR, which is associated with more aggressive tumor behavior and endocrine resistance. This has led to phase 1 trials with FGFR inhibitors such as lucitanib, dovitinib, pazopanib, and nindetanib, based on promising preclinical behavior.

NGS has also been important in identifying genomic alterations in melanoma patients. Numerous alterations in addition to BRAF have come to light recently, which suggests that resistance to BRAF inhibitors may be a result of activation or reactivation of various pathways such as MAPK. Other pathways commonly affected either directly or indirectly include the PI3K/AKT/mammalian target of rapamycin (mTOR) axis, the Wnt signaling pathway, as well as tumor suppressor pathways. Additional genomic alterations have also been detected including amplifications in BRAF, MET, and aurora kinase A.

With more and more data being generated with NGS on alterations in tumors of all types, a common thread is emerging, many tumors have multiple aberrations; however, that alone does not imply that patients will not respond well to targeted therapy. In a recent study, clinical proof of concept was achieved showing the utility of comprehensive genomic profiling in assigning therapy to patients with refractory malignancies [5]. What is important is identifying the action-

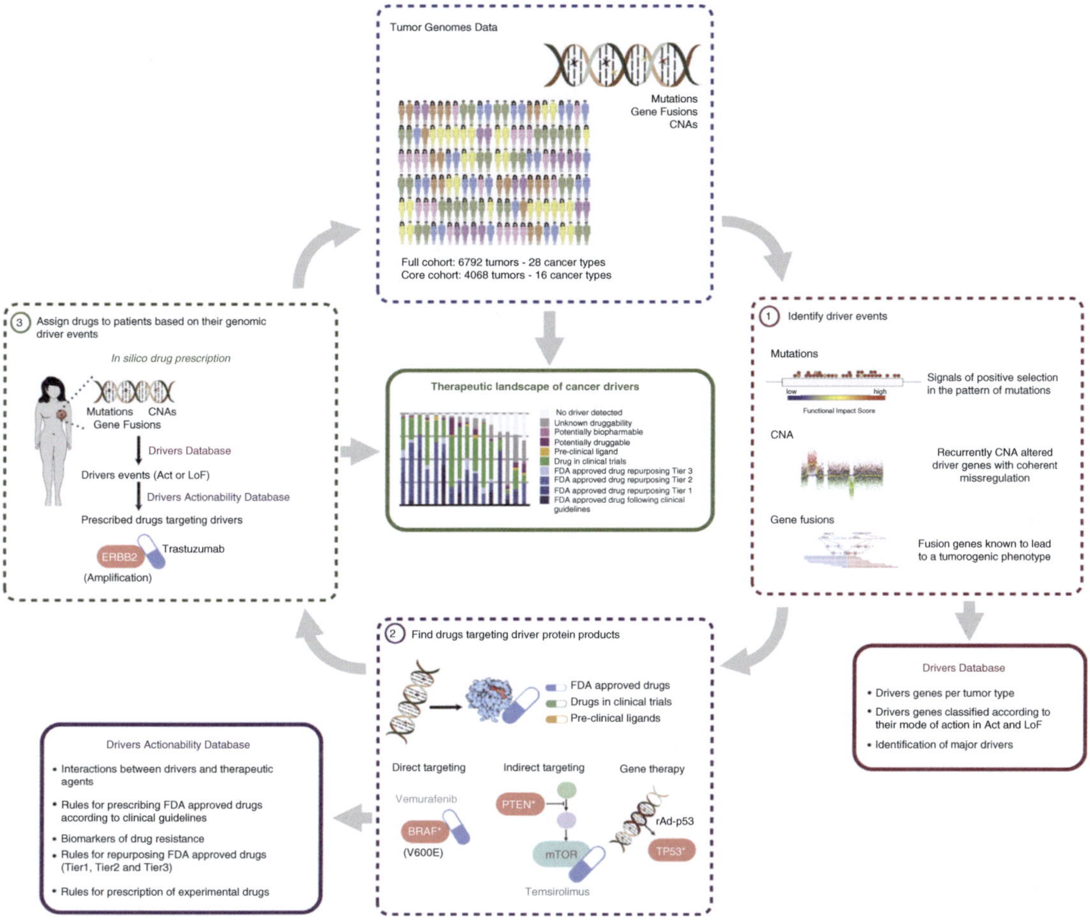

Fig. 10.2 Graphical summary of approach to understanding the therapeutic landscape of cancer drivers. Step 1 involves identifying the genomic driver events that are occurring in the tumor cohort. Step 2 consists of pinpointing the driver events that take place in the tumor cohort. Step 3 applies the information based on their particular genomic driver events in silico to select drugs to prescribe to those patients. The therapeutic landscape of cancer drivers is shown in the middle panel derived from all patients in the cohort. (Reprinted from Rubio-Perez et al. [3]. With permission from Elsevier)

able molecular alterations prior to commencement of treatment, and NGS is playing a greater and greater role in this process.

Importance of Mutations in Signaling Pathways in Carcinogenesis

Starting with the notion that most cancers are genetically complex, the importance of pathways rather than individual genes in carcinogenesis is becoming more evident. It is the driver pathways that need to be determined to help understand the molecular mechanisms that are at the heart of the cancer and contribute to the design of effective treatments for cancer patients. More and more, recent evidence suggests that in fact multiple pathways can function cooperatively in carcinogenesis as well as in other important biological processes. Given the heterogeneity of mutations in cancer genomes, it is the identification of the driver pathways and use of prior knowledge of those pathways and/or protein interaction networks that is coming to the forefront in helping understand cancer and how to better target it.

It has become clear that a number of driver pathways are required for cancer development and that these pathways operate in a cooperative manner in tumorigenesis [1]. Furthermore, it is also emerging that mutations occurring in the tumor of a patient usually function in different pathways, whereas those occurring in the same pathways are rarely mutated in the same sample. This type of information has been used to help in the detection of driver pathways. However, determining which are the driver pathways is a complex task. There is still a considerable amount of information that is missing regarding protein interactions and signal transduction pathways. What lies ahead is gathering more pathway information that will enable the systematic exploration of the cooperation between different biological pathways, which will hopefully enhance our understanding of the cellular mechanisms that are essential to carcinogenesis.

Identification of Acquired Resistance and Sensitivity Using NGS

Notwithstanding advances in targeted therapy, a common cause of cancer treatment failure is acquired drug resistance, and given the multiple molecular cancer types that evolve over time, mutational mechanisms contribute directly to acquired drug resistance. With the emergence of NGS that can yield complete molecular profiles of cancer genomes, elucidation of somatic genetic alterations associated with resistance to targeted therapies has materialized. Unlike traditional chemotherapy agents, for which establishing specific mechanisms of resistance have not met with success in large part because of the nonspecific nature of antitumor mechanisms associated with these types of drugs (e.g., mitotic inhibitors and alkylating agents), resistance mechanisms related to pathway-targeted drugs, such as the large class of clinically active kinase inhibitors, have been more readily elucidated and have enabled the development of drugs (e.g., in the case of chronic myeloid leukemia, dasatinib, a more potent inhibitor of the fusion gene BCR–ABL, can overcome imatinib-refractory clinical activity) that can overcome the resistance.

Fundamentally, mechanisms of acquired drug resistance fall into two major categories, genetic and epigenetic. Two main strategies have been used in the investigation of acquired resistance, namely, preclinical cell line modeling studies, in which paired samples of pretreatment drug-sensitive cells and posttreatment cells that become resistant are used to identify the underlying drug resistance mechanisms, and analysis of clinical biopsy specimens, comparing those collected prior to with posttreatment. This has led to the use of combination drug therapy, which has been found to overcome, and in some cases, prevent the acquisition of drug resistance. Described below are some noteworthy examples of genetic alterations that have been linked to resistance for several targeted therapies.

HER2, a receptor tyrosine kinase (RTK), is the target of the monoclonal antibody trastuzumab that binds to the extracellular domain of HER2 and is used in the treatment of breast cancer patients whose tumors have an amplified *ERBB2* gene. Blockade of HER2 is thought to result in inhibition of the downstream PI3K-AKT signaling pathway as well as HER2 shedding and activation of antibody-dependent cellular cytotoxicity. Acquired resistance occurs eventually in ~70% of HER2-postitive patients because of either compensatory activation of other RTKs or through activation of downstream signaling pathways. Use of NGS has helped identify activating alterations in the PI3K-AKT pathway and/or loss of the tumor suppressor PTEN as being implicated in resistance to trastuzumab. Two different approaches are currently being tested to overcome this acquired resistance to trastuzumab, namely, the use of trastuzumab-DM1, a conjugate of trastuzumab with a potent antimitotic drug designed to deliver tumor-targeted chemotherapy, and the co-administration of an HSP90 inhibitor that blocks HER2 trafficking to the cell membrane. Early clinical studies with the HSP90 inhibitor 17-AAG are encouraging showing clinical activity in patients who previously progressed on trastuzumab.

In the setting of non-small-cell lung cancer (NSCLC), which accounts for 80% of all lung cancers, NSCLCs have been identified in which the drivers are gene translocations resulting in targetable fusion oncokinases, the first of which was the echinoderm microtubule associated protein like 4 (EML4)-anaplastic lymphoma kinase (ALK) oncogene fusion. This oncokinase fusion has been identified in 4–6% of lung adenocarcinomas. Thus EML4-ALK exemplifies a novel molecular target in a small subset of NSCLCs. The FDA-approved TK inhibitor, crizotinib, is used to treat NSCLC patients harboring EML4-ALK rearrangements.

Recently, another clinically actionable oncokinase fusion that involves the TK ROS1, an orphan RTK that is evolutionarily related to ALK, has been detected in ~1.5% of NSCLCs. In 30% of ROS1 fusion-positive tumors, a recurrent translocation that creates the CD74 molecule, major histocompatibility complex, and class II invariant chain (CD74)-ROS fusion kinase has been detected. Crizotinib, a cMET/ALK/ROS1 TKI, has been found to inhibit the ALK fusion protein in a phase 1 trial of ROS1-postivive advanced stage NSCLC patients, and this has translated into an impressive objective response rate in treated patients.

NSCLCs treated with crizotinib eventually develop resistance due to novel acquired resistance mutations in the ROS1 kinase domain. Resistance, however, can be overcome by screening for inhibitors that are not affected by the newly identified secondary mutations, such as cabozantinib, a small molecule that inhibits the activity of multiple tyrosine kinases, including RET, MET, and VEGFR2. Cabozantinib, which is available for the treatment of refractory medullary thyroid cancer, can potentially be efficacious in the treatment of NSCLC patients who have become resistant to crizotinib treatment.

Malignant melanomas have been found to carry ~50% of BRAFV600E-activating mutations, which can be treated with BRAF inhibitors that have shown promise. But like other TKIs, following the initial beneficial therapeutic responses, acquired resistance becomes the issue. In the case of BRAF inhibitors, resistance mechanisms are either predominantly MAPK pathway-dependent or MAPK-independent. NGS has been used to identify mutated BRAF kinases, which include mutations in the gatekeeper residue, mutations that eliminate dimerization, or mutations that cause aberrant BRAF mRNA splicing, all of which can cause resistance. There are a number of additional MAPK-dependent mechanisms of BRAF inhibitor resistance that have been identified, including amplification of the *BRAF* gene, acquired mutations in NRAS, overexpression of CRAF or the MAPK COT1, as well as mutations in MEK that increase catalytic activity. Compensatory signaling by alternate pathways is usually implicated in MAPK-independent resistance to BRAF inhibitors.

Molecular Classification: Present Necessities and Future Directions

The way in which tumors are classified is undergoing important changes as a result of data generated using NGS. Previous transcriptional analyses led to the classification of breast cancer into four distinct molecular subtypes with diverse genomic signatures: luminal A, luminal B, HER2-enriched, and basal-like subtype. NGS has identified numerous additional genomic alterations in breast tumors that are further subdividing subtypes into additional molecular forms. For example, in luminal breast cancer, other genomic alterations that are frequently observed occur in PIK2CA and TP53 genes at a frequency of about 40% and 20%, respectively. Similarly, in lung cancer there are at least six subtypes that exist and have different genetic origins, which have enabled the identification of new driver oncogenes for which new drugs have been developed to treat these new subforms. In 2014, The Cancer Genome Atlas (TCGA) Network's NGS of lung adenocarcinomas in 2014 uncovered more than 15 different gene events (Fig. 10.3) that could be exploited for treatment and/or used for subclassifying patients into new taxa [6].

Similarly, this is occurring for all types of cancer whereby NGS data have yielded new ways to classify tumors and have pointed to previously unrecognized drug targets and carcinogens. In a study, Lawrence et al. [7] assessed the practicality of creating a comprehensive catalog

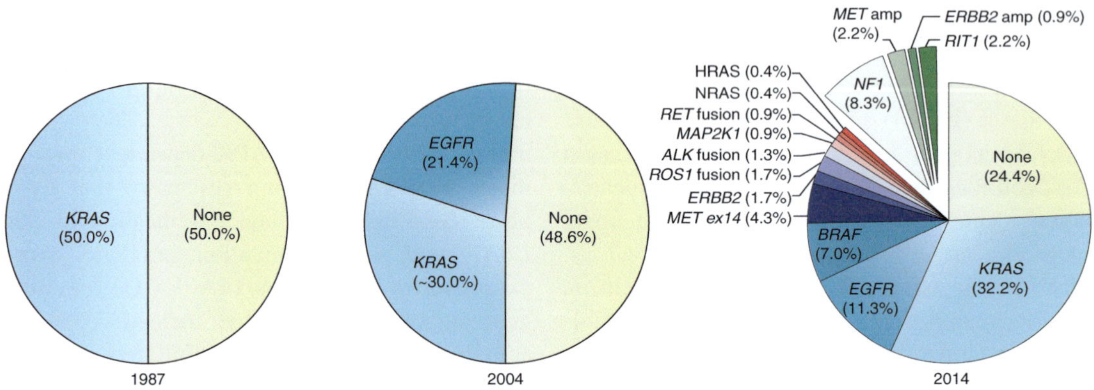

Fig. 10.3 In the past subclassification of lung cancer was based on histology grouping it into small-cell lung cancer and non-small-cell squamous cell carcinoma or adenocarcinoma. The utilization of next-generation sequencing through The Cancer Genome Atlas (TCGA) Network in 2014 has greatly expanded lung cancer subgroups to more than 15 based on different gene events that could be used in the treatment of patients. ALK anaplastic lymphoma kinase; amp, amplification; ex, exon; RIT1, Ras like without CAAX 1. (Reprinted from Vargas and Harris [6]. With permission Nature Publishing Group)

of cancer genes that contained point mutations in exome sequences from 4,742 tumor-normal pairs across 21 cancer types. They were able to identify all known cancer genes in these tumor types, but more importantly, they also identified 33 genes not previously known to be significantly mutated. The main takeaway from this NGS project is that there is only a minimal number of cancer genes that are mutated in a large proportion of a given tumor type (>20%); however, most are mutated at intermediate frequencies (2–20%). Therefore, Lawrence et al. fell far short of identifying all potential genes and estimated that this could only be achieved with 600–5,000 samples per tumor type, depending on the background mutation rate. This type of data clearly points out that subclassification of tumors based on earlier histological data and past tumor biomarker data only scratched the surface in terms of molecular classification of tumors. The NGS study by Lawrence and colleagues underscores the importance of using NGS in the classification process and emphasizes that thorough NGS of the required number of tumors of each type and subtype will greatly accelerate achieving a comprehensive molecular classification of all cancers.

Personalized Therapy for Lung Cancer and Melanoma

With the emergent knowledge that understanding cell signaling pathways in cancer is crucial in terms of optimizing therapy, it has also become clear that designing clinical trials that can effectively target patient populations more likely to benefit from a particular regimen has taken center stage. The groundbreaking Biomarker-integrated Approaches of Targeted Therapy for Lung Cancer Elimination (BATTLE) trial incorporated this forward thinking into the trial strategy in which mandatory biopsies and molecular profiling in real time were employed in an effort to match the patient to the right therapy [8].

The BATTLE trial was a biopsy-mandated, biomarker-based, adaptively randomized study in 255 pretreated NSCLC cancer patients enabling molecular profiling in real time, and analysis was performed, and treatment decisions made based on the findings. Following an initial equal randomization period, chemorefractory patients (97) were randomized equally to 4 treatment arms and 158 patients were assigned through adaptive randomization to erlotinib, vandetanib, erlotinib plus bexarotene, or sorafenib, based on relevant molecular biomarkers analyzed in fresh core needle biopsy specimens. Overall results included a 46% 8-week disease control rate (DCR) (primary end point), confirming prespecified hypotheses; among patients in the *KRAS/BRAF* marker group, sorafenib demonstrated an impressive 79% (11 of 14) DCR. BATTLE was the first study to incorporate mandated tumor profiling in real time, bringing PM in lung cancer therapy to the forefront by including molecular laboratory findings used in defining specific patient populations for individualized treatment.

In a phase 3 prospective, randomized trial in 230 patients with metastatic NSCLC and EGFR mutations who had not previously received chemotherapy, use of the EGFR inhibitor gefitinib resulted in progression-free survival that was twice as long as that obtained with the use of carboplatin-paclitaxel [9]. As an added benefit, the toxicity profile was more tolerable with less hematologic toxicity and neurotoxicity than was seen with chemotherapy. Gefitinib, however, was ineffective in patients with wild-type EGFR, clearly demonstrating that stratification of patients with EGFR mutations is critical for selecting those who will benefit from the drug.

The BRAF gene is the most commonly mutated protein kinase gene in human cancers. Melanomas, which are reliant on activation of the RAF/MEK/ERK pathway as the oncogenic driver, frequently have mutations in *BRAF*. Exploitation of this pathway as a target for blockade was thought would benefit melanoma patients. PLX4032 (RG7204), a potent inhibitor of oncogenic B-RAF kinase activity, was initially shown in preclinical experiments to selectively block the RAF/MEK/ERK pathway resulting in regression of BRAF mutant xenografts. A phase 1 clinical trial confirmed that blockade of >80% of ERK phosphorylation in tumors of patients correlated with a clinical response. The response rate seen was impressively high at 81% in metastatic melanoma patients with tumors that were highly dependent on B-RAF kinase activity.

Omics Assays in Oncology

Understanding molecular disease pathways in cancer is critical in improving tailoring and timing of preventative and therapeutic actions, thereby optimizing PM for the individual cancer patient. This will require obtaining biological information and identifying biomarkers by measuring transcripts, proteins, and small biological molecules, or metabolites, which define the fields of transcriptomics, proteomics, and metabolomics, respectively. Bioinformatics will be critical in deriving knowledge from the massive quantities of diverse biological, genetic, genomic, and gene expression data generated.

The ability to identify the genes/proteins that are part of a pathway or complex network will enable the evaluation of their association to cancer. Gathering the massive amounts of data necessary to accomplish this will require using high-throughput omics technologies that include NGS, mass spectrometry, nuclear magnetic resonance, and separation systems along with an integrated bioinformatics approach. The development of databases and knowledge bases and the implementation of computational modeling will support the integration of data from numerous fields. The challenge will be to derive meaningful information that can be translated into practical applications in the clinical setting and in the development of new targeted drugs.

There are a number of issues that must be overcome to allow this omics revolution to take hold. These include the high false-positive rate observed with candidate biomarkers identified using omics data, the limited understanding of the context in which biomarkers interact with each other within pathways or networks associated with cancer, the limited information available on biomarkers that are solely identified from omics data, and the inability to combine and integrate diverse omics data from several sources that can replicate signaling pathways and networks. To overcome these issues, pathway and network-centric approaches have come to the fore.

Omics technology and computational analytics are advancing rapidly with the large-scale integration of data generated from genomics, transcriptomics, proteomics, and metabolomics. This is allowing for a more effective means of discovering clinically usable cancer biomarkers (Fig. 10.4). More and more studies are focusing on unraveling pathways and networks by applying omics data to gain a more in-depth understanding of the underlying biological functions and processes, such as cell signaling and metabolic pathways, that are implicated in gene regulatory networks [10].

With this in mind, progress is being made in a number of areas related to pathway/network methodologies that will improve prediction of cancer outcomes, generate novel hypotheses for pathways implicated in tumor progression, and aid in the discovery of cancer-related biomarkers. Examples below highlight the progress being manifested using a diverse number of technical platforms.

Researchers are combining data from various sources to identify prognostic biomarkers. This includes gene expression data with physical protein-protein interaction data to identify subnetwork markers (Fig. 10.5) that can be used in the prognosis of metastasis in cancer patients. Gene co-expression networks are being applied to determine tumor-initiating genes in various cancers including breast and colorectal and in glioblastomas. Various new tools are being developed to analyze signaling pathways such as MAPIT (Multi Analyte Pathway Inference Tool), which aids in the identification of prognostic network markers that can predict patient survival time. Several emerging applications in systems biology are becoming prominent, including analysis of pathway-based biomarkers, generation of global genetic interaction maps, systems biology

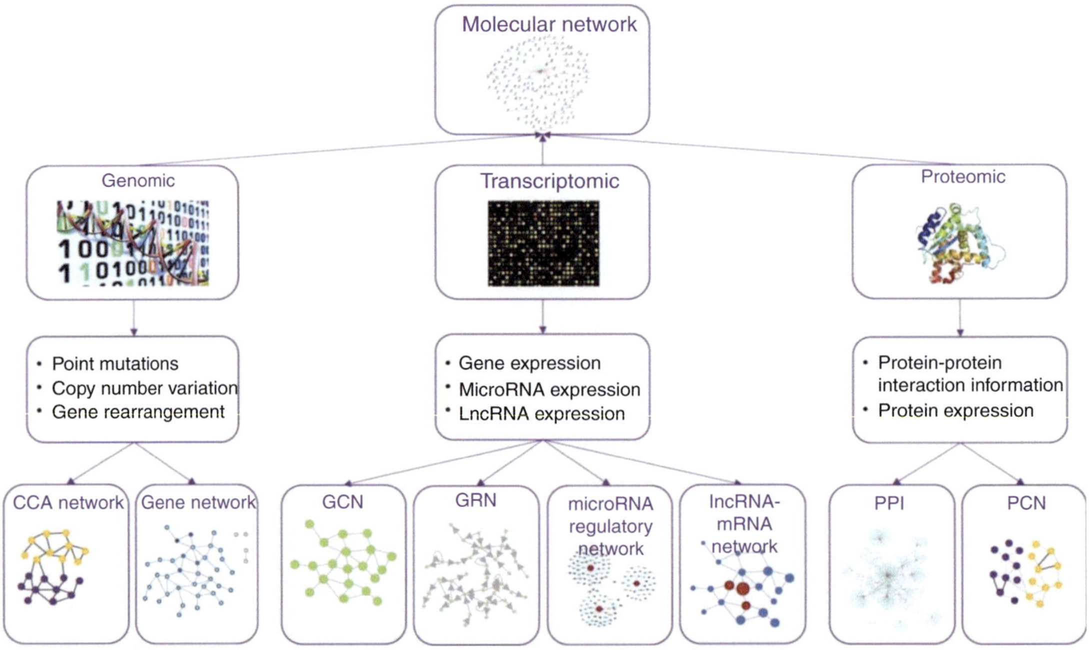

Fig. 10.4 There are a number of molecular networks that are being employed in biomarker discovery. These various networks are used in the analysis of data generated through genomics (CCA and Gene networks), transcriptomics (GCN, GRN, microRNA regulatory network and lncRNA–mRNA network) and proteomics (PPI network and PCN) that enable the identification of potential biomarkers. CCA, cancer genes with co-occurring and anti-co-occurring mutations; GCN, gene co-expression network; GRN, gene regulatory network; lncRNA, long noncoding RNA, PPI, protein-protein interaction; PCN, protein contact network. (Reprinted from Yan et al. [10]. With permission from SAGE Publications Ltd.)

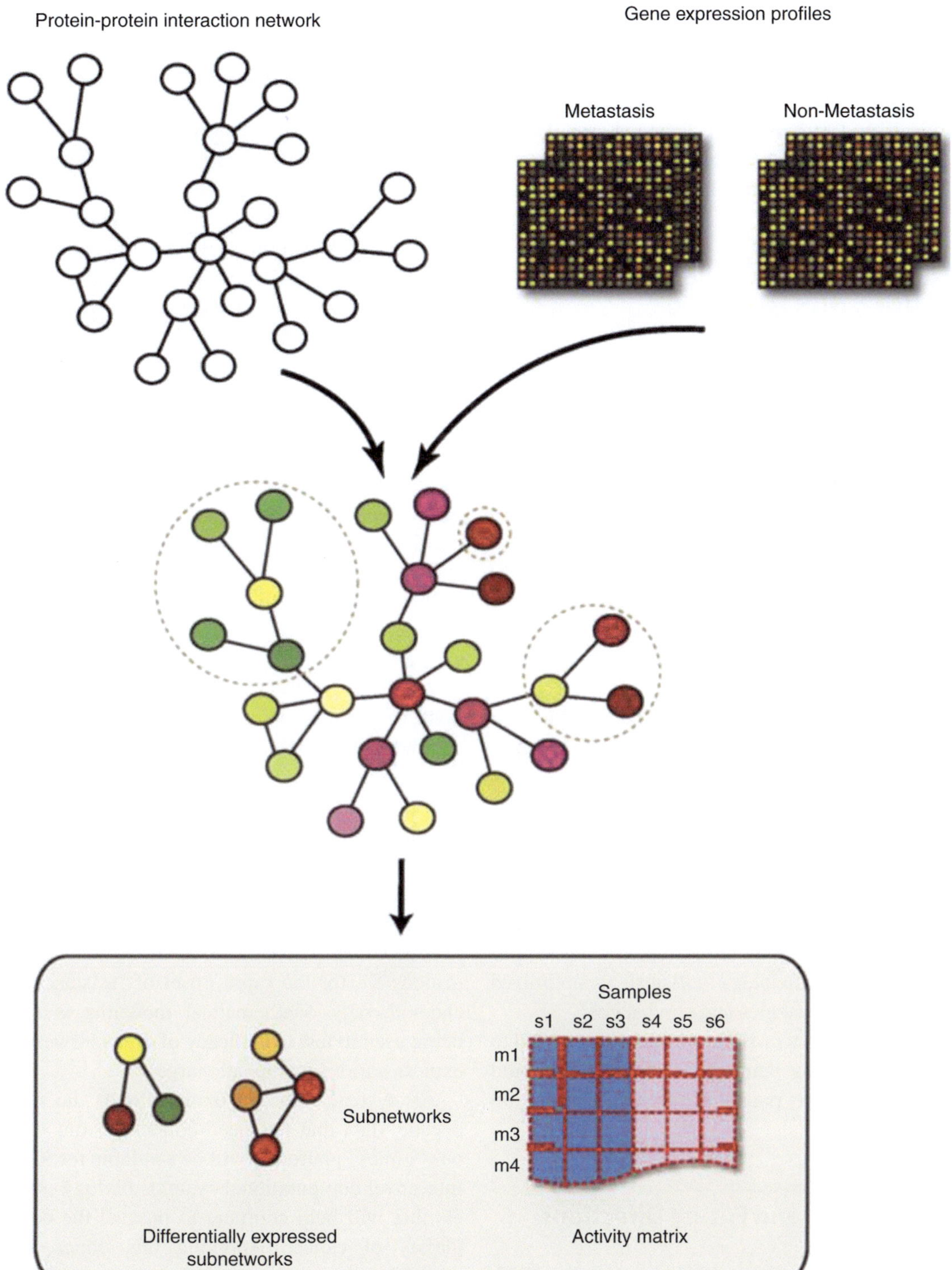

Fig. 10.5 Metastatic and nonmetastatic tumor samples that are subjected to gene expression profiling can be superimposed onto a protein-protein interaction network. This can generate information and an activity score for the various subnetworks in a patient's samples. Differentially expressed metastatic subnetworks can then be discerned from an activity matrix, which can aid in the identification of potential metastatic-related biomarkers. (Reprinted from Auffray [14]. With permission from EMBO and Nature Publishing Group under Creative Commons License 3.0: https://creativecommons.org/licenses/by/3.0/)

methodologies to find disease genes, and stem cell systems biology. Computational advances and powerful software tools are also contributing greatly to our ability to explore system-wide models and formulate novel hypotheses. Omics data and integrated bioinformatics analysis will help take PM to the next level.

Limitations and Challenges of Using NGS Technique

With the advent of NGS, a great deal of progress has been made in cancer research that could otherwise not have occurred. However, as with any new technology, there are several limitations and challenges still ahead of us:

- PM is based on NGS, but more evidence from prospective clinical trials is needed.
- The SHIVA trial did not demonstrate any improvement in PFS or OS (clinical trials involving NGS will be covered more thoroughly in Chap. 53).
- The absolute cost of NGS is too expensive on a per-patient basis compared with current standard molecular testing.
- NGS will encourage the use of off-label targeted agents, which may be less effective and more costly compared with standard evidence-based therapies.
- More supporting evidence is required to determine whether NGS data coupled with computational methodology will lead to optimized treatment strategies and at what cost.
- More rigorous prospective trials are needed to unequivocally demonstrate that NGS should be adopted as part of the standard of care in oncology.

Conclusions and Future Directions

The opportunities that PM, directed by NGS, omics-generated data, and molecular biomarkers can bring are expected to be far-reaching with respect to individualized treatment, improved quality of life, and major cost-efficiencies in the healthcare system. The vision is one of personalized oncology therapeutics, with seamless boundaries between omics data-driven research and optimized treatment regimens. But realizing these goals, given that cancer is a highly complex and heterogeneous disease, which involves a succession of genetic changes that eventually result in the conversion of normal cells into cancerous ones, will necessitate the integration and analysis of massive quantities of data as it is being collected from current omics platforms, as well as a comprehensive systems biology approach [11].

This will require a concerted effort reaching across many research fields. For example, current computational methods are being applied to transcriptomic and proteomic data to develop graphical models of gene-protein regulatory networks. Furthermore, several additional computational approaches are being applied to incorporate and connect experimental data into biological systems that can be simulated and used for hypothesis testing.

Although systems biology is still an emerging field, progress is taking place, and a number of computational approaches have been applied to the biological complexity of cancer models integrating vast amounts of data that include many interacting genes, proteins, and protein modifications. A simulation of a human cancer cell has been developed, and more recently, Waclaw and colleagues described a model for tumor evolution [12] in which mechanisms could be potentially responsible for the rapid onset of resistance to chemotherapy. Mathematical modeling is also being used to test the efficacy of drugs as well as explore various therapeutic targets.

As systems biology matures over the next decade, data that has been collected from various "omics" platforms will be available for input into novel computational systems biology models that will help continue to unravel the complexity of cancer. Applying this omics and biomarker-driven approach to cancer, in conjunction with algorithmic methods to infer the genomic evolution inherent to cancer, has the potential to more rapidly lead to early diagnosis,

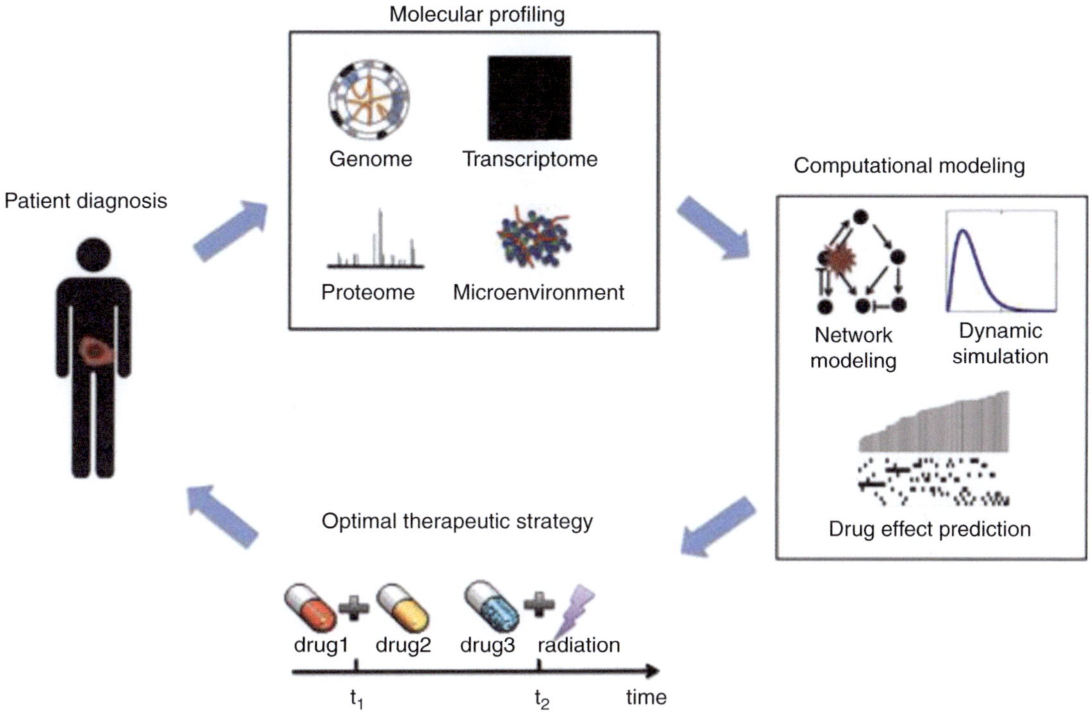

Fig. 10.6 Personalized cancer therapy strategy based on the integration of omics-generated data through the use of computational modeling. When patients are first diagnosed, tumor and blood samples are subjected to multidimensional experimental profiling to obtain a complete picture of the patient's specific cancer alterations. A computational model is then built, specific to the patient, and can be used to predict an optimized short-term therapeutic strategy. This process is then performed in an iterative manner to rapidly adapt to any potential resistance acquired due to continuous cancer evolution, resulting in the final eradication of the cancer. (Reprinted from Du and Elemento [11]. With permission from Nature Publishing Group)

to the individualization of treatment, and to overcoming acquired resistance [13].

Clinically, NGS has been used or is being developed for genetic screening, diagnostics, and clinical assessment. Though there are still many hurdles to overcome, clinicians are in the early stages of using genetic data to make treatment decisions for cancer patients. As integration of NGS in the study and treatment of cancer continues to mature, the field of cancer genomics will need to move toward more complete 100% genome sequencing. At present, technologies and methods are mainly limited to coding regions of the genome. Several recent studies have determined that mutations in noncoding regions may have direct tumorigenic effects or lead to genetic instability. Thus, noncoding regions denote a critical frontier in cancer genomics.

In the near future, PM will move in the direction of obtaining complete multidimensional profiles of a patient's cancer before and after drug treatment, particularly at the time of disease progression. This will enable serial pharmacodynamic assessment of tumor samples using panels of molecular assays that will become more standardized, which will aid in identifying acquired resistance mechanisms and selection of the most appropriate follow-on therapy (Fig. 10.6) (see Appendix 10.1).

Acknowledgments We would like to acknowledge the Avera Cancer Institute, Sioux Falls, SD, for providing support for our research.

Conflicts of Interest The authors have no conflicts of interest to declare.

Appendix 10.1: Next-Generation Sequencing (NGS)

"Massive Parallel Sequencing" or informally called next-generation sequencing (NGS) has greatly increased the speed of DNA sequencing, taking it from 84 kilobase (kb) per run in 1998 to greater than 1 gigabase (Gb) per run in 2005 to multiple Gb per run today. The ability to perform NGS, also known as massive parallel sequencing, has revolutionized throughput, heralding genomic science's "next generation."

Sequencing the human genome involves sequencing 3.2 billion bases at 30× coverage (on average each base in the genome is sequenced 30 times). In 2005, capacity was limited to 1.3 human genomes sequenced annually. This has risen exponentially to the point where as of 2014, approximately 18,000 genomes per year can be sequenced, which has come with a tremendous reduction in cost (approximately $1,000 per genome).

Since the introduction of NGS, major advances have focused on further increasing speed and accuracy, which has greatly reduced manpower and cost. The current bottleneck is storage, processing, and analysis of the voluminous amount of sequencing data generated.

The Nobel Prize in 1980 was awarded to Wally Gilbert and Fred Sanger for developing the first methods for DNA sequencing. Sanger sequencing became the gold standard in molecular diagnostics, but it has finally given way to NGS. While NGS is based on Sanger sequencing, which involves the incorporation of fluorescently labeled deoxyribonucleotide triphosphates (dNTPs) into a DNA template strand during sequential cycles of DNA synthesis that are identified using fluorophore excitation, the major difference is that in NGS, millions of fragments are being sequenced simultaneously. It is this massively parallel process that has brought sequencing into the twenty-first century.

There are several companies (e.g., Life Technologies and Applied Biosystems (Thermo Fisher Scientific), Illumina, Roche, and Pacific Biosciences) that have developed NGS systems, and while there are differences, four fundamental steps are shared: (1) DNA preparation of the sequencing library, (2) amplification, (3) sequencing, and (4) data analysis (see Fig. 10.7). Each of these is dealt with in turn:

1. *DNA Preparation of the Sequencing Library*
   Crucial to this step is the preparation of random DNA fragments, and size is dependent on the particular sequencing platform and application: whole-genome versus whole-exome sequencing (only exons of genes are sequenced or ~1% of the genome). The DNA sample is prepared using a process that involves either sonication or enzymes to generate random fragments. Adapters are then added to both ends of the fragments and this constitutes the sequencing library. This library can now be anchored and immobilized to a solid support on which the sequencing reaction will take place. Different types of adapters and support systems can be used.
2. *Amplification*
   In this next step, amplification of fragments takes place either in an emulsion or in solution. On the Illumina platform, for example, fragments are captured on a surface of bound oligos complementary to the library adapters. This allows each fragment to be amplified into distinct, clonal clusters through what is termed bridge amplification.
3. *Sequencing*
   Sequencing can be accomplished using different methodologies depending on the platform. In general, fluidic systems running on a microliter scale are involved in the sequencing reaction. The immobilized DNA reacts with the regulated flow of reagents. Life Technologies and Roche sequencing systems involve the addition of a single nucleotide, which, if complementary to the sequence, is incorporated. Any nucleotides that are not incorporated are washed away, and the DNA is mixed with another nucleotide-containing solution. If this additional nucleotide is incorporated, then the system registers the event. Detection can be based on light emission (GS FLX system, Roche) or emission of hydrogen ions released

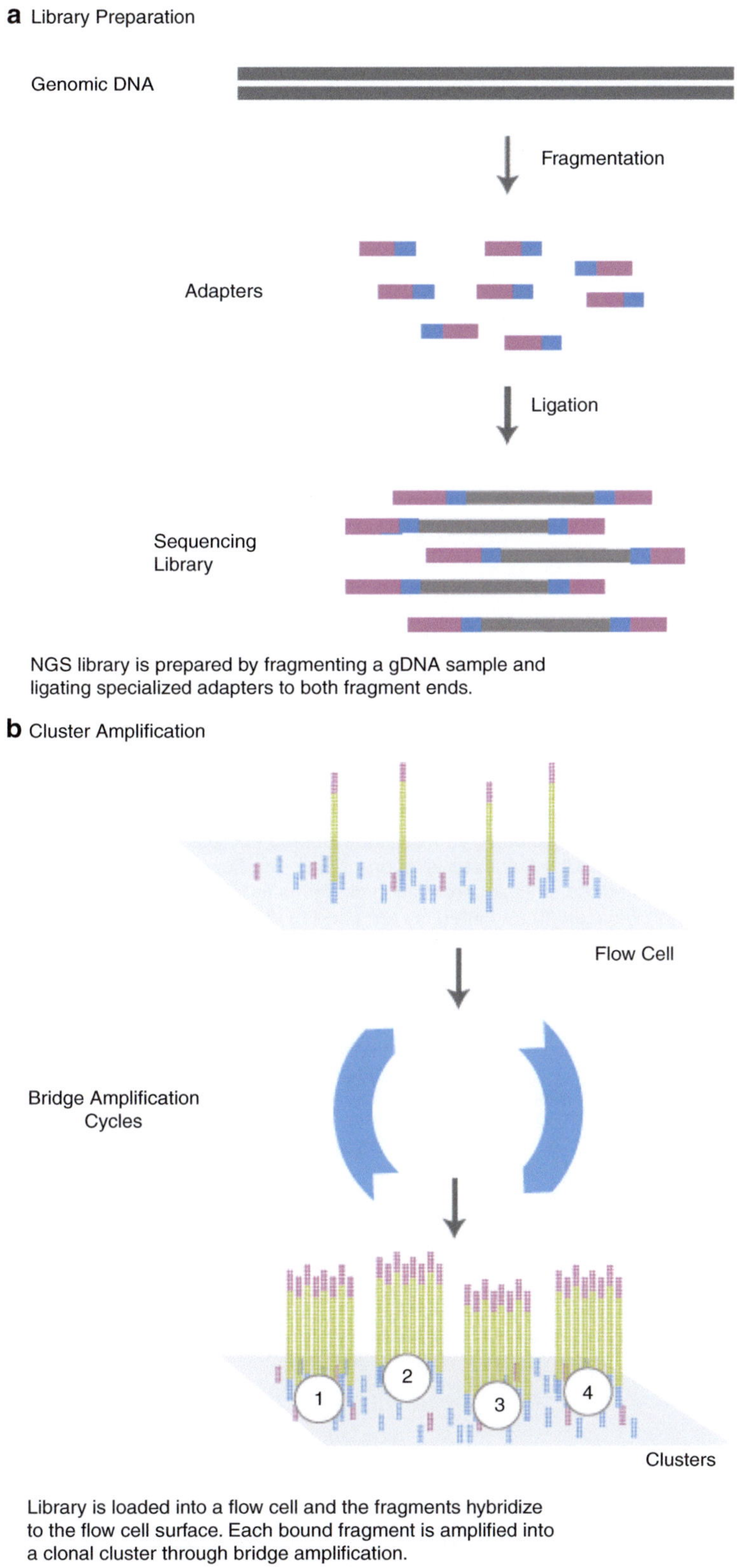

Fig. 10.7 (a–d) Next-generation sequencing steps. (Courtesy of Illumina, Inc.)

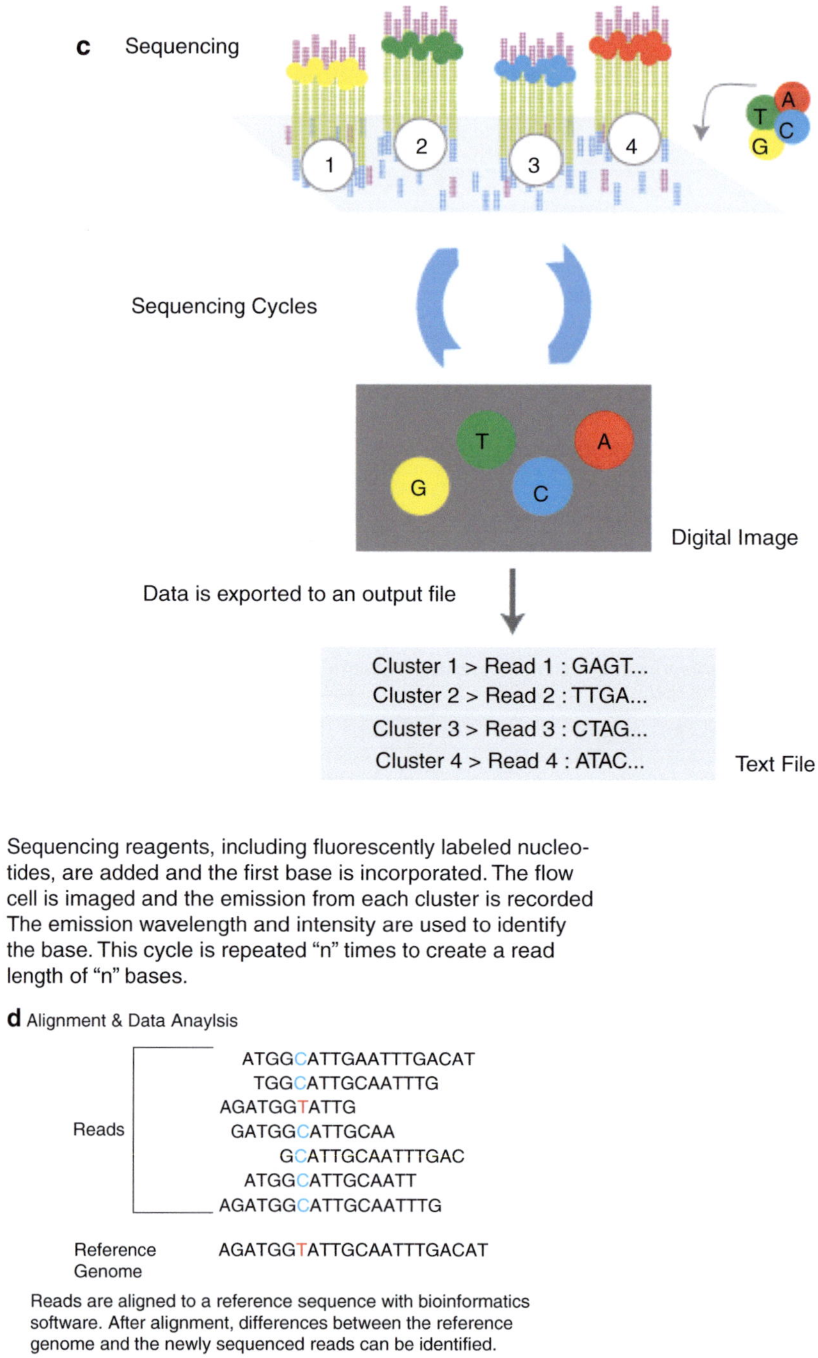

Fig. 10.7 (continued)

during the polymerization reaction (Ion Torrent, Life Technologies). On the Illumina platform, their proprietary sequencing by synthesis (SBS) system is used, in which all four reversible terminator-bound dNTPs are present in each sequencing cycle, resulting in natural competition that effectively minimizes incorporation bias and reduces raw error rates.

4. *Data Analysis*
   Data analysis systems are critical to the effective interpretation of the vast amounts of sequencing data generated and represent a potential bottleneck for going from raw output to aligned sequences. The "draft" sequencing data must first be aligned to a reference genome. Once processed, various analyses can be performed, including but not limited to identifying single nucleotide polymorphisms (SNPs), insertion-deletions (indels), performing read counting for RNA methods, as well as phylogenetic or metagenomic analysis.
   Increasing the speed of sequence data analysis and developing the necessary data storage capacity are important considerations moving forward. By some estimates, up to one billion people may have their genomes sequenced by 2025, producing an inordinate amount of data within the next decade. How this will be handled is a top priority as we continue to embrace PM.

References

1. Hanahan D, Weinberg RA. Hallmarks of cancer: the next generation. Cell. 2011;144(5):646–74.
2. Eifert C, Powers RS. From cancer genomes to oncogenic drivers, tumour dependencies and therapeutic targets. Nat Rev Cancer. 2012;12(8):572–8.
3. Rubio-Perez C, Tamborero D, Schroeder MP, et al. In silico prescription of anticancer drugs to cohorts of 28 tumor types reveals targeting opportunities. Cancer Cell. 2015;27(3):382–96.
4. Ledford H. Big science: the cancer genome challenge. Nature. 2010;464(7291):972–4.
5. Wheler JJ, Janku F, Naing A, et al. Cancer therapy directed by comprehensive genomic profiling: a single center study. Cancer Res. 2016;76(13):3690–701.
6. Vargas AJ, Harris CC. Biomarker development in the precision medicine era: lung cancer as a case study. Nat Rev Cancer. 2016;16(8):525–37.
7. Lawrence MS, Stojanov P, Mermel CH, et al. Discovery and saturation analysis of cancer genes across 21 tumour types. Nature. 2014;505(7484):495–501.
8. Kim ES, Herbst RS, Wistuba II, et al. The BATTLE trial: personalizing therapy for lung cancer. Cancer Discov. 2011;1(1):44–53.
9. Maemondo M, Inoue A, Kobayashi K, et al. Gefitinib or chemotherapy for non-small-cell lung cancer with mutated EGFR. N Engl J Med. 2010;362(25):2380–8.
10. Yan W, Xue W, Chen J, et al. Biological networks for cancer candidate biomarkers discovery. Cancer Informat. 2016;15(Suppl 3):1–7.
11. Du W, Elemento O. Cancer systems biology: embracing complexity to develop better anticancer therapeutic strategies. Oncogene. 2015;34(25):3215–25.
12. Waclaw B, Bozic I, Pittman ME, et al. A spatial model predicts that dispersal and cell turnover limit intratumour heterogeneity. Nature. 2015;525(7568):261–4.
13. Caravagna G, Graudenzi A, Ramazzotti D, et al. Algorithmic methods to infer the evolutionary trajectories in cancer progression. Proc Natl Acad Sci U S A. 2016;113(28):E4025–34.
14. Auffray C. Protein subnetwork markers improve prediction of cancer outcome. Mol Syst Biol. 2007;3:141.

11 Bioinformatic Methods and Resources for Biomarker Discovery, Validation, Development, and Integration

Júlia Perera-Bel, Andreas Leha, and Tim Beißbarth

Introduction

This chapter aims to give a brief overview of bioinformatic and biostatistical methods and tools used in biomarker research and discovery in the testing of biomarkers in clinical trials, up to the processing and reporting issues when used in clinical routine. Research and clinical applications of biomarker-based diagnostics usually require special knowledge and methods in bioinformatics and biostatistics, and the different applications of biomarkers pose very diverse challenges for the researchers in these areas. Different applications of biomarkers are, for example, the diagnosis of diseases (*diagnostic biomarkers*), prediction of disease risk (*preventive medicine*; screening for genetic diseases), prediction of the future onset of a disease (*prognostic biomarkers*), stratification of patient cohorts into different subgroups that respond to different treatments (*predictive biomarkers*), suggestion of personalized treatment (*personalized medicine*), suggestion of drug targets of an individual patient (*precision medicine*), and modeling the interaction effects of complex networks of biomarkers (*systems medicine*). Bioinformatics and statistics challenges include biomarker discovery in statistical learning approaches, testing of biomarker-based treatment strategies in clinical trials, data processing, bioinformatic pipelines, quality assurance, and reporting.

Section "Public Resources and Open-Source Tools" of this chapter gives an overview of resources that can be used for biomarker discovery and testing, as well as databases for cancer *omics* data and for clinically used biomarkers and drugs. With the growing availability of high-throughput technology, (e.g., *next-generation sequencing* to measure *single nucleotide variations*, *copy number variations*, gene expression, *microRNA expression* or *methylation* or *mass spectrometry* to measure *protein expression*, *protein phosphorylation*, or *metabolites*), it becomes increasingly a bioinformatic challenge in medical research to discover individual biomarkers or biomarker signatures.

Section "Bioinformatic and Statistical Methods" covers the bioinformatic and machine learning methods needed for biomarker discovery from *omics* data. In situations where there are many more measured potential biomarkers than there are patients in the training cohort, the task of discovering biomarkers is made difficult by the so-called curse of dimensionality. In the area of "systems medicine," the aim is to create mathematical models not using just one biomarker but often a complex network of interactions. This model should then be able to predict parameters important for patients' diagnosis or outcome.

J. Perera-Bel (✉) · A. Leha · T. Beißbarth
Department of Medical Statistics, University Medical Center Göttingen, Göttingen, Germany
e-mail: tim.beissbarth@ams.med.uni-goettingen.de

S. Badve, G. L. Kumar (eds.), *Predictive Biomarkers in Oncology*,
https://doi.org/10.1007/978-3-319-95228-4_11

Section "Clinical Evaluation" summarizes clinical trials and validation strategies to bring biomarkers into clinical practice. Once single biomarkers, biomarker signatures, or complex classifiers of mathematical models using biomarkers have been discovered in medical research, these biomarker-based diagnostics must be transferred into clinical practice and tested in clinical trials. *Omics* technologies like next-generation sequencing (NGS) are becoming increasingly more applicable in clinical routine. Therefore, means of quality control and standardized bioinformatics processing pipelines must be established to work with such data in clinics.

Section "Clinical Application" summarizes some issues of reporting and interpreting biomarkers in clinical routine. Methods of reporting and data visualization that should allow the treating doctor or possibly the individual patient to assess and interpret the results of biomarker-based diagnostics are still in development.

Public Resources and Open-Source Tools

Over the course of the past decade, there have been increasing numbers of large-scale, international efforts to generate, gather, and analyze cancer *omics* data. The success of such efforts is explained not only by direct publications of the groups involved but is also justified by thousands of secondary publications by independent research groups, facilitated by the public release of the involved datasets. Also, the tremendous amount of data (petabytes) has pushed researchers to generate metadatabases, to create tools for analyzing the data, and to develop smaller, curated databases [1]. In this section, we will review the existing public resources of cancer *omics* data which can be used in the process of biomarker discovery. A summary of the most important resources can be found in Table 11.1.

Public Data Repositories

Patient-Derived Omics and Clinical Data

Characterization of patient tumor samples is crucial for the identification of biomarkers, especially when clinical samples are coupled with clinical records. *The Cancer Genome Atlas* (TCGA) project is probably the largest, most renowned effort toward multi-*omics* cancer data generation. This dataset consists of paired normal and tumor tissue samples from more than 11,000 patients with 33 different cancer types, using 7

Table 11.1 List of tools and databases

	Resource	URL	Description
Repositories clinical data	TCGA	https://cancergenome.nih.gov	Largest project of multi-*omics* profiling of 33 tumor types (11,000 patients). Taken over by *GDC*
	ICGC	http://icgc.org	Collection of 55 cancer genomics projects (including *TCGA*) and tools for visualizing and analyzing the data
	GDC	https://gdc.cancer.gov	Includes *TCGA* and *TARGET* projects. Developed by National Institutes of Health (NIH) and National Cancer Institute (NCI)
Repositories cell line data	NCI-60	https://discover.nci.nih.gov/cellminer/home.do	Web tool to browse and analyze panel of 60 cell lines against 100,000 chemical compounds
	CCLE	http://www.broadinstitute.org/ccle	Largest screening of genetically characterized cell lines (~1000)
	GDSC	http://www.cancerrxgene.org	Characterization of 700 cell lines against 138 anticancer drugs performed by Sanger Institute
Expression repositories	GEO	https://www.ncbi.nlm.nih.gov/geo	Gene Omnibus Express. Functional genomics repository (microarrays and RNA-seq mainly)
	ArrayExpress	https://www.ebi.ac.uk/arrayexpress	Archive of high-throughput functional genomics experiments (includes GEO). Developed by the European Bioinformatics Institute

Table 11.1 (continued)

	Resource	URL	Description
Tools for data analysis and visualization	cBioPortal	http://www.cbioportal.org	Visualization, analysis and download of ~80 cancer genomics projects, developed by Memorial Sloan Kettering Cancer Center
	COSMIC	http://cancer.sanger.ac.uk/cosmic	Largest database of somatic mutations (curated list: *Cancer Gene Census*)
	IGV	http://software.broadinstitute.org/software/igv	Desktop application for genomic coordinates visualization
	Regulome Explorer	http://explorer.cancerregulome.org	Multivariate analysis methods and visualization for heterogeneous data types in TCGA data
	UCSC Genome Browser	http://genome.ucsc.edu	Genome browser, including vertebrate and model organism assemblies and annotations
	Bioconductor	https://www.bioconductor.org	R programming language. Provides tools for the analysis and comprehension of high-throughput genomic data
	Cytoscape	http://www.cytoscape.org	Desktop application. Visualization of molecular networks and biological pathways
	Gene Ontology	http://geneontology.org	Ontology linking genes to gene products. It allows functional interpretation of experimental data
Tools for learning biomarker signatures from *omics* data	limma	http://bioinf.wehi.edu.au/limma	R package to test for significantly differential genes
	blkbox	https://cran.r-project.org/package=blkbox	R package providing a unified interface to many binary classifiers
	caret	https://cran.r-project.org/package=caret	R package supporting many steps of predictive modeling
	netClass	https://cran.r-project.org/package=netClass	R package to train classifiers using pathway information
Somatic variants interpretation	MyCancer Genome	https://www.mycancergenome.org	Precision cancer medicine knowledge resource for physicians, patients, caregivers and researchers
	civic	https://civic.genome.wustl.edu	Knowledge database for clinical interpretation of somatic variants in cancer (~1700 gene-drug associations)
	TARGET	http://archive.broadinstitute.org/cancer/cga/target/	*TARGET* (tumor alterations relevant for genomics-driven therapy). Database of prognostic, diagnostic and predictive biomarkers in cancer (135 genes)
	CGI	https://www.cancergenomeinterpreter.org/biomarkers/	Cancer predictive biomarkers database (~1000 gene-drug associations)
	Clinicaltrials.gov	https://clinicaltrials.gov	Registry of publicly and privately supported clinical studies (>200,000 trials)
	EU-CTR	https://www.clinicaltrialsregister.eu	*The European Union Clinical Trials Register* (~50,000 trials)

different data types and up to 15 genomic assays. Overall, it consists of over 2.5 PB of data that are accessible via several methods. TCGA ends in 2017 with *Genomic Data Commons* (GDC) taking over this model of collaborative data generation together with other NCI initiatives. The GDC Data Portal[1] already collects TCGA and TARGET (T*umor Alterations Relevant for Genomics-driven Therapy*) projects, which have been made comparable and include almost

[1]Accessible through https://gdc-portal.nci.nih.gov

15,000 cases. Similarly, *The International Cancer Genome Consortium* (ICGC) was born with the aim of coordinating 55 research projects with the overall goal of characterizing the genome, transcriptome, and epigenome from 25,000 patients. The data can be accessed, analyzed, and downloaded in the ICGC data portal.

A common pitfall of these projects is the lack of comprehensive clinical data, such as follow-up or treatments. Without clinical covariates, it becomes very difficult to link genotypes to phenotypes and thus to perform the translation of research findings into real clinical outcomes. Bioinformatic tools such as *cBioPortal* (Memorial Sloan Kettering Cancer Center), *RTCGAToolbox*, *firehose_get* (Broad Institute), *UCSC Genome Browser*, *Synapse client*, and *Genomic Data Commons Data Portal* (National Cancer Institute) can help to link genotypes to phenotypes and thus perform the translation of research findings into real clinical outcomes.

Cell Line Databases to Predict Drug Responses

Drug and perturbation screens are crucial for candidate biomarker discovery in early research stages. Cell lines are a fast, commonly used tool to perform large screens to test which drugs affect cancer cell survival and which genotypes are predictive of drug response. These analyses consist of pharmacogenetics and genetic perturbation experiments. Of the most comprehensive publicly available resources, NCI-60, *Genomics of Drug Sensitivity in Cancer* (GDSC), and the *Cancer Cell Line Encyclopedia* (CCLE) are the largest.

NCI-60 contains 59 human tumor cell lines characterized by protein, RNA, DNA, and enzyme activity assays. Around 100,000 drugs have been tested. GDSC from Sanger Institute has tested 138 anticancer drugs on 700 cancer cell lines. Cell lines are characterized by gene expression and mutations (also copy number) in known cancer genes. CCLE from the Broad Institute has characterized the largest amount of human cancer cell lines (1036) and tested 24 drugs on 504 cell lines. Broad Institute also has the *Cancer Therapeutics Response Portal* (CTRP) and project *Achilles*.[2] CTPR provides sensitivity measurements of 481 small molecules and drugs on 860 cell lines molecularly characterized by CCLE. *Achilles* project focuses on genetic perturbations (using RNAi screens and CRISPR-Cas9) to identify the role of around 11,000 genes in cell survival. Finally, the *Library of Integrated Network-based Cellular Signals*[3] (LINCS) project has characterized 356 cell lines (gene and protein expression) after genetic and environmental perturbations.

Microarray/RNAseq Data Repositories

DNA microarray technologies had a profound impact on the examination of gene expression on a genomic scale in research and have been used widely for the identification of cancer biomarkers. They have demonstrated that levels of RNA transcripts stratify patients and predict outcomes in a variety of diseases (e.g., breast cancer), providing the basis for several important clinical tests. Similarly, the RNA-Seq technique allows transcriptome studies based on next-generation sequencing technologies. This technique is largely dependent on bioinformatics tools developed to support the different steps of the process. Both technologies generate tremendous amounts of data, and these data are usually made publicly available in *Gene Expression Omnibus* (GEO), a public repository that accepts array and sequencing-based genomic data comprising more than 4000 datasets. The European equivalent of GEO is *Array Express*, fr*om The European Molecular Biology Laboratory and The European Bioinformatics Institute (EMBL-EBI*), an archive of functional genomics data with more than 44 TB of stored data. Both repositories have different tools to access and download data. For R users, both have an interface to Bioconductor,

[2] *CTRP* accessible through: http://portals.broadinstitute.org/ctrp; *Achilles* accessible through: https://portals.broadinstitute.org/achilles

[3] Accessible through: http://www.lincsproject.org

the packages *ArrayExpress*[4] and *GEOquery*.[5] In both cases, data can be downloaded from the web browser, ftp sites or by programmatic access.

Tools for Data Analysis and Visualization

It is necessary to use bioinformatic tools and statistical methods to visualize, integrate, and analyze large datasets. The most popular tool to browse cancer genome studies is *cBioPortal* from the Memorial Sloan Kettering Cancer Center. It contains pre-calculated data from over 80 cancer projects (147 cancer studies) and allows integrative queries of somatic mutations, copy number changes, gene expression, methylation profiles, and protein phosphorylation. Specific to *cBioPortal* are the multi-*omic* networks and heatmap visualizations. Heatmap and network visualizations of pre-calculated public data are also provided by *Regulome Explorer*.[6] There are several genome browsers available as web tools or desktop applications that allow a visualization of genomic coordinates in linear or circular display including different data tracks, the most widely used of which are *UCSC Genome Browser* and *Integrative Genomics Viewer*. As for somatic mutations, *The Catalogue of Somatic Mutations in Cancer* (COSMIC) is the most comprehensive database for searching and analyzing all known somatic mutations in cancer. Also, tools for predicting the impact of somatic mutations in gene function exist, such as SIFT, *PolyPhen*, and *MutationAssessor*, and a combined functional score is computed in *IntoGen*[7] for recurrent mutations.

Bioconductor provides another concept for data analysis as an open-source environment for statistical analysis, data preprocessing, integration, and visualization of high-throughput genomic data. It is a highly extensible, open development platform that uses *R* programming language.[8] Other sources for R packages are CRAN or *GitHub*.[9]

Bioinformatic and Statistical Methods

Deriving predictive biomarkers from data entails the training of predictive models by the application of supervised learning methods to prelabeled data. The label is a measure of treatment success. The data are measurements of the biological condition prior to the treatment, such as potential predictors like gene expression or methylation status. The statistical task is then to train a classifier (or regression model, etc.) on this labeled training data (Fig. 11.1a). Depending on how treatment response is measured, different types of predictive models must be built (Table 11.2). The reader is referred to [2] for the theoretical background of these models and to [3] for a more practical introduction. As one example of a successful biomarker discovery process, Box 11.1 retraces the development of MammaPrint.

Learning biomarker signatures from high-dimensional data is usually comprised of the following steps: normalization, handling of missing values (e.g., imputation), handling of outliers, training, and validation of the classifier. There are several R packages that help with the preprocessing (e.g., *caret*, *vtreat*) and tools that help with the modeling and validation (e.g., *modelr*, *pipeliner*).

Special care is necessary to ensure the validity of any proposed biomarker. Therefore, the importance of multiplicity correction and validation methods must be stressed. These are crucial to reduce the risk of overfitting which any method working with high-dimensional data is prone to. The second focus of this section is the integration of prior knowledge in the predictive model, which potentially increases the reproducibility of suggested biomarkers.

[4]Accessible through: https://www.bioconductor.org/packages/release/bioc/html/ArrayExpress.html

[5]Accessible through: https://bioconductor.org/packages/release/bioc/html/GEOquery.html

[6]Accessible through: http://explorer.cancerregulome.org

[7]Accessible through: https://www.intogen.org/search

[8]*R* accessible through: https://www.r-project.org; *Bioconductor* accessible through: http://bioconductor.org

[9]*CRAN* accessible through: https://cran.r-project.org; *github* accessible through: https://github.com

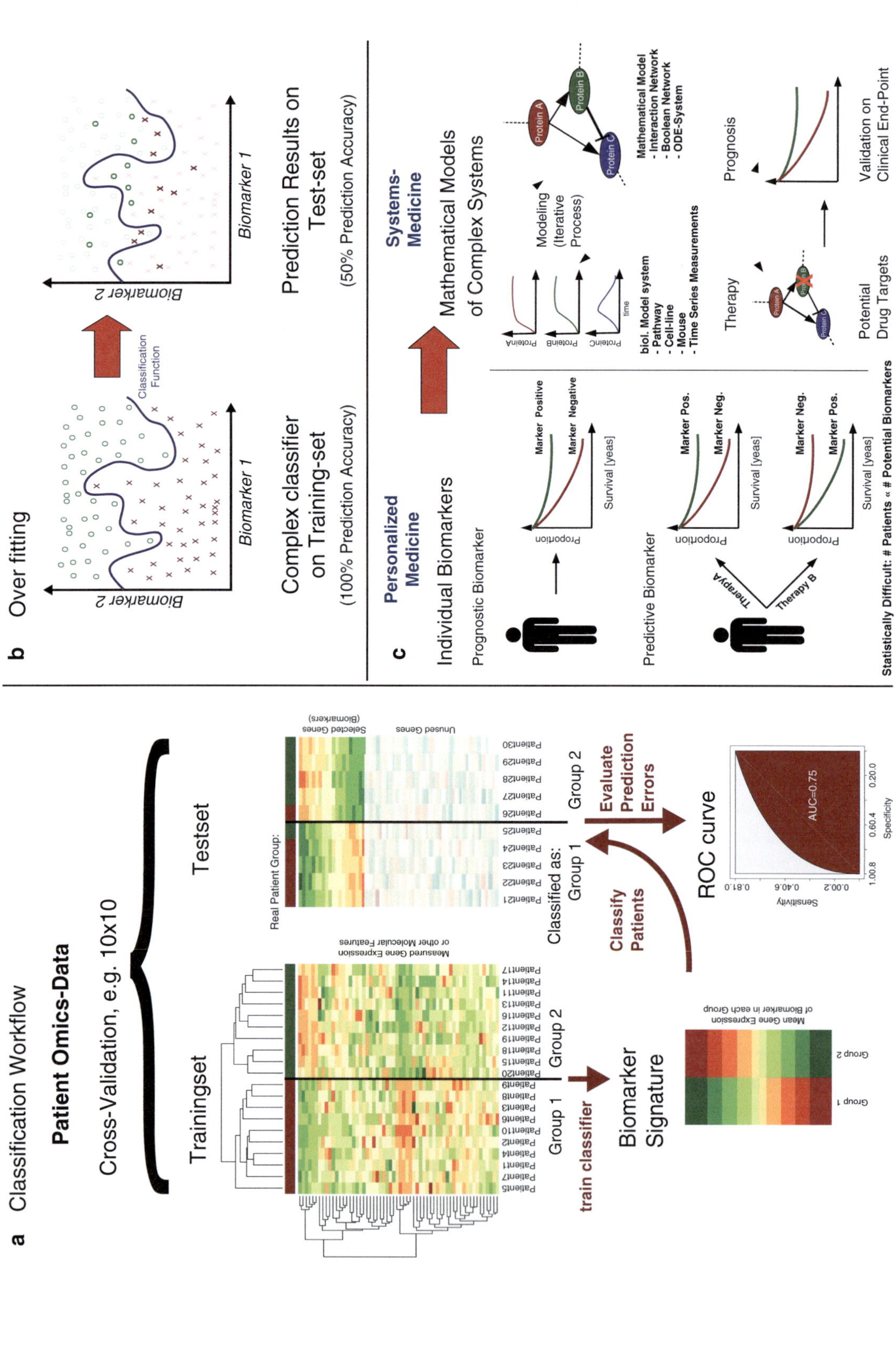
a Classification Workflow
Patient Omics-Data
Cross-Validation, e.g. 10x10
Trainingset
Testset
Measured Gene Expression or other Molecular Features
Real Patient Group:
Selected Genes (Biomarkers)
Unused Genes
Group 1
Group 2
Classified as: Group 1
Group 2
train classifier
Biomarker Signature
Mean Gene Expression of Biomarker in each Group
Classify Patients
Evaluate Prediction Errors
ROC curve
AUC=0.75
Sensitivity
Specificity
b Over fitting
Biomarker 2
Biomarker 1
Classification Function
Complex classifier on Training-set
(100% Prediction Accuracy)
Prediction Results on Test-set
(50% Prediction Accuracy)
c Personalized Medicine
Individual Biomarkers
Prognostic Biomarker
Predictive Biomarker
Therapy A
Therapy B
Proportion
Survival [yeas]
Marker Positive
Marker Negative
Marker Pos.
Marker Neg.
Statistically Difficult: # Patients « # Potential Biomarkers
Systems-Medicine
Mathematical Models of Complex Systems
Proteins
time
Modeling (Iterative Process)
Protein A
Protein B
Protein C
biol. Model system
- Pathway
- Cell-line
- Mouse
- Time Series Measurements
Mathematical Model
- Interaction Network
- Boolean Network
- ODE-System
Therapy
Potential Drug Targets
Prognosis
Validation on Clinical End-Point

Box 11.1 Cancer Biomarker Study (MammaPrint)

In 2002, van't Veer and colleagues (van't Veer et al. 2002, doi: https://doi.org/10.1038/415530a) suggested a 70-gene expression signature able to predict the risk of distant metastasis in patients with lymph node-negative breast cancer. They measured the expression of ~25.000 genes in 78 primary breast tumors using microarray technology. An unsupervised clustering showed that samples clustered into two main groups, "good prognosis" and "bad prognosis." They proceeded with a three-step approach to reduce the number of features (genes). First, 231 genes that correlated with disease outcome were selected. Then, these 231 genes were ranked according to the strength of the correlation coefficient. Finally, to optimize the number of features, predictions were made by sequentially adding sets of five genes; the performance was evaluated by using the leave-one-out cross-validation procedure. The peak of accuracy was reached at 70 genes, leading to the 70-gene signature.

This prognostic gene signature was further validated with another dataset of the same institution (van de Vijver et al., 2002, doi: https://doi.org/10.1056/NEJMoa021967). It was also proven that the signature was a more powerful predictor of disease outcome than other clinical and histological variables (e.g., histologic grade, estrogen-receptor status). After other studies had validated the signature with external datasets (Buyse et al. 2006, doi: https://doi.org/10.1093/jnci/djj329), a commercial test using this signature (MammaPrint test developed by Agendia) was approved for clinical use by FDA in February 2007.

Table 11.2 Statistical tools for different endpoints

Measurement of treatment success	Example (with reference)	Statistical model	R package
Two-class	Tumoral and stromal lymphocytic infiltration predict pathologic complete response (pCR) to neoadjuvant chemotherapy (Li et al. 2016, doi: https://doi.org/10.1093/ajcp/aqw045)	Logistic regression	Stats
Multi-class	Gene expression differentiates poor, mixed, and good outcome associated stroma subtypes (Finak et al. 2008, doi: https://doi.org/10.1038/nm1764)	Multi-class classification	SAMR
Continuous	Transthyretin predicts tumor size in breast cancer (Chung et al. 2014, doi: 10.1186/bcr3676)	Linear regression	Stats
Time-to-event	miRNA predicts recurrence-free survival after radical prostatectomy (Fredsøe et al. 2017, doi: https://doi.org/10.1016/j.euf.2017.02.018)	Proportional hazards regression	Survival

Fig. 11.1 (**a**) Workflow on how to train biomarker-based classification signatures based on genome-wide high-throughput data. Data consist of a set of features (e.g., expression levels of different genes) measured in a number of patients (see heatmap). Patients are labeled with different group labels (e.g., good prognosis, bad prognosis). Data has to be split into a training and a test set. A classifier (e.g., a biomarker signature) is derived from the training data and evaluated on the test data. The performance of the classifier can be evaluated by comparing the predicted labels with the real labels in terms of sensitivity and specificity (e.g., using ROC curves). (**b**) Visualization of the concept of overfitting. In the left panel patients of a training set (represented as circles and crosses in a two dimensional biomarker space) are separated by a complex classification function (blue line). In the right panel patients from a test set are highlighted. The data in these patients does not follow the complex function of the classifier since here biomarker 1 is not informative. Overfitting occurs especially when the dimensionality of the data is high or classification functions are complex. (**c**) Illustration on how a functional understanding of the interactions of individual biomarkers could be used to train predictive models. The idea here is to build a mathematical model that describes the functional relations in a cellular system (e.g., molecules in a signaling pathway) and that is able to make predictions, e.g., on patients prognosis or drug targets

Dealing with High Dimensionality

One problem often encountered within biomarker detection is the so-called curse of dimensionality. This means when searching for biomarkers in a training cohort, one has a high chance of discovering biomarkers that can predict the outcome of each patient in the training cohort, but which do not have any functional relevance or are not able to predict anything on an independent cohort (Fig. 11.1b). Unless strong prior knowledge guides the biomarker discovery process toward a few selected compounds, biomarker discovery mostly starts with screening experiments where multitudes of potential biomarkers are tested. Thus, modern statistical approaches focus on methods for penalization to reduce the number of features to include in a biomarker signature or on dimension reduction methods where linear combinations of several biomarkers can be used as predictors. Another approach is to guide the feature selection using prior knowledge and thus to transfer knowledge from basic research and functional understanding of biological networks and pathways into the process of selecting relevant biomarkers. In the new field of systems medicine, the aim is to construct mathematical models of the complex interactions of the molecular systems in order to improve prediction (Fig. 11.1c).

Multiple Testing Correction

Ignoring all possible interactions and correlations between the potential biomarkers, it is possible to perform one test for each potential biomarker. Care needs to be taken regarding the multiple testing that occurs here. When carried out naively, feature-wise testing leads to an increased risk of false-positive findings, as each test is at risk of producing a false-positive result. Typically, one allows this risk to be 5%. As an example, conducting gene-wise tests for 20,000 genes, each performed with a risk of $\alpha = 5\%$ of being a false positive, one must expect 1000 significant hits just by chance, even in the case that there is no true effect in any gene. To account for this issue, multiple testing correction methods must be applied to recalculate the probabilities obtained from performing a statistical test multiple times (Box 11.2).

> **Box 11.2 Multiple Testing Correction**
>
> **Significance level (α):** probability of a wrong test decision given the null hypothesis is true. It is often set to 0.05 (5% probability of making a false-positive error).
>
> **Bonferroni correction:** method to correct for multiple testing. It controls the probability of at least one false-positive result and is done by performing each individual test on an individual significance level α/ (number of tests). For biomarker screening, this is typically too stringent and does not leave enough power to discover any potential biomarkers.
>
> **FDR (*false discovery rate*):** expected proportion of false-positive results within all positive results. Methods that correct for multiple testing controlling the FDR are less stringent than the Bonferroni correction.

Feature Selection

Biomarker signatures based on a combination of many features (e.g., the expression of 20,000 genes) are typically undesirable. Instead, the subset of features (e.g., a gene panel) needed to fit the model is of at least the same importance as the model itself. Therefore, the feature selection (i.e., the removal of uninformative features from the model) is sometimes regarded as the most important component of predictive modeling.

The simplest method for feature selection is to only consider features that are predictive on their own. Such *filter methods* for feature selection are easy to implement and computationally light. On the downside, if two biomarkers have predictive power only when paired (which is called *interaction effect*), this will be missed by such simple filtering. Alternatively, so-called wrapper methods directly assess the prediction performance of each proposed subset of features. This yields better performing feature sets but is computationally very intensive.

Some methods for predictive modeling do not strictly depend on an externally performed feature selection but intrinsically perform their own *embedded* feature selection. Prominent representatives of such methods are regression methods that directly penalize the number of features in the model. Here, the model equation which is minimized during the fitting gets extended by an additional term which grows with the number of features in the model. Different methods to construct this term lead to different forms of penalization, LASSO and ridge regression methods being the most prominent ones.

Data Integration

Data integration plays a vital role in biomarker detection. It is crucial in the integration of several types of biomarkers (e.g., molecular layers), integration with clinical data, and integration with prior biological knowledge. Each of these overlap as the boundaries between the clinical parameters and the biomarkers are not fixed and the integration of different types of biomarkers is often done via prior knowledge.

Integration of Multiple Biomarker Types

Horizontal integration is the primary focus of screening studies. This is the integration of similar types of data, such as gene expression data. This is routinely performed using the techniques for high-dimensional data as discussed above. On the other hand, vertical integration involves the integration of data from several molecular levels (e.g., DNA, gene expression, protein expression). Clearly, the easiest targets to identify in the search for predictive biomarkers are those in which a single marker is associated with a phenotype with a detectable effect. Many of these targets have already been identified. Thus, vertical data integration has generated increasing interest as a tool to assist in the identification of more complex targets. Integrative analyses have the potential to detect interaction effects where each single effect is too small to exceed the noise level, but which have a large joint effect. In addition, many associations that have already been found are not biologically interpretable. Most genetic variants which are discovered in genome-wide association studies, for instance, fall within noncoding parts of the genome where it is not clear through which mechanism this variant influences phenotypic behavior. Integrating such analyses with other levels of molecular data (for instance, the combined presence of gene expression and the genome-wide association scan) can shed light on the signal mediation from one level to another. So, by making interaction effects subject to study, data integration can lead to new discoveries and can also serve as a tool to advance our biological knowledge.

The canonical example for vertical integration are eQTL studies, which integrate gene expression and DNA sequence data by scanning pairs of gene expression and genetic variant for significant correlations. Other examples include the observance of methylation around the transcription start sites of genes, or the analysis of mRNA-miRNA pairs on the basis of miRNA target predictions.

Several approaches to integrate different molecular layers have been devised (Box 11.3). For a detailed categorization of existing approaches to data integration methods, the reader is referred to [4].

Box 11.3 Methods for Integrating Molecular Layers

Concatenation-based: This is the easiest approach. The different molecular layers are analyzed together ignoring that they represent different layers (for instance, just concatenating the expression levels of 20,000 genes and 1000 miRNAs). While this is easily implemented, it enforces the problems induced by high dimensionality and additionally introduces problems with differently scaled data from different levels.

Model-based: Each layer is analyzed separately first. The resulting models (one for each layer) are then combined to give an integrated answer. Since model-based methods first train models on each data layer separately, they will only detect effects that are strong enough to be detected in one layer individually. Also, these methods do not lead to interpretable interactions.

Transformation-based: Each layer is transformed into a common abstract format (e.g., kernel matrices or graphs). Such abstract formats facilitate the integration and allow the use of prior knowledge.

Integration of Clinical Data

In many cases there exist already known biomarkers which exhibit a good predictive performance and are routinely used to guide the course of treatment. To date, these are often single markers such as ER and HER2 status. Such existing models should not only be used as benchmark models, but the known predictors should be made part of the new model to explicitly assess the information gained from adding new biomarkers and in order to benefit from their already known good predictive performance. To that end, the known predictors should be added as mandatory variables to the predictive model and excluded from feature selection. This is straightforward in linear/logistic models and also possible in other machine learning techniques, such as boosting for survival models (available in the R package *CoxBoost*).

Similarly, clinical parameters can be integrated into the predictive model. Biomarkers might show different behavior in women compared to men, for instance, so gender information is typically considered during the biomarker derivation. Other prime candidate parameters to include in any model are patient age or body mass index (BMI), which can heavily influence metabolism and thus might change the biomarker level.

Integration of Prior Knowledge

Integration of external knowledge is useful in different situations:

1. The signal from single biomolecules might be too low to exceed the noise level and to survive multiple testing correction. In these situations, the aggregated signal of several molecules summed into modules of known to be connected molecules (such as pathways) may be strong enough, especially given that there are typically much fewer modules than single molecules to be considered. Currently the Kyoto Encyclopedia of Genes and Genomes (KEGG) pathway database contains just over 500 pathways as compared to many thousands of annotated genes.
2. Alternatively, with too many molecules differentially abundant between the conditions, prior knowledge is equally beneficial as it aids the formation of a long list of molecules interpretable as one pathway might be overrepresented in this list of molecules. Additionally, detected lists of potential biomarkers are known to be very unstable so that even a slight change in the number of patients could lead to vastly different lists of biomarkers. The discovery can be stabilized by directing the process toward molecules with known importance, such as molecules with many connections in an interaction network or molecules upstream of the other parts of the network. Similarly, the discovery can be directed toward the most interesting targets, such as actionable molecules with known agents.
3. The integration of several molecular layers often uses external knowledge to emphasize biologically meaningful combinations. Prior knowledge can encode known interactions (protein-protein interaction (PPI), miRNA-mRNA pairs), gene coexpression networks, or gene regulation networks. The encoding is often in the form of networks (e.g., regulatory pathways). See reference [5] for a review of options how to import such network data from different formats into R.

Many different methods using prior knowledge during classification have been proposed in recent years. Reviewed here are some general properties (for details, refer to [6, 7]). To incorporate prior knowledge efficiently, the focus shifts from gene level to pathway level. In this setting, the objective shifts. No longer is the focus for predictive therapy response on a single gene; it is now relocated onto the entire pathway.

The question becomes the predictability of the expression of all genes within a pathway when prior knowledge of the pathway is introduced. This can be achieved in several ways. One can aggregate the expression signal from all the members of the pathway by simply taking the average expression or by other more sophisticated dimension reduction techniques as implemented in the R package *netClass*. One can also test each gene individually and aggregate the test results of all members of the pathway in gene set enrichment tests. Alternatively, one can directly formulate a global test against the null hypothesis that the class membership is independent from the biomarker data or similarly that the biomarker data distribution is the same in the classes (e.g., R packages globaltest and *RepeatedHighDim*).

All the methods above only take the grouping information of the genes into account and ask only if the gene is involved in the pathway or not. If prior knowledge can be encoded as a network, more sophisticated models can attempt to capture the topology of the network and consider the connections between the genes in addition to the mere membership information. The R package *PathNet* utilizes this method. A different approach is to use a feature selection which is biased to prefer features which are according to prior knowledge important or strongly connected features. As an example, "gene rank" is based on Google's page-rank algorithm and has been proposed as measure of gene importance in connection with an SVM classifier (R package *pathClass*). Prior knowledge encoded in networks lends itself for data visualization, as the biomarker data can be mapped onto the network, which greatly helps to understand the biological function [8].

Internal Validation (Prior to Clinical Evaluation)

Before testing potential biomarkers in an external validation cohort, internal validation methods must be applied to reduce the number of false-positive findings from the high-dimensional biomarker screening studies (see [2] for an introduction). Internal validation methods rely on splitting the available data into training and test sets so that the predictive model can be built using data from the training set and evaluated on different data from the test set. Box 11.4 introduces some available procedures.

Regardless of the method of choice, it is imperative not to utilize the validation data in any step prior to the performance estimation. While this seems trivial, it is easily violated. One commonly applied but flawed procedure is to perform a feature selection on the full dataset, to build a predictive model using the selected features, and to evaluate the prediction performance in a cross-validation scheme. This procedure will yield highly optimistic estimates of the prediction performance as the model building has at the feature selection stage seen the test data already. All stages of the model building need to be validated; in this example, the feature selection needs to be performed in each fold to get realistic performance estimates.

Box 11.4 Procedures for Internal Model Validation

Hold-out method: The data are split into one training set and one test set. This is a reasonable choice only if there are enough samples. The split ratio (commonly used: 66% training set and 33% test set) is a trade-off: if the test set is small, the estimation of the model performance might be poor; if the training set is small, the model fit might be poor.

K-fold cross-validation: The data are divided into *k* roughly equally sized sets. Each of these *k* sets in turn functions as the test set and the other $k-1$ sets are used as training set to build the model. The performance measure averaged across the folds is reported (Fig. 11.1a). This is a better choice if samples are scarce, because it provides better usage of the available samples.

Leave-one-out cross-validation: Special case of *k*-fold cross-validation, where $k = n$ (*n* being the total number of samples). In each fold, the test set consists of a single sample and the training set of the remaining n-1 samples.

Clinical Evaluation

After a biomarker has been analytically validated, its clinical performance must be established before it can reach a successful clinical application. For that, the clinical validity and clinical utility of the biomarker has to be tested in a clinical trial. These clinical trials are done using low-dimensional data with many samples for few variables. Such clinical trials will, thus, only test a few potential biomarkers and are very costly, which makes strong hypotheses from the earlier phases in the biomarker discovery process necessary. In this section, we will discuss genomically guided clinical trial designs as well as challenges and considerations for the clinical evaluation of biomarkers before being used in patient management.

Clinical validity is the ability of a biomarker to separate the population in two groups: those patients that will respond to a drug and those that will not. Of course, the definition also applies to drug resistance, toxicity, disease recurrence, etc. On the other hand, clinical utility determines whether testing for the biomarker leads to better outcome than the standard of care. Both clinical validity and utility need to be assessed in phase 2 and 3 clinical trials.

Drugs and Biomarkers Co-development

One of the purposes of using a predictive biomarker is to be included as a companion diagnostic for a drug, i.e., the status of the biomarker will determine if the patient will respond to the drug. Hence, the most efficient way of validating a predictive biomarker is validating it with the drug in a randomized clinical trial, also known as biomarker and drug co-development.

If the drug performs better than the standard treatment in the biomarker-positive arm, but not in the biomarker-negative arm, then the biomarker test will be included in the drug indication.

Genomically Guided Clinical Trial Designs

There are clinical trial designs that take into account the status of a biomarker to stratify patients. However, the complexity of such designs grows with the use of NGS techniques in biomarker identification. Whereas NGS provides a unique opportunity on genome-wide biomarker testing, there is a need for developing new trial designs accounting for several biomarkers and/or several drugs at the same time [9]. In an ideal world, we would perform a randomized clinical trial (RCT) for every biomarker that reaches clinical validation. But RCTs with biomarker arms need a large number of participants with biomarker-positive and biomarker-negative arms and experimental and control groups, making it complicated to reach a sufficient sample size. One way to overcome this problem is to explore predictive biomarkers in early clinical trials as part of the inclusion criteria or with *enrichment* strategies. If they show potential in predicting drug response, these early trials can be used as proof of concept for larger RCT trials.

Biomarker-driven clinical trial designs can be histology agnostic or histology specific. Histology-agnostic trials are based on the fact that many genes are mutated across cancer types; hence, there is the opportunity to discover pan-cancer predictive biomarkers. One example is the *Basket trial*, a nonrandomized approach that tests one drug in patients with the same genomic alteration regardless of the cancer type. On the other hand, an example of histology-specific design is the *Umbrella trial*, a nonrandomized strategy that tests for multiple biomarkers and matches biomarker-positive patients with targeted therapies (i.e., enrichment approach) under the umbrella of a common histology.

A more sophisticated modification of enrichment approach is the addition of a sequential step, as executed in MATCH trial (*NCT02465060*). In this trial, all patients are tested for several biomarkers and assigned to a drug. Each drug is a new treatment arm. However, if there is progression, a new drug can be selected. MATCH has now 24 arms and is a good design to study mechanisms of acquired resistance. A completely different approach is being used by MOSCATO trial, which follows a *N-of-1* sequential approach: the patient is used as its own control, meaning that the drug effect is compared to the earlier drug effect in terms of progression-free survival.

In a clinical trial, randomization is always recommended for establishing the clinical validity and utility of a biomarker-drug efficacy. *Biomarker* versus *control* designs follow randomization either before or after biomarker testing. Therefore, instead of comparing two treatment methods such as drug vs. control, two treatment strategies are compared, i.e., biomarker testing vs. non-biomarker testing. This strategy was followed in the SHIVA trial (*NCT01771458*), where they compared *personalized treatment* using molecular profiling versus conventional therapy. SAFIR02 uses the same strategy, but it is specific for some histologies (e.g., *NCT02299999* for breast, *NCT02117167* for lung). In general, randomized designs require large sample size which, for some genomic alterations, will not be feasible.

Clinical Application

Standards and Existing Methods

Routine testing of biomarkers must be performed in qualified laboratories having an accreditation such as ISO 15189 in the *European Union or Clinical Laboratory Improvement Amendments* (CLIA) in the USA. These certifications ensure certain assay precision, accuracy, sensitivity, specificity, and reproducibility among all clinical centers. Current clinical routine mainly uses individual biomarkers, often single molecules evaluated in immunohistological staining or individual genetic variations in single genes. For single biomarkers, the technologies for biomarker testing in clinics are quite standard: FISH, *Sanger sequencing* or PCR for DNA biomarkers, *immunohistochemistry* or ELISA for protein biomarkers, RT-PCR for RNA biomarkers, and ELISA or *chemical/colorimetric assays* for metabolites. Instead of using complex biomarker signatures from *omics* technologies, in clinical practice often single biomarkers are used which are measured on one of the established technology. Thus, there is a need to find proxies from single biomarkers, which can reproduce a similar patient stratification as a complex classifier. There is no systematic process to find such proxies, and results of different biomarker-based stratifications are often incomparable.

With rapid developments of NGS technologies, it is now possible to test biomarkers in a high-throughput way. However, NGS implies not only standardization regarding sample processing (DNA extraction, library preparation, barcoding), platform selection, or other analytical approaches such as targeted sequencing and exome or whole-genome sequencing but also standardization of bioinformatic workflows for data processing. For ensuring quality of raw data, high coverage is essential for clinical use. Standardization of aligning software can be tested by using reference sequences, benchmark data, control samples or parallel validation with other technologies. Regarding variant calling (identification of somatic variants), reproducibility and uniformity of calls are achieved through parameters that include filters for base quality, alignment mismatches, multi-mapped reads, and coverage at sites with variation.

Since 2011, several clinical and research organizations, as well as governments, have published guidelines and recommendations for dealing with NGS for diagnostic applications (e.g., *Food and Drug Administration*, *American College of Medical Genetics*, *Clinical and Laboratory Standards Institute*, *European Society of Human Genetics*). These guidelines cover issues such as ethical considerations, terminology, test quality, turnaround time, biobanking, bioinformatic pipelines, and interpretation and reporting of NGS data [10].

Interpretation and Reporting

When a tumor is sequenced with NGS, the output of bioinformatic pipelines are standard variant calling files (vcf, maf, or bed formats[10]) with a list from tens to hundreds of somatically altered genes (comprising point mutations, amplifications, deletions, insertions, fusions). For the clinician who is responsible of making sense of NGS results in a molecular tumor board, the difficulty relies on assigning clinical meaning to the identified genomic alterations, also referred as variant interpretation.

Clinicians will only prescribe a treatment based on a somatic variant if there is clinical evidence showing that the variant predicts response to a drug. For that, there are several databases that compile clinical information with varying levels of details, curation, and comprehensiveness. A clinically relevant selection would comprise databases of clinical trials (*ClinicalTrials.gov*, EU-CTR), predictive biomarker sites (*mycancergenome.org*, GKDB, TARGET, CIViC, CGI), and treatment guidelines (NCCN, FDA, ACMG) (more details in Table 11.1).

Also, searching in *PubMed* or *Google Scholar* for case reports and preclinical data becomes crucial for the interpretation of somatic variants. Another common approach to identify important somatic alterations, especially for new variants, is by using public *omic* resources or computational prediction tools reviewed in the section "Public Resources and Open-Source Tools" of this chapter. This is a broad field that addresses the following issues: identification of frequent mutations driving the tumor development; prediction of functional impact of the mutations; impact estimation on signaling pathways and networks; inference of synthetic lethal pairs of genes; and prediction of mutated fragments of DNA that trigger immune responses (i.e., neoantigens). Finally, when dealing with such amount of data, it is also common to make use of integrative visualization tools to see genomic alterations in different contexts (biological pathways, genome browsers, correlation between data types).

New platforms integrating the aforementioned databases with patient data will need to be developed and implemented in hospitals. Electronic medical records should incorporate biomarker status (probably coming from high-throughput technologies) and, in turn, link to biomarker knowledge databases and visualization tools. Also, strategies to prioritize biomarkers according to clinical evidence of biomarker-drug associations need to be defined. Figure 11.2 depicts the implementation of NGS data in molecular tumor boards. Finally, the standard way of transmitting, saving, and accessing patient information in the clinical environment is through reports. Treatment decisions are made based on the data shown in these reports. Therefore, it is crucial to determine how genomic findings need to be reported to clinicians.

It is clear that variant interpretation and reporting need automation, but efforts on bringing biomarkers knowledge together have still not been successful. Moreover, although there are some biomarkers present in a sizable fraction of cancer patients, most of them are rare events, and inferring the effect of a drug on off-label situations is never straightforward. For example, different mutations in the same gene can predict opposite reactions to a drug (e.g., exon 19,21 versus exon 20 mutations in EGFR), or the same mutation in the same gene in different cancer types can also predict opposite impact. A common thought is that if the target of a drug is mutated, the drug will have a positive effect; this is yet not true in most of the cases, nor it necessarily implies better performance than the standard of care. Facts show that most drugs fail in phase III trials because they don't show better performance than the control arm. And, if an off-label prescription manages to reduce the tumor in a patient, either it is published as a case report, or else this valuable information will be lost.

[10]Standard formats provided by variant calling/read counting software. File format descriptions can be found here: https://en.wikipedia.org/wiki/Variant_Call_Format, https://gdc-docs.nci.nih.gov/Data/File_Formats/MAF_Format, http://www.ensembl.org/info/website/upload/bed.html

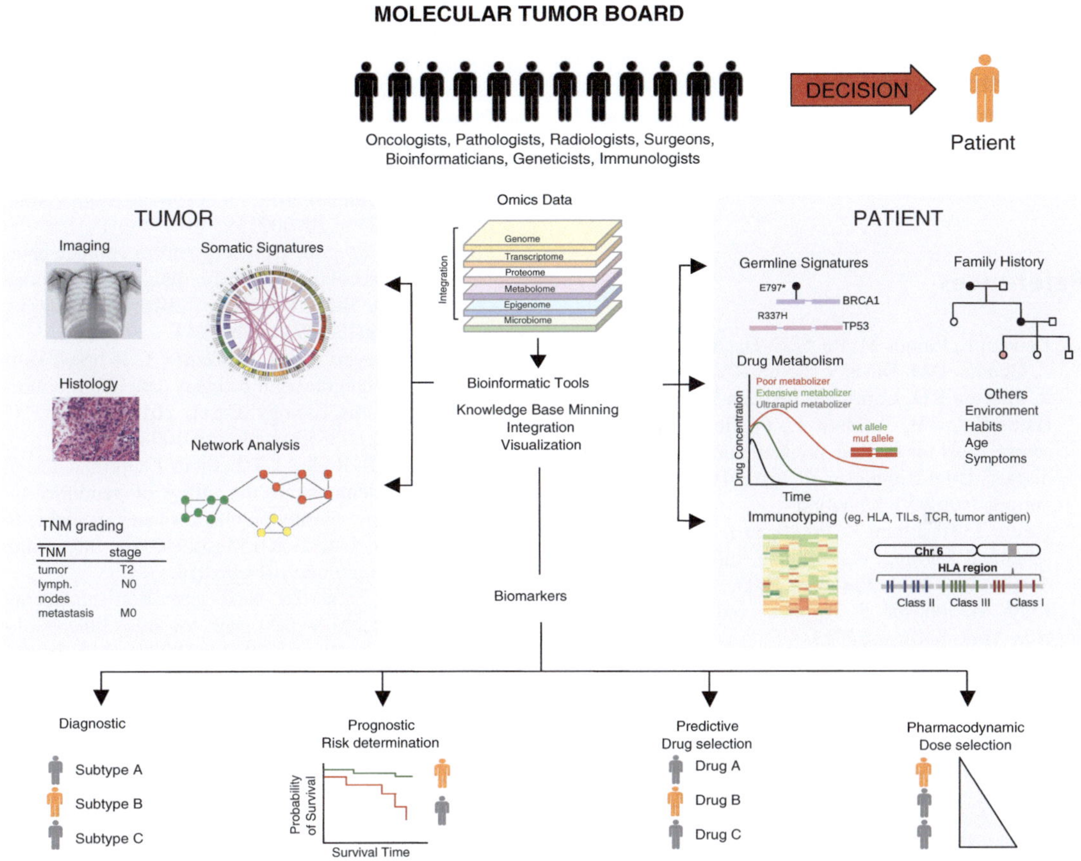

Fig. 11.2 Molecular tumor boards incorporating omics data. A multidisciplinary board of experts will decide the best therapy for the patient based on traditional clinical tests as well as on omics data results. Omics data requires bioinformatic tools to molecularly characterize the tumor and the patient. These tools will help to narrow down the data to informative biomarkers providing information on diagnosis, prognosis, drug targets and other therapeutic strategies

Summary and Future Directions

In this chapter, we give a brief overview over the various bioinformatic challenges in research related to predictive biomarkers and their translation into the clinic. This spans a wide range of different topics from resources, tools, and databases to the wide area of machine learning. Bioinformatics and biostatistics expertise are needed in the process of biomarker discovery as well as in the translation into the clinic in clinical trials. Standards of how to summarize and report the results will become increasingly important also in clinical routine and involve sophisticated bioinformatic tools and expertise. Thus, it seems evident that once we move to a more personalized or stratified biomarker-based medicine, bioinformatics will become an increasingly important part in medical research as well as in clinical routine. However, here we can only give a fleeting glimpse and touch the various topics rather than giving a comprehensive overview. Given its integral part in personalized medicine, bioinformatics should become an integral part also in medical education. We are now at a stage where most of the clinical research can only be performed by interdisciplinary teams involving bioinformaticians. It is yet unclear, how this expertise will be available to the patients and treating physicians in a routine clinical setting. Reporting standards and artificial intelligence might influence strongly how the diagnosis of complex diseases and the suggestion of an appropriate treatment course are carried out in the future. Also, here it will be crucial that interdisciplinary teams such as *molecular tumor boards* will have expertise on the medical side as well as in bioinformatics.

References

1. Kannan L, Ramos M, Re A, El-Hachem N, Safikhani Z, Gendoo DM, Davis S, Gomez-Cabrero D, Castelo R, Hansen KD, Carey VJ, Morgan M, Culhane AC, Haibe-Kains B, Waldron L. Public data and open source tools for multi-assay genomic investigation of disease. Brief Bioinform. 2016;17(4):603–15. https://doi.org/10.1093/bib/bbv080.
2. Hastie T, Tibshirani R, Friedman J. The elements of statistical learning: data mining, inference, and prediction. New York: Springer; 2003.
3. Kuhn M, Johnson K. Applied predictive modeling. New York: Springer; 2013.
4. Ritchie MD, Holzinger ER, Li R, Pendergrass SA, Kim D. Methods of integrating data to uncover genotype-phenotype interactions. Nat Rev Genet. 2015;16(2):85–97. https://doi.org/10.1038/nrg3868.
5. Kramer F, Beißbarth T. Working with ontologies. Methods Mol Biol. 2017;1525:123–13.
6. Porzelius C, Johannes M, Binder H, Beissbarth T. Leveraging external knowledge on molecular interactions in classification methods for risk prediction of patients. Biom J. 2011;53(2):190–201. https://doi.org/10.1002/bimj.201000155.
7. Glaab E. Using prior knowledge from cellular pathways and molecular networks for diagnostic specimen classification. Brief Bioinform. 2016;17(3):440–52. https://doi.org/10.1093/bib/bbv044.
8. Kramer F, Bayerlová M, Beißbarth T. R-based software for the integration of pathway data into bioinformatic algorithms. Biology (Basel). 2014;3(1):85–100. https://doi.org/10.3390/biology3010085.
9. Dienstmann R, Rodon J, Tabernero J. Optimal design of trials to demonstrate the utility of genomically-guided therapy: putting precision cancer medicine to the test. Mol Oncol. 2015;9(5):940–50. https://doi.org/10.1016/j.molonc.2014.06.014.
10. Bennett NC, Farah CS. Next-generation sequencing in clinical oncology: next steps towards clinical validation. Cancers (Basel). 2014;6(4):2296–312. https://doi.org/10.3390/cancers6042296.

Part II

Major Cell Signaling Pathways

Overview of Cell Signaling Pathways in Cancer

12

Amanda J. Harvey

Introduction to Cancer Cell Signaling

Cell signaling is the "catch-all" phrase that provides an overview of the communication system and is often linked to a single signaling pathway. In this one simple term, there is a sense of cells communicating with one another and changing their behavior as a result of such communication. This ability of cells to sense external signals and respond to them is a basic requirement for tissue development and repair, immunity, and homeostasis.

Signal transduction defines the precise series of molecular events that occur to convert an external stimulus into a cellular response. Most frequently these events involve phosphorylation of target molecules by enzymes with kinase activity. A signal transduction pathway is initiated when a ligand binds to its receptor resulting in a conformational change which then allows for activation of its kinase activity and receptor transphosphorylation, e.g., in the case of epidermal growth factor (EGF)-mediated signaling, binding of downstream substrates, and activation of the kinase activity. Often (but not always) the receptors cross the cell membrane allowing for ligand binding outside of the cell with the subsequent phosphorylation event occurring internally. This is a fundamental process by which cells can communicate with each other. One cell releases a ligand (e.g., growth factor or cytokine), which then binds to receptors on adjacent cells activating their internal signaling mechanisms.

Following receptor phosphorylation and binding of an adaptor molecule, a signaling cascade becomes activated allowing for a series of phosphorylation events to occur transmitting the signal from the cell membrane to other parts of the cells, most often the nucleus where, upon phosphorylation, transcription factors become activated. Transcription factor activation results in changes in gene expression, subsequent translation, and the production of a biological response by the cell.

Where nuclear receptors also act as transcriptional regulators, ligands diffuse into the cell and bind to the receptor in the cytoplasm resulting in a conformational change and subsequent nuclear translocation of the receptor. Once in the nucleus, these activated receptors are capable of binding to their respective consensus sequences within the promoter regions, altering gene transcription (Fig. 12.1).

A. J. Harvey
Department of Life Sciences, Brunel University London, Uxbridge, Middlesex, UK
e-mail: amanda.harvey@brunel.ac.uk

S. Badve, G. L. Kumar (eds.), *Predictive Biomarkers in Oncology*,
https://doi.org/10.1007/978-3-319-95228-4_12

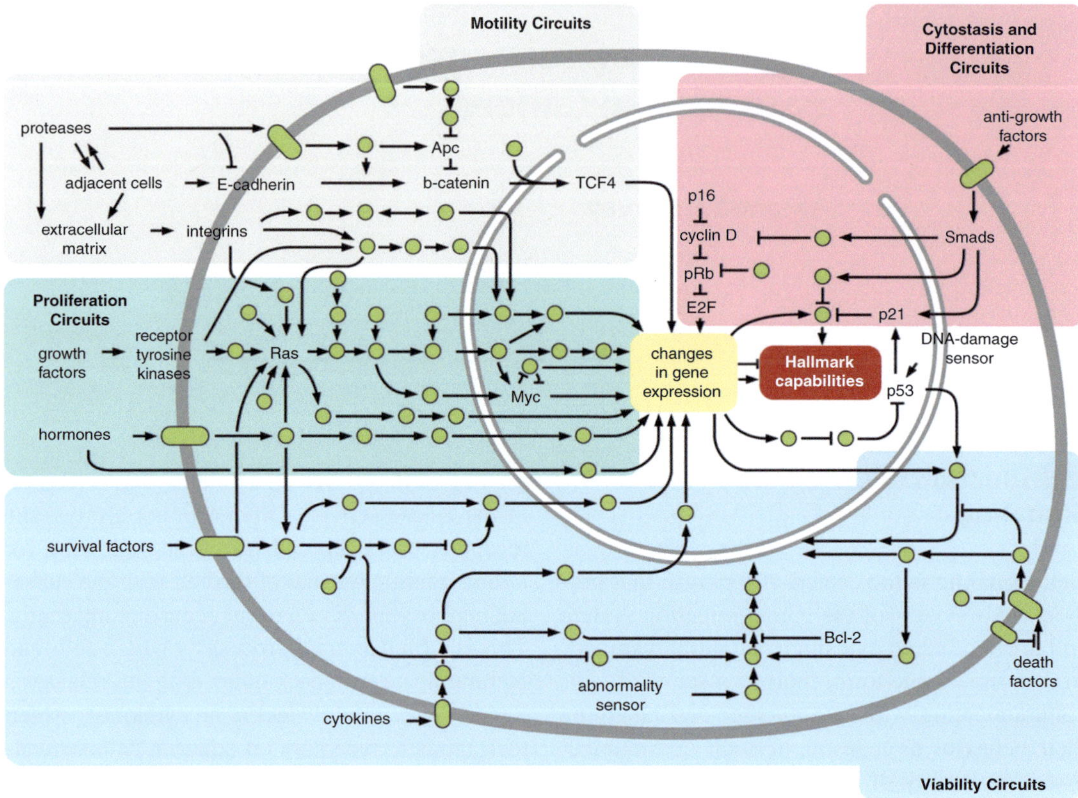

Fig. 12.1 Intracellular signaling networks regulate the operations of the cancer cell. An elaborate integrated circuit operates within normal cells and is reprogrammed to regulate hallmark capabilities within cancer cells. Separate subcircuits, depicted here in differently colored fields, are specialized to orchestrate the various capabilities. At one level, this depiction is simplistic, as there is considerable cross talk between such subcircuits. In addition, because each cancer cell is exposed to a complex mixture of signals from its microenvironment; each of these subcircuits is connected with signals originating from other cells in the tumor microenvironment, as outlined in Fig. 12.4. (Reprinted from Hanahan and Wienberg [1]. With permission from Elsevier)

Membrane Receptors

- *ErbB/HER Signaling Pathway*
- EGFR/HER family comprises four receptors and initiate signaling pathways (including PI3K/Akt, mTOR, and MAPK) involved in cell survival and proliferation. EGFR signaling is central to development.
- ***Roles in disease.*** These pathways have been implicated in several cancers (e.g., squamous-cell lung carcinomas, breast, colorectal, and epithelial head and neck cancers).
- EGFR and HER2 are targets for kinase inhibitors (e.g., lapatinib, gefitinib) and monoclonal antibody (biological) therapies (e.g., trastuzumab, pertuzumab). HER2 can also be targeted indirectly via inhibitors of heat shock protein 90 (Hsp90). (See Chap. 18 in this book.)
- *G Protein-Coupled Receptors (GPCRs) Signaling*
- GPCR signaling involves two principal signal transduction pathways: the cAMP signal pathway and the phosphatidylinositol signal pathway.
- GPCRs are the largest signaling receptor family; the receptors themselves are characterized by the seven transmembrane domains, and they have broad physiological functions including cell proliferation and invasion as well as immune cell-mediated functions and nervous system transmission. Canonical

signaling involves coupling with G proteins resulting in phosphorylation of the receptor.

- ***Roles in disease.*** GPCRs are involved in numerous cancers, especially at secondary sites such as the lung, bone, lymph nodes, and liver.
- GPCRs are potential targets for therapy but, currently, this has not been fully explored.
- *Fibroblast Growth Factor (FGF) Signaling Pathway*
- FGFs are considered to be either paracrine (locally acting) or endocrine (relating to hormones secreted into the blood) and signal through four receptors (FGFR1-2) to regulate several cell outcomes including survival, proliferation differentiation, and cell metabolism. They also regulate immunity, angiogenesis, and epithelial to mesenchymal transition (EMT). Downstream signaling components include PI3K/Akt, mTOR, MAPK, and phospholipase signaling.
- ***Roles in disease.*** FGF signaling is implicated in several cancers (e.g., gastric, lung, and breast cancers).
- FGF23 is a target for biological (monoclonal antibody) therapy (e.g., KRN23), while the receptors are targets for numerous antibodies or small molecule inhibitors (e.g., NVP-BGJ398).
- *Insulin Receptor (IR) and Insulin-Like Growth Factor Receptor (IGFR) Signaling Pathways*
- Insulin is critical for regulation of glucose and energy metabolism, while IGF plays an important role in growth, through adapter proteins, the insulin receptor substrate (IRS) family; both hormones mediate their effects via AMPK, PI3K/Akt, mTOR, and MAPK signaling pathways.
- ***Roles in disease.*** The IR and IGFR signaling pathways are widely implicated in many cancers (e.g., breast, prostate, ovarian, and colorectal cancers, Ewing's sarcoma, rhabdomyosarcoma, and non-small-cell lung carcinomas).
- IGFR1 can be targeted with both monoclonal antibodies (biological therapy) (e.g., cixutumumab) and small molecule tyrosine kinase inhibitors (linsitinib), and second-generation antisense oligonucleotides are in development. As with FGF23, IGF1, and IGF2 are targets for anti-ligand antibodies (e.g., MEDI-573 or BI836845). (See Chap. 22 in this book.)
- *Transforming Growth Factor-β (TGF-β)/ Smad Signaling Pathway*
- TGF-β signaling has opposing roles in different cellular contexts. It plays key roles in embryonic stem cell renewal, differentiation, proliferation, immune system suppression, and homeostasis of mature cells. The canonical pathway is well characterized, and signaling is carried out via the Smad signaling cascade which links the transmembrane receptors with the cell nucleus.
- ***Roles in disease.*** TGF-β signaling is implicated in pathologies such as benign prostatic hyperplasia as well in various cancers (e.g., colorectal, gastric, endometrial, breast liver and pancreatic cancers). TGF-β is a target for ligand traps (by antibodies such as lerdelimumab and metelimumab) or antisense oligonucleotides (e.g., trabedersen), but translation into the clinical has been disappointing. (See Chap. 25 in this book.)
- *Vascular Endothelial Growth Factor (VEGF) Receptor Signaling*
- VEGF signaling is crucial during embryonic development as it is required for the formation of new blood vessels (angiogenesis). It is also required to restore oxygen levels in tissues when blood supply is compromised and to create new blood vessels after injury. There are three receptors VEGFR1 (FLT-1), VEGFR2 (FLK-1), and VEGFR3 which homo- and heterodimerise.
- ***Roles in disease.*** VEGF signaling has been implicated in metastatic colorectal cancer (mCRC); metastatic renal cell carcinoma (mRCC); locally advanced, recurrent, or metastatic non-small-cell lung cancer (NSCLC); progressive glioblastoma; and breast cancer.
- VEGF receptors (VEGFRs) are targets for both kinase inhibitors (e.g., sorafenib) and biological (antibody-based) therapies (e.g., ramucirumab). VEGF is a target for ligand-blocking antibodies (e.g., bevacizumab).

Small oligonucleotides (such as Veglin) are also being tested to prevent expression of VEGF genes. (See Chap. 19 in this book.)

- *Toll-like Receptors (TLRs) Pathway*
- The TLR family belongs to the larger group of pattern recognition receptors (PRRs). They are present on antigen-presenting cells (APCs), and ligand binding results in maturation of the cell, cytokine induction, and the priming of naïve T cells to drive acquired immunity because of downstream signaling causing nuclear translocation of NF-κB. TLR ligands have potential as vaccine adjuvants and could be co-administered with protein subunit vaccines to boost immune responses.
- ***Roles in disease.*** TLR activation is linked to the pathology of immune diseases and cancer. Unlike other cancer targets where inhibition is key, agonists of TLR2, such as SMP105 and Sumitomo, have potential as anticancer agents.
- *B-Cell Receptor (BCR) Signaling Pathway*
- The BCR is central to regulating maturation and proliferation of, and antibody production by, B cells. Signaling from the receptor activates Src family members and PI3K with recruitment of Bruton tyrosine kinase (BTK), ultimately causing NF-κB to translocate to the nucleus inducing cytokine production.
- ***Roles in disease.*** B-cell receptor cascade is implicated in the development of B-cell malignancies as upregulated signaling modulates cell migration and adhesion through remodeling of the microenvironment. BTK signaling plays a role in a number of autoimmune and inflammatory diseases such as rheumatoid arthritis and multiple sclerosis.
- BTK is a B-cell-specific target for small molecule inhibitors, and compounds such as PRN2246, which readily crosses the blood-brain barrier, are in clinical trial.
- *T-Cell Receptor (TCR) Signaling Pathway*
- TCRs recognize fragments of antigens and function as complex whose signaling is enhanced through a co-receptor (e.g., CD4 or CD8). As with BCR, signaling from the TCR activates Src family members resulting in phospholipase activation; MAPK and NF-κB pathways are also triggered.
- ***Roles in disease.*** As well as being disease targets for drugs such as dasatinib that target the downstream elements of the pathway thereby inhibiting T-cell activation, T cells themselves are being engineered for use in immunotherapy.
- *Hepatocyte Growth Factor (HGF)/Met Receptor Signaling*
- MET is a cell surface receptor tyrosine kinase found in both epithelial and endothelial cells. Like other receptor tyrosine kinases, MET signaling positively regulates a number of key cellular functions including proliferation, survival, and cell migration; however, the MET receptor has a single ligand (HGF). There are several downstream pathways of MET signaling with the Ras-Raf-MAPK cascade and the PI3K-Akt axis being the most relevant to disease development.
- ***Roles in disease***. In normal cells MET expression and activity is low with activation in tumor cells arising from gene amplification or increased HGF levels. In glioblastoma, MET activation is associated with the higher-grade tumors. Potential therapeutic strategies target different target different aspects of MET function. C-Met peptides bind to the receptor preventing HGF from binding, whereas antibodies such as rilotumumab bind to HGF directly, although clinical trials showed adverse effects with this agent. As with other receptor tyrosine kinases, small molecules (such as) can target the kinase activity of c-MET [2]. (See Chap. 21 in this book.)
- *Platelet-Derived Growth Factor (PDGF) Signaling*
- Platelet-derived growth factors are important during embryonic development where oligodendrocyte precursor cells are stimulated to proliferate in response to PDGF.
- There are two receptor monomers that dimerize, resulting in three possible receptor dimer combinations, *and their kinase activity is activated by the binding of one of four ligand dimers. As with other receptors, downstream effectors include the MAPK cascades (via*

rash activation) and JAK/STAT signaling. During development negative feedback is limited, so signaling is controlled primarily through PDGF availability.

- ***Roles in disease****. PDGF receptors are often mutated, or expression is amplified in glioblastoma, and increased activation of PDGFRα* signaling may be a disease initiating event PDGF.
- Clinical trials with the signaling antagonist, imatinib, have not yielded the hoped-for results in glioblastoma although there has been more success with the same drug in some gastrointestinal tumors. Quinine derivatives (e.g., NSC13316) may prove to be more successful. The inhibitor nintedanib is used to target PDGFR (as well as VEGFR and FGFR) in non-small-cell carcinoma and pulmonary fibrosis [3].
- A nice animation of PDGFR activation can be accessed through the following link http://www.cellsignallingbiology.org/csb/001/csb001_mov016.htm.
- *Death Receptor Signaling*
- The growth factor superfamilies that directly regulate cell death are large with 19 ligands and 29 receptors and are predominantly expressed by immune cells. Members such as tumor necrosis factor (TNF) and Fas Ligand (FASL/CD95) bind to their receptors, TNF receptor (TNFR) and Fas/CD95, initiating cell death through recruitment of adaptor proteins such as TNF receptor-associated death domain (TRADD) and Fas-associated death domain (FADD), which both associate with their corresponding receptor death domain. This leads to the activation of caspases resulting in cell death. Receptors lacking in death domains recruit molecules such as TNF receptor-associated proteins (TRAF) to initiate cell death via signal transduction pathways and activation of transcription factors such as AP-1 and NF-κB.
- Ligands such as TNF-related apoptosis inducing ligand (Trail or Apo2L) and TNF-like weak inducer of apoptosis (Tweak or Apo3L) are also members of this superfamily with Trail binding to its own receptors and initiating cell death. Like TNF, Trail can also activate NF-κB, a pro-survival transcription factor, indicating the importance of signaling balance and activation of pro- and anti-apoptotic factors by this superfamily.
- ***Roles in disease***. Dysregulation of TNF occurs in rheumatoid arthritis and other inflammatory diseases such as ankylosing spondylitis, ulcerative colitis, and Crohn's disease. The main target for therapy in this superfamily is TNF with infliximab (an anti-TNF antibody).
- Treating Crohn's disease patient with combinatorial therapy that includes TNF inhibition can result in an increased risk of non-Hodgkin's lymphoma and skin and lung cancers, potentially highlighting the requirement for functioning death pathways in normal tissue homeostasis [4].

Cytoplasmic Signaling Molecules

- *Phosphatidylinositol-3-kinase (PI3K)/Akt Signaling Pathway*
- The PI3K/Akt pathway is downstream of several growth factor receptors, most notably the EGFR/HER family, and is upstream of mTOR. It plays an essential role in regulating growth, metabolism, and survival of normal cells, and its activity is negatively regulated by the phosphatase and tensin homologue, PTEN.
- ***Roles in disease.*** Activating mutations in this pathway are some of the most common mutations in cancer and human pathologies. PI3K/Akt activation results in conditions of clinical overgrowth disorders (e.g., Proteus syndrome) and Cowden's disease (due to inactivation of PTEN) as well as a range of solid tumors and hematological cancers (e.g., breast, colorectal, hepatocellular, and ovarian cancers and acute myeloid leukemia).
- PI3K is a target for inhibitors that either inhibit all PI3Ks (e.g., XL147) or are targeted to specific isoforms, and several are in phase II or phase III trials (e.g., CAL-101/idelalisib). Akt inhibitors (e.g., GSK2141795) are less selective but are also in clinical trials [5]. (See Chap. 20 in this book.)

- *mTOR Signaling Pathway*
- The mechanistic target of rapamycin (mTOR) pathway is a serine/threonine kinase belonging to the phosphoinositide-3-kinase-related kinase (PIKK) family. It forms two distinct complexes and is activated by PI3K/Akt signaling so is therefore critical in cell growth, metabolism, and survival, as well as protein synthesis. In addition, mTOR functions as a nutrient sensor so is central to the regulation of intracellular glucose and amino acids. In some animal models (e.g., *C. elegans* and *S. cerevisiae*), decreased mTOR activity is linked to an increase in life span.
- ***Roles in disease.*** mTOR signaling is implicated in central nervous system disorders and cancers. It is frequently upregulated in cancers including breast and renal cancers.
- mTOR is a target for inhibition in multiple cancers (by rapalogues such as everolimus, temsirolimus). Combined inhibitors that also target PI3K have also been designed (e.g., BEZ-235, XL765) [6]. (See Chap. 20 in this book.)
- *Protein Kinase C (PKC) Signaling*
- The PKC subgroup are a family of intracellular serine/threonine kinases, expressed in many different tissues types. They play a key role in many different signaling pathways contributing to the formation and degradation of focal adhesions, as well as regulating cell proliferation and invasion.
- ***Roles in disease.*** Because they act in a many different signaling pathways, PKCs have been implicated in a range of cancers including pancreatic cancers.
- PKCs are potential targets for small molecule inhibitors (e.g., UCN-01) and compounds such as bryostatin that induce membrane localization of PKC isoforms, but these have been unsuccessful in clinical trials. (See Chap. 14 in this book.)
- *MAPK/Erk in Growth and Differentiation Signaling Pathway*
- The mitogen-activated protein kinases (MAPK) and the extracellular signal-regulated kinases (Erk) are subfamilies of serine/threonine and tyrosine/threonine kinases which function in a canonical signaling cascade known as the MAPK cascade. MAPK/Erk signaling is downstream of several transmembrane receptors, including FGFR, IGFR, EGFR, VEGFR, and GPCR, and controls vital functions such as proliferation, differentiation, apoptosis, development, inflammation, and stress responses. MAPK also regulates the activities of transcription factors.
- ***Roles in disease.*** MAPK signaling is implicated in several pathologies including some neuropathologies and cancers (e.g., melanoma, renal cell carcinoma, and Hodgkin's disease), and elevated MAPK activity is common in all inflammatory diseases.
- RAF and MEK kinases are targets of FDA-approved small molecule inhibitors, and Erk is a current target for preclinical kinase inhibitors (e.g., AZD7624) [7].
- *Phospholipase Signaling*
- Phospholipases are widely occurring; they are a class of enzymes that cleave phospholipids, and it is likely that that they signal through MAPKs and other kinase pathways to regulate differentiation, programmed cell death, and immune cell activation.
- ***Roles in disease.*** Phospholipase signaling has a mixed role in tumor development. Some isoforms play key roles in cell migration and invasion so contribute to carcinogenesis, whereas others are linked to tumor suppression, especially in colorectal cancers.
- Phospholipases have potential as targets for inhibitors and molecules that target protein-protein interactions, but there are no compounds currently in clinical trial.
- *AMP-Activated Protein Kinase (AMPK) Signaling Pathway*
- AMPK is an intracellular serine/threonine kinase that is widely expressed as a nutrient sensor. It is phosphorylated in response to stress and subsequently activates its downstream substrates. It is a critical regulator of metabolic homeostasis, as well as having a role in cell proliferation and cell cycle regulation.
- ***Roles in disease.*** AMPK is implicated in the pathology of Peutz-Jeghers syndrome and

several cancers (e.g., lung, liver, and cervical cancers).
- It is a drug target in prostate cancer cell growth where metformin is believed to have both direct and indirect effects on AMPK activity.
- *Hedgehog Signaling Pathway*
- The hedgehog (Hh) pathway has a central role in segmental pattern formation and in development. Depending on the context, it can induce both cell proliferation and differentiation, and its signaling is cross-linked with the MAPK cascade and PI3K/Akt and mTOR signaling [8].
- ***Roles in disease.*** Hh is involved in developmental diseases such as abnormal tube development and cancers (e.g., medulloblastomas, neuroblastomas, gliomas, and breast cancers).
- Smoothened (SMO) is a target for natural inhibitors and vismodegib, the first Hh-targeting compound to get USFDA approval, entered clinical trial in 2017. (See Chap. 24 in this book.)
- *Glycogen Synthase Kinase-3 (GSK-3) Signaling*
- GSK-3 is a serine/threonine kinase central to many cellular processes such as metabolism, apoptosis, cell cycle progression, migration, differentiation, and embryogenesis. It interacts with multiple signaling pathways including PI3K/Akt, MAPK, Wnt/β-Catenin, Notch, and Hedgehog [8].
- ***Roles in disease.*** GSK-3 plays a role in several cancer types (e.g., breast, colorectal, pancreatic, and ovarian cancers and melanomas and glioblastomas) and is a target in Alzheimer's disease.
- GSK-3 can be therapeutically targeted by lithium and small molecule inhibitors (such as benzimidazoles and pyrimidines) and a potential target for miRNAs.

Signaling Molecules and Nuclear Receptors

- *Jak/STAT Signaling Pathway*
- The Janus kinase (JAK) family are non-receptor tyrosine kinases activated by cytokines. Cytokines phosphorylate the cell membrane cytokine receptors, causing binding and activation of the signal transducers and activators of transcription (STATs). STATs translocate to the nucleus where they regulate gene expression resulting in a wide range of biological effects that regulate T- and B-cell activities.
- ***Roles in disease.*** JAK/STATs play a role in numerous diseases including rheumatoid arthritis, colitis, and Crohn's disease, as well as in hematological malignancies such as leukemia and lymphoma and some solid tumors. JAKs are also targets for first- and second-generation small molecule inhibitors. A number of molecules targeting JAKS or STATS are in clinical trials such as sorafenib (STAT3 inhibitor in breast and thyroid cancer), WHI-P131, or WHI-P154 (JAK3 inhibitors in glioblastoma). (See Chap. 26 in this book.)
- *Wnt/β-Catenin Signaling Pathway*
- The Wnt/β-Catenin signaling pathway is important in normal cell growth and development. In the presence of Wnt, β-Catenin forms a complex with transcription factors to regulate gene expression. In the absence of Wnt, β-Catenin is phosphorylated and subsequently degraded by the proteasome.
- ***Roles in disease.*** Wnt/β-Catenin is involved in cancers such as medulloblastomas and ovarian and colorectal cancers. The most well-known genetic mutation in the pathway is in the *APC* gene resulting in familial adenomatous polyposis (FAP).
- Wnt/β-Catenin is a target for traditional compounds such as iron chelators and nonsteroidal anti-inflammatory drugs (NSAIDs). It is also a potential target for biological therapies (e.g., vantictumab) and small molecules (e.g., LGK974) as well as for natural inhibitors that degrade β-Catenin (e.g., flavonoids) [8]. (See Chap. 23 in this book.)
- *Notch Signaling Pathway*
- Notch is critical in many cellular processes and is activated in response to cell-cell interactions. Activation occurs through cleavage of Notch to form Notch intracellular domain (NCID) which is capable of nuclear translocation where it

regulates gene expression to control cell proliferation, survival, and differentiation [8].

- ***Roles in disease.*** Notch is involved in the development of gastrointestinal, gastric, colorectal, and pancreatic cancers. Notch is a target for gamma secretase inhibitors, a few which are in a clinical trial (e.g., RO4929097). (See Chap. 17 in this book.)
- *NF-κB Signaling Pathway*
- Nuclear factor kappa B (NF-κB) is a transcription factor that functions in a complex to regulate expression of genes involved in proliferation, apoptosis, inflammation, and immune responses. It is required at a low level for normal hematopoiesis [9].
- ***Roles in disease.*** NF-κB is implicated in leukemia (e.g., acute myeloid leukemia).
- NF-κB is a target for inhibitors, and some of its regulators such as IRAK1, TAK1, Bruton tyrosine kinase (BTK), and IKK are also considered potential targets (e.g., by PCI-32765/ibrutinib). (See Chap. 27 in this book.)
- *Nuclear Receptor Signaling*
- The retinoic acid-related orphan receptors (ROR α-γ or NR1F1-3), the orphan receptor TAK1 (TR4 or NR2C2), and the estrogen receptor (ER) are members of the nuclear receptor superfamily of ligand-dependent transcription factors. These receptors exhibit critical functions in regulating embryonic development and many other physiological processes and have been implicated in a variety of pathologies.
- ***Roles in disease.*** The RORs, TAK1/TR4, and ER have been implicated in a number of pathologies, including various cancers (e.g., breast cancer).
- The ROR, TAK1/TR4, and ER nuclear receptors are targets for endocrine disruptors and drug therapy (e.g., by tamoxifen). ER activity can also be indirectly targeted through inhibition of the aromatase enzyme (e.g., by letrozole, anastrozole). (See Chap. 13 in this book.)
- *Progesterone and Androgen Receptor Signaling*
- Like estrogen receptors, progesterone and androgen receptors are steroid hormone receptors. Progesterone and androgens (e.g., testosterone) bind to their respective receptors in the cytoplasm, initiating a conformational change and nuclear translocation. Once in the nucleus, the receptors predominantly function as DNA-binding transcriptional regulators.
- ***Roles in disease.*** The most notable examples of diseases involving these receptors are breast (progesterone) and prostate (androgen) cancers with the receptors being targets for drugs such as tamoxifen (progesterone receptors) and bicalutamide (testosterone receptors).
- *Aurora Kinases*
- Aurora kinases became a focus of interest over the last 20 years after they were discovered during screens for proteins involved in mitotic spindle dysfunction; their role is to regulate mitosis. They are located at the kinetochores, and their levels increase and decrease during the cell cycle, peaking between late S-phase and M-phase.
- ***Roles in disease***. All three human aurora kinases play roles in the development of both hematological malignancies and solid tumors (e.g., CML, AML, breast and colon cancer). They are targets for small molecule inhibitors such as danusertib and barasertib [10].

Common Signaling Components in Cancer

Most of the pathways discussed in section "Introduction to Cancer Cell Signaling" contribute to a more "active" cellular phenotype; therefore, they are all implicated in cancer development in some way. What is also clear is that several of these pathways contribute to the development of multiple cancer types and that few cancer types arise from dysregulation of only a single pathway. For example, breast cancer can arise due to elevated ER, EGFR/HER, or IGFR signaling, and, on many occasions, dysregulation of more than one of these pathways is involved.

Several of the cell membrane receptor families activate the same downstream intracellular pathways meaning there are common signaling components in the development of cancer. The MAPK cascade is activated by EGFR/HER,

FGFR, IGFR, VEGFR, PDGFR, and GPCR signaling (Fig. 12.2).

In the nucleus transcription of genes involved in cancer progression is increased; nuclear receptors are directly involved in mediating the transcriptional, whereas activation of the other cell signaling pathways results in phosphorylation of transcriptional activators (e.g., STATs) which in turn increases transcription.

This often means that there can be increased activity of MAPK signaling in the absence of specific genetic or expression abnormalities, purely because an upstream receptor is more active. This point is nicely demonstrated in non-small-cell lung carcinoma where, in 39 tumors with increased intracellular signaling due to activating mutations, 30% had mutations only in the EGFR/HER receptors and not in the Ras-Raf-MAPK cascade.

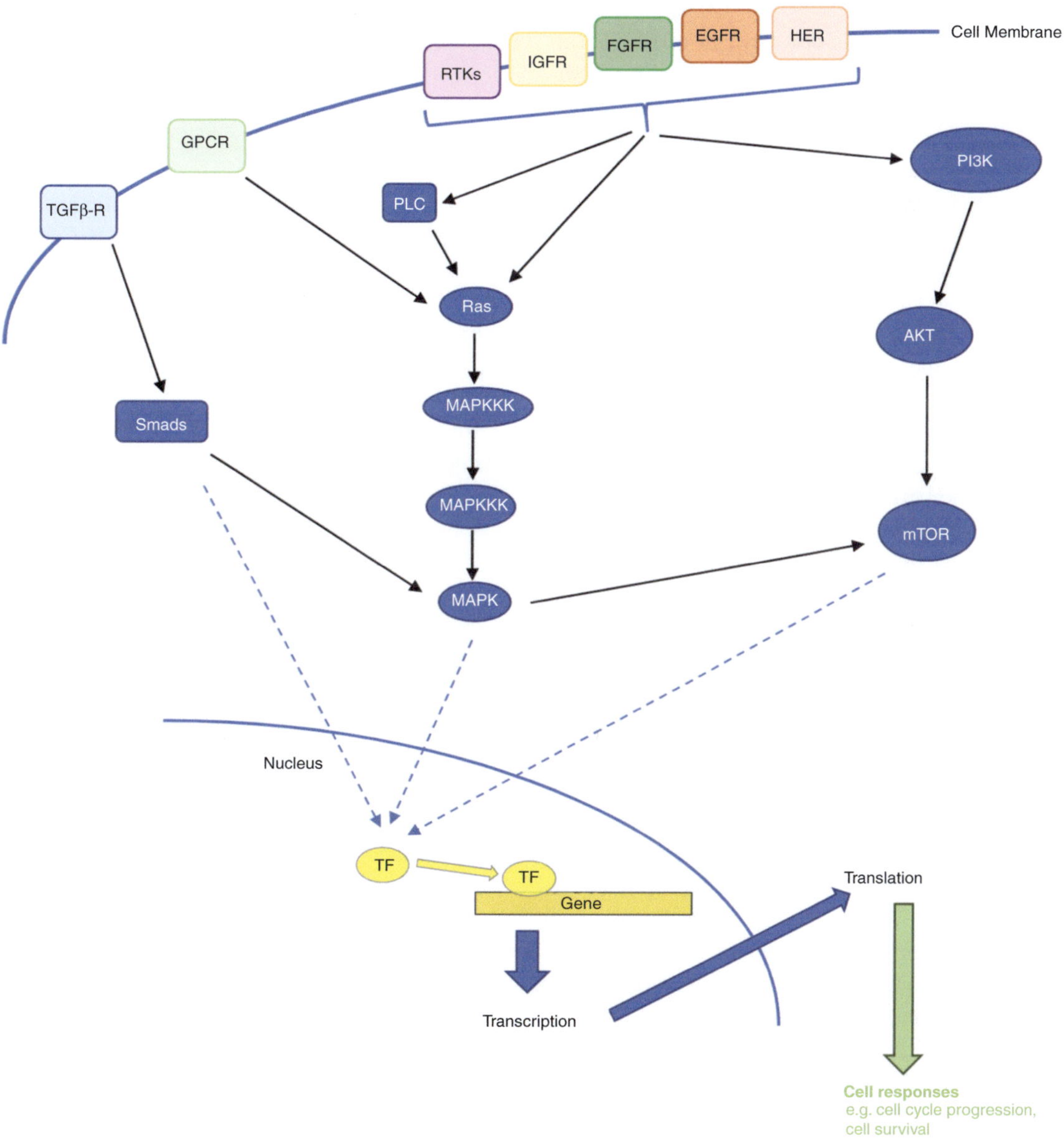

Fig. 12.2 Common signaling components in cancer. In response to increased signaling from cell surface receptors transcription of genes encoding of pro-survival proteins and positive regulators of cell cycle progression is increased resulting in the cell adopting a more cancerous phenotype. TF transcription factor

PI3K/Akt signaling has a long association with many types of cancer. Patients with Cowden's disease, characterized by PTEN mutations, have elevated PI3K/Akt signaling and are at a much-increased risk of developing cancers most notably, breast cancer. Seventy percent of breast cancers have gene mutations resulting in increased PI3K/Akt activity. PI3K/Akt signaling is crucial in tumor development as it links receptor signaling with downstream effects such as MAPK and mTOR.

mTOR signaling is downstream of PI3K/Akt, and therefore several upstream signaling pathways, including EGFR/HER, FGFR, and IGFR, converge at this focal point. mTOR signaling is often overactive as result of mutations in mTOR; however, in some cancers including breast cancer, activation of the EGFR/HER family of receptors and activating mutations in PI3K/Akt signaling also result in elevated mTOR activity [7].

JAK/STAT signaling tends to be more closely linked to the development of hematological malignancies, largely due to its involvement in cytokine signaling and the reliance of T and B cells on cytokines for their normal function. There is, however, a role for JAK/STAT signaling in the development of solid tumors as STAT5 can be activated by binding to EGFR and so could play a role in the signal cross talk (section "Cytoplasmic Signaling Molecules").

From a pharmacological perspective, activation of signaling pathways provides an opportunity for therapeutic intervention. The heterogeneity of signaling across cancers means drugs that are designed to inhibit specific signaling molecules have potential clinical benefit in more than one tumor type. The reality is, however, that some compounds are not as effective as predicted and this may well be due to the intricate balance of intracellular signaling required to maintain tumor growth that is potentially different to that required to establish initial tumor formation, development, and metastasis. For example, VEGF signaling plays a niche role in the development of solid tumors. The barrier to a microscopic tumor progressing to a larger mass is the requirement for oxygen, delivered by a blood supply. VEGF signaling is therefore critical early on in tumor development for neo-angiogenesis (formation of new blood vessels). Once a solid tumor is established, the reliance on VEGF signaling is likely to diminish; however, as a tumor becomes metastatic and cells disseminate to distant locations, VEGF signaling is once again required in the development of distant metastases. This potentially means that inhibition of VEGF signaling is maximal during the developmental stages or in treating tumor types where remodeling, and therefore angiogenesis, is a common occurrence.

Signaling Cross Talk

The commonality between the signaling pathways discussed in section "Introduction to Cancer Cell Signaling," and the fact that these provide many common signaling components in cancer development, also results in the biggest barrier to therapy, namely, signaling cross talk and compensatory signaling [11].

Signaling cross talk can occur via different mechanisms:

- A molecule in one pathway can affect the rate of activation of signaling molecules in a second pathway (signal flow cross talk).
- Two pathways can compete for common components (substrate availability cross talk).
- Receptors can have altered ability to detect ligands, or if receptors are overexpressed (as with HER2), signaling can happen in the absence of ligand (receptor function cross talk).
- Individual pathways could have opposing effects on transcription factor activation (gene expression cross talk).
- Ligand availability can be altered because of different mechanisms but often occurs in response to gene expression changes (intracellular communication cross talk).

The cross-talk mechanisms are not mutually exclusive and will often influence each other. For example, because signaling pathways converge at focal points, inhibiting one route to the focal point still allows signaling to that point to be rerouted via a different path and potentially free components to be activated via the second pathway (examples of signal flow and substrate availability cross talk).

Reducing PI3K/Akt or mTOR signaling, for example, through inhibition of membrane receptor activity, will initially achieve the desired outcome; however, overtime, tumor cells will adapt and find alternative mechanisms for increasing signaling. For example, if EGFR/HER is inhibited, more PI3K/Akt becomes available for IGFR signaling.

In the development of drug resistance EGFR/HER, inhibition could be mitigated through a compensatory increase in FGF, IGFR, or GPCR signaling, all of which would sustain elevated PI3K/Akt or mTOR activity. Indeed, IGF-1R signaling reduces the sensitivity of breast cancer cells to anti-HER2 monoclonal antibody therapy; sensitivity to trastuzumab is increased through inhibition of IGF-1R [11].

What is also perhaps most surprising is the promiscuity of receptors in drug resistant cells. It is easy to presume that receptors only dimerize with their designated partners and that they only signal within their discreet pathways. This is not always the case. Both IGF-1R/HER2 dimers and IGF-1R/HER2/HER3 trimers have been detected in trastuzumab resistant cells suggesting firstly that compensation for EGFR/HER signaling inhibition could be mediated through insulin-like growth factor signaling and, secondly, that there is a clinical rationale for combined EGFR/HER and IGF-1R targeting in tumors resistant to anti-HER2 or anti-EGFR therapy.

In addition to EGFR and IGFR, Wnt signaling also activates mTOR, where cross talk results in activation of both Notch and STAT signaling. Phosphorylation of EGFR/HER family receptors depends on the specific activating ligand. In some circumstances, phosphorylation of EGFR or HER4 will facilitate cross talk through STAT5 binding and activation which, under normal conditions, is an infrequent event; however, in breast cancer, STAT5b could contribute to an increased proliferative phenotype through enhanced transcriptional activation.

Although canonical TGF-β signaling occurs via the Smad proteins, there is signal flow and gene expression cross talk between TGF-β signaling and the MAPK pathways. MAPK signaling can activate expression of TGF-β target genes, and specific MAPK activity is central to breast cancer cell migration mediated by TGF-β [11].

Predictive Biomarkers and Therapeutic Targets

There is a very clear need for cancer biomarkers, both from a diagnostic and prognostic perspective. As our understanding of signaling has developed and the range of possible therapeutic options expands, it is vital to have reliable biomarkers that will predict which patients will benefit from specific treatment regimens. Many clinical trials now included evaluation of potential biomarkers as part of the study aims.

Unsurprisingly many prognostic biomarkers are also therapeutic targets, for example, the estrogen receptor (ER) predicts patient outcomes. Tumors lacking hormone receptors have worse outcomes, partly because triple-negative breast cancers are more aggressive in nature and less responsive to chemotherapy but also because the ER is itself a target for antihormone therapies such as tamoxifen.

Given the broad nature of cell signaling and the variety of signaling pathways outlined in section "Introduction to Cancer Cell Signaling," there are many potential biomarkers in cancer. The discussion in this section will focus on EGFR/HER signaling (Fig. 12.3), with other examples being illustrated in Table 12.1.

As discussed above in signaling cross talk, many patients do not respond to their targeted therapy, or they initially respond and then develop resistance. This is very evident in colorectal cancer, where patients with elevated EGFR signaling are offered anti-EGFR monoclonal antibody therapy. Elevated EGFR signaling in colorectal cancer can be categorized based on (i) increased upstream components, (ii) increased amount or aberrant EGFR, (iii) activation of downstream molecules, or (iv)

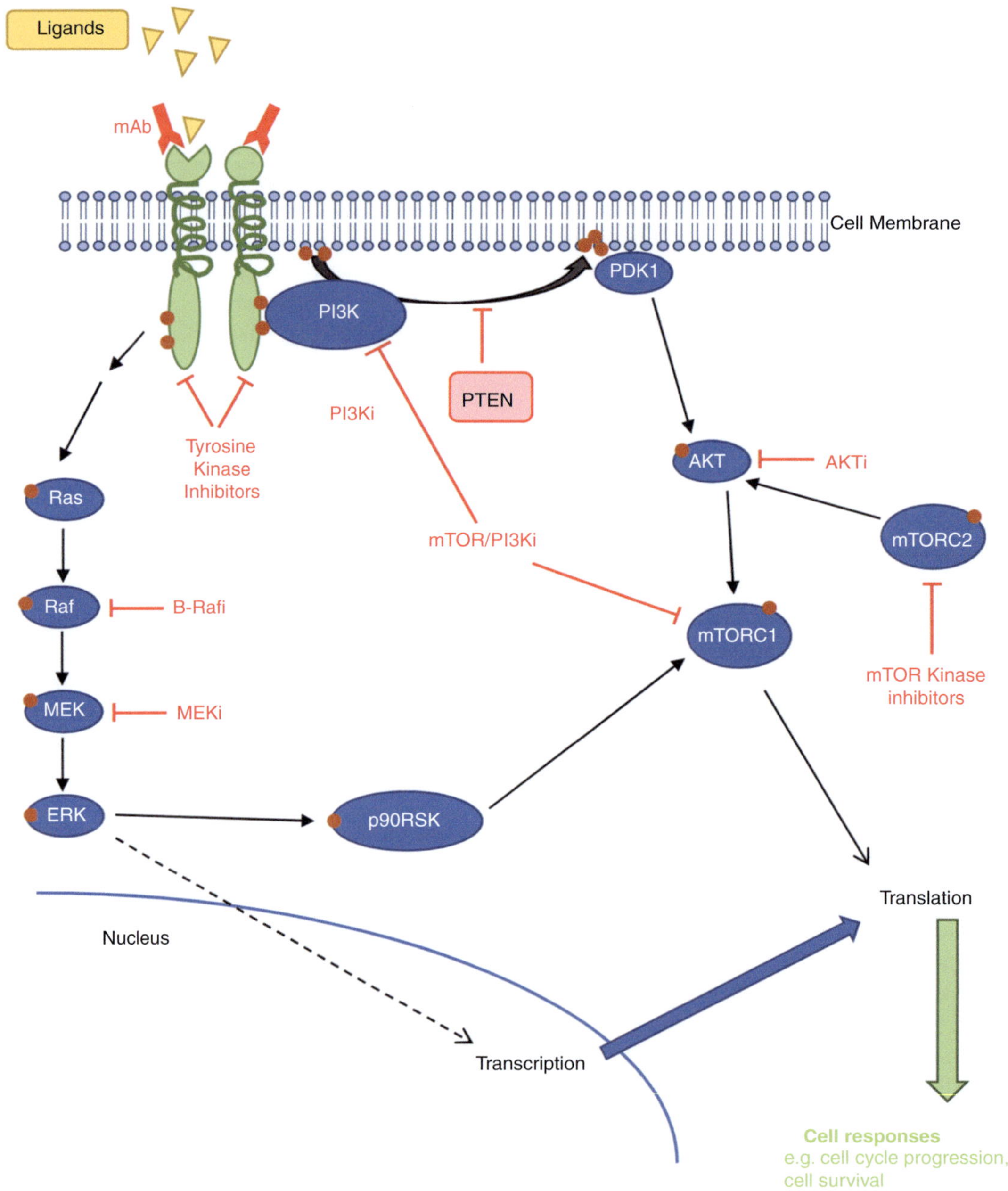

Fig. 12.3 Examples of biomarkers and therapeutic targets in the EGFR/HER signaling pathways. In response to increased Her2 signaling, transcription of genes encoding of pro-survival proteins and positive regulators of cell cycle progression is increased, resulting in the cell adopting a more cancerous phenotype in response to transcription of pro-survival genes. When EGFR/HER signaling is inhibited by some of the compounds listed in the above figure, the increase in transcription is ablated with a downregulation of the biological response. (Adapted from Montemurro and Scaltriti [12]. With permission from John Wiley & Sons)

activation of alternative bypass pathways. Only patients with tumors categorized in (i) or (ii) will respond to anti-EGFR monoclonal antibody therapy, so it is vital to have biomarkers that are predictive of response or resistance to treatment. So far, Ras has proved to be the most useful biomarker to predict resistance to anti-EGFR monoclonal antibodies in colorectal cancers as

Table 12.1 Examples of cancer biomarkers and therapeutic targets and their relationship to the hallmarks of cancer identified by Hanahan and Weinberg

Hallmarks of cancer	Signaling pathways	Example of biomarkers	Example of a major therapeutic target in signaling
Sustaining proliferative signaling	EGFR/HER	Breast cancer:	
	IGFR	ER	ER
	PKC	PR	
	MAPK	HER2	HER2
		p95HER2	
		IGF-1R/IRS-1	
		EREG (CRC)	
		IRS1 (BC)	
		IGF2 (CRC)	
		PTEN (BC)	
Activating invasion and metastasis	PKC	TGFα (CRC)	EGFR
	MAPK	TGFα/amphiregulin (NSCLC)	
	EGFR/HER		
	IGFR		
	TGF-β		
Evading growth suppressors	EGFR/HER	PTEN (BC)	EGFR
	MAPK		
Resisting cell death	IGFR	IGF2 (CRC)	EGFR
	EGFR/HER	PTEN (BC)	
Inducing angiogenesis	VEGF	VEGF	VEGFR
	EGFR/HER	EREG (CRC)	
	Ras		
Enabling replicative immortality	B-catenin	Telomerase length [13]	

Based on data from Ref. [1]
BC breast cancer, *CRC* colorectal carcinoma, *NSCLC* non-small-cell lung carcinoma

Ras mutations are linked to resistance to anti-EGFR therapy in colorectal cancer [14].

Epiregulin (EREG) is an EGFR ligand that is initially released as a transmembrane precursor. It regulates angiogenesis, and cell proliferation and increased levels are associated with a more aggressive tumor phenotype. Colorectal carcinoma patients with wild-type Ras tumors and high EREG gene expression have better outcomes in response to anti-EGFR therapy (cetuximab), with and without chemotherapy than those with low EREG expression. When serum levels of EREG were considered, the reverse was noted; both overall and progression-free survival times were shorter in patients with higher EREG levels than those with low. These inconsistencies are not surprising given the lack of correlation between protein levels and gene expression, but it does highlight the difficulties in identifying reliable prognostic biomarkers. More recent studies have indicated that BRAF mutations are more likely to serve as independent prognostic factors.

TGFα activates the EGFR, stimulating the MAPK pathway resulting in increased proliferation invasion and metastasis in both colorectal carcinoma and breast cancer patients. High tumor levels of TGFα are linked with resistance to anti-EGFR antibodies in colorectal carcinoma patients. In breast cancer, high TGFα expression is linked with poorer outcomes and resistance to chemotherapy, while high serum levels correlate with a more aggressive tumor in non-small-cell lung carcinoma (NSCLC). This illustrates that the same biomarker has potential in different tumor types, but it needs to be measured differently between the types; in some cases, tumor levels are required; in others it is serum levels that matter. In NSCLC, EGFR mutations are indicative of response to kinase inhibitors rather than absolute levels of EGFR.

In breast cancer, elevated HER2 is indicative of prognosis, and relative HER2 and HER3 levels are predictive of patient responses to trastuzumab and pertuzumab, respectively. This highlights that the complexities of signaling and receptor dimerization need to be considered alongside overexpression and mutations when considering biomarkers, therapeutic targets, and patient responses. A proportion of HER2 positive tumors also express a shorter form of HER2 (p95HER2). It lacks the extracellular domain, meaning it has no trastuzumab binding site and is hyperactive and very tumorigenic. In metastatic breast cancer, p95HER2 expression correlates with intrinsic resistance to trastuzumab [12].

Other potential biomarkers in breast cancer are linked to IGF-1R; however, measuring levels of IGF-1R alone is not enough to select breast tumors that maybe sensitive to IGF signaling inhibition. It is the combined levels of IGF-1R and IRS-1 that maybe more informative especially as IRS-1 is associated with reduced disease-free survival in breast cancers.

New Signaling Pathways and Future Strategies

When signaling molecule inhibitors were first developed, many lacked specificity and exhibited a variety of cross-reactivity. For this reason, they were not considered suitable for clinical use, and researchers were skeptical about their value in in vitro preclinical studies as it was difficult to determine whether data generated was a result of a desired inhibitory effect or as an artifact of an off-target. It is clear that single targeting has clinical benefit; however, it is also evident that cross talk and compensatory signaling result in therapeutic resistance such that targeting of sole signaling molecules might not be a fruitful long-term treatment strategy. There are several clinical trials examining the combinatorial effects of multiple inhibitors, and current thinking is that combined targeting strategies are likely to be the most successful for long-term patient survival.

In addition to multiple targeting, targeting adaptor molecules that link receptors to downstream effectors and signaling focal points are likely to have the most impact. To that end, multiple mTOR and dual mTOR/PI3K inhibitors are either undergoing clinical trials or are already in clinical use. Moreover, it is worth revisiting previous avenues that had previously been disregarded. The adaptor tyrosine kinases of the Src family were once perceived as potential drug targets. However, the amino acid homology between family members meant designing specific inhibitors was difficult and, when Src was inhibited, lack of activity was compensated for by signaling via other family members. A broad-spectrum approach to kinase inhibitor design could ameliorate these issues.

There is also scope for novel drug targets to be identified, and some, such as Brk/PTK6, may prove to be of therapeutic value as part of a combined therapeutic strategy especially in tumors for which there is currently no other viable signaling target (e.g., triple-negative breast cancers) [15].

So far, this chapter has largely focused on intracellular signaling and cross talk. To develop novel, more effective anticancer treatments, the effects of the tumor microenvironment and its interaction with tumor cells must be taken into consideration. The "seed and soil hypothesis" is not new, and it has long been known that certain tumor cell types "prefer" to colonize specific extracellular environments to form metastases.

To colonize the microenvironment, cancer cells must be attached to the extracellular matrix (ECM) and signal to the cells within it such as macrophages and fibroblasts which then become associated with the tumor and are referred to as tumor-associated macrophages (TAMs) and cancer-associated fibroblasts (CAFs). Expression of factors that regulate the ECM can promote tumor formation; in addition, many factors within the ECM can enhance the ability of tumor cells to be invasive and remodel the microenvironment through a process termed epithelial to mesenchymal transition (EMT). Understanding the interplay between the microenvironment and tumor cells is critical in developing novel therapies. Although enhanced CAF activity by tumor secreted growth factors is well documented, it is still not clear what initiates CAF activation [16].

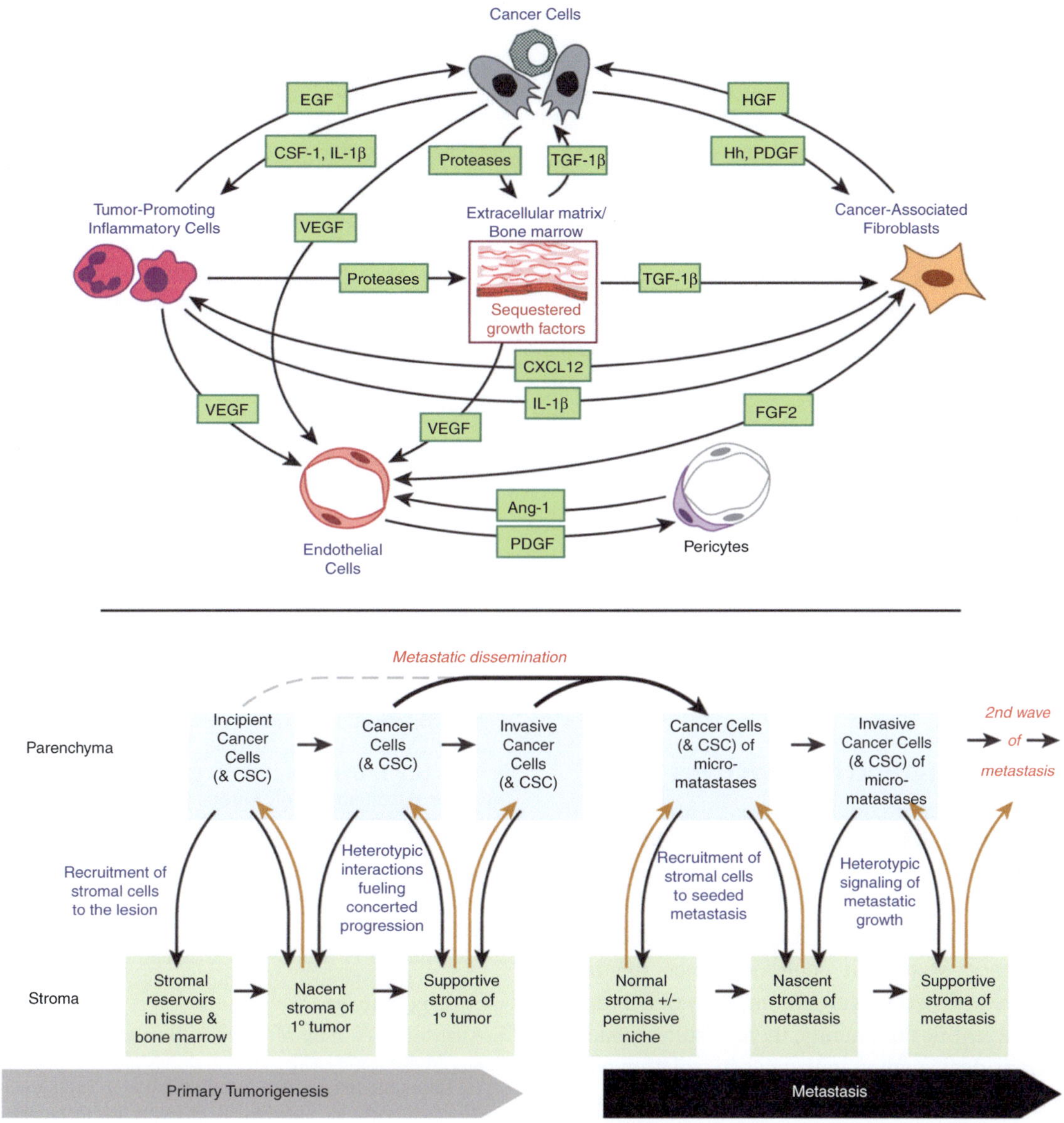

Fig. 12.4 Signaling interactions in the tumor microenvironment during malignant progression. (Upper) The assembly and collective contributions of the assorted cell types constituting the tumor microenvironment are orchestrated and maintained by reciprocal heterotypic signaling interactions, of which only a few are illustrated. (Lower) The intracellular signaling depicted in the upper panel within the tumor microenvironment is not static but instead changes during tumor progression as a result of reciprocal signaling interactions between cancer cells of the parenchyma and stromal cells that convey the increasingly aggressive phenotypes that underlie growth, invasion, and metastatic dissemination. Importantly, the predisposition to spawn metastatic lesions can begin early, being influenced by the differentiation program of the normal cell of origin or by initiating oncogenic lesions. Certain organ sites (sometimes referred to as “fertile soil” or “metastatic niches”) can be especially permissive for metastatic seeding and colonization by certain types of cancer cells,as a consequence of local properties that are either intrinsic to the normal tissue or induced at a distance by systemic actions of primary tumors. Cancer stem cells may be variably involved in some or all of the different stages of primary tumorigenesis and metastasis. (Reprinted from Hanahan and Wienberg [1]. With permission from Elsevier)

The link between tumor cells and the microenvironment could be mediated through NF-κB which, in addition to its role in proliferation and apoptotic control, can regulate the expression of pro-inflammatory cytokines that will initiate signaling required for ECM remodeling, thereby promoting tumor progression. Targeting the production of such cytokines could have enhanced clinical benefit in comparison to focusing solely on the tumor cells.

As a result of their interaction with the microenvironment, tumor cells are also capable of evading detection by the immune system. Immunotherapeutics is being developed to reactivate the immune system to recognize and destroy tumor cells. Products such as Sipuleucel-T, a therapeutic immunovaccine, and ipilimumab, a monoclonal antibody, both have FDA approval. At a cost of over $100,000 per individual treatment course, identifying patients who are most likely to benefit is crucial (Fig. 12.4).

Conclusions and Perspectives

There is no doubt that the wealth of knowledge relating to cell signaling in cancer has vastly improved in last 20 years. More is known about cross talk and how this could contribute to drug resistance or how it could influence treatment options and therapeutic combinations of the future. As a scientific community, there is still a tendency to consider signaling molecules in isolation and to teach students about individual pathways, largely for simplicity. There is a need to be much more aware of intracellular signaling networks and the cross talk between pathways, as well as the extracellular cross talk if the gains of the last two decades are to be continued in the next 20 years.

References

1. Hanahan D, Wienberg RA. Hallmarks of cancer: the next generation. Cell. 2011;144(5):646–74.
2. Cruickshanks N, Zhang Y, Yuan F, Pahuski M, Gibert M, Abdounder R. Role and therapeutic targeting of the HGF/MET pathway in glioblastoma. Cancer. 2017;9:87–102.
3. Westermark B. Platelet-derived growth factor in glioblastoma—driver or biomarker? Ups J Med Sci. 2014;119:298–305.
4. Slebioda TJ, Kmiec Z. Tumour necrosis factor superfamily members in the pathogenesis of inflammatory bowel disease. Mediat Inflamm. 2014;2014:325129.
5. Keppler-Noreuil KM, Parker VER, Darling TN, Martinez-Agosoto JA. Somatic growth disorders of the PI3K/AKT/mTOR pathways and therapeutic strategies. Am J Med Genet. 2016;172C:402–21.
6. Hare SH, Harvey AJ. mTOR function and therapeutic targeting in breast cancer. Am J Cancer Res. 2017;7(3):383–404.
7. Cristea S, Sage J. Is the canonical RAF/MEK/ERK signaling oath way a therapeutic target in SCLC. J Thorac Oncol. 2016;11(8):1233–41.
8. McCubrey JA, Rakus D, Gizak A, Steelman LS, Abrams SL, Lertpiriyapong K, et al. Effects of mutations in the Wnt/β-catenin, hedgehog, notch and PI3K pathways on GsK-3 activity – diverse effects on cell growth, metabolism and cancer. Biochim Biophys Acta. 2016;1863:2942–76.
9. Bosman MCJ, Schuringa JJ, Vellenga E. Constitutive NK-κB activation in AML: causes and treatment strategies. Crit Rev Oncol. 2016;98:35–44.
10. Choudary I, Barr PM, Friedberg J. Recent advances in the development of Aurora kinases inhibitors in hematological malignancies. Ther Adv Hematol. 2015;6(6):282–94.
11. Harvey AJ. Signalling cross talk. In: Harvey AJ, editor. Cancer cell signalling. Chichester: Wiley Blackwell; 2013. p. 193–206.
12. Montemurro F, Scaltriti M. Biomarkers of rugs targeting HER-family signaling in cancer. J Pathol. 2014;232:219–29.
13. Augustine TA, Baig M, Sood A, Budagov T, Atzmon G, Mariadason JM, Aparo S, Maitra R, Goel S. Telomere length is a novel predictive biomarker of sensitivity to anti-EGFR therapy in metastatic colorectal cancer. Br J Cancer. 2015;112:313–8.
14. Yang J, Li S, Wang B, Wu Y, Chen Z, Lv M, et al. Potential biomarkers for anti-EGFR therapy in metastatic colorectal cancer. Tumour Biol. 2016;37:11645–55.
15. Hussain HA, Harvey AJ. Evolution of breast cancer therapeutics: breast tumour kinase's role in breast cancer and hope for breast tumour kinase targeted therapy. World J Clin Oncol. 2014;5(3):299–310.
16. Martin M, Wei H, Tao L. Targeting the microenvironment in cancer therapeutics. Oncotarget. 2016;7(32):52272–583.

Steroid Hormone and Nuclear Receptor Signaling Pathways

13

Sunil Badve

Introduction

The role of steroid receptors in cancer has been well recognized for more than 100 years. This is partly due to the understanding of the importance of estrogen receptor in breast cancer and androgen receptor in prostate cancer. Experimental and molecular studies have contributed significantly to the understanding of the key players in this system. However, there are still large gaps in our knowledge. There are at least 48 steroid hormone and nuclear receptors (NRs) described in humans [1]. The major players, particularly with regard to cancer, include estrogen receptor (ERα and ERβ), progesterone receptor (PR), androgen receptor (AR), glucocorticoid receptor (GR), and related nuclear receptors such as retinoic acid receptors and vitamin D receptors. Apart from the steroid receptors that are well-known for their role in cancer, a number of other receptors such as mineralocorticoid receptors and thyroid receptors play a key role in metabolic responses of the body and could indirectly affect the behavior of cancer cells. The role of peroxisome proliferator-activated receptor (PPAR) is not only well studied in metabolic disorders such as diabetes but also in cancers. A quick search for the term "PPAR and cancer" identifies around 4000 articles in PubMed. A detailed discussion of all the members of this nuclear receptor family is beyond the scope of this chapter, and the discussion herein will be limited to a handful of key players that have established predictive roles in cancer.

S. Badve
Department of Pathology and Lab Medicine, Indiana University School of Medicine,
Indianapolis, IN, USA
e-mail: sbadve@iupui.edu

Steroid Hormone Receptors

All steroid hormones are derived from the same precursor, cholesterol, and many are initially secreted by the adrenal cortex and/or gonads (i.e., ovaries and testes) and diffuse into the bloodstream [2]. As they are lipid soluble, steroid hormones can freely diffuse through cellular membranes and bind to steroid hormone receptors in their target tissues and organs, where they exert a wide range of biological functions including cell homeostasis, differentiation, and regulation of proliferation, survival, and cell death [2]. In addition, they share amino acid homology and a common structure (Fig. 13.1) containing (1) amino (N)-terminal domain, (2) DNA-binding domain (DBD), and (3) hormone-/ligand-binding domain (LBD).

Within this broad outline, there are significant variations in the structure of these receptors and their splice variants that affect the functionality of the receptors. The functionality is broadly classified as "genomic pathway," which is involved in

S. Badve, G. L. Kumar (eds.), *Predictive Biomarkers in Oncology*,
https://doi.org/10.1007/978-3-319-95228-4_13

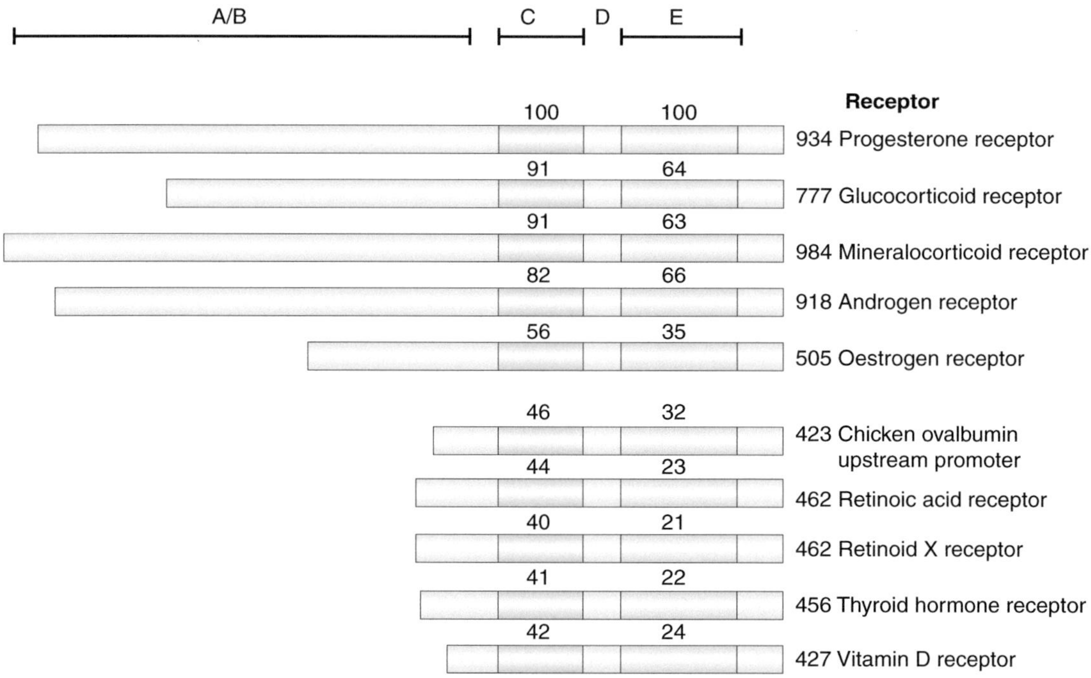

Fig. 13.1 Common structural framework of steroid nuclear receptors. Relative lengths of several members of the steroid/ nuclear hormone receptor superfamily. Variability between members of the steroid hormone receptor family primarily results from differences in the length and amino acid sequence of the amino (N)-terminal domain [3]. (Reprinted from Tata [45]. With permission from Springer Nature)

the binding of the activated receptor to the DNA, and "non-genomic pathway" that involves interaction with other predominantly cytoplasmic proteins [4, 5] (Fig. 13.2). The genomic pathway consists of the activation of the receptor by ligand binding and dimerization followed by translocation to the nucleus and directly binding to "hormone response elements" (HREs) on the DNA or indirectly through other transcriptional factors to regulate gene expression. This function can be modified by a host of factors including co-activators and co-repressors. A number of the co-regulators have been well characterized and belong to the steroid receptor co-activator (SRC) family, steroid receptor RNA activator (SRA), androgen receptor-associated proteins (ARAs), and the PIAS (protein inhibitor of activated signal transducer and activator of transcription) family [6]. In addition, other factors, termed "pioneer transcription factors," determine the occupancy of the receptor complex on the HRE. The mostly well described of these "pioneer transcription factors" is FOXA1 for estrogen receptor; whose expression has been well documented to be prognostic in breast cancer [7]. Currently, the non-genomic pathway actions are not well understood and consist of a variety of rapid intracellular signaling cascades that affect key cellular processes such as metabolism, proliferation, and apoptosis (Fig. 13.2). A number of posttranslational modifications of the receptor or co-factors have been described; these can modify the functionality of the pathway by altering the expression, protein stability, nuclear localization, hormone sensitivity, DNA binding, protein-protein interactions, and transcriptional activity [8, 9].

Estrogen Receptor (ESR1) and Its Signaling

The structure of the estrogen receptor (ESR1) follows the general architecture of the steroid receptor family with well-defined domains that

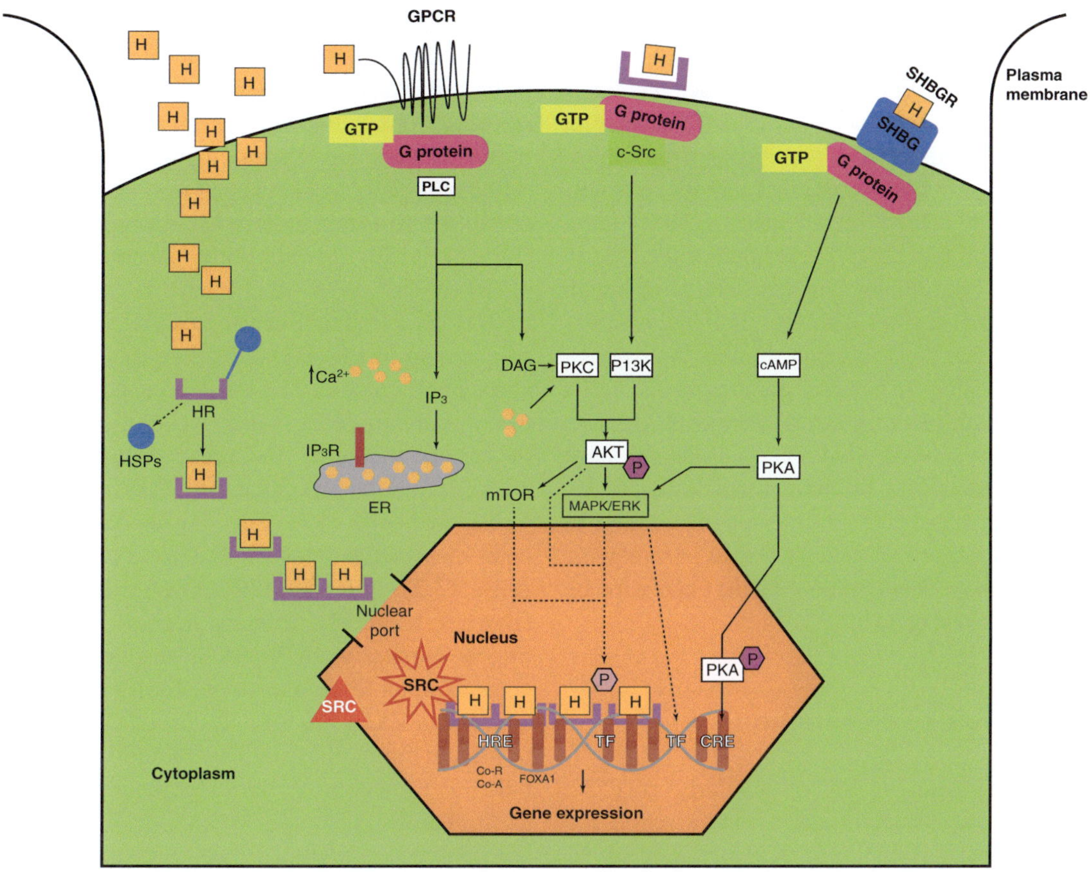

Fig. 13.2 Nuclear steroid signaling. Steroid hormones (H) work in conjunction with hormone receptors (HR). Hormone receptors for glucocorticoids (GCs) and androgen (A) are primarily in the cytoplasm as monomers bound to heat shock proteins (HSPs). Others, such as the estrogen receptors are located as monomers primarily in the nucleus, although a small percentage may also be bound to HSPs in the cytoplasm. In the case of GC and A, steroid binding to cytoplasmic receptors triggers release from the HSPs, receptor dimerization, alterations in receptor conformation, and nuclear localization. Estrogen binds to its nuclear receptors to promote dimerization and changes in receptor conformation. Dimerized receptors then bind to specific hormone response elements (HREs) and interact with various co-regulators to modulate gene transcription through either repression or activation. Nuclear steroid receptors can also modulate gene expression without direct DNA binding. In this case, they bind to other transcription factors (TFs) to either repress or activate transcription. Hormones can also bind G protein receptors and steroid hormone-binding globulin and get transported into the cell. These pathways result in activation of phosphoinositol 3-kinase (PI3K), protein kinase C (PKC), and cyclic adenosine monophosphate (cAMP) and downstream direct or indirect actions resulting in altered transcription factors (TFs) and/or cAMP response elements (CREs)

bind to specific regions of the DNA or to the ligand [10] (Fig. 13.1). Classical estrogen receptor (ERα or hERα-66) contains an amino-terminal region (AF-1), a central DNA-binding domain (DBD), and a carboxy-terminal hormone-binding domain (HBD). AF-1 domain function is ligand-independent, whereas AF-2 contains the ligand-dependent activation function. Binding of hormone to ERα facilitates "classical" genomic activities of the receptor, and it's binding to estrogen response elements (ERE) in target genes results in activation or repression of gene expression. Therefore, any mutations in these critical domains may alter the function of the ESR1 and its downstream signaling.

The importance of the domains in cancer lies in the differential response to tamoxifen, which might exert an agonist activity at AF-1 but inhibit

AF-2 domain. The ligand-binding domain consists of 12 helices arranged in an anti-parallel sandwich formation in which α-helical elements are linked by short loops [11–13]. The structure of helix 12 is thought to be critical for the ligand-dependent AF-2 transcriptional activity and interaction with other proteins such as SRC kinase [14, 15]. Mutations in this residue of this region have also been shown to disrupt interactions with co-activators such as RIP140, TIF1, and mSUG1 [16–18]. The DNA-binding domain is responsible for binding to specific regions in the DNA termed estrogen-responsive elements (EREs). These consist of inverted repeats of the sequence GGTCA separated by three variable bases [19, 20]. Interestingly, EREs are not only found in promoter regions of genes but also in intergenic regions; these could influence gene expression by epigenetic mechanisms.

Regulation of ER Function

Phosphorylation of ER, particularly at T537, seems to play an important role in the regulation of function. Mutations at this locus have been identified in metastatic carcinomas and thought to be associated with ligand-independent activity. Mutations in ER were thought to be uncommon in the days prior to next-generation sequencing but have now been well described in endocrine resistant recurrent/metastatic tumors (Fig. 13.3) [21–23]. Similarly, phosphorylation of S118 and S167 might have a significant impact on ER function [24] and an important mechanism of ER activation, particularly in the absence of a ligand. The binding of the receptor to downstream EREs is influenced by a number of factors including the presence of pioneer transcription factors such as FOXA1, PBX1, TLEs, and AP2 [25]. The impact of this has been classically demonstrated for FOXA1 [7] (Fig. 13.2). Silencing of FOXA1 has been shown to result in a dramatic decrease in levels of gene expression in ER+ cells in spite of exposure to similar levels of estrogen [26]. Alternative splicing of ESR1 has been described to give rise to variant forms that can have altered structure (such as truncation of the C-terminal function) and function.

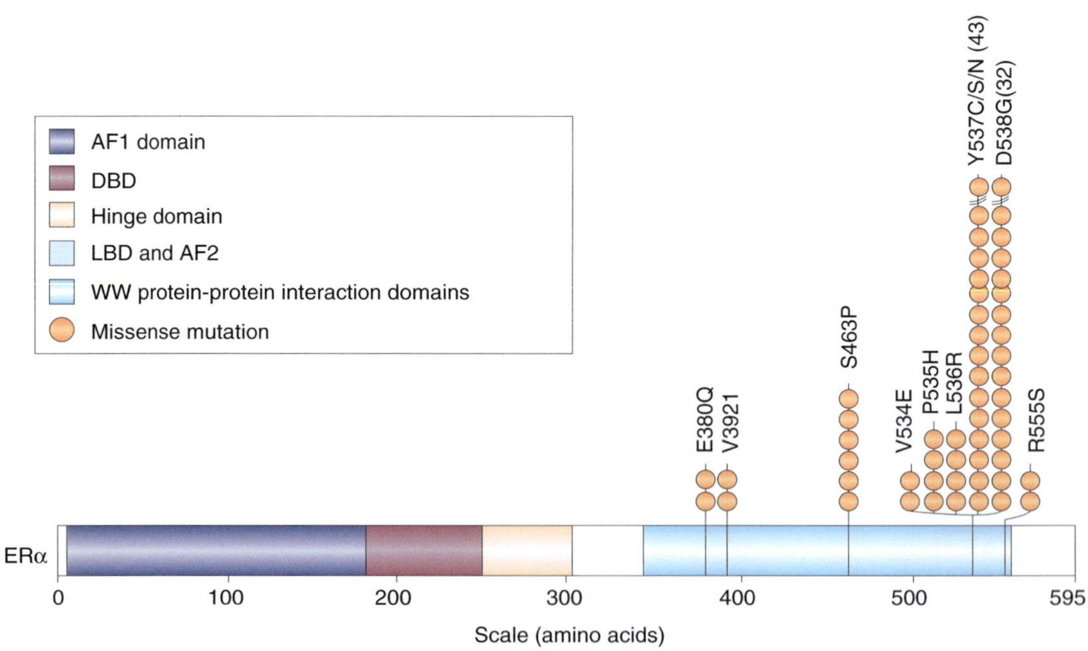

Fig. 13.3 Location and frequency of mutations in ER in breast cancers. (Adapted from Ma et al. [21]. With permission from Springer Nature)

Therapeutic Relevance

Targeting of ER has been in clinical use for decades using either agents that affect receptor activity or by decreasing estrogen synthesis. Selective estrogen receptor modulators (SERMs, e.g., tamoxifen) and selective estrogen receptor downregulators (SERDs, e.g., fulvestrant) have been in clinical use for decades and are very effective in controlling breast cancer. Numerous trials have documented the efficacy of tamoxifen in the treatment of ER+ breast cancer (Table 13.1). Tamoxifen has been used as a sole agent or in combination with (adjuvant or neo-adjuvant) chemotherapy. It is particularly of value in low-risk, low-grade ER-positive tumors. It has also been used in prevention trials to decrease the development of invasive carcinoma in women at high risk of cancer. The duration of therapy (in ATLAS (Adjuvant Tamoxifen, Longer Against Shorter) and aTTOM (adjuvant Tamoxifen—To offer more?) clinical trials have shown that increased duration of therapy (10 years) is associated with a decrease in the incidence of recurrence. Tamoxifen is metabolized to 4OH-tamoxifen by CYP 2D6 enzymes [27]. Concerns about the metabolism of tamoxifen by CYP 2D6 variants have resulted in increased use of toremifene, another SERM. Tamoxifen is not without toxicity; the patients are at increased risk for development of venous thrombosis and gynecological cancers in addition to sexual symptoms such as dryness and dyspareunia. Together, these result in high drop-out rates, which might be as high as 30–40%. Recent studies have documented significant efficacy for 500 mg fulvestrant, which had been previously used at a lower dose (250 mg). This has resulted in being used with increased frequency at some centers. The ATAC (Arimidex, Tamoxifen Alone or in Combination) clinical trial documented the superiority of aromatase inhibitors (AIs), which prevent estrogen synthesis in adipose tissues, over tamoxifen [28]. This has resulted in them being preferred agents, particularly in postmenopausal patients. The toxicity profile of AIs includes osteoporosis and joint pains, the latter being a major cause for discontinuation of therapy. Both steroidal and non-steroidal AIs are in clinical use, and resistance to one does not preclude the use of the other. Pre-receptor mechanisms such as use of gonadotrophin agonists (e.g., goserelin) have been successfully employed in Suppression of Ovarian Function Trial (SOFT) and Tamoxifen and Exemestane Trial (TEXT) clinical trials [29]. Similarly, post-receptor mechanisms of control of ER signaling have also been explored. The combination of AIs with everolimus, a mTOR inhibitor, has been documented to be effective in the treatment of breast cancer (BOLERO-2 trial [30]).

Estrogen Receptor Beta (ESR2)

The discovery of ER-beta in 1996 [31] significantly changed the thinking regarding the role of ER. Although smaller in size as compared to ERα, it has a similar structural organization with AF-1 and AF-2 domains that have ligand-independent and ligand-dependent activities. A number of splice variants of ERβ have been described; these have contributed to the increased difficulty in understanding the exact function of ERβ. Its function is also influenced by co-activators and co-repressors. ERβ has a major role in the immune, cardiovascular, and nervous system and in the prostate. It is thought to oppose the action of ERα and function as a tumor suppressor [32]. Its expression is lost in early stages of ductal breast cancer (DCIS) and in low Gleason score prostate cancer [33, 34]. Of note, it is expressed in lobular carcinomas and in 20% of TNBCs. Much of the current interest in ERβ is focused on its upregulation in prostate and breast cancers, particularly TNBCs, as a modality of cancer prevention and treatment [35]. A number of natural and synthetic agonist and antagonists of ERβ are available; they vary in their degree of specificity in modulation ERβ or both ERs.

Progesterone Receptor Signaling

The structure of progesterone receptor follows the general outline of the nuclear receptor family and exhibits both genomic and non-genomic

Table 13.1 Steroid receptor directed therapy

Disease	Target	Drug mechanism	Drug	Company	Clinical trials	Details
BRCA						
Breast cancer	Estrogen receptors	Estrogen receptor antagonists	Tamoxifen (Nolvadex®)	AstraZeneca	Standard of care	Combination with variety of other therapies in clinical trials
			Raloxifen (Evista ®)	Eli Lilly	Multiple trials	STAR trial; NCT00019500; in addition to other trials
			Toremifene (Fareston ®)	Kyowa Kirin, Inc.	Multiple trials	Used to circurmvent the impact of SNPs on tamoxifen metabolism
		Selective estrogen receptor downregulator (SERD)	Fulvestrant (Faslodex ®)	AstraZeneca	Standard of care	Combination with variety of other therapies in clinical trials
			Elacestrant (RAD1901)	Radius	Early trials	ClinicalTrials.gov Identifier:NCT01479946
	Pre-receptor targets	Aromatase imhibitors	Exemastane (Aromasin ®)	Pfizer Inc.	Standard of care	Combination with variety of other therapies in clinical trials
			Atamestane	Intarcia therapeutics	Multiple clinical trials. ClinicalTrials.gov Identifier: NCT00044291	In combination with other anti-estrogenic agents
			Anastrazole (Arimidex®)	AstraZeneca and ANI Pharmaceuticals, Inc	Multiple trials	Combination with variety of other therapies in clinical trials
			Letrozole (Femara®)	Novartis	Multiple trials	Combination with variety of other therapies in clinical trials
		GnRH agonists	Leuprorelin (LupronDepot®)	Abbvie	Multiple trials	In combination with variety of other therapies
			Goserelin (Zoladex®)	AstraZeneca and TerSera therapeutics LLC	Multiple trials	SOFT, TEXT clinical trials and in combination with variety of other therapies
	Androgen receptors	Androgen receptor antagonists	Bicalutamide (Casodex®)	AstraZeneca	Early trials for AR+ cancers	ClinicalTrials.gov Identifier:NCT03055312
			Enzalutamide (Xtandi®)	Astellas Pharma Inc. and Pfizer Oncology	Multiple trials	Combination with variety of other therapies in clinical trials
		Selective androgen receptor downregulator (SARD)	Enobosarm (Ostarine)	GTX, Inc	Phase I/II	NCT02971761

	Pre-receptor targets	Androgen synthesis inhibitors	Abiraterone acetate (Zytiga®)	Janssen Oncology	Multiple trials	ClinicalTrials.gov Identifier:NCT01842321; NCT00755885
		Anti-AR Oligo	AZD3512	AstraZeneca	Phase 1	ClinicalTrials.gov Identifier:NCT02144051
PRAD						
Prostate cancer	Androgen receptors	Androgen receptor antagonists	Bicalutamide (Casodex®)	AstraZeneca	Standard of care	In combination with variety of other therapies
			Nilutamide (Anandron®)	Sanofi-Aventis	Multiple trials	In combination with dasatinib, e.g., ClinicalTrials.gov Identifier:NCT00918385
			Hydroxyflutamide	Generic	Multiple trials	For example, NCT02341404
			Flutamide (Eulexin)	Generic	Multiple trials	In combination with variety of other therapies
			Enzalutamide (Xtandi®)	Astellas/Pfizer	Standard of care	In combination with variety of other therapies
		N-terminal domain antagonists	Ralaniten acetate (EPI-506)	ESSA Pharmaceutical Corp.	Study terminated by its developer in favor of next-generation androgen receptor N-terminal domain (AR-NTD) inhibitors with improved potency and tolerability	NCT02606123
		Selective androgen receptor downregulator (SARD)	Enobosarm (Ostarine)	GTx, Inc	Multiple trials	http://www.gtxinc.com/science/
			AZD3514	AstraZeneca	Multiple trials	In combination with variety of other therapies ClinicalTrials.gov Identifier: NCT01162395 ClinicalTrials.gov Identifier: NCT01351688
	Pre-receptor targets	Androgen synthesis inhibitors	Aminoglutethimide	Generic	Multiple trials	ClinicalTrials.gov Identifier: NCT00006371
			Abiraterone acetate (Zytiga®)	Janssen Oncology	Standard of care	In combination with variety of other therapies

(continued)

Table 13.1 (continued)

Disease	Target	Drug mechanism	Drug	Company	Clinical trials	Details
			Seviteronel (IN)-464)	Viamet Pharmaceuticals and Innocrin Pharmaceuticals	Multiple trials	ClinicalTrials.gov Identifier: NCT02445976
		5α-reductase inhibitors	Finasteride (Proscar®)	Merck	Multiple trials	ClinicalTrials.gov Identifier: NCT01342367; NCT00003323; NCT01296672
			Epristeride (Apuliete and Chuanliu)	Generic		Introduced for the treatment of enlarged prostate in China in 2000
			Alfatradiol (Avicis)	Generic		Used as a topical medication in the treatment of androgenic alopecia (pattern hair loss) in men and women
			Dutasteride (Avodart™)	Developed by GSK	Multiple trials around the world	Discountinued by GSK for treatment of prostrate cancer
		GnRH agonists	Leuprorelin (LupronDepot®)	Abbvie	Multiple trials	In combination with variety of other therapies
			Goserelin (Zoladex®)	AstraZeneca and TerSera Therapeutics LLC	Multiple trials	In combination with variety of other therapies
	Female hormones	Estrogens	Multiple agents		Multiple trials	In combination with variety of other therapies
		Progesterogens	Multiple agents		Multiple trials	In combination with variety of other therapies
	Others	Anti-receptor Oligos	AZD5312	AstraZeneca	Multiple trials	In combination with variety of other therapies
Retinoids						
Blood cancers	Retinoic acid receptor	Retinoic acid receptor pathway	Retinol	Generic	Multiple trials	In combination with variety of other therapies
Melanoma	Retinoic acid receptor	Retinoic acid receptor pathway	Retinol	Generic	Multiple trials	In combination with variety of other therapies
Solid tumors	Retinoic acid receptor	Retinoic acid receptor pathway	Retinol	Generic	Multiple trials	In combination with variety of other therapies

Acute promyelocytic leukemia	Retinoic acid receptor	Retinoic acid receptor pathway	Tretinoin (Vesanoid®)	Roche, OncBioMune Pharmaceuticals, Inc. and Cheplapharm Arzneimittel GmbH	Standard of care	FDA approved
Solid tumors	Retinoic acid receptor	Retinoic acid receptor pathway	Tretinoin	Generic	Multiple trials	In combination with variety of other therapies
Solid tumors	Retinoic acid receptor	Retinoic acid receptor pathway	Isotretinoin (Accutane™)	Roche	Multiple trials	In combination with variety of other therapies
Kaposi's sarcoma	Retinoic acid receptor	Retinoic acid receptor pathway	Alitretinoin (Panretin®)	Eisai Inc.	Standard of care	FDA approved
Rexinoids						
Lung	Retionic X receptor	Retionic X receptor pathway	IRX4204	Io Therapeutics, Inc.	ClinicalTrials.gov Identifier: NCT02991651	Advanced NSCLC
Breast	Retionic X receptor	Retionic X receptor pathway	9CUAB30	National Cancer Institute (NCI)	ClinicalTrials.gov Identifier: NCT02876640	Early-stage BRCA
Non-melanoma skin cancer	Retionic X receptor	Retionic X receptor pathway	9CUAB30	University of Alabama at Birmingham	ClinicalTrials.gov Identifier: NCT03327064	A biomarker evaluation trial of UAB30 in renal transplant recipients at high risk for non-melanoma skin
Cutaneous T-cell lymphoma, or CTCL	Retionic X receptor	Retionic X receptor pathway	Bexarotene (Targretin)	Ortho Dermatologics	FDA approved	In combination with variety of other therapies
Multiple blood and solid cancer	Retionic X receptor	Retionic X receptor pathway	Bexarotene (Targretin®)	Ortho Dermatologics	Multiple trials	In combination with variety of other therapies
Prostate	Retionic X receptor	Retionic X receptor pathway	NXR 194204	Io Therapeutics, Inc.	ClinicalTrials.gov Identifier:NCT01540071	Castration- and taxane-resistant prostate cancer
Lung	Retionic X receptor	Retionic X receptor pathway	NXR 194204	NuRx Pharmaceuticals, Inc.	ClinicalTrials.gov Identifier:NCT00964132	Advanced NSCLC

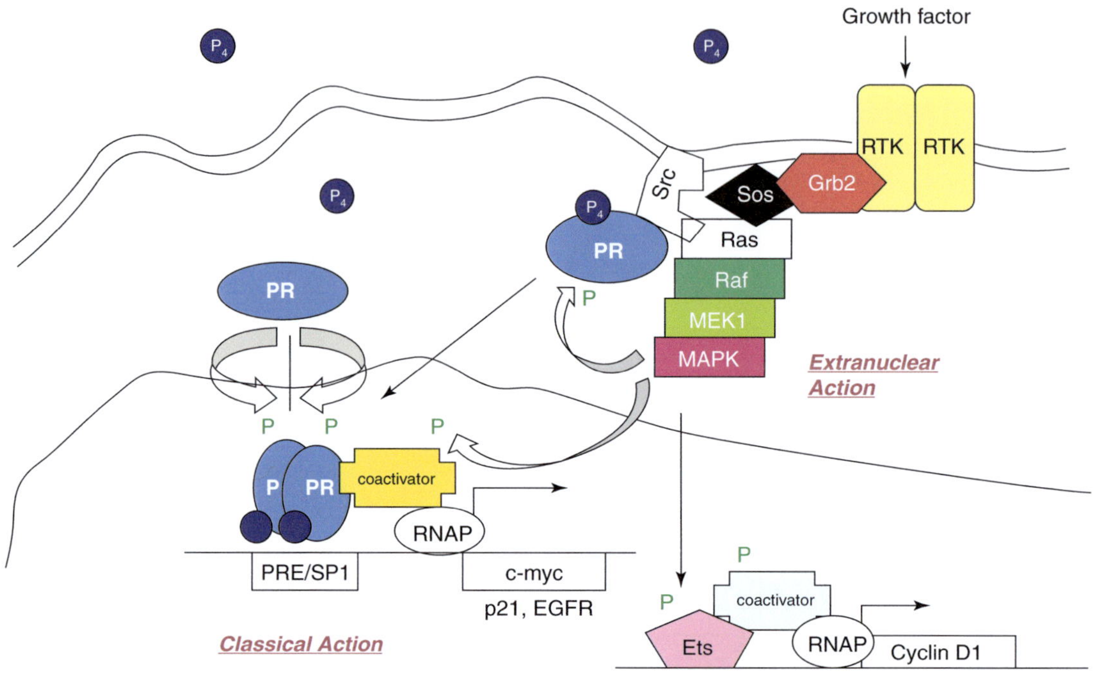

Fig. 13.4 Integration of PR rapid signaling and transcriptional activities. Progesterone (P4) binding to PR induces the rapid association of PR and c-Src. This interaction leads to a c-Src-dependent activation of the MAPK module through Ras/Raf signaling. This MAPK activation can lead to phosphorylation (P) of PR and transcriptional co-activators and/or activation of downstream MAPK target genes (i.e., cyclin D1). Phosphorylated PRs can activate transcription directly by binding to progesterone response elements (PREs) or indirectly though tethering interactions (i.e., SP1). Extranuclear and classical actions of PR are likely highly integrated actions, rather than separable events mediated by discrete populations of receptors. (Reprinted from Haga et al. [46]. With permission from Elsevier)

activity [36]. The function of PR has been predominantly studied in the breast. In knockdown animal models, there is severe impairment of the lobular alveolar development. Although there are multiple splice variants, PR-A, PR-B, and PR-C are most often the forms that are recognized. PR-B is the full form of the protein that contains the transcription activation function (TAF) region, while the PR-A and PR-C are shorter isoforms that antagonize the functions of PR-B. The ratio of the isoforms is thought to determine the outcome of PR activation. Testing for PR is routinely performed in breast cancer by either IHC or RT-PCR. The tests are however not isoform specific. High PR connotes a better prognosis. The vast majority of cases that express PR are also ER positive. Approximately 5% of cases are PR+/ER- by IHC and <1% by RT-PCR. The activation of PR in the absence of ER expression is thought to be due to non-genomic actions of ER and/or crosstalk with various other growth receptor pathways (Fig. 13.4). Synthetic progesterones (agonists) such as norethisterone and medroxyprogesterone acetate have been used in oral contraceptive pills. Both progesterone antagonists and selective progesterone receptor modulators are available and predominantly used in the treatment of gynecological cancers [37].

Androgen Receptor Signaling

Androgen receptor also has the basic structure of the nuclear receptor family with minor differences. It is located in the cytoplasm and when activated translocates to the nucleus to bind androgen-responsive elements (AREs). Splice variants and SNPs in AR have been reported to affect its localization, alter binding of AR to the co-regulators, and affect its activity. AR signaling is one of the primary pathways in prostate cancer, from which much of the known information has been obtained. AR has more recently been implicated in a subtype of triple-negative breast cancer, designated as luminal androgen receptor positive (LAR+); these patients also may respond to antiandrogenic agents. Androgens are the principal ligands of AR and primarily synthesized in the testis, but synthesis at other sites such as adrenal glands is possible. The traditional method for androgen deprivation had been castration; however, recent methods employ nonsurgical techniques. The current first line of therapy for recurrent prostate cancer after prostatectomy consists of the use of gonadotrophin-releasing hormone (GnRH). This can be supplemented by the addition of competitive androgen receptor antagonists to further impede AR signaling. In spite of this, recurrence of cancer is not infrequent; the recurrent cancer often retains active AR pathway. Genomic amplification of the AR locus is seen in up to 30% of the castration resistant prostate cancer. Similarly, mutations (such as L702H, W742C, H875Y, and T878A) involving the ligand-binding domain have been identified in multiple studies. Some of these mutations appear to be related specifically to the drugs used for the treatment.

Multiple splice variants of AR have been described in human tissues and cell lines (Fig. 13.5) and implicated in the development of prostate cancer. AR splice variant, AR-V7, is associated with nuclear localization. Its role in resistance to AR antagonist (enzalutamide) and CYP17A1 inhibitor (abiraterone) needs further analysis [38, 39]. More recently, detection of this variant in blood has been developed as a clinical assay.

Therapeutic Relevance

The control of AR signaling in cancer is achieved by pre-receptor, receptor, and post-receptor mechanisms. A number of agents that modify the GnRH pathway such as agonists (leuprolide and goserelin) and antagonists (degarelix) have been used for the treatment of prostate cancer. A number of agents that decrease the synthesis of androgens by modifying the activity of CYP17A1 are either FDA approved (abiraterone) or in phase III clinical trials (VT-464 and galeterone). Similarly, inhibitors of 5α-reductase which converts testosterone to dihydrotestosterone might have a role in controlling androgen production. At the receptor level, agents such as flutamide, bicalutamide, nilutamide, and enzalutamide are FDA approved, and additional agents (such as JNJ-56021927 and BAY1841788) are in phase III clinical trials. Crosstalk of AR pathway with glucocorticoid (GR) pathway has been implicated in resistance. The role these post-receptor mechanisms in resistance is being actively explored in clinical trials (GR antagonist mifepristone in combination with enzalutamide; NCT02012296).

Glucocorticoid Receptor Signaling

Glucocorticoids are steroid hormones secreted by the zona fasciculata of the adrenal gland. Their production is controlled by circadian rhythms and stress via the hypothalamic-pituitary-adrenal axis. Several splice variants of the glucocorticoid receptors exist of which GRα is the principal isoform. It is located primarily in the cytoplasm in a complex with HSP70/90 chaperone proteins. The GC-GR axis plays a critical role in metabolism and immune related pathways; the latter may be accomplished by multiple genomic mechanisms including transcriptional repression of NFkB. GR can also mediate activity by no-genomic competing with other transcription factors such as NFkB, IRF3, AP-1 for binding with essential co-activators such as CREB1-binding protein, nuclear receptor co-activator 1, GRIP1, and/or p53 (for details see Desmet et al. [40]). The importance of this pathway in cancer could be related to suppression of immune mechanisms and due to crosstalk with other nuclear receptors. As stated previously, combination of GR antagonist mifepristone in combination with enzalutamide is being analyzed for control of prostate cancer (NCT02012296). The ability of GC-GR pathway to cause growth suppression and apoptosis of immune cells has been put to good use in the treatment of lymphoid malignancies. The importance of GCs in treatment of leukemias is perhaps best highlighted by comparisons of patient

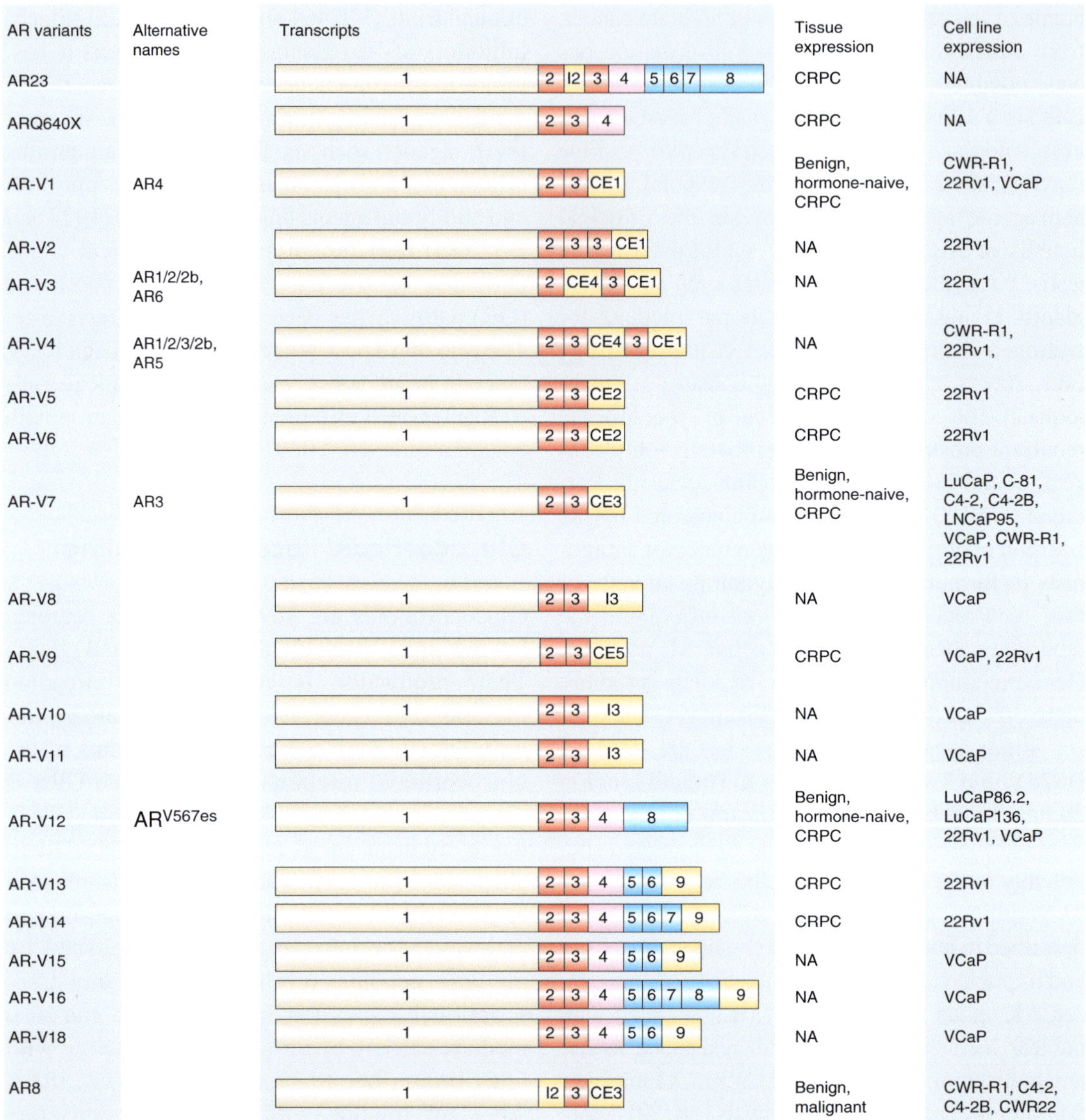

Fig. 13.5 Alternatively spliced AR-Vs in prostate cancer. Transcripts of AR-V functional domains are shown in the same colors as for AR-FL. Novel exons are shown in yellow. C-terminal sequences of AR-V8, AR-V10, and AR-V11 are derived from the DNA sequence within I3. Not to scale. (Adapted from Lu et al. [47]. With permission from Springer Nature)

response to dexamethasone and prednisone [41]. Dexamethasone (Dex) and prednisolone (Pred) are both derivatives of cortisol, with Dex differing by addition of a fluorine at the 9α and a methyl at C16. These two differences make Dex more specific for GR, with little to no MR activity, and about 10–16× more potent according to established indices [42]. In clinical trials, substituting Dex for Pred in high-risk ALL patients improves outcome by over 10% (81–94% overall survival) (Children's Oncology Group study COG AALL0232), despite each inducing indistinguishable MRD after induction [41].

Retinoic Acid Receptors

The term retinoids refers to natural and synthetic products that have structural and biological similarities to retinol or Vitamin A [43]. Dietary retinoids are predominantly taken up by the liver, but a small fraction binds to retinol-binding proteins or albumin and is found in the circulation. Retinol can be reversibly converted into isomers such as 9-*cis* retinoic acid (9-*cis*-RA) and all-trans retinoic acid (atRA). They exhibit differential binding affinities to the two receptors of RA, namely, retinoic acid receptors (RARs) and retinoid X receptors (RXRs). Each of these has three isoforms named α, β, and γ. RARs and RXRs play important roles in cell differentiation and are often silenced during cellular transformation. Both families have highly similar structure and often form heterodimers with each other or with other members of the nuclear receptor family. The make-up of the dimers not only determines which genes are regulated but also determines which of the many co-regulatory molecules the receptor can bind [43].

Therapeutic Relevance

Retinoic acid derivatives are in use for cancer prevention and in the treatment of several cancers. AtRA is used for the treatment of acute promyelocytic leukemia (APL), which is characterized by RXRα-PML fusion (Fig. 13.6). It has synergy with arsenic trioxide, which acts on PML protein, as well as with chemotherapy. It has also been used in cutaneous T-cell lympho-

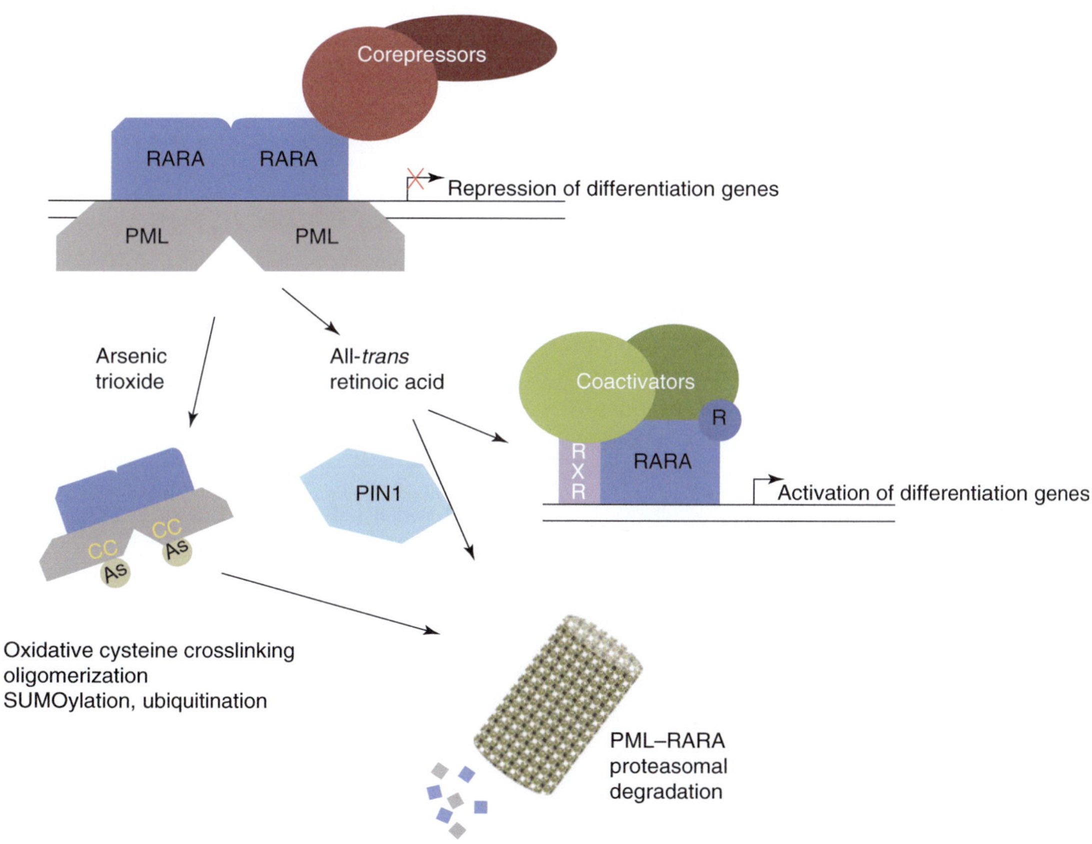

Fig. 13.6 Targeting of PML–RARA in acute promyelocytic leukemia. The PML–RARA fusion protein binds to RARA sites in the genome, recruiting co-repressors and repressing RAR-target genes. All-trans retinoic acid (ATRA, labeled as R) binds to PML–RARA and switches it from a repressor of myeloid differentiation genes to an activator. ATRA also induces degradation of PML–RARA through a direct effect on the PIN1 prolyl isomerase. Arsenic trioxide also binds directly to the fusion protein and induces its degradation interacting with the zinc finger of the PML moiety. Arsenic leads to ROS production and subsequent disulfide crosslinking of the cysteines resulting in oligomerization SUMOylation and subsequent ubiquitin-mediated proteolysis. *Abbreviations*: PIN1 peptidylprolyl cis/trans isomerase 1, PML promyelocytic leukemia, RARA retinoic acid receptor α, ROS reactive oxygen species, SUMO small ubiquitin-related modifier [44]. (Adapted from Bhagwat and Vakoc [44]. With permission from Elsevier)

mas and in Kaposi's sarcoma. Retinoids may also have a role in the treatment of basal and squamous cell carcinomas of the skin and head and neck region. The combination of bexarotene (rexinoid) and erlotinib was found to be efficacious in the BATTLE trial of non-small cell lung carcinoma.

Summary and Future Directions

The steroid nuclear receptor family is a large heterogeneous group of proteins, which clearly plays an important role in cancer. Crosstalk between the various receptors and co-activators and repressors is difficult to comprehend fully and may vary in individual tumors. Targeting ER and AR have proved to be challenging in breast and prostate cancers not only because of these complexities but also due to the dependence of bodily functions on these pathways. Novel strategies that intervene upstream or downstream of the receptors have been employed to circumvent these pathways and decrease the long-term toxicities. The implications of newer findings such as mutations and splice variants are still not clear, but clinical associations suggest that they will need to be addressed. Targeting pioneer factors such as FOXA1 could provide such a means; however, additional studies are necessary to establish the feasibility of such approaches.

References

1. Evans RM, Mangelsdorf DJ. Nuclear receptors, RXR, and the big bang. Cell. 2014;157(1):255–66.
2. Zheng Y, Murphy LC. Regulation of steroid hormone receptors and coregulators during the cell cycle highlights potential novel function in addition to roles as transcription factors. Nucl Recept Signal. 2016;14:e001.
3. Wahli W, Martinez E. Superfamily of steroid nuclear receptors: positive and negative regulators of gene expression. FASEB J. 1991;5(9):2243–9.
4. Contrò V, Basile JR, Proia P. Sex steroid hormone receptors, their ligands, and nuclear and non-nuclear pathways. AIMS Mol Sci. 2015;2(3):294–310.
5. Levin ER, Hammes SR. Nuclear receptors outside the nucleus: extranuclear signalling by steroid receptors. Nat Rev Mol Cell Biol. 2016;17(12):783–97.
6. Gao X, Loggie BW, Nawaz Z. The roles of sex steroid receptor coregulators in cancer. Mol Cancer. 2002;1:7.
7. Nakshatri H, Badve S. FOXA1 as a therapeutic target for breast cancer. Expert Opin Ther Targets. 2007;11(4):507–14.
8. Weigel NL, Moore NL. Kinases and protein phosphorylation as regulators of steroid hormone action. Nucl Recept Signal. 2007;5:e005.
9. Weigel NL, Moore NL. Steroid receptor phosphorylation: a key modulator of multiple receptor functions. Mol Endocrinol. 2007;21(10):2311–9.
10. Barone I, Brusco L, Fuqua SA. Estrogen receptor mutations and changes in downstream gene expression and signaling. Clin Cancer Res. 2010;16(10):2702–8.
11. Renaud JP, Rochel N, Ruff M, Vivat V, Chambon P, Gronemeyer H, et al. Crystal structure of the RAR-gamma ligand-binding domain bound to all-trans retinoic acid. Nature. 1995;378(6558):681–9.
12. Wagner RL, Apriletti JW, McGrath ME, West BL, Baxter JD, Fletterick RJ. A structural role for hormone in the thyroid hormone receptor. Nature. 1995;378(6558):690–7.
13. Bourguet W, Ruff M, Chambon P, Gronemeyer H, Moras D. Crystal structure of the ligand-binding domain of the human nuclear receptor RXR-alpha. Nature. 1995;375(6530):377–82.
14. Pearce ST, Liu H, Jordan VC. Modulation of estrogen receptor alpha function and stability by tamoxifen and a critical amino acid (Asp-538) in helix 12. J Biol Chem. 2003;278(9):7630–8.
15. Nieto L, Tharun IM, Balk M, Wienk H, Boelens R, Ottmann C, et al. Estrogen receptor folding modulates cSrc kinase SH2 interaction via a helical binding mode. ACS Chem Biol. 2015;10(11):2624–32.
16. Le Douarin B, Zechel C, Garnier JM, Lutz Y, Tora L, Pierrat P, et al. The N-terminal part of TIF1, a putative mediator of the ligand-dependent activation function (AF-2) of nuclear receptors, is fused to B-raf in the oncogenic protein T18. EMBO J. 1995;14(9):2020–33.
17. vom Baur E, Zechel C, Heery D, Heine MJ, Garnier JM, Vivat V, et al. Differential ligand-dependent interactions between the AF-2 activating domain of nuclear receptors and the putative transcriptional intermediary factors mSUG1 and TIF1. EMBO J. 1996;15(1):110–24.
18. L'Horset F, Dauvois S, Heery DM, Cavailles V, Parker MG. RIP-140 interacts with multiple nuclear receptors by means of two distinct sites. Mol Cell Biol. 1996;16(11):6029–36.
19. Klein-Hitpass L, Ryffel GU, Heitlinger E, Cato AC. A 13 bp palindrome is a functional estrogen responsive element and interacts specifically with estrogen receptor. Nucleic Acids Res. 1988;16(2):647–63.
20. Walker P, Germond JE, Brown-Luedi M, Givel F, Wahli W. Sequence homologies in the region preceding the transcription initiation site of the liver estrogen-responsive vitellogenin and apo-VLDLII genes. Nucleic Acids Res. 1984;12(22):8611–26.

21. Ma CX, Reinert T, Chmielewska I, Ellis MJ. Mechanisms of aromatase inhibitor resistance. Nat Rev Cancer. 2015;15(5):261–75.
22. Jeselsohn R, De Angelis C, Brown M, Schiff R. The evolving role of the estrogen receptor mutations in endocrine therapy-resistant breast cancer. Curr Oncol Rep. 2017;19(5):35.
23. Gu G, Fuqua SA. ESR1 mutations in breast cancer: proof-of-concept challenges clinical action. Clin Cancer Res. 2016;22(5):1034–6.
24. Anbalagan M, Rowan BG. Estrogen receptor alpha phosphorylation and its functional impact in human breast cancer. Mol Cell Endocrinol. 2015;418(Pt 3):264–72.
25. Jozwik KM, Carroll JS. Pioneer factors in hormone-dependent cancers. Nat Rev Cancer. 2012;12(6):381–5.
26. Hurtado A, Holmes KA, Ross-Innes CS, Schmidt D, Carroll JS. FOXA1 is a key determinant of estrogen receptor function and endocrine response. Nat Genet. 2011;43(1):27–33.
27. Maximov PY, McDaniel RE, Fernandes DJ, Korostyshevskiy VR, Bhatta P, Murdter TE, et al. Simulation with cells in vitro of tamoxifen treatment in premenopausal breast cancer patients with different CYP2D6 genotypes. Br J Pharmacol. 2014;171(24):5624–35.
28. Baum M, Budzar AU, Cuzick J, Forbes J, Houghton JH, Klijn JG, et al. Anastrozole alone or in combination with tamoxifen versus tamoxifen alone for adjuvant treatment of postmenopausal women with early breast cancer: first results of the ATAC randomised trial. Lancet. 2002;359(9324):2131–9.
29. Pagani O, Regan MM, Francis PA. Are SOFT and TEXT results practice changing and how? Breast. 2016;27:122–5.
30. Baselga J, Campone M, Piccart M, Burris HA 3rd, Rugo HS, Sahmoud T, et al. Everolimus in postmenopausal hormone-receptor-positive advanced breast cancer. N Engl J Med. 2012;366(6):520–9.
31. Kuiper GG, Enmark E, Pelto-Huikko M, Nilsson S, Gustafsson JA. Cloning of a novel receptor expressed in rat prostate and ovary. Proc Natl Acad Sci U S A. 1996;93(12):5925–30.
32. Heldring N, Pike A, Andersson S, Matthews J, Cheng G, Hartman J, et al. Estrogen receptors: how do they signal and what are their targets. Physiol Rev. 2007;87(3):905–31.
33. Huang B, Omoto Y, Iwase H, Yamashita H, Toyama T, Coombes RC, et al. Differential expression of estrogen receptor alpha, beta1, and beta2 in lobular and ductal breast cancer. Proc Natl Acad Sci U S A. 2014;111(5):1933–8.
34. Muthusamy S, Andersson S, Kim HJ, Butler R, Waage L, Bergerheim U, et al. Estrogen receptor beta and 17beta-hydroxysteroid dehydrogenase type 6, a growth regulatory pathway that is lost in prostate cancer. Proc Natl Acad Sci U S A. 2011;108(50):20090–4.
35. Warner M, Huang B, Gustafsson JA. Estrogen receptor beta as a pharmaceutical target. Trends Pharmacol Sci. 2017;38(1):92–9.
36. Mani SK, Oyola MG. Progesterone signaling mechanisms in brain and behavior. Front Endocrinol (Lausanne). 2012;3:7.
37. Chabbert-Buffet N, Meduri G, Bouchard P, Spitz IM. Selective progesterone receptor modulators and progesterone antagonists: mechanisms of action and clinical applications. Hum Reprod Updat. 2005;11(3):293–307.
38. To SQ, Kwan EM, Fettke HC, Mant A, Docanto MM, Martelotto L, et al. Expression of androgen receptor splice variant 7 or 9 in whole blood does not predict response to androgen-axis-targeting agents in metastatic castration-resistant prostate cancer. Eur Urol. 2018. Feb 2. pii: S0302-2838(18)30016-2; https://doi.org/10.1016/j.eururo.2018.01.007. [Epub ahead of print]
39. Antonarakis ES, Lu C, Wang H, Luber B, Nakazawa M, Roeser JC, et al. AR-V7 and resistance to enzalutamide and abiraterone in prostate cancer. N Engl J Med. 2014;371(11):1028–38.
40. Desmet SJ, De Bosscher K. Glucocorticoid receptors: finding the middle ground. J Clin Invest. 2017;127(4):1136–45.
41. Pufall MA. Glucocorticoids and Cancer. Adv Exp Med Biol. 2015;872:315–33.
42. Inaba H, Pui CH. Glucocorticoid use in acute lymphoblastic leukaemia. Lancet Oncol. 2010;11(11):1096–106.
43. Uray IP, Dmitrovsky E, Brown PH. Retinoids and rexinoids in cancer prevention: from laboratory to clinic. Semin Oncol. 2016;43(1):49–64.
44. Bhagwat AS, Vakoc CR. Targeting transcription factors in cancer. Trends Cancer. 2015;1(1):53–65.
45. Tata JT. Signalling through nuclear receptors. Nat Rev Mol Cell Biol. 2002;3:702–10.
46. Haga CR, Daniel AR, Dressing GE, et al. Role of phosphorylation in progesterone receptor signaling and specificity. Mol Cell Endocrinol. 2012;357:43–9.
47. Lu J, Van der Steen T, Tindall DJ. Are androgen receptor variants a substitute for the full-length receptor? Nat Rev Urol. 2015;12(3):137–44.

Protein Kinase C Signaling in Carcinogenesis

14

Thao N. D. Pham and Debra A. Tonetti

Abbreviations

AML	Acute myeloid leukemia
CSC	Cancer stem cell
DAG	Diacylglycerol
GPCR	G-protein-coupled receptor
IP3	Inositol triphosphate
PDK1	Phosphoinositide-dependent kinase-1
PKA	cAMP-dependent protein kinase
PKC	Protein kinase C
PKG	cGMP-dependent protein kinase
PLC	Phosphoinositide phospholipase C
PS	Phosphatidylserine
RTK	Receptor tyrosine kinase

Introduction

Protein kinase C (PKC) is a family of cytoplasmic serine-threonine kinases that belong to the highly conserved AGC protein kinase group.

Based on sequence alignment of the catalytic kinase domain, PKC is most closely related to the cAMP-dependent protein kinase (PKA or PKAC) and cGMP-dependent protein kinase (PKG or CGK1α) [1]. Twelve isoforms have been identified, which can be further divided into subgroups (classical, novel, atypical, and PKC-related PKN) based on their domain composition and cofactors for activation [1, 2]. The PKC-related PKN family members PKN1, PKN2, and PKN3 are not a focus of this discussion, but they all have the kinase domain homologous to PKC and a unique regulatory region containing antiparallel coiled-coil domains [1, 2]. The finding that phorbol ester, a potent tumor promoter, can activate PKC triggered an intensive examination of PKC function and regulation for several decades, prompting the discovery of a diversity of effectors and cellular consequences resulting from PKC activation. As a result, mutations and/or dysregulation of PKC have been identified as drivers in a number of cancers. The important contribution of PKC signaling in cancer initiation and progression has fueled both academic and clinical efforts toward the development of therapeutic PKC modulators.

Structure and Regulation

All PKC members are composed of an N-terminal regulatory and a C-terminal catalytic domain, separated by a hinge region [1, 2] (Fig. 14.1). The regulatory domain, housing the C1 and C2 regions, slightly varies among members. The C1 region contains a Cys-rich motif and forms a binding site for diacylglycerol (DAG)/phorbol

T. N. D. Pham · D. A. Tonetti (✉)
Department of Biopharmaceutical Sciences, College of Pharmacy, University of Illinois at Chicago, Chicago, IL, USA
e-mail: dtonetti@uic.edu

S. Badve, G. L. Kumar (eds.), *Predictive Biomarkers in Oncology*,
https://doi.org/10.1007/978-3-319-95228-4_14

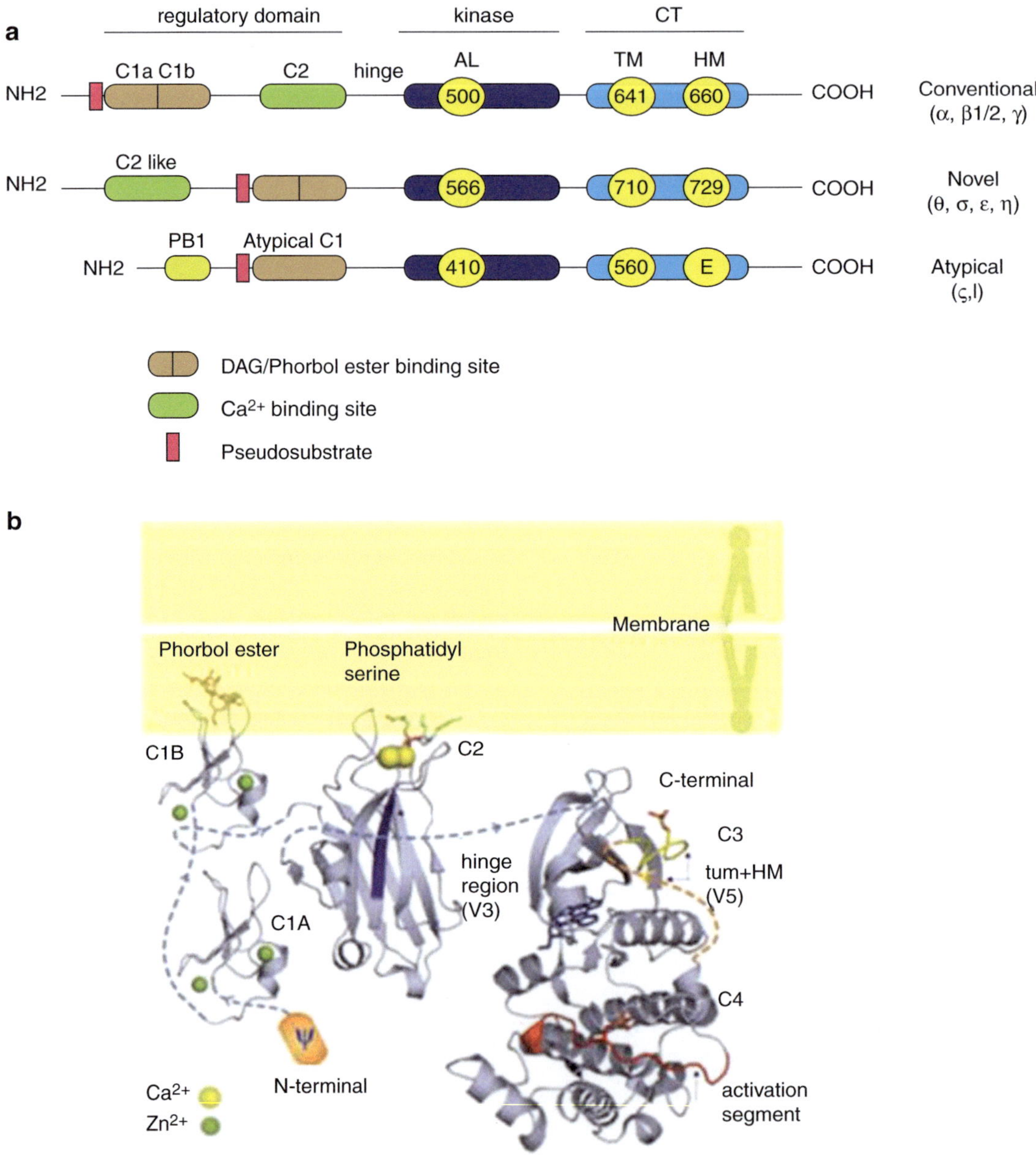

Fig. 14.1 PKC structure. (**a**) Schematic representation of the primary structure of conventional, novel, and atypical PKCs. With functional C1 and C2 regions, the conventional group can be activated by both diacylglycerol (DAG) and Ca^{2+}. The novel group can be activated by DAG alone. Activation of the atypical group does not require DAG or Ca^{2+}, but is activated by phospholipids. The PB1 domain on atypical isoforms is important for their interactions with other proteins. All PKC members undergo a series of phosphorylation events to gain catalytic competence, which occurs at the activation loop (AL), the turn motif (TM), and the hydrophobic motif (HM). Numbers indicate the specific amino acid residues that are phosphorylated: PKCβ as an example for the conventional group, PKCε for novel, and PKCζ for atypical [3]. Note that atypical PKC isoforms have a Glu residue at the phospho-acceptor position of the hydrophobic motif. (**b**) The tertiary structures for the C1A and C1B (PKCδ), C2 (PKCα), C3, and C4 domains (PKCθ) are illustrated by ribbon diagrams. (**a**) (Adapted from Gallegos and Newton [11]. With permission from John Wiley & Sons). (**b**) (Reprinted from Gomperts et al. [12]. With permission from Elsevier)

ester [1, 2], whereas C2 is the binding site for anionic phospholipids, such as phosphatidylserine (PS), and Ca^{2+} [2]. Based on these structures, PKC members are classified as classical (PKCα, PKCβ, and PKCγ), novel (PKCδ, PKCε, PKCη, and PKCθ), or atypical (PKCζ and PKCι). As seen in Fig. 14.1, activation of classical PKCs requires DAG, PS, and Ca^{2+}, of novel PKCs requires Ca^{2+} and PS, and atypical PKC do not require either DAG or Ca^{2+} but do require different lipid metabolite second messengers for activity. The catalytic domain is highly conserved among all PKC members and comprised of the C3 and C4 regions, responsible for binding ATP and phosphorylation substrate, respectively [2].

PKC activity is under tight regulation at both the structural and spatial level. Post-synthesis, PKC undergoes several phosphorylation events before reaching maturity [2]. The events take place at the activation domain, turn motif, and hydrophobic domain [2, 3] (Fig. 14.1). Among these, phosphorylation of the activation domain by phosphoinositide-dependent kinase-1 (PDK1) is critical for subsequent phosphorylation of the turn motif and hydrophobic domain [2]. In the absence of activating ligands, phosphorylated, mature PKC retains a closed, inactive conformation with the C1 domain masked and the substrate-binding cavity occupied by a pseudosubstrate (Fig. 14.2). In response to agonists, the binding of Ca^{2+} to the C2 domain targets PKC to the plasma membrane where it interacts with phospholipids and membrane-embedded DAG. The coordinated engagement of both the C1 and C2 domains on the membrane provides enough energy for the release of the pseudosubstrate, exposing the substrate binding site (Fig. 14.2) [2].

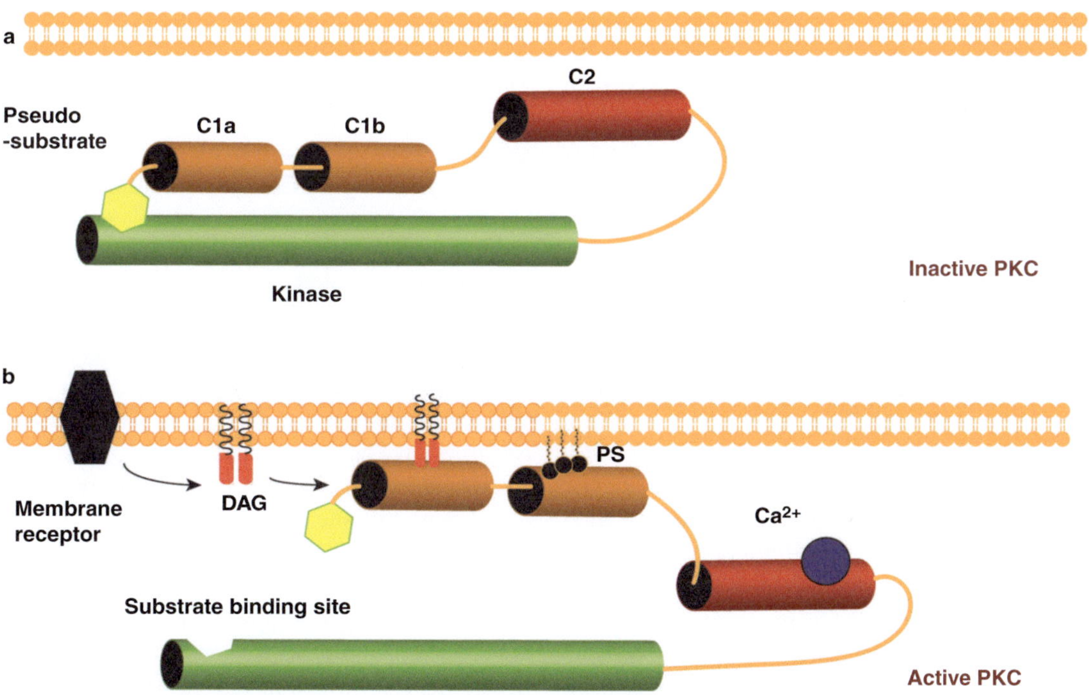

Fig. 14.2 Allosteric activation of PKC. (**a**) In the absence of activating ligands (such as diacylglycerol (DAG) and Ca^{2+}), mature PKC remains inactive, due to the binding of the pseudosubstrate to the substrate binding site. (**b**) Activation of membrane-bound receptors such as G-protein-coupled receptors (GPCRs) or receptor tyrosine kinases (RTKs) generates DAG and Ca^{2+}. These molecules bind to their respective binding sites, prompting PKC to undergo a conformation change that expels the pseudosubstrate and exposes the substrate-binding site. Binding of both DAG and Ca^{2+} to PKC is proposed to stabilize the enzyme active conformation

PKC Expression and Contribution in Cancers

Altered expression of the PKC family is a frequent event in cancer development and progression [3, 4]. However, as most PKC members are ubiquitously expressed in various tissues and can activate a myriad of signaling pathways resulting in numerous cellular outcomes, it remains challenging to unravel the precise contribution of each isoform in a specific tissue or disease. Nevertheless, vigorous studies employing both pharmacological and genetic means have helped flesh out our current understanding of this complex family of enzymes. It is now well accepted that PKC as a family can either promote or repress tumor initiation and development. Furthermore, some PKC isoforms may have opposite roles in different types of cancer. Therefore, PKC functions are not intrinsic but tissue specific. For instance, PKCα has been reported to promote progression of breast cancer but appear to be a tumor suppressor in colon cancer [3, 4]; PKCδ enhances cell survival in chronic lymphocytic leukemia but promotes cell death in acute myeloid leukemia (AML) [4].

Some isoforms, such as PKCα, β, δ, and ε, have been better studied in human cancer compared to others. In general, a number of studies support a tumor-promoting role of these isoforms in a majority of cancers [3, 4]. PKCα appears to be the most commonly altered isoforms in a majority of human malignancies, including adenocarcinoma (breast, lung, prostate, pancreatic), squamous cell carcinoma (head and neck, thyroid, bladder), and leukemia (acute lymphocytic leukemia and AML) [4]. In these cancers, PKCα supports cancer cell proliferation, drug resistance, and resistance to cell death caused by therapy [4]. Similarly, PKCε has been reported to promote tumor progression in various cancers (breast, lung, prostate, thyroid, melanoma, liver, head and neck) [4]. These observations, originally examined and validated using cancer cell lines and animal models, were subsequently validated in clinical samples where elevated expression levels of these isoforms (compared to healthy, normal individuals) are often correlated with poor clinical performance, including therapy resistance, early relapse, and occurrence of metastases [4]. Besides classical and novel isoforms, recent studies have also implied a tumor-promoting role for atypical PKC isoforms. For example, high expression of PKCη has been reported in metastatic human samples and can be correlated with lymph node metastases and drug resistance [4].

As a result, expression of PKC can potentially be used as a biomarker for patients' prognosis, and depletion of their expression and inhibition of their activity appear to be attractive goals in the field of cancer therapy. For example, high PKCα expression is predictive for early relapse and therapy resistance in breast cancer patients [4, 5]. Inhibition of PKCα expression using genetic means resulted in longer overall survival in several animal models of breast cancer [4, 5]. These outcomes may be attributed to reduced metastases and restored chemosensitivity [4, 5]. Similar positive results were observed in other aggressive diseases such as lung and pancreatic cancers [4].

PKC Signaling in Carcinogenesis

As PKC is often activated directly downstream of membrane-associated receptors, such as the G-protein-coupled receptors (GPCRs), receptor tyrosine kinases (RTKs), and integrin receptors, the family plays an important role in a number of signal transduction cascades [3]. As a result, PKC isoforms have been reported to participate in processes important for cancer cell survival and tumorigenicity, such as proliferation, invasion, migration, angiogenesis, and anticancer drug resistance [3].

PKC signaling is simplified and illustrated in Fig. 14.3. GPCRs or RTKs can activate phospholipase C (PLC), which generates inositol triphosphate (IP_3) and DAG. While DAG remains associated with the plasma membrane, IP_3 diffuses through the cytoplasm and interacts with IP_3 receptors on the endoplasmic reticulum, causing the release of calcium via the calcium channel. The generation of DAG and intracellular

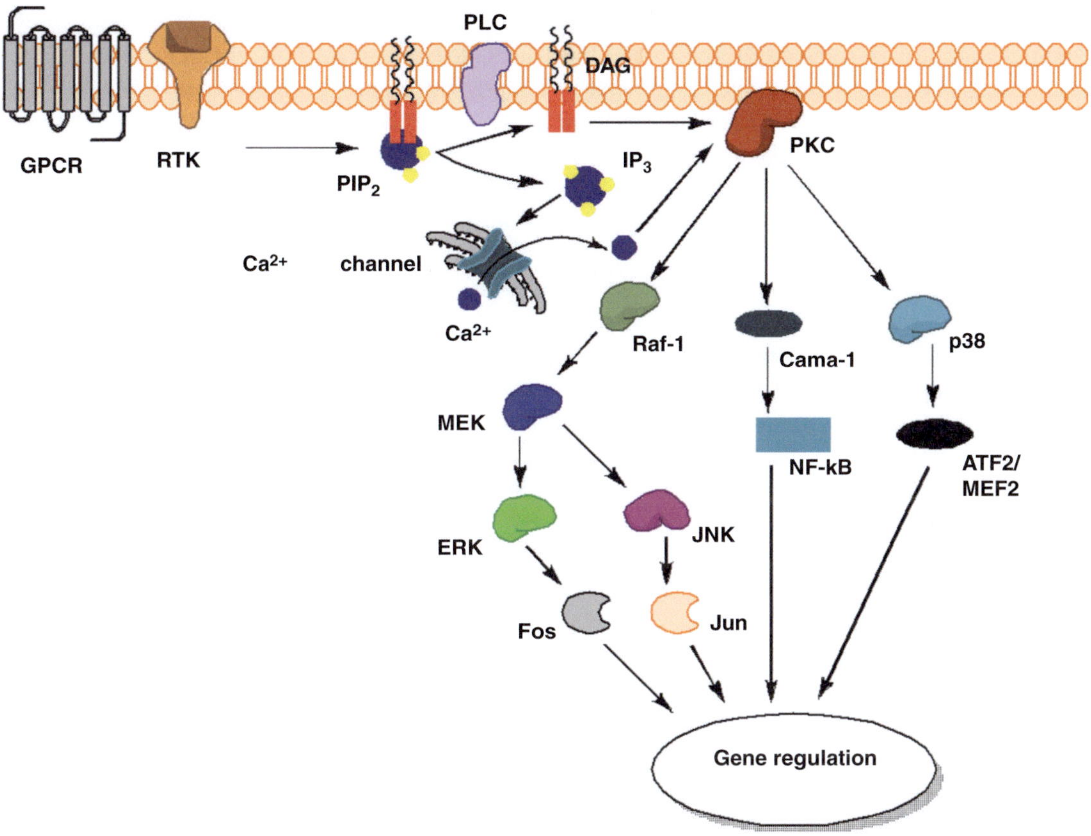

Fig. 14.3 PKC signaling in carcinogenesis. Following activation of membrane-bound receptors (GPCR, RTK), phospholipase C (PLC) generates inositol trisphosphate (IP_3) and diacylglycerol (DAG) from phosphatidylinositol 4,5-bisphosphate (PIP_2). IP_3 diffuses through the cytoplasm and interacts with the IP_3 receptors on the endoplastic reticulum (gray membrane) to prompt Ca^{2+} release. Binding of Ca^{2+} to the C2 domain of PKC increases the enzyme affinity for the plasma membrane, resulting in coordinated engagement of PKC with membrane-embedded DAG and phospholipids. Once fully activated, PKC phosphorylates a wide variety of substrates that activate MAPK/ERK, NF-κB, and p38 MAPK pathways to name a few. Major changes in gene expression take place as a result and play crucial roles in cancer development and progression

Ca^{2+} leads to the activation of PKC [2, 6]. One major phosphorylation target of PKC is Raf-1 [2, 4]. Activated Raf-1 by PKC triggers a protein kinase cascade that initiates the MAPK signaling pathway and is important for cell survival and proliferation [2, 4]. Several isoforms, such as PKCδ and PKCα, can activate the p38 MAPK signaling pathway, thereby promoting cancer cell migration and invasion [4]. PKC has also been shown to activate the NF-κB pathway [2]. It is noteworthy that the effect of a signaling molecule on NF-κB often strictly depends on the cell type or the microenvironment and that even opposite effects can occur in distinct cell types. In the context of cancer, activated NF-κB can not only enhance invasive properties but also exert an anti-apoptotic effect [2].

In addition, PKC can regulate cancer cell spreading and migration by influencing cytoskeleton morphology. For instance, it modifies integrins, which are surface molecules that mediate the interaction between cells and their extracellular matrix [7]. Alterations in integrin expression levels mediated by PKC can have a direct influence on cell movement. There are a number of reports on the direct association between PKC and integrins, such as between PKCα, PKCε with integrin β_1, and between PKCβ and β_3 integrin [7]. The interaction is believed for most cases to enhance transportation and distribution of integrins at the leading edge on the cell surface, facilitating movement and migration [7]. Atypical PKC, such as PKCλ and PKCι, can form complexes with members of the Rho family (Cdc42, Rac), PAR6, and PAR3, which then localize to the leading edge of cells and regulate locomotion [7].

Another area of PKC research has been the investigation of their contribution to cancer stem cell (CSC) survival and maintenance. CSCs have been demonstrated to be the underlying mechanism for therapy resistance, relapse, and metastasis of cancer cells [8]. Conventional therapy, such as chemotherapy and radiotherapy, tend to target proliferating non-stem cells and leave behind stem cells that have the ability to regenerate a new tumor bulk and can initiate cancer dissemination [8]. As a result, cellular regulation of CSCs became an intense area of investigation. Several recent studies have examined and demonstrated a role of PKC in controlling cellular signaling in CSCs. These examples include a potential role of PKCε phosphorylation of the stem cell marker Nanog in breast cancer cell lines, as well as PKCι in K-Ras-mediated bronchoalveolar stem cell expansion and lung cancer growth [4]. Recently, in breast cancer, PKCα was demonstrated to be critical for the phosphorylation of FRA-1, a member of the Fos transcription factors, which is required for the survival and function of the stem cell compartment [4, 5]. In agreement with this, PKCα has also been shown to enhance Notch4 signaling activity, which is responsible for estrogen-independent, endocrine-resistant growth and chemotherapy resistance in breast cancer [4, 5]. Targeting these isoforms therefore has the potential to halt cancer progression by depleting the cancer initiating cell population.

Current Therapies and Future Strategies

Efforts at targeting PKCshave mainly focused on disrupting the following regions or domains: ATP-binding site, C1, or C2. Overall, these inhibitors prevent PKC from achieving open catalytic conformation. Many of these compounds, however, do not have isoform specificity due to a high degree of structural conservation within the PKC family [3, 6, 9].

Small molecules blocking the ATP-binding site, such as staurosporine, midostaurin, and enzastaurin, are water-soluble bisindolylmaleimides and have slightly different specificity and efficacy. While staurosporine is fairly non-specific, its derivatives midostaurin and enzastaurin seem to be more specific for classical PKCs: midostaurin has a half maximal inhibitory concentration (IC_{50}) less than 30 nM for all conventional PKCs, and enzastaurin has an IC_{50} for PKCβ as low as 6 nM. However, clinical trials for these compounds so far have been disappointing, demonstrating little clinical benefit for patients [3, 6, 9]. Midostaurin has recently been recognized as a potential treatment for AML mainly due to its effect on other tyrosine kinases and is therefore classified as a potential broad-spectrum antineoplastic agent [3, 6, 9]. The April 2017 FDA approval of midostaurin (Rydapt®) in combination with standard cytarabine and daunorubicin induction and cytarabine consolidation is the first approved targeted treatment option for adult patients with AML as well as advanced systemic mastocytosis who have an FMS-like tyrosine kinase-3 (*FLT3*) mutation.

The second class of PKC inhibitors targets the C1 domain and competes for binding with DAG. Bryostatin is the most well-characterized and well-studied compound of this class in preclinical trials. Bryostatin can act in combination with other cancer therapies to be effective against a large number of cancers such as lung, prostate, and non-Hodgkin's lymphoma in animal models. However, bryostatin yielded disappointing results in clinical trials for a number of solid malignancies and has since been suspended [3, 6, 9] (Table 14.1).

A number of anchoring partners that bind to PKCs and regulate their activity have been described. Some of them bind PKC at the regulatory domain and some at the catalytic domain. For example, the actin-binding motif of PKCε is located in the regulatory domain of the isoform, between C1A and C1B; the anchoring protein PAR3 interacts with PKCζ at the kinase core [10]. Surprisingly, there is no defined consensus

Table 14.1 Summary on approved and potential therapies against PKC signaling pathway

Signaling pathway	Pathway active in cancer type	Affected biomarker	Method of detection	Target	Active drugs
Protein kinase C	Breast Lung Prostate Colon	FLT3 receptor tyrosine kinase	Internal tandem duplication detection	FLT3 receptor	Midostaurin (Rydapt®), Novartis
	Thyroid, bladder, pancreatic, liver, leukemia	PKC isoforms	Immunohistochemical staining	PKCβ	DB102 (formerly known as enzastaurin), Denovo Biopharma
	Head and neck melanoma			PKC isoforms	Bryostatin, Neurotrope Inc.

of PKC binding sites. Rather, each anchoring protein has been reported to bind to their respective PKC on a unique site. This knowledge opens up exciting opportunities for therapeutic intervention as one can identify isoform-specific binding partners and target them with minimal influence on other anchoring proteins. Depending on the specific situation and tissue, intervention can aim to either activate or inhibit PKC activity. Inhibitors that block the engagement between the anchoring protein with a specific PKC isoform have the potential to limit unwanted downstream effects from PKC activity. For instance, an inhibitor of PDK's docking on PKCδ was demonstrated to effectively inhibit the phosphorylation of PDK without affecting phosphorylation of the other PKCδ substrates. As a result, this inhibitor blocks PDK-mediated cardiac injury in the event of a heart attack [9]. Similarly, activation of PKC can be achieved by peptide mimics that have sequence similarity to the binding site on the anchoring protein. Interaction between these peptide mimics should activate PKC in the same manner as the anchoring protein [3, 9].

Another approach might take advantage of PKC-targeting miRNAs. As miRNA-mediated degradation of a transcript is based on their precise sequence alignment, these noncoding RNA molecules may be tools to develop inhibitors that have isoform-specific properties. For example, several miRNAs that can specifically target PKCα have been identified. In breast cancer, ectopic expression of miR-200b drastically reduced PKCα expression and was demonstrated to block tumor metastasis in animal models [5]. Further advancement in the understanding of PKC-mediating miRNAs and in vivo miRNA delivery will likely provide a new, useful reservoir of isoform-specific mediators.

Summary

In general, high expression of PKC isoforms is closely related to poor prognosis, therapy resistance, and poor patient survival. Lying at the crossroad of several major signaling pathways, this family represents a therapeutically meaningful target for the treatment of human malignancies and a potential cancer diagnostic marker. However, limited success has been achieved in terms of generating effective, PKC isoform-specific modulators mainly due to the highly conserved structural similarity within the family. In addition, there is an inadequate understanding of the role of each isoform in the early (initiation) and late (latency and/or progression) stages of the disease. Furthermore, stringent and clinically relevant evaluation of the significance and prevalence of PKC expression in human cancers is currently lacking. Moving forward, studies that identify interacting partners that exhibit isoform-specific properties will be essential for the development and evaluation of novel PKC modulators. Additionally, we need to improve our understanding of the interactions among PKC isoforms and their contribution in specific diseases and stages. Finally, surrogate biomarkers to predict for patient's prognosis and response to PKC therapy, besides PKC expression itself, will undoubtedly improve the likelihood of PKC modulators moving from bench to bedside.

References

1. Pearce LR, Komander D, Alessi DR. The nuts and bolts of AGC protein kinases. Nat Rev Mol Cell Biol. 2010;11(1):9–22.
2. Zeng L, Webster SV, Newton PM. The biology of protein kinase C. Adv Exp Med Biol. 2012;740: 639–61.
3. Dowling CM, Kiely PA. Targeting protein kinase C downstream of growth factor and adhesion signalling. Cancer. 2015;7(3):1271–91.
4. Kang JH. Protein kinase C (PKC) isozymes and cancer. New J Sci. 2014;231418–231456.
5. Pham TND, Tonetti DA. Protein kinase C alpha in breast cancer: a focus on endocrine resistant and triple negative breast cancer. J Cancer Biol Res. 2016;4(1):1076–87.
6. Mackay HJ, Twelves CJ. Targeting the protein kinase C family: are we there yet? Nat Rev Cancer. 2007;7(7):554–62.
7. Fogh BS, Multhaupt HA, Couchman JR. Protein kinase C, focal adhesions and the regulation of cell migration. J Histochem Cytochem. 2014;62(3):172–84.
8. Dawood S, Austin L, Cristofanilli M. Cancer stem cells: implications for cancer therapy. Oncology. 2014;28(12):1101–7, 10
9. Mochly-Rosen D, Das K, Grimes KV. Protein kinase C, an elusive therapeutic target? Nat Rev Drug Discov. 2012;11(12):937–57.
10. Newton AC. Protein kinase C: structural and spatial regulation by phosphorylation, cofactors, and macromolecular interactions. Chem Rev. 2001;101(8):2353–64.
11. Gallegos LL, Newton AC. Spatiotemporal dynamics of lipid signaling: protein kinase C as a paradigm. IUBMB Life. 2008;60(12):782–9.
12. Gomperts BD, Kramer IMIJ, Tatham PER. Phosphorylation and dephosphorylation: protein kinases A and C. In: Gomperts BD, Kramer IMIJ, Tatham PER, editors. Signal transduction. 2nd ed. Burlington: Academic; 2009. p. 243–72.

Roles of Rho/ROCK in Cancer Signaling

15

Yesim Gökmen-Polar

Introduction

The Rho family of small GTPases (Rho, Rac, and Cdc42), members of the Ras superfamily of small GTP-binding proteins, play important roles in the regulation of actin cytoskeleton organization and dynamics as well as other cellular functions such as polarity, motility, invasion, cell cycle progression, and survival [1]. A Rho-associated coiled-coil kinase (ROCK), belonging to the AGC (protein kinase A, G, and C) family of serine/threonine protein kinases, is the downstream effector of Rho signaling. Two mammalian ROCK isoforms have been identified, namely, ROCK1 (also known as ROCK I, ROKβ, Rho-kinase β, or p160ROCK) and ROCK2 (also known as ROCK II, ROKα, or Rho kinase) [2]. These molecules share a 65% overall homology and 92% homology in the kinase domain. Both kinases contain a catalytic kinase domain at the N terminus followed by a central coiled-coil domain, which includes the Rho-binding domain (RBD), and a C-terminal pleckstrin-homology (PH) domain [3] (Fig. 15.1). ROCK1 and ROCK2 share many downstream substrates due to the high degree of homology in their kinase domains important for controlling the dynamics of the actin cytoskeleton and cell morphology. They also share common substrates with the other members of AGC kinase family including myosin light chain (MLC) kinase. New studies are emerging to determine the functional differences of ROCK1 and ROCK2 isoforms in the regulation of adhesion, migration, cell cycle, and survival.

Several studies have reported the role of Rho/ROCK signaling in multiple biological processes that contribute to tumor progression [3–8]. Aberrant regulation of this pathway in cancer is dependent on the different level of alterations. Somatic mutations in Rho genes (RhoA, RhoB, and RhoC) and ROCK genes have been identified in cancer cell lines and human primary tumors. Rho is overexpressed at both mRNA and protein levels resulting in its hyperactivation in several cancers, including breast, colon, and lung cancer, as well as metastatic melanoma. In particular, overexpression or increased activation of RhoA protein is correlated with advanced stages of human cancer, including invasion and metastasis of testicular germ cell, urinary tract, and cervical cancers. Elevated expression of RhoC mRNA and protein has also been shown to correlate with an invasive phenotype in breast cancer cells and in human clinical samples. On the other hand, RhoB protein expression presented inhibitory effects on migration, invasion, and metastasis of human carcinoma cells through inhibition of the Ras/PI3 kinase/Akt pathway.

Y. Gökmen-Polar
Department of Pathology and Laboratory Medicine, Indiana University School of Medicine, Indianapolis, IN, USA
e-mail: ypolar@iu.edu

S. Badve, G. L. Kumar (eds.), *Predictive Biomarkers in Oncology*,
https://doi.org/10.1007/978-3-319-95228-4_15

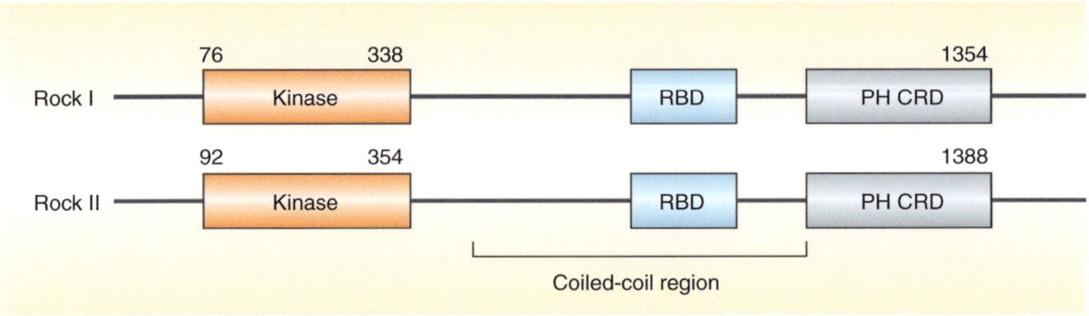

Fig. 15.1 Molecular structure of Rho kinase (ROCK) I and II. The two isoforms share an overall sequence identity at the amino-acid level of approximately 60%. Their kinase domains are more than 90% identical. The catalytic domain is located at the amino terminus, followed by a coiled-coil-forming region that encompasses the Rho-binding domain (RBD) and a pleckstrin-homology domain (PH) with a cysteine-rich repeat domain (CRD) at the carboxyl terminus10. (Reprinted from Mueller et al. [3]. With permission from Springer Nature)

Elevated protein levels of ROCKs have also been identified in several human cancers. Protein levels of both ROCKs were elevated in breast cancer. High ROCK1 expression also correlated with poor overall survival in breast and osteosarcoma. High expression of ROCK2 protein has been associated with aggressive behavior in hepatocellular carcinomas and colon and bladder cancers. However, further studies are necessary to understand whether ROCK1 and ROCK2 expression cause or effect for tumor progression.

Some studies have shown ROCK activation as an oncogenic process, whereas others show that ROCK functions as a negative regulator in cancer progression. This suggests that the role of ROCKs may be tissue-context-dependent, mainly on the cell type and the microenvironment surrounding the tumor. Another possibility may be due to the isoform-specific functions. Although they present overlapping functions, new isoform-specific partners have been identified suggesting their distinct functions in these processes [4, 5]. For example, the interaction of ROCK1 at the cell periphery in malignant melanoma is involved in the regulation of amoeboid cancer cell migration. On the other hand, other targets such as morgana/chp-1 and nucleophosmin/B23 (NPM/B23) compete for binding on ROCK2, regulating its activity in centrosome duplication and neoplastic transformation. These results emphasize the importance of isoform-specific functions in diseases including cancer. Although several studies report functional importance using rat and mouse cancer models, further cancer-specific studies are necessary to understand the complex regulation of Rho/ROCK isoforms.

ROCKs can be activated in a Rho-dependent and independent manner. Figure 15.2 shows the major players of Rho/ROCK signaling pathway. Upon activation by G-protein-coupled receptors (GPCRs), receptor tyrosine kinases (RTKs), and/or integrins, Rho GTPases undergo a conformational change in the effector-binding region of GTPase resulting in interaction with downstream targets. This activity is regulated by guanine nucleotide exchange factors (GEFs), GTPase-activating proteins (GAPs), and guanine nucleotide dissociation inhibitors (GDIs). GEFs activate Rho proteins by catalyzing the exchange of GDP for GTP. GAPs control the ability of the GTPase to hydrolyze GTP to GDP, controlling the transformation of the active conformation to the inactive conformation. GDI proteins, on the other hand, serve as an anchor and control Rho activation. ROCKs (ROCK1 and ROCK2), being the downstream effector molecules of Rho GTPases, phosphorylate their substrates and further regulate multiple processes in diseases including cancer. The well-known substrates of ROCKs for the regulation of cytoskeletal rearrangement, motility, and invasion comprise myosin light chain (MLC) phosphatase, myosin phosphatase 1 (MYPT1), and LIM kinases (LIMK1/2). These substrates further interact with their target proteins including ezrin/radixin/moesin (ERM) family proteins, Tau/MAP2, and cofilin [6]. The

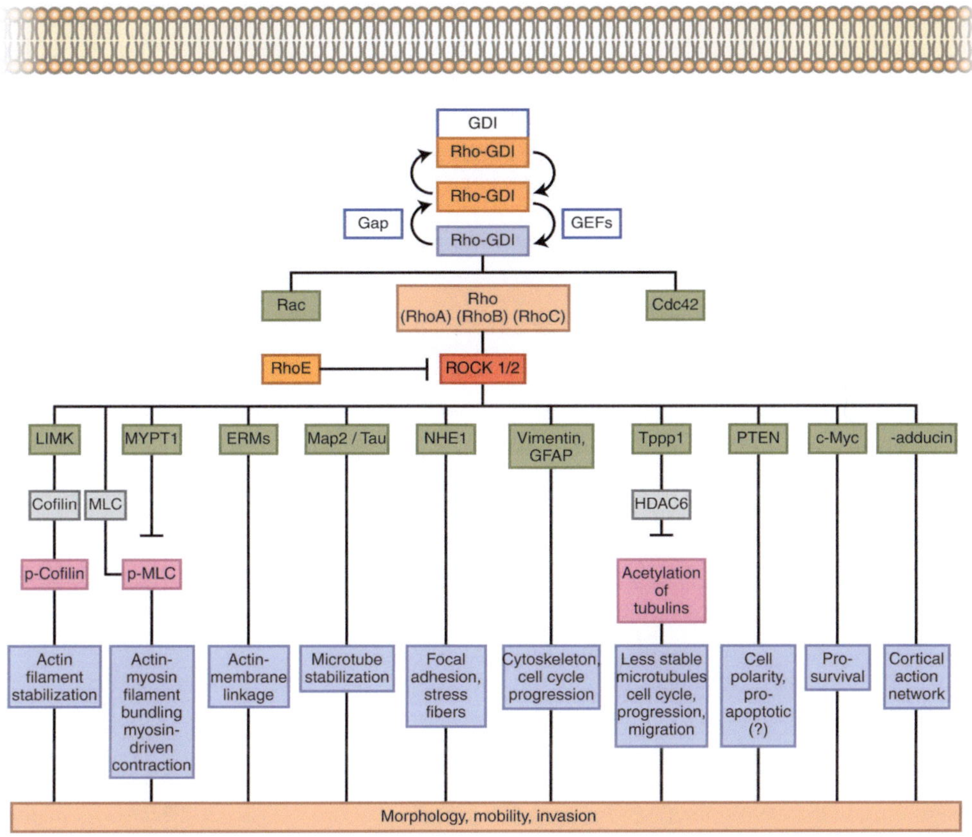

Fig. 15.2 Overview of Rho/ROCK signaling in cancer. Ligand-bound receptors such as G-protein-coupled receptors (GPCRs) or receptor tyrosine kinases (RTKs) activate guanine nucleotide exchange factors (GEFs), GDP-bound Rho, or Cdc42 to their respective GTP-bound states. GTP-bound RhoA and Cdc42 further activate ROCK (ROCK1/2) regulating multiple ROCK substrates that are responsible for diverse functions of Rho/ROCK pathway activation. Rock substrates include *LIMK* LIM domain kinase, *MYPT1* myosin phosphatase target subunit 1, *MLC* myosin light chain, *ERMs* ezrin, radixin, and moesins, *Map2/Tau* microtubule-associated protein 2/Tau, *NHE1* Na+/H+-exchanger 1, vimentin, *GFAP* glial fibrillary acidic protein, *Tppp1* tubulin polymerization promoting protein 1, *PTEN* phosphatase and tensin homolog c-Myc, and α-adducin

ROCK/MLC phosphatase/MLC and ROCK/LIM kinase/cofilin are major pathways that regulate actin filament dynamics affecting cell contractility, motility, and morphology. ROCK promotes actomyosin contractility through increasing MLC phosphorylation and stabilizes actin filaments through LIM kinase activation, resulting in cofilin phosphorylation and thereby inhibiting its actin-depolymerization activity. Rho/ROCK activation has been implicated to activate other substrates that are directly or indirectly involved in the regulation of cell cycle (p^{27kip1}), pro-survival/proliferation (c-Myc, ERK1/2), cell apoptosis and cell polarity (PTEN), and tumor invasion/metastasis [7]. In addition, ROCK can associate with nucleolar phosphoprotein nucleophosmin (NPM-1) after phosphorylation by cyclin-dependent kinase 2 (CDK2)/cyclin E, which is important for the initiation of centrosome duplication and the coupling of centrosome duplication and DNA replication during S-phase.

Therapeutic Targeting of Rho/ROCK Pathway

Inhibitors of the Rho/ROCK pathway can be grouped into three classes: ROCK inhibitors (RIs), geranylgeranyl transferase-1 (GGTIs), and 3-hydroxy-3-methylglutaryl-Coenzyme A (HMG-

CoA) reductase (also known as statins) [8]. Technically speaking, statins and GGTase inhibitors (GGTIs) are not specific inhibitors of the Rho/ROCK activity but rather decrease the formation of isoprenoid intermediates required for the activation of Rho/Rho kinase (ROCK) and inhibit the prenylation of oncogenic GTPases including K-Ras, N-Ras, RhoA, RhoC, Cdc42, RalA, RalB, and Rac1, respectively. Because of this reason, we will only focus on ROCK inhibitors in the following section.

ROCK inhibitors are in use or in clinical trials for the treatment of several clinical conditions. Fasudil (ERIL™, Asahi Kasei Corporation, Japan), a potent adenosine triphosphate (ATP) competitor for ROCK binding, has been used for the treatment of cerebral vasospasm after subarachnoid hemorrhage in Japan since 1995 [9]. Hydroxyfasudil, the main metabolite of fasudil and H-1152P, an analog of fasudil, are more potent than the originator substance fasudil [9, 10]. In cancer, fasudil has been effective in reducing tumor growth in various cancer cell lines and in vivo models, suggesting the importance of ROCK signaling in the development and progression of cancer [9, 11]. Other ROCK inhibitors such as Wf-536 [12], H1152 [13], and RKI-1447 [14] also reduced tumor progression in a number of cancers including hepatocellular, lung, melanoma, and breast cancers. These early ROCK inhibitors are non-isoform specific, as they target the ATP-dependent kinase domain of ROCK1 and ROCK2. They also inhibit other serine/threonine kinases of AGC protein kinase family such as protein kinase A (PKA) and protein kinase C (PKC) at higher concentrations [11]. This fact may result in off-target effects based on the isoform-specific functions. For example, ROCK inhibitors led to induction of pressure overload cardiac hypertrophy in mice resulting in elevated ROCK1, but not ROCK2 expression [10]. These off-target effects may be prevented by developing isoform-specific inhibitors rather than pan-inhibitors that target the common ATP-dependent kinase domain. Targeting ROCK2 may have less toxicity than inhibitors targeting both isoforms or ROCK1. Attempts to produce more specific and clinically suitable ROCK inhibitors are ongoing, with an increased focus on isoform-specific regulation and inhibition [10]. With the advent of medicinal chemistry and high-throughput drug screening, novel inhibitors are tested mostly in preclinical models [4]. These methods opened the avenue to isoform-specific inhibitors, particularly more potent inhibitors of ROCK2. These inhibitors are mostly tested for diseases other than cancer. Table 15.1 shows the inhibitors tested in various cancers and their method of detection as biomarkers. However, most preclinical studies measured the tumor efficacy and invasion rather than blocking the expression of Rho/ROCK expression/activity.

Despite these efforts in preclinical phase, only one clinical trial using ROCK inhibitors in cancer treatment has been reported in clinical Trials.gov: AT13148 in phase 1 clinical trial initiated in 2012 for the treatment of advanced solid tumors (ClinicalTrials.gov identifier NCT01585701). AT13148 also inhibits several members of AGC kinase family including AKT and PKA. This may work better in a wider group of cancer patients as a multiple kinase inhibitor. Alternatively, it may introduce more off-target effect which needs to be investigated. The study is recruiting, and results are not yet available.

Challenges and Future Directions

ROCK isoforms have overlapping as well as distinct functions. However, ROCK inhibitors are not isoform selective. This fact presents certain limitations for the current ROCK inhibitors. One limitation is that high concentrations of ROCK inhibitors inhibit serine/threonine kinases in other enzymes such as protein kinase C (PKC) and protein kinase A (PKA); the other limitation is that they are not specific to tumor cells. In particular, their effect on cardiac hypertrophy and development of cardiac fibrosis is a major concern. Furthermore, their functions are tissue-context-dependent. Therefore, it is necessary to test isoform specific inhibition and understand their mechanisms of action. Some novel isoform-selective inhibitors are becoming commercially available and will serve as valuable tools for fur-

Table 15.1 ROCK inhibitors tested in various cancers and their method of detection as biomarkers

Rho/ROCK signaling pathway	Pathway active in cancer type	Affected biomarker/target	Method of detection	Active drugs
	Bladder, brain leukemia, lung, liver, ovarian	ROCK1/ROCK2	Immunohistochemical staining and/or Western blotting	Fasudil (ERIL™, Asahi Kasei Corporation, Japan and CoTherix, Inc. USA)
	Breast leukemia, melanoma, ovarian, prostate	ROCK1/ROCK2	No biomarker-related approach	Y-27632 [(+)-(*R*)-*trans*-4-(1-aminoethyl)-*N*-(4-pyridyl) cyclohexanecarboxamidedihydrochloride]
				Yoshitomi Pharmaceutical Industries Ltd., Iruma, Saitama 358, Japan [4]
	Breast	Rho pathway	No biomarker-related approach	H-1152 (4-methyl-5-[[(2S)-2-methyl-1,4-diazepan-1-yl] sulfonyl]isoquinoline) [15]
	Lung	ROCK activity,	ROCK kinase activity assay	PT262 (7-Chloro-6-piperidin-1-yl-quinoline-5,8-dione) [16]
		Phospho-MLC	Western blotting	
	Breast	Phospho-MLC-2 and phospho-MYPT1	Western blotting	RKI-1447 1-[(3-hydroxyphenyl)methyl]-3-(4-pyridin-4-yl-1,3-thiazol-2-yl)urea [14]
	Breast	Phospho-MLC-2	Western blotting	RKI-18 [17]
	Lung	AGC kinases including ROCK1/ROCK2	ProfilerPro kinase selectivity	OXA-06 [18]
		Phospho-MYPT1 and phospho-cofilin	Western blotting	
	Melanoma	AGC kinases (phospho-Akt, phospho-GSK3β, phsopho-70S6K, and RSK)	Western blotting	AT13148
				(1S)-2-amino-1-(4-chlorophenyl)-1-[4-(1H-pyrazol-4-yl) phenyl]ethanol [19]
Phase 1 trial	Breast, prostate, and ovarian	AGC kinases	No biomarker-related approach	AT13148 Clinicaltrials.gov: NCT01585701
				Sponsor: Cancer Research UK
	Melanoma	ROCK activity	High-throughput kinase assay profiling	CCT129254 (AT11854) [20]

The references are based on the drug information rather than indication for the specific cancers ther dissecting the roles of ROCK1 and ROCK2 and the clinical utility of Rho/ROCK in diseases including clinical cancer management.

Conflict of Interest The author has no conflict of interest.

References

1. Porter AP, Papaioannou A, Malliri A. Deregulation of Rho GTPases in cancer. Small GTPases. 2016;7(3):123–38.
2. Julian L, Olson MF. Rho-associated coiled-coil containing kinases (ROCK): structure, regulation, and functions. Small GTPases. 2014;5:e29846.
3. Mueller BK, Mack H, Teusch N. Rho kinase, a promising drug target for neurological disorders. Nat Rev Drug Discov. 2005;4(5):387–98.
4. Wei L, et al. Novel insights into the roles of Rho kinase in cancer. Arch Immunol Ther Exp. 2016;64(4):259–78.
5. Morgan-Fisher M, Wewer UM, Yoneda A. Regulation of ROCK activity in cancer. J Histochem Cytochem. 2013;61(3):185–98.
6. Etienne-Manneville S, Hall A. Rho GTPases in cell biology. Nature. 2002;420(6916):629–35.
7. Kale VP, et al. The regulatory roles of ROCK and MRCK kinases in the plasticity of cancer cell migration. Cancer Lett. 2015;361(2):185–96.
8. Amin E, et al. Rho-kinase: regulation, (dys)function, and inhibition. Biol Chem. 2013;394(11):1399–410.
9. Olson MF. Applications for ROCK kinase inhibition. Curr Opin Cell Biol. 2008;20(2):242–8.
10. Hahmann C, Schroeter T. Rho-kinase inhibitors as therapeutics: from pan inhibition to isoform selectivity. Cell Mol Life Sci. 2010;67(2):171–7.
11. Ying H, et al. The Rho kinase inhibitor fasudil inhibits tumor progression in human and rat tumor models. Mol Cancer Ther. 2006;5(9):2158–64.
12. Nakajima M, et al. Wf-536 prevents tumor metastasis by inhibiting both tumor motility and angiogenic actions. Eur J Pharmacol. 2003;459(2–3):113–20.
13. Teiti I, et al. In vivo effects in melanoma of ROCK inhibition-induced FasL overexpression. Front Oncol. 2015;5:156.
14. Patel RA, et al. RKI-1447 is a potent inhibitor of the Rho-associated ROCK kinases with anti-invasive and antitumor activities in breast cancer. Cancer Res. 2012;72(19):5025–34.
15. Fagan-Solis KD, et al. The RhoA pathway mediates MMP-2 and MMP-9-independent invasive behavior in a triple-negative breast cancer cell line. J Cell Biochem. 2013;114(6):1385–94.
16. Tsai CC, et al. 7-Chloro-6-piperidin-1-yl-quinoline-5,8-dione (PT-262), a novel ROCK inhibitor blocks cytoskeleton function and cell migration. Biochem Pharmacol. 2011;81(7):856–65.
17. Patel RA, et al. Identification of novel ROCK inhibitors with anti-migratory and anti-invasive activities. Oncogene. 2014;33(5):550–5.
18. Vigil D, et al. ROCK1 and ROCK2 are required for non-small cell lung cancer anchorage-independent growth and invasion. Cancer Res. 2012;72(20):5338–47.
19. Xi Y, et al. AT13148, a first-in-class multi-AGC kinase inhibitor, potently inhibits gastric cancer cells both in vitro and in vivo. Biochem Biophys Res Commun. 2016;478(1):330–6.
20. Sadok A, et al. Rho kinase inhibitors block melanoma cell migration and inhibit metastasis. Cancer Res. 2015;75(11):2272–84.

Mitogen-Activated Protein Kinase (MAPK) Signaling

16

Andrei Zlobin, Jeffrey C. Bloodworth, and Clodia Osipo

Introduction to MAPK Signaling

One of the key mechanisms to transmit extracellular signals from the membrane to intracellular compartments and to the nucleus is through protein phosphorylation. Two groups of enzymes involved in this process are mitogen-activated protein (MAP) kinases and threonine-tyrosine dual-specificity phosphatases (DUSP) also known as MAPK phosphatase (MKP). MAPK phosphorylation events can be reversed by MKPs that dephosphorylate both phosphothreonine and phosphotyrosine residues on MAPKs. They have opposing roles on signaling, and their activation or downregulation is tightly coordinated to confer appropriate signaling. Their well-concerted activity is important in maintaining an appropriate signaling balance inside cells. "In particular, mitogen-activated protein kinases (MAPKs; ERK1/2, p38, JNK, and ERK5) transduce environmental and developmental signals (growth factors or stress) into adaptive and programmed responses such as survival, proliferation, differentiation, inflammation, and apoptosis" [1, 2] (Fig. 16.1). The MAP kinases lie downstream of a wide variety of effectors including receptor tyrosine kinases, G-protein-coupled receptors, receptor serine/threonine kinases, Src family kinases, and cytokine receptors among many others. Activation of these effectors is typically induced by a ligand-binding event which initiates a cascade of enzymatic reactions. The activation of MAP kinases includes a core of three-kinase cascades consisting of a MAP kinase kinase kinase (MAP 3K or MAPKKK) which phosphorylates and activates a MAP kinase kinase (MAP 2K, MEK, or MKK) which subsequently phosphorylates and increases the activity of one or more MAP kinases [1]. Upon activation, MAPKs can phosphorylate a variety of intracellular targets including transcription factors, nuclear pore proteins, membrane transporters, cytoskeletal elements, and protein kinases. "Mitogen-activated protein kinases are protein kinases that phosphorylate their own dual serine and threonine residues (auto-phosphorylation), or those found on their substrates, to activate or de-activate their targets. MAPKs are ubiquitously expressed and evolutionarily conserved in eukaryotes" [3]. Some authors describe up to six different groups of MAPKs in mammalian cells, namely, extracellular signal-regulated

A. Zlobin
Oncology Research Institute, Loyola University Medical Center, Maywood, IL, USA

J. C. Bloodworth
Cardinal Bernardin Cancer Center, Loyola University Chicago, Maywood, IL, USA

C. Osipo (✉)
Department of Microbiology and Immunology, Stritch School of Medicine, Cardinal Bernardin Cancer Center of Loyola University Chicago, Maywood, IL, USA
e-mail: cosipo@luc.edu

S. Badve, G. L. Kumar (eds.), *Predictive Biomarkers in Oncology*,
https://doi.org/10.1007/978-3-319-95228-4_16

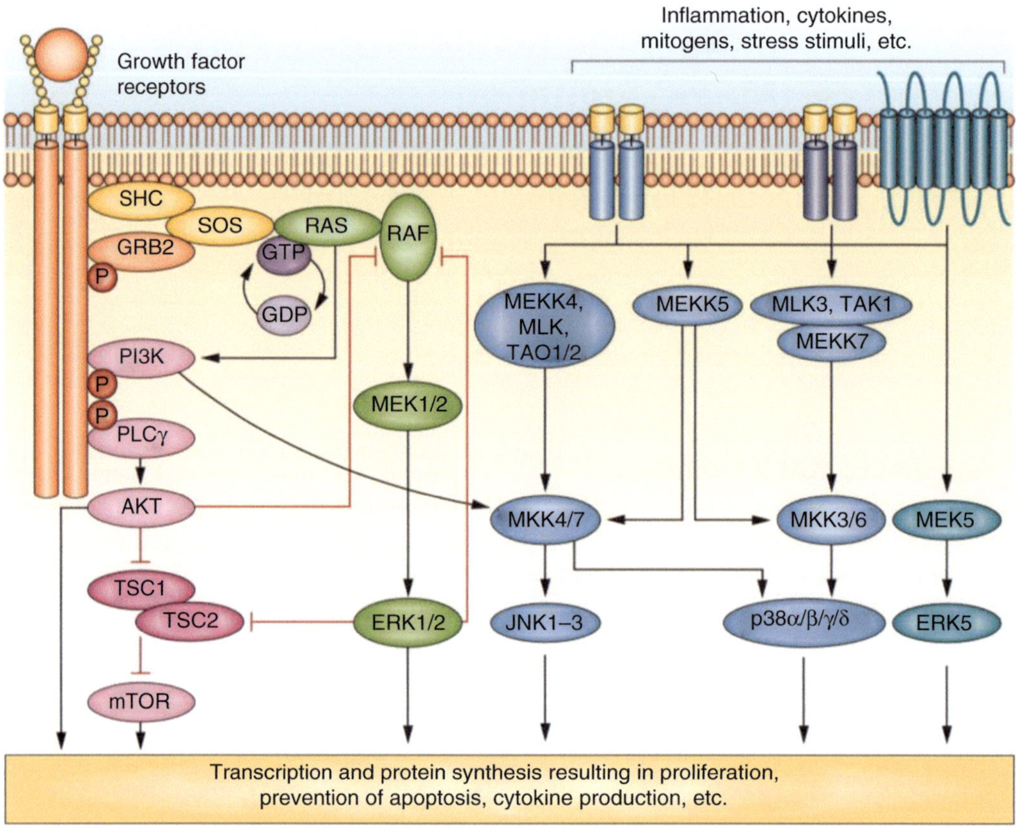

Fig. 16.1 Simplified schematic of MAPK signaling. Receptor tyrosine kinase activation leads to signaling via the RAS-RAF-MEK1/2-ERK1/2 MAPK pathway, which triggers a number of cellular responses that are relevant to cancer cell survival and proliferation, including transcription and protein synthesis, protection from apoptosis, and DNA replication. Other parts of MAPK pathways are associated with cell proliferation, motility, invasiveness, and angiogenesis, and some of these pathways might also be activated downstream of the RAS-RAF pathway. In addition, other signal cascades—such as the PI3K-AKT pathway—are activated, and cross talk with the RAS-RAF-MEK1/2-ERK1/2 pathway exists. Indeed, co-inhibition of these pathways might be required to optimize treatment efficacy. *Abbreviations*: GRB2 growth-factor-receptor-bound protein 2, mTOR mammalian target of rapamycin, PLCγ phospholipase Cγ, SHC Src homology two domain-containing-transforming protein, SOS Son of Sevenless, TSC1/2 tuberous sclerosis protein 1/2. (Reprinted from Zhao and Adjei [19]. With permission from Springer Nature)

kinase (ERK) 1/2, ERK3/4, ERK5, ERK7/8, Jun N-terminal kinase (JNK1, JNK2, JNK3), and the p38 isoforms α/β/γ, (ERK6)/δ [1], while others group them as four "well-known MAPK pathways: ERK1/2; the c-JUN N-terminal kinase 1, 2 and 3 (JNK1/2/3); the p38 MAPK α, β, δ and γ pathways and ERK5 (Figs. 16.1 and 16.2). In this later system, ERK, JNK, and p38 isoforms are grouped according to their activation motif, structure, and function" [3]. ERK1/2 is activated in response to growth factors, hormones, and pro-inflammatory stimuli, while JNK1/2/3 and p38 MAPK α, β, δ, and γ are activated by cellular and environmental stresses, in addition to pro-inflammatory stimuli (Figs. 16.1 and 16.2). Commonly, "these kinases are designated from upstream to downstream, toward to the nucleus, for example MAPK kinase-kinase (MAPKKK), MAPK kinase (MAPKK) and MAPK. The most basic MAPK/ERK pathway is composed of three types of MAPKKK: A-RAF, B-RAF and RAF-1 or C-RAF kinases (Fig. 16.1). Interestingly, *BRAF* is the most commonly mutated gene at this level in human cancers. One level below are the MAPKKs, which are composed of MEK1 and MEK2. Further downstream are ERK1 and

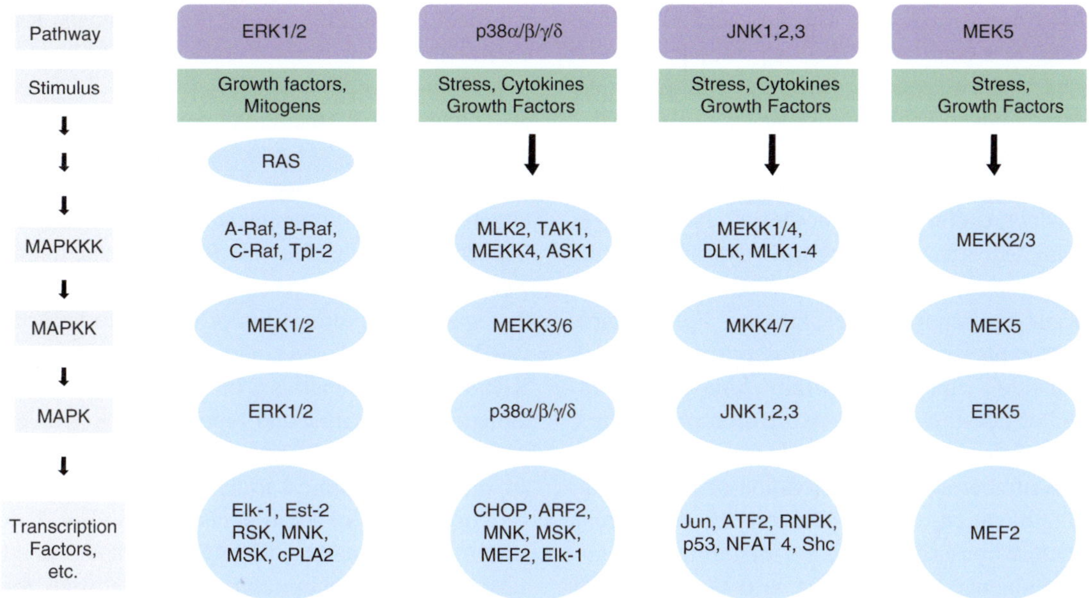

Fig. 16.2 MAPK signaling showing four major mammalian protein kinase cascades: (I) ERK1/2; (II) p38 MAPK α, β, γ, δ; (III) c-JUN N-terminal kinases 1, 2, and 3 (JNK1/2/3); and (IV) MEK5

ERK2, and transcription factors Elk-1, Est-2, RSK, MNK, MSK and cPLA2 which are the final effectors of the MAPK pathway" [4] (Figs. 16.1 and 16.2). The transcription factors described above bind to specific DNA sequence motifs and activate transcription of genes such as *CCND1* (coding for cyclin D1). The biological effect of expression of these genes is proliferation of a cell. Another important feature of MAPK signaling is its cross talk between molecules and feedback or forward loops actively engaged in an exchange of different signals (Fig. 16.2) [5].

ERK Pathway

ERK (extracellular signal-regulated kinase), also known as MAPK, is involved in the regulation of meiosis, mitosis, and post-mitotic functions in differentiated cells. Disruption of the ERK pathway is common in many human cancers, particularly Ras, c-Raf, and tyrosine kinase receptors such as HER2. Multiple extracellular signals such as growth factors or mitogens are able to activate ERK signaling. For instance, binding of epidermal growth factor (EGF) to one of its cognate receptors causes the receptor to dimerize with another identical or closely related EGF receptor. Upon dimerization, the intracellular tyrosine kinase domains of these receptors phosphorylate each other, thus creating phosphotyrosine sites which can be recognized by the SH2 domain of the scaffolding protein GRB2. GRB2 recruits the guanine nucleotide exchange factor denoted as Son of Sevenless (SoS) via SH3 domain interaction, which effectively localizes SoS to the cell membrane. Membrane localization brings SoS into proximity to its target, Ras [6]. SoS activates Ras by the hydrolysis of GTP. The GTP-bound Ras adopts an active conformation which can in turn bind to and induce conformational activity of its target, the MAP kinase kinase kinase, Raf. Raf phosphorylation initiates a cascade of phosphorylation events wherein Raf activates the MAP kinase kinase MEK and MEK activates the MAP kinases ERK1 and ERK2. ERK1/2, in turn, phosphorylates a number of substrates that generally act to promote cell survival and proliferation. One well-described substrate of ERK1/2 is c-Fos. c-Fos is a member of the AP-1 transcription factor. ERK1/2 phosphorylates c-Fos, thereby stabilizing it and encouraging dimerization with other

AP-1 family members. The activated AP-1 dimer binds to a consensus DNA sequence proximal to genes which generally favor survival and proliferation phenotypes. In a classic example of AP-1-mediated control of proliferation, AP-1 activates the transcription of the CCND1, and the protein product, cyclin D1, promotes cell cycle progression. "Most cancer-associated dysfunctions that lead to constitutive activation of ERK signaling occur upstream of the pathway. For example, over-expression of receptor tyrosine kinases, activating mutations within kinase domains, and/or constitutive over-expression of ligands to name a few. Ras and B-Raf mutations are examples of aberrant hyper-activation of the pathway. Amplification or deregulation of downstream nuclear transcription factors including MYC and components of AP-1 have also been reported" [3]. The high rate of Ras-Raf mutations deserves particular attention in terms of carcinogenesis. For instance, whereas the Ras→Raf→ERK1/2 cascade is the accepted mode of ERK1/2 activation, the typical mode of JNK activation is TAK1→MKK4→JNK. Though these putative pathways have been described, it is very basic, as extensive cross talk, feedback, forward loops, and transactivation in between many of signaling cascades have been reported [7].

RAS

The three Ras genes that exist in mammals (H-Ras, K-Ras, and N-Ras) are the most common oncogenes in human cancer. Mutations in K-Ras and N-Ras are activating by nature and are present in many types of malignancies (e.g., pancreatic cancer) [8, 9]. For this reason, Ras inhibitors are being investigated as a treatment for cancer and other diseases with Ras hyperactivation and/or overexpression. Ras GTPases control the activity of many signaling pathways. Ras as a small GTPase hydrolyzes bound GTP to GDP and in its GDP form is inactive. When it is mutated, Ras remains in its GTP-bound state and is constitutively active and enhances activation of downstream effectors. Subcellular localization of Ras oncogenes has been shown to play a critical role in the transduction of signals. Ras creates numerous complexes with scaffold proteins, namely, KSR and SUR-8/SHOC-2, which also regulate the activation of downstream targets such as Raf.

There is active research in exploring Ras for targeted cancer treatments. Some examples include reoviruses that kill Ras-activated tumor cells [10], type II herpes simplex virus (HSV-2) that specifically targets tumor cells with an activated Ras pathway [11], and siRNA anti-mutated K-RAS targeted treatment. In fact, Reolysin, a formulation of reovirus, and type II herpes simplex virus (HSV-2)-based agent, designated FusOn-H2, are currently in clinical trials for the treatment of various cancers. In addition, a treatment based on siRNA anti-mutated K-RAS (G12D) called siG12D LODER is currently in clinical trials for the treatment of locally advanced pancreatic cancer (NCT01188785, NCT01676259).

RAF

RAF is an acronym for **r**apidly **a**ccelerated **f**ibrosarcoma. The first *raf* gene was described in 1983 as a retroviral oncogene, *v-raf*, transduced by the murine sarcoma virus isolate 3611. This gene encodes for the Raf protein which is an essential connector between Ras and the MEK-ERK pathway. Raf kinases are the target of Ras by direct protein-protein interaction that leads to a cascade of phosphorylation events, ultimately activating ERK. In addition, all Raf proteins share MEK1/2 kinases as substrates. In general three Raf paralogs (A-Raf, B-Raf, and c-Raf aka Raf-1) are similar, but there are significant differences in structure, and they have different mechanisms of activation [12]. Particularly among the *Raf* genes, *B-Raf* is the most well studied, and approximately 20% of all examined human tumor samples display a mutated B-Raf gene [13]. The overwhelming majority of these mutations involve the exchange of a single amino acid: Val 600 into Glu, and this aberrant gene product (BRAF-V600E) can be visualized by immunohistochemistry for clinical molecular diagnostics. There are several Raf inhibitors that selectively target the BRAF-V600E mutant that have been developed to combat cancer. Examples include sorafenib, vemurafenib, regorafenib, and dabrafenib.

MEK and ERK

Mitogen-activated protein kinase/extracellular signal-regulated kinase pathway or MAPK/ERK pathway (also known as the Ras-Raf-MEK-ERK pathway) is a chain of proteins in the cell that communicates a signal from a receptor on the membrane to the DNA in the nucleus. When one of the proteins in the MEK/ERK pathway is mutated, it can become stuck in the "on" or "off" position, resulting in the development of many cancers. The first drug licensed to act on this pathway is sorafenib—a Raf kinase inhibitor. Other Raf prominent inhibitors include dabrafenib and vemurafenib. Some MEK inhibitors include cobimetinib, binimetinib (MEK162), selumetinib, and trametinib (see Table 16.1).

B-Raf is by far the most active MEK. Phosphorylation of MEK at position S298 promotes interaction with Raf, while phosphorylated S212 has a repressive effect. Upon activation, members of the ERK group regulate a number of various processes by phosphorylating regulatory components of cells with different localization. A number of genes with different functions involved in regulation of ERK signaling are Fos, Jun, Myc, Egr-1, p21, and p27. Also, MAP kinase phosphatases (MKPs) and Sprouty group members play an important role. Regulation of kinase activity is achieved by the counterbalance of different positive and negative feedback loops, cross talk with other pathways, including oncogenes and tumor suppressors [14].

The JNK Pathway

The c-Jun N-terminal protein kinases (JNK) are responsible for phosphorylating c-Jun at Ser-63 and Ser-73. These master protein kinases regulate many physiological processes, including inflammatory responses, morphogenesis, cell proliferation, differentiation, survival, and death. However, the physiological and pathological functions of JNK signaling have been very difficult to predict because of the contradictory role of JNK in promoting cell survival and proliferation on one hand and cell death on the other. This could be due to cell context dependency.

The p38 MAPK Pathway

One of the important components of MAPK pathway is p38. It is mainly activated by JNK. However, it can also be regulated by MEK3, MEK4, and MEK6. There are four isoforms of p38, α, β, γ which is also described as (ERK6), and γ form. Their distinctive feature is a specific phosphorylation TGY (Thr-Gly-Tyr) motif. The p38 MAPKs are responsive to environmental stresses and inflammatory cytokines. Upon activation, p38 MAPK proteins can translocate into the nucleus where they, in turn, phosphorylate serine/threonine residues on multiple substrates. The p38 MAPK is also involved in the regulation of HSP27, MAPKAPK-2 (MK2), MAPKAPK-3 (MK3), and several transcription factors including ATF-2, Stat1, the Max/Myc complex, MEF-2, Elk-1, and indirectly CREB via activation of MSK1 leading to cytokine production and apoptosis. In addition to their function in stress responses, p38 pathway MAPK proteins play critical roles in cell cycle, growth, and differentiation. Analysis of the function of the p38 α gene or its activators MEK3 and MEK6 has suggested that p38 may be a tumor suppressor [15]. Repression of p38 MAPK activity enhances apoptosis in response to cisplatin as well as taxol, vincristine, and vinblastine. Many chemotherapeutic agents require p38 MAPK activity for the induction of apoptosis [16].

MEK4 (MAP 2K4)/MKK4

MEK4/MKK4 (or *JNKK1*, *MAP 2K4*, and *SEK1*) is a dual-specificity kinase gene on chromosome 17p11. MKK4 protein is activated by over ten kinases, and active MKK4 cooperates with

Table 16.1 Approved therapeutic agents for MAP kinase

MAP kinase	Pathway active in cancer type	Affected biomarker	Method of detection	Target	Approved drug(s)/company ClinicalTrials.gov Identifier
MAP kinase	Melanoma	MEK1/2, MEK1/2	Western blot analysis	MEK1/2, MEK1/2	Trametinib (Mekinist ®) Novartis
	Non-small cell lung cancer	p15^{INK4b} and/or p27^{KIP1}	NanoPro™1000 technology (ProteinSimple, San Jose, CA) for signaling molecule phosphorylation ELISA and dot blot	p15^{INK4b} and/or p27^{KIP1}	ClinicalTrials.gov Identifier
		ERK		ERK	NCT01723202
					NCT02672358
					NCT02939846
					NCT02447939
					NCT02645149
					NCT02967692
MAP kinase	Stage III or IV differentiated thyroid cancer (DTC)	ERK1/2 phosphorylation, p90^{RSKc} caspase-3 caspase-7, cyclin D1	Immunohistochemistry (IHC)	MEK1	Selumetinib (AZD6244)
	Solid tumors and colorectal cancer locally advanced	Cdc-2	NanoPro™1000 technology (ProteinSimple, San Jose, CA) for signaling molecule phosphorylation	ERK1/2	AstraZeneca
	Adenocarcinoma of the rectum	CDK2			ClinicalTrials.gov Identifier
	Neurofibromatosis type 1 and cutaneous neurofibroma	CDK4			NCT01287130
		Cyclin B1			NCT01134601
		c-Myc			NCT01306045
					NCT03040986
					NCT03004105
					NCT02839720
MAP kinase	Advanced melanoma	BRAF V600E non-V600E	Immunohistochemistry (IHC)	MEK1	Cobimetinib (GDC-0973, RG7420)
	Chronic lymphocytic leukemia		NanoPro™1000 technology (ProteinSimple, San Jose, CA) for signaling molecule phosphorylation		(Cotellic ®)
	Hodgkin lymphoma				Genentech (Roche)
	Non-Hodgkin lymphoma				NCT02908672
					NCT03005639
					NCT02968303
					NCT02902029
					NCT02876224
					NCT02721459

MAP kinase	Gastrointestinal stromal tumor	RET	Enzyme-linked immunosorbent assay	VEGFR1, VEGFR2, VEGFR3, PDGFRβ, Kit	Regorafenib (BAY 73-4506)
	Colorectal cancer	Raf-1	Immunohistochemistry (IHC)	RET and Raf-1	(Stivarga®),
	Hepatocellular carcinoma	VEGFR2	In situ hybridization, immunohistochemistry		Bayer HealthCare Pharmaceuticals Inc
		Kit	RT-PCR		NCT02795156
		VEGFR1			NCT03042689
					NCT02940223
					NCT02955940
					NCT02889328
MAP kinase	Melanoma, colon cancer	B-Raf^{V600E}	Enzyme-linked immunosorbent assay	B-Raf^{V600E} SRMS	Vemurafenib (PLX4032, RG7204), RO5185426 (Zelboraf®)
	Non-small cell lung cancer	MEK1/2 and ERK1/2	Immunohistochemistry (IHC)	ACK1	Hoffmann-La Roche Inc.
	Solid tumorl lymphoma, stage IIIB–C melanoma		In situ hybridization, immunohistochemistry	C-Raf	NCT01585415
			RT-PCR	MAP4K5 (KHS1)	NCT01843738
					NCT02314481
					NCT03013491
					NCT02908672
					NCT03005639
MAP kinase	Primary kidney cancer	C-Raf	Enzyme-linked immunosorbent assay	B-Raf	Sorafenib
	Advanced primary liver cancer	B-Raf	Immunohistochemistry (IHC)	(V599E)	(Nexavar®)
	Radioactive iodine resistant advanced thyroid carcinoma	CYP3A4 UGT1A9	In situ hybridization, immunohistochemistry	VEGFR	Bayer
		(Flk-1)	RT-PCR	mVEGFR2 (Flk-1)	NCT00727233
		mVEGFR3mPDGFRβ		mVEGFR3, mPDGFRβ, Flt3 c-Kit	NCT02989870
		Flt3		PDGFR	NCT01445080
		c-Kit		VEGFR2	NCT01434602
		PDGFR VEGFR2		Flt-3	NCT02988440
		Flt-3		eIF4E	NCT03037437
		eIF4E		Mcl-1	
		Mcl-1			

(continued)

Table 16.1 (continued)

MAP kinase	Pathway active in cancer type	Affected biomarker	Method of detection	Target	Approved drug(s)/company ClinicalTrials.gov Identifier
MAP kinase	Hepatocellular carcinoma	C-Raf	Enzyme-linked immunosorbent assay	Raf-1	Sorafenib
	Renal cell carcinoma	B-Raf	Immunohistochemistry (IHC)	VEGFR2/Flk1B-Raf	Tosylate
	Thyroid cancer	CYP3A4 UGT1A9	In situ hybridization, immunohistochemistry	B-Raf (V599E)	Bayer
		(Flk-1)	RT-PCR	PDGFRβ	NCT02728050
		mVEGFR3mPDGFRβFlt3			NCT02779283
					NCT02143401
					NCT02035527
					NCT02066181
					NCT02050919
MAP kinase	Melanoma	B-Raf(wt)	Enzyme-linked immunosorbent assay	BRAF V600	Dabrafenib (GSK2118436)
	Non-small cell lung cancer	C-Raf, v	Immunohistochemistry (IHC)	B-Raf(wt)	(Tafinlar®)
			In situ hybridization, immunohistochemistry		
			RT-PCR	C-Raf	GlaxoSmithKline
					NCT01723202
					NCT02672358
					NCT02447939
					NCT02967692
					NCT02858921
					NCT03026517

MKK7 to phosphorylate and thereby activate Jun NH_2-terminal kinase (JNK) in the stress-activated cascade. MKK4 is commonly disrupted by a mutation in cancers of the breast, pancreas, bile ducts, colon, lung, and testis [17]. In some ovarian cancers, MEK4 expression has been shown to be suppressed in 75% of cases, leading to the possibility that *MKK4* is either a tumor-suppressor or genome-maintenance gene. The exact mechanism how MEK4 can drastically suppress tumorigenesis is far from being understood [18].

Summary

MAPK pathway is one of the most important signaling mechanisms involved in the regulation of gene expression, cellular growth, and survival. Abnormal MAPK signaling may lead to increased or uncontrolled cell proliferation (cancer) and resistance to apoptosis, chemotherapy, radiotherapy, and targeted therapies. One of the most studied and understood mechanisms of MAPK pathway is the overactivation of MAPK signaling by oncogenic BRAF in multiple malignancies such as melanoma tumors, papillary thyroid tumors, serous ovarian tumors, and colorectal tumors, making it a potential target in oncology. Based on these findings, a number of companies are developing anti-MAPK drugs to target MAPK signaling.

References

1. Avruch J. MAP kinase pathways: the first twenty years. Biochim Biophys Acta. 2007;1773(8):1150–60.
2. Escos A, et al. p38gamma and p38delta mitogen activated protein kinases (MAPKs), new stars in the MAPK galaxy. Front Cell Dev Biol. 2016;4:31.
3. Dhillon AS, et al. MAP kinase signalling pathways in cancer. Oncogene. 2007;26(22):3279–90.
4. Burotto M, et al. The MAPK pathway across different malignancies: a new perspective. Cancer. 2014;120(22):3446–56.
5. Rauch N, et al. MAPK kinase signalling dynamics regulate cell fate decisions and drug resistance. Curr Opin Struct Biol. 2016;41:151–8.
6. Segal RA, Greenberg ME. Intracellular signaling pathways activated by neurotrophic factors. Annu Rev Neurosci. 1996;19:463–89.
7. Wagner EF, Nebreda AR. Signal integration by JNK and p38 MAPK pathways in cancer development. Nat Rev Cancer. 2009;9(8):537–49.
8. Sebolt-Leopold JS, Herrera R. Targeting the mitogen-activated protein kinase cascade to treat cancer. Nat Rev Cancer. 2004;4(12):937–47.
9. Yan C, Theodorescu D. RAL GTPases: biology and potential as therapeutic targets in cancer. Pharmacol Rev. 2018;70(1):1–11.
10. Lal R, et al. Reovirus: rationale and clinical trial update. Curr Opin Mol Ther. 2009;11(5):532–9.
11. Thirukkumaran C, Morris DG. Oncolytic viral therapy using reovirus. Methods Mol Biol. 2009;542:607–34.
12. Wellbrock C, Karasarides M, Marais R. The RAF proteins take centre stage. Nat Rev Mol Cell Biol. 2004;5(11):875–85.
13. Forbes SA, et al. COSMIC: mining complete cancer genomes in the catalogue of somatic mutations in cancer. Nucleic Acids Res. 2011;39(Database issue):D945–50.
14. Mirza AM, et al. Cooperative regulation of the cell division cycle by the protein kinases RAF and AKT. Mol Cell Biol. 2004;24(24):10868–81.
15. Bulavin DV, Fornace AJ Jr. p38 MAP kinase's emerging role as a tumor suppressor. Adv Cancer Res. 2004;92:95–118.
16. Cicenas J, et al. JNK, p38, ERK, and SGK1 inhibitors in cancer. Cancers (Basel). 2017; 10(1).
17. Cunningham SC, et al. Targeted deletion of MKK4 in cancer cells: a detrimental phenotype manifests as decreased experimental metastasis and suggests a counterweight to the evolution of tumor-suppressor loss. Cancer Res. 2006;66(11):5560–4.
18. Krishnan V, et al. Using MKK4's metastasis suppressor function to identify and dissect cancer cell-microenvironment interactions during metastatic colonization. Cancer Metastasis Rev. 2012;31(3–4):605–13.
19. Zhao Y, Adjei AA. The clinical development of MEK inhibitors. Nat Rev Clin Oncol. 2014;11:385–400.

Notch Signaling Pathway in Carcinogenesis

17

Andrei Zlobin, Jeffrey C. Bloodworth, Andrew T. Baker, and Clodia Osipo

Introduction

The *NOTCH* gene owes its name to the notched phenotype observed by J.S. Dexter in 1914 at Olivet College in Olivet, Michigan, in the wingtips of mutant fruit flies *Drosophila melanogaster*. Later, the alleles of the gene were identified by T.H Morgan in 1917 at Columbia University in New York City, New York. Since then, the research on *NOTCH* performed on a number of species from worms to humans has led to the establishment of an evolutionarily conserved Notch signaling pathway [1]. The Notch signaling cascade is critical for development, cell proliferation, differentiation, and homeostasis. Aberrant Notch signaling is found in various cancers, such as breast, prostate, lung, colorectal, T-cell leukemia, as well as central nervous system (CNS) malignancies.

In mammals, there are four Notch receptors, Notch1–4, and five ligands, delta-like ligand 1 (DLL1), delta-like ligand 3 (DLL3), delta-like ligand 4 (DLL4), Jagged-1 (JAG1), and Jagged-2 (JAG2). Notch proteins span a cell's plasma membrane, extend outward into extracellular space, and function as receptors to receive signals from neighboring cells (see Fig. 17.1).

Activation of Notch signaling involves binding of the Notch ligand to the Notch receptor, followed by endocytosis of the extracellular portion of the receptor into the ligand bearing, signal-sending cell. The pulling force on the Notch receptor generated by endocytosis enables the LAG-12 domain of the Notch receptor to unfold. This exposes the S2 cleavage site making it vulnerable to proteolysis by either ADAM-10 or ADAM-17, a member of metalloproteinase (ADAM) group of proteinases. The S2 cleavage yields two fragments: the N-terminal fragment (N^{ECD}) and the C-terminal fragment (NEXT). The N-terminal fragment bound to the ligand is internalized into the signal sending cell. The C-terminal fragment, NEXT domain, undergoes S3/4 cleavage by the γ-secretase complex. For many proteins, the transmembrane domain is viewed as an inert linker between extracellular and intracellular domains. However, this is not the case for substrates of the γ-secretase complex, including Notch, for which the transmembrane domain

A. Zlobin
Oncology Research Institute, Loyola University Medical Center, Maywood, IL, USA

J. C. Bloodworth
Cardinal Bernardin Cancer Center, Loyola University Chicago, Maywood, IL, USA

A. T. Baker
Department of Integrative Cell Biology, Loyola University Chicago, Maywood, IL, USA

C. Osipo (✉)
Department of Microbiology and Immunology, Stritch School of Medicine, Cardinal Bermardin Cancer Center of Loyola University Chicago, Maywood, IL, USA
e-mail: cosipo@luc.edu

S. Badve, G. L. Kumar (eds.), *Predictive Biomarkers in Oncology*,
https://doi.org/10.1007/978-3-319-95228-4_17

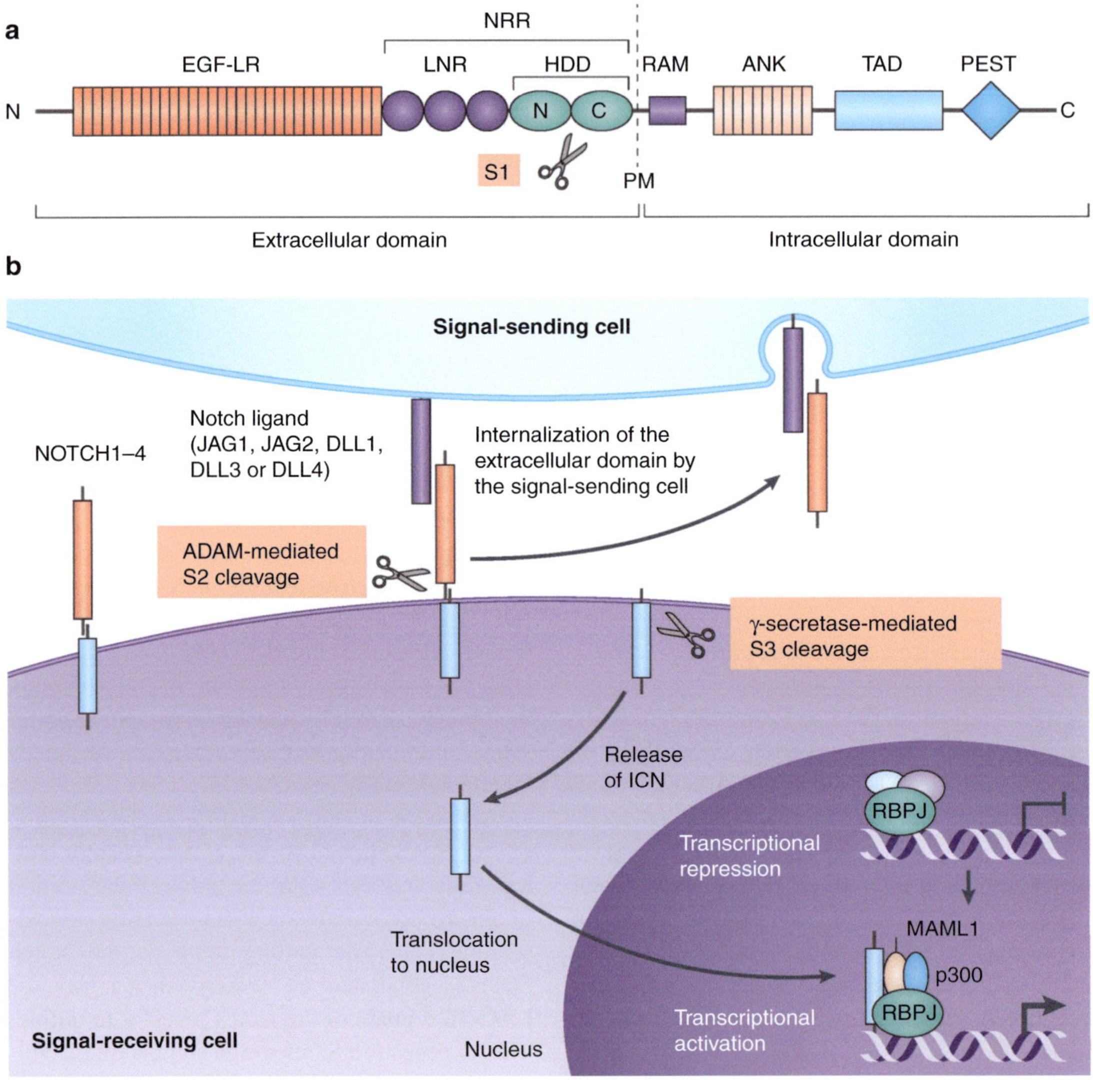

Fig. 17.1 Structure of Notch receptors and the Notch signaling cascade. (**a**) The Notch receptors (NOTCH1–4) are heterodimers consisting of an extracellular ligand-binding domain and an intracellular domain that mediates target gene transcription. The extracellular portion consists of epidermal growth factor (EGF)-like repeats (EGF-LR), which are essential for ligand binding, and LIN12-NOTCH repeats (LNR) and N- and C-terminal heterodimerization domains (HDDs) that together form the negative regulatory region (NRR). The NRR prevents ligand-independent Notch activity. Before transport to the cell surface, an S1 cleavage mediated by furin-like convertase occurs in the Golgi within the HDD, which is thought to be important for the generation of the mature receptor, although the precise function is unclear. The intracellular domain contains the recombining binding protein suppressor of hairless (RBPJ) association molecule (RAM), ankyrin repeats (ANK) that contain nuclear localization signals (NLSs) in NOTCH1–3 (the NLS domain in NOTCH4 is not well defined), a transactivation domain (TAD), and a polypeptide enriched in proline, glutamate, serine, and threonine (PEST) sequence important for degradation of the receptor. (**b**) Interaction between NOTCH receptors 1–4 and ligand jagged1 (JAG1), JAG2, or delta-like ligand 1 (DLL1), DLL3, or DLL4 results in an S2 cleavage in the extracellular portion close to the transmembrane domain mediated by disintegrin and metalloproteinase domain-containing protein (ADAM) metalloproteinases. This removes the ligand-binding domain, leaving the intracellular portion anchored to the plasma membrane. Subsequently, an S3 cleavage in the transmembrane domain mediated by the γ-secretase complex liberates the intracellular domain of Notch (ICN) from the plasma membrane, facilitating translocation to the nucleus, where it associates with the transcriptional repressor RBPJ. This displaces transcriptional repressors and recruits coactivators, such as Mastermind-like 1 (MAML-1) and p300, resulting in the transcription of genes bound by the ICN–RBPJ complex. (Reprinted from Nowell and Radtke [15]. With permission from SpringerNature)

serves as a substrate for a third proteolytic cleavage S3. The γ-secretase complex is a multi-pass transmembrane protein composed of a four subunits: Presenilin, Nicastrin, APH1, and PEN2. Presenilin, the catalytic subunit of the complex, is an aspartyl proteinase which has a vast number of transmembrane protein substrates. Though the other members of the γ-secretase complex lack enzymatic activity, they are critical cofactors for enzyme activity and substrate specificity. Nicastrin is able to specifically interact with the N-terminal cleavage product formed after S2 cleavage. Nicastrin binding brings the NEXT domain into position within the γ-secretase complex allowing Presenilin to perform cleavage at S3 in the NEXT domain. Cleavage by γ-secretase releases the Notch intracellular domain (NICD) from the membrane, enabling NICD to be translocated to the nucleus and carry out its function as a transcriptional activator.

Notch Nuclear Activity

The extracellular structure of the four Notch paralogs is similar, although not identical, including the different numbers of EGF-like repeats. Also, the cytoplasmic portions of the four Notch intracellular domains (NICDs) differ in terms of their structure and function. These different structures confer diversity and complexity to the Notch signaling pathway. All of the Notch paralogs harbor nuclear localization signals, which facilitate translocation to the nucleus after the S3 cleavage. Once in the nucleus, Notch facilitates the formation of an active transcription complex, which is centered about the CSL (CBF-1, suppressor of hairless, Lag-2 after its orthologues, respectively) DNA-binding protein, also known as RBPκJ (recombination signal-binding protein for immunoglobulin kappa J). In the absence of nuclear NICD, CSL inhibits transcription by recruiting transcriptional repressors such as NCoR or SMRT. Upon NICD recruitment to CSL, NICD displaces the transcriptional repressors and recruits transcriptional coactivators such as the p300/CAF histone acetyltransferase. The NICD/CSL interaction is stabilized by another protein called Mastermind-like 1 (MAML-1), a factor that is necessary for recruitment of additional coactivators. The domains required for these protein-protein interactions have been described in detail by Bray et al. The N-terminus of NICD harbors the RAM domain, which is the site responsible for the NICD/CSL interaction. MAML binds to NICD at the ankyrin repeat domain, which consists of seven highly conserved ankyrin motifs. Targeting transcription factors or activators has been challenging for drug discovery. The structural identification of the Notch1/MAML1 interface has allowed for development of novel MAML1-specific inhibitors [2].

Notch in Cancer

Notch facilitates conversion of the CSL transcriptional repressor to a transcriptional activator and results in a complex cascade of events. Among these events is primary event represented by transcription of the primary Notch target genes HEY (hairy/enhancer-of-split related with YRPW motif protein) and HES (hairy enhancer of split). In turn, Notch signaling regulates a number of genes. The most notable of these genes include p21, c-Myc, cyclinD1, $p27^{kip1}$, Slug, NF-kB, and a comprehensive list of genes that can be found in reviews devoted to Notch signaling [1]. Many of the Notch target genes are implicated in targeted combination therapy of several different types of cancer. Notch signaling functions in a synergistic fashion with a number of different oncogenes. Notch1 is required for the transforming activity of H-Ras and TGF-α during pancreatic tumorigenesis [3]. The formation of adenocarcinomas and their metastases in transgenic mouse models demonstrates the synergistic effects between Notch and MYC.

Notch1 is involved in cross talk with the PI3K-AKT pathway. It has been shown that malignant transformation by Notch requires signals from the ERK/MAP kinase and PI-3 kinase (PI3K) pathways downstream of Ras [4]. Cross talk between Notch and the estrogen receptor in breast cancer has also been extensively investigated as it was shown that estrogen inhibits Notch signaling, while estrogen deprivation reactivates the Notch pathway. Another important cross talk mechanism between Notch and HER2 receptors has been shown in a breast cancer model. HER2 overexpression inhibits

Notch signaling, while repression of HER2 transcription, or HER2 pharmacological inhibition, results in increased Notch signaling [4, 5]. Also of significance is the correlation between Notch1, Notch4, and PEA3. PEA3 is a transcription factor whose expression has been connected with tumorigenesis possibly through activation of Notch1 and Notch4 [6].

Notch signaling contributes to epithelium-mesenchymal transition (EMT), a normal process during embryogenesis as well as organ development in which epithelial cells adopt mesenchymal cell characteristics. EMT is a critical step of oncogenesis and is involved in promoting metastasis of the cancer cell. Notch signaling is activated in the EMT process, and this occurs through cross talk with TGF-β, Wnt, and/or Hedgehog pathways. Notch has been shown to promote EMT by upregulating several transcription factors involved in the EMT process such as Snail, Slug, and Twist.

An important function of Notch signaling is its ability to stimulate stem cell proliferation, differentiation, and self-renewal. Notch signaling has recently been established as a driver of stemness and tumorigenicity of esophageal adenocarcinoma [6]. Notch activity has been implicated in promoting stemness and EMT in colorectal cancer (CRC). In CRC cells, Notch has been shown to promote the expression of CD44, Slug, Smad-3, and Jagged-1 resulting in increased CRC cell migration and anchorage-independent growth, as well as stimulating the metastatic spread of the CRC cells. Pan Notch inhibitors specifically repress Notch-mediated activation of CD44, Slug, and Smad-3 via a cascade of Notch receptors through induction of Jagged-1 [7]. The ability of Notch to promote cancer stem cell maintenance, EMT regulation, and drug resistance can occur through regulated expression of HES1, a canonical Notch target gene.

Therapeutic Targeting of Notch Pathways

A role for Notch signaling in the development and progression of cancer stem cells (CSCs) or tumor initiating cells (TICs) has been addressed in studies ranging from early preclinical to clinical trials [8]. Accordingly, targeting the Notch pathway with γ-secretase inhibitors (GSIs) or humanized monoclonal antibodies (mAbs) targeting Notch ligands or receptors are in clinical development. However, therapeutic advancements of these trials has been challenging as drug toxicity is difficult to overcome, thus significantly limiting the usefulness of treatments. Of all the possible nodes that could be therapeutically targeted along the Notch pathway, the majority of anti-Notch drugs aim to inhibit the release of NICD from the membrane. Most attention regarding targeting of Notch focuses on γ-secretase inhibitors. However, other attractive targets such as the ADAM proteinases as well as the receptors and ligands themselves have been investigated for targeted anti-Notch therapy [9] .

The γ-Secretase Inhibitors (GSIs)

The γ-secretase inhibitors (GSIs) are the most extensively studied modulators of Notch signaling. They were first developed as potential therapy for Alzheimer's disease. Currently, numerous clinical trials have been conducted in patients with a wide variety of cancer diagnoses. Table 17.1 cites some trials on the ClinicalTrials.gov site.

Monoclonal Antibodies

A phase 1a trial for a fully human delta-like ligand 4 (Dll4) monoclonal antibody in patients with advanced solid tumors, enoticumab (SAR153192) (REGN421) from Regeneron Pharmaceuticals, was concluded in 2014 [13]. Enoticumab was tolerated well, and of the 53 patients enrolled, 2 exhibited partial tumor response, while 16 showed disease stability. OncoMed Pharmaceuticals currently have four monoclonal antibodies in their pipeline that target multiple components of the Notch signaling pathway. Two of their antibodies target the Notch ligand DLL4. The anti-DLL4 antibody, demcizumab, is labeled as an anticancer stem cell drug.

Table 17.1 Approved therapeutic agents for Notch and related clinical trails

Notch signaling pathway	Pathway active in cancer type	Affected biomarker	Method of detection	Target	Drugs in clinical trials Sponsor ClinicalTrials.gov Identifier:
	Breast	NOTCH	Histology or cytology	Gamma secretase inhibitors	Drug: MK-0752
	Central nervous system				Sponsor: Loyola University
	Pancreatic cancer				ClinicalTrials.gov Identifier: NCT00756717
	T-cell acute lymphoblastic leukemia/lymphoma				There are eight more studies of MK-0752 ongoing on the ClinicalTrials.gov
	Non-small cell lung Cancer				
	Recurrent fallopian tube carcinoma	NOTCH	Response was assessed by response evaluation criteria in solid tumors (RECIST 1.1)	Gamma secretase/ notch signaling pathway inhibitor	Drug: RO4929097
	Recurrent ovarian carcinoma				Sponsor: National Cancer Institute (NCI)
					ClinicalTrials.gov Identifier: NCT01175343
	Recurrent primary peritoneal carcinoma				There are 34 more studies of RO4929097 ongoing on the ClinicalTrials.gov
	Colon mucinous adenocarcinoma				
	Colon signet ring cell adenocarcinoma				
	Rectal mucinous adenocarcinoma				
	Adenoid cystic carcinoma	NOTCH	Notch immunohistochemistry	Notch signaling pathway inhibitor	Drug: LY3039478
	Advanced solid tumor				Eli Lilly and Company
					ClinicalTrials.gov Identifier: NCT02836600
					There are six more studies of LY3039478 ongoing on the ClinicalTrials.gov [10]

(continued)

Table 17.1 (continued)

Notch signaling pathway	Pathway active in cancer type	Affected biomarker	Method of detection	Target	Drugs in clinical trials Sponsor ClinicalTrials.gov Identifier:
	Advanced solid malignancies		Immunohistochemistry on archival tumor samples for expression of Dll4, Notch1 and 3	Membrane-bound Notch ligand: delta-like (Dll4)	Enoticumab REGN421z (SAR153192)
					Sponsor: Regeneron pharmaceuticals
					ClinicalTrials.gov identifier: NCT00871559 [11]
	Malignant melanoma		Immunohistochemical (IHC) testing of archived tumor tissue or on-study biopsy	Delta-like protein 3 (DLL3)	Rovalpituzumab Tesirine (Rova-T)
	Medullary thyroid Cancer				Sponsor: Stemcentrx
	Glioblastoma		mRNA and/or protein expression		ClinicalTrials.gov Identifier: NCT02709889 [12]
	Large cell neuroendocrine carcinoma				
	Neuroendocrine prostate Cancer				
	High-grade gastroenteropancreatic neuroendocrine carcinoma				
	Other neuroendocrine carcinoma				
	Other solid tumors				

Demcizumab is currently undergoing phase 1b (active) and phase 2 (recruiting) clinical trials as a combination therapy for treating non-small cell lung, pancreatic, and ovarian cancers. Results from the phase 1b trial demonstrate that 89% of pancreatic cancer patients exhibited clinical benefit (50% partial response and 39% disease stability) from the demcizumab+gemcitabine +Abraxane combination. These patients exhibited improved survival metrics over those who receive gemcitabine+Abraxane alone. Due to the promising preclinical investigation, patients with solid tumors are being recruited for a phase 1a trial of another anti-DLL4 antibody, OMP-305B83. Tarextumab has similar antitumor effects but targets the receptors, Notch2 and Notch3, instead of the DLL4 ligand. Results from the phase 1b clinical trial of tarextumab demonstrate tolerability as well as desirable antitumor effects. The fourth antibody from OncoMed, OMP-52M51, targets Notch1 and is currently recruiting for phase 1a trials in solid and lymphoid malignancies [14].

ADAM Inhibitors

The ADAM family of proteins is comprised of over 30 members, and ADAM family members contain a metalloproteinase-like domain. Proteinase activity is absent in all except for approximately ten of the ADAM proteins. One function of this class of proteinases is to facilitate signal transduction by shedding the inhibitory regulatory domains of certain transmembrane receptors, as in Notch signaling. ADAM proteins that contain a functional proteinase domain carry out similar roles to the matrix metalloproteinases (MMPs). Isoform-specific MMP inhibitors are elusive, and broad-spectrum MMP inhibitors provide no benefit to cancer patients. However, inhibitors that can target specific ADAM isoforms have been developed and tested in the preclinical phase. Examples include INCB3619, a dual inhibitor of ADAM10 and ADAM17. ADAM10 and ADAM17 have shown antitumor activity in lung adenocarcinoma xenografts in mice. However, as per our knowledge, there are no human clinical trials on clinicaltrials.gov site.

Conclusion and Future Directions

In this review, we have described the role of Notch in cancer and characterized existing therapeutic strategies targeting the Notch pathway. Activation, or in some cases repression, of Notch signaling has been shown to promote tumorigenesis, EMT, and stimulate the formation of cancer stem cells leading to metastasis as well as enhancing drug resistance. The vast majority of Notch-targeted therapies focus on exploitation of GSIs, which, together with monoclonal antibodies, present an attractive approach to personal cancer treatment, yet more prognostic and diagnostic markers are needed. Notch signaling is underscored in a number of other cancer hallmarks such as recurrence, drug resistance, tumor angiogenesis, and cancer stem cell initiation and proliferation. Notch is strongly implicated as an important and viable target in cancer patients. As of 2017, at least seven notch inhibitors are in clinical trials, and some of them (MK-0752) has shown promising results in an early clinical trial for breast cancer.

Conflict of Interest The authors have no conflict of interest.

References

1. Yamamoto S, Schulze KL, Bellen HJ. Introduction to notch signaling. Methods Mol Biol. 2014;1187:1–14.
2. Braune EB, Lendahl U. Notch – a goldilocks signaling pathway in disease and cancer therapy. Discov Med. 2016;21(115):189–96.
3. Espinoza I, et al. Notch signaling: targeting cancer stem cells and epithelial-to-mesenchymal transition. Onco Targets Ther. 2013;6:1249–59.
4. Fitzgerald K, Harrington A, Leder P. Ras pathway signals are required for notch-mediated oncogenesis. Oncogene. 2000;19(37):4191–8.
5. Rizzo P, et al. Rational targeting of notch signaling in cancer. Oncogene. 2008;27(38):5124–31.
6. Espinoza I, Miele L. Notch inhibitors for cancer treatment. Pharmacol Ther. 2013;139(2):95–110.
7. Fender AW, et al. Notch-1 promotes stemness and epithelial to mesenchymal transition in colorectal cancer. J Cell Biochem. 2015;116(11):2517–27.
8. Panneerselvam J, Munshi A, Ramesh R. Molecular targets and signaling pathways regulated by interleukin (IL)-24 in mediating its antitumor activities. J Mol Signal. 2013;8(1):15.
9. Pannuti A, et al. Targeting notch to target cancer stem cells. Clin Cancer Res. 2010;16(12):3141–52.

10. Even C, et al. Notch pathway inhibition with LY3039478 in adenoid cystic carcinoma (ACC). J Clin Oncol. 2017;35(suppl; abstr 6024).
11. Chiorean EG, et al. A phase I first-in-human study of enoticumab (REGN421), a fully human delta-like ligand 4 (Dll4) monoclonal antibody in patients with advanced solid tumors. Clin Cancer Res. 2012;21(12):2695–703.
12. Kavalerchik E, et al. An open-label study of rovalpituzumab tesirine in patients with DLL3-expressing advanced solid tumors. J Clin Oncol. 2017;35(suppl; abstr TPS2597).
13. Caffo M, et al. Innovative therapeutic strategies in the treatment of brain metastases. Int J Mol Sci. 2013;14(1):2135–74.
14. Takebe N, Nguyen D, Yang SX. Targeting notch signaling pathway in cancer: clinical development advances and challenges. Pharmacol Ther. 2014;141(2):140–9.
15. Nowell CS, Radtke F. Notch as a tumor suppressor. Nat Rev Cancer. 2017;17:145–59.

Signaling of the ErbB Receptor Family in Carcinogenesis and the Development of Targeted Therapies

18

Zheng Cai, Payal Grover, Zhiqiang Zhu,
Mark I. Greene, and Hongtao Zhang

Abbreviations

4E-BPs	eIF-4E-binding proteins
ADC	Antibody-drug conjugate
AP-1	Activator protein-1
AR	Amphiregulin
BBB	Blood-brain barrier
BTC	Betacellulin
CAR-T	Chimeric antigen receptor T cells
CEP17	*HER2/neu*-to-chromosome 17 centromere
CR	Complete response
CR	Cysteine-rich domain
DAG	Diacylglycerol
EGF	Epidermal growth factor
EGFR	Epidermal growth factor receptor
EP	Epigen
ErbB	Avian erythroblastosis oncogene B
ERK	Extracellular signal-regulated kinases
FFPET	Formalin-fixed, paraffin-embedded tissue
FISH	Fluorescent in situ hybridization
Grb2	Growth factor receptor-bound protein 2
HB-EGF	Heparin-binding EGF
HER	Human epidermal growth factor receptor
IHC	Immunohistochemistry
JAK	Janus kinase
L	Large EGF-binding domain
MAPK	Mitogen-activated protein kinase
mCRC	Metastatic colorectal cancer
mTOR	Mammalian target of rapamycin
NRG	Neuregulins
NSCLC	Non-small cell lung cancer
ORR	Objective response rate
OS	Overall survival
$p70^{S6K}$	p70 ribosomal S6 kinase
PH	Pleckstrin-homology
PI3K	Phosphatidylinositol 3-kinase
PIP3	Phosphatidylinositol (3,4,5) trisphosphate
PKB	Protein kinase B
PKC	Protein kinase C
PLC	Phospholipase C
PTEN	Phosphatase and tensin homolog deleted on chromosome 10
Raf	Rapidly accelerated fibrosarcoma
Ras	Rat sarcoma
RTKs	Receptor tyrosine kinases
scFv	Single-chain variable fragment
Shc2	Src homology 2
SHP	SH2 domain-containing inositol 5′-phosphatase
SOS	Son of sevenless

Z. Cai · P. Grover · Z. Zhu · H. Zhang (✉)
Department of Pathology and Lab Medicine, University of Pennsylvania, Philadelphia, PA, USA
e-mail: zhanghon@pennmedicine.upenn.edu

M. I. Greene
Department of Pathology and Laboratory Medicine, The Perelman School of Medicine at the University of Pennsylvania, Philadelphia, PA, USA

S. Badve, G. L. Kumar (eds.), *Predictive Biomarkers in Oncology*,
https://doi.org/10.1007/978-3-319-95228-4_18

SRS	Stereotactic radiosurgery
STAT	Signal transducer and activator of transcription
T-DM1	Ado-trastuzumab emtansine
TGF-α	Transforming growth factor-α
TKI	Tyrosine kinase inhibitor
WBRT	Whole brain radiation therapy

Introduction

The epidermal growth factor receptor (EGFR) family of tyrosine kinases (RTKs), also known as the HER or ErbB receptor family, plays crucial roles in the proliferation of many types of cells—notably epithelial—as well as in the pathogenesis and progression of a variety of carcinomas. The activation of the ErbB receptors, either by their ligands or by genetic amplification/mutations, has been associated with many aspects of transformation. As a result, many therapeutic agents have been developed which target distinct receptors or receptor complexes within this family.

Discovery of the ErbB Receptor Family

The ErbB RTK family is comprised of four members: EGFR (ErbB1, HER1), ErbB2 (HER2/neu, p185 $^{erbB2/neu}$), ErbB3 (HER3), and ErbB4 (HER4). Epidermal growth factor (EGF) was initially isolated and characterized in 1962 as a salivary-gland protein that induced eyelid opening and tooth eruption in the newborn animal but was later called EGF for its activity to stimulate the proliferation of epithelial cells. In 1972, the complete amino acid sequence of EGF was determined.

The receptor for EGF (EGFR) was later discovered using ^{125}I-labeled EGF as a phosphorylated protein on the cell surface of the squamous cell carcinoma cell line A431. In 1984, the cDNA sequence of human EGFR was elucidated and found to be similar to the v-erbB oncogene [1].

The second ErbB receptor, ErbB2, was initially discovered as a 185 kDa membrane protein encoded by the *neu* oncogene [2]. In humans, ErbB2 is encoded by the *ERBB2* gene. It is also frequently called *HER2* (for human epidermal growth factor receptor 2) or *HER2/neu*. The third and fourth members, ErbB3 and ErbB4, were found to share structural domain and sequence similarity with EGFR and ErbB2. All receptors consist of four functional domains: an extracellular ligand-binding domain, a hydrophobic transmembrane domain, an intracellular cytoplasmic tyrosine kinase domain, and a C-terminal tyrosine-containing regulatory domain. The extracellular domain is further divided into four domains (L1, CR1, L2, and CR2; L, large EGF-binding domain; CR, cysteine-rich domain) (Fig. 18.1). The cytoplasmic tyrosine kinase domain consists of two lobes (N and C), and the cleft between the two lobes constitutes the ATP-binding site.

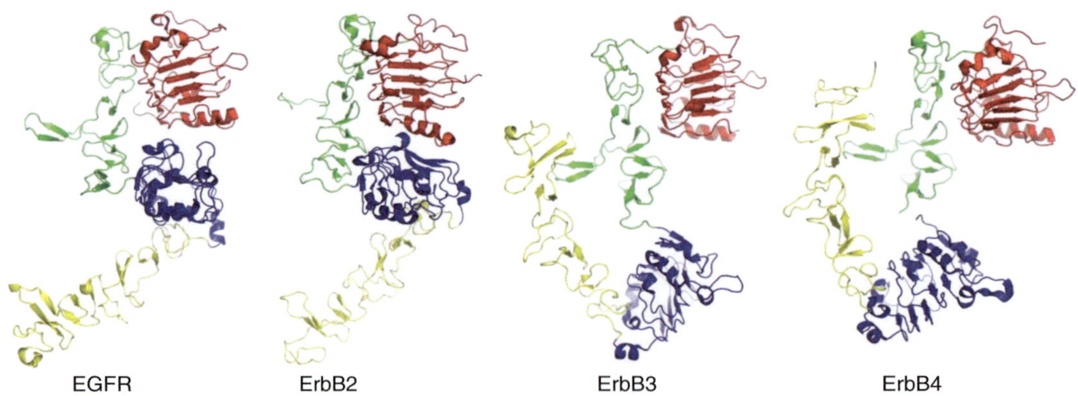

Fig. 18.1 Cartoon models of ErbB extracellular domain structures. Their three-dimensional structures are extracted from PDB files 3njp (EGFR), 3n85 (ErbB2), 4leo (ErbB3), and 2ahx (ErbB4). The four subdomains, L1, CR1, L2, and CR2, are shown in red, green, blue, and yellow, respectively

With the exception of ErbB3, all ErbB receptors are active receptor tyrosine kinases. Upon binding to ligands, ErbB receptors form dimers that enhance tyrosine kinase activity. In ErbB3, the three critical residues that are highly conserved throughout the protein kinases are substituted with other amino acids, and thus ErbB3 is known to possess little intrinsic kinase activity. However, ErbB3 has been shown to form ligand-induced heterodimers with other ErbB receptors, and overexpression of ErbB3 is considered a mechanism by which drug-resistant tumor cells arise during targeted therapies.

The extracellular domains of ErbB receptors are less conserved, thereby accounting for the selectivity of ligand binding [3]. ErbB family receptors are known to bind to 11 ligands which can be classified into three groups: (a) ligands which bind specifically to EGFR, EGF, transforming growth factor-α (TGF-α), and amphiregulin (AR); (b) ligands specific for both EGFR and ErbB4, betacellulin (BTC), heparin-binding EGF (HB-EGF), and Epogen (EP); and (c) neuregulins (NRG) which bind to both ErbB3 and ErbB4 (NRG-1 and NRG-2) or only to ErbB4 (NRG-3, NRG-4, and tomoregulin) [4].

Signal Transduction by ErbB RTK Family

Among ErbB receptors, ErbB2 is the preferred dimerization partner for all other ErbB receptors owing to its open and extended conformation. EGFR represents a typical ErbB receptor; it stays in a closed conformation with the ectodomain lying close to the membrane in the absence of ligand. Once it binds to its ligand and forms a receptor dimer, the ectodomain adopts an extended conformation. The ligand-induced homo- and heterodimerization of ErbB receptors is the main natural mechanism to activate ErbB signaling. Ligand-induced receptor dimerization is followed by subsequent activation of the intrinsic kinase domain and autophosphorylation of tyrosine residues in the intracellular domain. Alternatively, mutations in the receptors that lead to easy dimerization or elevated kinase activity will also initiate signaling pathways. Under certain conditions, ErbB receptors can interact with other membrane proteins to initiate signaling. It has been shown that such interactions further induce phosphorylation of the ErbB receptor dimers [5].

Various adaptor proteins bind to the phosphorylated residues in the intracellular domain of ErbB receptors and orchestrate downstream signaling events that affect cell proliferation, adhesion, migration, invasion, apoptosis, and metastasis. There are four major activation pathways that are initiated upon activation of this set of receptors [6] (Fig. 18.2).

The Ras/Raf/MAPK Pathway

Tyrosine phosphorylation of ErbB receptors creates docking sites for Src homology 2 (Shc2) and growth factor receptor-bound protein 2 (Grb2) proteins that activate the Ras/Raf/MAPK pathway through son of sevenless (SOS). SOS then activates RAS leading to the activation of Raf, which through phosphorylation leads to activation of the ERK1/2 or MAPK (mitogen-activated protein kinase) pathway. ERK further regulates the transcription of early genes such as c-fos and c-jun that further drive the transcription of AP-1, which is responsible for regulating various cellular responses such as proliferation and apoptosis.

The PI3K/Akt Cell Survival Pathway

The PI3K/Akt/mTOR signaling is triggered by ErbB3 and regulates cell survival and growth. This pathway is activated through the recruitment of the p85 adaptor subunit of PI3K (phosphatidylinositol 3-kinase) to the activated ErbB receptors, which contain p85 docking sites. This leads to the production of phosphatidylinositol (3,4,5) trisphosphates (PIP3) by PI3K. These lipids serve as plasma membrane docking sites for proteins that harbor pleckstrin-homology (PH) domains, such as Akt or protein kinase B (PKB).

Akt, a serine/threonine kinase, has emerged as a critical enzyme in several signal transduction pathways involved in cell proliferation, apoptosis,

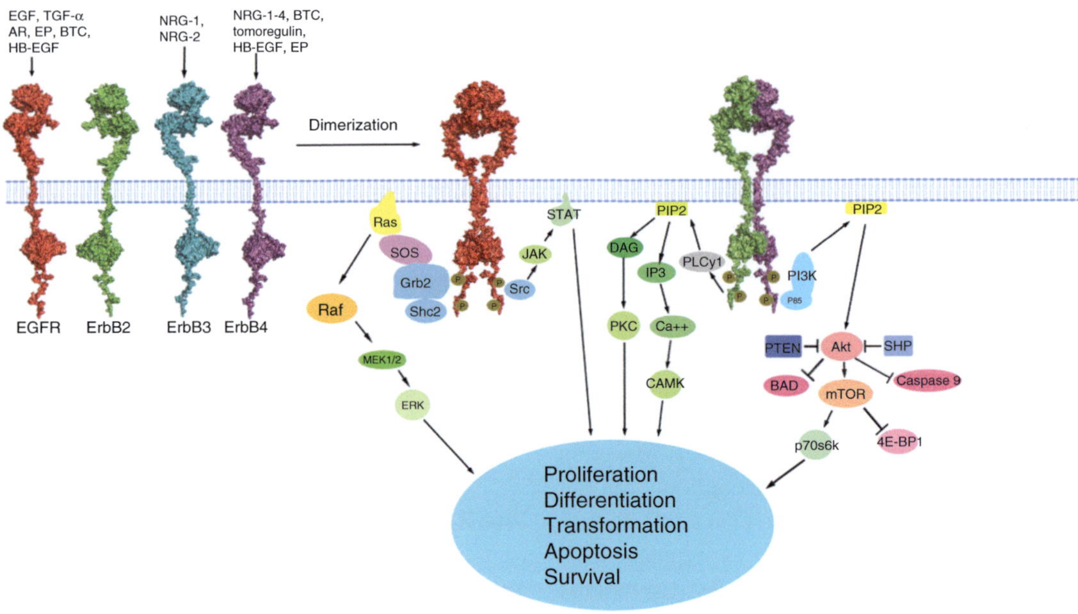

Fig. 18.2 ErbB signaling pathways

and angiogenesis. Akt is fully activated following its phosphorylation at two regulatory residues: a threonine residue in the kinase domain and a serine residue in the hydrophobic motif. The activity of Akt is itself negatively regulated by PTEN (phosphatase and tensin homolog deleted on chromosome 10, a PIP3-specific phosphatase) and SHP (SH2 domain-containing inositol 5′-phosphatase).

An important role of Akt is to facilitate cell survival and block apoptotic cell death. This process is accomplished by phosphorylating and inactivating pro-apoptotic factors such as caspase-9, Bad, and forkhead transcription factors. Akt also activates the mammalian target of rapamycin (mTOR), which further regulates the cap-dependent translation machinery through the phosphorylation of its downstream substrates p70 ribosomal S6 kinase ($p70^{S6K}$) and eIF-4E-binding proteins (4E-BPs).

The PLCγ1/PKC Pathway

Activation of phospholipase C gamma (PLCγ) hydrolyses phosphatidylinositol 4,5-biphosphate to diacylglycerol (DAG) and inositol triphosphate (IP3). DAG is a cofactor for the activation of the serine/threonine kinase protein kinase-C (PKC), whereas IP3 mediates calcium release from intracellular stores to affect $Ca2^{+}$-dependent enzymes. The activation of PKC results in cell-cycle progression, transformation, differentiation, and apoptosis.

The JAK/STAT Pathway

Both EGFR and HER2 hyperactivate these pathways in various cancers. The JAK/STAT pathway is also involved in cell survival. Recruitment of Src to phosphorylated tyrosine residues leads to the phosphorylation of JAK, which further phosphorylates STAT proteins located at the plasma membrane. Translocation of STAT to the nucleus leads to the activation of genes associated with cellular survival.

Clinical Applications of Anti-EGFR/Anti-Her2 Therapies

As the ErbB receptors are closely related to cancer development and progression, therapeutic agents have been developed against these targets.

Many of these agents are already in clinical use, which include antibodies targeting the extracellular domains of EGFR or ErbB2 and small-molecule tyrosine kinase inhibitors (TKIs) targeting the intracellular kinase domain (Table 18.1).

Anti-ErbB2 Monoclonal Antibodies

In 1984, Drebin et al. reported that the anti-ErbB2/anti-neu monoclonal antibody 7.16.4 was able to induce downregulation of cell surface expression of ErbB2 receptors and reverse transform the malignant phenotype of tumor cells growing in soft agar. This was followed by studies demonstrating that 7.16.4 can inhibit the growth of the *her2/neu*-overexpressing tumors in athymic mice and syngeneic rats of the BDIX strain. This was the first time that targeted therapy was used to down modulate an oncoprotein in a syngeneic system. Shortly after this discovery, an antihuman $p185^{her2/neu}$ antibody, 4D5, was developed and demonstrated tumor-inhibiting features similar to the 7.16.4 mAb. The 4D5 was engineered and humanized to reduce immunogenicity for clinical use. The humanized antibody 4D5, or rhumAb4D5, was later named trastuzumab/Herceptin™ and was approved by FDA for use in combination with first-line chemotherapeutic agents in patients with *her2/neu*-overexpressing metastatic breast cancer. Trastuzumab was also used to prevent tumor emergence for patients with *her2/neu*-expressing breast cancer [4]. Since ErbB2 is more commonly referred to as HER2 in the clinical field, in the following sections, we will use HER2 when this receptor is mentioned as the target for clinical treatment.

Greene's laboratory also showed that combinations of anti-ErbB2/anti-neu monoclonal antibodies which bound to separate domains of the ectodomain of ErbB2/neu exerted synergistic antitumor activity in vitro in soft agar cultures and dramatically limited ErbB2/neu tumor growth in vivo [7].

Roche recognized this important and novel effect of combining ectodomain-binding monoclonals and developed pertuzumab (Perjeta™, 2C4), which is a recombinant humanized IgG1**κ** mAb, targeting the extracellular dimerization domain II of HER2. This antibody, like other anti-Her2/anti-neu antibodies [8], can diminish heterodimer formation and ligand-dependent signaling. Its activity does not depend on the level of *her2* expression. The epitope of pertuzumab is distinct from that of trastuzumab and 7.16.4, which binds to subdomain IV of HER2 extracellular domain and inhibits ligand-independent signaling. Recently, clinical studies have demonstrated that the combination of trastuzumab and pertuzumab, targeting different epitopes of HER2 ectodomain, produced a more complete and effective blockade of *her2*-driven tumors in patients. Based on this, the trastuzumab/pertuzumab combination has been approved to be used together with docetaxel to treat metastatic breast cancer patients as well as early-stage breast cancer patients in the neoadjuvant setting.

Trastuzumab emtansine (also named ado-trastuzumab emtansine, Kadcyla™, T-DM1) is an antibody-drug conjugate (ADC) in which a cytotoxic chemical agent emtansine is covalently linked to the antibody trastuzumab. After binding to HER2 overexpressed on the surface of cancer cells, T-DM1 can deliver emtansine to cancer cells to target tubulin and to disrupt microtubule stability. In clinical trials, T-DM1 demonstrated significant clinical benefit for patients whose breast cancers progressed after trastuzumab-based therapy. T-DM1 was approved by FDA in 2013 for late-stage breast cancer.

Anti-EGFR Monoclonal Antibodies

Two EGFR-specific mAbs (225, IgG1 and 528, IgG2a) were found to bind EGFR with affinity comparable to the natural ligand. These antibodies compete with the binding of EGF to its receptor and inhibit EGF-induced activation of tyrosine kinase activity and cell proliferation. These mAbs were also capable of substantially inhibiting the growth of EGFR-overexpressing human tumor xenografts of the vulvar squamous carcinoma A431 and the breast adenocarcinoma

Table 18.1 Approved therapeutic agents for ErbB receptors

ErbB signaling pathway	Pathway active in cancer type	Affected biomarker	Method of detection	Target	Active drugs
Overexpression of EGFR	Colorectal	EGFR	PCR	EGFR	Cetuximab, Panitumumab, Necitumumab (Mendelsohn et al., *JCO*, 2003, 21:2787–2799; Herbst and Shin, *Cancer*, 2002, 94:1593–1611; Thatcher et al., *Lancet Oncol*, 2015, 16:763–774)
	Head and neck		NGS		
	Lung (squamous non-small cell lung cancer)		NGS		
Amplification/ overexpression of HER2	Breast,	HER2	HER2 IHC or FISH	HER2	Trastuzumab Pertuzumab T-DM1 Lapatinib Neratinib (Slamon et al., *N Eng J Med*, 2001, 344:783–792; Baselga et al., *JCO*, 2010, 28:1138–1144; Verma et al., *N Engl J Med*, 2012, 367:1783–1791) (Cameron et al., *Breast Cancer Res Treat*, 2008, 112:533–543; Chan et al., *Lancet Oncol*, 2016, 17:367–377)
	Esophageal				
	Gastric				
EGFR mutations	Non-small cell lung cancer	EGFR	*therascreen*® EGFR PCR test	EGFR exon 19 deletions or exon 21 (L858R)	Gefitinib, Erlotinib Afatinib Osimertinib (Zhang et al., *J Clin Invest*, 2007, 8:2051–2058; Soria et al., *Lancet Oncol*, 2015, 16:897–907; Mok et al., *N Engl J Med*, 2017, 376:629–640)
	Pancreatic		cobas EGFR mutation test v2, real time PCR		
	Advanced medullary thyroid cancer	VEGF, EGFR, RET			Vandetanib (Thornton et al., *Clin Cancer Res*, 2012, 18:3722–3730)

MDA-MB-468 in vivo [9]. To reduce the immunogenicity of mAb 225, a chimeric antibody consisting of the variable domain of mAb 225 and the human IgG1 constant region was generated. The chimeric antibody, named cetuximab (formerly called C225; also known as Erbitux™), has a higher affinity for EGFR than the parent antibody and is more effective at inhibiting tumor growth. The antigenic structures recognized by cetuximab have been identified as a broad structural surface on domain III of EGFR. Cetuximab has been approved to treat patients with advanced metastatic colorectal cancer.

Another EGFR-specific antibody, panitumumab (ABX-EGF or Vectibix™), is a fully humanized antibody generated from transgenic mice (XenoMouse) which express human Ig genes. Panitumumab also has sub-nanomolar affinity for EGFR, prevents ligand binding, and eradicates A431 xenograft tumors in nude mice. The epitope recognized by panitumumab remains poorly defined, but it is clear that the panitumumab epitope is not identical to that of cetuximab and that panitumumab is effective in some patients who have developed resistance to cetuximab treatment. In contrast to cetuximab, which requires a combination of chemotherapeutic agents, panitumumab functions as monotherapy. It was granted accelerated approval by the FDA in 2006 as a single agent for the treatment of metastatic colorectal carcinoma with disease progression on or following fluoropyridine, oxaliplatin, and irinotecan chemotherapy regimens.

Most recently, necitumumab (Portrazza™), a recombinant human IgG1 monoclonal antibody that also blocks the ligand-binding site of human EGFR, was approved for the first-line treatment of patients with metastatic squamous non-small cell lung cancer when used in combination with gemcitabine and cisplatin.

Small-Molecule Inhibitors

Small-molecule EGFR kinase inhibitors have been developed for more than 20 years. Gefitinib and erlotinib were the first-generation FDA-approved EGFR kinase inhibitors that showed effectiveness in treating non-small cell lung cancer (NSCLC) driven by somatic mutations in the EGFR kinase domain. A subset of NSCLC is caused by these somatic mutations that abnormally activate the kinase activity without ligand binding. Several EGFR mutations have been reported in NSCLC, including deletion in exon 19, insertions in exon 20, and point mutations in exons 18 and 21. The most common point mutation L858R (Leu^{858} to Arg) accounts for about 40% of EGFR mutations in NSCLC. The clinical responsiveness of these first-generation EGFR inhibitors in NSCLC patients greatly correlates with the presence of these activating mutants, especially L858R and the exon 19 deletion.

Unfortunately, patients with somatic EGFR mutations who initially have good clinical responses to first-generation inhibitors usually develop resistance in 6–12 months. In approximately 50% of cases, this acquired resistance is due to a secondary somatic mutation T790 M (Thr^{790} to Met) in the EGFR kinase domain. Another 25% of drug resistance to first-generation TKIs involves an amplification of the c-Met tyrosine kinase gene, which in turn activates a parallel signaling pathway, bypassing EGFR.

A bulky gatekeeper residue Met^{790} structurally impedes the accessibility of inhibitors to the hydrophobic pocket lying at the back of the ATP-binding site. However, newer crystal structures revealed that an alternate side-chain rotamer of Met^{790} allows for binding of the compound with little loss of affinity as compared with the wild-type EGFR kinase. Thus, it is suggested that other mechanisms may determine the sensitivity of tumor cells harboring EGFR mutants to TKIs.

Somatic EGFR mutations not only change the affinity for inhibitors but also affect their affinity binding to ATP. While the sensitizing mutations (e.g., L858R, deletion in exon 19, etc.) compromise affinity for ATP as compared to the wild-type EGFR, the secondary mutation T790 M actually rescues the affinity for ATP and demonstrates a reduced Michaelis constant $K_{m[ATP]}$ that is comparable to that of wild-type EGFR [10].

To overcome EGFR T790 M mutation-mediated resistance, a second generation of

TKIs were developed to covalently bind to Cys^{797}. Nevertheless, most of these inhibitors had to be used at reduced doses in patients due to their equally strong reactivity with the wild-type EGFR. The third generation of TKIs (e.g., WZ4002, CO-1686, AZD9291) were then developed to address the on-target toxicity to wild-type EGFR. The FDA recently granted an accelerated approval of AZD9291 (osimertinib/Tagrisso®) to treat patients with the resistant T790M EGFR mutation after the use of first-generation EGFR inhibitors. However, the emergence of a new mutation C797S after AZD9291treatment renders resistance to irreversible TKIs (both second- and third-generation inhibitors).

One small-molecule TKI, lapatinib, can inhibit both EGFR and HER2 kinase activity. Lapatinib was approved by FDA as a second-line therapy to treat HER2-positive advanced or metastatic breast cancer, in combination with the chemotherapeutic agent capecitabine. In postmenopausal hormone-positive and HER2-positive breast cancer patients, lapatinib is approved for use in combination with letrozole. One unique feature for lapatinib is its ability to pass through the blood-brain barrier and mediate CNS activity against metastatic lesions. In patients with HER2-positive breast cancer with progressive brain metastases after trastuzumab and cranial radiotherapy, the CNS objective response rate (ORR) for the lapatinib plus capecitabine treatment was 38% (95% confidence interval [CI] 13.9–68.4) [11]. The CNS objective response was defined as a $\geq$50% volumetric reduction of CNS lesion(s) in the absence of new or progressive CNS or non-CNS lesions or increasing steroid requirements. In contrast, no responses were observed, but excess toxicity was noted in patients treated with lapatinib in combination with topotecan.

In 2017, FDA approved a second pan-ErbB small-molecule inhibitor, neratinib, for the extended adjuvant treatment of early-stage, HER2-positive breast cancer. Major side effects of neratinib include diarrhea, which can be managed by medication. Neratinib adds marginal efficacy to the current trastuzumab-based treatment, but patients at higher risk of recurrence may benefit most from this newly approved drug.

Diagnostic Markers for EGFR/HER2 Therapies

As targeted therapies are developed to best treat cancers that carry the "target," patients usually have to be screened first for corresponding target biomarkers. For HER2-targeted therapies, such as the trastuzumab-based therapy, the HER2 status is primarily evaluated using immunohistochemistry (IHC) and fluorescent in situ hybridization (FISH) in the clinical setting. According to the most recent guidelines published by the American Society of Clinical Oncology and the College of American Pathologists, only patients with more than 10% of invasive tumor cells showing a uniform, intense membrane staining on IHC or a *HER2/neu*-to-chromosome 17 centromere (CEP17) ratio of greater than 2.0 on FISH are considered HER2/neu-positive and eligible for trastuzumab treatment [12].

Serum tests for circulating HER2 ECD have also been developed. Although many studies have evaluated the correlation between abnormal HER2 ECD in serum and HER2 positivity in tissue, the serum assay is unfortunately not used as often as IHC and FISH due to assay sensitivity issues. Monitoring serum HER2 has been shown to have clinical value to follow tumor progression in breast cancer patients and as a means to guide further treatments when targeting mAbs fail [13].

For anti-EGFR antibody treatments, diagnostic assays for EGFR expression in tumor tissue have not been widely used. Since the *KRAS* mutation is predictive of nonresponse and shorter survival in metastatic colorectal cancer (mCRC) patients treated by the anti-EGFR antibody, the European Medicines Agency approved the use of cetuximab in patients with wild-type *KRAS* tumors. Unfortunately, fewer than 50% of these patients show clinical response to cetuximab treatment.

In contrast, companion EGFR mutation tests are required to screen NSCLC patients for targetable mutations. The first generation of EGFR TKIs, such as erlotinib (Tarceva®) and gefitinib (Iressa®), are effective on tumors with exon 19 deletions and L858R mutations. Patients with the EGFR T790M first-generation inhibitors but sensitive to the third-generation TKI osimertinib. Tests for EGFR mutations can be performed on plasma specimens (liquid biopsies) or on formalin-fixed, paraffin-embedded tissue (FFPET) samples. Although EGFR mutations are predictive for TKI activity, activation of the EGFR pathway as the driver for cancer cells is critical. Increased EGFR gene copy number is statistically significantly associated with survival in advanced NSCLC patients after gefitinib therapy (hazard ratio = 0.44, 95% CI = 0.23–0.82), especially in patients displaying elevated phospho-Akt protein levels [14].

Future Therapies

Although targeted therapies against EGFR and HER2 have successfully changed the way we treat cancer patients, there are still unmet medical needs for additional therapeutic agents targeting this family of receptor tyrosine kinases. One such need is the treatment of brain metastases, which occurs frequently in cancer patients. Another area for drug development is the treatment of patients with acquired resistance that occurs during targeted therapies. See Table 18.2 for EGFR receptor family-related clinical trials that have been undertaken in various indications.

Table 18.2 EGFR-/HER2-related active clinical trials (2018)

ErbB signaling pathway	Pathway active in cancer type	Drugs in clinical trials[a]
Amplification	Brain tumors (glioblastoma: prevalence 50%)	Phase II or III clinical trials: Tesevatinib (a multiple target TKI, NCT02844439) HER2-CAR T (NCT01109095) ABT-414 (anti-EGFR ADC, NCT02573324) Cetuximab (anti-EGFR, NCT02861898) Sym004 (anti-EGFR, NCT02540161)
Amplification mutation	Ovarian cancer (prevalence 17–73%)	Phase II or III clinical trials Gefitinib (NCT00317772) HER-2/neu peptide vaccine (for HER2 positive, NCT00194714) A166 (anti-HER2 ADC, NCT03602079) HER2 CAR-T (NCT02713984) Cetuximab
Overexpression and mutation	Cervical and uterine cancer (prevalence 54–71%)	Phase II or III clinical trials Cetuximab (NCT00997009, NCT02979977) Nimotuzumab (anti-EGFR, NCT03469531) A166 (anti-HER2 ADC, NCT03602079) Afatinib (NCT02979977) Lapatinib
Mutation and overexpression		Phase II clinical trials Gefitinib Erlotinib Karp and Falchook [31]
Tumor promotion	Hepatocellular carcinoma	Erlotinib (NCT02273362) Neratinib (NCT01953926)
	Thymoma and thymic carcinoma (prevalence 69%)	Gefitinib Erlotinib Cetuximab
		Karp and Falchook [31]
	Uterine (endometrial cancer)	Gefitinib Erlotinib Cetuximab
		Karp and Falchook [31]
Amplification for HER2	Biliary tract adenocarcinoma (gallbladder, bile duct) (prevalence 24%)	Erlotinib
Activation	Pancreatic cancer	Clinical trial registry number: NCT01013649

[a]General Reference: Karp and Falchook [31]

Brain Metastases

With current standard therapies for treating advanced HER2-positive breast cancers, approximately one-third of advanced HER2-positive patients develop brain metastases. Currently, there is a lack of effective targeted therapeutics for this indication. In general, antibody molecules owing to their large size are unable to cross the blood-brain barrier (BBB) and thus are ineffective treating brain metastases. However, there are some data [15] to suggest that perturbation in the BBB occurs during metastases and allows some penetration of T-DM1.

In addition, most small-molecule TKIs possess modest activity for brain lesions [16]. For example, lapatinib and neratinib, both dual inhibitors of HER2 and EGFR kinases, have been tested in patients with HER2-positive brain metastases. In a multicenter phase II open-label trial, three women achieved a partial response with a CNS objective response rate of 8% [17]. The CNS activity of neratinib was enhanced when combined with capecitabine and showed 49% CNS ORR. TKIs with improved BBB permeability are also being developed [18].

NSCLC patients with activating EGFR mutations have a longer overall survival, due to current available targeted therapies, and a higher incidence of brain metastasis. While whole brain radiation therapy (WBRT) has been the foundation of management of brain metastasis, multiple approaches, such as higher dosages, next-generation TKIs, or local controlled surgery and stereotactic radiosurgery (SRS) for symptomatic, accessible lesions, have been tried to delay the use of WBRT as much as possible [18].

Resistance to Current Therapies

Trastuzumab-Resistant HER2 Tumors

Trastuzumab-based targeted therapies significantly improve the average overall survival (OS) of metastatic breast cancer patients [19]. However, greater than 30% of patients with HER2-positive breast cancers still develop resistance over time and succumb to the disease [20].

Successful activation of the immune system is thought to boost the clinical efficacy of anti-HER2 antibody [21, 22]. Recently, we have shown that the combination of anti-HER2 antibody and IFN-γ leads to a much more effective inhibition of tumor growth in vivo [23]. Treatment of IFN-γ polarized tumor microenvironment to pro-inflammatory agents changed tumor cell features relevant to epithelial-mesenchymal transition. We have also developed a HER2-targeted antibody-like protein that carries the IFN-γ function and displays much better in vivo activity than the antibody alone [24].

EGFR Tumors with Acquired Resistance Mutations

In NSCLC patients treated with the third generation of EGFR TKIs, the resistance mutation C797S can emerge and lead to the failure of irreversible inhibitors that rely on the cysteine residue at this position. Thus, inhibitors with activity against EGFR C797S mutants have been under development to rescue these patients. Jia et al. reported a rationally developed allosteric inhibitor, EAI045, which binds to a site created by the displacement of the regulatory C-helix in an inactive conformation of the kinase domain [25]. Although EAI045 inhibits the EGFR L858R/T790M mutant with a low-nanomolar potency in biochemical assays, this compound is not effective as a single agent, and it requires the synergistic activity of cetuximab for efficacy in mouse models of lung cancer driven by EGFR mutations (L858R/T790M or L858R/T790M/C797S). Clearly this set of targeted and mutated surfaces requires novel pharmaceuticals.

CAR-T Therapy for the EGFR Family of Receptors

Engineered T cells expressing chimeric antigen receptor (CAR) genes have recently demonstrated success in treating chemotherapy-resistant hematologic cancers [26].

The CAR-T approach has been adopted for HER2 and EGFR, with HER2 CAR-T cells being tested in sarcoma patients. However, the phase I results have been less than impressive, with responses in only 4 out of 17 patients [27]. This is in sharp contrast with the 90% complete response (CR) rate for CD19 CAR-T therapies in leukemia. In addition, HER2 CAR-T therapy in a colon cancer patient led to the patient's death due to cytokine storm which was triggered by the engineered T cells binding to HER2 in lung epithelial cells [28]. To reduce on-target toxicity, Liu et al. reduced the affinity of anti-HER2 scFv and created CAR-T cells that could discriminate HER2-overexpressing tumors from normal tissues that express physiologic levels of HER2 [29].

CAR-T therapy has also been attempted for EGFR-positive tumors. In a phase I clinical trial in 11 NSCLC patients, EGFR-targeted CAR-T treatment led to partial responses in 2 patients and stable disease in 5 patients [30]. In tumor biopsies, tumor-infiltrating CAR-T cells could be identified, and pathological eradication of EGFR-positive tumor cells was observed. No severe toxicity was observed in this phase I trial. Also undergoing clinical assessment is a CAR-T therapy targeting the EGFR variant III (EGFRvIII) mutant, which occurs in 40–70% of glioblastomas.

Acknowledgments This work was supported by grants from the Breast Cancer Research Foundation and the National Institutes of Health to M.I.G. (R01CA089481, R01CA149425).

References

1. Downward J, Yarden Y, Mayes E, Scrace G, Totty N, Stockwell P, et al. Close similarity of epidermal growth factor receptor and v-erb-B oncogene protein sequences. Nature. 1984;307(5951):521–7. PubMed PMID: 6320011.
2. Drebin JA, Shilo BZ, Weinberg RA, Greene MI. Preliminary evidence of an association between an activated cellular transforming gene and a tumor specific transplantation antigen. In: Vitetta E, editor. ICN-UCLA Symposia. New York: Academic Press; 1982.
3. Yarden Y, Sliwkowski MX. Untangling the ErbB signalling network. Nat Rev Mol Cell Biol. 2001;2(2):127–37.
4. Zhang H, Berezov A, Wang Q, Zhang G, Drebin J, Murali R, Greene MI. ErbB receptors: from oncogenes to targeted cancer therapies. J Clin Invest. 2007;117(8):2051–8. Epub 2007/08/03. https://doi.org/10.1172/JCI32278. PubMed PMID: 17671639; PMCID: 1934579.
5. Qian X, LeVea CM, Freeman JK, Dougall WC, Greene MI. Heterodimerization of epidermal growth factor receptor and wild-type or kinase-deficient Neu: a mechanism of interreceptor kinase activation and transphosphorylation. Proc Natl Acad Sci U S A. 1994;91(4):1500–4.
6. Nyati MK, Morgan MA, Feng FY, Lawrence TS. Integration of EGFR inhibitors with radiochemotherapy. Nat Rev Cancer. 2006;6(11):876–85. https://doi.org/10.1038/nrc1953.
7. Drebin JA, Link VC, Greene MI. Monoclonal antibodies specific for the neu oncogene product directly mediate anti-tumor effects in vivo. Oncogene. 1988;2(4):387–94. Epub 1988/04/01. PubMed PMID: 2896329.
8. Cai Z, Zhang G, Zhou Z, Bembas K, Drebin JA, Greene MI, Zhang H. Differential binding patterns of monoclonal antibody 2C4 to the ErbB3-p185her2/neu and the EGFR-p185her2/neu complexes. Oncogene 2008;27(27):3870–4. Epub 2008/02/12. https://doi.org/10.1038/onc.2008.13. PubMed PMID: 18264138; PMCID: 2819401.
9. Mendelsohn J, Baselga J. Status of epidermal growth factor receptor antagonists in the biology and treatment of cancer. J Clin Oncol: Off J Am Soc Clin Oncol. 2003;21(14):2787–99. Epub 2003/07/16. https://doi.org/10.1200/JCO.2003.01.504. PubMed PMID: 12860957.
10. Yun CH, Mengwasser KE, Toms AV, Woo MS, Greulich H, Wong KK, Meyerson M, Eck MJ. The T790M mutation in EGFR kinase causes drug resistance by increasing the affinity for ATP. Proc Natl Acad Sci U S A. 2008;105(6):2070–5. Epub 2008/01/30. https://doi.org/10.1073/pnas.0709662105. PubMed PMID: 18227510; PMCID: 2538882.
11. Lin NU, Eierman W, Greil R, Campone M, Kaufman B, Steplewski K, et al. Randomized phase II study of lapatinib plus capecitabine or lapatinib plus topotecan for patients with HER2-positive breast cancer brain metastases. J Neuro-Oncol. 2011;105(3):613–20. https://doi.org/10.1007/s11060-011-0629-y.
12. Wolff AC, Hammond ME, Hicks DG, Dowsett M, McShane LM, Allison KH, Allred DC, Bartlett JM, Bilous M, Fitzgibbons P, Hanna W, Jenkins RB, Mangu PB, Paik S, Perez EA, Press MF, Spears PA, Vance GH, Viale G, Hayes DF, American Society of Clinical O, College of American P. Recommendations for human epidermal growth factor receptor 2 testing in breast cancer: American Society of Clinical Oncology/College of American Pathologists clinical practice guideline update. J Clin Oncol. 2013;31(31):3997–4013. https://doi.org/10.1200/JCO.2013.50.9984. PubMed PMID: 24101045.

13. Lam L, Czerniecki BJ, Fitzpatrick E, Xu S, Schuchter L, Xu X, Zhang H. Interference-free HER2 ECD as a serum biomarker in breast cancer. J Mol Biomark Diagn 2014;4(3):151. Epub 2014/08/05. https://doi.org/10.4172/2155-9929.1000151. PubMed PMID: 25089226; PMCID: 4114390.
14. Cappuzzo F, Hirsch FR, Rossi E, Bartolini S, Ceresoli GL, Bemis L, et al. Epidermal growth factor receptor gene and protein and gefitinib sensitivity in non-small-cell lung cancer. J Natl Cancer Inst. 2005;97(9):643–55. https://doi.org/10.1093/jnci/dji112. PubMed PMID: 15870435.
15. Keith KC, Lee Y, Ewend MG, Zagar TM, Anders CK. Activity of Trastuzumab-Emtansine (Tdm1) in Her2-positive breast cancer brain metastases: a case series. Cancer Treat Commun 2016;7:43–6. https://doi.org/10.1016/j.ctrc.2016.03.005. PubMed PMID: 27114895; PMCID: PMC4840897.
16. Baik CS, Chamberlain MC, Chow LQ. Targeted therapy for brain metastases in EGFR-mutated and ALK-rearranged non-small-cell lung cancer. J Thorac Oncol. 2015;10(9):1268–78. https://doi.org/10.1097/JTO.0000000000000615. PubMed PMID: 26107553.
17. Freedman RA, Gelman RS, Wefel JS, Melisko ME, Hess KR, Connolly RM, et al. Translational breast cancer research consortium (TBCRC) 022: a phase II trial of neratinib for patients with human epidermal growth factor receptor 2-positive breast cancer and brain metastases. J Clin Oncol. 2016;34(9):945–52. https://doi.org/10.1200/JCO.2015.63.0343. PubMed PMID: 26834058; PMCID: PMC5070554 online at http://www.jco.org. Author contributions are found at the end of this article.
18. Zeng Q, Wang J, Cheng Z, Chen K, Johnstrom P, Varnas K, et al. Discovery and evaluation of clinical candidate AZD3759, a potent, oral active, central nervous system-penetrant, epidermal growth factor receptor tyrosine kinase inhibitor. J Med Chem. 2015;58(20):8200–15. https://doi.org/10.1021/acs.jmedchem.5b01073. PubMed PMID: 26313252.
19. Berghoff AS, Bago-Horvath Z, Dubsky P, Rudas M, Pluschnig U, Wiltschke C, et al. Impact of HER-2-targeted therapy on overall survival in patients with HER-2 positive metastatic breast cancer. Breast J. 2013;19(2):149–55. Epub 2013/01/29. https://doi.org/10.1111/tbj.12070.
20. Merry CR, McMahon S, Forrest ME, Bartels CF, Saiakhova A, Bartel CA, et al. Transcriptome-wide identification of mRNAs and lincRNAs associated with trastuzumab-resistance in HER2-positive breast cancer. Oncotarget. 2016;7(33):53230–44. https://doi.org/10.18632/oncotarget.10637. PubMed PMID: 27449296; PMCID: PMC5288181.
21. Park S, Jiang Z, Mortenson ED, Deng L, Radkevich-Brown O, Yang X, Sattar H, Wang Y, Brown NK, Greene M, Liu Y, Tang J, Wang S, Fu YX. The therapeutic effect of anti-HER2/neu antibody depends on both innate and adaptive immunity. Cancer Cell 2010;18(2):160–70. Epub 2010/08/17. https://doi.org/10.1016/j.ccr.2010.06.014. PubMed PMID: 20708157; PMCID: 2923645.
22. Bianchini G, Gianni L. The immune system and response to HER2-targeted treatment in breast cancer. Lancet Oncol. 2014;15(2):e58–68. Epub 2014/02/01. https://doi.org/10.1016/S1470-2045(13)70477-7. PubMed PMID: 24480556.
23. Nagai Y, Tsuchiya H, Runkle EA, Young PD, Ji MQ, Norton L, Drebin JA, Zhang H, Greene MI. Disabling of the erbB pathway followed by IFN-gamma modifies phenotype and enhances genotoxic eradication of breast tumors. Cell Rep 2015;12(12):2049–59. Epub 2015/09/15. https://doi.org/10.1016/j.celrep.2015.08.044. PubMed PMID: 26365188; PMCID: PMC4591220.
24. Zhang H, Lam L, Nagai Y, Zhu Z, Chen X, Ji MQ, Greene MI. A targeted immunotherapy approach for HER2/neu transformed tumors by coupling an engineered effector domain with interferon-γ. Oncoimmunology. 2018;7(4):e1300739. PMCID: PMC5889208.
25. Jia Y, Yun CH, Park E, Ercan D, Manuia M, Juarez J, Xu C, Rhee K, Chen T, Zhang H, Palakurthi S, Jang J, Lelais G, DiDonato M, Bursulaya B, Michellys PY, Epple R, Marsilje TH, McNeill M, Lu W, Harris J, Bender S, Wong KK, Janne PA, Eck MJ. Overcoming EGFR(T790M) and EGFR(C797S) resistance with mutant-selective allosteric inhibitors. Nature. 2016;534(7605):129–32. https://doi.org/10.1038/nature17960. PubMed PMID: 27251290; PMCID: PMC4929832.
26. Newick K, O'Brien S, Moon E, Albelda SM. CAR T cell therapy for solid tumors. Annu Rev Med. 2017;68:139–52. https://doi.org/10.1146/annurev-med-062315-120245. PubMed PMID: 27860544.
27. Ahmed N, Brawley VS, Hegde M, Robertson C, Ghazi A, Gerken C, et al. Human epidermal growth factor receptor 2 (HER2) -specific chimeric antigen receptor-modified T cells for the immunotherapy of HER2-positive sarcoma. J Clin Oncol. 2015;33(15):1688–96. https://doi.org/10.1200/JCO.2014.58.0225. PubMed PMID: 25800760; PMCID: PMC4429176.
28. Morgan RA, Yang JC, Kitano M, Dudley ME, Laurencot CM, Rosenberg SA. Case report of a serious adverse event following the administration of T cells transduced with a chimeric antigen receptor recognizing ERBB2. Mol Ther. 2010;18(4):843–51. https://doi.org/10.1038/mt.2010.24. PubMed PMID: 20179677; PMCID: PMC2862534.
29. Liu X, Jiang S, Fang C, Yang S, Olalere D, Pequignot EC, et al. Affinity-tuned ErbB2 or EGFR chimeric antigen receptor T cells exhibit an increased therapeutic index against tumors in mice. Cancer Res. 2015;75(17):3596–607. https://doi.org/10.1158/0008-5472.CAN-15-0159. PubMed PMID: 26330166; PMCID: PMC4560113.
30. Feng K, Guo Y, Dai H, Wang Y, Li X, Jia H, et al. Chimeric antigen receptor-modified T cells for the immunotherapy of patients with EGFR-expressing advanced relapsed/refractory non-small cell lung cancer. Sci China Life Sci. 2016;59(5):468–79. https://doi.org/10.1007/s11427-016-5023-8.
31. Karp DD, Falchook GS, editors. Handbook of targeted cancer therapy. Philadelphia, PA: Wolters Kluwer; 2015.

Angiogenic Signaling Pathways and Anti-angiogenic Therapies in Human Cancer

19

Aejaz Nasir

Abbreviations

BRC	Breast carcinoma
CIN	Chromosomal instability
CRC	Colorectal carcinoma
DV	Draining vein
ECOG	Eastern Cooperative Oncology Group
EGFR	Epidermal growth factor receptor
FA	Feeder artery
FDA	Food and Drug Administration
FGFR	Fibroblast growth factor receptor
FISH	Fluorescent in situ hybridization
FLK1	Fetal liver kinase 1
FOLFIRI	Folinic acid, 5-fluorouracil, and irinotecan
GEJ	Gastroesophageal junction
GIST	Gastrointestinal stromal tumor
GMP	Glomeruloid microvascular proliferation
HCC	Hepatocellular carcinoma
HER2	Human epidermal growth factor receptor 2
HGF	Hepatocyte growth factor
IgG	Immunoglobulin G
IHC	Immunohistochemistry
ILF	Irinotecan, bolus fluorouracil, and leucovorin
INF	α Interferon-α
MC	Mast cell
MDSCs	Myeloid-derived suppressor cells
MET	Mesenchymal epithelial transition factor
MTC	Medullary thyroid carcinoma
MVD	Microvascular density
NGS	Next-generation sequencing
NRP1	Neuropilin 1
NRP2	Neuropilin 2
NSCLC	Non-small cell lung carcinoma
OC	Ovarian cancer
OS	Overall survival
PACA	Pancreatic cancer
PDGFR	Platelet-derived growth factor
PDGF-Rb	Platelet-derived growth factor receptor-beta
PFS	Progression-free survival
PlGF	Placental growth factor
RCC	Renal cell carcinoma
RNA	Ribonucleic acid
TAM	Tumor associated macrophage
TAMs	Tumor-associated macrophages
TCGA	The Cancer Genome Atlas
TGF-b	Transforming growth factor-beta
TKIs	Tyrosine kinase inhibitors
Treg	Regulatory T cells
VEGF	Vascular endothelial growth factor
VEGFR	Vascular endothelial growth factor receptor
VPF	Vascular permeability factor

A. Nasir
Diagnostic & Experimental Pathology, Tailored Therapeutics, Eli Lilly & Co., Indianapolis, IN, USA

BJ's Diagnostic & Precision Oncology, Tampa, FL, USA

S. Badve, G. L. Kumar (eds.), *Predictive Biomarkers in Oncology*,
https://doi.org/10.1007/978-3-319-95228-4_19

Introduction

Several drugs targeting VEGF or its receptors (VEGFRs) have been approved for the treatment of various malignancies, and many more are in clinical trials [1]. Unfortunately, these agents, used as monotherapy or in combination with chemotherapy, have only provided modest survival benefits in some tumor types and have not been efficacious at all in others [1]. For example, bevacizumab (Avastin®, Roche/Genentech), a humanized antibody against VEGF-A, prolongs the life of patients with advanced colon cancer by 4–5 months when combined with triple chemotherapy (irinotecan, bolus fluorouracil, and leucovorin) [2]. In a recent meta-analysis of 9 randomized trials with 3710 mCRC patients, the OS/PFS benefit was observed only in the presence of irinotecan-based regimen (ILF or FOLFIRI) [3]. A recent editorial has questioned whether bevacizumab was "boon or bust" because of its limited effectiveness, serious (though uncommon) side effects, and high cost [4]. Other US Food and Drug Administration-approved drugs that bind VEGF-A such as aflibercept (Zaltrap), or that target VEGF receptors such as various tyrosine kinase inhibitors, have fared no better [5].

Tumors acquire blood supply via multiple mechanisms: angiogenesis (sprouting new vessels from existing vessels), cooption (tumor cells engulf host vessels in the normal surrounding tissue as the tumor invades), intussusception (new vessels are generated by the fission of existing vessels), vasculogenic mimicry (tumor cells directly form vascular channels that are perfused via connection to the host vasculature), and transdifferentiation of cancer cells into endothelial cells [6]. The original concept of anti-angiogenic therapy aimed to destroy ("starve") tumor vessels was put forward by Judah Folkman [7]. It turned out that, in reality, anti-angiogenic drugs "normalize" tumor vasculature and as a result offer an improved delivery of chemotherapeutic agents to the tumor tissues [1, 8]. Furthermore, the initial idea that anti-angiogenic therapy would be resistance-free failed to materialize, and currently, we are faced with resistance to anti-angiogenic therapy as one of the major clinical challenges. Also, an increasing number of preclinical and clinical observations have shown that the process of angiogenesis is far from clearly understood. Apart from targeting the VEGF pathway, novel therapeutic strategies aim to influence other molecular factors that are involved in tumor angiogenesis.

Here I will review the clinically relevant aspects of biology of pathologic (aberrant) angiogenesis in human cancer, especially with reference to the principal angiogenic (VEGF-VEGFR) signaling pathway, cross talk between the main and alternative angiogenic pathways including an overview of the recent scientific and clinical advances in the field, and some of the challenges that we face in tailoring these agents to the right patients and also with reference to accurate prediction of response or resistance to these therapies, when administered in unselected cancer patient populations. In the end, I will summarize key patho-biologic and clinical learnings about tumor angiogenesis and anti-angiogenesis and outline some strategies to develop predictive biomarkers to improve clinical efficacy and the overall value of these drugs for cancer patients.

Angiogenic Signaling Pathways in Human Cancer

VEGF Signaling Pathway

Vascular endothelial growth factor (VEGF, VEGF-A) is the major player in the VEGF-driven angiogenic signaling pathway and principal regulator of physiological and pathological angiogenesis [9]. Discovery of VEGF (aka vascular permeability factor, VPF) as the primary tumor angiogenesis factor prompted the development of a number of drugs (e.g., bevacizumab, aflibercept, ramucirumab, and small-molecule tyrosine kinase inhibitors) that targeted this ligand or its receptors. These agents have often been successful in halting tumor angiogenesis and in regressing rapidly growing mouse tumors [10]. However, results in human cancer have been less impressive. Furthermore, while tumors induce their heterogeneous vasculature by secreting vascular endothelial growth factor (VEGF)-A, the underlying mechanisms how anti-VEGF/VEGF receptor (VEGFR) drugs treat cancer still remain unclear [11].

VEGF Ligands, Receptors, and Co-receptors

The complex process of angiogenesis is predominantly regulated by a single growth factor, VEGF (also known as VEGF-A), which is overexpressed in many human cancers. VEGF family consists of five members (VEGF-A, VEGF-B, VEGF-C, VEGF-D, and PlGF (placental growth factor)), which transmit signals via three receptors (VEGFR-1, VEGFR-2, VEGFR-3) (Fig. 19.1).

The most important factor is VEGF-A, which has been shown to stimulate endothelial cell mitogenesis and cell migration, leading to cancer progression and metastasis via binding to VEGFR-2 (also known as fetal liver kinase 1 (FLK1)). VEGF-B plays a role in the maintenance of newly formed blood vessels via VEGFR-1. VEGF-C and VEGF-D bind to VEGFR-3, predominately expressed in lymphatic vessels and play a role in lymphangiogenesis and metastatic spread to lymph nodes. PlGF is a multitasking cytokine that stimulates angiogenesis by direct or indirect mechanisms and also activates bone-marrow-derived endothelial progenitor and myeloid cells, as well as stromal cells, to create a nurturing "soil" for tumor cells, in addition to activating tumor cells [12]. By skewing the polarization of tumor-associated macrophages (TAMs), the loss

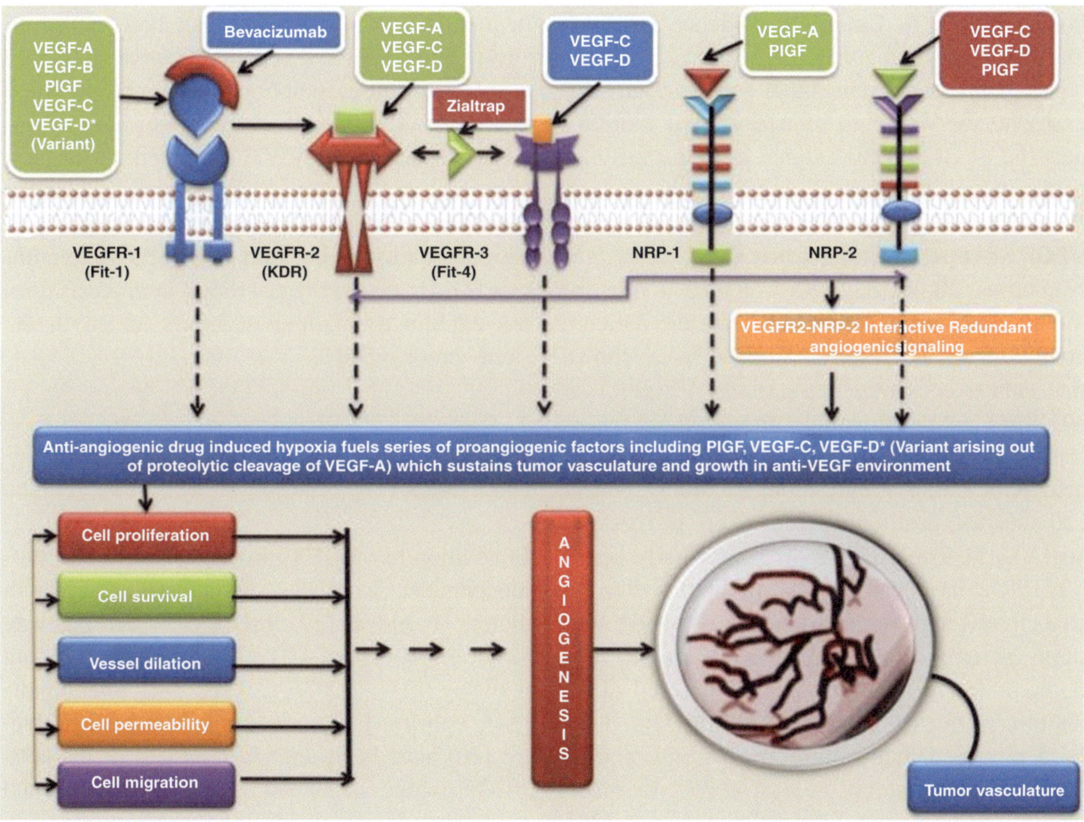

Fig. 19.1 Schematic representation of the mechanism of action of VEGF in regulating pathologic angiogenesis in human cancer. The VEGF signaling pathway with the three principal VEGF receptors (VEGFR1, VEGFR2, VEGFR3), co-receptors (neuropilin 1 (NRP1) and neuropilin 2 (NRP2)), and various ligands are illustrated. Redundancy of VEGF signaling in the form of cross talk among various receptors and ligands is obvious. In response to anti-VEGF therapies, hypoxia plays a major role in driving various proangiogenic factors to sustain proliferation and growth of tumor vessels and cells – forming the basis of acquired resistance to anti-VEGF therapy [14]. (Reprinted from Gacche and Assaraf [14]. With permission from Elsevier)

of PlGF improves vessel perfusion and maturation and enhances responses to chemotherapy [13].

Several VEGF co-receptors have also been identified, including heparan sulfate proteoglycans, neuropilin 1 (NRP1), neuropilin 2, and CD146. Moreover, VEGF receptors can cross talk with additional cell surface molecules, including integrins and other growth factor receptors. Neuropilins such as NRP1 and NRP2 are VEGF co-receptors, which enhance the activity of VEGFR-2, but also signal independently [15].

Soluble VEGF isoforms promote vessel enlargement, whereas matrix-bound isoforms stimulate branching. Paracrine VEGF, released by tumor, myeloid, or other stromal cells, increases vessel branching and renders tumor vessels abnormal [16], whereas autocrine VEGF, released by endothelial cells, maintains vascular homeostasis [17]. Emerging evidence indicates that the biological effect of VEGFR-2 signaling depends on its subcellular localization – for example, for VEGF to induce arterial morphogenesis, VEGFR-2 must signal from intracellular compartments [18].

VEGF Receptors and Tumor Cells

Numerous studies have documented a role for VEGF signaling in tumor cells, but the data are conflicting [19]. Several studies have shown that cancer cell lines can express VEGFR1 or VEGFR2 and that signaling through these receptors in cancer cells can promote events associated with tumor progression, including cancer cell survival, proliferation, invasion, or metastasis [20–23]. The presence of functionally active VEGFR-2 has been shown on human ovarian cancer cells and suggests that the observed antitumor activity of VEGF-targeted therapies may be mediated by both anti-angiogenic and direct antitumor effects [23]. Based on these data, it has been proposed that inhibition of VEGF signaling in tumor cells may, at least in part, be mediated by direct activity against tumor cells [24]. In preclinical studies inhibition of VEGF signaling in CRC and glioblastoma cells made these cells more invasive [25, 26]. Further studies are required to determine the clinical significance of tumor cell expression of VEGF/VEGFR2.

Targeting VEGF Versus VEGFR2 May Have a Different Clinical Outcome

Because VEGFR2 is thought to be the main receptor conveying the proangiogenic signals downstream of VEGF, it is generally assumed that targeting VEGFR2 would have similar biological effects as targeting the ligand. However, this is not the case in some malignancies [1]. For example, although bevacizumab monotherapy has not improved overall survival in any phase III trial, the anti-VEGFR2 antibody ramucirumab led to an OS advantage of 1.4 months in advanced gastric or gastroesophageal junction (GEJ) adenocarcinomas. Interestingly, when added to paclitaxel, ramucirumab also increased OS by 2.3 months in patients with GEJ tumors. When combined with chemotherapy, both bevacizumab and ramucirumab failed to improve OS in metastatic breast cancer, but both improved survival in non-small cell lung cancer (NSCLC) (Table 19.1). It is tempting to assume that blood vessels of GEJ tumors are highly or even exclusively dependent on VEGFR2 signaling for their survival, and, hence, ramucirumab's benefits result from starving these tumors, which is in support of the original anti-angiogenesis hypothesis [1]. However, the starvation hypothesis does not explain the failure of bevacizumab in the same tumor type [1].

VEGF-Independent Signaling Pathways

In addition to VEGF pathway, a series of VEGF-independent pathways like fibroblast growth factors 1 and 2 [27, 28], HGF/cMet pathway [29], angiopoietins [30], Delta-Notch signaling pathway [31], PDGF-C [32, 33], interleukins [34], ephrins [35], and epidermal growth factor [36] have been described as part of the anti-VEGF escape mechanisms. The abovementioned angiogenic factors and their redundant angiogenic signaling pathways are summarized in Fig. 19.2. These anti-VEGF/VEGFR escape mechanisms may contribute to acquired resistance to anti-angiogenic therapies and may contribute to subsequent recurrence and/or metastases.

Table 19.1 FDA approved anti-angiogenic drugs [14]

Name of the approved drug (trade name)	Therapeutic targets/ potential biomarker (s)	Year of approvals	Types of cancer treated	Guidelines for treatment
Monoclonal antibodies/chimeric fusion proteins				
Bevacizumab (Avastin®, Roche/ Genentech)	VEGF-A	2004	Metastatic colorectal cancer (MCC)	First- and second-line treatment for MCC, first line for NSCLC, with interferon for RCC, with chemotherapy for OC
		2006		
		2009	Non-small cell lung cancer (NSCLC)	
		2014	Renal cell carcinoma (RCC) platinum-resistant recurrent ovarian cancer (OC)	
		2009	Approval withdrawn for treating breast cancer (BRCA)	
Aflibercept (Zaltrap®, Sanofi Genzyme): a chimeric VEGF/PlGF neutralizing receptor	VEGFA, VEGFB, PLGF	2012	colorectal cancer (CRCA), pancreatic cancer (PACA), NSCLC	Second-line metastatic treatment for CRCA, with chemotherapy for PACA and NSCLC
Ramucirumab (Cyramza®, Eli Lilly)	VEGFR2	2014	Gastric or gastroesophageal junction adenocarcinoma (GOAC)	Refractory with or without chemotherapy for GOAC; refractory with chemotherapy for NSCLS and MCC
		2014	NSCLC	
		2015	MCC	
Small-molecule tyrosine kinase inhibitors with anti-VEGFR activity				
Axitinib (Inlyta®, Pfizer)	VEGFR 1–3	2012	RCC	Second-line single drug therapy
Cabozantinib (Cabometyx™, Exelixis)	All VEGFRs	2012	Progressive metastatic medullary thyroid cancer	Second-line therapy with chemotherapy
Pazopanib (Votrient™)	All VEGFRs	2009	Renal cell carcinoma	Second-line treatment with chemotherapy
		2012	Soft tissue sarcoma, recommended treatment for RCC, NSCLC	
Regorafenib (Stivarga®, Bayer)	All VEGFRs	2013	Resistant metastatic colorectal cancer	Single-drug treatment for resistant advanced gastrointestinal stromal tumors, second-line treatment for MCC
Sorafenib (Nexavar®, Bayer/Onyx)	All VEGFRs	2005	Renal cell carcinoma	Second-line treatment for metastatic or recurrent thyroid carcinoma and advanced renal cell carcinoma
		2007	HCC	
		2013	Differentiated thyroid cancer	
			Recommended treatment for melanoma and NSCLC	
Sunitinib (Sutent®, Pfizer)	All VEGFRs	2006	RCC	Single-drug, first-line treatment for RCC, single drug for treatment of progressive well-differentiated pancreatic neuroendocrine tumors
		2011	Pancreatic neuroendocrine tumors; also recommended for RCC, gastrointestinal stromal tumor (GIST), BRCA, HCC, CRCA	

(continued)

Table 19.1 (continued)

Name of the approved drug (trade name)	Therapeutic targets/ potential biomarker (s)	Year of approvals	Types of cancer treated	Guidelines for treatment
Lenvatinib (Lenvima®, Eisai)	All VEGFRs	2015	Thyroid cancer	Treatment of locally recurrent or metastatic, progressive, radioactive iodine-resistant differentiated thyroid cancer
Vandetanib (Caprelsa®, Sanofi Genzyme)	All VEGFRs	2011	NSCLC, medullary thyroid carcinoma (MTC)	Unresectable, locally advanced, or metastatic MTC

Reprinted from Gacche and Assaraf [14]. With permission from Elsevier

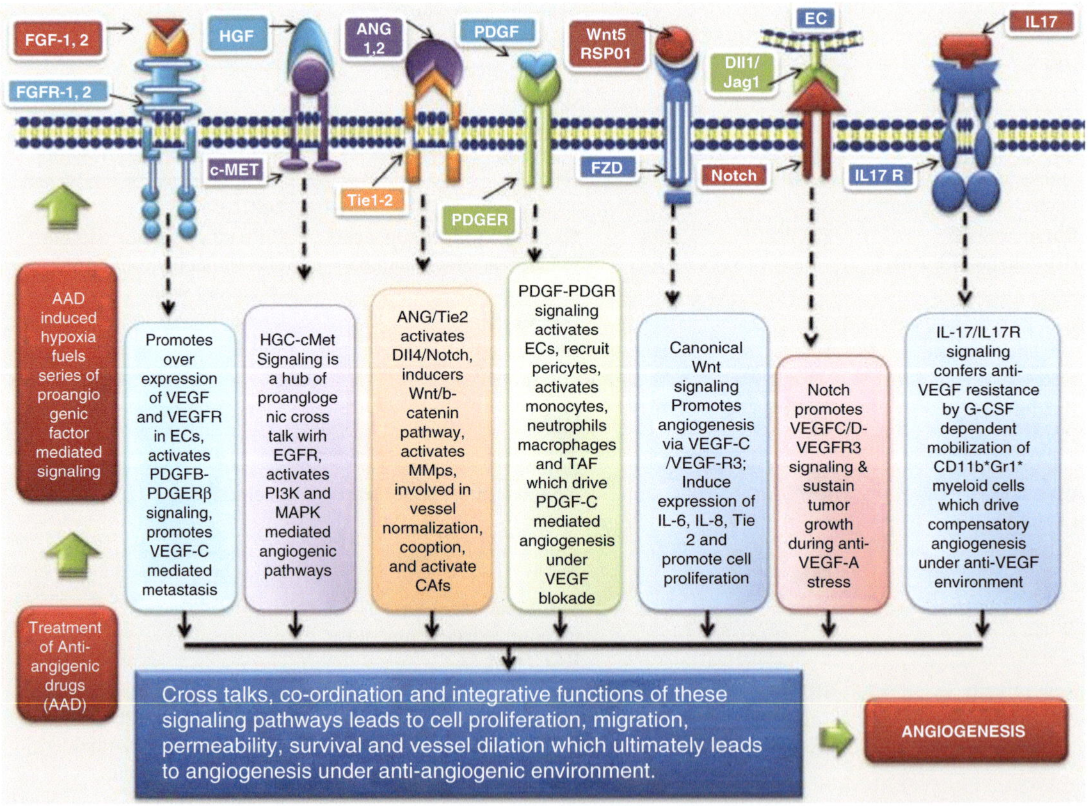

Fig. 19.2 VEGF-independent signaling pathways. In response to treatment of cancer patients with anti-angiogenic agents, cross talk and redundancy among various VEGF-independent compensatory proangiogenic signaling pathways can drive continued tumor angiogenesis and progression [14]. (Reprinted from Gacche and Assaraf [14]. With permission from Elsevier)

Cross Talk Between VEGFR2 and Other RTKs

VEGFR2 can also form direct complexes with other receptor tyrosine kinases. For example, stimulation of vascular smooth muscle cells with VEGF promotes the formation of a complex between VEGFR2 and the receptor tyrosine kinase PDGF-Rb [37]. This results in suppression of PDGF-Rb signaling and decreased pericyte coverage in tumors and may explain the observation that, in some experimental systems, inhibition of VEGF signaling leads to increased pericyte coverage of tumor vessels and increased maturation/normalization of the tumor vasculature [38]. In glioblastoma cells, VEGF stimulates the formation of a complex between VEGFR2 and the receptor tyrosine kinase, MET, which results in suppression of MET signaling and reduced tumor cell invasion [25]. Consequently, inhibition of VEGF has been shown to release MET from this inhibitory mechanism and allows for increased tumor invasion. Cross talk between VEGF and HER2 signaling pathways has been demonstrated in human breast cancer tissues in the form of higher VEGFR2 expression in HER2+ breast cancer [39]. Therefore, cross talk between VEGF/VEGF receptor signaling and other receptors may offer (1) plausible explanation for the diversity of clinical responses observed with VEGF-targeted therapies and (2) new opportunities to combine anti-angiogenic therapies with other targeted therapies to treat specific cancer types more effectively.

Pathologic Angiogenesis

Aberrant (pathologic) angiogenesis is one of the hallmarks of cancer. The imbalance of pro- and anti-angiogenic signaling within tumors creates an abnormal vascular network that is characterized by dilated, tortuous, and hyperpermeable vessels, which were elegantly described in Ad-VEGFA mouse model by Harold Dvorak as six different vessel types [40]. The physiological consequences of these vascular abnormalities include temporal and spatial heterogeneity in tumor blood flow and oxygenation and increased tumor interstitial fluid pressure. These abnormalities and the resultant microenvironment fuel tumor progression and also lead to a reduction in the efficacy of chemotherapy, radiotherapy, and immunotherapy. With the discovery of vascular endothelial growth factor (VEGF) as a major driver of tumor angiogenesis, efforts have focused on novel therapeutics aimed at inhibiting VEGF activity, with the goal of regressing tumors by starvation. Unfortunately, clinical trials of anti-VEGF monotherapy in patients with solid tumors have been largely negative or resulted in marginal clinical benefit. Intriguingly, the combination of anti-VEGF therapy with conventional chemotherapy has improved survival in cancer patients compared with chemotherapy alone. These seemingly paradoxical results could be explained by the concept of "normalization" of the tumor vasculature by anti-VEGF therapy. Preclinical studies have shown that anti-VEGF therapy changes tumor vasculature toward a more "mature" or "normal" phenotype. This "vascular normalization" is characterized by attenuation of hyperpermeability, increased vascular pericyte coverage, and more normal vascular basement membrane, resulting in reduced tumor hypoxia and interstitial fluid pressure. These, in turn, can lead to an improvement in the metabolic profile of the tumor microenvironment, the delivery and efficacy of exogenously administered therapeutics, the efficacy of radiotherapy and of effector immune cells, and a reduction in number of metastatic cells shed by tumors into circulation in mice. These findings are consistent with data from clinical trials of anti-VEGF agents in patients with various solid tumors [41].

Evaluation of Pathologic Angiogenesis in Human Cancer

Although during last few decades there has been substantial research that contributed a great deal to our understanding of biology of tumor angiogenesis, including primary and acquired resistance mechanisms, due to several different factors, oncology biomarker research community has not been so successful in the development, optimization, and technical and clinical validation of biomarkers of pathologic angiogenesis in human cancer, especially in the context of histopathologic and molecular heterogeneity of cancers in various parts of the human body. This continues to be an area of unmet need, which if addressed appropriately by the biomarker and clinical trial teams, has the potential to further refine the current level of success of both the single agent and combinatorial anti-angiogenic therapies in clinical trials. Despite the urgent need for greater focus and investment in biomarker research and practice to support current and future trials of anti-angiogenic therapies, some of the ongoing efforts to investigate and advance tissue angiogenic biomarkers are outlined in the following sections.

Heterogeneity of Tumor Vessels in Experimental Models and Human Tumors

Therapies directed against VEGF-A and its receptors are effective in treating many mouse tumors but have been less so in treating human cancer patients. Such variation has been attributed to the nature of blood vessels that appear in human and mouse cancers and the tumor "surrogate" blood vessels that develop in immunodeficient mice in response to an adenovirus expressing the VEGF-A^{164} protein [40]. Both tumor and tumor surrogate blood vessels are heterogeneous and form by two distinct processes, angiogenesis and arterio-venogenesis [40].

The first new angiogenic blood vessels to form are mother vessels (MV); MV arise from preexisting venules and capillaries and evolve over time into glomeruloid microvascular proliferations (GMP) and subsequently into capillaries and vascular malformations (VM). Arterio-venogenesis results from the remodeling and enlargement of preexisting arteries and veins, leading to the formation of feeder arteries (FA) and draining veins (DV) that supply and drain angiogenic vessels. Among these, only MVs and GMPs were highly responsive to anti-VEGF therapy, whereas "late"-formed capillaries, VMs, FAs, and DVs, were relatively unresponsive. These findings were further supported by results of immunohistochemistry: early-forming MVs and GMPs, in which the lining endothelial cells expressed high levels of VEGFR-2, were highly susceptible to anti-VEGF blockade by VEGF-Trap (ziv-aflibercept, Zaltrap®, Sanofi-Aventis). In contrast, late-forming VMs, FAs, and DVs that expressed low levels of VEGFR-2 were largely resistant. Taken together, these findings may explain, at least in part, the relatively poor response of human cancers to anti-VEGF/VEGFR therapies, because human cancers, present for months or years prior to discovery, are expected to contain a large proportion of late-formed blood vessels [40]. Translating VEGFR2 IHC findings from Ad-VEGFA^{164} model to human cancer tissues will be an important consideration. As such high VEGFR2 expression levels in human tumor vessels are likely to correlate with tumor response to anti-VEGF/VEGFR therapies. Of course, in cancer patients the overall complexity of the angiogenic process and the redundancy of various signaling pathways can make such correlations less than straightforward and may need systematic evaluation of multiple biomarkers of pathologic angiogenesis.

VEGF/VEGF Receptor Expression in Human Tumor Tissues

In recent years progress has been made with regard to the evaluation of the clinical significance of VEGF/VEGF receptors in human cancer tissues by the development of robust methodologies with appropriate controls to accurately and reproducibly determine vascular and tumor cell expression levels of various angiogenic ligands and receptors. We and others have developed technically robust immunohistochemical assays [42, 43] to evaluate VEGFR2 and other VEGF receptors on archival tumor tissues. As part of the technical validation of the above assay, we carried out extensive optimization experiments and demonstrated comparable levels of VEGFR2 protein and in situ VEGFR2 RNA levels in serial sections of human (bladder) cancer and H441 (non-small cell lung cancer) xenograft tissues. In our solid tumor analyses, a frequent finding has been the heterogeneity of vascular and tumor cell expression of VEGFR2 in different areas of the same tumor (intra-tumor heterogeneity) and among different tumors (inter-tumor heterogeneity) (Fig. 19.3) and a degree of variation in subcellular localization of VEGFR2 in tumor cells [43, 45]. In our experience with different human cancer tissues, we came across many of the heterogeneous vessels described by Dvorak in Ad-VEGFA^{164} mouse model [44]. In survival analyses, tumor cell expression of VEGFR2 was found to be associated with adverse outcome in non-small cell lung

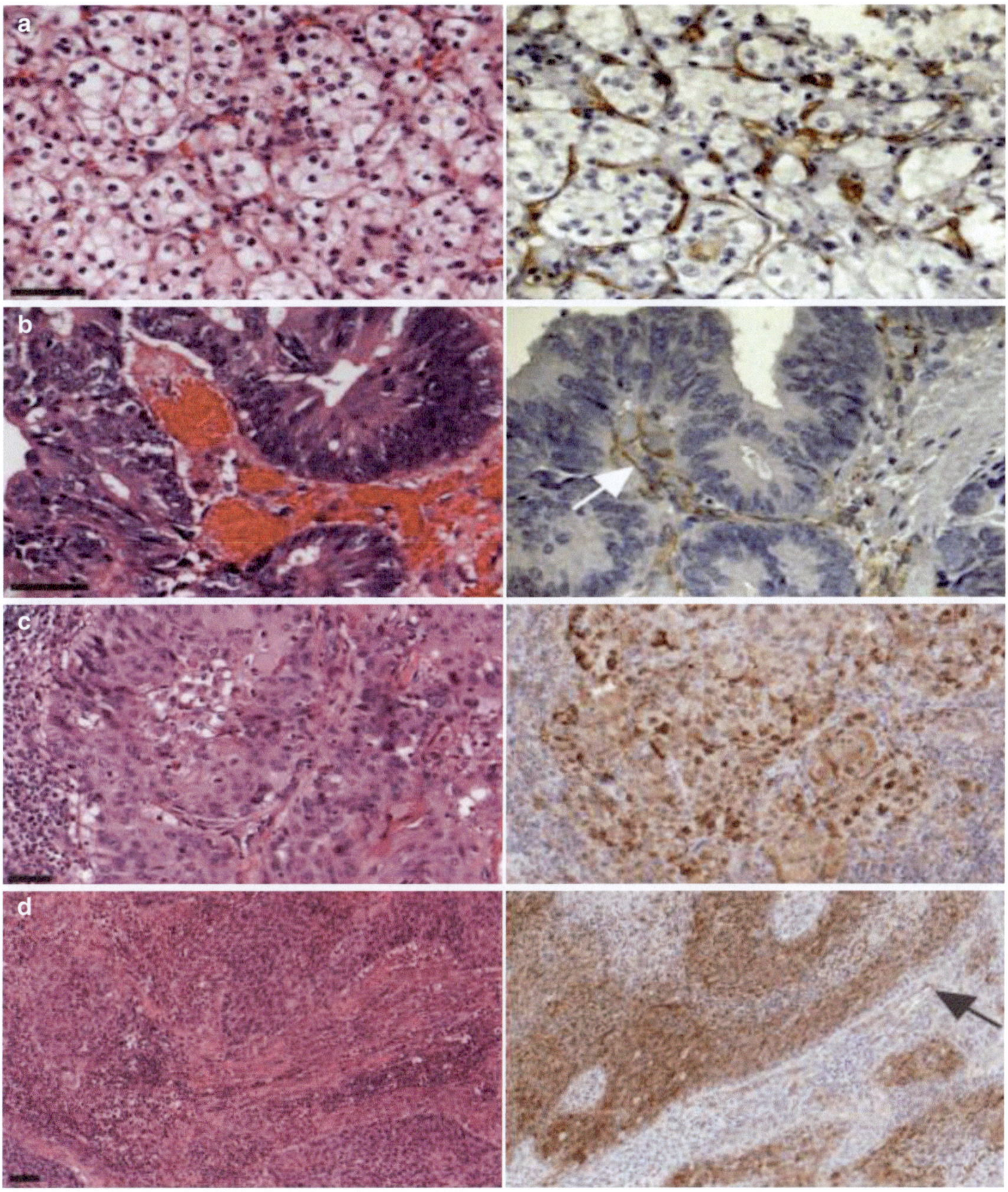

Fig. 19.3 Vascular endothelial cell and tumor cell expression of VEGFR2 on representative cases from a multi-tumor survey. Left panels H&E; right panels VEGFR2 IHC. (**a**) VEGFR2 IHC on renal cell carcinoma of the kidney showing endothelial cell immunoreactivity (X400). (**b**) VEGFR2 IHC on ADC of the colon showing endothelial cell immunoreactivity in the stromal blood vessels. Tumor cells are negative for VEGFR2 (X400). (**c**) VEGFR2 IHC on SCC of the lung showing endothelial cell and a range of tumor cell (nuclear cytoplasmic membranous) immunoreactivity (X200). (**d**) VEGFR2 IHC is showing vascular endothelial cell immunoreactivity and a range of tumor cell cytoplasmic and nuclear immunoreactivity on SCC of the cervix (X200). Immunoreactivity in endothelial cells lining vessels (*white* and *black arrows*). Slides were counterstained with hematoxylin. Scale bars: 50 μm. In renal and colonic cancer tissues, the VEGFR2 expression is restricted to tumor vessels. In squamous cell cancers, however, VEGFR2 expression was found both in tumor stromal vessels and tumor cells. Such variation in distribution and localization of VEGFR2 may result in part account for the differences in sensitivity of these cancer tissues to anti-VEGF/VEGFR2 therapies. (Reprinted from Holzer et al. [43]. With permission from Creative Commons License)

cancer (NSCLC) [43] but favorable prognosis in bladder cancer [46]. In a more recent disease state characterization analysis, we have demonstrated significantly higher expression of the VEGFR2 protein in HER2+ breast cancer compared to other BRC subtypes [39]. Based on these findings, we hypothesized that compared to hormone receptor positive or triple negative subsets, HER2+ human breast cancers with high VEGFR2 expression might respond differently to anti-angiogenic therapies. Utilizing high-quality reagents, stringent method development strategies, well-optimized laboratory protocols, and "fit-for-purpose" yet sufficiently informative interpretation and scoring approaches, such methodologies are being evaluated in early- and late-phase clinical trials of anti-angiogenic agents.

Microvascular Density (MVD)

In breast tumors, high baseline microvascular density (MVD) is considered a positive response to vascular normalization index induced by bevacizumab [47]. In the case of high MVD, the effect of anti-angiogenic drugs would be to remove some vessels and increase the functions of others, by inducing their normalization. In the case of low MVD, the anti-angiogenic therapy would reduce tumor microvasculature further and prevent their normalization. This can make the tumor insensitive to anti-angiogenic therapy. Therefore, baseline MVD can be a key factor in predicting the success of treatment with anti-angiogenic drugs [48].

Predictive Biomarkers

High variability in patient response to anti-angiogenic therapy across different indications exists, and this is coupled with the development of therapy resistance [49]. As with other targeted compounds, reliable biomarkers to identify patients with cancer who will benefit from anti-angiogenic therapy are still needed. One of the main challenges in identifying potential biomarkers for anti-angiogenic therapy is the complex nature of the angiogenic signaling process, which is characterized by multiple pathways that not only overlap but that continuously cross talk, making it difficult to eliminate an angiogenic stimulus [50]. Several types of biomarkers are being investigated across different indications: circulating biomarkers (e.g., concentrations of soluble angiogenic receptors/ligands), genetic biomarkers (e.g., single nucleotide polymorphisms), tissue biomarkers (e.g., immunohistochemical staining of angiogenic receptors), and physiologic biomarkers (e.g., hypertension) [49].

Elevated levels of soluble VEGFR1 (sVEGFR1) prior to treatment were associated with a poor outcome from bevacizumab in rectal carcinoma, hepatocellular carcinoma (HCC), and metastatic colorectal carcinoma patients [51–54]. A retrospective analysis has shown that a genetic polymorphism in the VEGFR1 gene correlates with increased VEGFR1 expression and a poor outcome of bevacizumab treatment in metastatic renal cell carcinoma and pancreatic ductal adenocarcinoma patients [55]. Similarly, elevated levels of NRP1 were associated with a poor outcome in some trials [56]. It is possible that VEGFR1 and NRP1 function as endogenous VEGF-Traps. Therefore, adding an external anti-VEGF agent may not have significant biologic effects in patients with high sVEGFR1/NRP1 levels. Additionally, increased VEGFR1 levels may induce increased proangiogenic signaling by PlGF when VEGF is blocked [55].

In recent years, our laboratory developed technically robust immunohistochemical assays for localization of VEGFR2, VEGFR1, and VEGFR3 in archival human tissues. Large-scale biomarker prevalence and disease characterization analyses have shown significant variation in VEGF receptor profiles (VEGFR1, VEGFR2, VEGFR3) among NSCLC, BRC, and CRC tissues [43, 45, 57]. Since various anti-angiogenic therapies target one (VEGFR2 in case of ramucirumab) or multiple VEGF receptors (VEGFR1, VEGFR2, VEGFR3 in case of small-molecule tyrosine kinase inhibitors), the clinical significance of various VEGF receptor profiles may be determined by retrospective-prospective (VEGFR receptor profiling) analyses on tissue specimens from positive anti-angiogenic therapy trials. Just as assessments of circulating VEGF receptor levels are frequently performed in clinical trials of anti-angiogenic therapies (more for logistical reasons than true science, given the frequent uncertainty about the source of circulating receptors), it will also be prudent to evaluate all three VEGF receptors along with NRPs in pathologically well-characterized archival tumor tissues in order to determine their value in predicting response or resistance to anti-VEGF/VEGFR2/VEGFR1,2,3 (bevacizumab, ramucirumab, TKIs) agents.

Despite initial reports on predictive biomarkers, overall reproducibility of candidate biomarkers across indications is limited, and there is a paucity of studies comparing the same biomarkers for the same indication. The appropriate use of genomic and proteomic technologies will be key in improving our ability to match a target pathology with the efficacy of a specific anti-angiogenic therapy, although a lot of cross-platform validation work will be required to implement newly discovered candidate predictive biomarkers into clinical practice. Another area that needs urgent attention in clinical trials is an accurate diagnostic evaluation of clinical trial tissues, which can result in significant misclassification of clinical trial tissue specimens – with obvious negative impact on the quality of biomarker data from clinical trials. Central pathology review and even sub-specialty level histopathologic characterization of cancer tissue specimens from clinical trials and implementation of comprehensive angiogenic biomarker approaches in analyzing the clinical trial data sets need to be serious considerations in future trials of anti-angiogenic therapies.

Molecular Cancer Subtyping and Response to Anti-angiogenic Therapies

A recent analysis of patients with gastroesophageal carcinoma demonstrated that the ratio of progression-free survival (PFS) on the molecular profile (MP)-based treatment to PFS on treatment prior to molecular profiling exceeds 1.3, suggesting the potential value of MP in guiding selection of individualized therapy [58]. Biologic rationale for this clinical finding is evident by the presence of four molecular subtypes of human gastric cancer by TCGA [59]. The TCGA subgroup labeled "chromosomal instability (CIN)" is characterized by amplifications of several therapeutic targets including HER2, VEGF-A, MET, and others. In the current and future trials of anti-VEGF/VEGFR2 therapies, it will be interesting to see if this molecular subset responds better to such therapies compared to the other molecular subsets of GC. In a recent analysis of stromal gene signature in GC, we identified differentially expressed genes in various molecular subtypes of GC [60]. Such analyses can provide invaluable biologic insights and can help with the selection of promising predictive biomarker candidates for subsequent clinical validation in clinical trials of anti-angiogenic therapies. An important technical challenge in tailoring anti-angiogenic therapies to various molecular subgroups of human cancers will be that we still do not have robust, clinical grade tissue-based methodologies with clinically validated scoring cutoffs for pertinent solid tumors for many of the newer therapeutic targets, so that various biologically relevant combinatorial therapeutic approaches can be tested in appropriately selected cancer patient subsets.

Anti-angiogenic Therapies to Treat Human Cancer

Currently, there are four main approaches targeting angiogenesis in human cancer, which have been tested in clinical trials and approved for clinical practice: (1) neutralizing monoclonal antibody that binds circulating VEGF (bevacizumab, Avastin®, Roche/Genentech), (2) recombinant protein called decoy receptor or "VEGF-Trap" (Aflibercept, Zaltrap®, Sanofi Genzyme) that binds more than one proangiogenic growth factor, (3) small-molecule tyrosine kinase inhibitors (like sunitinib (SUTENT®, Pfizer), sorafenib (Nexavar®, Bayer)) that block tyrosine kinase activity of VEGFRs, and (4) therapeutic monoclonal antibodies targeting VEGF receptor 2 (ramucirumab, Cyramza®, Eli Lilly).

Anti-VEGF Therapy (Bevacizumab)

One of the first anti-angiogenic therapies was the monoclonal antibody neutralizing circulating VEGF. In 2004, the first phase III trial results showed that bevacizumab (Avastin®, Roche/Genentech), a humanized monoclonal antibody binding specifically to VEGF-A alone, when combined with chemotherapy in metastatic colorectal cancer improved progression-free survival (PFS) (10.6 vs. 6.2 months) and overall survival (OS) (23 vs. 15.3 months) compared to chemotherapy arm [2]. An improvement in PFS for the combination of bevacizumab plus chemotherapy was next shown in two phase III trials in non-squamous non-small cell lung cancer (NSCLC) [61–63], but only one study reported an improvement in OS [61]. Within the next few years, bevacizumab was approved as a monotherapy in second-line treatment of glioblastoma and in combination with interferon-α (INF- α) for renal cell carcinoma. There were some controversies in cases of using bevacizumab in the treatment of metastatic breast cancer. The ECOG-2100 trial showed that adding bevacizumab to paclitaxel improved PFS (11.8 vs. 5.9), as well as OS rates (36.9% vs. 21.2%) compared to paclitaxel alone. Based on those results, the US Food and Drug Administration (FDA) accelerated in 2008 approval of bevacizumab in combination with paclitaxel in metastatic breast cancer. Further trials, AVADO and RIBBON-1, confirmed the improvement of PFS by bevacizumab, but neither demonstrated any improvement of OS. In addition, bevacizumab-induced hypertension was reported as a clinically relevant adverse event in a phase III breast cancer trial (E5103) [64]. In 2011, FDA withdrew approval for bevacizumab in metastatic breast cancer. In 2014, bevacizumab was approved for the treatment of patients with platinum-resistant recurrent epithelial ovarian, fallopian tube, or primary peritoneal cancer in combination with paclitaxel, pegylated liposomal doxorubicin, or topotecan [65, 66] and for recurrent or metastatic cervical cancer in combination with paclitaxel and cisplatin or paclitaxel and topotecan [67, 68].

Aflibercept (Human Recombinant Fusion Protein)

Aflibercept (Zaltrap ®, Sanofi Genzyme) is a human recombinant fusion protein that acts as a decoy receptor of angiogenic factors. Unlike bevacizumab, it targets not only VEGF-A but also VEGF-B and placental growth factor (PIGF). This is a fusion protein of the second immunoglobulin domain of VEGFR1, third immunoglobulin domain of VEGFR2, and constant region Fc of human IgG1. In 2012, FDA approved aflibercept in the treatment of metastatic colorectal cancer (CRC) with infusional fluorouracil, leucovorin, and irinotecan, based on phase III trial results [69].

Anti-VEGFR2 Therapy (Ramucirumab)

Ramucirumab (Cyramza® Eli Lilly) is a human monoclonal antibody that inhibits angiogenesis by blocking binding of VEGF to the extracellular domain of VEGFR2. It is advantageous due to its receptor selectivity with minimal off-target activity. Preclinical studies showed that ramucirumab binds selectively to VEGFR2 with a greater efficacy than its natural ligand VEGF-A. It is approved for second-line treatment in gastric cancer, NSCLC, and colon cancer. Based on the RAISE study, ramucirumab was approved in combination with FOLFIRI (folinic acid, 5-fluorouracil, and irinotecan) in metastatic CRC patients, if disease progressed after therapy with bevacizumab, oxaliplatin, and fluoropyrimidine. In NSCLC, ramucirumab was approved in combination with docetaxel after platinum-based chemotherapy. In gastric cancer patients, FDA approved ramucirumab as a monotherapy in advanced or metastatic disease or in gastroesophageal junction carcinoma patients for whom first-line chemotherapy had failed [70, 71]. The FDA guidelines for the therapeutic applications of the approved anti-angiogenic drugs are summarized in Table 19.1 [11].

Small-Molecule Tyrosine Kinase Inhibitors

Tyrosine kinase inhibitors (TKIs) are small-molecular-weight drugs that inhibit the kinase activity of different receptors. The mechanism of action of TKIs relies on binding around the ATP-binding site of a given receptor, thus hindering phosphorylation of the tyrosine residue of that receptor and downstream signaling. There are several small-molecule kinase inhibitors, tyrosine kinase, serine/threonine kinase, or dual protein kinase inhibitors, approved by the FDA (sunitinib, sorafenib, axitinib, and pazopanib), some of which target VEGF receptors (VEGFRs) and are used to treat a number of different types of cancer (Table 19.1) [72, 73]. Compared to VEGF neutralizing antibodies, TKI does not interfere with the binding of VEGF to its receptors, and they usually target not only VEGFR but other kinases as well like PDGFR, FGFR, and c-KIT [41].

Bispecific Antibodies Targeting Both Tumor Cells and Angiogenesis

In recent years, monospecific antibodies targeting cell surface receptors have achieved remarkable success with cancer treatment. However, redundant signaling and cross talk between different pathways within tumor cells and between tumor cells and their microenvironment can limit the efficacy of receptor-targeted monospecific-based therapies [74]. During tumor progression, hypoxia and acidosis are known to induce angiogenesis within the tumor. Both tumor cells and tumor-associated endothelial cells express growth factors and their corresponding receptors, such as EGFR and VEGFR. In a mouse model of colon cancer [75], dual inhibition of EGFR and VEGFR by kinase inhibitors reduced tumor growth and metastasis, suggesting that the EGFR and VEGFR pathways have important roles in regulating tumor progression and neovascularization. More strikingly, an increase in EGFR expression and loss of ErbB3 expression has been identified in tumor vasculature and provides the rationale to target EGF-induced endothelial cell proliferation in tumor vasculature [76]. Co-inhibition of PDGFRβ and VEGFR has been shown to prevent new blood vessel growth better than VEGFR alone [77]. A number of tyrosine kinase inhibitors (axitinib, sorafenib, and sunitinib) targeting both VEGFR and PDGFR are effective in renal cell carcinoma and other types of human cancer [78]. Bispecific antibodies – such as single-chain variable fragments dual-targeting PDGFRβ and VEGF-A [79] and Ang-2-VEGF-A CrossMab, which targets angiopoietin 2 and VEGF-A simultaneously [80] – inhibit two distinct pathways targeting tumor angiogenesis. Bispecific antibodies against those targets that are shared by tumor cells and tumor-associated endothelial cells have the potential for an enhanced therapeutic efficacy.

Potential Mechanisms of Resistance to Anti-angiogenic Therapies

As illustrated by Vasudev and Reynolds [19] (Fig. 19.4), several different mechanisms are involved in lack of clinical response (resistance) to anti-angiogenic therapies. These include:

A. *Heterogeneity of tumor vessels* in the form of therapy-sensitive and therapy-insensitive vessels. For example, some of the tumor vessels may be destroyed by the therapy, while others may survive.
B. *Alternative signaling pathways* that can regulate the sensitivity of vessels to therapy.
C. *Stromal cells*, such as immature myeloid cells or fibroblasts that can infiltrate the tumor and mediate resistance either by releasing pro-angiogenic growth factors or by physically incorporating into the tumor vessels.
D. *Tumor cells that can survive conditions of stress*. For example, some tumor cells may have survived the loss of a vascular supply because they are adapted to survive conditions of hypoxia or nutrient shortage.
E. Tumors may use *alternative mechanisms of vascularization* besides sprouting angiogenesis. In *intussusceptive microvascular growth,* new vessels are generated by the fission of existing vessels. *Glomeruloid angiogenesis* is characterized by tight nests of vessels that resemble the renal glomerulus. In *vasculogenic mimicry*, tumor cells directly form vascular channels (blue cells) that are perfused via connection to the host vasculature (red cells). In *looping angiogenesis*, contractile myofibroblasts (green) pull host vessels out of the normal surrounding tissue (pink region). In *vessel co-option* tumor cells engulf host vessels in the normal surrounding tissue (pink region) as the tumor invades.

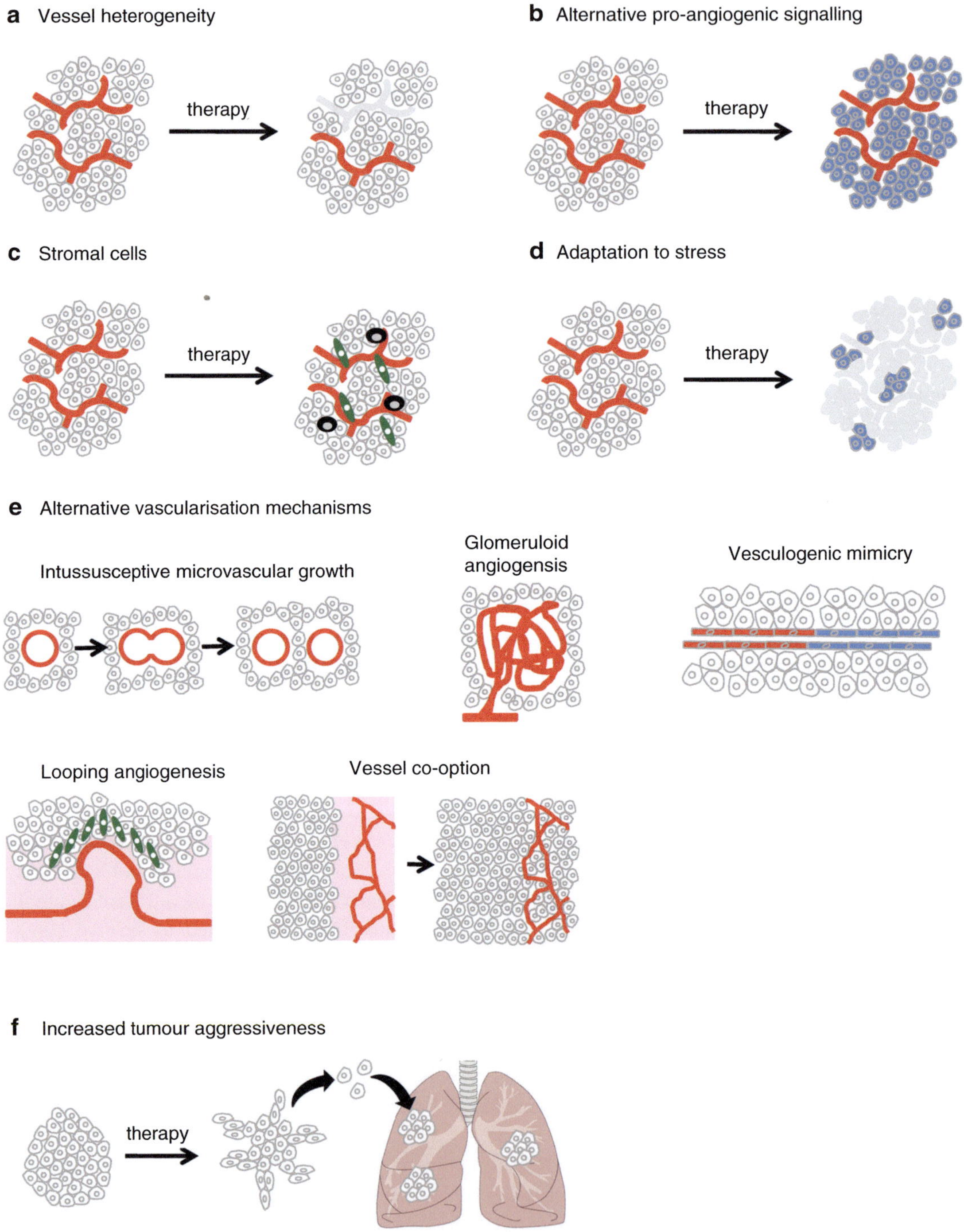

Fig. 19.4 Potential mechanisms involved in resistance to VEGF-targeted therapy; see text. (Reprinted from Vasudev and Reynolds [19]. With permission from Springer Nature)

F. *Increased tumor aggressiveness*, i.e., therapy causes the tumor to become more invasive and/or accelerates the growth of metastases.

Emerging Trends in Anti-angiogenic Therapies

The anti-angiogenic (anti-VEGF/VEGFR) therapies, intended to block tumors' blood supply, may cause hypoxia, which may fuel tumor progression and treatment resistance. Emerging clinical data suggest that patients whose tumor perfusion or oxygenation increases in response to these agents may actually survive longer. Hence, strategies aimed at alleviating tumor hypoxia while improving perfusion may enhance the outcome of radiotherapy, chemotherapy, and immunotherapy [1].

Vascular Normalization and Pruning After Anti-angiogenic Therapy

In normal tissue, the blood vessels have normal structure and function due to the balance of the signals downstream of the pro-angiogenic molecules (e.g., VEGF, Ang2) and anti-angiogenic molecules (e.g., sVEGFR1, thrombospondins, semaphorins). In contrast, tumor vessels are structurally and functionally abnormal due to an imbalance between pro- and anti-angiogenic signals. This creates an abnormal microenvironment in tumors – characterized by hypoxia, acidosis, and elevated fluid pressure – which fuels tumor progression and treatment resistance via multiple mechanisms [1]. Inhibiting pro-angiogenic signaling or enhancing anti-angiogenic signaling can prune some abnormal vessels and remodel the rest resulting in a "normalized vasculature." Depending upon the extent of normalization versus pruning, tumor perfusion/oxygenation may increase, remain unchanged, or decrease. Some tumors might be intrinsically resistant to a given AA agent, and others may switch to non-sprouting mechanisms of vessel recruitment (e.g., vessel cooption) that are refractory to the given AA agent and continue to make abnormal vessels again.

Combining Anti-angiogenic Agents with Drugs That Target Oncogenic Pathways

Combining AA agents with agents targeting oncogenic pathways, similar to chemotherapeutic agents, has led to some unexpected results.

Combining Anti-VEGF and Other Targeted Therapies

Despite promising preclinical results from combining VEGF- and EGFR-targeted agents in colorectal and NSCLC models, all phase III trials combining these targeted agents have failed [81]. Similarly, phase III trials combining VEGF- and HER2-targeted therapies in HER2+ breast cancer patients also failed [82]. A potential mechanism for these failures is that the dose of bevacizumab used may have decreased the size of pores in the tumor vessel walls and compromised the delivery of antibodies [83]. This hypothesis is consistent with elevated baseline plasma VEGF concentrations being associated with a greater bevacizumab benefit. It is also consistent with the recent randomized phase II trial showing the benefit of combining bevacizumab with a smaller drug, erlotinib, in EGFR-mutant NSCLC patients [84].

Combining Anti-VEGFR2 and Anti-HER2 Agents

Treatment of HER2+ breast tumors in the mouse brain with trastuzumab leads to increased VEGF production by host cells in the brain [85]. To this end, we combined HER2-targeted drugs (trastuzumab and lapatinib) with an anti-VEGFR2 antibody and demonstrated a significant improvement in survival of mice bearing HER2+ tumors in the brain [86]. Moreover, a phase II clinical trial with dual HER2 blockade and bevacizumab showed encouraging results in heavily pretreated HER2+ breast cancer patients with brain metastases [87]. Some of these clinical results are in line with our recent finding of higher VEGFR2 protein levels by IHC in HER2+ breast cancer [39], pointing toward potential clinical relevance of combining anti-VEGFR2 therapy (Ramucirumab: Cyramza®) with anti-HER2 therapies in HER2+ breast cancer.

Combining Anti-angiogenic and Immunotherapeutic Agents

Vascular Normalization Can Improve Benefit from Immunotherapy

The abnormal tumor vasculature can impede T effector cell infiltration into tumors and create a hypoxic and acidic tumor microenvironment that upregulates PD-L1 on myeloid-derived suppressor cells (MDSCs), dendritic cells, and cancer cells; increases the accumulation of regulatory T cells (Tregs); impairs T effector cells; and polarizes TAMs to the immune inhibitory M2-like phenotype to suppress T effector cell function.

Hypoxia can also upregulate multiple immune-suppressive growth factors and cytokines (e.g., VEGF and TGF-b). Vascular normalization with an appropriate dose and schedule of anti-angiogenic treatment can normalize the tumor vasculature and generate a more homogeneous distribution of perfused tumor vessels, facilitating the infiltration of T effector cells while reducing MDSC and regulatory T cell (Treg) accumulation. In addition, alleviation of hypoxia and acidity by improved vascular perfusion polarizes TAMs to an immunostimulatory M1-like phenotype [88].

Conclusions and Perspective

In recent years, significant advances in cancer treatment have been made with anti-angiogenic therapies, many of which have focused on inhibition of the vascular endothelial growth factor/VEGFR receptor (VEGF/VEGFR) pathway. VEGF/VEGFR targeting alone, however, has not been as efficacious as originally hoped. Based on recent advances in demystifying the complex biology of tumor angiogenesis, it has become clear that there are many redundant, compensatory signaling pathways that can overcome VEGF/VEGFR-targeted inhibition of tumor angiogenesis and may contribute to subsequent tumor progression. Therefore, refinement of the efficacy of various anti-angiogenic therapies will, at one end, require more focused approach rationalized by predictive biomarkers and, on the other, a rather broader therapeutic approach using biologically relevant combinatorial strategies or multitargeted anti-angiogenic agents, based on diverse molecular pathologic profiles of various human cancer types and subtypes.

Advanced histopathologic characterization and molecular classification of human cancer tissues from clinical trials will enable clinical trial teams to accurately interpret clinical efficacy data emerging from ongoing trials of anti-angiogenic therapies. As high-throughput technologies like NGS and more targeted sequencing approaches are becoming less and less cost-prohibitive, in addition to gold standard single marker methodologies like IHC and FISH, a great deal of progress can be made in terms of discovery and analytical and clinical validation of appropriate targeted panels of molecular biomarkers of response or resistance to AA-Rxs. Anatomic pathologists with sub-specialty expertise in molecular oncologic pathology will have a key role in designing and advancing predictive and prognostic biomarker science on well-characterized human cancer tissues. Acquisition of high-quality human tissue specimens and relevant clinicopathologic data will facilitate exploratory biomarker analyses at earlier stages of clinical development of AA-Rxs.

Although a great deal of scientific knowledge has accumulated about the highly complex biology of VEGF/VEGFR and non-VEGF/VEGFR signaling pathways with frequent cross talk and redundant mechanisms of primary and acquired resistance to anti-angiogenic therapies, there is an urgent need to incorporate and translate that scientific knowledge into patient-tailoring strategies in the current and future clinical trials of anti-angiogenic therapies, so that these agents can be offered to the right patients in order to maximize and sustain clinical benefit at the individual patient level. Because of its complexity, it will also be important to develop reliable methodologies to more fully characterize pathobiology of angiogenesis in the context of molecular pathology of various human cancer types. Such efforts will benefit from well-integrated, interdisciplinary clinical, translational, and basic research teamwork including the industry, academia, diagnostic, and biotechnology companies. Prioritization of well-established and innovative technologies to develop and standardize predictive biomarker assays, the definition of optimal scoring strategies/cutoffs, and accurate diagnostic classification of clinical trial cancer tissues will be important considerations for next-generation clinical trials of anti-angiogenic therapies.

Future Directions

To Improve Efficacy of Anti-angiogenic Therapies

While approved anti-angiogenic therapies have become a notable advance in targeted therapeutic options for patients with several different cancer types, in order to further improve clinical efficacy of these agents in the future, there is an urgent need to:

1. Discover, validate, and qualify clinically relevant predictive biomarkers for various anti-angiogenic therapies.
2. Develop and standardize technically robust tissue-based molecular methodologies that can quantify relevant molecular targets at protein, RNA, or DNA level, and provide an objective measure of the tumor biology to allow correlations with circulating angiogenic biomarker levels, which in isolation may not be uniformly representative of the biology of primary tumor or various metastases in an individual patient at a given time.
3. Further develop and advance technically robust and cost-effective tissue-based methodologies for broader evaluation of the elegant concept of vascular heterogeneity put forward by Harold Dvorak from Harvard University ("early" and "late" tumor vessel phenotypes) and to systematically evaluate the clinical relevance of tumor vessel phenotypes as potential predictors of response or resistance to anti-VEGF/VEGFR therapies in various cancer indications.
4. Carry out further clinical evaluation of tissue-based VEGF receptor profiling to determine its clinical significance and utility in the context of various anti-angiogenic therapies targeting one or more VEGF receptors.
5. Consider implementing central sub-specialty level human tumor pathology reporting in clinical trials of anti-angiogenic therapies, so as to minimize (ideally eliminate) histologically misclassified tumor data submitted by global clinical trial sites and to improve reliability of tissue biomarker data analyses from clinical trials.
6. Develop rational patient-tailoring hypotheses, based on patho-biologically relevant predictive biomarker/drug target profiles of tumor subsets, to be tested and refined in future clinical trials of anti-angiogenic therapies.
7. Build effective collaborations among basic, translational, and clinical research teams in the industry led by experienced oncologists and pathologists, so that tailoring biomarker research can be implemented early along the drug development process.
8. Generate biologically relevant retrospective biomarker data sets using optimal technologies to interrogate and advance promising single markers and carefully designed targeted marker panels representing broader biologic profile of a given human cancer type.

Using high-quality tumor tissues as full sections or leveraging tissue microarray technology can help generate data-driven hypotheses that can provide clinical teams with the rationale to design monotherapy or combinatorial trials of anti-angiogenic therapies with other promising oncology drugs like anti-HER2 or immunotherapy agents.

Acknowledgments The author would like to acknowledge the dedication and hard work of Timothy Holzer, Angie Fulford, Beverly Falcon, Drew Nedderman, Leslie O'Neal Reising, James Alston, and Mia Chen in the lab in developing and optimizing technically robust immunohistochemical and bright-field in situ hybridization technologies for VEGFR2, VEGFR3, and VEGFR1 for archival human tissues. Thanks to Jeff Hanson for supporting pathologic angiogenesis and oncologic disease state characterization analyses. Thanks also to Andrew Schade, Kelly Credille, Aafia Chaudhry, Katherine Marie Bell-McGuinn, Mayukh Das, Richard Walgren, Laura Benjamin, Bronislaw Pytowsky, Mark Uhlik, and Jeremy Graff for great scientific and clinical collaboration on tumor angiogenesis and anti-angiogenesis projects over the years.

Disclaimer *Expert scientific and clinical views expressed in this chapter are those of the author.*

References

1. Jain RK. Antiangiogenesis strategies revisited: from starving tumors to alleviating hypoxia. Cancer Cell. 2014;26(5):605–22.
2. Hurwitz H, Fehrenbacher L, Novotny W, Cartwright T, Hainsworth J, Heim W, et al. Bevacizumab plus irinotecan, fluorouracil, and leucovorin for metastatic colorectal cancer. N Engl J Med. 2004;350(23):2335–42.
3. Jang HJ, Kim BJ, Kim JH, Kim HS. The addition of bevacizumab in the first-line treatment for metastatic colorectal cancer: an updated meta-analysis of randomized trials. Oncotarget. 2017;8(42):73009–16.
4. Hayes DF. Bevacizumab treatment for solid tumors: boon or bust? JAMA. 2011;305(5):506–8.
5. Jain RK. Lessons from multidisciplinary translational trials on anti-angiogenic therapy of cancer. Nat Rev Cancer. 2008;8(4):309–16.
6. Carmeliet P, Jain RK. Molecular mechanisms and clinical applications of angiogenesis. Nature. 2011;473(7347):298–307. https://doi.org/10.1038/nature10144.
7. Folkman J. Tumor angiogenesis: therapeutic implications. N Engl J Med. 1971;285(21):1182–6.
8. Jain RK. Normalizing tumor microenvironment to treat cancer: bench to bedside to biomarkers. J Clin Oncol. 2013;31(17):2205–18.
9. Ferrara N. Vascular endothelial growth factor. Arterioscler Thromb Vasc Biol. 2009;29(6):789–91.
10. Sitohy B, Nagy JA, Dvorak HF. Anti-VEGF/VEGFR therapy for cancer: reassessing the target. Cancer Res. 2012;72(8):1909–14.
11. Sitohy B, Chang S, Sciuto TE, Masse E, Shen M, Kang PM, et al. Early actions of anti-vascular endothelial growth factor/vascular endothelial growth factor receptor drugs on angiogenic blood vessels. Am J Pathol. 2017;187(10):2337–47.
12. Fischer C, Mazzone M, Jonckx B, Carmeliet P. FLT1 and its ligands VEGFB and PlGF: drug targets for anti-angiogenic therapy? Nat Rev Cancer. 2008;8(12):942–56.
13. Rolny C, Mazzone M, Tugues S, Laoui D, Johansson I, Coulon C, et al. HRG inhibits tumor growth and metastasis by inducing macrophage polarization and vessel normalization through downregulation of PlGF. Cancer Cell. 2011;19(1):31–44.
14. Gacche RN, Assaraf YG. Redundant angiogenic signaling and tumor drug resistance. Drug Resist Updat. 2018;36:47–76.
15. Neufeld G, Kessler O. The semaphorins: versatile regulators of tumour progression and tumour angiogenesis. Nat Rev Cancer. 2008;8(8):632–45.
16. Stockmann C, Doedens A, Weidemann A, Zhang N, Takeda N, Greenberg JI, et al. Deletion of vascular endothelial growth factor in myeloid cells accelerates tumorigenesis. Nature. 2008;456(7223):814–8.
17. Lee S, Chen TT, Barber CL, Jordan MC, Murdock J, Desai S, et al. Autocrine VEGF signaling is required for vascular homeostasis. Cell. 2007;130(4):691–703.
18. Lanahan AA, Hermans K, Claes F, Kerley-Hamilton JS, Zhuang ZW, Giordano FJ, et al. VEGF receptor 2 endocytic trafficking regulates arterial morphogenesis. Dev Cell. 2010;18(5):713–24.
19. Vasudev NS, Reynolds AR. Anti-angiogenic therapy for cancer: current progress, unresolved questions and future directions. Angiogenesis. 2014;17(3):471–94.
20. Fan F, Wey JS, McCarty MF, Belcheva A, Liu W, Bauer TW, et al. Expression and function of vascular endothelial growth factor receptor-1 on human colorectal cancer cells. Oncogene. 2005;24(16):2647–53.
21. Dales JP, Garcia S, Bonnier P, Duffaud F, Carpentier S, Djemli A, et al. Prognostic significance of VEGF receptors, VEGFR-1 (Flt-1) and VEGFR-2 (KDR/Flk-1) in breast carcinoma. Ann Pathol. 2003;23(4):297–305.
22. Guo P, Fang Q, Tao HQ, Schafer CA, Fenton BM, Ding I, et al. Overexpression of vascular endothelial growth factor by MCF-7 breast cancer cells promotes estrogen-independent tumor growth in vivo. Cancer Res. 2003;63(15):4684–91.
23. Spannuth WA, Nick AM, Jennings NB, Armaiz-Pena GN, Mangala LS, Danes CG, et al. Functional significance of VEGFR-2 on ovarian cancer cells. Int J Cancer. 2009;124(5):1045–53.
24. Ellis LM, Hicklin DJ. VEGF-targeted therapy: mechanisms of anti-tumour activity. Nat Rev Cancer. 2008;8(8):579–91.
25. Lu KV, Chang JP, Parachoniak CA, Pandika MM, Aghi MK, Meyronet D, et al. VEGF inhibits tumor cell invasion and mesenchymal transition through a MET/VEGFR2 complex. Cancer Cell. 2012;22(1):21–35.
26. Fan F, Samuel S, Gaur P, Lu J, Dallas NA, Xia L, et al. Chronic exposure of colorectal cancer cells to bevacizumab promotes compensatory pathways that mediate tumour cell migration. Br J Cancer. 2011;104(8):1270–7.
27. Ronca R, Benkheil M, Mitola S, Struyf S, Liekens S. Tumor angiogenesis revisited: regulators and clinical implications. Med Res Rev. 2017;37(6):1231–74.
28. Welti JC, Gourlaouen M, Powles T, Kudahetti SC, Wilson P, Berney DM, et al. Fibroblast growth factor 2 regulates endothelial cell sensitivity to sunitinib. Oncogene. 2011;30(10):1183–93.
29. Shojaei F, Lee JH, Simmons BH, Wong A, Esparza CO, Plumlee PA, et al. HGF/c-Met acts as an alternative angiogenic pathway in sunitinib-resistant tumors. Cancer Res. 2010;70(24):10090–100.
30. Eklund L, Saharinen P. Angiopoietin signaling in the vasculature. Exp Cell Res. 2013;319(9):1271–80.
31. Li JL, Sainson RC, Oon CE, Turley H, Leek R, Sheldon H, et al. DLL4-Notch signaling mediates tumor resistance to anti-VEGF therapy in vivo. Cancer Res. 2011;71(18):6073–83.
32. Crawford Y, Kasman I, Yu L, Zhong C, Wu X, Modrusan Z, et al. PDGF-C mediates the angiogenic and tumorigenic properties of fibroblasts associated with tumors refractory to anti-VEGF treatment. Cancer Cell. 2009;15(1):21–34.

33. di Tomaso E, London N, Fuja D, Logie J, Tyrrell JA, Kamoun W, et al. PDGF-C induces maturation of blood vessels in a model of glioblastoma and attenuates the response to anti-VEGF treatment. PLoS One. 2009;4(4):e5123.
34. Huang D, Ding Y, Zhou M, Rini BI, Petillo D, Qian CN, et al. Interleukin-8 mediates resistance to anti-angiogenic agent sunitinib in renal cell carcinoma. Cancer Res. 2010;70(3):1063–71.
35. Salvucci O, Tosato G. Essential roles of EphB receptors and EphrinB ligands in endothelial cell function and angiogenesis. Adv Cancer Res. 2012;114:21–57.
36. Cascone T, Herynk MH, Xu L, Du Z, Kadara H, Nilsson MB, et al. Upregulated stromal EGFR and vascular remodeling in mouse xenograft models of angiogenesis inhibitor-resistant human lung adenocarcinoma. J Clin Invest. 2011;121(4):1313–28.
37. Greenberg JI, Shields DJ, Barillas SG, Acevedo LM, Murphy E, Huang J, et al. A role for VEGF as a negative regulator of pericyte function and vessel maturation. Nature. 2008;456(7223):809–13.
38. Carmeliet P, Jain RK. Principles and mechanisms of vessel normalization for cancer and other angiogenic diseases. Nat Rev Drug Discov. 2011;10(6):417–27.
39. Nasir A, Holzer TR, Chen M, Man MZ, Schade AE. Differential expression of VEGFR2 protein in HER2 positive primary human breast cancer: potential relevance to anti-angiogenic therapies. Cancer Cell Int. 2017;17:56.
40. Nagy JA, Dvorak HF. Heterogeneity of the tumor vasculature: the need for new tumor blood vessel type-specific targets. Clin Exp Metastasis. 2012;29(7):657–62.
41. Goel S, Duda DG, Xu L, Munn LL, Boucher Y, Fukumura D, et al. Normalization of the vasculature for treatment of cancer and other diseases. Physiol Rev. 2011;91(3):1071–121.
42. Smith NR, Baker D, James NH, Ratcliffe K, Jenkins M, Ashton SE, et al. Vascular endothelial growth factor receptors VEGFR-2 and VEGFR-3 are localized primarily to the vasculature in human primary solid cancers. Clin Cancer Res. 2010;16(14):3548–61.
43. Holzer TR, Fulford AD, Nedderman DM, Umberger TS, Hozak RR, Joshi A, et al. Tumor cell expression of vascular endothelial growth factor receptor 2 is an adverse prognostic factor in patients with squamous cell carcinoma of the lung. PLoS One. 2013;8(11):e80292.
44. Sitohy B, Nagy JA, Jaminet SC, Dvorak HF. Tumor-surrogate blood vessel subtypes exhibit differential susceptibility to anti-VEGF therapy. Cancer Res. 2011;71(22):7021–8.
45. Nasir A, Reising LO, Nedderman DM, Fulford AD, Uhlik MT, Benjamin LE, et al. Heterogeneity of vascular endothelial growth factor receptors 1, 2, 3 in primary human colorectal carcinoma. Anticancer Res. 2016;36(6):2683–96.
46. Nasir A, Falcon B, Wang D, et al. Vascular and tumor cell expression of VEGFR2 and molecular subtyping: an innovative biomarker approach in bladder cancer. ASCO-GU Cancers Symposium. San Francisco; 2018.
47. Tolaney SM, Boucher Y, Duda DG, Martin JD, Seano G, Ancukiewicz M, et al. Role of vascular density and normalization in response to neoadjuvant bevacizumab and chemotherapy in breast cancer patients. Proc Natl Acad Sci U S A. 2015;112(46):14325–30.
48. Jeong HS, Jones D, Liao S, Wattson DA, Cui CH, Duda DG, et al. Investigation of the lack of angiogenesis in the formation of lymph node metastases. J Natl Cancer Inst. 2015;107(9):699.
49. Wehland M, Bauer J, Magnusson NE, Infanger M, Grimm D. Biomarkers for anti-angiogenic therapy in cancer. Int J Mol Sci. 2013;14(5):9338–64.
50. Pilotto S, Bonomi M, Massari F, Milella M, Ciuffreda L, Brunelli M, et al. Anti-angiogenic drugs and biomarkers in non-small-cell lung cancer: a 'hard days night'. Curr Pharm Des. 2014;20(24):3958–72.
51. Duda DG, Willett CG, Ancukiewicz M, di Tomaso E, Shah M, Czito BG, et al. Plasma soluble VEGFR-1 is a potential dual biomarker of response and toxicity for bevacizumab with chemoradiation in locally advanced rectal cancer. Oncologist. 2010;15(6):577–83.
52. Meyerhardt JA, Ancukiewicz M, Abrams TA, Schrag D, Enzinger PC, Chan JA, et al. Phase I study of cetuximab, irinotecan, and vandetanib (ZD6474) as therapy for patients with previously treated metastastic colorectal cancer. PLoS One. 2012;7(6):e38231.
53. Willett CG, Duda DG, di Tomaso E, Boucher Y, Ancukiewicz M, Sahani DV, et al. Efficacy, safety, and biomarkers of neoadjuvant bevacizumab, radiation therapy, and fluorouracil in rectal cancer: a multidisciplinary phase II study. J Clin Oncol. 2009;27(18):3020–6.
54. Zhu AX, Ancukiewicz M, Supko JG, Sahani DV, Blaszkowsky LS, Meyerhardt JA, et al. Efficacy, safety, pharmacokinetics, and biomarkers of cediranib monotherapy in advanced hepatocellular carcinoma: a phase II study. Clin Cancer Res. 2013;19(6):1557–66.
55. Lambrechts D, Claes B, Delmar P, Reumers J, Mazzone M, Yesilyurt BT, et al. VEGF pathway genetic variants as biomarkers of treatment outcome with bevacizumab: an analysis of data from the AViTA and AVOREN randomised trials. Lancet Oncol. 2012;13(7):724–33.
56. Lambrechts D, Lenz HJ, de Haas S, Carmeliet P, Scherer SJ. Markers of response for the anti-angiogenic agent bevacizumab. J Clin Oncol. 2013;31(9):1219–30.
57. Holzer TR, Fulford AD, Reising LO, Nedderman DM, Zhang X, Benjamin LE, et al. Profiling of vascular endothelial growth factor receptor heterogeneity identifies protein expression-defined subclasses of human non-small cell lung carcinoma. Anticancer Res. 2016;36(7):3277–88.
58. Kankeu Fonkoua L, Yee NS. Molecular characterization of gastric carcinoma: therapeutic implications for biomarkers and targets. Biomedicines. 2018;6(1).
59. Cancer Genome Atlas Research Network. Comprehensive molecular characterization of gastric adenocarcinoma. Nature. 2014;513(7517):202–9.
60. Uhlik MT, Liu J, Falcon BL, Iyer S, Stewart J, Celikkaya H, et al. Stromal-based signatures for

the classification of gastric cancer. Cancer Res. 2016;76(9):2573–86.
61. Sandler A, Gray R, Perry MC, Brahmer J, Schiller JH, Dowlati A, et al. Paclitaxel-carboplatin alone or with bevacizumab for non-small-cell lung cancer. N Engl J Med. 2006;355(24):2542–50.
62. Reck M, von Pawel J, Zatloukal P, Ramlau R, Gorbounova V, Hirsh V, et al. Phase III trial of cisplatin plus gemcitabine with either placebo or bevacizumab as first-line therapy for nonsquamous non-small-cell lung cancer: AVAil. J Clin Oncol. 2009;27(8):1227–34.
63. Reck M, von Pawel J, Zatloukal P, Ramlau R, Gorbounova V, Hirsh V, et al. Overall survival with cisplatin-gemcitabine and bevacizumab or placebo as first-line therapy for nonsquamous non-small-cell lung cancer: results from a randomised phase III trial (AVAiL). Ann Oncol. 2010;21(9):1804–9.
64. Schneider BP, Li L, Shen F, Miller KD, Radovich M, O'Neill A, et al. Genetic variant predicts bevacizumab-induced hypertension in ECOG-5103 and ECOG-2100. Br J Cancer. 2014;111(6):1241–8.
65. Poveda AM, Selle F, Hilpert F, Reuss A, Savarese A, Vergote I, et al. Bevacizumab combined with weekly paclitaxel, pegylated liposomal doxorubicin, or topotecan in platinum-resistant recurrent ovarian cancer: analysis by chemotherapy cohort of the randomized phase III AURELIA trial. J Clin Oncol. 2015;33(32):3836–8.
66. Liu JF, Matulonis UA. Bevacizumab in newly diagnosed ovarian cancer. Lancet Oncol. 2015;16(8):876–8.
67. Krill LS, Tewari KS. Integration of bevacizumab with chemotherapy doublets for advanced cervical cancer. Expert Opin Pharmacother. 2015;16(5):675–83.
68. Crafton SM, Salani R. Beyond chemotherapy: an overview and review of targeted therapy in cervical cancer. Clin Ther. 2016;38(3):449–58.
69. Ciombor KK, Berlin J. Aflibercept – a decoy VEGF receptor. Curr Oncol Rep. 2014;16(2):368.
70. Aprile G, Rijavec E, Fontanella C, Rihawi K, Grossi F. Ramucirumab: preclinical research and clinical development. Oncol Targets Ther. 2014;7:1997–2006.
71. Ramucirumab TP. Boon or bane. J Egypt Natl Canc Inst. 2016;28(3):133–40.
72. Wu P, Nielsen TE, Clausen MH. FDA-approved small-molecule kinase inhibitors. Trends Pharmacol Sci. 2015;36(7):422–39.
73. Wu P, Nielsen TE, Clausen MH. Small-molecule kinase inhibitors: an analysis of FDA-approved drugs. Drug Discov Today. 2016;21(1):5–10.
74. Zhu Y, Choi SH, Shah K. Multifunctional receptor-targeting antibodies for cancer therapy. Lancet Oncol. 2015;16(15):e543–e54.
75. Yokoi K, Thaker PH, Yazici S, Rebhun RR, Nam DH, He J, et al. Dual inhibition of epidermal growth factor receptor and vascular endothelial growth factor receptor phosphorylation by AEE788 reduces growth and metastasis of human colon carcinoma in an orthotopic nude mouse model. Cancer Res. 2005;65(9):3716–25.
76. Amin DN, Hida K, Bielenberg DR, Klagsbrun M. Tumor endothelial cells express epidermal growth factor receptor (EGFR) but not ErbB3 and are responsive to EGF and to EGFR kinase inhibitors. Cancer Res. 2006;66(4):2173–80.
77. Erber R, Thurnher A, Katsen AD, Groth G, Kerger H, Hammes HP, et al. Combined inhibition of VEGF and PDGF signaling enforces tumor vessel regression by interfering with pericyte-mediated endothelial cell survival mechanisms. FASEB J. 2004;18(2):338–40.
78. Hojjat-Farsangi M. Small-molecule inhibitors of the receptor tyrosine kinases: promising tools for targeted cancer therapies. Int J Mol Sci. 2014;15(8):13768–801.
79. Mabry R, Gilbertson DG, Frank A, Vu T, Ardourel D, Ostrander C, et al. A dual-targeting PDGFRbeta/VEGF-A molecule assembled from stable antibody fragments demonstrates anti-angiogenic activity in vitro and in vivo. MAbs. 2010;2(1):20–34.
80. Kienast Y, Klein C, Scheuer W, Raemsch R, Lorenzon E, Bernicke D, et al. Ang-2-VEGF-A CrossMab, a novel bispecific human IgG1 antibody blocking VEGF-A and Ang-2 functions simultaneously, mediates potent antitumor, antiangiogenic, and antimetastatic efficacy. Clin Cancer Res. 2013;19(24):6730–40.
81. Tol J, Koopman M, Cats A, Rodenburg CJ, Creemers GJ, Schrama JG, et al. Chemotherapy, bevacizumab, and cetuximab in metastatic colorectal cancer. N Engl J Med. 2009;360(6):563–72.
82. Gianni L, Romieu GH, Lichinitser M, Serrano SV, Mansutti M, Pivot X, et al. AVEREL: a randomized phase III trial evaluating bevacizumab in combination with docetaxel and trastuzumab as first-line therapy for HER2-positive locally recurrent/metastatic breast cancer. J Clin Oncol. 2013;31(14):1719–25.
83. Chauhan VP, Stylianopoulos T, Martin JD, Popovic Z, Chen O, Kamoun WS, et al. Normalization of tumour blood vessels improves the delivery of nanomedicines in a size-dependent manner. Nat Nanotechnol. 2012;7(6):383–8.
84. Seto T, Kato T, Nishio M, Goto K, Atagi S, Hosomi Y, et al. Erlotinib alone or with bevacizumab as first-line therapy in patients with advanced non-squamous non-small-cell lung cancer harbouring EGFR mutations (JO25567): an open-label, randomised, multicentre, phase 2 study. Lancet Oncol. 2014;15(11):1236–44.
85. Izumi Y, Xu L, di Tomaso E, Fukumura D, Jain RK. Tumour biology: herceptin acts as an anti-angiogenic cocktail. Nature. 2002;416(6878):279–80.
86. Kodack DP, Chung E, Yamashita H, Incio J, Duyverman AM, Song Y, et al. Combined targeting of HER2 and VEGFR2 for effective treatment of HER2-amplified breast cancer brain metastases. Proc Natl Acad Sci U S A. 2012;109(45):E3119–27.
87. Falchook GS, Moulder SL, Wheler JJ, Jiang Y, Bastida CC, Kurzrock R. Dual HER2 inhibition in combination with anti-VEGF treatment is active in heavily pretreated HER2-positive breast cancer. Ann Oncol. 2013;24(12):3004–11.
88. Huang Y, Goel S, Duda DG, Fukumura D, Jain RK. Vascular normalization as an emerging strategy to enhance cancer immunotherapy. Cancer Res. 2013;73(10):2943–8.

Role of PI3K/AKT/mTOR in Cancer Signaling

20

Nicci Owusu-Brackett, Maryam Shariati, and Funda Meric-Bernstam

PI3K/AKT/mTOR Signaling Pathway

Phosphatidylinositol 3-kinases (PI3Ks) are a conserved family of lipid kinases, divided into three classes based on their structures and specific lipid substrates. Class I PI3Ks are heterodimeric molecules which are further divided into class IA and class IB. Class IA is comprised of p110 catalytic subunit containing three isoforms of p110α, p110β, and p110δ and p85 regulatory subunit with five variants of p85α, p55α, p50α, p85 β, and p55γ. Class IB PI3Ks are heterodimers containing p110γ catalytic subunit and p101 regulatory subunit [1]. The physiological role of class II and III in signal transduction remains enigmatic.

In response to extracellular growth signals, receptor tyrosine kinase (RTK) binds to PI3K directly or indirectly through insulin receptor substrate (IRS). PI3K activation, which can also be induced by G protein-coupled receptor (GPCR), recruits class I PI3K to plasma membrane relieving p85 inhibition of p110 and resulting in phosphorylation of phosphatidylinositol (PtdIns) 4,5-bisphosphate (PIP2) to generate PtdIns(3,4,5)-triphosphate (PIP3) [1]. This triggers translocation of AKT (serine/threonine protein kinase B) to the membrane where it binds to PIP3 through its pleckstrin homology (PH) domain allowing phosphoinositide-dependent kinase 1 (PDK1) to access and phosphorylate AKT at Thr308. Full activation of AKT requires phosphorylation of Ser473 which can be catalyzed by PDK2, mechanistic target of rapamycin complex2 (mTORC2), and DNA-dependent protein kinase (DNA-PK).

Once active, AKT leads to substrate-specific phosphorylation in both cytosol and nucleus including inhibitory phosphorylation of proline-rich AKT substrate 40kD (PRAS40), glycogen synthase kinase 3 β (GSK-3β), and pro-apoptotic proteins of BAD, a BCL-2 family member, and FOXO [2]. Additionally AKT phosphorylates and inhibits tuberous sclerosis protein 2 (TSC2) leading to activation of mammalian target of rapamycin (mTOR), a serine/threonine

N. Owusu-Brackett
Department of Surgical Oncology, The University of Texas MD Anderson Cancer Center, Houston, TX, USA

M. Shariati
Department of Investigational Cancer Therapeutics, Graduate School of Biomedical Sciences, The University of Texas MD Anderson Cancer Center, Houston, TX, USA

F. Meric-Bernstam (✉)
Investigational Cancer Therapeutics, The University of Texas MD Anderson Cancer Center, Houston, TX, USA

Sheikh Khalifa Bin Zayed Al Nahyan Institute for Personalized Cancer Therapy, The University of Texas MD Anderson Cancer Center, Houston, TX, USA

Department of Breast Surgical Oncology, The University of Texas MD Anderson Cancer Center, Houston, TX, USA
e-mail: fmeric@mdanderson.org

S. Badve, G. L. Kumar (eds.), *Predictive Biomarkers in Oncology*,
https://doi.org/10.1007/978-3-319-95228-4_20

kinase. mTOR is present in two structurally distinct complexes: mTOR complex 1 (mTORC1) and mTOR complex 2 (mTORC2). Activated mTORC1 phosphorylates eukaryotic translation initiation factor 4E-binding protein 1 (4EBP1), and ribosomal protein S6 kinase beta-1 70kD (S6K1) which, in turn, phosphorylates ribosomal protein S6 (S6), leading to protein synthesis.

Mitogen signals such as insulin-like growth factor 1 (IGF1) activate MAPK/ERK pathway resulting in TSC1/TSC2 complex inhibition, thus stimulating mTORC1 [3]. In addition, Wnt pathway, a major regulator of cell growth, is involved in mTORC1 activation through GSK-3β phosphorylation and subsequent TSC2 inhibition. The main negative regulator of the pathway is PTEN lipid phosphatase which converts PIP3 to PIP2 by removing the 3′ phosphate from PIP3. PI3K/AKT pathway negatively controls PTEN at the transcriptional level through regulation of transcription factor NF-kB and tumor necrosis factor α (TNFα), which in turn repress PTEN expression. The pathway is also controlled by protein phosphatase 2A (PP2A) and phosphatase PHLPP which dephosphorylate AKT at Thr308 and Ser473, respectively. mTORC1 triggers the most prominent negative feedback regulation of PI3K signaling. During chronic insulin-mediated PI3K stimulation, mTORC1 and S6K activation can lead to downregulation of IRS-1 adapter protein through transcriptional regulation and proteasomal degradation [3]. In another negative feedback control mechanism, S6K1 is able to phosphorylate IRS-1 at multiple Ser residues preventing its binding to RTKs and initiation of PI3K signaling (Fig. 20.1).

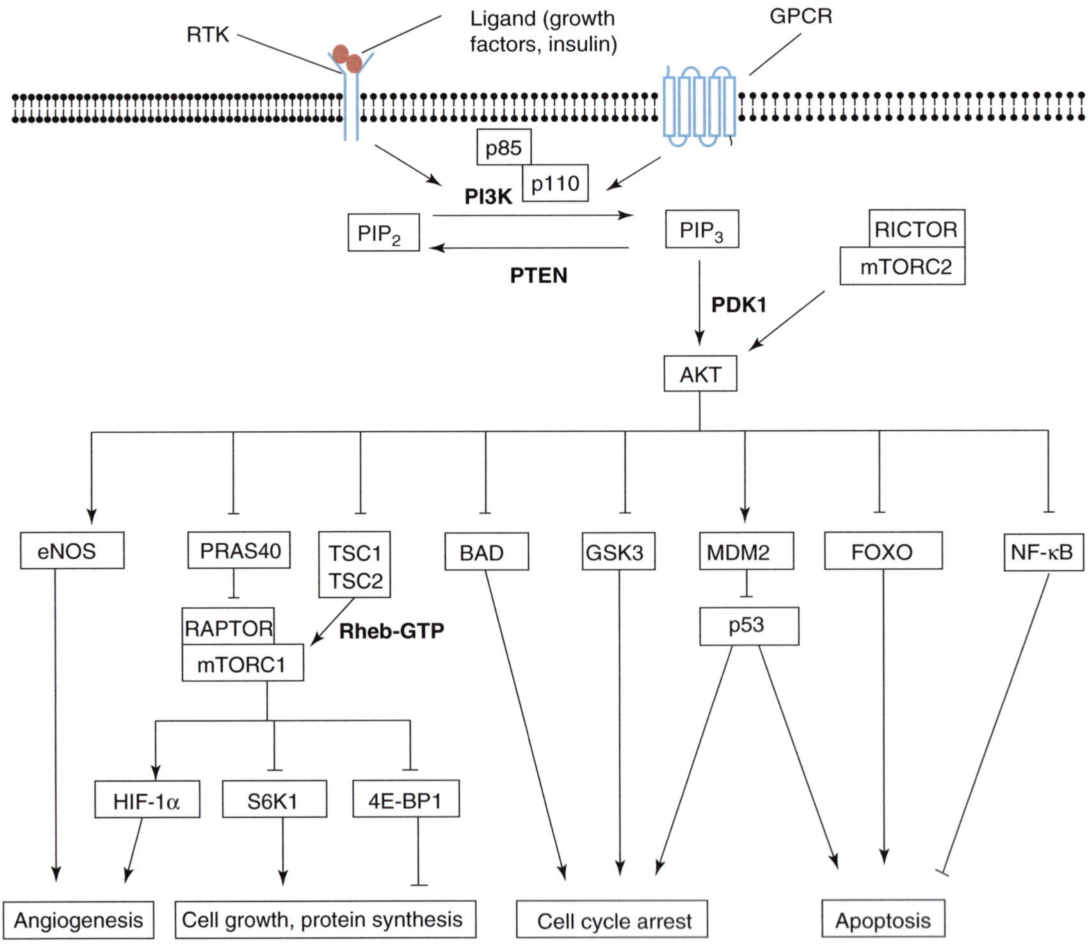

Fig. 20.1 Key molecules involved in PI3K/AKT/mTOR signaling

PI3K/AKT/mTOR Signaling in Carcinogenesis

The PI3K/AKT/mTOR signaling network is an important pathway that integrates extracellular and environmental stimuli translating them into intracellular signals which drive cellular functions such as cell growth, proliferation, motility, survival, and vesicular trafficking.

Aberrant activation of PI3K signaling contributes to most hallmarks of cancer including increased cell cycle, metabolism, survival, and motility. Molecular alterations of the major nodes of this signaling axis have been found in numerous tumor types (Table 20.1). PI3K activity is regulated by growth factor receptors, and therefore any modulation in RTKs upstream of PI3K can lead to increases in the pathway activity. Mutation activation of epidermal growth factor receptor (EGFR) as an activator of PI3K is one of the major alterations in the carcinogenesis of non-small cell lung cancer (NSCLC). *EGFR* gene amplification and overexpression are observed frequently in glioblastoma, while not very common in other cancer types. HER2 (human epidermal growth factor receptor 2) is another member of the EGFR family which has been overexpressed and amplified in invasive breast and gastric cancer and less frequently in other tumor types such as lung, colon, biliary, ovarian, and salivary cancer. *PIK3CA* gene encoding p110α catalytic subunit of PI3K is often mutated in many human cancer types including breast, endometrial, colorectal, and ovarian tumors. These mutations cluster *predominantly* in three regions of helical (E542K and E545K) and catalytic (H1047R) domains known as "hot spot." *PTEN* tumor suppressor gene that antagonizes PI3K signaling is the second most frequently mutated gene in human cancer. PTEN loss or mutation has been identified in both spontaneous and heritable cancers including glioblastoma and endometrial sarcoma. Amplification and activating somatic mutation in the PH domain (E17K) of AKT1 have been reported in breast, colorectal, pancreatic and ovarian cancers [4].

Table 20.1 Common PI3K pathway genetic alterations in cancer

Pathway elements	Genetic variants	Common tumor type
RTK class I (EGFR, HER2, etc.)	Amplification, mutation	NSCLC, glioblastoma, breast, gastric
PIK3CA	Amplification, mutation	Breast, endometrial, colorectal, ovarian
AKT	Amplification, mutation	Breast, colorectal, pancreatic, ovarian
PTEN	Mutation, loss of heterozygosity	Glioblastoma, endometrial

The pathway affects different aspects of tumorigenicity including angiogenesis, cell cycle progression, and metastasis. PI3K/AKT/mTOR signaling promotes tumor angiogenesis by shifting the homeostatic balance toward pro-angiogenic factors. Through induction of NOS and inhibition of GSK-3β and FOXO, this signaling pathway increases HIF-1α expression, which induces vascular endothelial growth factor (VEGF) transcriptional activation [5]. In addition, AKT suppresses endogenous angiogenic inhibitor TSP-1 [5]. Thus, the PI3K pathway promotes angiogenesis via overexpression of pro-angiogenic factors, inhibition of antiangiogenic factors, and induction of factors that promote the stability of developing vasculature.

PI3K pathway promotes tumor metastasis through several mechanisms. PI3K pathway promotes tumor metastasis via activation of matrix metalloproteinases and urokinase-type plasminogen activator, which degrade the extracellular matrix (ECM) [5]. In addition, the production of chemokines such as C-X-C motif ligand 1 (CXCL-1), cyclooxygenase-2 (COX-2), and interleukin-8 (CXCL-8) in conjunction with promotion of epithelial-mesenchymal transition (EMT) by NF-kB and repression of E-cadherin allows for the cell motility integral to tumor metastasis [5].

In addition to regulating metastasis, PI3K/AKT/mTOR signaling promotes cell cycle progression by increasing transcription of S-phase entry genes and initiation of translation of messenger RNAs while inhibiting protein regulators of cell cycle progression and pro-apoptotic factors. AKT inhibition of GSK3β and retinoblastoma protein through cyclin-dependent kinase 4/6 (CDK4/6) results in E2F-mediated transcription of S-phase entry genes [6]. TORC1-mediated activation of S6K1 and inhibition of 4E-BP1, which allows translation of messenger RNAs, is required for progression to S phase [6]. In addition, transition to the S phase occurs by inhibiting negative regulators of cell cycle progression such p27 and p21 [6]. Finally, inhibition of pro-apoptotic factors such as forkhead box O3 (FOXO3) proteins, mouse double minute 2 homolog (MDM2), and BAD promotes cell cycle progression [6].

Adenosine Monophosphate-Activated Protein Kinase (AMPK) Signaling in Cancer

AMPK is a major regulator of cellular metabolism and energy homeostasis and has been studied extensively in diabetes. AMPK is activated by decreased cellular energy levels especially during stresses such as hypoxia, ischemia, and glucose deprivation. As a result, AMPK signaling inhibits pathways that consume energy favoring pathways that generate ATP. The accelerated growth rate of cancer cells demands an increased energy requirement and creates an environment of metabolic stress that is dependent on deregulating cellular energetics. AMPK acts as a tumor suppressor by opposing these metabolic changes that occur during tumorigenesis. AMPK signaling plays a significant role in a variety of cancers such as lung, colorectal, liver, prostate, breast, and melanoma. Mutations in the upstream protein kinases responsible for activation of the AMPK such as liver kinase B1 (LKB1), calcium/calmodulin-dependent protein kinase (CaMKK), and transforming growth factor β (TGF-β)-activated kinase (TAK1) allow unsuppressed cell proliferation due to an inability to activate AMPK. In addition, AMPK interacts with many molecules and tumor suppressors implicated in tumorigenesis and cancer cell proliferation such as AKT, mTOR, and p53. Given the importance of AMPK in tumorigenesis and cancer cell proliferation, activation of AMPK is an intriguing potential therapeutic target. Potential treatments being investigated include metformin, flavones, nonsteroidal anti-inflammatory drugs, and AICAR.

Current Therapies and Future Strategies for Treatment

Given the critical role that PI3K pathway plays in carcinogenesis, this signaling cascade is an important therapeutic target. Of the drugs targeting the PI3K/AKT/mTOR pathway, everolimus and temsirolimus are currently FDA approved. Everolimus is approved for five indications, which include adults with advanced renal cell carcinoma after failure of treatment with sunitinib or sorafenib; adults with renal angiomyolipoma and tuberous sclerosis complex not requiring immediate surgery; and postmenopausal women with advanced hormone receptor-positive, HER2-negative breast cancer in combination with exemestane after failure of treatment with letrozole or anastrozole. The final two indications are adults with progressive neuroendocrine tumors of pancreatic origin and adults with progressive, well-differentiated, nonfunctional neuroendocrine tumors of gastrointestinal or lung origin that are unresectable, locally advanced, or metastatic [7]. Temsirolimus is approved for treatment of advanced renal cell carcinoma [7]. Currently several PI3K/AKT/mTOR pathway inhibitors are in development with a focus on development in tumors with known PI3K pathway activation (e.g., breast cancer) as well as tumors with genomic alteration in the pathway such as *PIK3CA* mutations for PI3Kα inhibitors (Table 20.2) [7].

Table 20.2 Current therapies targeting the PI3K pathway

Target	Name	Company
PI3K isoforms		
PIK3CA	ACP-319	Acerta Pharma
	BYL719	Novartis
	Serabelisib	Takeda
PIK3CB	GSK2636711	GlaxoSmithKline
	SAR260301	Sanofi
PIK3CA/PIK3CB	BAY1082439	Bayer
PIK3CB/PIK3CD	AZD8186	AstraZeneca
	KA2237	Karus
PI3KD/PI3KG	Duvelisib	Verastem
PIK3CA/PIK3CD/PIK3CG	Taselisib (GDC0032)	Genentech
PIK3CA/PIK3CB/PIK3CD/PIK3CG (pan-PI3K)	AZD8835	AstraZeneca
	Buparlisib	Novartis
	CLR457	Novartis
	Copanlisib	Bayer
	CUDC-907	Curis
	GDC0077	Genentech
	GDC0941	Piramed
	PA799	Chugai
	Pilaralisib	Exelixis and Sanofi
	WX-037	UCB Pharma
	ZSTK474	Zenyaku Kogyo
AKT inhibitors	ARQ092	Arqule
	AZD5363	Astex
	Afuresertib	Novartis
	Archexin	Rexahn Pharmaceutical Inc
	BAY1125976	Bayer
	GSK690693	Novartis
	Ipatasertib	Genentech
	LY2780301	Lilly
	MK2206	Merck
	M2698	EMD Serono
	Perifosine	Aeterna Zentaris
	SR13668	SRI International
	TAS-117	Taiho
	Triciribine (PTX-200)	Prescient Therapeutics
	Uprosertib	Novartis
mTOR inhibitors	BI860585	Boehringer
	CC-223	Celgene
	DS-3078a	Daiichi
	GDC-0349	Genentech
	ME-344	Novogen
	OSI-027	OSI
	P529	Diffusion Pharmaceuticals Takeda
	Sapanisertib	AstraZeneca
	Vistusertib	

(continued)

Table 20.2 (continued)

Target	Name	Company
Rapamycin and its analogues rapalogues	Nanoparticle albumin-bound rapamycin	Wyeth
	Ridaforolimus	Ariad
	Temsirolimus	Wyeth
	Everolimus	Novartis
	Sirolimus	Pfizer
PI3K/mTOR inhibitors	Apitolisib	Genentech
	DCBCI0901	Development Center for Biotechnology
	DS7423	Daiichi
	Dactolisib	Novartis
	GDC0084	Genentech
	Gedatolisib	Pfizer
	LY294002	Lilly
	LY3023414	Lilly
	PF04691502	Pfizer
	PQR309	PIQUR Therapeutics AG
	PWT33597	Pathway Therapeutics, Inc.
	SF1126	Semafore Pharmaceuticals
	VS-5584	S*Bio Pte Ltd
	Voxtalisib	Exelixis

Monotherapies targeting PI3K pathway have had lower clinical response rates than hoped. Mechanisms of this observed suboptimal drug response or resistance to pathway inhibition include reactivation of the PI3K pathway and/or activation of a parallel pathway. Drug-induced reactivation of the PI3K pathway occurs by loss of AKT feedback regulation of RTK expression due to AKT inhibitors resulting in activation of the PI3K pathway [6]. In addition, mTOR inhibition promotes AKT activation via loss of negative feedback on IRS1 (insulin receptor) and relief of GRB10-mediated PI3K suppression [6]. Furthermore, loss of PTEN was noted as an acquired resistance mechanism to PI3Kα inhibitors [8]. Resistance can also be mediated by activation of parallel signaling pathways due to common upstream adaptor proteins. For instance, mTORC1 inhibition can activate ERK signaling [6].

Given the numerous resistance mechanisms, rationally derived drug combinations are the key to future strategies for treatment. Four key strategies utilized in constructing combination therapies to overcome resistance include inhibition of proximal and distal targets in the PI3K pathway, inhibition of parallel pathways, inhibition with targeted therapies, and inhibition with non-targeted therapies [6]. As a result, a variety of dual inhibitors are currently being investigated in clinical trials with majority focusing on dual PI3K/mTOR inhibition (Table 20.2) [7].

Clinical trials are also underway which combine two inhibitors of the PI3K pathway in addition to combination therapies targeting parallel signaling pathways such as UCN-01, ASN003, and ONC201 [7]. PI3K pathway inhibitors combined with distinct targeted agents such as metformin and inhibitors of VEGF, EGFR, CDK4/6, and PARP are also being evaluated. The final combination strategy focuses on the addition of non-targeted inhibitors such as chemotherapy, hormonal agents, immunotherapy, and biological therapies [6].

Another key component of creating successful therapeutics is identifying predictive markers of sensitivity or resistance to therapies. Current markers of interest include PTEN loss, activating mTOR mutations, and *TSC2* mutations in addition to markers of intrinsic and acquired resistance to inhibitors of this pathway (Table 20.3).

Table 20.3 Potential biomarkers targeting the PI3K pathway

Actionable genomic alterations	Potential therapeutic implications
AKT1, AKT2, AKT3 – activating mutation or gene amplification	AKT1 or mTOR inhibitors
mTOR – activating mutation or gene amplification or selected binding domain mutation	mTOR inhibitors
NF2 – inactivating mutations or deletions [9]	PI3K/AKT/mTOR inhibitors
PIK3CA – activating mutation or gene amplification	PI3K/AKT/mTOR inhibitors
PIK3CB – activating mutation or gene amplification	PI3KCB inhibitors
PIK3CD – activating mutation or gene amplification	PI3KCD inhibitors
PIK3R1– inactivating mutations	PI3K/AKT/mTOR inhibitors
PIK3R2 – inactivating mutations	PI3K/AKT/mTOR inhibitors
PTEN – inactivating mutations or deletions	P110β, AKT or mTOR inhibitors
RICTOR – activating mutation or gene amplification [9]	AKT or MTORC2 inhibitors
STK11 – inactivating mutations or deletions [10]	mTOR inhibitors
TSC1, TSC2- inactivating mutations or deletions [10]	mTOR inhibitors
IHC based biomarkers	
PTEN loss [11]	AKT/mTOR inhibitors
ER+/PR+	mTOR inhibitors

For instance, in preclinical studies catalytic mTOR inhibitors were found to be effective in tumors with intrinsic and acquired rapamycin-resistance [12]. In addition, activating mTOR mutations and loss of function TSC1, TSC2, or *STK11* mutations sensitize tumors to mTOR inhibition in a variety of cancer types; however, acquired mTOR mutations in the FKBP domain result in resistance to allosteric mTOR inhibition despite initial sensitivity but remain sensitive to direct ATP-competitive mTOR kinase inhibitors [10, 12]. Therefore, an understanding of the mechanisms of clinical resistance and sensitivity to available PI3K pathway inhibitors will improve patient selection and the success of rationally derived combination therapies.

Summary

The PI3K-mTOR pathway serves a wide range of normal physiological functions in addition to contributing to carcinogenesis. In a variety of cancers, the PI3K-mTOR pathway is commonly dysregulated due to genomic alterations resulting in hyperactivity of the signaling cascade from overactivation of the positive regulators and loss of the negative regulators. Hyperactivity of this pathway plays a significant role in all aspects of carcinogenesis including angiogenesis and metastasis. Therefore, the PI3K-mTOR pathway is an important therapeutic target. Clinical responses to PI3K-mTOR inhibitors as monotherapy have been modest. As a result, efforts have turned to combination therapies to overcome mechanisms of resistance. Drug combinations focusing on dual inhibition of the PI3K-mTOR pathway, inhibition of parallel pathways, and targeted and non-targeted combinations including immunotherapy are currently being investigated in clinical trials.

References

1. Vanhaesebroeck B, et al. The emerging mechanisms of isoform-specific PI3K signalling. Nat Rev Mol Cell Biol. 2010;11(5):329–41.
2. Vivanco I, Sawyers CL. The phosphatidylinositol 3-Kinase AKT pathway in human cancer. Nat Rev Cancer. 2002;2(7):489–501.
3. Laplante M, Sabatini DM. mTOR signaling in growth control and disease. Cell. 2012;149(2):274–93.
4. Yuan TL, Cantley LC. PI3K pathway alterations in cancer: variations on a theme. Oncogene. 2008;27(41):5497–510.
5. Jiang BH, Liu LZ. PI3K/PTEN signaling in angiogenesis and tumorigenesis. Adv Cancer Res. 2009;102:19–65.
6. Dey N, De P, Leyland-Jones B. PI3K-mTOR in cancer and cancer therapy. 2016. Humana Press/Springer, New York.

7. John Mendelsohn GM, Meric-Bernstam F. Personalized cancer therapy. Knowledge base for precision oncology. The MD Anderson Cancer Center Sheikh Khalifa Bin Zayed Al Nahyan Institute for Personalized Cancer Therapy University of Texas MD Anderson Cancer Center. 2015. https://pct.mdanderson.org.
8. Juric D, et al. Convergent loss of PTEN leads to clinical resistance to a PI(3)Kalpha inhibitor. Nature. 2015;518(7538):240–4.
9. Meric-Bernstam F, et al. A decision support framework or genomically informed investigational cancer therapy. J Natl Cancer Inst. 2015;107(7):djv098.
10. Wagle N, et al. Response and acquired resistance to everolimus in anaplastic thyroid cancer. N Engl J Med. 2014;371(15):1426–33.
11. Andre F, et al. Everolimus for women with trastuzumab-resistant, HER2-positive, advanced breast cancer (BOLERO-3): a randomised, double-blind, placebo-controlled phase 3 trial. Lancet Oncol. 2014;15(6):580–91.
12. Hassan B, et al. Catalytic mTOR inhibitors can overcome intrinsic and acquired resistance to allosteric mTOR inhibitors. Oncotarget. 2014;5(18):8544–57.

Met Signaling in Carcinogenesis

21

Dinuka M. De Silva, Arpita Roy, Takashi Kato, and Donald P. Bottaro

Abbreviations

CRC	Colorectal carcinoma
EGFR	Epidermal growth factor receptor
GC	Gastric carcinoma
HGF	Hepatocyte growth factor
IHC	Immunohistochemistry
IPT	Ig-like, plexins, transcription factors
JM	Juxtamembrane
MM	Multiple myeloma
MST1R	Macrophage stimulating 1 receptor
OS	Overall survival
PD	Pharmacodynamic
PFS	Progression-free survival
PRC	Papillary renal cell carcinoma
PSI	Plexins, semaphorins, and integrins
RCC	Renal cell carcinoma
SCLC	Small lung cell cancer
SH2	Src homology-2
TK	Tyrosine kinase
TKI	TK inhibitor
TPR	Translocated promoter region

Introduction

Hepatocyte growth factor (HGF) was discovered by several research groups working independently [reviewed in 1]. HGF is the only known ligand for the Met receptor tyrosine kinase (TK) [1]. HGF signaling is primarily paracrine: secretion by mesenchymally derived cells in a variety of tissues and organs reaches target cells locally or through the systemic circulation, where it stimulates cell motility, proliferation, survival, and morphogenesis. Signaling through the HGF/Met pathway is critical for normal embryonic development, postnatal maturation, and adult homeostasis [1].

Oncogenic HGF/Met pathway activation occurs through locally induced HGF overproduction, co-expression of HGF and Met in the same cells leading to autocrine pathway activation, constitutive Met TK activation associated with *MET* gene amplification, and *MET* gene mutation [reviewed in 2, 3]. As in embryogenesis, pathway

D. M. De Silva · A. Roy · T. Kato · D. P. Bottaro (✉)
Urologic Oncology Branch, National Cancer Institute, Bethesda, MD, USA
e-mail: don.bottaro@nih.gov

S. Badve, G. L. Kumar (eds.), *Predictive Biomarkers in Oncology*,
https://doi.org/10.1007/978-3-319-95228-4_21

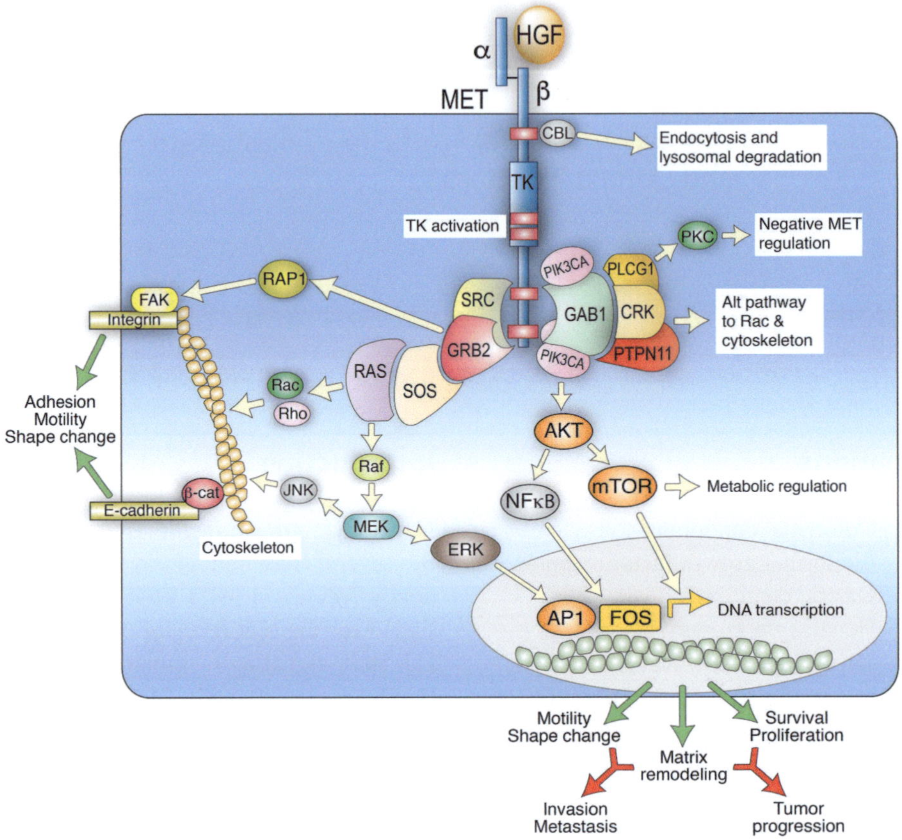

Fig. 21.1 Simplified schematic of the HGF/Met signaling pathway. Pleiotropic signaling through HGF-induced Met tyrosine kinase activity begins with autophosphorylation (red bars) at sites in kinase A-loop and c-terminus (positive regulators and adaptor binding sites, respectively) and in the juxtamembrane domain (negative regulation). Proximal positive affectors include the GRB2/SOS1/RAS/RAF/MEK/ERK and GAB1/ PIK3CA/AKT1/mTOR cascades which impact cycle progression, metabolism, and cytoskeletal change (green arrows), as well as the CRK/PTPN11, MEK/JNK, RAS/RAC/RHO1, and GRB2/RAP1/FAK cascades which impact cell shape change, motility, and morphogenesis (green arrows). Intracellular negative regulation occurs through CBL-mediated vesicular guidance of activated Met to lysosomal degradation and through PKC activation. Red arrows indicate pathways and resulting cellular activities subverted in cancer cells

activation can drive transitions between epithelial and mesenchymal phenotypes in tumor cells: cell dissociation and increased cell motility, protease production, and extracellular matrix turnover promote tumor invasiveness and metastasis. In the tumor microenvironment, HGF/Met signaling in vascular endothelial cells stimulates tumor angiogenesis, supporting the progression of cancers that are growth limited by hypoxia and further enabling tumor metastasis. Hypoxia, or the pseudohypoxic state associated with *VHL* loss of function, upregulates *MET* expression and enhances pathway signaling, motility, and invasiveness in cultured cells and mouse tumor models [4–6]. Silencing overexpressed *MET* gene in tumor cells suppresses tumor growth and metastasis and induces regression of established metastases in mouse models [7] (Fig. 21.1).

Oncogenic *MET* Gene Alterations and Met Overproduction

Missense *MET* mutations in the TK domain were first identified in connection with hereditary and sporadic forms of papillary renal cell carcinoma (PRC) [reviewed in 8] and also occur at low frequency in other cancers. Tumor regression in

PRC patients treated with a Met TK inhibitor [reviewed in 9] confirms that these are targetable disease-driving alterations.

MET sequence alterations in the Sema and JM domains occur in some lung cancer-derived cell lines and patient tumor samples [reviewed in 10]. *MET* gene amplification can result in dramatic Met overproduction and HGF-dependent or HGF-independent TK activation. *MET* gene amplification has been found at low but measurable frequency (<10%) in patients with acquired resistance to TK inhibitors (TKIs) of the epidermal growth factor receptor (EGFR) [11, 12 and references therein]. *MET* amplification also occurs at low but measurable frequency in glioblastoma [13].

RNA sequencing of 42 pediatric glioblastoma tumors revealed fusion transcripts resulting from structural rearrangements in 64% samples involving *FGFR2*, *NTRK2* and *PIK3R2*, and *MET*, where *MET* was the most frequently affected (10%) [14]. Gene fusion events in pediatric glioblastomas involving *MET* include *TFG-MET* and *CLIP2-MET*, where products retain only the Met TK domain, and *PTPRZ1-MET*, that encodes full-length Met but is driven by the highly active *PTPRZ1* promoter, leading to Met overproduction [14, 15].

Oncogenic *HGF* Expression

Lung adenocarcinoma patients harboring the *EML4-ALK* gene rearrangement and treated with alectinib also eventually acquire drug resistance [16]. Among several different routes to resistance, *EML4-ALK*-positive lung adenocarcinoma cells can acquire *HGF* expression, leading to autocrine activation of Met [17]. Autocrine-driven oncogenesis, where tumor cells produce both HGF and Met, has been characterized preclinically in several settings [reviewed in 2, 3, 9]. Establishing that autocrine HGF/Met signaling is an oncogenic driver in the clinical setting is more challenging, although many early studies reported suggestive evidence in various cancer types. Compelling evidence of widespread oncogenic autocrine HGF/Met signaling has been reported for acute myeloid leukemia (AML) cell lines and clinical samples [18]. In that study, genetic suppression of Met or pharmacological inhibition by crizotinib significantly reduced phospho-Met (pMet) and induced apoptosis in AML cells within 12 h, but was accompanied by dramatic upregulation of HGF relative to control cells, ultimately leading to drug resistance through restored HGF/Met signaling [18]. In cases where Met was coactivated with other TKs, concomitant inhibition of both RTKs blocked this compensatory HGF upregulation, resulting in sustained cell killing in vitro and in xenograft models [18].

In addition to *MET* gene amplification in lung adenocarcinoma patients with acquired resistance to EGFR TKIs, HGF overabundance was detected in 61% and 29% of patients with acquired and intrinsic resistance, respectively [reviewed in 19]. The hypothesis that innate drug resistance might be caused by factors secreted by the tumor microenvironment, and the extent to which this might occur generally, was addressed experimentally by measuring the sensitivity of cancer cell lines to anticancer drugs in the absence or presence of stromal cell types [20]. Stromal cells profoundly influenced sensitivity to anticancer drugs, and HGF was a predominant factor that conferred resistance to molecularly targeted drugs such as RAF inhibitors [20].

Other Aberrant HGF/Met Signaling Routes in the Tumor Microenvironment

HGF functions as a stromal cell-derived factor that strongly influences cancer cell invasiveness in the tumor microenvironment. In a genetically engineered murine model of hepatocellular carcinoma and in human patients with this cancer, Horwitz et al. [17] found that amplification of the gene encoding vascular endothelial cell growth factor-A (*VEGFA*) in tumor cells was associated with increased HGF production in tumor-associated macrophages and that HGF, not VEGF, directly drove tumor cell proliferation. A subset of patients harboring *VEGFA* amplification showed significantly better survival after treatment with sorafenib than patients without amplification [17].

Leung et al. [21] found that HGF produced by endothelial cells was essential for directional breast cancer tumor cell migration, blood vessel infiltration, and metastasis. Most often it is pro-HGF that is secreted into the microenvironment, and thrombin or other proteases (e.g., matriptase and/or hepsin) present locally can activate HGFA to process inactive pro-HGF to active HGF [reviewed in 22]. Conversion of pro-HGF to mature HGF at target cell surfaces is essential for signaling [reviewed in 1]. Indeed, some early commercial HGF/Met inhibitor development programs included strategies to disrupt this critical step, although such agents have not advanced to human clinical trials.

Met can also be transferred between cells in the tumor microenvironment through exosomes [23, 24]. Met in exosomes promotes metastatic microenvironment formation in metastatic melanoma [23, 24]. Exosomes from highly metastatic mouse and human melanoma cells contained high levels of Met, and exosomes in the circulation localized to sites of metastatic tissues and increased vascular permeability and, in turn, the migration of tumor cells [23, 24]. Circulating exosomes also increased Met activation in bone marrow-derived cells, thereby reprogramming these cells to a proangiogenic phenotype that mobilized to the lungs where they could facilitate angiogenesis, invasion, and metastasis [23]. Experimental administration of exosomes that contained high levels of Met facilitated metastasis of melanoma cells with lower metastatic capability [24]. Met in circulating serum exosomes accurately predicted melanoma progression [23], and circulating bone marrow-derived cells isolated from patients with advanced melanoma had increased *MET* transcript levels and phosphorylated Met protein [23].

Pharmacological Inhibitors of the HGF/Met Pathway: An Update

The prevalence of HGF/Met pathway activation in human malignancies has driven the rapid development of HGF/Met pathway inhibitors, which can be broadly subdivided into biological (protein-based) agents and low-molecular-weight synthetic compounds (Tables 21.1 and 21.2). We highlight below a subset of HGF/Met pathway inhibitors whose clinical trial results have been particularly informative and/or have progressed in clinical development since our last comprehensive review of this subject [9].

At least 230 human clinical trials of 28 different HGF/Met pathway antagonists have been conducted to date. These clinical trials can be found on the ClinicalTrials.gov site. Due to space restrictions, Tables 21.1 and 21.2 show only one representative trial for every US FDA-approved drug and experimental drug, respectively. For details of all clinical trials under each drug, refer to Cecchi et al. [9] or search the ClinicalTrials.gov site with the drug's name and the indicated cancer.

Rilotumumab [25] has been evaluated as monotherapy in phase Ib/II trials for ovarian and renal cancer and in combination with Avastin in glioma, erlotinib in lung adenocarcinoma, and platinum-based chemotherapy in SCLC, mesothelioma, and gastric cancer, as well as mitoxantrone in prostate cancer. Rilotumumab monotherapy did not show significant antitumor activity in patients with recurrent glioblastoma who had previously received bevacizumab compared with bevacizumab-naive patients. Rilotumumab combined with panitumumab in patients with wild-type *KRAS* metastatic colorectal cancer showed an increased response rate over panitumumab alone. Rilotumumab in combination with epirubicin, cisplatin, and capecitabine (ECX) as a first-line treatment for metastatic gastric cancer showed improved progression-free (PFS) and overall survival (OS) over ECX alone, especially in patients with high Met expression (NCT00719550), prompting randomized, placebo-controlled phase III studies in this indication (RILOMET-1, NCT01697072, and RILOMET-2, NCT02137343). Both trials were terminated in November 2014 based on a planned safety review by the RILOMET-1 independent data monitoring committee that found an increase in the number of deaths in the rilotumumab and chemotherapy treatment arm when compared to the chemotherapy treatment-only arm.

Table 21.1 US FDA-approved drugs targeting the HGF/Met signaling pathway

Signaling pathway	Pathway active in cancer type	Affected predictive Biomarker/ target	Methods of detection[a]	Active drug(s) Representative ClinicalTrials.gov Identifier
HGF/ Met pathway	Metastatic castration-resistant prostate cancer treated with docetaxel and abiraterone or MDV3100	MET	Immunohistochemistry Fluorescent in situ hybridization (FISH) Immunoblotting	Cabozantinib (Cometriq®) Exelixis, Inc. (XL184) ClinicalTrials.gov Identifier NCT01605227 21 clinical trials on the ClinicalTrials.gov Identifier site
	Altered ALK or MET in advanced or metastatic anaplastic large cell lymphoma, inflammatory myofibroblastic tumor, PRCC type 1, alveolar soft part sarcoma, clear cell sarcoma, alveolar rhabdomyosarcoma	MET	Immunohistochemistry Fluorescent in situ hybridization (FISH) Immunoblotting	Crizotinib (Xalkori®) (PF02341066) Pfizer Oncology and EMD Serono ClinicalTrials.gov Identifier NCT01524926 13 clinical trials on the ClinicalTrials.gov Identifier site

[a]Methods of detection are subject to change while trial is active

Table 21.2 Experimental drugs targeting the HGF/Met signaling pathway

Signaling pathway	Pathway active in cancer type	Affected potential predictive Biomarker/target	Method of detection[a]	Active drug(s) Representative ClinicalTrials.gov Identifier
HGF/Met pathway	Japanese subjects with advanced solid tumors or advanced or metastatic gastric or GEJ	HGF	Quantitative plasma immunoassay	Rilotumumab (AMG 102) Amgen ClinicalTrials.gov Identifier NCT01105390 2 more clinical trials on the ClinicalTrials.gov Identifier site
	Advanced solid tumors, gastric esophageal adenocarcinoma	MET	Immunohistochemistry Fluorescent in situ hybridization (FISH) Immunoblotting	AMG337 Amgen ClinicalTrials.gov Identifier NCT02096666
	Advanced or metastatic solid tumors	MET	Immunohistochemistry Fluorescent in situ hybridization (FISH) Immunoblotting	BMS-777607 (ASLAN002) Aslan Pharmaceuticals ClinicalTrials.gov Identifier NCT01721148
	Genomic subpopulations of NSCLC	MET	Immunohistochemistry Fluorescent in situ hybridization (FISH) Immunoblotting	Foretinib (GSK1363089) GlaxoSmithKline ClinicalTrials.gov Identifier NCT02034097

(continued)

Table 21.2 (continued)

Signaling pathway	Pathway active in cancer type	Affected potential predictive Biomarker/target	Method of detection[a]	Active drug(s) Representative ClinicalTrials.gov Identifier
	Metastatic gastric cancer, HER2−, Met+	MET	Immunohistochemistry Fluorescent in situ hybridization (FISH) Immunoblotting Quantitative plasma immunoassay	Onartuzumab (MetMab) Genentech/Roche ClinicalTrials.gov Identifier NCT01662869 2 more clinical trials on the ClinicalTrials.gov Identifier site
	Recurrent/metastatic squamous cell carcinoma of the head and neck (SCCHN)	HGF	Quantitative plasma immunoassay	Ficlatuzumab (AV-299) Aveo Oncology ClinicalTrials.gov Identifier NCT02277197 2 more clinical trials on the ClinicalTrials.gov Identifier site
	Healthy participants	MET	Immunohistochemistry Fluorescent in situ hybridization (FISH) Immunoblotting	LY2801653 Eli Lilly and Company ClinicalTrials.gov Identifier NCT01981408
	Glioblastoma multiforme, gliosarcoma, colorectal cancer, renal cell carcinoma	MET	Immunohistochemistry Fluorescent in situ hybridization (FISH) Immunoblotting	Capmatinib (INCB28060) (INC280) Novartis Pharmaceuticals ClinicalTrials.gov Identifier NCT02386826 13 more clinical trials on the ClinicalTrials.gov Identifier site
	Japanese participants with advanced cancer	MET	Immunohistochemistry Fluorescent in situ hybridization (FISH) Immunoblotting	LY2875358 Eli Lilly and Company ClinicalTrials.gov Identifier NCT01602289 4 more clinical trials on the ClinicalTrials.gov Identifier site
	Metastatic breast cancer, triple-negative breast cancer	MET	Immunohistochemistry Immunoblotting	Met RNA CAR-T cells University of Pennsylvania ClinicalTrials.gov Identifier NCT01837602
	Metastatic breast cancer, triple-negative breast cancer	MET	Immunohistochemistry Fluorescent in situ hybridization (FISH) Immunoblotting	MGCD265 Mirati Therapeutics Inc. ClinicalTrials.gov Identifier NCT01930006
	Asian subjects with HCC	MET	Immunohistochemistry Fluorescent in situ hybridization (FISH) Immunoblotting	MSC2156119J (EMD1214063) Merck KGaA ClinicalTrials.gov Identifier NCT01988493 3 more clinical trials on the ClinicalTrials.gov Identifier site

Table 21.2 (continued)

Signaling pathway	Pathway active in cancer type	Affected potential predictive Biomarker/target	Method of detection[a]	Active drug(s) Representative ClinicalTrials.gov Identifier
	Advanced c-MET-positive (+) solid tumors followed by expansion in selected tumor types	MET	Immunohistochemistry Fluorescent in situ hybridization (FISH) Immunoblotting	SAIT301 Young Suk Park, Samsung Medical Center ClinicalTrials.gov Identifier NCT02296879
	Asian advanced malignant solid tumor patients	MET	Immunohistochemistry Fluorescent in situ hybridization (FISH)	SAR125844 Sanofi ClinicalTrials.gov Identifier NCT01657214
	Locally advanced or metastatic kidney cancer	MET	Immunohistochemistry Fluorescent in situ hybridization (FISH) Immunoblotting	Volitinib (Savolitinib) (AZD6094) (HMPL-504) (HMP-504) National Cancer Institute (NCI) ClinicalTrials.gov Identifier NCT02761057

[a]Methods of detection are subject to change while trial is active

TAK-701 and ficlatuzumab [26, 27] are humanized anti-HGF mAbs now in early-phase clinical trials; both potently block HGF-Met binding. These mAbs are well tolerated, show dose proportional pharmacokinetics, and like rilotumumab, reduce free (unbound) plasma HGF to undetectable levels. At present, studies remain open for ficlatuzumab in head and neck squamous cell carcinoma (NCT02277197) and acute myeloid leukemia (NCT02109627; Table 21.1).

Onartuzumab is a monovalent anti-Met mAb that blocks HGF binding [reviewed in 9]. A phase II study evaluating onartuzumab or placebo in combination with erlotinib showed improved outcome in patients with Met-positive, advanced stage lung adenocarcinoma. Patients whose tumors had high levels of Met protein as determined by IHC that were treated with onartuzumab plus erlotinib showed significantly improved PFS, OS, and nearly threefold reduction in risk of death over those treated with erlotinib alone, prompting a phase III trial for Met-positive advanced lung adenocarcinoma patients (NCT01456325); that trial was terminated after interim review for lack of efficacy, and onartuzumab is not currently in clinical development [reviewed in 28].

Cabozantinib, now in phase III trials, is a multikinase inhibitor that targets primarily Met, KDR, and RET. Cabozantinib was US FDA approved for the treatment of progressive metastatic medullary thyroid cancer; its primary target in that indication is perceived to be RET [29]. Median PFS was 11.2 vs 4.0 months for the cabozantinib and placebo arms, respectively. Although a phase II study of patients with metastatic castration-resistant prostate cancer showed that cabozantinib treatment reduced or stabilized soft tissue lesions, bone metastases, bone pain, and narcotic use (NCT01599793), two follow-up phase III trials were halted when interim review showed failure to meet efficacy endpoints. Cabozantinib is also in trials for the treatment of breast, hepatocellular, melanoma, lung adenocarcinoma, ovarian [30], brain [31], and kidney cancers [32]. Cabozantinib was approved by the US FDA for treatment for renal cell carcinoma after prior anti-angiogenic therapy failure (NCT01865747) in April 2016.

Crizotinib potently inhibits Met and anaplastic lymphoma kinase (ALK) TKs [reviewed in 9]. Crizotinib is highly effective against activated products of *ALK* gene translocations (most frequently *EML4-ALK*) that occur in a subset of lung adenocarcinoma patients [33] and has been approved by the US FDA to treat that group on the basis of a companion diagnostic test for *ALK* rearrangement. Other active efficacy trials of crizotinib target gastric carcinoma and lung adenocarcinoma in the third-line setting (not restricted to *ALK* translocations), urothelial carcinoma of the bladder, anaplastic large cell lymphoma, PRC type 1, inflammatory myofibroblastic tumor, alveolar rhabdomyosarcoma, and glioma, where Met and/or ALK pathways are thought to be involved (Table 21.1).

The multikinase inhibitor foretinib targets Met, VEGFR2, AXL, MST1R, and TEK with high affinity. In the largest clinical trial to date devoted to papillary renal cell carcinoma, foretinib demonstrated antitumor activity, modulation of several target indicator plasma proteins, and a manageable toxicity profile [reviewed in 9]. A recently published phase I/II study of foretinib showed promising antitumor activity and good tolerability in the first-line setting in Asian patients with advanced HCC [34], and a recent phase II study examines efficacy in genomically defined lung adenocarcinoma subpopulations (NCT02034097; Table 21.1).

Other Met TKIs in phase II trials for safety and efficacy include MK8033, golvatinib, amuvatinib, BMS777607, MGCD265, and MK246170 (Table 21.1). Many of these agents are more Met-selective than their predecessors, and phase I trials indicate that some of these are well tolerated, e.g., golvatinib, which targets Met and KDR [35], and the Met-selective agent capmatinib (INC280), which in combination with the EGFR agent EGF816, effectively treated drug-resistant activating mutations in the EGFR and *MET* gene amplification (NCT02335944) [36]. Other Met-selective agents continue clinical development, primarily in drug-resistant lung adenocarcinoma and other cancers where recent genomic and proteomic evidence reinforces older data supporting a driver role in specific individuals.

By far, the biggest hurdle facing further development of Met inhibitors has become patient selection. While several promising phase II clinical trials appeared to have succeeded in identifying indications where HGF/Met signaling was a frequent and critical driver, this was not often borne out in subsequent phase III trials. It has become clear that the finding of any *single* suspicious pathway-related aberration – overexpression, gene amplification, or mutation – is, with the exception of hereditary PRC, rarely sufficient basis for identifying patients most likely to benefit from a targeted therapeutic approach. As anticipated, the most direct means of identifying pathway activation, e.g., quantitative detection of activated Met protein in flash-frozen biopsies of tumor and metastases [37], also present risk and technical difficulty. Moreover, even this information is not in itself direct evidence supporting a "driver" role. Reaching that conclusion reliably is likely to require a complex profile composed from multiple biomarker tests; such a battery can compensate for, and may benefit from, the poor positive predictive value of certain markers that are sensitive and readily measured, such as plasma HGF level. The summary of biomarker discovery related to the HGF/Met pathway that follows is not a comprehensive overview of this subject. It does intentionally include many studies showing simple associations between basic pathway components, such as HGF and clinical outcome in specific cancers, to help provoke thought along fundamental lines in the conceptual development of multiplexed assays of HGF/Met pathway functionality for specific cancer types.

Representative Results from the Cancer Genome Atlas (TCGA) Glioma Project and the International Cancer Genome Consortium PedBrain Tumor Project

An integrative analysis by The Cancer Genome Atlas (TCGA) program revealed primary sequence alterations and gene copy number changes for components of the RTK/PI(3)K pathways in 86% of glioblastoma samples [13]. *MET* showed frequent alteration (4%), in addition to alterations in *EGFR*, *ERBB2*, and *PDGFRA*, in 13% of samples [13]. Of nucleotide sequence variations in 91 of 206 glioblastomas, 10 had either amplifications or point mutations in at least 2 of the RTKs *EGFR*, *ERBB2*, *PDGFRA*, and *MET*, suggesting that co-activation of RTKs may be due to genomic activation [13]. The *EGFR* variant *EGFR*vIII, which lacks exons 2–7 and is thus unable to bind ligand, signals constitutively and forms a complex with Met, thereby promoting Met activation [38]. RNA sequencing of 42 pediatric glioblastoma tumors revealed fusion transcripts resulting from structural rearrangements in 64% samples involving *FGFR2*, *NTRK2* and *PIK3R2*, and *MET*, where *MET* was the most frequently affected (10%) [14]. Gene fusion events in pediatric glioblastomas involving *MET* include *TFG-MET* and *CLIP2-MET*, where products retain only the Met TK domain, and *PTPRZ1-MET* that encodes full-length Met but is driven by the highly active *PTPRZ1* promoter, leading to Met overproduction [14, 15]. An 8-year-old patient with recurrent *PTPRZ1-MET*-positive glioblastoma was treated with crizotinib and initially experienced tumor shrinkage, but eventually drug-resistant lesions resulted in death, highlighting the growing need for combination treatments designed to overcome resistance [15].

Conclusions and Future Prospects

Evidence of aberrant HGF/Met signaling has been found in a variety of human cancers, including evidence of driving oncogenesis, tumor progression and metastasis in animal models, and human cancer patients. This evidence, and our growing understanding of HGF/Met biology, has driven rapid and extensive development of HGF/Met antagonists as anticancer therapy candidates. At least 230 human clinical trials of 28 different HGF/Met pathway antagonists have been conducted to date [39]. The development of clinically useful biomarkers, however, has fallen behind rapid progress in medicinal chemistry. As a class, these agents are generally well tolerated, and clinical responses have been reported in lung, gastric, prostate, and PRC patients treated with HGF and Met inhibitors. However, widespread efficacy has not been seen in many completed phase II and III clinical studies of prevalent cancer types where the pathway was believed to be a frequent disease driver.

In hindsight, although many details play into negative clinical trial outcomes, three related problems are apparent: (1) an incomplete understanding of the molecular heterogeneity of major cancer types (lung cancer, breast cancer, colon cancer, etc.), (2) the relatively delayed development of reliable clinical tests to detect and characterize oncogenic HGF/Met signaling, and (3) the failure to stratify or select patients identified by such tests for efficacy trials of targeted experimental agents. The negative impact of two other factors, related more to the business than the science of drug development, grew over the course of clinical drug development: (4) a historic, widely held expectation of success for single-agent therapies on large cancer populations and (5) the reflection of poor trial outcome on the drug target as well as the drug, and its impact on resource allocation. Together these five factors strongly curtailed or ended several commercial drug development programs for HGF/Met that once drew upon vast financial and intellectual resources.

Despite these setbacks in scope, progress in the field continues to narrow the target spectrum of small molecule kinase inhibitors, improve methods to identify clinically relevant pathway alterations, and identify complementary or synergistic treatment combinations. These efforts should contribute to better clinical trial design and more informative outcomes. For example, small-molecule kinase inhibitors such as AMG337, tepotinib (EMD1214063), capmatinib (INC280), and others target Met with far greater selectivity than early agents such as foretinib (GSK1363089) and cabozantinib (XL184). To date, more than 60 high-resolution structures of the Met kinase domain in combination with ATP-competitive inhibitors have been deposited to the Protein Data Bank. These include several variants arising from *MET* missense mutations, which have been shown to impact inhibitor binding and potency. In addition to revealing an unusual level of conformational plasticity in the Met TK domain, these structures have shown that the ATP-binding pocket accommodates two inhibitor classes: type I ligands, which displace ATP in the active kinase, and type II ligands, which bind to the inactive kinase [40]. Early type II ligands were less Met-selective than type I inhibitors. Within a few years, studies identified ways to conformationally restrain key pharmacophores, enabling the development of selective type II inhibitors, some of which overcome the resistance to type I compounds displayed by specific Met variants [41]. Achieving high potency and singular target selectivity may not translate directly to improved efficacy, but they should help minimize off-target toxicities in combination therapies and thereby better inform about the functional role of the target in carefully selected patient groups.

Assay methods for relevant biomarkers in clinical samples have also become more reliable and more sensitive in detecting Met alterations. Tissue sampling from specific sites, though not without risk, has seen more widespread practice in recent clinical trials. Analysis of circulating DNA has become more sensitive for both sequence variations and copy number variation (from read-depth data) and has distinct advantages in being low risk, systemically informative, and able to identify relevant genomic changes over short time intervals. Future developments in the analysis of circulating cell-free DNA have the potential to provide low-cost longitudinal assessment of the patient's genomic and transcriptomic profiles. Combining evidence of receptor and/or ligand overexpression, amplification, or mutation to correlate with clinical responses to selective pathway inhibitors will improve our ability to use these agents to their full potential. The observation that foretinib was more active against PRC patients with germline *MET* mutations than in those without, but with otherwise histologically indistinguishable tumor phenotype [reviewed in 9], illustrates the need to include the best available molecular biomarkers in therapeutic clinical trials. The inability to detect some small nucleotide deletions using high-throughput, low-cost assays has led to underestimates of their frequency, e.g., lung cancer with *MET* exon 14 skipping alterations. Clinically tractable assays for detecting HGF and *MET* aberrations underlying drug resistance, and for improving our estimates of the prevalence of these alterations, will also improve clinical studies and the effective use of existing high-quality HGF/Met pathway inhibitors. Software-aided IHC and in situ hybridization to assess HGF and Met protein and mRNA abundance in tumor sections, and implementation of available immunoassays of detergent extracts from flash-frozen tissue, also warrant further development. Combining markers to create a profile may also improve predictive reliability. For example, using TCGA and patient-derived xenograft data, Johnson et al. [42] identified a group of 20 genes highly associated with HGF overexpression in GBM that were modulated only in tumors sensitive to Met inhibitors. The Met inhibitors used affected tumor (human) and host (mouse) cells within the tumor through distinct mechanisms to impede tumor growth. *EGFR* amplified tumors displaying erlotinib resistance responded to a combination of Met and EGFR inhibitors [42].

Growing evidence that HGF/Met pathway inhibitors will be more effective in combination with agents targeting other pathways further underscores the need for diagnostic and pharmacodynamic (PD) biomarker development. A recent case report [43] described a patient with lung adenocarcinoma harboring a mutation in *EGFR* and amplification of *MET*, who progressed on erlotinib and responded dramatically to combined Met and EGFR inhibition with savolitinib and osimertinib. When resistance later developed to this combination, a newly acquired *MET* mutation, D1228V, was detected by monitoring circulating plasma cell-free DNA [43]. Modeling studies indicated that the D1228V mutation provided resistance to type I Met TKIs, including savolitinib, through impaired drug binding, while sensitivity to type II Met TKIs, e.g., cabozantinib, was likely to be maintained. Based on these findings, the patient was treated with erlotinib combined with cabozantinib (for which safety data on the combination was available) and exhibited a dramatic response [43]. This case illustrates the superior medical decision-making afforded by strong basic knowledge of oncogenesis for both pathways, advanced inhibitor development for both targets, and reliable assays for additional oncogenic alterations, which were fortunately at hand. Other promising combinations for HGF/Met-targeted agents on the horizon, e.g., with PARP1 inhibitors in ovarian cancer [44], immunotherapy in metastatic RCC [45], radiation in head and neck cancers [46], and VEGF inhibition in lung adenocarcinoma [47], will benefit from concerted parallel biomarker development efforts. A recent patient-derived xenograft study to identify new combination therapies for metastatic melanoma [48] found that single agent capmatinib (targeting Met), encorafenib (targeting BRAF), or binimetinib (targeting MEK) were ineffective, whereas the triple combination resulted in complete and sustained tumor regression in all animals. The most effective use of this knowledge in cancer patients will require reliable and clinically tractable diagnostic and PD biomarkers for all targeted drivers, as well as clear patient selection criteria based on their results.

Acknowledgments This work was supported by the Intramural Research Program of the NIH (ZIA BC011124 awarded to DPB), National Cancer Institute, Center for Cancer Research.

Disclosure Statement The authors disclose no conflicts of interest, financial or otherwise.

References

1. Rubin JS, Bottaro DP. UCSD molecule pages: HGF. 2011. URL: http://www.signaling-gateway.org/molecule/query?afcsid=A004032.
2. Peschard P, Park M. From Tpr-Met to Met, tumorigenesis and tubes. Oncogene. 2007;26:1276–85.
3. Gherardi E, Birchmeier W, Birchmeier C, et al. Targeting MET in cancer: rationale and progress. Nat Rev Cancer. 2012;12:89–103.
4. Pennacchietti S, Michieli P, Galluzzo M, et al. Hypoxia promotes invasive growth by transcriptional activation of the met protooncogene. Cancer Cell. 2003;3:347–61.
5. Peruzzi B, Athauda G, Bottaro DP. The von Hippel-Lindau tumor suppressor gene product represses oncogenic beta-catenin signaling in renal carcinoma cells. Proc Natl Acad Sci U S A. 2006;103:14531–6.
6. Lee YH, Morrison BL, Bottaro DP. Synergistic signaling of tumor cell invasiveness by hepatocyte growth factor and hypoxia. J Biol Chem. 2014;289:20448–61.
7. Corso S, Migliore C, Ghiso E, et al. Silencing the MET oncogene leads to regression of experimental tumors and metastases. Oncogene. 2008;27:684–93.
8. Dharmawardana PG, Giubellino A, Bottaro DP. Hereditary papillary renal carcinoma type I. Curr Mol Med. 2004;4:855–68.
9. Cecchi F, Rabe DC, Bottaro DP. Targeting the HGF/Met signaling pathway in cancer therapy. Expert Opin Ther Targets. 2012;16:553–72.
10. Salgia R. MET in lung cancer: biomarker selection based on scientific rationale. Mol Cancer Ther. 2017 Apr;16(4):555–65.
11. Cappuzzo F, Varella-Garcia M, Finocchiaro G, et al. Primary resistance to cetuximab therapy in EGFR FISH-positive colorectal cancer patients. Br J Cancer. 2008;99:83–9.
12. Kim JY, Welsh EA, Fang B, Bai Y, Kinose F, Eschrich SA, et al. Phosphoproteomics reveals MAPK inhibitors enhance MET- and EGFR-driven AKT signaling in KRAS-mutant lung cancer. Mol Cancer Res. 2016;14(10):1019–29.
13. Cancer Genome Atlas Research Network. Comprehensive genomic characterization defines human glioblastoma genes and core pathways. Nature. 2008;455:1061–8.
14. International Cancer Genome Consortium PedBrain Tumor, Project. Recurrent MET fusion genes represent a drug target in pediatric glioblastoma. Nat Med. 2016;22:1314–20.
15. Bao ZS, Chen HM, Yang MY, et al. RNA-seq of 272 gliomas revealed a novel, recurrent PTPRZ1-MET fusion transcript in secondary glioblastomas. Genome Res. 2014;24:1765–73.
16. Isozaki H, Ichihara E, Takigawa N, et al. Non-small cell lung cancer cells acquire resistance to the ALK inhibitor alectinib by activating alternative receptor tyrosine kinases. Cancer Res. 2016;76:1506–16.
17. Horwitz E, Stein I, Andreozzi M, et al. Human and mouse VEGFA-amplified hepatocellular carcinomas are highly sensitive to sorafenib treatment. Cancer Discov. 2014;4:730–43.
18. Kentsis A, Reed C, Rice KL, et al. Autocrine activation of the MET receptor tyrosine kinase in acute myeloid leukemia. Nat Med. 2012;18:1118–22.
19. Yano S, Takeuchi S, Nakagawa T, et al. Ligand-triggered resistance to molecular targeted drugs in lung cancer: roles of hepatocyte growth factor and epidermal growth factor receptor ligands. Cancer Sci. 2012;103:1189–94.
20. Straussman R, Morikawa T, Shee K, et al. Tumour micro-environment elicits innate resistance to RAF inhibitors through HGF secretion. Nature. 2012;487:500–4.
21. Leung E, Xue A, Wang Y, et al. Blood vessel endothelium-directed tumor cell streaming in breast tumors requires the HGF/C-Met signaling pathway. Oncogene. 2017;36:2680–92.

22. Kawaguchi M, Kataoka H. Mechanisms of hepatocyte growth factor activation in cancer tissues. Cancer. 2014;6:1890–904.
23. Peinado H, Alečković M, Lavotshkin S, et al. Melanoma exosomes educate bone marrow progenitor cells toward a pro-metastatic phenotype through MET. Nat Med. 2012;18:883–91.
24. Adachi E, Sakai K, Nishiuchi T, et al. Cell-autonomous changes in Met receptor expression regulate the growth and metastatic characteristics in malignant melanoma. Oncotarget. 2016;43:70779–93.
25. Jun HT, Sun J, Rex K, et al. AMG 102, a fully human anti-hepatocyte growth factor/scatter factor neutralizing antibody, enhances the efficacy of temozolomide or docetaxel in U-87 MG cells and xenografts. Clin Cancer Res. 2007;13:6735–42.
26. Jones SF, Cohen RB, Bendell JC, et al. Safety, tolerability, and pharmacokinetics of TAK-701, a humanized anti-hepatocyte growth factor (HGF) monoclonal antibody, in patients with advanced nonhematologic malignancies: first-in-human phase I dose-escalation study. J Clin Oncol. 2010;28:15s. (suppl abstr 3081), American Society of Clinical Oncology (ASCO) Annual Meeting; Chicago, IL
27. Tan K, Park K, Lim M, et al. Phase Ib study of ficlatuzumab (formerly AV-299), an anti-hepatocyte growth factor (HGF) monoclonal antibody (MAb) in combination with gefitinib (G) in Asian patients (pts) with NSCLC. J Clin Oncol. 2011;29 (Suppl Abstract 7571), American Society of Clinical Oncology (ASCO) Annual Meeting; Chicago, IL, 2011;7571–7571.
28. Rolfo C, Van Der Steen N, et al. Onartuzumab in lung cancer: the fall of Icarus? Expert Rev Anticancer There. 2015;15:487–9.
29. U.S. Food and Drug Administration. Cabozantinib. URL: http://www.fda.gov/Drugs/InformationOnDrugs/ApprovedDrugs/ucm330213.htm.
30. Gordon MS, Vogelzang NJ, Schoffski P, et al. Activity of cabozantinib (XL184) in soft tissue and bone: results of a phase II randomized discontinuation trial (RDT) in patients (pts) with advanced solid tumors. J Clin Oncol. 2011;29:(Suppl; Abstract 3010) 3010–3010.
31. Wen PY, Prados M, Schiff D, et al. Phase II study of XL184 (BMS 907351), an inhibitor of MET, VEGFR2, and RET, in patients (pts) with progressive glioblastoma (GB). J Clin Oncol. 2010;28 (Suppl Abstract 2006). American Society of Clinical Oncology (ASCO) Annual Meeting; Chicago, IL.
32. Vaishampayan U. Cabozantinib as a novel therapy for renal cell carcinoma. Curr Oncol Rep. 2013;15: 76–82.
33. Kwak EL, Bang YJ, Camidge DR, et al. Anaplastic lymphoma kinase inhibition in non-small-cell lung cancer. N Engl J Med. 2010;363:1693–703.
34. Yau TC, Lencioni R, Sukeepaisarnjaroen W, et al. A phase I/II multicenter study of single-agent foretinib as first-line therapy in patients with advanced hepatocellular carcinoma. Clin Cancer Res. 2017;23: 2405–13.
35. Molife LR, Dean EJ, Blanco-Codesido M, et al. A phase I, dose-escalation study of the multitargeted receptor tyrosine kinase inhibitor, golvatinib, in patients with advanced solid tumors. Clin Cancer Res. 2014;20:6284–94.
36. Jia Y, Juarez J, Li J, Manuia M, et al. EGF816 exerts anticancer effects in non-small cell lung cancer by irreversibly and selectively targeting primary and acquired activating mutations in the EGF receptor. Cancer Res. 2016;76:1591–602.
37. Srivastava AK, Hollingshead MG, Weiner J, et al. Pharmacodynamic response of the MET/HGF receptor to small-molecule tyrosine kinase inhibitors examined with validated, fit-for-clinic immunoassays. Clin Cancer Res. 2016;22:3683–94.
38. Li L, Puliyappadamba VT, Chakraborty S, et al. EGFR wild type antagonizes EGFRvIII-mediated activation of Met in glioblastoma. Oncogene. 2015;34:129–34.
39. Cecchi F, Wright C, Bottaro DP. Experimental therapeutics targeting the hepatocyte growth factor/Met signaling pathway. 2017;URL: https://ccrod.cancer.gov/confluence/display/CCRHGF/Home.
40. Bellon SF, Kaplan-Lefko P, Yang Y, Zhang Y, et al. c-Met inhibitors with novel binding mode show activity against several hereditary papillary renal cell carcinoma-related mutations. J Biol Chem. 2008;283(5):2675–83.
41. Norman MH, Liu L, Lee M, et al. Structure-based design of novel class II c-Met inhibitors: 1. Identification of pyrazolone-based derivatives. J Med Chem. 2012;55(5):1858–67.
42. Johnson J, Ascierto ML, Mittal S, et al. Genomic profiling of a hepatocyte growth factor-dependent signature for MET-targeted therapy in glioblastoma. J Transl Med. 2015;13:306.
43. Bahcall M, Sim T, Paweletz CP, et al. Acquired METD1228V mutation and resistance to MET inhibition in lung cancer. Cancer Discov. 2016;6:1334–41.
44. Du Y, Yamaguchi H, Wei Y, et al. Blocking c-Met-mediated PARP1 phosphorylation enhances anti-tumor effects of PARP inhibitors. Nat Med. 2016;22:194–201.
45. Kammerer-Jacquet SF, Medane S, Bensalah K, et al. Correlation of c-MET expression with PD-L1 expression in metastatic clear cell renal cell carcinoma treated by sunitinib first-line therapy. Target Oncol. 2017;12:487–494.
46. Chen GZ, Dai WS, Zhu HC, et al. Foretinib enhances the radiosensitivity in esophageal squamous cell carcinoma by inhibiting phosphorylation of c-Met. J Cancer. 2017;8:983–92.
47. Cascone T, Xu L, Lin HY, et al. The HGF/c-MET pathway is a driver and biomarker of VEGFR-inhibitor resistance and vascular remodeling in non-small cell lung cancer. Clin Cancer Res. 2017;23:5489-5501.
48. Krepler C, Xiao M, Sproesser K, et al. Personalized preclinical trials in BRAF inhibitor-resistant patient-derived xenograft models identify second-line combination therapies. Clin Cancer Res. 2016;22: 1592–602.

22 Role of Insulin-Like Growth Factor Receptors in Cancer Signaling

Douglas Yee

Role of Insulin-Like Growth Factor Receptors in Cancer Signaling

The insulin-like growth factor (IGF) signaling pathway is composed of multiple ligands and receptors. As their name implies, the key components of the IGF pathway are highly homologous to insulin and its receptor. In fact, insulin can best be considered a key member of this pathway as insulin also plays an important role in cancer signaling [1].

Ligands – The three ligands of the IGF system are IGF-I, IGF-II, and insulin. Due to its importance in maintaining glucose homeostasis, insulin was the first peptide hormone purified and sequenced. Subsequent cloning of the IGF-I and IGF-II genes revealed a high degree of sequence and amino acid homology. They share a common structure composed of intra-chain disulfide cross-links. Insulin, but not IGF-I or IGF-II, has its internal domain (C-peptide) proteolytically cleaved to form a two-chain ligand [2].

Regulation of ligand expression is well studied. For insulin, the expression is restricted to the beta cells of the pancreas and is tightly linked to serum glucose levels. IGF-I and IGF-II are expressed in many tissues. During puberty, the pulsatile release of growth hormone (GH) results in increased IGF-I production by the liver. IGF-I, originally named somatomedin C, then interacts with receptors in essentially all normal tissues and is important during the linear growth phase experienced during puberty. In mice, IGF-II is a fetal somatomedin. After birth, rodents experience a decline in serum IGF-II levels. In humans, IGF-II levels persist during life. However, in adult humans, a clear physiologic role for IGF-II has still not yet been identified. It is noteworthy humans with defective GH signaling have both low IGF-I and IGF-II serum levels and are relatively resistant to developing cancer [3]. In serum and extracellular fluids, IGFs are found to bind to high-affinity IGF-binding proteins (IGFBPs). Six well-characterized binding proteins have been identified, and each has the ability to affect release of ligand to the receptor [4].

Receptor Structure

Just as the ligands share homology, the receptors are also similar to each other in structure. Transcribed from a single gene, they are processed into two separate chains. The alpha subunit is extracellular and covalently bonded to the

D. Yee
Department of Medicine, Masonic Cancer Center, University of Minnesota, Minneapolis, MN, USA
e-mail: yeexx006@umn.edu

S. Badve, G. L. Kumar (eds.), *Predictive Biomarkers in Oncology*,
https://doi.org/10.1007/978-3-319-95228-4_22

beta subunit which contains a short extracellular domain, the transmembrane domain, and an intracellular tyrosine kinase domain. An alpha-beta subunit is bound to a partner; thus the functional receptor complex is a heterodimeric structure [5].

Since the insulin receptor (InsR) and type I IGF receptor (IGF1R) have similar structures, this allows for heterodimerization between a single chain of Ins and IGF1R. Additionally, cancer cells express a fetal isoform of Ins, known as Ins-A, while normal tissues express a splice variant Ins-B. This variation in receptor expression allows for the formation of receptor subtypes composed of either homodimers (InsR-A/InsR-A or InsR-B/InsR-B or IGF1R/IGF1R) or heterodimers (InsR-A/IGF1R or InsR-B/IGF1R or InsR-A/InsR-B). Since most cancer cells express both IGF-1R and InsR, the cell surface receptor complexes are themselves complex [6].

The InsR fetal and adult isoforms differ in the alpha subunit and affect ligand binding. The tyrosine kinase domains of InsR isoforms are identical. The tyrosine kinase domain of IGF1R has a high degree of homology with InsR. Thus, a functional receptor may have both different ligand binding and tyrosine kinase domains (Fig. 22.1).

Healthy humans have low levels of insulin in the absence of feeding. However, patients with insulin resistance (type II diabetes mellitus, metabolic syndrome, and prediabetes) all have elevated serum insulin levels to maintain glucose homeostasis. In these states of insulin resistance, the elevated insulin levels are thought to compensate for InsR insensitivity at the major target organs: the liver, muscle, and fat. Healthy humans have very high circulating levels of both IGF-I and IGF-II.

However, these ligands are not free to intact with receptors as they are complexed to high-affinity binding proteins in extracellular fluids [8]. In blood, IGF-I and IGF-II are found in a ternary complex composed of the ligand, IGF-binding protein-3, and an acid-labile subunit. While this ternary complex prevents IGF-I interaction with its receptors, it does provide a reservoir of IGF ligands available for release in times of stress. For example, major surgery, burns, and pregnancy all result in the proteolytic cleavage of the ternary complex allowing the IGFs to interact with tissues. Some cancer-specific proteases, such as prostate-specific antigen, have also been found to give rise to ~~result in~~ proteolytic cleavage of IGF-binding proteins.

Receptor Activation

It is well established the first step in IGF1R/InsR signaling is autophosphorylation. The two tyrosine kinase domains are physically constrained from interaction until ligand binding. Once ligands bind, the receptors undergo a conformational change which results in interaction between the two tyrosine kinase domains. Unlike single-chain transmembrane tyrosine kinase receptors, which may auto-activate if the receptor is overexpressed, HER2, for example, this family of IGF receptors cannot "auto-activate" and requires ligand binding. This has therapeutic implications as outlined below.

The ability of ligands to bind the receptors is similarly complex. While the insulin interaction with InsR-A and InsR-B is well characterized for its role in glucose homeostasis, other ligands will also bind InsRs. InsR-A has high affinity for IGF-II. The hybrid receptors composed of one chain of IGF1R and one of InsR tend to have higher affinity for the IGF ligands. However, supraphysiologic concentrations of any of the ligands may activate all the isoforms. Since there are a substantial number of people with elevated insulin levels (type II diabetes mellitus, metabolic syndrome, and prediabetes), the availability of ligands, including IGF-I and -II, makes it possible that activation of this receptor system occurs in many cancers [6].

Once autophosphorylated, the receptors now allow docking of adaptor proteins to the intracellular domain of the receptor. Many adaptor proteins have been identified including the insulin receptor substrates (IRS-1 through 4) Shc, Crk, etc. [9]. These adaptor proteins serve as scaffolds for other intermediate signal transduction pathways including activation of PI3K and MAPK. In normal cells, activation of PI3K is important for glucose transporter replication and glucose

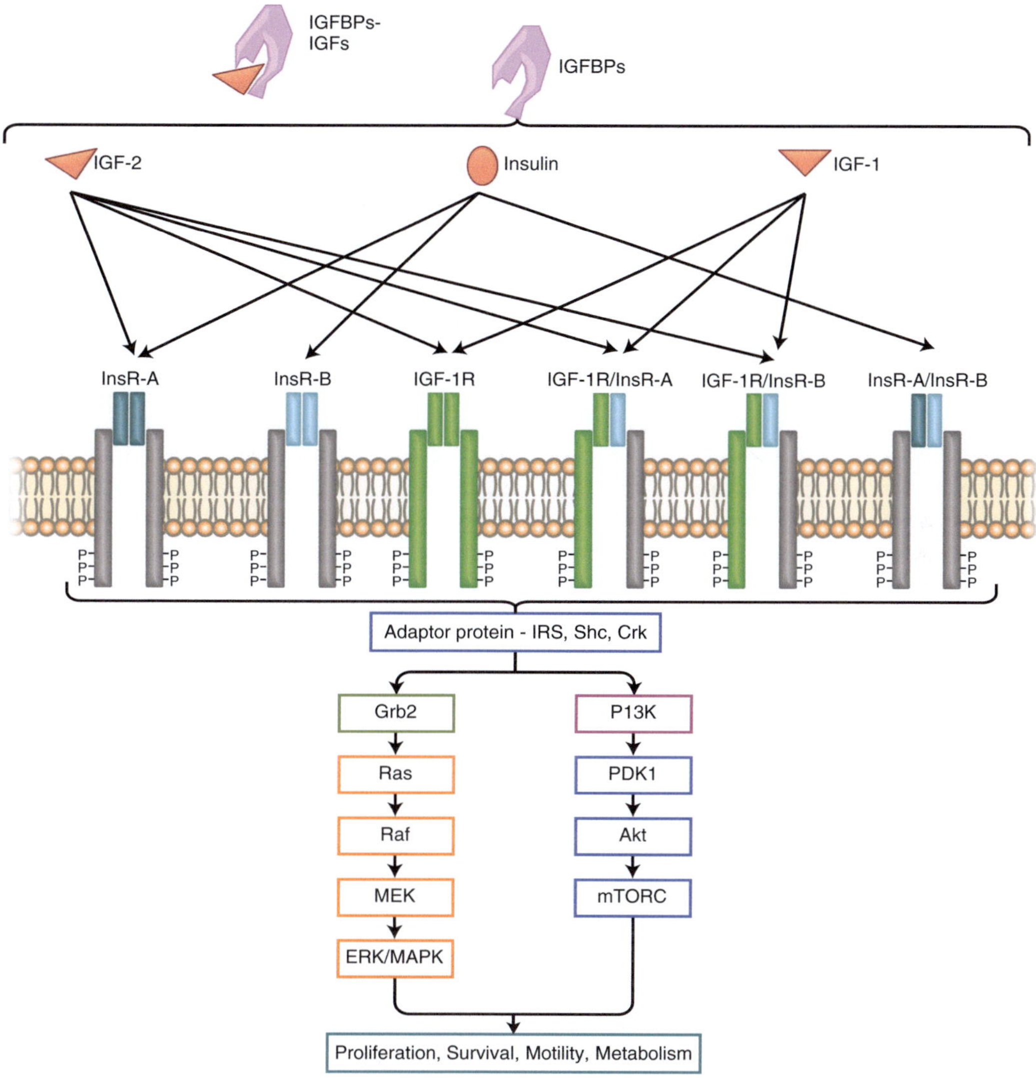

Fig. 22.1 IGF and insulin signaling pathways. The three IGF ligands (IGF-1, IGF-2, and insulin) interact with multiple cell surface receptors. In extracellular fluids, the IGF ligands, except insulin, are bound to high-affinity IGF-binding proteins (IGFBPs). In serum, IGF-I, IGFBP-3, and the acid-labile subunit form a complex which serves as a reservoir for the ligand. Holo-receptors (IGF1R, InsR-A, and InsR-B) and hybrid receptors expressing one chain of either IGF1R and InsR all bind ligands. Once ligand bound, receptors undergo a conformational change to activate their intracellular tyrosine kinase domains. After receptor autophosphorylation (signified by -P), the intracellular domain of the receptor serves as a docking site, and additional substrates will be phosphorylated. For example, the IRS proteins serve as scaffolds to activate multiple pathways including MAPK and PI3K. These secondary signaling pathways result in enhanced cancer cell proliferation, survival, motility, and metabolism. (Adapted from Ekyalongo and Yee [7]. With permission from Nature Publishing Group)

uptake. In cancer cells, this enhancement of glucose uptake has also been identified. In addition, activation of signaling pathways downstream of IGF1R and InsR stimulates cell survival, proliferation, and motility. While InsR is primarily linked to glucose uptake in normal host tissues and IGF1R was identified as an important growth regulatory signaling molecule, data suggest both pathways can stimulate either function. Further, distinct specific signaling pathways downstream of the receptors have not been identified. Rather, the signaling pathways activated downstream of this receptor family have significant shared properties. Preclinical models have shown both receptor families affect cancer cell biology.

Targeting IGF1R and InsR Signaling

There have been three strategies to target IGF1R and InsR signaling in cancer: monoclonal antibodies (moAbs) directed only against IGF1R, tyrosine kinase inhibitors targeting both IGF1R and InsR, and IGF-I and IGF-II ligand neutralization moAbs. Many clinical trials have been reported with the most extensive being that of the IGF1R moAbs [7].

The moAbs were all designed to specifically interact with IGF1R and had no binding affinity for InsR. Given the known function of InsR, it was felt a specific IGF1R inhibitor could be effective without disrupting glucose homeostasis. This premise proved to be incorrect; the IGF1R moAbs all disrupted glucose homeostasis, most likely due to the interruption of a negative feedback system between GH and IGF-I. Early phase I trials showed GH increased when patients received an IGF1R moAb thus showing the negative feedback of IGF-I on the hypothalamus and pituitary was disrupted by this class of drugs. GH elevation results in decreased insulin sensitivity most likely due to increased free fatty acid output by the liver as seen in conditions of GH excess. As insulin levels rise, they may effectively counteract the potential benefits of IGF1R inhibition by allowing an insulin-InsR interaction to proceed. Despite this potential negative outcome, several phase I clinical trials demonstrated responses to single agent IGF1R moAbs. However, phase III clinical trials failed to show a benefit for this class of drugs when compared to placebo. In fact, in endocrine-resistant breast cancer, there was a suggestion of harm when the IGF1R moAb ganitumab was combined with either exemestane or fulvestrant. While the mechanism of harm was not documented in this trial, it is tempting to speculate elevated insulin levels could stimulate InsR. Indeed, recent careful examination of IGF1R and InsR mRNA expression in hormone-refractory breast cancer suggests InsR, both A and B isoforms, are expressed at much higher levels than IGF1R [10].

The tyrosine kinase inhibitors were also reported in phase I clinical trials. For this class of drugs, it was evident they had little specificity for IGF1R; they were equipotent inhibitors of InsR. While the phase I trials suggested an activity, there was little enthusiasm to pursue these drugs due to their disruption of glucose homeostasis [6].

One remaining strategy exists. Since IGF1R and InsR require ligand binding, an attempt to inhibit ligand interaction with the receptors could be successful. Two neutralizing antibodies (xentuzumab and MEDI-573) have been described and are currently being pursued in a clinical trial (see Table 22.1) [11]. It remains to be seen if these drugs also induce GH elevation and subsequent hyperglycemia, but these potential toxicities will be monitored in ongoing clinical trials.

Table 22.1 Targeting strategies for IGF1R and InsR

Signaling pathway	Pathway active in cancer type	Affected biomarker	Method of detection	Target	Active drugs
Type I IGF receptor (IGF1R)	Most	IGF1R	Western blot to phospho-IGF1R/InsR	IGF1R	IGF-1R moAbs – figitumumab, ganitumab, cixutumumab, dalotuzumab
IGF ligands	Most	Serum IGF-I, IGF-II	ELISA for IGFs, Western blot for phospho-IGF1R/InsR	IGF1R	IGF ligand-neutralizing abs – MEDI-573, xentuzumab (BI 836845)
Type I IGF receptor (IGF1R) and insulin receptor (InsR)	Most	IGF1R and InsR	Western blot to phospho-IGF1R/InsR	IGF1R/InsR	BMS-754807, linsitinib (OSI-906)

Biomarkers for IGF1R and InsR Inhibitors

Compared to development of other drugs (such as trastuzumab) with known targets (HER2), none of the clinical trials conducted *for IGF1R and InsR inhibitors* were stratified for target expression. Early preclinical data suggested only a weak link between receptor expression and IGF sensitivity, and as a result, none of the clinical trials required measurement of either IGF1R or InsR expression. Since the receptors require ligand activation, there is evidence to show that serum ligand levels may be associated with the outcome. In addition, several gene expression profiles have suggested specific "IGF-activated" breast cancer can be identified [12].

The failure in the use of predictive biomarkers may explain the lack of benefit for IGF1R moAbs in a clinical trial. Since it is possible only a subset of cancers may have activation of this pathway, applying these therapies to the "activated" cancers would likely result in a demonstration of benefit. Beyond identifying IGF1R-activated tumors, the role for InsR has also not been studied. Given the disruption of insulin and glucose homeostasis, the measurement of InsR could be equally important. Finally, identifying the conformation of receptor subtypes on the cell surface (holo- versus hybrid receptors) may also identify cancers vulnerable to specific inhibitors. For example, cancer cells expressing equal amounts of IGF1R and InsR could have different cell surface conformations. A cell expressing equal amounts of holo-IGF1R and InsR may not be inhibited by a IGF1R moAb as InsR could serve as an alternative growth regulatory pathway. In contrast, if the cell only contained hybrid receptor (IGF1R and InsR dimers), then a moAb to IGF1R could be effective in blocking all downstream signaling for this receptor subtype.

Finally, measurement of downstream activated signaling pathways, such as IRS molecules, could also play a role in determining sensitivity to IGF1R antagonists.

Future Perspectives

Despite the abundant preclinical and clinical data suggesting that the IGF1R system plays an important role in regulating cancer biology, none of the current drugs have shown success in phase III trial designs. While many of the studies have shown rare, extraordinary responses to IGF1R inhibitors as single agents [13], phase III trials have not shown benefit for a combination of IGF1R inhibitors with conventional chemotherapy, endocrine therapy, or targeted therapy. The reasons for this lack of success may be due to several flaws in the design of the clinical trials including lack of patient selection on the expression of predictive biomarkers and failure to suppress reflex hyperinsulinemia with subsequent InsR activation.

Currently, strategies to neutralize the IGF ligands remain in a clinical trial, most notably with the monoclonal antibody xentuzumab. This approach has appeal as high circulating levels of IGF-I and IGF-II are not required for adult life as shown in patients with deficiencies in the growth hormone signaling pathway [3].

As with many other validated cancer targets, such as K-ras, the first attempt at inhibition of signaling may not result in a clinical benefit [14]. This inability to demonstrate the clinical relevance of a target with a specific therapy does not mean that the target is invalid. Rather, the drug was unable to effectively inhibit the target. This is likely the case with the first generation of IGF1R inhibitors. Because the drugs, especially the monoclonal antibodies, did not inhibit the highly related InsR, these drugs were insufficient to inhibit the entire IGF ligand signaling pathway. Further drug development and rational combinations will be needed to determine if the preclinical data translate into clinical benefit for patients.

References

1. Pollak MN, Schernhammer ES, Hankinson SE. Insulin-like growth factors and neoplasia. Nat Rev Cancer. 2004;4(7):505–18.
2. Leroith D, Kavsan VM, Koval AP, Roberts CT. Phylogeny of the insulin-like growth factors (IGFs) and receptors – a molecular approach. Mol Reprod Dev. 1993;35(4):332–8.
3. Guevara-Aguirre J, Balasubramanian P, Guevara-Aguirre M, Wei M, Madia F, Cheng CW, et al. Growth hormone receptor deficiency is associated with a major reduction in pro-aging signaling, cancer, and diabetes in humans. Sci Transl Med. 2011;3(70):70ra13.
4. Perks CM, Holly JM. IGF binding proteins (IGFBPs) and regulation of breast cancer biology. J Mammary Gland Biol Neoplasia. 2008;13(4):455–69.
5. Krywicki RF, Yee D. The insulin-like growth factor family of ligands, receptors, and binding proteins. Breast Cancer Res Treat. 1992;22(1):7–19.
6. Yee D. A tale of two receptors: insulin and insulin-like growth factor signaling in cancer. Clin Cancer Res. 2015;21(4):667–9.
7. Ekyalongo RC, Yee D. Revisiting the IGF-1R as a breast Cancer target. NPJ Precis Oncol. 2017;1:14.
8. Baxter RC. IGF binding proteins in cancer: mechanistic and clinical insights. Nat Rev Cancer. 2014;14:329–41.
9. White MF. IRS2 integrates insulin/IGF1 signalling with metabolism, neurodegeneration and longevity. Diabetes Obes Metab. 2014;16(Suppl 1):4–15.
10. Gradishar WJ, Yardley DA, Layman R, Sparano JA, Chuang E, Northfelt DW, et al. Clinical and translational results of a phase II, randomized trial of an anti-IGF-1R (cixutumumab) in women with breast cancer that progressed on endocrine therapy. Clin Cancer Res. 2016;22(2):301–9.
11. Iams WT, Lovly CM. Molecular pathways: clinical applications and future direction of insulin-like growth factor-1 receptor pathway blockade. Clin Cancer Res. 2015;21(19):4270–7.
12. Becker MA, Ibrahim YH, Oh AS, Fagan DH, Byron SA, Sarver AL, et al. Insulin receptor substrate adaptor proteins mediate prognostic gene expression profiles in breast cancer. PLoS One. 2016;11(3):e0150564.
13. Tolcher AW, Sarantopoulos J, Patnaik A, Papadopoulos K, Lin CC, Rodon J, et al. Phase I, pharmacokinetic, and pharmacodynamic study of AMG 479, a fully human monoclonal antibody to insulin-like growth factor receptor 1. J Clin Oncol. 2009;27(34):5800–7.
14. Macdonald JS, McCoy S, Whitehead RP, Iqbal S, Wade JL 3rd, Giguere JK, et al. A phase II study of farnesyl transferase inhibitor R115777 in pancreatic cancer: a Southwest oncology group (SWOG 9924) study. Investig New Drugs. 2005;23(5):485–7.

Role of Wnt/β-Catenin Pathway in Cancer Signaling

23

Casey D. Stefanski and Jenifer R. Prosperi

Abbreviations

APC	Adenomatous Polyposis Coli
CBP	CREB-binding protein
CK1	Casein Kinase 1
COSMIC	Catalogue of somatic mutations in cancer
CSC	Cancer stem cell
Dkk1	Dickkopf1
Dvl	Disheveled
EMT	Epithelial-to-mesenchymal transition
FAP	Familial adenomatous polyposis
FZD	Frizzled
FZD8CRD	Fusion protein containing the Fc region IgG fused to the cysteine-rich domain of FZD8
GSK3β	Glycogen Synthase Kinase 3β
HCC	Hepatocellular carcinoma
HNSCC	Head and neck squamous cell carcinoma
LGR 4–6	Leucine-rich repeat containing G-protein receptors
LRP5/6	Lipoprotein receptor-related protein
Prcn	Porcupine
SFRPs	Secreted frizzled-related proteins
TCF/LEF	T-cell factor/lymphoid enhance-binding factor
TNKS	Tankyrase
WIFs	WNT inhibitory factors

C. D. Stefanski
Department of Biological Sciences, University of Notre Dame, South Bend, IN, USA

J. R. Prosperi (✉)
Department of Biological Sciences, University of Notre Dame, South Bend, IN, USA

Department of Biochemistry and Molecular Biology, Indiana University School of Medicine – South Bend, South Bend, IN, USA
e-mail: jprosper@nd.edu

Introduction

The Wnt/β-catenin pathway is a highly conserved pathway essential for development and is one of the most commonly dysregulated signaling pathways associated with tumorigenesis. The discovery that activation of *int1* (or *Wnt1*) could promote mammary tumor development [1] gave rise to the importance of Wnt signaling in tumorigenesis. This notion was further substantiated when *Adenomatous Polyposis Coli* (*APC*), the gene responsible for the cancer syndrome familial adenomatous polyposis (FAP), was shown to interact with β-catenin, a key component of the Wnt pathway. APC, in addition to Axin, Glycogen Synthase Kinase 3β (GSK3β), and Casein Kinase 1α (CK1α) combine to form the so-called destruction complex. When the Wnt pathway is inactive, this complex, which will be described in detail below, is responsible for the sequestration, phosphorylation,

S. Badve, G. L. Kumar (eds.), *Predictive Biomarkers in Oncology*,
https://doi.org/10.1007/978-3-319-95228-4_23

and targeted degradation of β-catenin. Studies have demonstrated aberrant Wnt signaling in many diverse cancers, and recent efforts have focused on targeting the Wnt pathway to improve patient outcome. Regulation of the Wnt pathway is maintained through multiple proteins that positively and negatively influence the pathway status (Fig. 23.1). The specific components, their role in tumorigenesis, and current literature in targeting them will be described throughout this chapter.

The Wnt Pathway

Wnts are secreted glycoproteins capable of activating both canonical (β-catenin-dependent) and noncanonical (β-catenin-independent) Wnt pathways. This chapter will focus solely on canonical Wnt signaling due to its well-regarded role in tumorigenesis. Wnt signaling begins through Porcupine (Prcn), a membrane-bound O-acyl transferase, which acetylates Wnt ligands allowing them to bind to the seven-transmembrane receptor frizzled (FZD) and the co-receptor lipoprotein receptor-related protein (LRP5/6). Concurrently, R-spondin ligands bind to the seven-transmembrane receptor leucine-rich repeat containing G-protein receptors (LGR4-6), most commonly LGR5. This interaction then inhibits the cell surface transmembrane E3 ubiquitin ligases ZNRF3/RNF43 from targeting FZD receptors for degradation [1, 2]. Accumulated FZD receptors are therefore able to interact with disheveled (Dvl), which is phosphorylated by PAR-1 (Fig. 23.1) and CK1ε (not shown) [3, 4]. By binding to FZD, activated Dvl recruits Axin away from the β-catenin destruction complex. In addition, CK1γ and GSK3β are recruited by Dvl-Axin complex and are capable of phosphorylating LRP5/6 to promote Axin binding. Another level of regulation is tankyrase (TNKS) that phosphorylates Axin, allowing it to

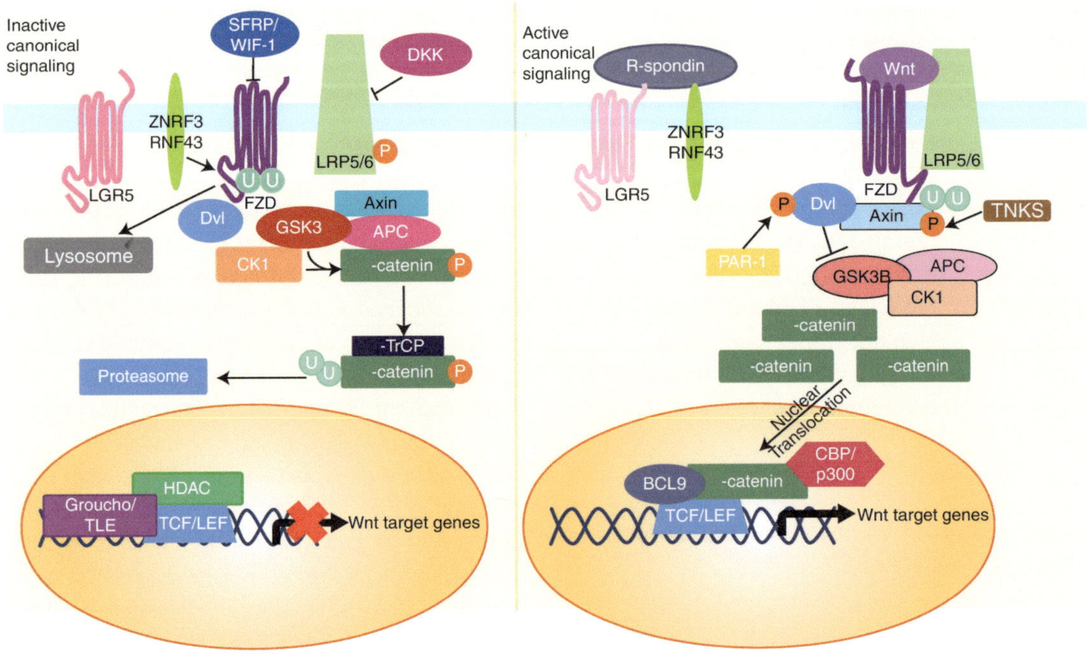

Fig. 23.1 The Wnt/β-catenin signaling pathway. The canonical Wnt pathway is essential for cell development, proliferation, and stem cell homeostasis. On the left, inactive Wnt signaling stabilizes the β-catenin destruction complex comprising Adenomatous Polyposis Coli (APC), Axin, Glycogen Synthase Kinase 3β (GSK3β), and Casein Kinase 1α (CK1α). The β-catenin destruction complex phosphorylates and targets β-catenin for proteasomal degradation. Without nuclear translocation, β-catenin is unable to activate transcription. On the right, active Wnt signaling destabilizes the β-catenin destruction complex. This allows β-catenin to be translocated into the nucleus and thereby activates transcription of Wnt target genes. Proteins that are inactive are in a lighter shade and outlined in black compared to their active counterpart. P phosphorylated, U ubiquitinated

be ubiquitinated for proteasomal degradation. Any of these methods to disrupt the destruction complex prevent GSK3β from phosphorylating β-catenin and targeting it for degradation. The stabilized β-catenin is translocated into the nucleus where it binds to the T-cell factor/lymphoid enhance-binding factor (TCF/LEF) transcription factors and coactivators, such as CREB-binding protein (CBP), p300, or BCL9 to initiate transcription. The interaction between nuclear β-catenin and TCF/LEF transcriptionally activates Wnt target genes, such as c-myc and cyclin D1, to promote proliferation [1, 5, 6]. β-catenin binds other transcription factors, such as hypoxia-inducible factor 1α and microphthalmia-associated transcription factor, to regulate other processes including epithelial-to-mesenchymal transition (EMT), oxidative stress, and pluripotency [4, 7, 8].

Due to the wide range of cellular impacts that aberrant β-catenin signaling can have, Wnt/β--catenin signaling is highly controlled with many regulatory mechanisms as mentioned above. First, Wnt ligands can be blocked by secreted frizzled-related proteins (SFRPs) or WNT inhibitory factors (WIFs). These bind competitively with Wnt ligands to FZD receptors preventing Wnt ligand-receptor interaction and downstream signaling [7]. Additionally, Dickkopf1 (Dkk1) inhibits Wnt signaling by binding to the co-receptor LRP5/6. With upstream inactivation of the Wnt pathway, Dvl cannot recruit Axin away from the destruction complex. Cytosolic β-catenin is phosphorylated allowing it to bind to the E3 ubiquitin ligase β-TrCP, which then leads to β-catenin being ubiquitinated for proteasomal degradation. Wnt target genes are therefore not transcriptionally activated because β-catenin is not translocated into the nucleus to bind to TCF/LEF transcription factors or to modify the chromatin by displacing the transcriptional repressor Groucho/TLE.

Wnt Signaling in Tumorigenesis

Wnt/β-catenin signaling is currently recognized as one of the most prominent signaling pathways in cancer development and progression. Although activation of Wnt signaling is long-recognized as important in tumor development, newer data suggest that inactivation of this pathway could also promote tumorigenesis. We have focused this chapter on the activation but briefly have addressed inactivation (below) as one of the challenges of targeting the pathway. Typically associated with colorectal cancer, the Wnt/β-catenin pathway has been implicated in the tumorigenesis of multiple other epithelial cancers including breast, lung, and prostate cancers (Table 23.1) [8].

Table 23.1 Biomarkers of Wnt signaling

Signaling pathway	Pathway active in cancer type	Affected biomarker	Method of detection
Wnt/β-catenin signaling	Breast	Nuclear β-catenin	Electrophoretic mobility shift assays
	Colorectal	β-Catenin/TCF transcriptional activation of Wnt target genes	Immunohistochemistry for nuclear β-catenin
	Lung		TOP-Flash luciferase assay for β-catenin/TCF transcriptional activation
	Prostate		Real-time PCR of Wnt target genes
	Gastric		
	Pancreatic		
	Liver		
	Skin		
	Musculoskeletal		
	Gynecological		
	CNS		
	Kidney		
	Bladder		
	Head and neck		
	Esophageal thyroid		

Wnt/β-catenin signaling can promote tumor progression, chemoresistance, and epithelial-to-mesenchymal transition (EMT) [5, 7]. The critical role of Wnt/β-catenin signaling in promoting tumor progression is best demonstrated in cancer stem cell (CSC) populations. For instance, increased Wnt signaling in LGR5-expressing breast cancer cells was shown to promote CSC-like characteristics [5]. CSCs are associated with tumorigenesis, chemoresistance, and poor patient outcome. The Wnt pathway can be activated either through enhanced positive regulators or decreased negative regulators, both of which will be discussed further.

Two broad examples of *enhanced positive regulators* are (1) upstream Wnt ligands and (2) β-catenin and the associated transcriptional regulation. Wnt activation allows for the transcription of target genes necessary for supporting hyperproliferation and increased metabolism, known hallmarks of cancer [5]. Increased expression of Wnt ligands and receptors have been observed in different cancer types. Interestingly, depending on the type of cancer, different Wnt ligands are upregulated. For example, WNT1 is highly expressed in prostate cancer, whereas WNT14 is highly expressed in malignant head and neck squamous cell carcinoma (HNSCC). Both prostate cancer and HNSCC are associated with WNT5A overexpression. In addition, stabilizing mutations and enhanced nuclear accumulation of β-catenin are biomarkers of Wnt signaling and are observed in many cancers including colon, breast, and lung cancers (Table 23.1). Mutations to the β-catenin gene, *CTNNB1*, which prevent the phosphorylation of β-catenin and subsequent ubiquitin-mediated degradation have been observed in liver, kidney, pancreatic, and soft tissue cancers [7]. According to the Catalogue of Somatic Mutations in Cancer (COSMIC) database, 42% of samples contained a mutated *CTNNB1* [7]. In addition, multiple cancer types exhibit overexpression of the transcription factors bound by nuclear β-catenin, which in turn enhances their expression in a positive feedback loop. For instance, increased expression of LEF-1 has been observed in colon, pancreatic, and liver cancers [8].

Moreover, *decreased Wnt pathway inhibitors* can also promote Wnt signaling and tumorigenesis. Loss of pathway inhibitors such as WIF1 have been associated with poor patient prognosis in hepatocellular carcinoma (HCC) [7]. One of the most notable mutations leading to tumorigenesis, especially in colorectal cancer, is loss of the tumor suppressor and the negative regulator of Wnt signaling, APC [8]. APC has also been shown to have decreased protein expression either through mutation or promoter hypermethylation in many epithelial cancers including lung, gastric, prostate, and breast cancers. Interestingly, in breast cancer, the APC-deficient phenotype resembles the more aggressive triple-negative subtype, which would benefit from a novel targeted therapy [8]. Also, mutated RNF43 incapable of directing FZD degradation has also been associated with tumor subtypes within colorectal cancer, endometrial cancer, and pancreatic ductal adenocarcinoma [1]. Overall, upregulation of Wnt signaling has been established to promote tumorigenesis in diverse cancer subtypes and corresponds to poor patient outcome. This further supports Wnt-targeted therapy as a potential therapeutic approach for diverse cancer types.

Targeted Wnt Therapies

Due to prevalence of enhanced Wnt signaling activation promoting tumor progression, therapies have focused on inhibition of the Wnt pathway. The key targeting sites for Wnt signaling inhibition have been Prcn, FZD, Dvl, the β-catenin destruction complex, nuclear β-catenin, and TNKS [9]. Two therapeutic methods have demonstrated antitumor effects in targeting Wnt signaling, namely, small molecule inhibitors and monoclonal antibodies [7]. Several *small molecule inhibitors* that are in the preclinical and clinical stages of drug development are β-catenin inhibitors, including PRI-724, CWP232228, and BC2059, which repress TCF/LEF target genes (Table 23.2) [10]. Previous studies have demonstrated that inhibiting the Wnt pathway, including blocking the β-catenin/TCF interaction, can arrest cancer cell growth [11]. In addition, there are two TNKS inhibitors, XAV939 and E7449, which reduce Wnt signaling. By inhibiting TNKS, Axin is not targeted for degradation and therefore stabilizes the destruction complex to promote β-catenin proteasomal degradation. Recent studies demonstrate XAV939 can reduce Wnt signaling *in vitro*; however, *in vivo* studies still need to validate this reduced signaling. Due to toxicity observed in XAV939, the less toxic E7449 inhibitor should be investigated [9]. Two small molecule Prcn inhibitors, WNT974 and ETC-159, have also shown promising results *in vitro* and *in vivo* [9]. Currently, WNT974 is in phase I/II clinical trials for treatment of metastatic colorectal cancer, melanoma, and pancreatic adenocarcinoma, while ETC-159 is in phase I for treatment of colorectal and renal solid tumors [1, 6, 12]. While inhibiting Wnt signaling through small molecule inhibitors is a potential therapeutic strategy to prevent cancer progression, more research needs to be conducted to determine how these inhibitors can affect other signaling pathways to minimize toxicity to non-cancerous cells as seen with XAV939 [9]. Overall there needs to be an emphasis on increasing the efficacy of these inhibitors while ensuring safety.

Another avenue being explored is the use of *monoclonal antibodies* to regulate Wnt signaling. There are many antibody-based therapies currently being studied (Table 23.3). Preclinical trials have shown the potential therapeutic utility of antibodies. The fusion protein, OMP-54F28, contains the Fc region IgG fused to the cysteine-rich domain of FZD8, which scavenges for FZD8-binding Wnt ligands. OMP-54F28 is in phase I clinical trials for treatment of ovarian,

Table 23.2 Wnt-targeted small molecule inhibitors

Small molecule inhibitors			
Target	Active drugs	Stage	Refs.
Porcupine	ETC-159	Phase I	[1–3]
	WNT974 (LGK974)	Phase I	[1, 2, 4, 5]
	WNTC59	Preclinical	[1, 2]
Tankyrase	XAV939	Preclinical	[1, 2, 5]
	E7449	Preclinical	[6]
	AZ1366	Preclinical	[1, 2]
β-Catenin-protein interactions	PRI-724	Phase I	[1, 4, 5]
	CWP232228	Preclinical	[1]
	BC2059	Preclinical	[1, 2]
	CGP049090	Preclinical	[1, 2, 5]
	LF3	Preclinical	[1, 2]
	MSAB	Preclinical	[1, 2]
	PKF115-584	Preclinical	[1, 2, 5]
	SAH-BCL9	Preclinical	[1, 2]

Table 23.3 Wnt-targeted monoclonal antibodies

Monoclonal antibodies that affect upstream Wnt signaling			
Target	Active drugs and sponsoring company/organization	Stage	Refs.
R-spondin3	OMP-131R10 (Rosmantuzumab. OncoMed Pharmaceuticals, Inc., USA)	Phase I	[1, 5]
		A study of OMP-131R10 in subjects with locally advanced or metastatic solid tumors	ClinicalTrials.gov identifier: NCT02482441
FZD 1,2,5,7,8	OMP-18R5 (Vantictumab. OncoMed Pharmaceuticals, Inc., USA)	Phase I	[1, 4, 5, 7]
		A study of vantictumab (OMP-18R5) in combination with paclitaxel in locally recurrent or metastatic breast cancer	ClinicalTrials.gov identifier: NCT01973309
FZD10	OTSA101 Centre Leon Berard, France OncoTherapy Science, Inc.	Phase I (terminated)	[1, 2]
		First-in-man study investigating the biodistribution, the safety, and the optimal recommended dose of a new radiolabelled monoclonal antibody targeting frizzled homolog 10 (SYNFRIZZ)	ClinicalTrials.gov identifier: NCT01469975
Wnt	OMP-54F28 (Ipafricept. OncoMed Pharmaceuticals, Inc., USA)	Phase I	[1, 4, 5, 7]
		Dose escalation study of OMP-54F28 in combination with sorafenib in patients with hepatocellular cancer	ClinicalTrials.gov identifier: NCT02069145

liver, and pancreatic cancer [6, 10, 11, 13]. Clinical trials are also investigating the use of antibodies as a potential therapy. For instance, the monoclonal antibody OMP-18R5 that targets five FZD receptors is in phase I clinical trials for solid tumors [13]. In addition, phase I clinical trials are ongoing for OMP-131R10, an anti-R-spondin3 antibody, for treatment of solid tumors and metastatic colorectal cancer [1]. The use of monoclonal antibodies to target Wnt signaling has shown promise, but further clinical investigation is needed to determine clinical relevance.

Addressing the Challenges in Targeting the Wnt Pathway

While great strides have been made in modulating Wnt signaling, the multifactorial regulation of the Wnt pathway complicates treatment. One issue that has arisen is the need for Wnt signaling in cellular development and during injury repair. The monoclonal antibody, OMP-18R5, was shown to reduce tumor xenograft growth, but data from clinical studies discovered off-target effects on skeletal composition. This adverse effect is likely due to the essential role of Wnt signaling in bone development [1]. Ectopic expression of Dkk1, the negative regulator of Wnt, also shows off-target effects in the intestine, where Wnt signaling is essential for crypt formation [9]. Yet, studies have demonstrated that acute loss of the Wnt-dependent LGR5+ stem cells in the intestine could be repopulated by a local quiescent stem cell population [9]. Therefore, potential side effects from Wnt perturbation could be potentially minimized by drug optimization.

Another confounding issue is the primary focus on Wnt pathway activation being the driver in cancer development. However, as mentioned above, studies have shown a reduction in Wnt signaling can promote tumorigenesis. Elevated nuclear β-catenin, a biomarker of elevated Wnt signaling, has been correlated with better patient prognosis in malignant melanoma, prostate, and ovarian cancer [7]. Likewise, overexpression of WNT3A is associated with less aggressive melanoma [4]. This suggests that the effects of Wnt signaling aberrations are likely cancer subtype and stage specific. The question of whether it would be more therapeutically beneficial to antagonize or agonize Wnt signaling has been a topic of recent debates. A study by Duffy et al. showed a bidirectional vulnerability of cancer cells, including neuroblastoma, melanoma, and colorectal cancers within a single tumor. By targeting the Wnt pathway for either activation or inhibition within cancer cells, they found both methods could be a potential therapeutic approach [14]. This study demonstrates the clinical importance of targeting the Wnt/β-catenin pathway, despite the complex signaling within cancer cells. Current studies are trying to determine which cancer subtypes and stages are best candidates for Wnt inhibition or activation.

The cross talk with other signaling pathways also demonstrates an obstacle in targeting Wnt signaling. While perturbing Wnt as a combinational therapy has shown promise, both activating and inhibiting Wnt can have beneficial effects. For instance, targeting Wnt signaling has also been shown to increase chemosensitivity in tumor cells. In prostate cancer cells, inhibiting the pathway through overexpression of WIF1 increased sensitivity to paclitaxel; however, activation of Wnt signaling in melanoma can sensitize cells to BRAF-MAPK inhibitors [7].

Summary

In summary, targeting the Wnt signaling pathway as a therapeutic mechanism has potential clinical use, but further studies need to elucidate how Wnt targeting affects tumor progression. It has been argued that a threshold for Wnt signaling, which is cancer and stage specific, exists where changes in either direction can be detrimental to the cell [14]. Understanding this threshold, and further acquisition of this information for individual patients, would facilitate understanding whether inhibition or activation of Wnt signaling would be more efficacious. Within the last 20 years, Wnt signaling has become a potential therapeutic target, and many clinical trials using Wnt inhibitors or activators are currently in progress. To utilize these Wnt modulators to improve patient prognosis, a biomarker needs to be identified to determine which direction, and to what level, Wnt signaling needs to be altered. The future practice of Wnt-targeted therapy is possible, but further research is needed to optimize this treatment.

References

1. Zhan T, Rindtorff N, Boutros M. Wnt signaling in cancer. Oncogene. 2017;36(11):1461–73.
2. de Lau W, Peng WC, Gros P, Clevers H. The R-spondin/Lgr5/Rnf43 module: regulator of Wnt signal strength. Genes Dev. 2014;28(4):305–16.
3. Karimaian A, Majidinia M, Bannazadeh Baghi H, Yousefi B. The crosstalk between Wnt/β-catenin signaling pathway with DNA damage response and oxidative stress: implications in cancer therapy. DNA Repair. 2017;51:14–9.
4. Kahn M. Can we safely target the WNT pathway? Nat Rev Drug Discov. 2014;13(7):513–32.
5. Mohammed MK, Shao C, Wang J, Wei Q, Wang X, Collier Z, et al. Wnt/β-catenin signaling plays an ever-expanding role in stem cell self-renewal, tumorigenesis and cancer chemoresistance. Genes Dis. 2016;3(1):11–40.
6. Blagodatski A, Poteryaev D, Katanaev VL. Targeting the Wnt pathways for therapies. Mol Cell Ther. 2014;2:28.
7. Anastas JN, Moon RT. WNT signalling pathways as therapeutic targets in cancer. Nat Rev Cancer. 2013;13(1):11–26.
8. Prosperi JR, Luu HH, Goss KH. Dysregulation of the Wnt pathway in solid tumors. In: Goss KH, Kahn M, editors. Targeting the Wnt pathway in cancer. New York: Springer; 2011. p. 81–128. Available from: http://link.springer.com/chapter/10.1007/978-1-4419-8023-6_5.
9. Tran FH, Zheng JJ. Modulating the wnt signaling pathway with small molecules. Protein Sci. 2017;26(4):650–61.
10. Katoh M, Katoh M. Molecular genetics and targeted therapy of WNT-related human diseases (review). Int J Mol Med. 2017;40(3):587–606.
11. Polakis P. Wnt signaling in cancer. Cold Spring Harb Perspect Biol. 2012;4(5):a008052.
12. Zhang X, Hao J. Development of anticancer agents targeting the Wnt/β-catenin signaling. Am J Cancer Res. 2015;5(8):2344–60.
13. Katoh M. Canonical and non-canonical WNT signaling in cancer stem cells and their niches: cellular heterogeneity, omics reprogramming, targeted therapy and tumor plasticity (review). Int J Oncol. 2017;51(5):1357–69.
14. Duffy DJ, Krstic A, Schwarzl T, Halasz M, Iljin K, Fey D, et al. Wnt signalling is a bi-directional vulnerability of cancer cells. Oncotarget. 2016;7(37):60310–31.

Hedgehog Signaling in Carcinogenesis

24

Victor T. G. Lin, Tshering D. Lama-Sherpa, and Lalita A. Shevde

Introduction

Hedgehog (Hh) signaling serves a central role in embryogenesis and tissue homeostasis. The Hh pathway was initially defined in *Drosophila*, where it was shown to play an important part in polarity and patterning of body segments in larval development. The name of the pathway derives from the characteristic short and hairy appearance of Hh-mutated *Drosophila* larvae, which resemble hedgehogs. While the signaling pathway in invertebrates is distinct from the mammalian one, they both consist of complex networks of proteins that ultimately drive the nuclear translocation of zinc finger transcription factors, promoting expression of specific target genes. Vertebrate Hh signaling is evolutionarily conserved and promotes tissue remodeling, patterning, differentiation, and vascularization. In development, Hh signaling is precisely controlled, and normal patterning relies on differences in ligand concentration and duration of exposure. However, abnormal engagement of the Hh pathway has also been implicated in tumor initiation and progression. Cancer is known to recapitulate normal development, and neoplastic cells co-opt ontogenic pathways such as Hh signaling in order to promote growth, metastasis, and resistance to therapy [1]. Aberrant activation of Hh signaling has been demonstrated in a wide variety of malignancies, including basal cell carcinoma (BCC), medulloblastoma, breast cancer, pancreatic ductal adenocarcinoma, glioma, rhabdomyosarcoma, ovarian cancer, and hepatocellular carcinoma, underlining its importance in cancer biology. Accordingly, there has been a great interest in targeting Hh signaling to treat cancer. Here, we introduce our current understanding of the Hh pathway, discuss its role in carcinogenesis, and examine the current state and future potential of Hh-directed therapies in the clinic.

An Overview of Hedgehog Signaling

Much of our current understanding of the Hh signaling pathway stems originally from studies in *Drosophila*. While there are evolutionarily conserved elements, key differences exist between the invertebrate and mammalian Hh pathways. For instance, in contrast to the invertebrate Hh pathway, the primary cilium plays an important role in the mammalian Hh pathway [2]. These specifics are beyond the scope of this text, which will focus only on vertebrate signaling (Fig. 24.1) and its relevance to human disease.

V. T. G. Lin
Division of Hematology and Oncology, Department of Medicine, The University of Alabama at Birmingham, Birmingham, AL, USA

T. D. Lama-Sherpa · L. A. Shevde (✉)
Department of Pathology, The University of Alabama at Birmingham, Birmingham, AL, USA
e-mail: lsamant@uab.edu

S. Badve, G. L. Kumar (eds.), *Predictive Biomarkers in Oncology*,
https://doi.org/10.1007/978-3-319-95228-4_24

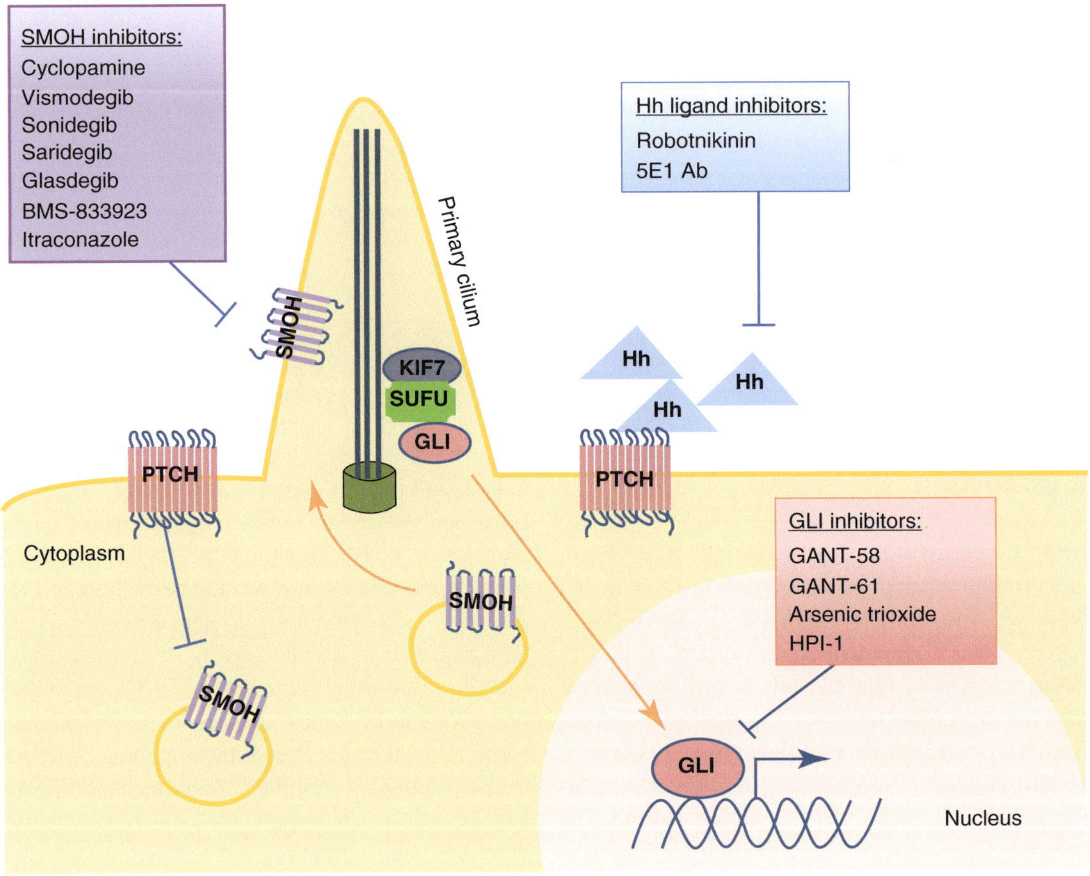

Fig. 24.1 Schematic of the classical, canonical vertebrate Hh signaling pathway. In the absence of Hh ligand, PTCH represses SMOH, which is sequestered in vesicles. After binding of Hh ligand to PTCH, SMOH translocates to the primary cilium. This releases GLI from a repressor complex that includes SUFU and KIF7. GLI is then able to translocate to the nucleus and function as a transcriptional activator of target genes. Inhibitors of this pathway can work at the level of Hh ligands, SMOH, and GLI

The ***classical**, ligand-initiated* Hh pathway is activated by the binding of Hh ligands—sonic hedgehog (SHH), desert hedgehog (DHH), and Indian hedgehog (IHH)—to the transmembrane protein receptor Patched (PTCH). PTCH spans the membrane 12 times and, in the absence of Hh ligand, constitutively represses vesicle-bound Smoothened (SMOH), a G-protein-coupled signal transduction molecule. PTCH forms a co-receptor complex with cell adhesion molecule (CAM) related/downregulated by oncogenes (CDO), brother of CDO (BOC), and growth arrest-specific 1 (GAS1). Ligand binding to this multimolecular co-receptor complex leads to the internalization of PTCH and consequently relieves its inhibitory effect on SMOH, which is then able to translocate to the primary cilium. This ultimately starts a signaling cascade that results in the release of *glioma-associated oncogene* (GLI) transcription factors from a repressor complex that includes suppressor of Fused (SUFU), kinesin family member 7 (KIF7), protein kinase A (PKA), glycogen synthase kinase 3β (GSK3β), and casein kinase 1 (CK1). GLI transcription factors are then free to translocate to the nucleus where they promote target genes. In the absence of Hh ligand binding, this macromolecular complex is sequestered at the microtubules of the primary cilium [3]. The Hh pathway can also be activated in a ***nonclassical**, Hh ligand-independent* manner. Signaling mediated by tumor-associated cytokines such as osteopontin (OPN), transforming growth factor β (TGF-β), platelet-derived growth factor (PDGF), and stromal-derived factor 1 (SDF-1) via their respective receptors can potentiate GLI activity independent of Hh ligands and SMOH [4].

Canonical, *GLI-mediated* Hh signaling requires GLI-initiated transcription of different genes, particularly those involved in cellular differentiation, stem cell maintenance, and tissue development. There are three GLI family members: GLI1, GLI2, and GLI3. Full-length GLI proteins act as transcriptional activators of downstream targets, and truncation via proteasomal processing changes their function from activation to repression. GLI2 and GLI3 carry a repressor domain in the N-terminus that GLI1 lacks, and thus GLI1 is thought to function only as a transactivator. Conversion of GLI2 or GLI3 from full-length activator to truncated transcriptional repressor involves phosphorylation catalyzed by PKA, GSK3β, and CK1, followed by ubiquitination by the Skp/Cullin/F-box (SCF)-βTrCP E3 ligase, resulting in proteasomal processing to remove the C-terminal transactivation domain [5]. Hh signaling can also proceed via the *GLI-independent,* ***noncanonical*** pathways, with or without direct SMOH involvement. SMOH-independent noncanonical Hh signaling leads to cell proliferation and survival, whereas SMOH-dependent signaling modulates intracellular Ca^{2+} balance and the actin cytoskeleton through activation of the Rac small GTPase [4].

Given the complex nature of Hh signaling, it is not surprising that a variety of inhibitors have been developed to study these pathways. Some agents act at the level of Hh ligands, such as the small-molecule inhibitor robotnikinin and 5E1, a monoclonal antibody directed against SHH. SMOH inhibitors represent the largest class of Hh inhibitors. These include cyclopamine, a natural compound derived from wild corn lilies, and its synthetic small-molecule derivatives, such as vismodegib and sonidegib (formerly known as erismodegib). Finally, there are direct GLI antagonists, which include GLI antagonists 58 and 61 (GANT-58 and GANT-61) and Hh pathway inhibitor 1 (HPI-1). In addition to the agents listed above, several drugs approved for other indications have also been identified as Hh inhibitors. These include the antifungal itraconazole, which inhibits SMOH through a site distinct from the cyclopamine derivatives, and arsenic trioxide, an agent used to induce differentiation in acute promyelocytic leukemia (APL) that has also been shown to inhibit Hh signaling at the level of GLI [6].

Hedgehog Signaling in Carcinogenesis

The most compelling demonstration of the importance of Hh signaling in carcinogenesis is the nevoid basal cell carcinoma syndrome (NBCCS), also known as basal cell nevus syndrome (BCNS) or Gorlin syndrome. This is an inherited cancer predisposition disorder that typically results from loss of function mutations in *PTCH1*, the gene encoding PTCH. In the absence of PTCH function, tonic repression of SMOH is relieved, and ligand-independent downstream signaling proceeds unabated. Patients with NBCCS are particularly susceptible to developing BCCs and medulloblastomas. As discussed in more detail later, targeting Hh signaling in these two tumor types has yielded the most encouraging clinical results.

Subsequently, multiple unrelated groups of patients meeting clinical criteria for NBCCS, but who lacked the expected mutation in *PTCH1*, underwent exome sequencing of lymphocyte DNA in an effort to identify other causative mutations. In this fashion, mutations in *SUFU* were also determined to cause NBCCS [7]. As SUFU acts to negatively regulate GLI function, loss of function mutations in *SUFU* would be predicted to similarly result in constitutive Hh pathway activation autonomous of ligand binding. However, because SUFU acts downstream of SMOH, it is expected tumors arising in SUFU-related NBCCS would be refractory to SMOH inhibitors, unlike PTCH-related NBCCS.

Aside from BCC and certain medulloblastomas, Hh signaling has also been implicated in a wide variety of cancers, including glioma, lymphoma, multiple myeloma, and carcinomas of the breast, colon, ovaries, pancreas, and prostate. The possible routes of activation are equally varied. In addition to mutations of proteins that result in ligand-independent pathway activation as described above, aberrant Hh signaling can also result from the overexpression of Hh ligands, either by the cancer cells themselves in an autocrine loop or from the surrounding stroma via paracrine signaling. Finally, nonclassical activation can arise from dysregulation of other pathways involved in carcinogenesis, including K-Ras, NF-κB, mTOR/S6K, and c-Jun [4].

Functionally, inappropriate Hh activation has been linked to multiple hallmarks of cancer [3]. It has been shown to drive cell proliferation through upregulation of Myc and cyclin D1 and immortalizes cancer cells by upregulating human telomerase reverse transcriptase (hTERT). Hh signaling can also allow cancer cells to resist apoptosis by upregulating the antiapoptotic protein B-cell lymphoma 2 (Bcl-2) and downregulating Bcl-2-associated death promoter (BAD). The Hh pathway potentially mediates immune evasion as well, with evidence suggesting that Hh inhibition enhances antigen presentation by increasing the expression of major histocompatibility complex (MHC) class I proteins in cancer cells and enhancing the number of tumor infiltrating lymphocytes (TILs). Finally, Hh activation can drive tumor invasion and metastasis by triggering the epithelial-to-mesenchymal transition (EMT) while stimulating angiogenesis through expression of vascular endothelial growth factor (VEGF) and fibroblast growth factor (FGF). It can also predispose to osseous metastases by preparing the metastatic niche through the expression of osteoclast-promoting cytokines, including receptor activator of NF-κB ligand (RANKL) and OPN.

Targeting Hedgehog in the Clinic

While there is a wide variety of Hh inhibitors used in scientific studies, the clinically relevant armamentarium is limited. At present, SMOH inhibitors are the only class of agents specifically targeting Hh signaling that are available for use in the clinic. There are two FDA-approved drugs on the market, vismodegib and sonidegib. Currently, the only labeled indication for Hh-directed therapy is in non-resectable BCC. The ideal treatment of BCC is with local therapies: resection with adjuvant radiation therapy if there are any positive surgical margins that cannot be re-excised. However, on the basis of the ERIVANCE, STEVIE, and BOLT trials, SMOH inhibitors are now an option in cases where local therapies are not possible, either because of unresectable lesions or metastasis. Recent clinical trial data also indicate SMOH inhibitors can reduce the frequency of recurrent BCCs in patients with NBCCS [8].

Although there is an abundance of data in the preclinical setting demonstrating the importance of Hh signaling in a variety of different cancer types, clinical trials assessing the addition of SMOH inhibitors to standard treatment in advanced colon cancer, small cell lung cancer, and pancreatic cancer have all yielded negative results. Clinical trials have also been negative for the use of SMOH inhibitors in the maintenance setting for ovarian cancer in second or third complete remission [9] and for unselected recurrent medulloblastomas [10]. However, while these last two clinical trials yielded negative results, they do illustrate important concepts critical for the future success of Hh-directed therapy in cancer treatment.

In the CONSORT trial, the intention was to use vismodegib maintenance to disrupt tumor-stroma interactions to prolong the progression-free interval in ovarian cancer in complete remission after prior relapse. However, the expression of Hh ligands—SHH and IHH, as measured by quantitative RT-PCR—was lower in comparison to prior studies of banked ovarian cancer tissue, suggesting that the eligibility criteria for this trial may have selected out the population likely to benefit from this targeted treatment approach [9]. In the PBTC-025B and PBTC-032 studies, patients with recurrent medulloblastoma were treated with vismodegib. When patients were stratified into SHH-driven and non-SHH-driven tumors by IHC, it was clear that the responses were confined to the SHH-driven tumors, as would be expected.

Interestingly, further analysis indicated that SHH-driven tumors that had concurrent strong diffuse staining of p53, which is associated with dominant-negative mutations of the p53 DNA-binding domain, were also insensitive to vismodegib [10]. Taken in combination, these data suggest that careful selection of patients will be a primary determinant for the clinical success of Hh-directed therapy, not only to identify those with tumors driven by the Hh pathway but also potentially to stratify their likelihood of response based on other factors, such as p53 status.

The Future of Hedgehog-Targeted Therapies

Going forward, the viability of Hh targeting as a cancer treatment strategy hinges on several critical factors. First and foremost is the development of other Hh inhibitors. As described above, Hh signaling is comprised of a collection of complex pathways that can be either dependent on or independent of Hh ligands, SMOH, and GLI. While a diverse collection of inhibitors is available in the laboratory setting, only SMOH-dependent pathways can be targeted in the clinic. Because of this significant limitation, there are two major challenges to overcome. First, mutations of SMOH that abrogate binding to cyclopamine-derived SMOH inhibitors have already been identified and render all currently available agents ineffective. Secondly, SMOH inhibitors will not be useful in cases where GLI is activated independent of SMOH. Development of direct GLI inhibitors in particular will help to bridge the gap that currently exists between the laboratory and the clinic. Alternatively, to bypass the long development cycle required for new drugs, another approach has been to evaluate drugs already approved for other indications that also function as Hh inhibitors. While preclinical testing suggests that itraconazole and arsenic trioxide can both be used to target Hh signaling and to bypass resistance mutations, these findings require confirmation in ongoing clinical trials before they can be used to treat patients [6].

Based on the available evidence, the future success of Hh-directed therapies will clearly be dependent on the appropriate selection of patients. Thus, another priority going forward must be the identification of candidates most likely to benefit from Hh-targeted therapies. Accordingly, the development and validation of biomarkers for Hh-driven cancers are of paramount importance (Table 24.1).

As the capability to profile cancers in the clinic using advanced molecular testing and liquid biopsies becomes more prevalent, it is also critical to determine the utility of Hh-directed therapy in tumors determined to be driven by Hh (Table 24.2). Because of this, ongoing large-scale precision oncology trials, such as NCI-MATCH and ASCO's TAPUR, may hold the key. These studies will screen large numbers of patients for Hh-driven cancers based on the specific tumor genomic profile and funnel these patients to Hh-directed therapies, unlike prior trials which were either unselected or too small to have a significant cohort of Hh-driven cancers. They may also help to identify other malignancies in which a role for Hh signaling has not yet been demonstrated. With the recent tumor agnostic approval of the immune checkpoint inhibitor pembrolizumab, there are likely to be more drug approvals based on specific tumor biology irrespective of cancer type. While Hh-targeted therapy is currently only approved for BCC, the hope is that these large-scale precision oncology trials will eventually justify a similar expansion of the role of Hh inhibitors.

Table 24.1 Examples of potential predictive biomarkers for the Hh signaling pathway

Hh signaling pathway	Cancer type	Affected biomarker	Method of detection	Target	Active drugs/company
Loss of PTCH1 function	BCC	PTCH1	Not routinely tested	SMOH	Vismodegib (Erivedge®) Genentech/Roche
					Sonidegib (Odomzo®) Novartis
					[1]
Loss of PTCH1 function	NBCCS	PTCH1	Molecular genetic testing	SMOH	Vismodegib (Erivedge®) Genentech/Roche [8]
Loss of SUFU function	NBCCS	SUFU	Molecular genetic testing	SUFU	None [7]
PTCH1 deletion	Medulloblastoma	GAB1*	IHC	SMOH	Vismodegib (Erivedge®) Genentech/Roche [10]
PTCH1 deletion	Medulloblastoma	PTCH1	FISH	SMOH	Vismodegib (Erivedge®) Genentech/Roche [10]
SHH/IHH amplification	Ovarian cancer	SHH/IHH	qRT-PCR	SHH/IHH	None [9]
PTCH1 deletion or inactivating mutations	Advanced solid tumors	PTCH1	NGS	SMOH	Vismodegib (Erivedge®) Genentech/Roche (Currently being evaluated in active clinical trials, such as ASCO TAPUR)

*Cytosolic expression of GRB2-associated-binding protein 1 (GAB1) has been identified as a surrogate for *PTCH1* deletion [11]
SUFU suppressor of Fused, *PTCH1* Patched 1, *SHH* sonic hedgehog, *IHH* Indian hedgehog, *SMOH* Smoothened

Summary

A large body of evidence has demonstrated the importance of Hh signaling in a diverse range of cellular processes involved in carcinogenesis and tumor progression. Numerous preclinical studies have shown that inhibition of Hh signaling is an effective strategy to target cancer cells. Unfortunately, with a few notable exceptions, preclinical successes have not yet translated to clinical efficacy. However, this seems likely to be the sequelae of suboptimal patient selection and a limited arsenal of Hh inhibitors available for use in human subjects. With recent improvements in technology that allow the heretofore unprecedented ability to rapidly and cost-efficiently analyze the genomic profiles of individual tumors, more precise identification of patients likely to benefit from Hh-targeted therapy will soon be possible. Furthermore, as new agents targeting different elements of the Hh pathway become available, advanced molecular testing may also help guide treatment choices, allowing clinical oncologists to stratify patients into subpopulations likely to benefit from specific inhibitors. While the enthusiasm for Hh-directed therapies has been somewhat dampened due to a number of negative clinical trials to date, with continuing technological advances in precision oncology, there is now reason for renewed optimism.

Table 24.2 Examples of ongoing trials of Hh inhibitors in cancer

Hh inhibitor	Disease	ClinicalTrials.gov identifier / Sponsor	Biomarker-selected
Arsenic trioxide and itraconazole	Basal cell carcinoma	NCT02699723 Stanford University, Palo Alto, CA, USA	No
Glasdegib (PF-04449913)	AML	NCT01841333 University of Colorado, Denver, USA	No
Glasdegib (PF-04449913)	AML	NCT02038777 Pfizer, USA	No
Glasdegib (PF-04449913)	Myelofibrosis	NCT02226172 Pfizer, USA	No
Glasdegib (PF-04449913)	MDS and AML	NCT02367456 Pfizer, USA	No
Sonidegib (Odomzo®)	Multiple myeloma	NCT02086552 Mayo Clinic	No
Vismodegib (Erivedge®)	Medulloblastoma	NCT01601184 Centre Leon Berard, France	Yes
Vismodegib (Erivedge®)	Medulloblastoma	NCT01878617 St. Jude Children's Research Hospital, Memphis, TN, USA	Yes
Vismodegib (Erivedge®)	AML	NCT02073838 Jewish General Hospital McGill University, Montreal, Canada	No
Vismodegib (Erivedge®)	Advanced solid tumors	NCT02091141 Genentech, Inc., USA	Yes
Vismodegib (Erivedge®)	Advanced solid tumors	NCT02465060 (NCI-MATCH) National Cancer Institute (NCI), USA	Yes
Vismodegib (Erivedge®)	Meningioma	NCT02523014 Alliance for Clinical Trials in Oncology	No
Vismodegib (Erivedge®)	Myelofibrosis	NCT02593760 Hoffmann-La Roche	No
Vismodegib (Erivedge®)	Advanced solid tumors	NCT02693535 (ASCO's TAPUR) American Society of Clinical Oncology	Yes
Vismodegib (Erivedge®)	Triple-negative breast cancer	NCT02694224 Clinica Universidad de Navarra, Universidad de Navarra, Madrid, Spain	No
Vismodegib (Erivedge®)	Urothelial carcinoma	NCT02788201 National Cancer Institute (NCI)	Yes
Vismodegib (Erivedge®)	Advanced solid tumors	NCT02925234 The Netherlands Cancer Institute, The Netherlands	Yes
Vismodegib (Erivedge®)	Advanced gastric adenocarcinoma	NCT03052478 Samsung Medical Center, S. Korea	Yes
Vismodegib (Erivedge®)	Glioblastoma	NCT03158389 University Hospital Heidelberg, Germany	Yes
Vismodegib (Erivedge®)	Advanced solid tumors	NCT03297606 Canadian Cancer Trials Group, Canada	Yes

AML acute myeloid leukemia, *MDS* myelodysplastic syndrome, *BCC* basal cell carcinoma

References

1. Lin VTG, Pruitt HC, Samant RS, Shevde LA. Developing cures: targeting ontogenesis in cancer. Trends Cancer. 2017;3(2):126–36.
2. Briscoe J, Therond PP. The mechanisms of hedgehog signalling and its roles in development and disease. Nat Rev Mol Cell Biol. 2013;14(7):416–29.
3. Hanna A, Shevde LA. Hedgehog signaling: modulation of cancer properties and tumor microenvironment. Mol Cancer. 2016;15:24.
4. Shevde LA, Samant RS. Nonclassical hedgehog-GLI signaling and its clinical implications. Int J Cancer. 2014;135(1):1–6.
5. Pan Y, Wang B. A novel protein-processing domain in Gli2 and Gli3 differentially blocks complete protein degradation by the proteasome. J Biol Chem. 2007;282(15):10846–52.
6. Kim J, Aftab BT, Tang JY, Kim D, Lee AH, Rezaee M, et al. Itraconazole and arsenic trioxide inhibit hedgehog pathway activation and tumor growth associated with acquired resistance to smoothened antagonists. Cancer Cell. 2013;23(1):23–34.
7. Smith MJ, Beetz C, Williams SG, Bhaskar SS, O'Sullivan J, Anderson B, et al. Germline mutations in SUFU cause Gorlin syndrome-associated childhood medulloblastoma and redefine the risk associated with PTCH1 mutations. J Clin Oncol. 2014;32(36):4155–61.
8. Tang JY, Ally MS, Chanana AM, Mackay-Wiggan JM, Aszterbaum M, Lindgren JA, et al. Inhibition of the hedgehog pathway in patients with basal-cell nevus syndrome: final results from the multicentre, randomised, double-blind, placebo-controlled, phase 2 trial. Lancet Oncol. 2016;17(12):1720–31.
9. Kaye SB, Fehrenbacher L, Holloway R, Amit A, Karlan B, Slomovitz B, et al. A phase II, randomized, placebo-controlled study of vismodegib as maintenance therapy in patients with ovarian cancer in second or third complete remission. Clin Cancer Res. 2012;18(23):6509–18.
10. Robinson GW, Orr BA, Wu G, Gururangan S, Lin T, Qaddoumi I, et al. Vismodegib exerts targeted efficacy against recurrent sonic hedgehog-subgroup medulloblastoma: results from phase II pediatric brain tumor consortium studies PBTC-025B and PBTC-032. J Clin Oncol. 2015;33(24):2646–54.
11. Ellison DW, Dalton J, Kocak M, Nicholson SL, Fraga C, Neale G, et al. Medulloblastoma: clinicopathological correlates of SHH, WNT, and non-SHH/WNT molecular subgroups. Acta Neuropathol. 2011;121(3):381–96.

TGF-β and the SMAD Signaling Pathway in Carcinogenesis

25

Wendy Greenwood and Alejandra Bruna

Introduction

Transforming growth factor beta (TGF-β) is a member of the TGF-β superfamily of cytokines important in organ development and tissue homeostasis. The TGF-β signaling pathway was one of the first pathways controlling multicellular life to emerge with the appearance of the animal species [1]. TGF-β is a pleiotropic cytokine, being instrumental in the regulation of numerous cellular processes including cell proliferation, differentiation, apoptosis, mobility, invasion, angiogenesis, immune response, and extracellular matrix (ECM) production. Through its ability to activate specific transcription factors, TGF-β is known to regulate hundreds of TGF-β target genes, thereby influencing a variety of cellular processes [2]. Because of the crucial role of TGF-β in controlling cellular programs regulating proliferation, differentiation, and tissue regeneration and its importance in the evolution of early life, it comes as no surprise that diseases, including cancer, result from malfunctions of this pathway [3].

The TGF-β Signaling Pathway

In this chapter, we will focus on canonical TGF-β signaling through ligands TGF-β1, TGF-β2, and TGF-β3 and the intracellular SMAD proteins. For an excellent review on signaling via all TGF-β superfamily members, we recommend the article by Massague et al. [4].

The TGF-β signaling pathway is activated by the binding of the TGF-β ligand to a heterodimeric complex of type I and type II membrane-bound serine/threonine protein kinase receptors, TGF-βRI and TGF-βRII, respectively. Upon ligand binding, TGF-βRII phosphorylates the glycine- and serine-rich TGF-βRI GS domain creating a binding site for the phosphorylation of TGF-β-specific transcription factor SMADs (Fig. 25.1).

The designation SMAD is a portmanteau of the homologous proteins: *Drosophila* homolog mothers against decapentaplegic (MAD) and *Caenorhabditis elegans* protein SMA [8]. SMAD proteins are classified, depending on their structure and function, into three groups: (1) The receptor-regulated SMADs (R-SMAD), which contain a C-terminal Ser-Ser-X-Ser motif, are directly phosphorylated by type I receptor

W. Greenwood · A. Bruna (✉)
Cancer Research UK Cambridge Institute, University of Cambridge, Cambridge, Cambridgeshire, UK
e-mail: Alejandra.bruna@cruk.cam.ac.uk

S. Badve, G. L. Kumar (eds.), *Predictive Biomarkers in Oncology*,
https://doi.org/10.1007/978-3-319-95228-4_25

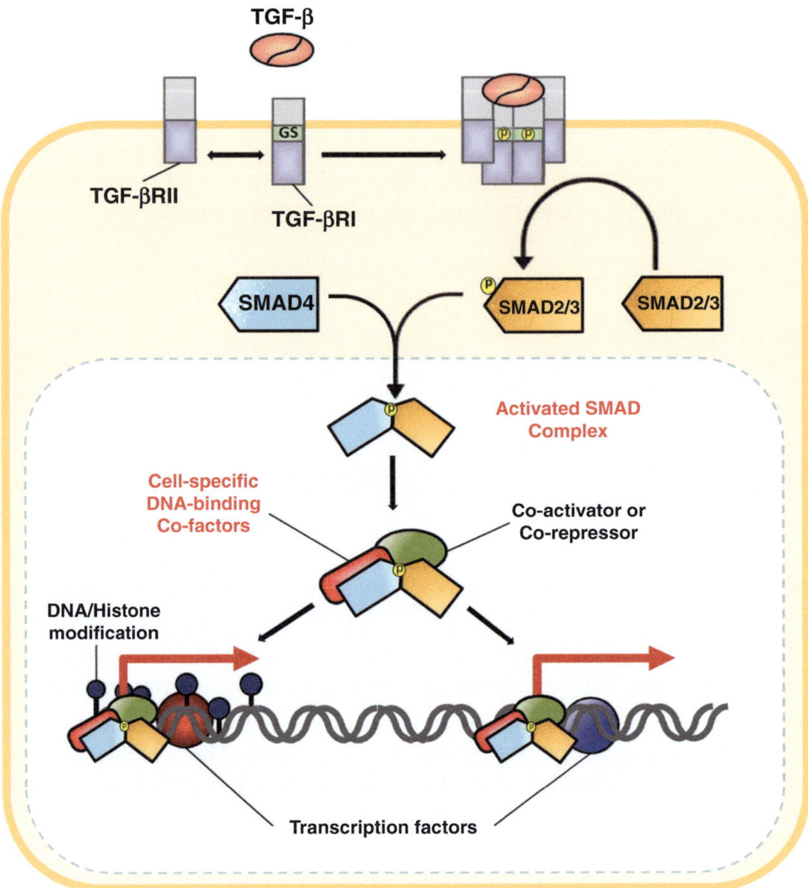

Fig. 25.1 A simplified schematic diagram of canonical TGF-β signaling. (Adapted from Siegel and Massague [5]). Upon ligand binding, TGF-βRII phosphorylates TGF-βRI at the GS domain. Activated TGF-βRI phosphorylates R-SMADs SMAD2/SMAD3 which complexes with SMAD4 and translocates to the nucleus. Here the complex interacts with DNA-binding cofactors to mediate the regulation of TGF-β-specific gene programs. This differential expression of TGF-β target gene programs is dependent on distinct cofactor binding, transcription factor binding, and DNA/histone modifications [6, 7]. Alternative binding of different transcription factors is here represented in red or blue

kinases. (2) The common mediator SMADs (co-SMAD), which complex with R-SMADs, translocate to the nucleus to mediate the regulation of TGF-β-specific target genes. (3) The inhibitory SMADs (I-SMADs) interfere with receptor activation of R-SMADs or complex formation with co-SMADs [9].

Structurally, SMAD proteins comprise two globular domains, MH1 and MH2, joined by a variable linker region. The amino-terminal MH1 domain acts as a DNA-binding region, and the C-terminal MH2 domain contains a series of hydrophobic regions, which mediate SMAD interactions with both cytoplasmic and nuclear pore proteins and with DNA-binding cofactors and signaling regulators [8]. The SMAD variable linker domain contains multiple phosphorylation sites for protein kinases including mitogen-activated protein kinases (MAPK) [10] and cyclin-dependent kinases (CDK), both of which are closely involved in the regulation of TGF-β/SMAD signaling [11]. These phosphorylation domains allow SMADs to act as integration hubs for multiple regulatory processes, for example, cell cycle progression and growth factor signaling.

Although the focus of this chapter is primarily on canonical TGF-β signaling through the TGF-β/SMAD pathway, TGF-β receptors have the ability to directly activate many other non-SMAD signaling pathways, designated as noncanonical TGF-β signaling. These noncanonical pathways include multiple components of the phosphatidylinositol 3-kinase (PI3K)/AKT, MAPK, and Rho-like GTPase signaling pathways [12].

In the instance of canonical TGF-β signaling, the TGF-β type I receptor phosphorylates the R-SMADs, SMAD2, and SMAD3, leading to their translocation to the nucleus and complex formation with the co-SMAD, SMAD4. Once in the nucleus, this complex recruits DNA-binding transcription factors resulting in transcriptional regulation and expression of TGF-β target genes. Signaling through TGF-β/SMADs is a tightly regulated process, and the SMAD protein complex only translocates to the nucleus, while TGF-β receptor activation is taking place. This localization is maintained through repeated phosphorylation/dephosphorylation and nuclear-cytoplasmic shuttling cycles. Following transcriptional activation, SMADs undergo rapid phosphorylation at the linker region by CDK8 and the cyclinT1/CDK9 complex. This phosphorylation primes the linker region for further phosphorylation by glycogen synthase kinase 3 (GSK3) and targets SMADs for polyubiquitylation and degradation in the proteasome [4].

TGF-β is crucially involved in the maintenance of tissue homeostasis. At the cellular level, TGF-β effects are well recognized for their pleiotropic nature. Embryonic stem (ES) cells and lineage-committed progenitors exhibit elevated levels of TGF-β signaling, as it functions to regulate pluripotency and differentiation. In mature epithelial, hematopoietic, and neural cells, TGF-β negatively regulates the cell cycle through a number of mechanisms, including the expression of the cyclin-dependent kinase inhibitors CDKN1A, CDKN1C, and CDKN2B and repression of the proto-oncogene MYC [5]. Because of the crucial role of TGF-β signaling in organ development and in maintaining tissue homeostasis, deregulation of the pathway is commonly seen in disease, including cancer.

TGF-β and Cancer

TGF-β signaling is strongly implicated in many aspects of cancer progression. The TGF-β dogma supports that TGF-β functions mainly as a potent tumor suppressor in early stages of cancer development through the potent cytostatic effects observed in the normal epithelium. Inactivation or loss of TGF-β pathway components in various cancer types strongly supports this hypothesis. For example, inactivating mutations in TGF-βRII are frequently observed in cancers associated with defects in the DNA mismatch repair system, such as gastric, colorectal, biliary, and lung adenocarcinoma. Loss of SMAD2 and SMAD4 is commonly observed in pancreatic and colon cancers due to mutation or loss of chromosome 18q [13], and CDKN2B loss is frequent in melanoma [4]. In sharp contrast, however, other cancers progress with an intact TGF-β/SMAD pathway. Here, evasion of tumor-suppressive TGF-β activity occurs downstream of SMAD signaling. In this context, most commonly seen in breast cancer, melanoma, and glioma, the TGF-β/SMAD signaling pathway remains intact, but tumors are no longer growth suppressed by TGF-β signaling. Consistently, increased TGF-β activity is a marker of malignancy in non-small cell lung carcinoma (NSCLC); colorectal, prostate, and gastric carcinoma; glioblastoma; and breast cancer [6, 13, 14].

Our work over the years in glioma and breast cancer supports opposing roles for TGF-β signaling in cancer, both tumor suppressive and tumor promoting. Additionally, our most recent work in the laboratory of Professor Caldas has provided further evidence that in breast cancer, the paradoxical role of TGF-β activity is not necessarily linked to the stage of breast cancer progression. We demonstrate that oncogenic TGF-β signaling occurs in a molecularly defined subset of breast cancer known as Claudin^{Low}. In Claudin^{Low} cells, TGF-β/SMADs converge with the Rho-MRTF-SRF pathway to transcriptionally regulate a stem cell-like gene expression program which is associated with worse outcome in a large breast cancer clinical cohort, irrespective of ER status [6, 7]. Other studies have failed to find a clear association between TGF-β-specific gene signatures and clinical outcomes. We hypothesize this is because they did not consider isolating TGF-β-specific gene signatures from the opposing tumor-promoting and tumor-suppressive branches.

The effects of TGF-β should not be taken in the context of the cell autonomous compartment alone; a significant number of TGF-β effects in cancer are mediated by the interaction between epithelial cells and the tumor microenvironment (stromal cells). Tumor stroma is a complex tissue comprising numerous cell types together with extracellular matrix proteins, all of which express TGF-β receptors and respond differently to TGF-β signaling in various manners. In contrast to the paradoxical role of TGF-β signaling in tumor epithelial cells, TGF-β signaling in tumor stromal cells mainly contributes toward promoting a favorable environment for tumor growth [14]. TGF-β signaling is also fundamental in immune regulation of the tumor microenvironment by inducing tumor immune evasion, allowing tumor cells to evade host immune surveillance. TGF-β signaling suppresses the host immune system by a number of mechanisms including the inhibition of M1 macrophages and expansion of tumor-promoting M2 macrophages, the suppression of cytotoxic T cells and NK cells, and an increase in the population of T-helper 2 cells, which possess humoral but no cytotoxic activity [5].

All of the above observations highlight the complex nature of TGF-β regulation and its role in cancer and the importance of fully understanding the pleiotropy of cellular response to TGF-β signaling.

Cancer Therapy

As discussed, the deregulation of TGF-β signaling is frequent in cancer and plays crucial roles in tumor initiation progression and metastasis. This has highlighted TGF-β signaling as an increasingly interesting target for drug development. However, because of the pleotropic effects of TGF-β signaling in normal epithelial cells and its paradoxical role in different cancer types, along with its complex roles in immune regulation, and the tumor microenvironment, TGF-β inhibition as a cancer therapy is not a straightforward prospect.

The development of anti-TGF-β therapy to date is focused in three main areas: (1) inhibition of TGF-β synthesis by antisense oligonucleotides either delivered intravenously or delivered as an allogeneic cancer cell vaccine via genetically engineered immune cells, (2) targeting TGF-β signaling at the ligand-receptor level using monoclonal antibodies or peptides, and (3) TGF-β receptor kinase inhibitors, which function to prevent signal transduction [15]. A list of current drugs in development can be seen in the table below adapted from Neuzillet et al. [15] (Table 25.1).

Although the prospect of TGF-β inhibition seems attractive considering the role of TGF-β tumor progression and the results of early drug trials appear promising, it is crucial to consider the opposing effect of TGF-β signaling in different tumor types to ensure patients are correctly stratified for treatment. Perhaps more importantly, current research should focus on identifying molecular markers of TGF-β tumor-promoting activity, which is crucial for appropriate patient stratification and treatment monitoring.

Table 25.1 TGF-β pathway inhibitor drugs and stage of clinical development

Compound/drug company	Biomarker/target	ClinicalTrials.gov identifier trial ID	Status
TGF-β ligand inhibitors			
Fresolimumab (GC1008) Sanofi/Aventis	TGF-β1, TGF-β2, TGF-β3	NCT00356460 NCT00923169 NCT01472731 NCT01112293 NCT01401062	Phase I study completed in RCC, melanoma, and glioma Phase II study completed in mesothelioma and breast cancer
Trabedersen (AP12009) Antisense Pharma	TGF-β2	NCT00844064 NCT00431561 NCT00761280	Phase I study in melanoma, pancreatic, and CRC completed Phase II study in glioma completed Phase III study in astrocytoma and secondary glioblastoma terminated
Lucanix (belagenpumatucel-L) NovaRx Corporation	TGF-β2	NCT01058785 NCT00676507	Phase II and phase III study in NSCLC completed
FANG™ vaccine (rhGMCSF/shRNAfurin) Gradalis	TGF-β1, TGF-β2	NCT01061840 NCT01309230 NCT01505166 NCT01453361	Phase I study in NSCLC and liver cancer ongoing Phase II study in melanoma, ovarian, and CRC ongoing
Disitertide (P144) Digna Biotech®	TGF-β1		Progress outside oncology Preclinical development in glioma
TGF-β receptor inhibitors			
Galunisertib (LY157299) Eli Lilly	TGF-βRI	NCT01246986 NCT01373164 NCT01220271 NCT02178358 NCT01582269	Phase I study in metastatic cancer completed Phase II study in HCC, glioma, and pancreatic cancer ongoing
LY3200882 Eli Lilly	TGF-βRI	NCT02937272	Phase I study recruiting
TEW-7197 MedPacto	TGF-βRI	NCT02160106	Phase I study in progress in solid tumors
PF-03446962 Pfizer	TGF-βRI	NCT00557856 NCT01337050 NCT01911273 NCT01486368 NCT01620970 NCT02116894	Phase I study in CRC completed Phase II study in mesothelioma and urothelial cancer completed Phase II study in HCC terminated Phase I study in combination with regorafenib in CRC completed
IMC-TR1 (LY3022859) Eli Lilly	TGF-βRII	NCT01646203	Phase I study in solid tumors completed

Adapted from Neuzillet et al. [15]. With permission from Elsevier
RCC renal cell carcinoma, *CRC* colorectal carcinoma, *NSCLC* non-small cell lung carcinoma, *HCC* hepatocellular carcinoma

References

1. Huminiecki L, et al. Emergence, development and diversification of the TGF-β signalling pathway within the animal kingdom. BMC Evol Biol. 2009;9:28.
2. Mullen AC, et al. Master transcription factors determine cell-type-specific responses to TGF-β signaling. Cell. 2011;147:565–76.
3. Massagué J. TGFβ in cancer. Cell. 2008;134:215–30.
4. Massagué J. TGFβ signalling in context. Nat Rev Mol Cell Biol. 2012;13:616–30.
5. Siegel PM, Massagué J. Cytostatic and apoptotic actions of TGF-β in homeostasis and cancer. Nat Rev Cancer. 2003;3:807–21.
6. Bruna, et al. TGFβ induces the formation of tumour-initiating cells in claudin^{low} breast cancer. Nat Commun. 2012;3:1055.

7. Tufegdzic Vidakovic A, et al. Context-specific effects of TGF-β/SMAD3 in cancer are modulated by the epigenome. Cell Rep. 2015;13:2480–90.
8. Massagué J, et al. Smad transcription factors. Genes Dev. 2005;19:2783–810.
9. Miyazono K. TGF-β signaling by Smad proteins. Cytokine Growth Factor Rev. 2000;11:15–22.
10. Kretzschmar M, et al. Opposing BMP and EGF signalling pathways converge on the TGF-β family mediator Smad1. Nature. 1997;389:618–22.
11. Matsuura I, et al. Cyclin-dependent kinases regulate the antiproliferative function of SMADs. Nature. 2004;430:226–31.
12. Zhang YE. Non-Smad pathways in TGF-beta signalling. Cell Res. 2009;19:128–39.
13. Padua D, Massagué J. Roles of TGFβ in metastasis. Cell Res. 2009;19:89–102.
14. Bruna, et al. High TGFbeta-Smad activity confers poor prognosis in glioma patients and promotes cell proliferation depending on the methylation of the PDGF-B gene. Cancer Cell. 2007;11(2): 147–60.
15. Neuzillet C, et al. Targeting the TGFβ pathway for cancer therapy. Pharmacol Ther. 2015;147:22–31.

Role of JAK-STAT Pathway in Cancer Signaling

26

Na Luo and Justin M. Balko

Pathway and Key Molecules

The *Ja*nus *k*inase/*s*ignal *t*ransducer and *a*ctivator of *t*ranscription (JAK-STAT) pathway plays a central role in cell proliferation, differentiation, survival, and developmental processes. The JAK-STAT pathway senses extracellular signals and responds in kind by controlling the expression of its target genes. In this regard, the JAK-STAT pathway transfers extracellular signals to the cell and, via a series of phosphorylation events, initiates the transcription of genes that are involved in cell proliferation and differentiation [1, 2].

The JAK-STAT pathway is primarily involved in cytokine signaling, such as the signaling induced by erythropoietin, thrombopoietin, interferons, interleukins, and granulocyte-colony-stimulating factors. JAKs initially interact with cytokine receptors in the inactive form. Upon ligand binding to the corresponding cytokine receptor, JAK transphosphorylation and activation are induced. Activated JAKs trigger a conformational change in the cognate cytokine receptor to provide a STAT docking site via the SH2 domain, thereby recruiting STAT family members to the JAK-cytokine receptor complex. After recruitment, the STATs become activated and form homo- or heterodimers. Once in dimeric form, STATs translocate to the nucleus and bind the promoter of target genes that participate in various cell processes as mentioned above [1, 2] (Fig. 26.1).

JAK family member-associated cytokine receptors are type I transmembrane proteins consisting of an extracellular domain, a transmembrane domain, and a cytoplasmic domain. The extracellular domain functions not only through cytokine recognition/binding but also in the aggregation of multiple subunits to form a receptor complex. The cytoplasmic domain functions in tyrosine phosphorylation and signal transduction. The cytokine receptor usually contains two different subunits. The α-subunit defines ligand specificity, whereas the β-subunit converts low-affinity α-chains into high-affinity receptors and is usually shared among related subfamilies of receptors.

The JAK family consists of four members (i.e., JAK1, JAK2, JAK3, and TYK2), each of which contains seven conserved *J*AK *h*omology domains (i.e., JH1–7) (Fig. 26.2a). The N-terminus of JAKs (JH5–7) constitutes a "4.1 protein, ezrin, radixin, moesin" (FERM) domain which functions

N. Luo
Department of Medicine, Vanderbilt-Ingram Cancer Center, Vanderbilt University, Nashville, TN, USA

Department of Anatomy and Histology, School of Medicine, Nankai University, Tianjin, China

J. M. Balko (✉)
Department of Medicine and Cancer Biology, Vanderbilt University Medical Center, Nashville, TN, USA
e-mail: Justin.balko@vanderbilt.edu

S. Badve, G. L. Kumar (eds.), *Predictive Biomarkers in Oncology*,
https://doi.org/10.1007/978-3-319-95228-4_26

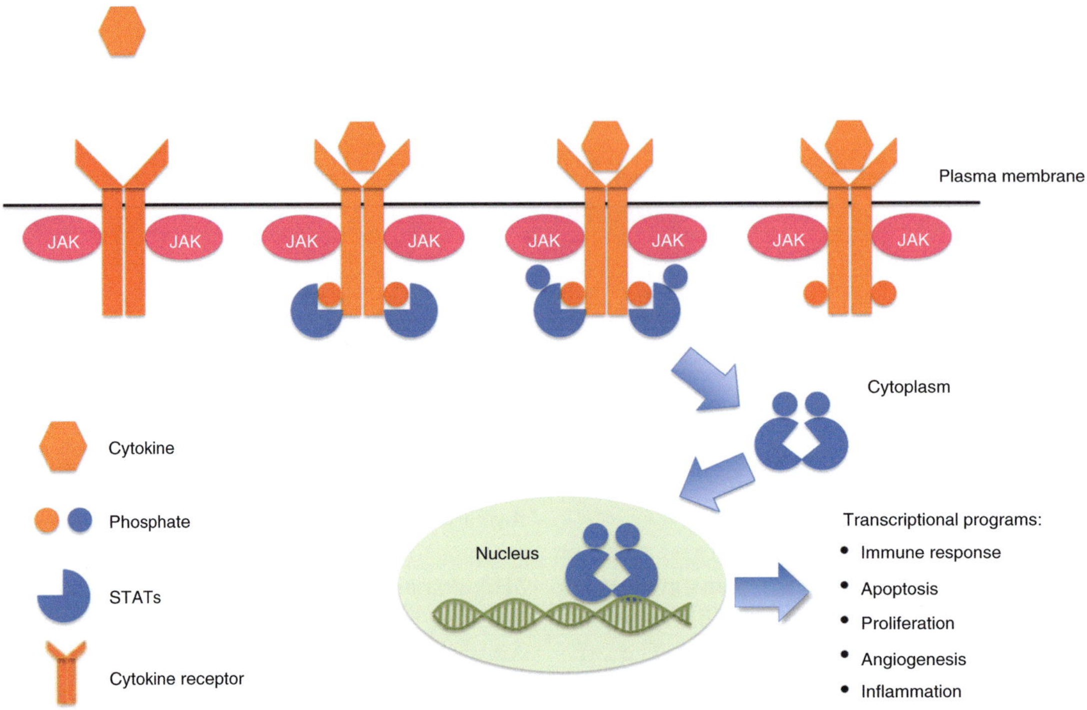

Fig. 26.1 The JAK-STAT signaling pathway. JAK family members initially interact with cytokine receptors in the inactive form. Upon ligand binding to the corresponding cytokine receptor, JAK transphosphorylation and activation are induced. The activated JAKs trigger the cytokine receptor to undergo a conformational change, exposing the SH2 domain thereby recruiting STATs to the JAK-cytokine receptor complex. After recruitment, the STATs are activated, inducing the formation of homo- or heterodimers. Dimeric STATs translocate to the nucleus and bind the promoter of cognate transcriptional targets to activate a variety of cell processes

in the association between JAKs and cytokine receptors as well as other kinases [3]. The JH3 and JH4 domains, which possess structural similarities with Src-homology-2 (SH2) domains, link the C-terminal and N-terminal domains and also aid in binding cognate cytokine receptors. The C-terminal of JAKs includes a JH1 kinase domain and a JH2 pseudo-kinase domain. The JH1 domain is essential for the kinase activity of JAKs and contains conserved tyrosine residues necessary for the activation of JAKs. Importantly, the JH2 domain has a predilection for gain-of-function mutations in cancer, which lead to constitutive activation of JAKs.

The STAT family consists of seven members (STAT1, STAT2, STAT3, STAT4, STAT5A, STAT5B, and STAT6). All STAT family members contain seven domains which include the oligomerization (OG) domain, coiled-coil (CC) domain, DNA-binding (DB) domain, linker (LK) domain, SH2 domain, phosphotyrosine tail (Y), and the transcriptional activation (TA) domain (Fig. 26.2b). The CC and SH2 domains act in the association between STATs and cytokine receptors. The TA, Y, SH2, and OG domains are involved in STAT dimerization. The DB and LK domains act in STAT binding to DNA. Finally, all domains (except the Y domain) participate either in STAT nuclear import or export [4]. The 3D ribbon structure of STAT1 dimers is shown in (Fig. 26.2c).

There are a number of negative regulators of the JAK-STAT pathway that suppress signaling at multiple levels which include (1) tyrosine phosphatases (SHP1 and SHP2) which dephosphorylate JAKs, (2) the *s*uppressors *o*f *c*ytokine *s*ignaling (SOCS) which compete with STAT binding to the cytokine receptor, and (3) the family of *p*rotein *i*nhibitors of *a*ctivated *S*TATs (PIAS) which interferes with STAT binding to DNA. Negative regulators of the JAK-STAT

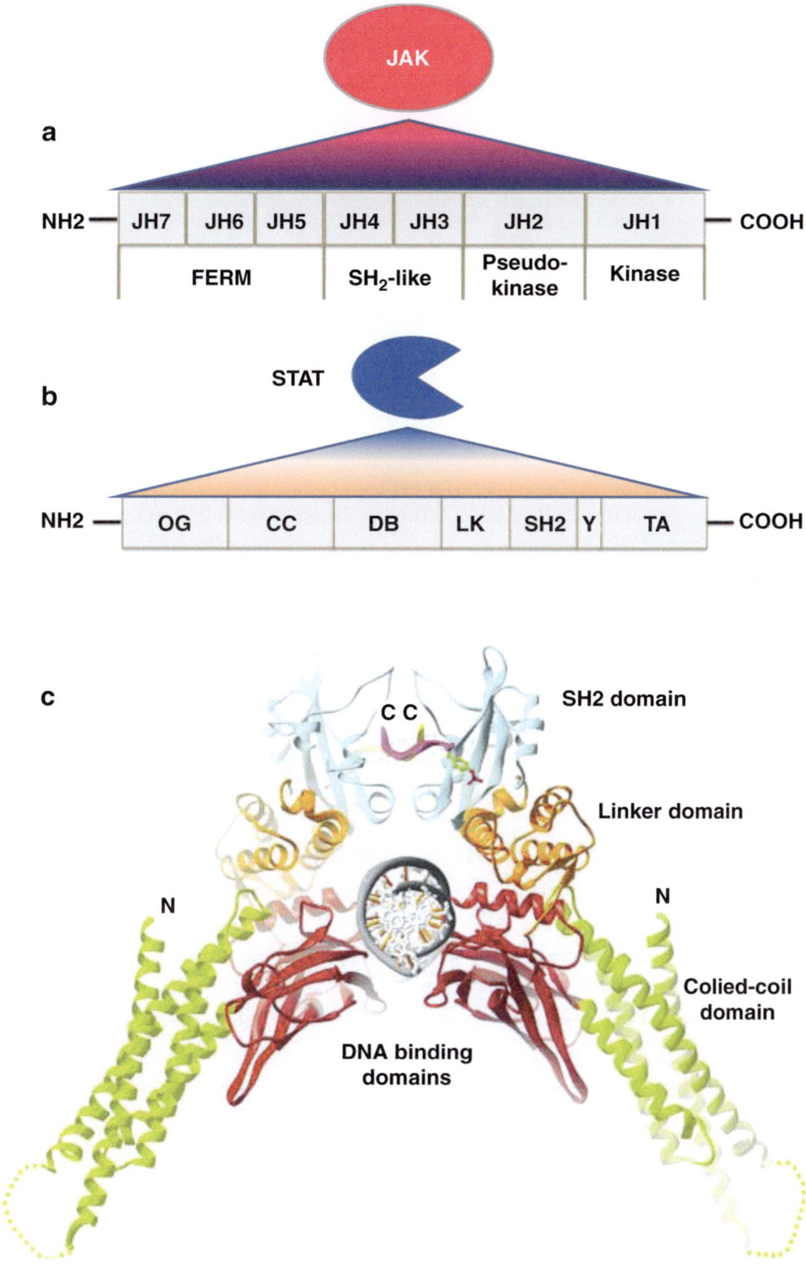

Fig. 26.2 JAK and STAT molecular structure. (**a**) The JAK family members (JAK1, JAK2, JAK3, and TYK2) contain seven conserved *J*AK *h*omology domains (JH1–7). The C-terminal of JAKs includes a JH1 kinase domain and a JH2 pseudo-kinase domain. The JH3 and JH4 domains function as a link between the C-terminal and N-terminal domains. In addition, the JH3 and JH4 domains possess structural SH2-like domain. The N-terminal of JAKs consists of 4.1 protein, ezrin, radixin, moesin (FERM) domain. (**b**) The STAT family members contain seven domains which include the oligomerization (OG) domain, coiled-coil (CC) domain, DNA-binding (DB) domain, linker (LK) domain, SH2 domain, phosphotyrosine tail (Y), and the transcriptional activation (TA) domain. (**c**) Ribbon diagram of the STAT-1 core dimer on DNA. The component domains are colored green (coiled-coil domain), red (DNA-binding domain), orange (linker domain), and cyan (SH2 domain). The tail segments are shown in magenta and yellow. Disordered loops (one in the coiled-coil domain and one connecting the SH2 domain to the tail segment) are shown as dotted lines. The phosphotyrosine residue is shown in a stick representation. The N- and C-termini of STAT-1 core are indicated by "N" and "C". (Reprinted from Chen et al. [11]. With permission from Elsevier)

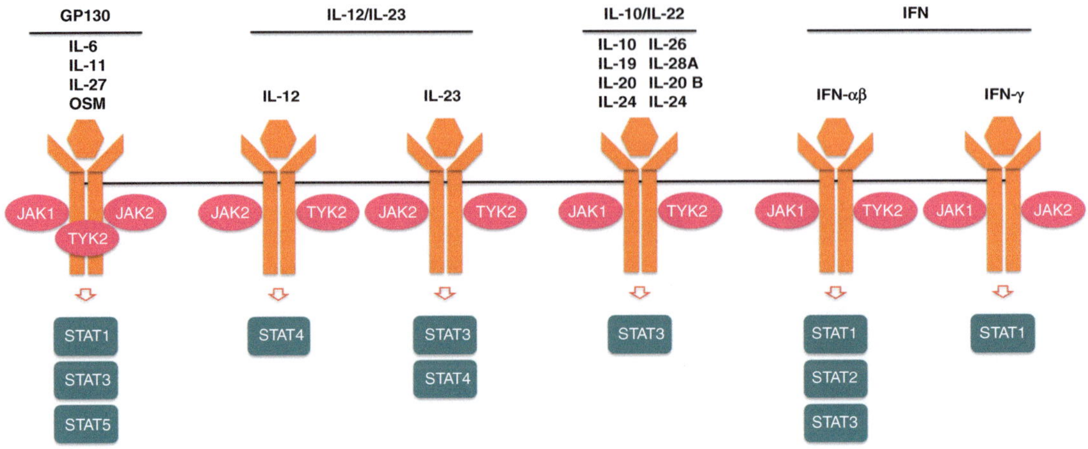

Fig. 26.3 Types of tyrosine receptors that associate with various combinations of JAK-STATs. (1) GP130 co-receptor family, (2) IL-12/IL-23 receptor family, (3) IL-10/IL-22 receptor family, and (4) IFN receptor family

pathway switch off the signaling cascade thereby inhibiting the JAK-STAT pathway, thereby regulating amplitude and temporal control of pathway signaling [2, 3].

There are four types of tyrosine receptors that associate with various combinations of JAK-STATs which include the (1) GP130 co-receptor family, (2) IL-12/IL-23 receptor family, (3) IL-10/IL-22 receptor family, and (4) interferon (IFN) receptor family [5] (Fig. 26.3).

1. The cytokines that employ the GP130 co-receptor family are the IL-6 family of cytokines (e.g., IL-6, IL-11, IL-27, OSM, and others). The JAK-STAT members involved in the GP130 co-receptor family signaling are JAK1, JAK2, and TYK2 along with downstream STAT1, STAT3, and STAT5 although to a lesser extent.
2. The cytokines that employ the IL-12/IL-23 receptor family are IL-12 and IL-23. The JAK-STAT members involved in the IL-12/IL-23 receptor family signaling are JAK2 and TYK2 along with downstream STAT4 for IL-12 versus STAT3 and STAT4 for IL-23.
3. The cytokines that employ the IL-10/IL-22 receptor family are IL-10, IL-19, IL-20, IL-24, IL-26, IL-28A, IL-28B, and IL-29. The JAK-STAT members involved in the IL-10/IL-22 receptor family signaling are JAK1 and TYK2 along with downstream STAT3 as the only transcription factor for signaling via the IL-10/IL-22 receptor family.
4. The cytokines that employ the IFN receptor family are IFNα/β and IFNγ. The JAK-STAT members involved in the IFN receptor family signaling are JAK1 and TYK2 along with downstream STAT1 and STAT2.

Signaling Pathways in Carcinogenesis

In general, JAKs typically interact with tyrosine receptors and remain in an inactive state until ligand binding occurs. The aberrant activation of the JAK-STAT pathway due to genetic mutations or polymorphisms leads to a persistent activation of JAKs in the absence of cytokine signaling, which may then result in tumorigenesis or oncogenic activity.

Role of the JAK-STAT Pathway in Hematological Malignancies

The JAK-STAT pathway plays a prominent role in hematological malignancies, particularly in myeloproliferative neoplasms (MPNs) which include polycythemia vera (PV), essential thrombocythemia (ET), and idiopathic myelofibrosis (IMF). MPNs are hallmarked by the enhanced proliferation of myeloid-derived cells (e.g., erythrocytes, platelets, and granulocytes). The occurrence of PV, ET, and IMF is almost always due to cytokine hypersensitivity or cytokine independence which results in an augmented or persistent activation of certain signaling pathways. In addition, the occurrence of PV, ET, and IMF is related to various mutations in the JAK-STAT pathway, of which the most notable alteration is JAK2 V617F. This mutation occurs in a majority of PV patients and in ~50% of ET and IMF patients and causes constitutive JAK2 activation and hypersensitivity leading to persistent activation of the JAK-STAT pathway. The JAK2 V617F mutation may also set up a competition between mutant JAK2 and wild-type JAK2 for cytokine receptor binding. Interestingly, the JAK1 V658F mutation and the TYK2 V678F mutation which are homologous to the JAK2 V617F mutation also lead to constitutively active proteins, suggesting that these alterations may likewise be involved in hematological malignancies [1, 6].

In addition, MPNs may occur in conjunction with less frequent mutations in the JAK-STAT pathway. These mutations have been identified by advanced next-generation sequencing techniques and can result in augmented activation of the JAK-STAT pathway. These mutations include [1, 6]:

1. A gain-of-function mutation in exon 12 of JAK2 covering the region F537–E543 that occurs in 3–5% of PV patients
2. A JAK2 T875N activation mutation that occurs in a megakaryoblastic myeloid leukemia cell line derived from a child with trisomy 21
3. A JAK2 ΔIREED activation mutation with a five-amino acid in-frame deletion that occurs in trisomy 21 and B-cell precursor acute lymphoblastic leukemia
4. JAK3 A572V, JAK3 V722I, and JAK3 P132T gain-of-function mutations that occur in trisomy 21 and acute megakaryoblastic myeloid leukemia

Role of the JAK-STAT Pathway in Solid Tumors

The noteworthy JAK2 V617F mutation associated with myeloproliferative neoplasms (MPNs) has not been found in solid tumors even though a wide range of solid tumors exhibit aberrant JAK-STAT signaling. However, the amplification of the 9p24 locus which includes the JAK2 gene has been identified in residual triple-negative breast cancers (TNBCs) after neoadjuvant chemotherapy using next-generation sequencing. The JAK2/9p24 locus amplification occurs at a higher rate in chemotherapy-treated TNBCs than in either untreated TNBCs, in basal-like breast cancers, or in other subtypes. This leads to a lower recurrence-free rate and a lower overall survival rate. Most importantly, the JAK2/9p24 locus amplification was selected during neoadjuvant chemotherapy and metastatic dissemination suggesting a role in chemotherapy resistance and possibly metastasis [7].

Receptor tyrosine phosphorylation and STAT nuclear immunostaining, which are indicative of JAK-STAT pathway activation, have been identified in a large cohort of human tumors. pSTAT immunostaining of patient tumor samples shows that STAT activation associates with patient outcomes across various tumor types in different ways. Oftentimes, this is dependent on the tumor context and specific downstream STATs that are activated. For example, in non-small cell lung cancer (NSCLC) and glioblastoma, the presence of STAT3 and pSTAT3 immunostaining correlates with a lower overall survival rate. In malignant melanoma patients, a prominent pSTAT3 immunostaining correlates with a higher recurrence rate of lymph node metastases, whereas a prominent pSTAT1 immunostaining correlates with a lower recurrence rate of lymph node and brain metastases. In contrast, in breast cancer patients undergoing adjuvant chemotherapy, the presence of pSTAT3 immunostaining correlates with a higher overall survival rate. In rectal/colorectal cancer patients, the correlation of pSTAT3 immunostaining with overall survival rate remains conflicting.

In prostate cancer patients, the presence of pSTAT5 immunostaining correlates with a lower cancer-specific survival rate and an early recurrence rate, while in breast cancer patients, the absence of pSTAT5 immunostaining correlates with a lower cancer-specific survival rate. In light of the variability in the abovementioned findings, further studies are needed to clarify the specific role of individual JAK-STAT activation patterns in cancer [2]. Although these findings illustrate the association of pSTATs with patient outcomes among different tumor types, this association does not suggest STAT activation as the primary causative factor. Therefore, further study of the roles of JAK-STATs across tumor types is needed.

The JAK-STAT pathway also demonstrates substantial cross talk through interaction with other signaling pathways involved in oncogenesis. For example, STAT activation can occur downstream of the epidermal growth factor receptor (EGFR) and androgen receptor (AR) signaling, which may occur independently of JAKs. Therefore, deciphering the ways different signaling pathways interact with each other and thereby cause JAK-dependent or JAK-independent activation of STATs should provide essential information for new clinical therapies and could help explain the failures of JAK inhibitors in clinical trials of solid tumors [2].

In addition to the abovementioned cellular processes of genetic mutation, gene amplification, and aberrant signaling that cause JAK-STAT pathway activation and lead to tumorigenesis, there are some noncanonical pathways that may contribute to JAK-STAT activity. These noncanonical pathways include noncoding RNAs (e.g., miRNAs, long noncoding RNAs) and heterochromatin stabilization, both of which modulate JAK-STAT pathway activity.

Oncogenic microRNAs (miRNAs) negatively regulate SOCS or PIAS (inhibitors of JAK-STAT signaling) expression and cause JAK-STAT pathway activation. These miRNAs include the following [3]:

1. miRNA-155 overexpression leads to STAT3 activation by downregulating SOCS3 in breast cancer and anaplastic large cell lymphoma.
2. miRNA-30 overexpression leads to elevated JAK-STAT3 signaling by targeting the 3′ UTR of SOCS3 in primary glioma tissue.
3. miRNA-221 limits SOCS3 expression in PCa cell lines.
4. miRNA-18a overexpression leads to elevated STAT3 expression by negatively modulating PIAS3 (a repressor of JAK-STAT signaling) expression.

Long noncoding RNAs (LncRNAs) may also contribute to JAK-STAT pathway regulation. These LncRNAs include the following [3]:

1. lncRNA OLA1P2 directly interacts with STAT3 tyrosine phosphorylation and thereby inhibits STAT3 nuclear translocation and activation.
2. lncRNA-DC binds to cytoplasmic STAT3 and prevents STAT3 dephosphorylation by SHP1 and thereby activates STAT3 signaling.

Lastly, the JAK-STAT pathway has been found to modulate heterochromatin stabilization in *Drosophila*. JAK activation causes heterochromatin destabilization, whereas un-phosphorylated STAT causes heterochromatin stabilization. STAT overexpression causes increased amounts of heterochromatin, increasing cellular resistance to DNA damage. Thus, un-phosphorylated STAT acts as a tumor suppressor by promoting heterochromatin stabilization [8], while activated STAT signaling may promote oncogenesis by converse effects.

Current Therapy and Future Strategy

The JAK-STAT pathway is an attractive therapeutic target due to its involvement across cancers. In addition, the discovery of the JAK2 V617F mutation in MPNs motivated the development of JAK2 inhibitors. The design of JAK inhibitors focuses on two basic mechanisms, which include either acting as an ATP-binding site competitor or targeting the pseudo-kinase domain, both of which will inhibit JAK activity. Herein, we will review the JAK-STAT inhibitors that are FDA-approved and summarize the JAK-STAT inhibitors that are in clinical trials and under investigation in Table 26.1 [9, 10].

Table 26.1 JAK inhibitors under clinical investigation in cancer

Drug	JAK targets (IC_{50})	Non-JAK targets (IC_{50})	FDA-approved	Trial identifier(s)	Indication	Phase
Ruxolitinib (INCB018424)	JAK2 (2.8 nM) JAK1 (3.3 nM) TYK2 (18 nM)	unknown	Yes	NCT02876302	Triple-negative or inflammatory breast cancer	II
				NCT01594216	Estrogen receptor-positive breast cancer	II
				NCT01877005, NCT02164500	Hodgkin's lymphoma	II
				NCT02723994	ALL	II
				NCT01914484	TKI-resistant Ph-leukemia	I/II
				NCT01776723, NCT01751425	CML	I/II
				NCT01712659	ATL	II
				NCT02117479	Metastatic pancreatic adenocarcinoma	III
				NCT01693601, NCT02718300, NCT01375140, NCT01790295, NCT02251821, NCT01732445	Myelofibrosis (MF)	I–III
				NCT02493530	MF/PV	I
				NCT00509899	MF/PV/thrombocytosis	I/II
				NCT02076191	MPN	I/II
				NCT01243944	PV	III
				NCT02577926	PV/thrombocythemia	III
Lestaurtinib (CEP-701)	JAK2 (0.9 nM)	FLT3 (3 nM)	No	NCT00494585	MF	II
		TrkA (<25 nM) Aurora kinase A/B (8.1 nM, 2.3 nM)		NCT00668421		I/II
Momelotinib (CYT387)	JAK1 (11 nM) JAK2 (18 nM) JAK3 (155 nM)	unknown	No	NCT02251821	MF	II
Pacritinib (SB-1518)	$JAK2^{V617F}$ (19 nM) JAK2 (23 nM) TYK2 (50 nM)	$FLT3^{D835Y}$ (6 nM) FLT3 (22 nM)	No	NCT00719836	Advanced myeloid malignancies	I/II
				NCT00745550, NCT03165734	MF	I/II
				NCT02532010	AML	II
AT9283	JAK3 (1.1 nM) JAK2 (1.2 nM)	Aurora kinase A/B (3.0 nM) Abl1 (4 nM)	No	NCT01431664	Relapsed or refractory acute leukemia	I
BMS-911543	JAK2 (1.1 nM) JAK1 (360 nM) JAK3 (75 nM) TYK2 (66 nM)	SET-2 (60 nM)	No	NCT01236352	MF	I/II

FDA-Approved Drugs

- Ruxolitinib (INCB018424) was the first FDA-approved JAK inhibitor used for treating intermediate- and high-risk myelofibrosis and for PV. Ruxolitinib is a selective JAK1 and JAK2 inhibitor and binds to the kinase domain of JAK in order to block JAK-STAT signaling. Therefore, ruxolitinib can be employed in tumorigenic conditions involving the JAK2 V617F mutation. Based on the phase III COMFORT-I study involving patients with intermediate- or high-risk myelofibrosis, ruxolitinib significantly improved splenomegaly, key symptoms, and quality-of-life measures. Ruxolitinib treatment resulted in an improved overall survival rate for patients with intermediate- and high-risk myelofibrosis. In addition, ruxolitinib provided rapid and long-lasting improvement for patients with myelofibrosis. However, any discontinuity in ruxolitinib treatment leads to a return of symptoms within 1 week. A long-term efficacy and safety analysis of ruxolitinib treatment for patients with myelofibrosis showed an improvement in splenomegaly, quality-of-life measures, and overall survival rate compared to placebo. Current clinical trials are directed toward analyzing the efficacy and safety of ruxolitinib in certain hematological malignancies and also in solid tumors (e.g., colorectal cancer, prostate cancer, and breast cancer) [1, 9, 10].
- Tofacitinib was the first selective JAK inhibitor tested in humans. Tofacitinib has a broad spectrum inhibition involving many JAK members but shows a greater specificity for JAK3. Since JAK3 primarily affects immune-associated organs and tissues, current clinical trials are directed toward analyzing the efficacy and safety of Tofacitinib in autoimmune disorders (e.g., rheumatoid arthritis, psoriasis, and ulcerative colitis) [1]. However, this drug has not been approved in cancer treatment.

Non-approved and Preclinical JAK Inhibitors

There are a number of drugs currently in preclinical development and/or clinical trials testing their efficacy and safety in hematological malignancies (e.g., PV, ET, myelofibrosis, and AML) and solid tumors (e.g., non-small cell lung cancer and metastatic pancreatic ductal adenocarcinoma). These drugs either target multiple JAK family members, such as momelotinib (a selective JAK1 and JAK2 inhibitor), pacritinib (a selective JAK2 and FLT3 inhibitor), and AZD1480 (a selective JAK3 and TYK2), or selectively target JAK2, such as BMS-911543 and XL019. Description of these drugs, their on-target and off-target specificities, and a brief description of the associated clinical trials and indications are shown in Table 26.1.

Summary

In this chapter, the JAK-STAT signaling pathway was introduced, and the major signaling components therein were discussed. This pathway plays a central role in a variety of cellular processes pertinent to cancer, but each of the JAK-STAT pathway members can be differentially activated in different types of malignancy, and cellular and molecular context can lead to varying effects on cancer phenotypes. As it stands, the role of the JAK-STAT signaling pathway is the most well-established in hematological malignancies, but there is increasing awareness for a role of this pathway in affecting solid tumor behavior. In hematological malignancies, the JAK2 V617F mutation induces JAK-STAT pathway activation and appears to be targetable with JAK-specific inhibitors. However, it remains to be seen whether JAK-STAT pathway inhibitors will be efficacious in solid tumors and how they should be combined with other therapies to elicit maximal patient benefit. It is also becoming increasingly clear that noncanonical JAK-STAT pathway activation may also be involved in carcinogenesis. Finally, there is currently only one FDA-approved drug for use in cancer, although there are a number of ongoing clinical trial testing agents that target the JAK-STAT pathway for cancer therapy. An improved context-specific understanding of the role of the JAK-STAT pathway in cancer should help elucidate which cancers are best targeted by JAK-STAT inhibitors, which patients are most likely to benefit, and what combinations of therapy can synergize to best target cancer-specific mechanisms.

References

1. O'Shea JJ, Schwartz DM, Villarino AV, Gadina M, McInnes IB, Laurence A. The JAK-STAT pathway: impact on human disease and therapeutic intervention. Annu Rev Med. 2015;66:311–28.
2. Thomas SJ, Snowden JA, Zeidler MP, Danson SJ. The role of JAK/STAT signalling in the pathogenesis, prognosis and treatment of solid tumours. Br J Cancer. 2015;113(3):365–71.
3. Pencik J, Pham HT, Schmoellerl J, Javaheri T, Schlederer M, Culig Z, et al. JAK-STAT signaling in cancer: from cytokines to non-coding genome. Cytokine. 2016;87:26–36.
4. Lim CP, Cao X. Structure, function, and regulation of STAT proteins. Mol BioSyst. 2006;2(11):536–50.
5. Buchert M, Burns CJ, Ernst M. Targeting JAK kinase in solid tumors: emerging opportunities and challenges. Oncogene. 2016;35(8):939–51.
6. Constantinescu SN, Girardot M, Pecquet C. Mining for JAK-STAT mutations in cancer. Trends Biochem Sci. 2008;33(3):122–31.
7. Balko JM, Schwarz LJ, Luo N, Estrada MV, Giltnane JM, Davila-Gonzalez D, et al. Triple-negative breast cancers with amplification of JAK2 at the 9p24 locus demonstrate JAK2-specific dependence. Sci Transl Med. 2016;8(334):334ra53.
8. Silver-Morse L, Li WX. JAK-STAT in heterochromatin and genome stability. Jak Stat. 2013;2(3):e26090.
9. Kontzias A, Kotlyar A, Laurence A, Changelian P, O'Shea JJ. Jakinibs: a new class of kinase inhibitors in cancer and autoimmune disease. Curr Opin Pharmacol. 2012;12(4):464–70.
10. Sonbol MB, Firwana B, Zarzour A, Morad M, Rana V, Tiu RV. Comprehensive review of JAK inhibitors in myeloproliferative neoplasms. Ther Adv Hematol. 2013;4(1):15–35.
11. Chen X, Vinkemeier U, Zhao Y, et al. Crystal structure of a tyrosine phosphorylated STAT1 dimer bound to DNA. Cell. 1998;93:827–39.

NF-κB Signaling Pathways in Carcinogenesis

27

Harikrishna Nakshatri

Abbreviations

COX2	Cyclooxygenase 2
CSN5	COP9 signalosome 5
CXCL	Chemokine (C-X-C) ligand
CXCR	Chemokine (C-X-C) receptor
EGFR	Epidermal growth factor receptor
ERBB2	ERB-B2 receptor tyrosine kinase 2
FAS	Fas cell death receptor
FASL	FAS ligand
IAP	Inhibitor of apoptosis
IKK	IκB kinase
IL	Interleukin
IκB	Inhibitor of kappaB
MLL	Mixed lineage leukemia
MMP	Matrix metalloproteinase
MyoD	Myogenic differentiation 1
NF-κB	Nuclear factor-kappaB
PD-L1	Programmed death ligand 1
PIK3CA	Phosphatidylinositol-4,5-bisphosphate 3-kinase catalytic subunit alpha
SASP	Senescence-associated secretory phenotype
TNF	Tumor necrosis factor
ZEB	Zinc finger E-box-binding homeobox

H. Nakshatri
Departments of Surgery, Biochemistry and Molecular Biology, Indiana University, Indianapolis, IN, USA
e-mail: hnakshat@iupui.edu

Introduction

NF-κB in Tumor Initiation

The potential role of NF-κB in cancer was first investigated in hematologic malignancies, particularly multiple myeloma and leukemia [1]. The focus subsequently shifted to solid tumors after discovery of its aberrant activity in breast cancers by others and us in 1998 [2]. Aberrant NF-κB activity was subsequently reported in prostate, bladder, lung, head and neck, and pancreatic cancers. Mechanistic studies, however, revealed dichotomous role of NF-κB in cancer. NF-κB has been shown to promote senescence and function as a master regulator of senescence-associated secretory phenotype (SASP) [3]. Soluble factors in SASPs reinforce senescence arrest, alter the microenvironment in the tumor, and trigger immune surveillance. In genetically engineered Kras-induced mouse model of pancreatic cancer, the p65 subunit of NF-κB triggers CXCL1 (part of SASP)-/CXCR2-dependent senescence and inhibits initial steps of carcinogenesis [4]. DNA damage also triggers NF-κB-dependent senescence. Switch to oncogenic function occurs once cells are immortalized. Other genomic aberrations in cancer cells will also determine tumor suppressor function of NF-κB. For example, in tumors where the expression of the anti-apoptotic BCL-2 is dependent on NF-κB, it functions as a survival factor. By

S. Badve, G. L. Kumar (eds.), *Predictive Biomarkers in Oncology*,
https://doi.org/10.1007/978-3-319-95228-4_27

contrast, in tumors with elevated BCL-2 independent of NF-κB, therapy-induced cell death requires NF-κB-mediated induction of SASPs [5]. In general, senescence is a double-edged sword in cancer because while senescent cells themselves rarely progress into cancer, SASPs from these cells can promote neoplastic progression of nearby preneoplastic cells by providing pro-inflammatory molecules and inducing epithelial-to-mesenchymal transition. Therefore, timing of NF-κB-induced senescence could have profound influence on cancer development. From the practical and clinical angle, this also poses challenge to ascertain whether concurrent presence of activated NF-κB (as determined by nuclear p65 and phosphorylated p65) and SASP is a good or bad prognostic marker.

NF-κB in Tumor Progression/Metastasis

Overcoming senescence barrier either through inactivation of cell cycle inhibitors such as p16 or overexpression of cell cycle protein cyclin D1 and inactivation of p53 or through telomerase overexpression leads to cellular immortalization. In immortalized cells, NF-κB cooperates with other oncogenes such as RAS or functions downstream of oncogenes such as KRAS, mutant PIK3CA, activated EGFR, or ERBB2 to promote cancer progression. Cancer-promoting functions include upregulation of anti-apoptotic proteins such as BCL-2, BCL-XL, GADD45β, XIAP, cIAP1, and cIAP2; cell cycle proteins such as Cyclin D1; pro-invasion molecules such as MMP9; pro-metastatic molecules such as CXCR4; pro-inflammatory molecules such as IL-1α, IL-1β, IL-6, IL-8, TNFα, and COX2 and metabolic pathway genes such as GLUT3; and genes linked to epithelial-to-mesenchymal transition such as ZEB1 and ZEB2 [2, 6]. Additionally, there is evidence for NF-κB playing a significant role in self-renewal and maintenance of cancer stem cell phenotype, particularly in glioblastoma. NF-κB is essential for lung tumor development upon p53 mutation and KRAS (G12D) expression [7]. Inhibition of NF-κB sensitizes EGFR-mutant lung tumors to EGFR-targeted therapies. MLL fusion proteins that typically cause leukemia are dependent on NF-κB for transformation. NF-κB is also a central player in epigenetic switch that links chronic inflammation to cell transformation and subsequent metastasis [8].

Cancer Cell Non-autonomous Functions of NF-κB

Although initial studies on NF-κB were primarily focused on cell autonomous functions, several recent reports highlight its non-cell autonomous roles. NF-κB-inducible cytokines such as IL-6 and IL-8 not only alter tumor microenvironment by attracting different immune cells, but also these cytokines are associated with systemic effects of cancer. IL-6 is a major contributor to cancer cachexia. NF-κB itself contributes to cancer cachexia by blocking myogenic differentiation by affecting skeletal muscle transcription factor-microRNA circuitry and by repressing differentiation factor MyoD [9].

Recent studies have demonstrated a role for NF-κB in resistance to immune therapy. PD-L1, expressed mainly by cancer cells, plays a significant role in creating antitumor immunity. Therefore, several antibodies targeting PD-L1 have entered clinic. Although correlation between PD-L1 levels and response to therapy is yet to be established, NF-κB has been shown to increase PD-L1 at both transcriptional and posttranscriptional level. Pro-inflammatory cytokines such as TNFα induce the expression of COP9 signalosome 5 (CSN5), which deubiquitinates PD-L1 and stabilizes the protein.

By increasing PD-L1 protein, NF-κB could reduce the effectiveness of PD-L1-targeting antibodies (Fig. 27.1) [10].

Current NF-κB-Targeted Therapies and Future Strategies for Treatment

Based on multiple functions ascribed to NF-κB in cancer and therapeutically targetable signaling cascades involved in NF-κB activation,

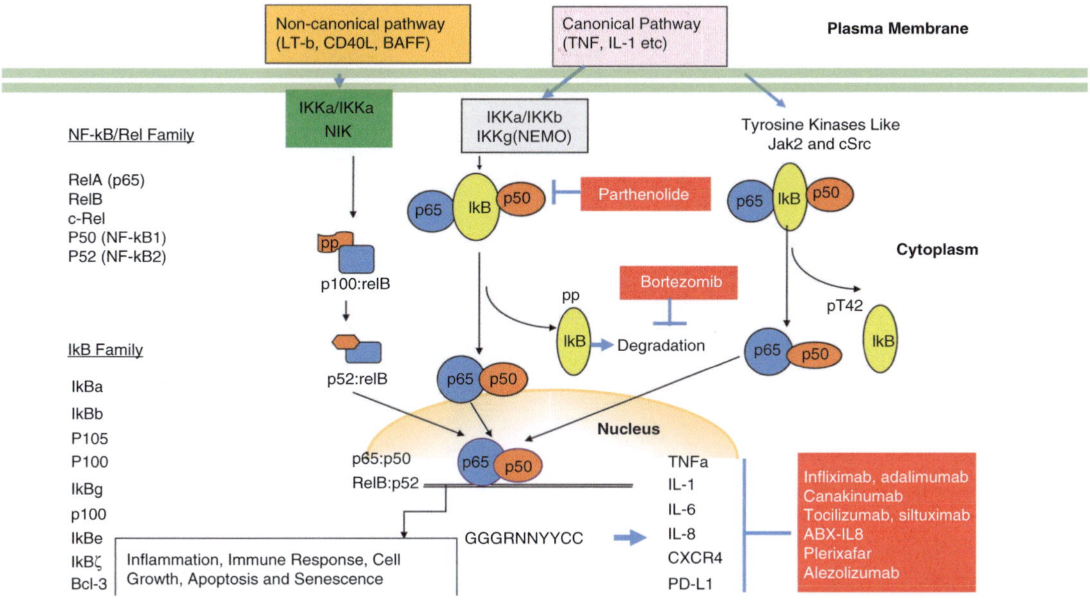

Fig. 27.1 Canonical and noncanonical pathways of NF-κB activation. Steps associated with NF-κB activation in response to extracellular signals are shown. Different members of NF-κB and IκB family are listed. In mammals, the NF-κB family is composed of five related transcription factors: p50, p52, RelA (p65), c-Rel, and RelB. IKK complex is the main signaling hub that integrates extracellular/membrane-activated signals to nuclear events by activating NF-κB. Potential therapeutic agents that may reduce NF-κB activation or reduce the activities of downstream targets of NF-κB are also indicated. For example, ABX-IL8 may inhibit IL8-dependent signaling, and Plerixafar prevents CXCR4 activation by inhibiting CXCR4/CXCL12 interaction. Infliximab is a TNFα antagonist, which can block NF-κB activation by TNFα as well as block the effects of TNFα downstream of NF-κB activation

there have been several attempts to develop drugs targeting NF-κB [11]. Commonly used drugs such as aspirin and food ingredient curcumin inhibit NF-κB. Several groups including us have shown anti-NF-κB activity of parthenolide, an active ingredient in the herb *Tanacetum parthenium* or feverfew used for migraine. Clinically, bortezomib used to treat multiple myeloma is a proteasome inhibitor that prevents IκB degradation and restricts NF-κB activation. However, for most of these drugs, NF-κB is one of their targets, and their clinical activity cannot be solely due to NF-κB inhibition. A highly specific NF-κB inhibitor is yet to enter clinic, although such an inhibitor may never be discovered. An ideal NF-κB inhibitor should dampen overactive NF-κB rather than completely eliminate its activity because basal NF-κB is necessary for normal function of the immune system and to prevent infection. Therefore, there is still considerable interest in both academia and industry to develop drugs that reduce but not eliminate NF-κB activity. Because of the role of NF-κB in upregulating anti-apoptotic proteins, these inhibitors will likely work as chemosensitizing agents rather than displaying single agent activity. However, chemosensitizing function NF-κB inhibitors needs to be tested with individual chemotherapeutic drugs because in certain instances, therapy-induced NF-κB, particularly therapies that promote replication stress, could augment cell death by activating extrinsic pathway of cell death through expression/activation of FAS-FASL-dependent cell death machinery [12].

Several of downstream targets of NF-κB can be targeted therapeutically (Table 27.1). In fact, drugs targeting TNF, IL-1, IL-6, IL-8, CXCR4, and PD-L1 are already in clinical use and can be exploited to treat cancers in which NF-κB pathway is essential for cancer cell survival.

Table 27.1 Aberrations and targeting NF-kB pathways in cancer

NF-κB signaling pathway	Pathway active in cancer type	Affected biomarker	Method of detection	Target	Active drugs
Overexpression of growth factors and receptors Growth factor receptor mutation	Breast Lung Pancreatic Prostate Bladder Leukemia Multiple myeloma	Nuclear p65 Phosphorylated IKKβ	Western blotting for nuclear p65 Electrophoretic mobility shift assays Western blotting with phospho-specific antibodies Immunohistochemistry for nuclear p65	IKK complex	Erlotinib, lapatinib Trastuzumab, pertuzumab (Blakely et al., *Cell Reports* 11:98–110; Merkhofer et al., *Oncogene* 29:1238–48; Pianetti et al., *Oncogene* 20:1287–99)
Increased production of cytokines such as TNFα, IL-1	Breast, lung, pancreatic, prostate Bladder Ovarian	Nuclear p65 Phosphorylated IKKβ	Western blotting for nuclear p65, electrophoretic mobility shift assays, Western blotting with phospho-antibodies, immunohistochemistry for nuclear p65	IKK complex	Infliximab, adalimumab Canakinumab (Nikolaus et al., *Lancet* 356:1475-9; Urbano et al., *Biologics* 8:211–20)
IKK family member mutation/amplification	Pancreatic, neuroendocrine prostate cancer, ovarian, uterine	Nuclear p65	Immunohistochemistry for nuclear p65	IKK complex and IκBα	Bortezomib (Sartore-Bianchi et al., *Clin Cancer Res*: 5942–51)
Amplification of IKKε	Breast cancer	Phosphorylation of p65	Western blotting or immunohistochemistry for p65 phosphorylated at Ser468 and Ser536	P65 and c-Rel	
Amplification of BCL3	Pancreatic, uterine Neuroendocrine prostate cancer Hodgkin lymphoma	Elevated cIAP1, cIAP2, IL-8, IL-17	qRT-PCR for pro-inflammatory genes	P50 homodimers, p65/p50 heterodimers	Bortezomib ABX-IL8 (Chang and Vancurova *Biochim Biophys Acta* 1843:2620–30; Mian et al., *Clin Cancer Res* 9:3167–75)

Summary

Although NF-κB was discovered more than 30 years ago, its regulation and function still remain at the forefront of research not only from the oncology point of view but also with other diseases such as autoimmune disorders, microbial infections, and neurodegenerative diseases. While constitutive activation of NF-κB has been reported in advanced stages of many cancers, causes of this activation vary widely and may involve both genomic and non-genomic events. Despite early-stage-specific tumor suppressor role described for this transcription factor complex, this observation should not hinder clinical development of NF-κB inhibitors because NF-κB has already transitioned to oncogenic role at the time of clinical manifestation of the disease. Thus, studies focusing on NF-κB, both at regulatory and functional level, will continue to unlock mysteries surrounding cancer progression and potentially to new cancer therapies. Reliable biomarkers of constitutive NF-κB activity in cancer are yet to be identified because of close link between NF-κB activation and inflammatory process. Although drugs that directly target NF-κB are yet to enter clinic, several drugs that inhibit the function of proteins overexpressed as a consequence of increased NF-κB activity in cancer are showing promising results.

References

1. Hayden MS, Ghosh S. NF-kappaB, the first quarter-century: remarkable progress and outstanding questions. Genes Dev. 2012;26(3):203–34.
2. Karin M, Cao Y, Greten FR, Li ZW. NF-kappaB in cancer: from innocent bystander to major culprit. Nat Rev Cancer. 2002;2(4):301–10.
3. Salminen A, Kauppinen A, Kaarniranta K. Emerging role of NF-kappaB signaling in the induction of senescence-associated secretory phenotype (SASP). Cell Signal. 2012;24(4):835–45.
4. Lesina M, Wormann SM, Morton J, Diakopoulos KN, Korneeva O, Wimmer M, et al. RelA regulates CXCL1/CXCR2-dependent oncogene-induced senescence in murine Kras-driven pancreatic carcinogenesis. J Clin Invest. 2016;126(8):2919–32.
5. Klein U, Ghosh S. The two faces of NF-kappaB signaling in cancer development and therapy. Cancer Cell. 2011;20(5):556–8.
6. Perkins ND, Gilmore TD. Good cop, bad cop: the different faces of NF-kappaB. Cell Death Differ. 2006;13(5):759–72.
7. Meylan E, Dooley AL, Feldser DM, Shen L, Turk E, Ouyang C, et al. Requirement for NF-kappaB signalling in a mouse model of lung adenocarcinoma. Nature. 2009;462(7269):104–7.
8. Kuo HP, Wang Z, Lee DF, Iwasaki M, Duque-Afonso J, Wong SH, et al. Epigenetic roles of MLL oncoproteins are dependent on NF-kappaB. Cancer Cell. 2013;24(4):423–37.
9. Bakkar N, Guttridge DC. NF-kappaB signaling: a tale of two pathways in skeletal myogenesis. Physiol Rev. 2010;90(2):495–511.
10. Lim SO, Li CW, Xia W, Cha JH, Chan LC, Wu Y, et al. Deubiquitination and stabilization of PD-L1 by CSN5. Cancer Cell. 2016;30(6):925–39.
11. DiDonato JA, Mercurio F, Karin M. NF-kappaB and the link between inflammation and cancer. Immunol Rev. 2012;246(1):379–400.
12. Perkins ND. The diverse and complex roles of NF-kappaB subunits in cancer. Nat Rev Cancer. 2012;12(2):121–32.

Immune Signaling in Carcinogenesis

28

Mahesh Yadav, Marcin Kowanetz, and Hartmut Koeppen

The immune system is designed to maintain homeostasis by controlling immune responses against pathogens and foreign antigens. It is regulated by activating/stimulatory and inhibitory pathways; the latter is also known as immune checkpoints. Under normal physiological conditions, immune checkpoints are crucial for maintaining self-tolerance to prevent autoimmunity and to protect tissues from damage during infections. However, in cancer, this balance is dysregulated, and immune checkpoints are often activated to suppress the ongoing antitumor immune responses. T cells are primary mediators of antitumor immune responses due to their ability to directly recognize and kill antigen-expressing cells (CD8 or cytotoxic T cells) or help other immune cells to boost immune functions (CD4 or helper T cells).

For a cytotoxic CD8 T cell to get activated or helper T cell to proliferate and differentiate into an effector cell, two kinds of signals are required (Fig. 28.1). Signal 1 is provided by a peptide, following which it is bound and presented by an MHC protein on the surface of the antigen-presenting cell (APC). This peptide-MHC complex signals through the T cell receptor on the T cells and its associated proteins. Signal 2 is provided by co-stimulatory proteins, especially the B7 proteins (CD80 and CD86), which are recognized by the co-receptor protein CD28 on the surface of the T cell. Co-stimulatory signal 2 is thought to amplify the intracellular signaling process triggered by signal 1. Many other T cell co-signaling receptors have been identified and are broadly defined as cell-surface molecules that can transduce signals into T cells to positively (co-stimulatory receptors) or negatively (co-inhibitory receptors or immune checkpoints) modulate TCR signaling.

In most cases, the immune system can recognize tumor antigens expressed by each patient's unique and frequently changing population of cancer cells, and activated T cells can kill tumor cells directly or indirectly. However, tumors escape this immune destruction by blunting the immune system through exhaustion/inactivation of T cell functions caused by increased expression of inhibitory receptors on tumor and immune cells. It is possible to overcome that defect and restore T cell function in cancers by boosting activation through agonist co-stimulatory signals or by blocking inhibitory signal through antagonists of immune checkpoints. Indeed, clinical evidence has shown that immune manipulation of T cells can yield remarkable outcomes in metastatic melanoma as well as other indications, including non-small cell lung cancer (NSCLC),

M. Yadav · M. Kowanetz
Department of Oncology Biomarker Development, Genentech, South San Francisco, CA, USA

H. Koeppen (✉)
Research Pathology, Genentech, South San Francisco, CA, USA
e-mail: hkoeppen@gene.com

S. Badve, G. L. Kumar (eds.), *Predictive Biomarkers in Oncology*,
https://doi.org/10.1007/978-3-319-95228-4_28

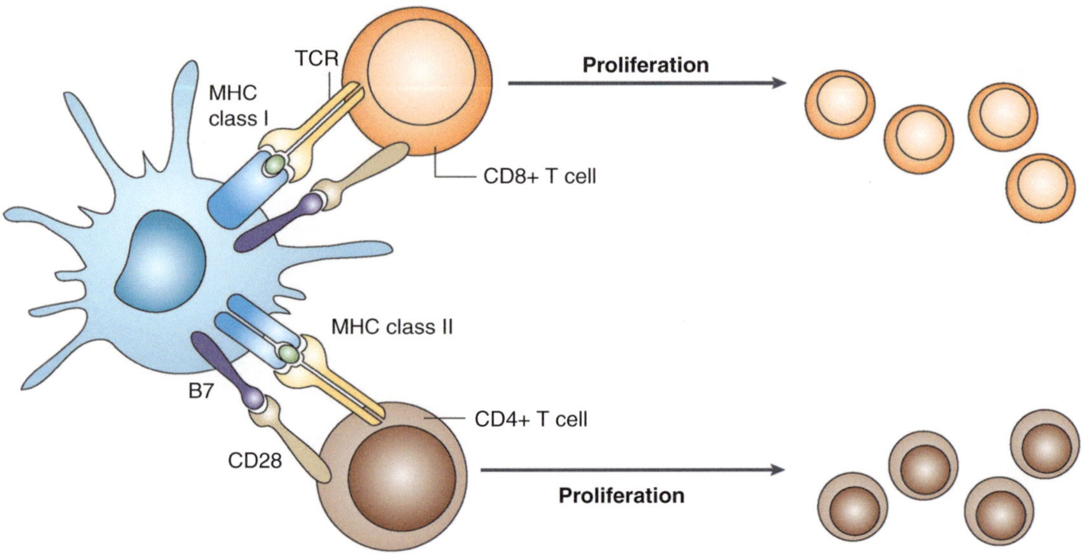

Fig. 28.1 T cell activation is initiated with the presentation of the antigen-derived peptide in association with major histocompatibility complex (MHC) class I or II on the antigen-presenting cell such as dendritic cell and its recognition by T cell receptor on CD8 or CD4 T cell, respectively. Co-stimulatory signal or signal 2, mediated through B7-CD28 interactions, is needed in most instances for full activation leading to a proliferation of CD4 and CD8 T cells. Additional signals mentioned later in the chapter may further promote survival and proliferative capacity of the activated T cells. (Adapted from Gilboa [17]. With permission from Springer Nature)

kidney cancer, bladder cancer, head and neck squamous cell carcinoma, Merkel cell carcinoma, and Hodgkin's lymphoma [1].

Multiple checkpoints have been implicated in regulating T cell function (shown in Fig. 28.2), but the two that have been most actively studied in the context of clinical cancer immunotherapy are cytotoxic T-lymphocyte-associated antigen 4 (CTLA-4, also known as CD152) and programmed cell death protein-1 (PD-1, also known as CD279). Both receptors regulate immune responses at different levels and by different mechanisms. CTLA-4 down-regulates the amplitude of T cell activation upon initial activation, whereas PD-1 limits the effector phase of T cell function in tissues [2]. Other inhibitory receptors on T cells include T cell immunoglobulin mucin domain 3 (TIM-3), lymphocyte-activation gene 3 (LAG-3), B- and T-lymphocyte attenuator (BTLA), V-domain immunoglobulin suppressor of T cell activation (VISTA), B7-H4, T cell immunoglobulin, and ITIM domain (TIGIT), all of which have been shown to modulate T cell responses to chronic infections and tumors by binding to their respective ligands. Some of the genes that encode proteins associated with immunosuppression, including PD-L1 and IDO-1 (indoleamine 2,3-dioxygenase), could also be induced on tumor cells through amplification or enhanced transcription and prevent tumor cells from cytotoxic T cells. Notably, the signaling pathways utilized by these checkpoint proteins are unique and nonredundant [2].

A number of tumor necrosis factor receptors (TNFRs) are expressed on T cells and directly influence the T cell response. Among these receptors are CD27, OX40 (Tnfrsf4, CD134), 4-1BB (Tnfrsf9, CD137), and GITR (Tnfrsf18). Co-stimulatory TNFRs further boost the T cell response when T cells are activated by TCR engagement. One of the key functions of co-stimulatory TNFRs is to keep activated T cells alive. They also complement the T cell response either through direct signaling or indirectly through activation of antigen-presenting cells (APCs) or by inducing inflammation and innate immunity.

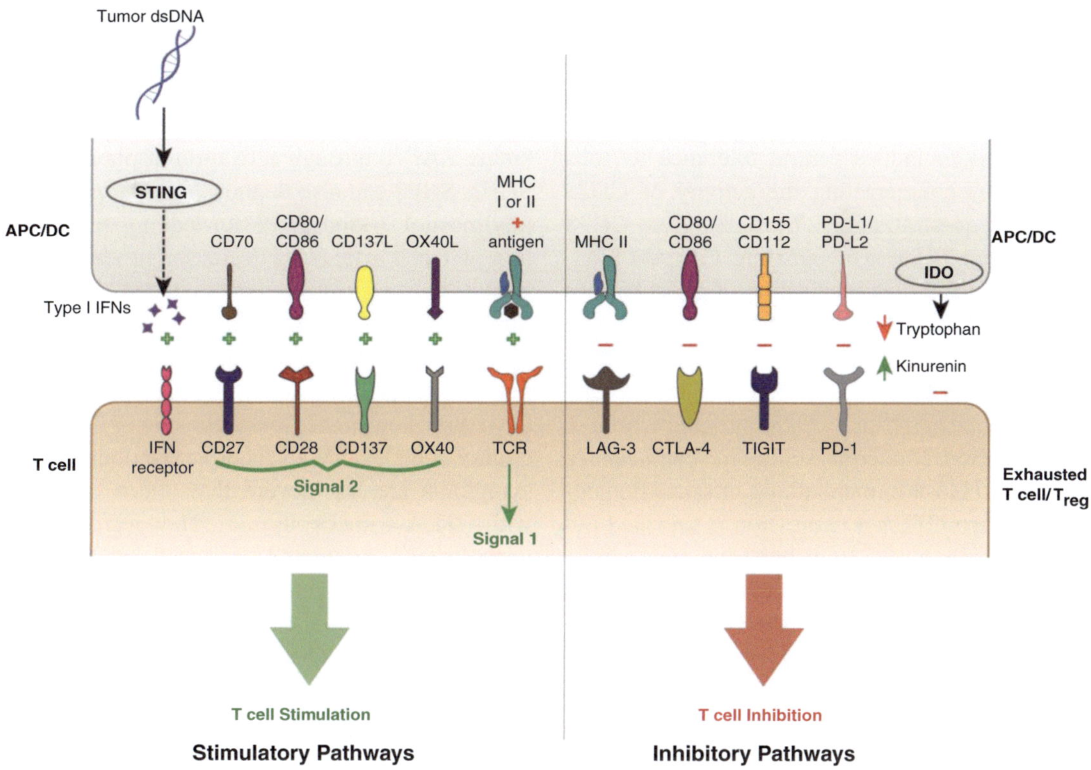

Fig. 28.2 Activating and inhibitory pathways in the regulation of T cell activation and functions. Multiple stimulatory and inhibitory receptors are involved in regulation of T cell responses. Shown is a schematic overview of various ligand-receptor interactions between T cells and antigen-presenting cells (APCs) involved in mediating T cell responses. Once the cognate antigen is recognized by T cell receptor on T cells, additional signals through activating or co-stimulatory receptors (shown by ± sign) are required for full activation of T cells. Once activated T cells upregulate expression of inhibitory receptors (shown by – sign) such as TIGIT D-1 and CTLA-4 on T cells which function to limit T cell activation. Many activating and inhibitory receptors have common ligands. Example CD28 is a co-stimulatory receptor on T cell and binds to CD80/CD86 or B7.1/2 which are also recognized by inhibitory CTLA-4. The functional outcome of these receptor ligand interactions is determined by kinetics of receptor expression and their affinity for the ligand. For example, CTLA-4 binds to a much higher affinity to CD80/CD86 but is not normally expressed on resting T cell. Many of the ligands for inhibitor receptors can also be expressed on tumor cells besides antigen-presenting cells thus providing another barrier to limit T cell activity in cancers. And finally, interactions between APCs and T cells are bidirectional and lead to initiation of immune signaling in both T cells and APCs

Emerging data also suggest that activation of the innate immune system could help boost T cell responses in tumors to counteract tumor-induced immunosuppression, which includes triggering of immune targets including Toll-like receptors (TLRs) and stimulator of interferon genes (STING).

In this chapter, we will summarize some of the inhibitory and stimulatory immune signaling pathways that are being targeted for cancer immunotherapy.

Inhibitory Pathways

CTLA-4

CTLA-4 is a transmembrane glycoprotein that is a homolog of the co-stimulatory protein CD28, which is expressed exclusively on T cells where it primarily regulates the amplitude of the early stages of T cell activation. Both CD28 and CTLA-4 share and compete for identical ligands: CD80 (also known as B7.1) and CD86 (also known as B7.2), which are expressed on APCs. However, CTLA-4 binds to B7 proteins with approximately 20 times greater affinity and hence outcompetes CD28. The primary function of CTLA-4 is to induce central tolerance to self-proteins by counteracting the activity of CD28 through sequestration of B7 proteins from CD28 engagement. In terms of signaling, CTLA-4 binding to ligands induces activation of phosphatases SHP2 or SH2 tyrosine phosphatase (also known as PTPN11) and protein phosphatase 2 (PP2A), which diminish the kinase signaling induced by the T cell receptor and CD28 and hence limit T cell activation. Numerous studies have indicated a role for CTLA-4 in maintaining immune homeostasis. When CTLA-4 expression is knocked out in mice, mice suffer from hyperimmune activation and succumb to multi-organ autoimmunity due to systemic T cell infiltration in tissues [3].

Regulatory T cells (Tregs), a subset of T cells that suppresses the function of CD4 and CD8 T cells, constitutively express CTLA-4 at levels higher than normal T cells. CTLA-4 is employed by Tregs as an immunosuppressive mechanism. Blocking CTLA-4 with monoclonal antibodies (mAbs) disrupts the ligation between CTLA-4 and CD80/CD86 and leads to depletion of Tregs in the tumor microenvironment, thereby reducing the immunosuppressive function of Tregs [4].

Targeting CTLA-4 in cancers leads to improved T cell functions through enhancement of cytotoxic T cell activity in addition to inhibiting or depleting immunosuppressive Tregs. An antibody targeting CTLA-4, namely, ipilimumab, was approved by the FDA for the treatment of metastatic melanoma in 2011. A second antibody targeting CTLA-4, tremelimumab, has a longer half-life compared to ipilimumab and is currently being investigated in cancer therapy. Many clinical trials are currently testing the combination of CTLA-4-blocking antibodies in combination regimens to further enhance the initial activity observed with the antibodies [1].

PD-1

In contrast to CTLA-4, the major role of PD-1 is to limit the activity of T cells that have gained effector functions and migrated to the tissues. PD-1 interaction with its ligands PD-L1 (B7-H1 or CD274) and PD-L2 (CD273) leads to dephosphorylation of the cytoplasmic protein tyrosine kinase ZAP70 through activation of phosphatase SHP2. SHP2 can also dephosphorylate phosphatidylinositol 3-kinase (PI3K) leading to inhibition of T cell activation, decreased production of inhibitory cytokines, and reduced survival. Similar to CTLA-4, PD-1 also is highly expressed by Tregs and is known to enhance their proliferation and suppressive activity [5]. In studies with PD-1 knockout mice, there are signs of autoimmunity and elevated levels of autoantibodies, but symptoms are less severe than those observed with CTLA-4 knockout mice. However, PD-1 is more broadly expressed on T cell subsets than CTLA-4. In addition, it can also be expressed on other immune cells including B cells and natural killer (NK) cells, whereby it can limit antibody production and lytic activity, respectively [6], thus making it an excellent target for cancer immunotherapy. Indeed, therapies using a blocking antibody against PD-1 have been very successful in treating cancers such as melanoma, lung cancer, renal cell cancer, and bladder cancer. The PD-1 pathway can also be targeted using antibody targeting its ligand PD-L1, which is widely expressed on immune cells as well as on tumor cells in different cancers. Expression of PD-L1 on tumor cells is largely stimulated by interferon-gamma (IFN-γ), although in some cases such as NSCLC, tumor cells can express PD-L1 in the absence of immune infiltration, suggesting alternative mechanisms for PD-L1 expression. The expression of PD-L1 on tumor cells leads to inhibition of T cell expressing PD-1 and hence provides a mechanism for the tumor to evade T cell killing. In addition to PD-1, PD-L1 can also bind to CD80 as a second mechanism of T cell suppression. Thus, PD-L1 expression in tumors can suppress T cell activity through interactions with both PD-1 and CD80. Consequently, anti-PD-1 and anti-PD-L1 therapies might have distinct antitumor effects, with PD-1 blockade leaving the CD80-PD-L1 interaction intact and PD-L1 blockade leaving the PD-L2-PD-1 interaction intact. Therapies targeting PD-1 and PD-L1 have shown remarkable efficacy in the clinic and are FDA approved for the treatment of multiple cancer types [7].

TIGIT

TIGIT (also known as WUCAM, Vstm3, VSIG9) is a member of the poliovirus receptor (PVR)/ nectin family, a subset of the immunoglobulin superfamily. It consists of an extracellular immunoglobulin variable-set (IgV) domain, a type 1 transmembrane domain, and an intracellular domain possessing a canonical immunoreceptor tyrosine-based inhibitory motif (ITIM) and an immunoglobulin tyrosine tail (ITT) motif. Similar to PD-1 and other immune checkpoints, TIGIT is upregulated on exhausted T cells in both chronic viral infections and cancer. In addition, TIGIT is highly expressed on NK cells and Tregs. TIGIT binds primarily to PVR (CD155) and with lower affinity to CD112 (also known as PVRL2/ nectin 2). PVR also binds to CD226, which is an activating receptor expressed on T and NK cells. The relationship between TIGIT and CD226 on T cells is analogous to the CTLA-4/CD28 receptor pair, with TIGIT fulfilling the role of co-inhibitory receptor to counterbalance the co-stimulatory function of CD226 [8].

TIGIT expression on Tregs defines an activated subset of Tregs with superior suppressive capacity in both humans and mice. TIGIT-expressing Tregs express higher amounts of PD-1, CTLA-4, LAG-3, and TIM-3 and produce more IL-10 compared to TIGIT-negative Tregs. TIGIT signaling has been primarily defined in NK cells, where binding of TIGIT to PVR results in phosphorylation of tyrosine residues in ITIM domains leading to recruitment of phosphatase SHP1, inhibition of PI3K, and initiation of MAPK signaling. TIGIT is co-expressed with PD-1 on exhausted CD8 T cells infiltrating lung cancer and colorectal carcinoma as well as several mouse tumor models. Combined blockade of TIGIT and PD-L1 dramatically improves effector cytokine production by CD8 T cells in tumors and leads to complete tumor rejection in mice, suggesting that co-blockade of TIGIT and PD-1 pathways could be an effective strategy to improve antitumor responses in human cancer. Indeed, anti-TIGIT mAb is currently being tested in combination with anti-PD-1/PD-L1 antibodies in clinical trials [9].

LAG-3

LAG-3 (CD223) is expressed on activated CD4 and CD8 T cells, activated Tregs, and a subset of NK and B cells. The interaction between LAG-3 and its major ligand, class II MHC expressed by APCs and DCs, is thought to play a role in modulating T cell function. LAG-3 blockade slightly improves the proliferation and function of T cells. LAG-3 associates with the TCR complex following TCR stimulation and negatively regulates downstream activation. Although the exact mechanism is not completely understood, a single lysine residue (K468) in the cytoplasmic tail of LAG-3 is essential for interaction with downstream signaling molecules [10]. Higher expression of LAG-3 on tumor-infiltrating T cells is observed in human cancers and also in preclinical mouse models, suggesting an important role for LAG-3 [10]. Antibodies targeting LAG-3 in combination with anti-PD-1 antibodies are currently being tested in the clinic to treat different cancers.

IDO

IDO (indoleamine 2, 3 dioxygenase) is an IFN-inducible enzyme that controls inflammation and suppresses adaptive T cell immunity. The IDO family consists of IDO1 and IDO2, the latter being less studied. IDO exerts its immunoregulatory properties by catabolizing the essential amino acid tryptophan into downstream factors including soluble kynurenine. Kynurenine is known to bind and activate aryl hydrocarbon receptors, which in turn promotes Treg differentiation and directs macrophages and DCs to an immunosuppressive state. Upregulation of IDO is also known to change the entire milieu of APCs from immunogenic to tolerogenic. IDO can also be expressed by tumor cells, which results in aggressive tumor growth and an increasingly immunosuppressive microenvironment in the tumor and hence resistance to T cell-targeting immunotherapies [11].

IDO inhibitors are one of the latest players in immunotherapy treatments. In a recent clinical trial of patients with advanced melanoma, a combination of IDO inhibitor indoximod with anti-PD-1 resulted in tumor shrinkage in greater than 50% of the tested patients. Other IDO inhibitors when used as combination products with anti-PD-1 drugs have showed effective responses in many different cancer trials, including those of the skin, kidney, and lung. In contrast to other combination drugs used with PD-1, IDO inhibitors do not display toxicity, thus, making them potent candidates for immunotherapy [12].

Stimulatory Pathways

OX40

OX40 (CD134) is a member of the TNF receptor superfamily, which is expressed on T cells upon activation and has co-stimulatory functions. On T cells, OX40 binding to ligand OX40L induces expression of survivin through sustained activation of P3K and PKB promoting their proliferation, survival, and the secretion of cytokines associated with both type 1 and type 2 T helper cell responses. OX40 co-stimulation also lowers the threshold for T cell activation leading to enhanced IFN-γ production upon TCR engagement. In contrast, OX40 signaling in Tregs inhibits their suppressive functions. However, agonistic anti-OX40 alone is not sufficient to work as a single agent in cancer immunotherapy. OX40 agonist complements the activity of PD-1/PD-L1 blockade by boosting the effector function of PD-1-expressing CD4 and CD8 T cells. Therefore, combination immunotherapy incorporating both OX40 and checkpoint inhibition through PD-1/PD-L1 or CTLA-4 blockade may be able to augment antitumor immunity [13].

4-1BB

4-1BB (CD137) is a member of the TNF receptor superfamily and works as a co-stimulatory target on NK and T cells. CD137 expression is induced upon TCR engagement, and if bound by its natural ligand (CD137L), it induces polyubiquitination-mediated signals that inhibit apoptosis while enhancing proliferation and effector functions. Targeting 4-1BB with agonistic mAb therapy demonstrated potent antitumor effects in murine tumor models. Due to selective expression of CD137 on antigen-experienced T cells, it can be used as a marker to select the T cells that are reactive against the tumor. CD137 is also being employed as a key component of anti-CD19 chimeric antigen receptors (CARs) that are used to redirect T cells against leukemia and lymphoma in the clinic [14].

CD27

Unlike other TNFRs which are mostly expressed upon T cell activation, CD27 is already present on naïve CD4 and CD8 T cells. Upon activation, expression of CD27 is upregulated and may act as a marker for T cell activation. The expression of its ligand, CD70, is tightly regulated as it is only transiently expressed on activated T and B cells, on subsets of professional antigen-presenting dendritic cells (DCs), and on NK cells. Ligation of CD27 to CD70 counteracts apoptosis in activated T cells by downregulating expression of FasL (on CD4 T cells) or decreasing the sensitivity of CD8 T cells to FasL-induced apoptosis. CD27 signaling activates NFκB, promotes cell survival, enhances T and B cell receptor-mediated proliferative signals, and induces cytokine production such as CXCL10. The cytokine induction, in turn, enhances the interaction of CD8 T cells with DCs for effector T cell generation. Interestingly, transient CD27/CD70 co-stimulation promotes the T cell response, but constitutive CD27/CD70 co-stimulation as found in chronic viral infection leads to T cell exhaustion and death. Induction of IL-2 through CD27 signaling can also lead to increased frequency of intratumoral Tregs. Thus, CD27/CD70 is an attractive target to improve antitumor immunity but may need to be regulated tightly to achieve the appropriate outcome [15].

STING

The stimulator of interferon genes (STING, also known as TMEM173, MPYS, MITA, ERIS) is encoded by the TMEM173 gene in the host. STING is an endoplasmic reticulum adaptor protein, which plays a pivotal role in host defense and innate immune responses by detecting the cytosolic nucleic acid ligands in response to bacterial, viral, and eukaryotic pathogens. Following the detection of cytosolic DNA, the STING pathway gets activated and results in type 1 IFN responses to alert the immune system and recruit T cells to the site during infections or cancer. Activation of the STING-dependent pathway in cancer cells can result in tumor infiltration with immune cells and modulation of the antitumor immune response. The function of the STING pathway in cancer cells is not completely understood. Recent studies indicate that STING signaling is repressed in many different types of cancers including colorectal cancer. In the absence of STING signaling, damaged cells in cancers are able to evade the immune system. Some of the common checkpoint inhibitors, namely, anti-PD-1 and anti-PD-L1, do not work as well in controlling tumor growth in mice lacking STING gene expression [16]. A further understanding of the STING pathways in human tumors might provide important insights into the development of effective therapeutic strategies. In this regard, agents that stimulate the STING pathway are being evaluated as potential cancer therapeutics.

Concluding Remarks

Cancer immunotherapies targeting immune checkpoints have the potential to generate robust antitumor responses, and clinical trials using different checkpoint inhibitors to treat patients with several tumor types have yielded unprecedented results. However, despite numerous advances, most patients do not respond favorably to T cell-based therapies. This may be due to additional mechanisms that can influence the enhancement of T cell function at play, in addition to multiple mechanisms used by tumors to evade the immune system. Recent studies have further provided insight into the immune signaling pathways that control T cell activation in the tumor and that could be targeted to improve tumor destruction by immune cells. However, because of heterogeneity of human cancers, the mechanism of immune escape is highly complex and dictated by the cellular and molecular features of the tumor microenvironment. Tumors called "inflamed or hot tumors" have preexisting immune responses and are infiltrated by immune cells, while tumors devoid of immune cell infiltration are called "cold tumors." This classification highlights the mechanism by which tumors can escape detection/destruction by the immune system. Hence, depending on the type of tumor, the two major phenotypes of tumor microenvironment may require distinct immunotherapeutic interventions for maximal therapeutic effect. Understanding immune signaling in tumors that can define the underlying mechanisms of immune resistance will help in finding the appropriate immunotherapeutic approach for different tumor types. This could be achieved by defining biomarkers that can identify the immune state of the tumor. Tumors that display the T cell-inflamed phenotype are characterized by infiltrating T cells, PD-L1 expression chemokine profile, and an elevated IFN- γ signature profile suggesting ongoing immune responses. These tumors resist immune attack by the dominant inhibitory effects of suppressive pathways of the immune system and are likely to respond to checkpoint inhibitors or to combination immunotherapies. Tumors lacking immune infiltration usually lack PD-L1 expression or IFN-γ signature and resist immune attack through immune system exclusion or ignorance. These tumors need agents that can initiate or enhance the detection of tumors by the immune system or augment trafficking and infiltration of T cells into the tumor. The lack of T cell responses in tumors could be due to the poor recognition of tumor antigens by T cells. However, which tumor antigens are recognized by T cells have been difficult to identify. Recent advances in next-generation sequencing have helped identify mutations in tumors that are potentially antigenic (called neoantigens) and are targeted by T cells.

Identifying and targeting such neoantigens to boost T cell responses using a personalized cancer vaccine could be another way of boosting the efficacy of the existing approaches. Hence, future effective immunotherapies will likely involve novel combinations of different immunotherapeutic approaches and other cancer targets that are personalized for maximal benefit.

References

1. Callahan MK, Postow MA, Wolchok JD. Targeting T cell co-receptors for cancer therapy. Immunity. 2016;44(5):1069–78. https://doi.org/10.1016/j.immuni.2016.04.023.
2. Sharma P, Allison JP. The future of immune checkpoint therapy. Science. 2015;348(6230):56–61. https://doi.org/10.1126/science.aaa8172.
3. Ise W, Kohyama M, Nutsch KM, Lee HM, Suri A, Unanue ER, et al. CTLA-4 suppresses the pathogenicity of self antigen-specific T cells by cell-intrinsic and cell-extrinsic mechanisms. Nat Immunol. 2010;11(2):129–35. https://doi.org/10.1038/ni.1835.
4. Khalil DN, Smith EL, Brentjens RJ, Wolchok JD. The future of cancer treatment: immunomodulation, CARs and combination immunotherapy. Nat Rev Clin Oncol. 2016;13(5):273–90. https://doi.org/10.1038/nrclinonc.2016.25.
5. Pardoll DM. The blockade of immune checkpoints in cancer immunotherapy. Nat Rev Cancer. 2012;12(4):252–64. https://doi.org/10.1038/nrc3239.
6. Nishimura H, Minato N, Nakano T, Honjo T. Immunological studies on PD-1 deficient mice: implication of PD-1 as a negative regulator for B cell responses. Int Immunol. 1998;10(10):1563–72.
7. Alsaab HO, Sau S, Alzhrani R, Tatiparti K, Bhise K, Kashaw SK, et al. PD-1 and PD-L1 checkpoint signaling inhibition for cancer immunotherapy: mechanism, combinations, and clinical outcome. Front Pharmacol. 2017;8:561. https://doi.org/10.3389/fphar.2017.00561.
8. Manieri NA, Chiang EY, Grogan JL. TIGIT: a key inhibitor of the cancer immunity cycle. Trends Immunol. 2017;38(1):20–8. https://doi.org/10.1016/j.it.2016.10.002.
9. Johnston RJ, Comps-Agrar L, Hackney J, Yu X, Huseni M, Yang Y, et al. The immunoreceptor TIGIT regulates antitumor and antiviral CD8(+) T cell effector function. Cancer Cell. 2014;26(6):923–37. https://doi.org/10.1016/j.ccell.2014.10.018.
10. Anderson AC, Joller N, Kuchroo VK. Lag-3, Tim-3, and TIGIT: co-inhibitory receptors with specialized functions in immune regulation. Immunity. 2016;44(5):989–1004. https://doi.org/10.1016/j.immuni.2016.05.001.
11. Holmgaard RB, Zamarin D, Li Y, Gasmi B, Munn DH, Allison JP, et al. Tumor-expressed IDO recruits and activates MDSCs in a treg-dependent manner. Cell Rep. 2015;13(2):412–24. https://doi.org/10.1016/j.celrep.2015.08.077.
12. Ledford H. (2017). Next-generation cancer drugs boost immunotherapy responses. Early clinical trial data suggest that combining medicines improves treatment. Nat News. https://doi.org/10.1038/nature.2017.22092.
13. Linch SN, McNamara MJ, Redmond WL. OX40 agonists and combination immunotherapy: putting the pedal to the metal. Front Oncol. 2015;5:34. https://doi.org/10.3389/fonc.2015.00034.
14. Bartkowiak T, Curran MA. 4-1BB agonists: multi-potent potentiators of tumor immunity. Front Oncol. 2015;5:117. https://doi.org/10.3389/fonc.2015.00117.
15. van de Ven K, Borst J. Targeting the T-cell co-stimulatory CD27/CD70 pathway in cancer immunotherapy: rationale and potential. Immunotherapy. 2015;7(6):655–67. https://doi.org/10.2217/imt.15.32.
16. Woo SR, Fuertes MB, Corrales L, Spranger S, Furdyna MJ, Leung MY, et al. STING-dependent cytosolic DNA sensing mediates innate immune recognition of immunogenic tumors. Immunity. 2014;41(5):830–42. https://doi.org/10.1016/j.immuni.2014.10.017.
17. Gilboa E. The promise of cancer vaccines. Nat Rev Cancer. 2004;4(5):401–11.

Predictive Biomarkers and Targeted Therapies in Immuno-oncology

29

Hartmut Koeppen, Mark L. McCleland, and Marcin Kowanetz

Introduction

Immunotherapy exploits the body's immune system to fight cancer. The explosion in the number of ongoing cancer immunotherapy (CIT) trials reflects the great enthusiasm and potential for a cancer cure the field is believed to hold. Strikingly, more than 800 different immunotherapy clinical trials are currently underway to evaluate anti-PD-L1/PD-1 and other immunomodulatory agents alone or in combination with existing or new molecular entities in cancer patients [1–3]. To date, immunotherapy has shown efficacy in numerous tumor types including melanoma, non-small cell lung cancer (NSCLC), renal cancer, bladder cancer, colon cancer, and head and neck cancers [4]. These successes are only just the beginning. In the years ahead, new drug combinations will allow us to unlock patient populations and new indications that are currently unresponsive to the existing CIT treatments. As additional therapeutic options come to the market, it will be critical to identify and co-develop novel predictive biomarkers that will facilitate matching the right patient with the right drug combination.

The first wave of immunotherapy trials compared CIT monotherapy versus standard-of-care chemotherapy and established efficacy in cancer patients with advanced and metastatic disease that failed previous lines of therapy. The current focus (second wave) of clinical trials is to augment the success seen with monotherapy CIT through the combination of immunotherapy with other modalities, such as chemotherapy, radiation, or with other immunomodulatory agents. Promising results and positive trials have now been reported in NSCLC and renal cell carcinoma, where anti-PD-L1 or anti-PD-1 inhibitors were combined with chemo- and/or anti-angiogenic therapy or immune doublet therapies such as the combination of PD-1 targeted agents with CTLA4 inhibitors [5, 6]. A third wave of trials will undoubtedly combine novel multiple immunomodulatory agents with anti-PD-L1/PD-1 with the intention to replace cytotoxic chemotherapy.

While clinical success has been observed across multiple tumor types and with multiple agents, only a minority of patients have benefited from durable and long-lasting responses. It will be critical in the coming years to develop a deeper understanding of this phenomenon in order to facilitate smarter and science-driven drug combinations. Biomarkers have been and will continue to define patient populations that respond best to

H. Koeppen (✉)
Research Pathology, Genentech, South San Francisco, CA, USA
e-mail: hkoeppen@gene.com

M. L. McCleland · M. Kowanetz
Department of Oncology Biomarker Development, Genentech, South San Francisco, CA, USA

S. Badve, G. L. Kumar (eds.), *Predictive Biomarkers in Oncology*,
https://doi.org/10.1007/978-3-319-95228-4_29

therapy (predictive biomarker), that are prone to poor clinical outcome (prognostic biomarker), or that are at risk of higher drug-associated toxicity (safety biomarker). Predictive biomarkers have garnered the most attention, as they are commonly used to guide treatment decisions for patients. Importantly, predictive biomarkers have increased clinical trial success rate, accelerated market access, decreased clinical development costs, and ultimately saved patients from receiving drug from which they would not benefit. To ensure biomarker results are robust and reproducible, biomarker assay development requires extensive analytical validation. As discussed in other chapters of this book, the emergence of novel high throughout technologies, such as mass spectrometry and gene expression profiling, provides an exciting opportunity for comprehensive biomarker profiling and co-development of companion diagnostics with new therapies [7, 8]. The remainder of this chapter focuses on predictive biomarkers, both approved as well as those still under development, for currently approved CIT drugs.

Immune checkpoint blockade therapies have transformed our approach to cancer treatment. The checkpoint inhibitors currently approved for clinical use are listed in Table 29.1. Based on the consistent clinical activity seen with anti-PD-L1/PD-1 inhibitors, the first-generation diagnostic tests focused on detecting PD-L1 expression in the tumor tissue. Tissue-based, immunohistochemical (IHC) tests were developed and employed in clinical development programs. PD-1, the receptor for PD-L1, is a member of the B7-CD28 super family and is expressed on numerous cell types, including activated T cells, B cells, and NK cells [28]. The interaction of PD-1 with its two ligands, PD-L1 and PD-L2, which can be expressed on various cell types within the tumor microenvironment (including tumor and various types of immune cells), may lead to downregulation of a potential antitumor immune response [29]. PD-L1 IHC assays have been granted regulatory approval as companion or complementary diagnostic tests (see details below) for PD-1/PD-L1 targeted therapies. Additional biomarkers such as those evaluating gene expression signatures in tissues, tumor mutational burden, and others have been evaluated in an exploratory fashion and may achieve regulatory approval at some point in the future [8, 30].

PD-L1 Immunohistochemistry (IHC)

IHC assays that are currently approved to detect PD-L1 in patients for anti-PD-L1/PD-1 therapies differ from each other in numerous ways, ranging from cutoffs defining the PD-L1 positivity to the type of cells (i.e., tumor vs immune) used to determine PD-L1 expression. Four tests currently being used in the clinic (22C3, 28-8, SP142, and SP263) are discussed below. These tests rely on formalin-fixed, paraffin-embedded (FFPE) tissue sections as source material, which is typically available in pathology laboratories for diagnostic and biomarker testing. The different assays are summarized in Table 29.1.

22C3

The PD-L1 IHC 22C3 pharmDx test from Agilent-Dako is a qualitative immunohistochemical companion diagnostic test to identify patients eligible for treatment with pembrolizumab (KEYTRUDA®; Merck). It is approved for patient identification in first-line (1L) and second-line and beyond (2L+) NSCLC and third-line gastroesophageal cancer. In 1L and 2L+ NSCLC, this test was approved based on the results of phase III randomized studies KEYNOTE-024 and KEYNOTE-010, respectively [21]; patients with PD-L1 expression on at least 50% (1L) and 1% (2L+) of tumor cells showed clinical benefit from pembrolizumab. While PD-L1 assessment is limited to tumor cells in NSCLC, it includes both tumor and stromal cells in gastroesophageal cancer. In the KEYNOTE-059 study, PD-L1 expression as determined by 22C3 was categorized as the PD-L1 combined positive score (CPS) defined as the percentage of PD-L1-expressing tumor and infiltrating immune cells relative to the total number of viable tumor cells.

Table 29.1 Currently approved checkpoint inhibitor therapies and associated diagnostic tests

Therapy target	Target expression	Drug/company	Approved indication	Patient selection	Trial	References	Dx assay
CTLA-4	Regulatory and activated T cells	Ipilimumab (Yervoy®) Bristol-Myers Squibb	Metastatic melanoma	None	Randomized phase III	[9]	
			Melanoma, adjuvant (stage III)	None	Randomized phase III	[10]	
PD-1	T cells, B cells, and NK cells	Nivolumab (Opdivo®)	Advanced melanoma				
		Bristol-Myers Squibb	Advanced NSCLC	None	Randomized phase III	CheckMate-017, 057 [11]	28–8 (compl.)[a]
			Renal cell carcinoma	None	Randomized phase III	[12]	
			Refractory Hodgkin's lymphoma (a)[b]		Phase II, single-arm, phase I	CheckMate-205 and -039 [13]	
			SCCHN	None	Randomized phase III	CheckMate-141 [14]	
			2L urothelial carcinoma (a)[b]	None	Phase II, single-arm	CheckMate-275 [15]	
			2L CRC	MSI-H, dMMR	Phase II, single-arm	CheckMate-142 [16]	
			2L HCC (a)[b]	None	Phase I/II	CheckMate -040 [17]	
			Melanoma, adjuvant	None	Phase III	CheckMate-238 [18]	
		Pembrolizumab (Keytruda®)	1L+ advanced melanoma	None	Randomized phase III	KEYNOTE-006 [19]	
		Merck	2L+ advanced NSCLC	$\geq$1% tumor cells positive for PD-L1		KEYNOTE-010 [20]	22C3 (cDx)[c]
			1L advanced NSCLC	$\geq$50% tumor cells positive for PD-L1	Phase III	KEYNOTE-024 [21]	22C3 (cDx)[c]
			Refractory Hodgkin's lymphoma (a)[b]	None	Phase II, single arm	KEYNOTE-087 [22]	
			1L advanced NSCLC-non-squamous	None	Randomized phase II	KEYNOTE-021G [5]	
			2L urothelial carcinoma	None	Randomized phase III	KEYNOTE-045 [23]	
			1L urothelial, platinum-ineligible	None	Single-arm phase II	KEYNOTE-052 [24]	
			2L advanced SCCHN (a)[b]	None	Single-arm phase II	KEYNOTE-055 [25]	
			Advanced solid tumors (a)[b]	MSI-H, dMMR	Single-arm trials	KN-158, KN-164 ascopubs.org/doi/abs/10.1200/JCO.2017.35.15_suppl.3071	MSI (cDx)
			2L+ GC (a)[b]	$\geq$1% PD-L1-positive		KEYNOTE-059 ascopubs.org/doi/abs/10.1200/JCO.2017.35.15_suppl.4003	

(continued)

Table 29.1 (continued)

Therapy target	Target expression	Drug/company	Approved indication	Patient selection	Trial	References	Dx assay
PD-L1	T, B, NK, DC, macrophages, tumor cells	Atezolizumab (Tecentriq ®)	2L+ metastatic NSCLC	None	Phase III	OAK; [26]	SP142 (compl.)[a]
		Genentech/ Roche	Locally advanced or metastatic urothelial carcinoma (a)[b]	None	Single-arm phase II	IMvigor 210 [27]	
		Durvalumab (IMFINZI™)	2L urothelial carcinoma (a)[b]	None	Single-arm phase II		SP263 (compl.)[a]
		AstraZeneca	2L advanced NSCLC	None	Randomized phase III		
		Avelumab (Bavencio®)	Urothelial carcinoma	None	Phase I	Javelin	

[a]Complementary diagnostic
[b]Accelerated approval
[c]Companion diagnostic

Patients with a CPS of ≥1% derived clinical benefit from treatment with pembrolizumab.

28-8

The PD-L1 IHC 28-8 pharmDx test from Agilent-Dako is a qualitative immunohistochemical test approved as complementary diagnostic assay to identify patients, who may derive clinical benefit from treatment with nivolumab (OPDIVO®; Bristol-Myers Squibb) in second-line NSCLC and first-line metastatic melanoma based on the results of the phase III randomized studies, CheckMate-057 and CheckMate-067, respectively. Both trials showed improved survival of patients receiving nivolumab compared to patients on the control arm; the survival benefit was more pronounced in patients with PD-L1-positive tumors (defined as tumors with ≥1% of tumor cells showing complete or incomplete membranous staining of any intensity). PD-L1 protein expression in this assay is defined as the percentage of tumor cells exhibiting positive membrane staining at any intensity; staining on stromal/immune cells is not considered. Unlike the companion diagnostic test 22C3, which is required for patient eligibility for pembrolizumab, 28-8 is a complementary diagnostic test for nivolumab and is intended to aide in the clinical decision-making process [31].

SP142

The SP142 PD-L1 IHC assay developed by Ventana is approved as a complementary diagnostic test for the use of atezolizumab (TECENTRIQ®; Genentech/Roche) in patients with NSCLC and urothelial carcinoma. Both tumor cells (TC) and tumor-infiltrating immune cells (IC) are evaluated for PD-L1 expression in this assay. PD-L1 TC results are expressed as the percentage of tumor cells staining positive at any intensity, and IC results are expressed as the percentage of the area of viable tumor occupied by PD-L1-positive immune cells. The complementary diagnostic approval for NSCLC was based on a randomized phase III trial (OAK) comparing atezolizumab with standard-of-care chemotherapy in patients who had failed first-line therapy; a survival benefit to atezolizumab was observed across all levels of PD-L1 expression but was greatest for patients with the highest PD-L1 expression (defined as expression on ≥50% TC or ≥10% IC) [26]. For urothelial carcinoma, approval as a complementary diagnostic was based on a single-arm, two-cohort phase II trial for patients who had failed platinum-based chemotherapy. Treatment benefit from atezolizumab was greatest in patients with tumors expressing PD-L1 on ≥5% of IC [27]. Most recently, atezolizumab in combination with bevacizumab showed positive results in a phase II trial for metastatic renal cell carcinoma. Patients with PD-L1-positive tumors on ≥1% of IC derived greater benefit than the intention-to-treat population (https://cancerletter.com/articles/20180209_7/).

SP263

The SP263 PD-L1 IHC assay developed by Ventana was first approved as a complementary diagnostic for treatment of patients with advanced urothelial carcinoma with durvalumab (IMFINZI®; AstraZeneca) based on results of single-arm phase II study. The SP263 scoring algorithm captures PD-L1 expression on both TC and IC. A tumor sample is scored as positive if ≥25% of tumor cells exhibit membrane staining of any intensity, if ≥25% of immune cells are positive and occupy >1% of the viable tumor area, or if 100% of IC are PD-L1-positive and occupy 1% of the tumor area (https://www.accessdata.fda.gov/cdrh_docs/pdf16/P160046C.pdf). SP263 has also been CE-marked in EU and can be used to identify NSCLC patients for treatment with nivolumab or pembrolizumab. This approval was based on comparability studies of SP263 with the 28-8 and the 22C3 IHC assays, respectively. In this application only TC are evaluated for PD-L1 expression with SP263 assay.

Limitations of PD-L1 IHC Biomarker Data

The variation in assay platforms, primary antibody clones, and secondary detection reagents among the various PD-L1 assays may yield different results when testing the same tumor for PD-L1 expression. Furthermore, the scoring algorithms used to identify the PD-L1-positive cell types and cutoffs differ among the assays (i.e., tumor cells only vs tumor and immune cells). To evaluate the analytical performance and address market harmonization of the approved PD-L1 IHC assays, a collaborative project called "The Blueprint Programmed Death Ligand 1 (PD-L1) Immunohistochemistry (IHC) Assay Comparison Project" was established [32]. Analytical comparison demonstrated good concordance between the 22C3, 28-8, or SP263 assays with respect to proportion of PD-L1-positive tumor cells; the SP142 assay showed lower proportion of positive tumor cells. Greater variability across the four assays was seen for immune cell staining but without a consistent pattern. Based on the limited sample set, the authors concluded that misclassification of a proportion of tumors with respect to PD-L1 status could occur when using assays and/or scoring algorithms interchangeably [32].

There are several limitations of this first phase of the Blueprint project that could be addressed in subsequent iterations. The cohort of NSCLC cases analyzed was small and not associated with clinical outcome; therefore, the predictive value of each individual assay and associated algorithm could not be evaluated. Pathologist readers ($n = 3$) did not receive training for interpretation of the assay they were not expert in. Lastly, an orthogonal methodology to verify degree and pattern of PD-L1 expression in the tissues was not attempted. Analysis of larger NSCLC cohorts has now been published by other investigators [33, 34]. It should also be noted that each of the four assays has demonstrated predictive value in pivotal clinical studies and is FDA approved. A recent study compared the performance and predictive value for two PD-L1 assays, 22C3 and SP142, on patients treated with an anti-PDL1 agent atezolizumab in second-line NSCLC; the results demonstrated equivalent survival benefit in patient populations defined as positive for PD-L1 by either assay [35].

Each of the four assays described above performs robustly and reliably when used as prescribed and can enrich for the appropriate patient population. However, the use of an IHC assay in general and PD-L1 as a predictive biomarker for anti-PD-L1/PD-1 targeted therapies specifically comes with challenges [36]. First of all, the tumor microenvironment is a dynamic space with interactions of multiple cell types and assessment of a single marker most likely represents an oversimplification. IHC tends to perform well for binary observations but is much less reliable for the readout of continuous variables; gradients of expression are most likely critical for activation or inhibition of a tumor immune response and might be measured more appropriately through alternate technologies (see below). Common to most cancer immunotherapy trials is the observation that biomarker (PD-L1)-negative patients may respond suggesting that – in the presence of an adequately performing assay – the tissue sample is inadequate (time, location) or the biomarker is imperfect. Below we focus on technologies which characterize the tumor microenvironment more globally and are tested in ongoing clinical trials for their value in identifying patients for checkpoint inhibitor treatment.

Gene Expression Signatures

Clinical benefit to CIT therapies has been observed in patients that lack PD-L1 expression (i.e., patients negative by PD-L1 IHC tests). This observation suggests a complex biology underlies the immune response and that a single-parameter assay may be insufficient to predict patient outcome to CIT. Furthermore, it highlights the need for continued biomarker discovery to more accurately identify patients that benefit most from CIT therapy [26]. Gene signature profiles are being studied as predictive biomarkers in trials using checkpoint inhibitors for a variety of indications. At the foundation of many of these signatures are T-cell- and immune biology-related genes that are thought to represent T-cell infiltration and pre-existing immunity within the tumor microenvironment. Gene expression assay outputs have the upside of a quantitative continuous variable and are not limited by subjective assessment that accompanies pathological assessment by IHC. In a phase II trial evaluating ipilimumab in advanced melanoma, increased numbers of tumor-infiltrating lymphocytes (TILS) and increased expression of FoxP3 and IDO by IHC were associated with clinical activity. Gene expression analysis in these samples also found increases in immune-related genes such as granzyme B, perforin-1, and T-cell receptor subunits in samples during treatment; however, such changes did not reach levels of statistical significance [37]. Likewise, a recent neoadjuvant ipilimumab melanoma study discovered that high expression of immune-related genes in baseline tumor samples predicted clinical benefit [38].

While there has not been a gene signature assay yet approved by regulatory agencies, favorable associations and outcomes have been similarly reported in a wide range of anti-PD-L1/PD-1 therapies. The phase II POPLAR and phase III OAK studies, both evaluating efficacy of atezolizumab vs docetaxel in 2L+ NSCLC patients, revealed a novel association between a T-effector gene signature profile and clinical benefit (PFS and OS) to atezolizumab [39, 40]. A similar association between the T-effector gene signature and clinical benefit to atezolizumab in combination with bevacizumab and chemotherapy was observed in a recently reported phase III trial, IMpower150 [41]. Similar results were also reported for durvalumab, whereby a baseline IFN-γ gene expression signature was associated with improved clinical outcomes in durvalumab-treated advanced NSCLC cancer patients [42]. Yet another independent study identified and validated gene signatures related to IFN-γ signaling and activated T-cell biology for pembrolizumab across multiple distinct tumor types including melanoma, head and neck squamous cell carcinoma (HNSCC), and gastric cancer [43]. This T-cell-inflamed gene expression profile (GEP) suggested that a tumor microenvironment characterized by antigen presentation, active IFN-γ signaling, and cytotoxic effector activity is potentially responsive to PD-1 checkpoint blockade [43]. Altogether, gene expression profiling affords promise that more accurate and sensitive RNA-based next-generation diagnostic assays will be available in the near future for CIT patients.

Tumor Mutational Burden

Tumor mutational burden (TMB) is yet another biomarker that has demonstrated potential as a predictive biomarker for CIT patients [44]. TMB measures the number of mutations per coding area within a tumor genome. All cancers are caused by somatic mutations, which are attributed to a number of factors including malfunctioning of the DNA replication machinery, exposure to various mutagens (e.g., tobacco, asbestos, UV light), DNA modification, defects in DNA replication, etc. As the adaptive immune system requires foreign antigen presented on MHC to initiate a T-cell-directed immune response, patients with highly mutated tumors are more likely to generate and present a neoantigen that can be subsequently recognized by immune cells to mount an effective antitumor response.

Alexandrov et al. conducted an exhaustive study of the mutation data set across thousands of tumors from 30 cancer types with the goal of evaluating average mutation burden across indications. Interestingly, the actual number of mutations per megabase varied on average across tumor type with the highest mutation burden observed in CIT responsive indications, including melanoma, NSCLC, and bladder cancer [45] (Fig. 29.1). Together with subsequently reported clinical correlation between TMB and efficacy of CIT, this study opened a novel diagnostic opportunity to test mutation burden as a predictor of clinical response to CIT therapy [45].

The commonly used method to measure TMB includes comprehensive genomic profiling using whole-exome sequencing, whereby all genes in the protein-coding region of the tumor genome are sequenced. As a robust alternative, targeted cancer genome panels, such as the FoundationOne® assay (Foundation Medicine), have provided reliable results. This method measures the somatic mutations occurring in selected genes, instead of sequencing the entire exome [27].

Multiple clinical studies have now demonstrated that TMB can predict responses to checkpoint inhibitor immunotherapies across different cancer types, including NSCLC, SCLC, and bladder cancer [46–49]. TMB was also associated with higher response rate and PFS, but not OS, to nivolumab vs chemotherapy in the exploratory analysis of the recently reported phase III study CheckMate-026 in PD-L1-selected 1L NSCLC patients [50]. The main caveat of TMB analysis in this study was that patients had already been selected based on PD-L1 IHC expression. Nevertheless, these results further solidified the importance of TMB as an important CIT biomarker. Importantly, PD-L1 IHC and TMB appear to be independent predictors of CIT efficacy, further suggesting that multiple biomarkers may be needed to most accurately identify patients benefiting from CIT. TMB can also be reliably measured in blood of cancer patients through NGS approaches on circulating tumor DNA. NSCLC

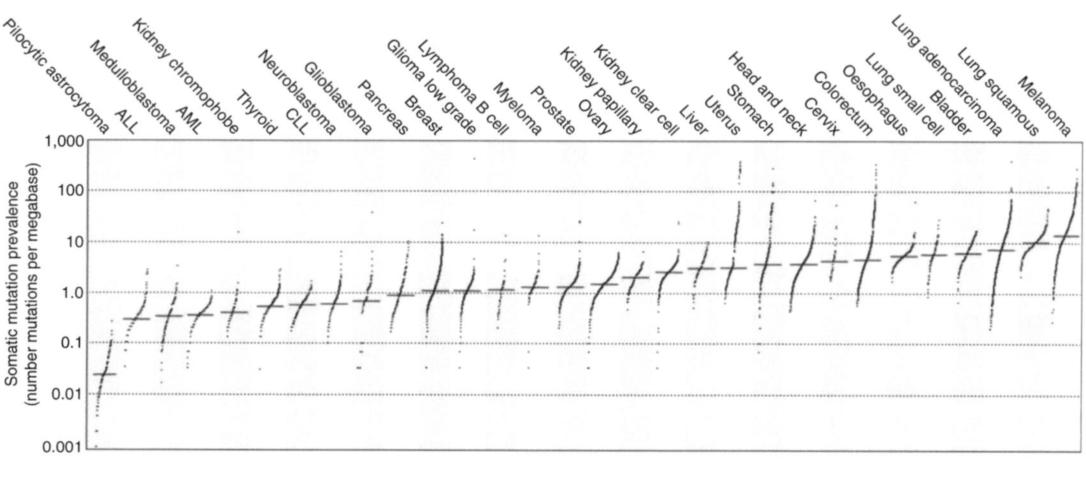

Fig. 29.1 Number of somatic mutations (per megabase of genome) across human cancer types. Dots represent individual samples. Red lines correspond to median numbers of mutations in each cancer type. (Reprinted from Alexandrov et al. [45]. With permission from Springer Nature)

patients with high TMB in their plasma derived an improved PFS benefit to atezolizumab vs docetaxel as demonstrated in an exploratory analysis of studies POPLAR and OAK [51]. Together these data warrant development of blood-based biomarkers for CIT in the near future.

In 2017, FDA approved pembrolizumab for treating solid tumors that are microsatellite instability-high (MSI-H) or mismatch repair deficient (dMMR). This was a seminal approval, as it was the first pan-tumor/pan-tissue agnostic FDA approval of a CIT drug. Approval was based on pembrolizumab efficacy across 15 cancer types, including colon cancer, renal cell cancer, and pancreatic cancer [52]. The efficacy observed in these cancer patients most likely reflects the high level of somatic mutations and mutation-associated neoantigens that trigger the immune system to mount a response against the tumor.

Other Biomarkers

In addition to the three platforms discussed, several other platforms have emerged that aim to profile the tumor immune interface in order to predict response to immunotherapy. These platforms are based on multiplexed transcriptome analysis, protein expression, and genomic variability. Examples include multiparametric flow cytometric immunophenotyping of peripheral blood, T-cell receptor (TCR) sequencing for clonality assessments of tumor-infiltrating lymphocytes (TILs), and assessing presence of CD8+ T cells and TILs within the tumor. These approaches are preliminary and require further investigation but have shown promise in predicting response to CIT checkpoint blockade [8, 53].

Summary

While many patients derive long-term clinical benefit from various CITs alone or in combination with other modalities, a substantial number of patients do not derive such benefit. Therefore, there is a great need to develop and validate predictive biomarkers of response to CIT. Here, we discussed the major biomarker platforms which are FDA approved or being actively pursued for CIT. It must be highlighted that the development of a biomarker test for clinical application is a highly regulated process that involves proper clinical trial design for clinical validation. The regulatory aspects of submission of biomarker assays to the FDA in the USA, as well as regulatory considerations in the European Union and other regions, must be well thought out during the planning and implementation phases of clinical and biomarker development. Ultimately, the approval of well-validated clinical biomarkers can maximize the benefits of CIT while reducing cost and toxicity [54]. Finally, a complex immune biology underlying responses to CIT suggests that using a single biomarker, such as PD-L1 IHC, to identify all patients benefiting from CIT is not possible and that most likely multiple approaches are needed to increase our precision in identifying patients benefiting from CIT [3]. Identifying such more complex biomarkers will be increasingly important to help to select a personalized treatment regimen for patients, as numerous CIT mono- and combination therapies will be becoming available in the future.

References

1. Farkona S, Diamandis EP, Blasutig IM. Cancer immunotherapy: the beginning of the end of cancer? BMC Med. 2016;14:73.
2. Chen DS, Mellman I. Oncology meets immunology: the cancer-immunity cycle. Immunity. 2013;39(1):1–10.
3. Chen DS, Mellman I. Elements of cancer immunity and the cancer-immune set point. Nature. 2017;541(7637):321–30.
4. Dempke WCM, et al. Second- and third-generation drugs for immuno-oncology treatment-the more the better? Eur J Cancer. 2017;74:55–72.
5. Langer CJ, et al. Carboplatin and pemetrexed with or without pembrolizumab for advanced, non-squamous non-small-cell lung cancer: a randomised, phase 2 cohort of the open-label KEYNOTE-021 study. Lancet Oncol. 2016;17(11):1497–508.
6. Jotte RM, et al. PS01.53: first-line atezolizumab plus chemotherapy in chemotherapy-naive patients with advanced NSCLC: a phase III clinical program: topic: medical oncology. J Thorac Oncol. 2016;11(11S):S302–3.

7. Gulley JL, et al. Immunotherapy biomarkers 2016: overcoming the barriers. J Immunother Cancer. 2017;5(1):29.
8. Yuan J, et al. Novel technologies and emerging biomarkers for personalized cancer immunotherapy. J Immunother Cancer. 2016;4:3.
9. Hodi FS, et al. Improved survival with ipilimumab in patients with metastatic melanoma. N Engl J Med. 2010;363(8):711–23.
10. Eggermont AM, et al. Prolonged survival in stage III melanoma with ipilimumab adjuvant therapy. N Engl J Med. 2016;375(19):1845–55.
11. Horn L, et al. Nivolumab versus docetaxel in previously treated patients with advanced non-small-cell lung cancer: two-year outcomes from two randomized, open-label, phase III trials (CheckMate 017 and CheckMate 057). J Clin Oncol. 2017;35(35):3924–33.
12. Motzer RJ, et al. Nivolumab versus everolimus in advanced renal-cell carcinoma. N Engl J Med. 2015;373(19):1803–13.
13. Younes A, et al. Nivolumab for classical Hodgkin's lymphoma after failure of both autologous stem-cell transplantation and brentuximab vedotin: a multicentre, multicohort, single-arm phase 2 trial. Lancet Oncol. 2016;17(9):1283–94.
14. Ferris RL, et al. Nivolumab for recurrent squamous-cell carcinoma of the head and neck. N Engl J Med. 2016;375(19):1856–67.
15. Sharma P, et al. Nivolumab in metastatic urothelial carcinoma after platinum therapy (CheckMate 275): a multicentre, single-arm, phase 2 trial. Lancet Oncol. 2017;18(3):312–22.
16. Overman MJ, et al. Nivolumab in patients with metastatic DNA mismatch repair-deficient or microsatellite instability-high colorectal cancer (CheckMate 142): an open-label, multicentre, phase 2 study. Lancet Oncol. 2017;18(9):1182–91.
17. El-Khoueiry AB, et al. Nivolumab in patients with advanced hepatocellular carcinoma (CheckMate 040): an open-label, non-comparative, phase 1/2 dose escalation and expansion trial. Lancet. 2017;389(10088):2492–502.
18. Weber J, et al. Adjuvant nivolumab versus ipilimumab in resected stage III or IV melanoma. N Engl J Med. 2017;377(19):1824–35.
19. Robert C, et al. Pembrolizumab versus ipilimumab in advanced melanoma. N Engl J Med. 2015;372(26):2521–32.
20. Herbst RS, et al. Pembrolizumab versus docetaxel for previously treated, PD-L1-positive, advanced non-small-cell lung cancer (KEYNOTE-010): a randomised controlled trial. Lancet. 2016;387(10027):1540–50.
21. Reck M, et al. Pembrolizumab versus chemotherapy for PD-L1-positive non-small-cell lung cancer. N Engl J Med. 2016;375(19):1823–33.
22. Chen R, et al. Phase II study of the efficacy and safety of pembrolizumab for relapsed/refractory classic Hodgkin lymphoma. J Clin Oncol. 2017;35(19):2125–32.
23. Bellmunt J, et al. Pembrolizumab as second-line therapy for advanced urothelial carcinoma. N Engl J Med. 2017;376(11):1015–26.
24. Balar AV, et al. First-line pembrolizumab in cisplatin-ineligible patients with locally advanced and unresectable or metastatic urothelial cancer (KEYNOTE-052): a multicentre, single-arm, phase 2 study. Lancet Oncol. 2017;18(11):1483–92.
25. Bauml J, et al. Pembrolizumab for platinum- and cetuximab-refractory head and neck cancer: results from a single-arm, phase II study. J Clin Oncol. 2017;35(14):1542–9.
26. Rittmeyer A, et al. Atezolizumab versus docetaxel in patients with previously treated non-small-cell lung cancer (OAK): a phase 3, open-label, multicentre randomised controlled trial. Lancet. 2017;389(10066):255–65.
27. Rosenberg JE, et al. Atezolizumab in patients with locally advanced and metastatic urothelial carcinoma who have progressed following treatment with platinum-based chemotherapy: a single-arm, multicentre, phase 2 trial. Lancet. 2016;387(10031):1909–20.
28. Okazaki T, Honjo T. PD-1 and PD-1 ligands: from discovery to clinical application. Int Immunol. 2007;19(7):813–24.
29. Blank C, Gajewski TF, Mackensen A. Interaction of PD-L1 on tumor cells with PD-1 on tumor-specific T cells as a mechanism of immune evasion: implications for tumor immunotherapy. Cancer Immunol Immunother. 2005;54(4):307–14.
30. Gnjatic S, et al. Identifying baseline immune-related biomarkers to predict clinical outcome of immunotherapy. J Immunother Cancer. 2017;5:44.
31. Scheerens H, et al. Current status of companion and complementary diagnostics: strategic considerations for development and launch. Clin Transl Sci. 2017;10(2):84–92.
32. Hirsch FR, et al. PD-L1 immunohistochemistry assays for lung cancer: results from phase 1 of the blueprint PD-L1 IHC assay comparison project. J Thorac Oncol. 2017;12(2):208–22.
33. Hendry S, et al. Comparison of four PD-L1 immunohistochemical assays in lung cancer. J Thorac Oncol. 2018;13(3):367–76.
34. Rimm DL, et al. A prospective, multi-institutional, pathologist-based assessment of 4 immunohistochemistry assays for PD-L1 expression in non-small cell lung cancer. JAMA Oncol. 2017;3(8):1051–8.
35. Gadgeel S, et al. 1296OClinical efficacy of atezolizumab (Atezo) in PD-L1 subgroups defined by SP142 and 22C3 IHC assays in 2L+ NSCLC: results from the randomized OAK study. Ann Oncol. 2017;28(suppl_5):mdx380.001–mdx380.001.
36. Kerr KM. The PD-L1 immunohistochemistry biomarker: two steps forward, one step back? J Thorac Oncol. 2018;13(3):291–4.
37. Hamid O, et al. A prospective phase II trial exploring the association between tumor microenvironment biomarkers and clinical activity of ipilimumab in advanced melanoma. J Transl Med. 2011;9:204.

38. Tarhini AA, et al. Expression profiles of immune-related genes are associated with neoadjuvant ipilimumab clinical benefit. Oncoimmunology. 2017;6(2):e1231291.
39. Fehrenbacher L, et al. Atezolizumab versus docetaxel for patients with previously treated non-small-cell lung cancer (POPLAR): a multicentre, open-label, phase 2 randomised controlled trial. Lancet. 2016;387(10030):1837–46.
40. Gadgeel S, et al. PL04a.02: OAK, a randomized Ph III study of atezolizumab vs docetaxel in patients with advanced NSCLC: results from subgroup analyses. J Thorac Oncol. 2017;12(1):S9–S10.
41. Reck M, et al. LBA1_PRPrimary PFS and safety analyses of a randomized phase III study of carboplatin + paclitaxel +/− bevacizumab, with or without atezolizumab in 1L non-squamous metastatic nsclc (IMPOWER150). Ann Oncol. 2017;28(suppl_11):mdx760.002–mdx760.002.
42. Higgs BW, et al. Relationship of baseline tumoral IFNγ mRNA and PD-L1 protein expression to overall survival in durvalumab-treated NSCLC patients. J Clin Oncol. 2016;34(15_suppl):3036–3036.
43. Ayers M, et al. IFN-gamma-related mRNA profile predicts clinical response to PD-1 blockade. J Clin Invest. 2017;127(8):2930–40.
44. Gibney GT, Weiner LM, Atkins MB. Predictive biomarkers for checkpoint inhibitor-based immunotherapy. Lancet Oncol. 2016;17(12):e542–51.
45. Alexandrov LB, et al. Signatures of mutational processes in human cancer. Nature. 2013;500(7463):415–21.
46. Rizvi NA, et al. Cancer immunology. Mutational landscape determines sensitivity to PD-1 blockade in non-small cell lung cancer. Science. 2015;348(6230):124–8.
47. Powles T, et al. Atezolizumab versus chemotherapy in patients with platinum-treated locally advanced or metastatic urothelial carcinoma (IMvigor211): a multicentre, open-label, phase 3 randomised controlled trial. Lancet. 2018;391(10122):748–57.
48. Kowanetz M, et al. OA20.01 tumor mutation burden (TMB) is associated with improved efficacy of atezolizumab in 1L and 2L+ NSCLC patients. J Thorac Oncol. 2017;12(1):S321–2.
49. Mutation load offers predictive biomarker in SCLC. Cancer Discov. 2017. [Epub ahead of print]. https://doi.org/10.1158/2159-8290.CD-NB2017-154.
50. Carbone DP, et al. First-line nivolumab in stage IV or recurrent non-small-cell lung cancer. N Engl J Med. 2017;376(25):2415–26.
51. Gandara DR, et al. 1295OBlood-based biomarkers for cancer immunotherapy: tumor mutational burden in blood (bTMB) is associated with improved atezolizumab (atezo) efficacy in 2L+ NSCLC (POPLAR and OAK). Ann Oncol. 2017;28(suppl_5):mdx380-mdx380.
52. Le DT, et al. PD-1 blockade in tumors with mismatch-repair deficiency. N Engl J Med. 2015;372(26):2509–20.
53. Masucci GV, et al. Validation of biomarkers to predict response to immunotherapy in cancer: volume I – pre-analytical and analytical validation. J Immunother Cancer. 2016;4:76.
54. Dobbin KK, et al. Validation of biomarkers to predict response to immunotherapy in cancer: volume II – clinical validation and regulatory considerations. J Immunother Cancer. 2016;4:77.

Role of Protein Tyrosine Phosphatases in Cancer Signaling

30

Elie Kostantin, Yevgen Zolotarov, and Michel L. Tremblay

PTEN

PTEN (phosphatase and tensin homolog) was identified as a tumor suppressor 20 years ago and has since been studied extensively. It contains a domain with high sequence similarity to protein tyrosine phosphatases. PTEN protein phosphatase activity is much lower than that of protein tyrosine phosphatase 1B (PTP1B). Recombinant PTEN can dephosphorylate proteins on serine, threonine, and tyrosine residues. Lipid dephosphorylation is the major result of PTEN activity that is important in cancer development [1]. PTEN genetic lesions are found in multiple cancers, and together with p53 mutations, it is one of the hallmarks of cancer development. Mutations in PTEN are often observed in prostate, skin, uterine, and central nervous system cancers [2].

In uterine cancer, mutations of PTEN are found in more than 60% of cases, and in metastatic prostate cancer, PTEN is deleted in almost 40% of cases (cBioPortal data). PTEN acts as a tumor suppressor by counteracting the activity of phosphatidylinositol 3-kinase (PI3K). PI3K phosphorylates phosphatidylinositol-4,5-bisphosphate (PIP_2) to generate phosphatidylinositol-3,4,5-trisphosphate (PIP_3), which brings protein kinase AKT to the membrane where it is phosphorylated and activated by mTORC2. Activated AKT acts on multiple downstream pathways involved in cell growth, cell cycle, apoptosis, and DNA repair. PTEN dephosphorylates PIP_3 to produce PIP_2, which prevents AKT activation [2]. PTEN interacts with another tumor suppressor, p53. p53 can bind to the promoter of PTEN and activate its expression, whereas PTEN can interact with p53 and stabilize it [1]. In prostate cancer patients, PTEN can be detected in exosomes circulating in the blood, whereas it is undetectable in normal subjects [3]. Therefore, PTEN status has a potential to be used as a biomarker in a noninvasive method (Fig. 30.1).

Four clinical studies directly targeting PTEN mutations/loss in neoplasms have been completed to date, and many more are ongoing. For example, an active phase 1 trial is targeting PTEN loss in solid tumors using two drugs: pazopanib (tyrosine kinase inhibitor) and everolimus (mTOR inhibitor) (see Table 30.1). However, there are no FDA-approved drugs for PTEN-targeted therapies.

PRLs

The phosphatases of regenerating liver 1, 2, and 3 (PRL 1–3), also known as *PTP4A1–3*, are members of the PTPs family. This phosphatase

E. Kostantin · Y. Zolotarov · M. L. Tremblay (✉)
Department of Biochemistry, Goodman Cancer Research Center, McGill University, Montreal, QC, Canada
e-mail: michel.tremblay@mcgill.ca

S. Badve, G. L. Kumar (eds.), *Predictive Biomarkers in Oncology*,
https://doi.org/10.1007/978-3-319-95228-4_30

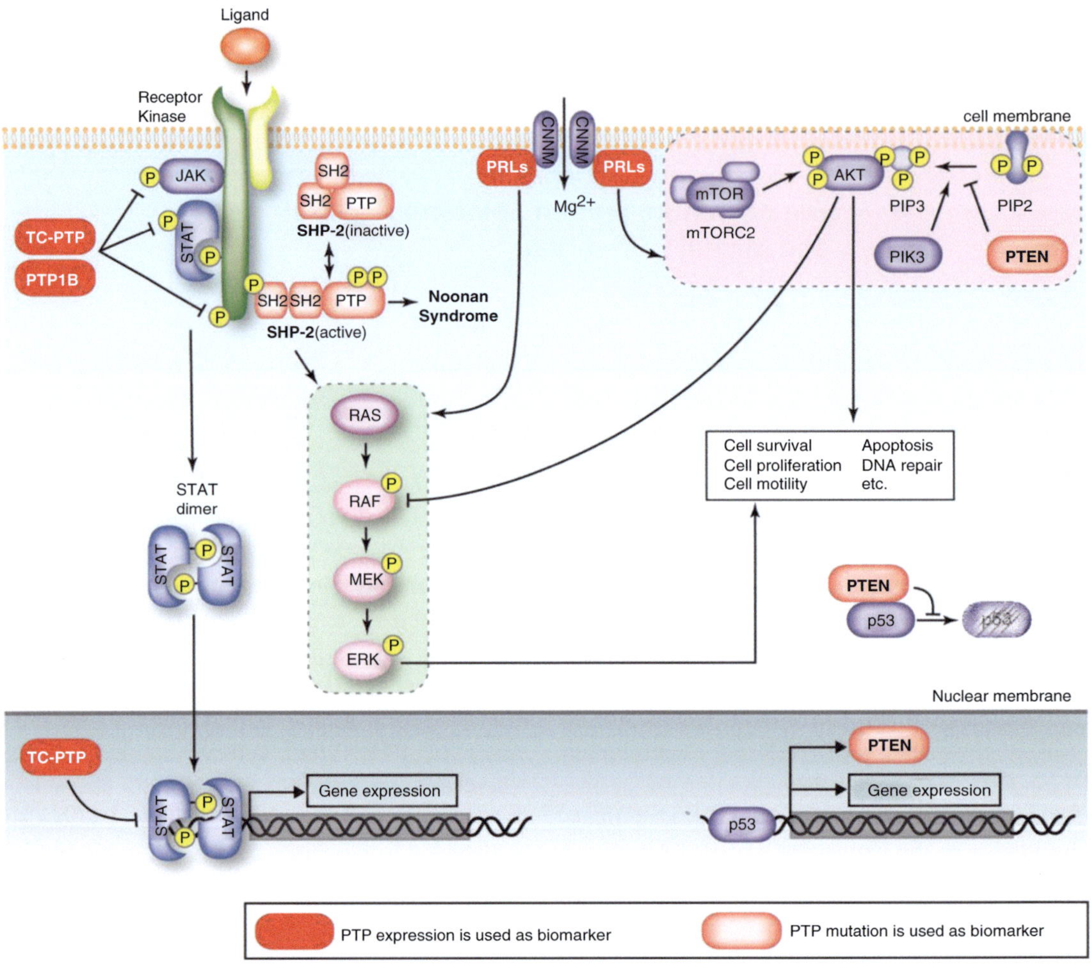

Fig. 30.1 Protein tyrosine phosphatases as biomarkers in cancer. Expression of PTPs and their mutations can be used as biomarkers in different cancers. TC-PTP and PTP1B are involved in regulation of the JAK-STAT signaling pathway. TC-PTP affects the pathway both in the cytoplasm and the nucleus. SH2 is involved in Ras/ERK signal cascade. PTEN counteracts the activity of PI3K, which prevents the activation of AKT. Additionally, PTEN stabilizes p53, which in turn activates PTEN expression. In contrasts with other PTPs, PRLs act not by dephosphorylating a substrate but rather by interacting with a family of magnesium transporters (CNNMs), which increases magnesium concentration within the cell. All these pathways are involved in the malignant transformation that results in increased cell survival, proliferation, and deregulation of apoptosis. (We thank Dr. Noriko Uetani, McGill University, for providing the figure)

subfamily is unique in its structure and its oncogenic properties. Indeed, it is the only phosphatase subfamily to have a prenylation motif at its C-terminus, which localizes it to the plasma membrane and early endosomes. Moreover, the PRL phosphatases are highly expressed in the majority of human tumors, as well as hematological cancers. Increasing the levels of those phosphatases correlates with increased proliferation and invasion of cancer cells through activation of ERK1/ERK2 and PI3K/AKT pathways, among others. Despite increasing cancer-related literature, the normal physiological function of PRLs is not well understood. They are poorly active in *in vitro* enzymatic assays with synthetic substrates, and their physiological substrates are still unknown [4].

Like protein phosphorylation, intracellular magnesium balance is altered in transformed cells and has been linked to modifications of

Table 30.1 PTPs in current clinical trials

Signaling pathway	Pathway active in the cancer type	Affected biomarker/ target	Method of detection	Drugs in clinical trials
PI3K/AKT	Advanced cancers	PTEN PI3KCA	N/A	Phase 1 Pazopanib (Votrient®: Novartis), Everolimus (Afinitor®: Novartis) ClinicalTrials.gov Identifier: NCT01430572
PI3K/AKT	Refractory solid tumors	PTEN PIK3CA	CT scan	Phase 2 Everolimus (Afinitor®: Novartis) ClinicalTrials.gov dentifier: NCT02449538
PI3K/AKT	Advanced cancers	PTEN	IHC	Phase 1/2a GSK2636771 ClinicalTrials.gov Identifier: NCT01458067
PI3K/AKT	Advanced cancers	PTEN AKT PI3K	N/A	Phase 1 ARQ 751 ClinicalTrials.gov Identifier: NCT02761694
Unknown	Advanced solid tumors	PRL-3	Plasma concentration quantification	Phase 1 PRL3-zumab ClinicalTrials.gov Identifier: NCT03191682
Ras/ERK	Advanced solid tumors	SHP2	IHC and plasma concentration quantification	Phase 1 TNO155 ClinicalTrials.gov Identifier: NCT03114319
Ras/ERK	Refractory juvenile Myelomonocytic leukemia (JMML)	SHP2	Mass spectrometry and plasma concentration quantification	Phase 2 Trametinib (Mekinist®: Novartis) ClinicalTrials.gov Identifier: NCT03190915

several hallmarks of cancer. Importantly, high intracellular levels of this cation seem to confer a metabolic advantage to the cells and promote the acquisition of a transformed phenotype. Magnesium is an essential intracellular cation, which regulates numerous cellular functions and enzymes, including ion channels, metabolic cycles, and important intracellular signaling molecules. This includes the human cyclin M (CNNM) gene family that comprises four isoforms (CNNM1-4) that are differentially expressed in human tissues. Interestingly, genome-wide association studies showed that CNNM2, CNNM3, and CNNM4 gene expression correlates with serum magnesium concentrations, supporting the role of these proteins in human magnesium homeostasis under physiological conditions. Furthermore, CNNM2 was proposed to function as a homodimer at the plasma membrane, hypothesized to sense intracellular magnesium concentrations and to regulate other magnesium transporters [5].

Recently, the discovery that PRLs bind to CNNMs and promote cancer progression through upregulation of intracellular magnesium was a milestone in the field [6]. Indeed, studies among patients with either breast or colorectal cancer have shown increased expression of PRLs and CNNMs by immunohistochemistry and gene expression quantification, which could be of interest for future clinical trials.

There is currently one ongoing phase 1 clinical trial using a PRL-3 antibody named PRL3-zumab. This antibody was previously shown targeting the secreted form of that protein which decreased tumor burden in mice [7] (see Table 30.1). However, there are still no FDA approved drugs targeting PRL phosphatases directly.

SHP2

Src homology region 2 (SH2)-containing protein tyrosine phosphatase 2 (Shp2) is one of the two SH2 domain-containing PTPs, with the other member being Shp1. This non-receptor PTP is encoded by the *PTPN11* gene. Shp2 is ubiquitously expressed in various tissues of vertebrates. Its role is to transduce mitogenic, pro-survival, cell-fate, and/or promigratory signals from numerous growth factor, cytokine, and extracellular matrix receptors. Shp2 possess two tandem SH2 domains (N-SH2 and C-SH2) in its N-terminus and a classical PTP catalytic domain followed by two tyrosine phosphorylation sites (Y542 and Y580) and a proline-rich region in its C-terminus. Shp2 toggles between active and inactive state, where in its inactive (closed) state, the catalytic domain is blocked by the N-SH2 domain. Upon growth factor or cytokine stimulation, binding of phosphorylated tyrosine residues of upstream substrates, such as receptor tyrosine kinases (RTKs), cytokine receptors, and scaffolding adaptors, to the N-SH2 domain disrupts autoinhibition and thus switching the enzyme to its active confirmation [8]. Alternatively, Y542 or Y580, phosphorylated by protein tyrosine or another kinase, can also intramolecularly interact with N-SH2 and C-SH2 domains, respectively, to relieve the basal inhibition of Shp2, although this role of the C-terminal tyrosine phosphorylation on Shp2 activity and function is controversial.

In most RTK signaling pathways, Shp2 is required for full activation of the Ras/ERK cascade. The precise target(s) that Shp2 must dephosphorylate is somewhat controversial, as the receptor and/or cell type specificity may determine how Shp2 regulates the Ras/ERK pathway. Shp2 mediates RTK- and integrin-evoked Src family kinase (SFK) activation or by controlling RasGAP recruitment through dephosphorylation of RasGAP binding sites on some RTK (e.g., PDGFR and Torso). Other candidates for Shp2-mediated Ras/Erk activation include the regulatory tyrosyl phosphorylation sites on SPROUTY and SPRED proteins [8]. Alternatively, regulation of Src kinase activity by Shp2 through either direct dephosphorylation of Src or through indirect regulation of the Src inhibitor Csk can enhance activation of the ERK pathway [9]. In addition to Ras/ERK activation, Shp2 has also been described as a regulator of PI3K, Fak, and the Rho family GTPase, RhoA, as well as NFAT signaling. In addition to RTK signaling, Shp2 has been implicated downstream of cytokine signaling in the regulation of JAK-STAT signaling pathways and in the activation of NF-κB [8].

Positioned at the apex of many cell signaling cascades, it is no surprise that this protein is involved in various types of cancers. Germline *PTPN11* mutations are commonly found in patients with Noonan syndrome, a common autosomal dominant developmental disorder. These patients are often at a higher risk for leukemia. Somatic mutations are commonly found in juvenile myelomonocytic leukemia and, to a lesser extent, in acute myelogenous leukemia, and several types of solid tumors. Studies have shown Shp2 to be upregulated in breast cancer. It is also involved in gastric, lung, cervical and laryngeal cancers, as well as oral squamous cell carcinoma. Past investigations correlate Shp2 expression negatively with patient prognosis [8].

There are currently two drugs undergoing clinical trials with patients harboring *PTPN11* mutations: TNO155 for solid tumors and trametinib for myelomonocytic leukemia (see Table 30.1).

TC-PTP

Human T-cell protein tyrosine phosphatase (TC-PTP) belonging to the family of classical non-receptors PTPs is encoded by the gene

PTPN2 [10]. Its name is a misnomer since it is ubiquitously expressed in T lymphocytes and hematopoietic tissues. TC-PTP possesses a conserved catalytic domain and a variable length C-terminus. This phosphatase is also closely related to another classical PTP, PTP1B, by sharing ~72% sequence identity in their catalytic domains and having similar enzymatic activities. In human and rodents, *PTPN2* encodes two different isoforms due to alternative splicing at its carboxyl end: a 45 kDa TC-PTP (TC45) and 48.5 kDa TC-PTP (TC48). Those differences at their C-terminus provided TC45 with a nuclear localization signal (NLS) that localizes the protein to the nucleus. Interestingly, TC45 shuttles between the cytoplasm and the nucleus in response to different cellular stimuli. On the other hand, TC48 also harbors an NLS but is localized to the endoplasmic reticulum (ER) due to hydrophobic sequences in its elongated carboxyl end that inhibit its NLS activity. TC45 is actively transported to the nucleus, but its small size allows it to passively diffuse to the cytoplasm [11].

The nucleocytoplasmic shuttling of TC45 was proposed to be regulated in part by AMPK. More specifically, patients with Peutz-Jeghers syndrome harbor an inactivating mutation in LKB1 leading to decreased activity of AMPK and increased activity of c-Src oncogene. The Peutz-Jeghers syndrome is characterized with the growth of noncancerous polyps in the gastrointestinal track and a greatly elevated risk of cancer. In addition, activated AMPK is known to inhibit nuclear import, hence TC45 is mostly in the cytoplasm. On the other hand, inactive LKB1 will lead to impaired activity of AMPK, and TC45 would be mostly nuclear. Furthermore, TC-PTP was shown to dephosphorylate and inactivate c-Src family kinases. Hence, in the context of Peutz-Jeghers syndrome and gastrointestinal tumors where LKB1 is mutated, and c-Src activity is enhanced, TC45 is proposed to be unable to inactivate c-Src due to its nuclear localization. In cervical adenocarcinoma cells, TC45, but not TC48, was also shown to be acting as a tumor suppressor due to its localization differences [11].

Among many TC-PTP substrates, several members of the JAK-STAT pathway are present. In brief, this pathway serves as a communication relay between extracellular chemical signal binding to surface receptor and transcription changes in the nucleus. The binding of a ligand to its appropriate receptor will result in Janus kinase (JAK) proteins to phosphorylate each other. Those new phosphorylated sites will recruit a multitude of proteins that will in turn get phosphorylated. Among them are the signal transducer and activator of transcription (STAT) proteins which will activate specific gene transcription in the nucleus upon phosphorylation. More specifically, the loss of TC-PTP was shown in mice to enhance phosphorylation of JAK1 and STAT5 in primary leukemic T cells. In humans, 6% of patients with T-cell acute lymphoblastic leukemia (T-ALL) harbor a loss of *PTPN2* and overexpression of TLX1 transcription factor [12]. TLX1 aberrant expression promotes genetic lesion that causes fusion of NUP214 to ABL1 gene and the subsequent expression of the NUP214-ABL1 oncogene. Furthermore, TC-PTP expression was rescued in cellular models and showed a decreased NUP214-ABL1 activity through direct dephosphorylation. Most importantly, the genomic levels of *PTPN2* increased after treatment when patients were in remission and were lost again at relapse of the disease. Mechanistically, the contribution of TC-PTP in TLX1+ T-ALL patients was attributed to its negative regulation of the JAK-STAT signaling pathway. Taken together, the tumor suppressor role of TC-PTP in TLX1+ T-ALL makes it a potential predictive biomarker for recurrence of the disease. Nonetheless, there are still neither active clinical trials nor FDA-approved drugs to therapeutically target TC-PTP to date.

PTP1B

Protein tyrosine phosphatase 1B (PTP1B) (encoded by *PTPN1* gene) is one of the most studied phosphatases and has also been shown to act either as a tumor suppressor or promoter depending on the cellular context [10]. In metastatic prostate cancer, a region of chromosome 20, 20q13, containing *PTPN1* is co-amplified with androgen receptor, which has been linked with increased rates of biochemical relapse. Involvement of PTP1B in prostate cancer is complicated by the fact that in mice with deletion of both *PTPN1* and *PTEN*, the progression of the disease is exacerbated by high-fat diet, but it is not in PTEN-null mice. Amplification of *PTPN1* is often observed in colorectal, breast and ovarian cancer. However, in hepatocellular carcinoma downregulation of *PTPN1* expression is frequently observed, and it is associated with poor prognosis. Overexpression of PTP1B promotes oncogenesis by activating Src [13]. This is thought to be accomplished by dephosphorylating Tyr314 of phosphoprotein associated with glycosphingolipid-enriched microdomains (PAG), which releases C-terminal Src kinase (CSK) and prevents Src inactivation through phosphorylation of its C-terminus.

In breast cancer studies of a mouse model, knockout of PTP1B in MMTV (mouse mammary tumor virus)-Erbb2 mice resulted in delayed tumorigenesis. A small molecule inhibitor of PTP1B had a similar effect on tumor development [14]. In human breast epithelial cell line MCF-10A, ERBB2 activation results in PTP1B overexpression which activates Src, as described above, and that results in the induction of Src-dependent transformed phenotype [15]. PTP1B plays a complex role in hematological malignancies such as classical Hodgkin lymphoma and primary mediastinal B cell lymphoma (PMBCL). *PTPN1* mutations found in these malignancies with appreciable frequency result in increased phosphorylation of members of JAK-STAT pathway, which lead to upregulation of genes related to drug resistance and increased survival [16]. One of the genes whose expression was upregulated upon *PTPN1* knockdown in Hodgkin lymphoma cell line is a proto-oncogene Myc, which is very often upregulated in various cancers [17]. It has been shown that not only mRNAs but also double-stranded DNA that covers the whole genome in an unbiased manner can be found in tumor-derived exosomes [18]. For example, in prostate cancer, urinary exosomes have been used to detect transcripts linked with increased oncogenicity [19]. Therefore, it should be possible to detect genetic aberrations of PTP1B together with identification of PTEN status in prostate cancer to use as a biomarker guiding personalized treatment decisions. However, there is still neither FDA-approved drug for PTP1B nor ongoing clinical trials at the moment.

Conclusion and Future Perspectives

PTPs have long been overlooked as the simple counterparts to kinases, yet they have been found to have a wide array of functions, and some were shown to be oncogenic. It is then without surprise that putative PTP targets in cancer therapies have been increasing in number and are expected to play a central role in the years to come. Recently, the clinical relevance of TC-PTP-and PTP1B-targeted immunotherapies was demonstrated by showing that their inhibition enhanced dendritic cell function [20]. This *ex vivo* approach could boost the antitumoral immune response and be combined with several currently used therapies. Finally, as drug resistance and treatment unresponsiveness are still the main deficiencies of cancer therapeutic regimen, studying PTPs diversity of action and their targeting is now becoming an exciting direction to unveil new therapeutic strategies.

References

1. Worby CA, Dixon JE. PTEN. Annu Rev Biochem. 2014;83(1):641–69.
2. Chalhoub N, Baker SJ. PTEN and the PI3-kinase pathway in cancer. Ann Rev Pathol Mech Dis. 2009;4(1):127–50.
3. Gabriel K, Ingram A, Austin R, Kapoor A, Tang D, Majeed F, et al. Regulation of the tumor suppressor PTEN through exosomes: a diagnostic potential for prostate cancer. PLoS One. 2013;8(7):e70047.
4. Rubio T, Kohn M. Regulatory mechanisms of phosphatase of regenerating liver (PRL)-3. Biochem Soc Trans. 2016;44(5):1305–12.
5. de Baaij JH, Hoenderop JG, Bindels RJ. Magnesium in man: implications for health and disease. Physiol Rev. 2015;95(1):1–46.

6. Hardy S, Uetani N, Wong N, Kostantin E, Labbe DP, Begin LR, et al. The protein tyrosine phosphatase PRL-2 interacts with the magnesium transporter CNNM3 to promote oncogenesis. Oncogene. 2015;34(8):986–95.
7. Thura M, Al-Aidaroos AQ, Yong WP, Kono K, Gupta A, Lin YB, et al. PRL3-zumab, a first-in-class humanized antibody for cancer therapy. JCI Insight. 2016;1(9):e87607.
8. Chan G, Neel BG. Role of PTPN11 (SHP2) in cancer. In: Protein tyrosine phosphatases in cancer. New York: Springer; 2016. p. 115–43.
9. Grossmann KS, Rosario M, Birchmeier C, Birchmeier W. The tyrosine phosphatase Shp2 in development and cancer. Adv Cancer Res. 2010;106:53–89.
10. Julien SG, Dube N, Hardy S, Tremblay ML. Inside the human cancer tyrosine phosphatome. Nat Rev Cancer. 2011;11(1):35–49.
11. Tiganis T. The Role of TCPTP in cancer. In: Protein tyrosine phosphatases in cancer. New York: Springer; 2016. p. 145–68.
12. Kleppe M, Lahortiga I, El Chaar T, De Keersmaecker K, Mentens N, Graux C, et al. Deletion of the protein tyrosine phosphatase gene PTPN2 in T-cell acute lymphoblastic leukemia. Nat Genet. 2010;42(6):530–5.
13. Labbé DP, Tremblay ML. PTP1B: from metabolism to cancer. 263. New York: Springer; 2016. p. 169–99.
14. Feldhammer M, Uetani N, Miranda-Saavedra D, Tremblay ML. PTP1B: a simple enzyme for a complex world. Crit Rev Biochem Mol Biol. 2013;48(5): 430–45.
15. Arias-Romero LE, Saha S, Villamar-Cruz O, Yip S-C, Ethier SP, Zhang Z-Y, et al. Activation of Src by protein tyrosine phosphatase 1B is required for ErbB2 transformation of human breast epithelial cells. Cancer Res. 2009;69(11):4582–8.
16. Pike KA, Tremblay ML. TC-PTP and PTP1B: regulating JAK–STAT signaling, controlling lymphoid malignancies. Cytokine. 2016;82:52–7.
17. Gunawardana J, Chan FC, Telenius A, Woolcock B, Kridel R, Tan KL, et al. Recurrent somatic mutations of PTPN1 in primary mediastinal B cell lymphoma and Hodgkin lymphoma. Nat Genet. 2014;46(4):329–35.
18. Thakur BK, Zhang H, Becker A, Matei I, Huang Y, Costa-Silva B, et al. Double-stranded DNA in exosomes: a novel biomarker in cancer detection. Cell Res. 2014;24(6):766–9.
19. Truong M, Yang B, Jarrard DF. Toward the detection of prostate cancer in urine: a critical analysis. J Urol. 2013;189(2):422–9.
20. Penafuerte C, Feldhammer M, Mills JR, Vinette V, Pike KA, Hall A, et al. Downregulation of PTP1B and TC-PTP phosphatases potentiate dendritic cell-based immunotherapy through IL-12/IFNgamma signaling. Oncoimmunology. 2017;6(6):e1321185.

Part III

Predictive Biomarkers in Specific Organs

31 Predictive and Prognostic Biomarkers in Myeloid Neoplasms

Raju K. Pillai

Overview

The major categories of myeloid neoplasms include myeloproliferative neoplasms, myelodysplastic syndromes, myelodysplastic/myeloproliferative neoplasms, acute myeloid leukemia, mastocytosis, blastic plasmacytoid dendritic cell neoplasms, and myeloid/lymphoid neoplasms with eosinophilia.

Myeloproliferative neoplasms (MPNs) are hematopoietic stem cell disorders characterized by proliferation of cells of one or more of the myeloid lineages (granulocytic, erythroid, and megakaryocytic) and a tendency to transform to acute myeloid leukemia. Chronic myeloid leukemia (CML) is defined by the presence of a specific gene fusion, BCR-ABL1. BCR-ABL1-negative MPNs – polycythemia vera (PV), essential thrombocythemia (ET), and primary myelofibrosis (PMF) – are referred to as classic MPNs. Chronic neutrophilic leukemia (CNL) and chronic eosinophilic leukemia (CEL) are uncommon MPNs, referred to as non-classic MPNs (Fig. 31.1).

Myelodysplastic syndromes (MDS) are clonal hematopoietic neoplasms characterized by simultaneous proliferation and apoptosis of hematopoietic cells that results in a normocellular or hypercellular marrow with peripheral blood cytopenias and a tendency to evolve into acute myeloid leukemia. The incidence of MDS in the United States is estimated at 75 cases per 100,000 in individuals 65 years of age and older. Genomic studies over the past decade have shown that sequential acquisition of somatic mutations in a set of genes involved in hematopoiesis leads to dysregulation of cellular processes leading to asymptomatic clonal hematopoiesis and later to MDS [2].

The myelodysplastic/myeloproliferative neoplasms (MDS/MPNs) include clonal myeloid neoplasms, which at the time of initial presentation are associated with features that support the diagnosis of MDS and other findings more consistent with an MPN. MDS/MPNs are characterized by a hypercellular bone marrow due to proliferation in one or more of the myeloid lineages. Often the proliferation is effective in some lineages with increased numbers of circulating cells and ineffective in other lineages leading to concurrent cytopenia.

Acute myeloid leukemia (AML) results from the clonal expansion of myeloid blasts in the peripheral blood, bone marrow, or tissues. The worldwide annual incidence is approximately three cases per 100,000 of the population per

R. K. Pillai
Department of Pathology, City of Hope National Medical Center, Duarte, CA, USA
e-mail: rpillai@coh.org

S. Badve, G. L. Kumar (eds.), *Predictive Biomarkers in Oncology*,
https://doi.org/10.1007/978-3-319-95228-4_31

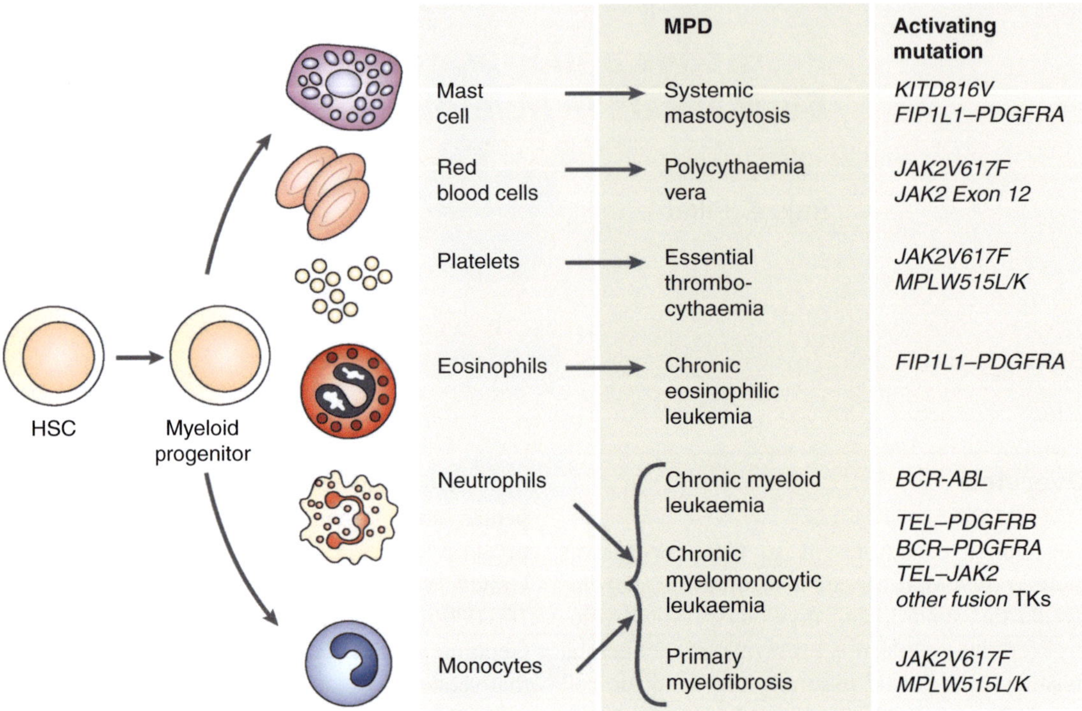

Fig. 31.1 Myeloproliferative disorders can be classified by predominant terminally differentiated myeloid cell involved in the disorder. The most common mutations that lead to constitutive receptor tyrosine kinase signaling are shown. HSC hematopoietic stem cell, JAK2 Janus kinase 2, MPL thrombopoietin receptor, PDGFR platelet-derived growth factor receptor. (Reprinted from Levine et al. [1]. With permission from Springer Nature)

year and is highest in Australia, Western Europe, and the United States. AML constitutes 15–20% of all cases of acute leukemia in children less than 15 years with a peak incidence in the first 3–4 years of life. The WHO classification requires 20% blasts in the peripheral blood or bone marrow for a diagnosis of AML, except in cases associated with t(8;21) (q22; q22.1), inv(16) (p13.1q22), t(16;16)(p13.1;q22), or the t(15;17) *promyelocytic leukemia protein-retinoic acid receptor alpha* (PML-RARA) translocation.

Predictive Biomarkers and Approved Targeted Therapy in Myeloid Neoplasms

The currently approved targeted therapies and their predictive biomarkers are shown in Table 31.1.

Pathogenetic Mechanisms in Myeloid Neoplasms

Myeloproliferative Neoplasms

Dysregulation of JAK2 signaling by direct or indirect mechanisms has emerged as the central theme in classic MPNs. Figure 31.2 depicts the principal driver mutations in MPNs. JAK2 V617F mutation is seen in the majority of classic MPNs [3–5].

JAK2 is a non-receptor tyrosine kinase that plays an important role in signal transduction from class I cytokine receptors such as erythropoietin receptor (EPOR), thrombopoietin receptor (TPOR encoded by MPL gene), and the granulocyte colony-stimulating factor receptor. The mutation burden of JAK2 V617F is usually higher in PV and PMF compared with

Table 31.1 Currently approved targeted therapies in myeloid neoplasms

Gene/RNA protein biomarkers	Function/pathogenic process	Patient population	Clinical use and limitations	Approved Drug(s) generic/ trade name
JAK2	JAK/STAT pathway	PMF	Intermediate or high-risk MF	Ruxolitinib Jakafi® (Incyte)
FLT3	Receptor tyrosine kinase	Patients with FLT3-positive AML	Used in combination with chemotherapy	Midostaurin (RYDAPT®) Novartis
IDH2	Intermediary metabolism and energy production	Patients with IDH2-positive AML	Promotes maturation of blasts	Enasidenib (IDHIFA®) Celgene/Agios
CD33	Targeted immunotherapy	CD33-positive AML	Used in combination with chemotherapy	Gemtuzumab Ozogamicin (MYELOTARG™) Pfizer oncology
t(15;17)(PML/ RAR translocation		APL	Acute promyelocytic leukemia	All-trans retinoic acid (ATRA)

PMF primary myelofibrosis, *AML* acute myeloid leukemia, *APL* acute promyelocytic leukemia

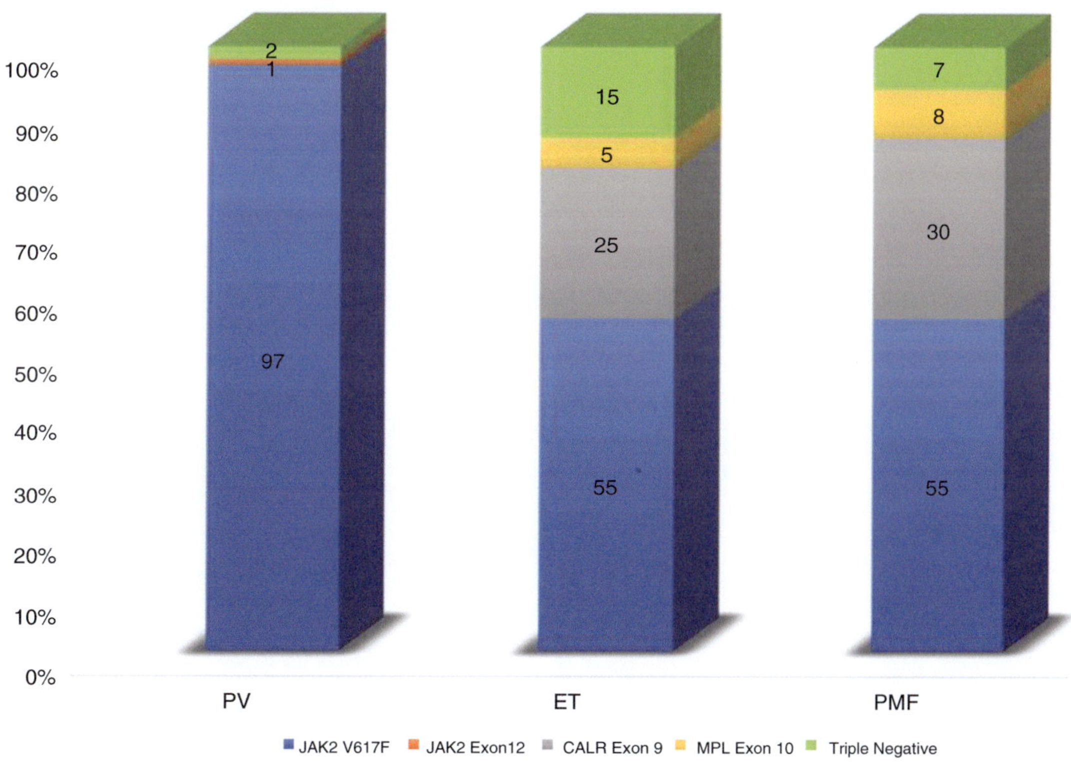

Fig. 31.2 Distribution of molecular drivers in classic MPN's. PV polycythemia vera, ET essential thrombocythemia, PMF primary myelofibrosis. (Courtesy of Raju K. Pillai)

ET. Many patients with PV and PMF demonstrate a homozygous JAK2 mutant clone. One third of patients with JAK2 V617F-negative PV have JAK2 exon 12 mutations which are mostly complex insertion/deletion events. Activating mutations in MPL or CALR gene results in indirect activation of JAK/STAT signaling. MPL mutations commonly involve W515 of the TPOR, which is located at the junction between the transmembrane and cytoplasmic domains and is required for maintaining TPOR in its inactive confirmation. Mutations in W515 constitutively activate JAK2 signaling [6]. ET and PMF cases negative for the classical driver mutations in JAK2, CALR, and MPL are referred to as triple-negative MPNs (Fig. 31.3).

Myelodysplastic Syndromes

Although a large number of driver genes have been implicated in MDS, these can be organized into a limited number of categories: RNA splicing factors (SF3B1, SRSF2, U2AF1, ZRSR2), epigenetic regulators (TET2, DNMT3A, IDH1, and IDH2), cohesin components (STAG2,

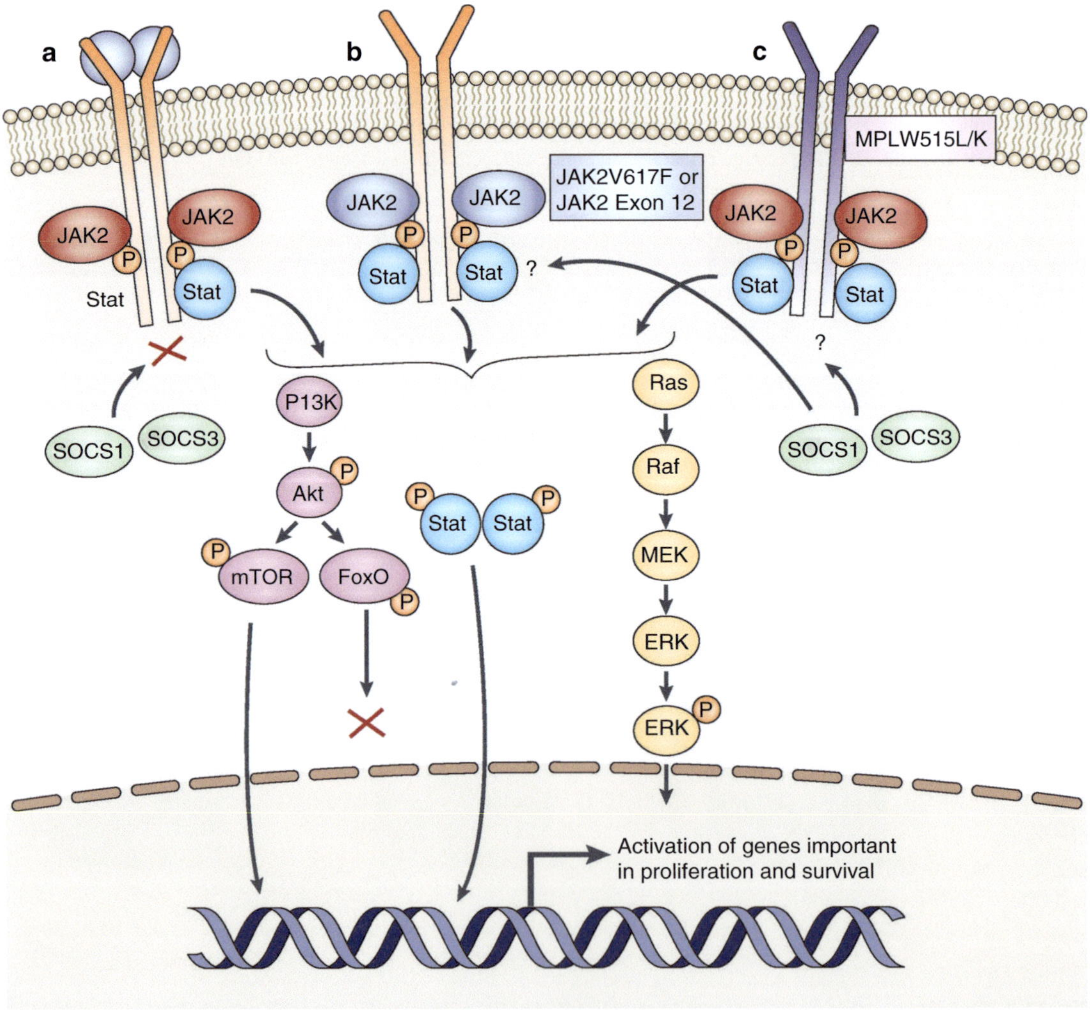

Fig. 31.3 (**a**) Binding of cytokine ligands to cognate receptors results in JAK2 phosphorylation, recruitment of Stat (signal transducer and activator of transcription) proteins and activation of downstream pathways. (**b**) The presence of JAK2 mutations leads to constitutive phosphorylation in the absence of cytokine ligands. (**c**) Mutated MPL proteins can phosphorylate wild -type JAK2 in the absence of thrombopoietin ligands. (Reprinted from Levine et al. [1]. With permission from Springer Nature)

RAD21, SMC1A, SMC3), transcription factors (RUNX1, ETV6, CUX1, GATA2), DNA damage response (TP53, PPM1D), and signal transduction molecules (CBL, JAK2, NRAS, KRAS, MPL, NF1, PTPN11, KIT, FLT3) [2]. RNA splicing factors are mutated in up to 60% of patients with MDS [7, 8], are heterozygous, and are generally mutually exclusive. Mutant splicing factors result in altered patterns of splicing. However, the precise pathogenetic mechanisms are not fully understood. SF3B1 mutations are seen in up to 80% of MDS cases with ringed sideroblasts.

The epigenome is composed primarily of cytosine modifications, histone modifications, and expression of noncoding RNA molecules. In a normal somatic cell, approximately 70–80% of cytosines that occur as a CpG dinucleotide are methylated, but CpG islands are largely unmethylated. Methylation status is maintained by DNA methyl transferases. DNA methyl transferase 1 (DNMT1) targets hemi-methylated DNA in order to maintain methylation patterns during DNA replication, whereas DNA methyl transferases 3A and 3B (DNMT3A and DNMT3B) primarily target unmethylated CpGs. Nucleosomes are made up of histone octamers composed of H2A, H2B, H3, and H4 proteins that form the scaffold upon which DNA is wound. Histones can be chemically modified by acetylation, methylation, phosphorylation, sumoylation, or ubiquitinylation. Trimethylation of H3 lysine 27 (H3K27me3) by the polycomb repressive complex 2 (PRC2) which includes EZH2, SUZ12, and EED is associated with transcriptional repression.

Mutations in genes involved in DNA methylation and histone modification frequently occur in MDS. Recurrent loss of function mutations has been identified in DNM3TA and TET2, an enzyme that initiates the process of DNA demethylation [9]. Mutations in isocitrate dehydrogenase (IDH1 and IDH2) alter their enzymatic activity resulting in the generation of 2-hydroxy butyrate that inhibits the activity of numerous targets including TET2 [10]. Loss of function mutations in ASXL1 and EZH2, which are involved in histone modification, occurs in approximately 20% and 5% of MDS, respectively [11].

Transcription factors such as GATA2 and RUNX1 play an important role in hematopoietic differentiation. Loss of function mutations in GATA2 and RUNX1 is seen in MDS. Such mutations can also be inherited as a germline alteration leading to familial bone marrow failure syndromes [7, 12]. The tumor suppressor gene TP53 is induced and activated by a number of stress signals, including DNA damage, oxidative stress, telomere shortening, hypoxia, and activated oncogenes leading to cell cycle arrest, cellular senescence, or apoptosis. Missense mutations in TP53 are particularly common in MDS patients who have undergone chemotherapy (up to 40%) [13, 14]. Loss of the second TP53 allele due to 17p deletions is associated with thrombocytopenia, complex karyotype, and a poor prognosis [15].

Pathogenic mutations seen in MDS can be detected in apparently normal individuals with a clonal frequency of up to 10–20% of all circulating nucleated cells [16–18]. Clonal hematopoiesis of indeterminate potential (CHIP) is defined by the presence of somatic mutations of driver genes in the blood or bone marrow present at greater than 2% variant allele frequency in individuals without a diagnosed hematologic disorder [19]. Based on exome sequencing studies, the incidence of CHIP increases with age and exceeds 15% in individuals over 70 years of age. Mutations are most frequent in epigenetic regulators such as DNMT3A, TET2, and ASXL1. The estimated rate of progression is 1% per year and typically occurs in individuals with involvement of more than 20% of their peripheral blood cells. Idiopathic cytopenia of undetermined significance (ICUS) is used to refer to patients with cytopenias with no evidence of clonal hematopoiesis or evidence of involvement by myelodysplastic syndrome. Clonal cytopenia of undetermined significance (CCUS) refers to patients with cytopenias with somatic mutations at greater than 2% variant allele frequency that do not meet diagnostic criteria for myelodysplastic syndrome.

Acute Myeloid Leukemia

Acute myeloid leukemia is characterized by somatic mutations affecting multiple cellular pathways and has a complex clonal architecture. Mutational profiling of 200 de novo AML cases in the "The Cancer Genome Atlas" (TCGA) project revealed an average number of 13 coding mutations, one somatic copy number variant and less than one gene fusion event per patient [20]. Recurrent mutations were seen in 23 genes which were grouped into 7 functional categories (Table 31.2). Mutations in the epigenetic pathway including genes such as DNMT3A, ASXL1, TET2, and IDH1/IDH2 are acquired early in the disease process. Mutations involving the signal transduction pathway or NPM1 are typically secondary events that occur later during the evolution of the disease.

Representative Prognostic Biomarkers in Myeloid Neoplasms

Classical cytogenetics and molecular genetic studies are of great importance in prognostic evaluation of acute myeloid leukemia. Classical cytogenetics studies have been used for risk stratification to favorable, intermediate, and adverse prognostic groups. Integration of molecular studies is used to further refine prognostic categorization. Examples are listed in Table 31.3.

Table 31.2 Categories of genes commonly affected in acute myeloid leukemia (AML)

Pathway	Selected genes	Incidence (TCGA study – %)
Signal transduction	FLT3, KIT, PTPN11, KRAS, NRAS	59
DNA methylation	DNMT3A, TET2, IDH1, IDH2	44
Chromatin modification	ASXL1, EZH2, KMT2A	30
Transcription factors	RUNX1, CEBPA, CBF	40
Tumor suppressors	TP53, WT, PHF6	16
Splice factor	SRSF2, SF3B1, U2AF1, ZRSR2	14
Cohesin complex	STAG2, RAD21	13

Table 31.3 Prognostic significance of molecular genetic alterations

Gene/RNA protein biomarkers	Cytogenetic group	Prognostic significance
FLT3-ITD	Normal or abnormal karyotype	Significantly shorter DFS and OS [21]
Biallelic CEBPA mutation	Normal or abnormal karyotype	Significantly higher DFS and OS [22]
NPM1 mutation	Normal karyotype	In some studies, significantly higher CRR and DFS [23]
RUNX1 mutation	Normal or abnormal karyotype	Higher resistant disease rate, shorter DFS and OS [24]
WT1 mutation	Normal or abnormal karyotype	Signet significantly worse RR and OS [25]
TET2 mutation	Normal or abnormal karyotype	In younger patients (less than 60 years), no significant difference in RFS or OS [26]
ASXL1 mutation	Normal or abnormal karyotype	Significantly worse RFS and OS [27]
IDH1 mutation	Normal karyotype	No significant difference in DFS and OS in some studies; among patients with NPM1 or CEBPA mutation without FLT3-ITD, higher RR, and shorter OS [28]
TP53 mutation	Abnormalities of chromosome 5, 7, or 17 and/or complex karyotype (≥5 abnormalities)	Significantly shorter OS [29]

CRR complete remission rate, *DFS* disease-free survival, *OS* overall survival, *RFS* relapse free survival, *RR* risk of relapse

Summary and Future Directions

For many years, classical cytogenetics was used to determine the prognosis of AML by dividing patients into favorable, intermediate, or adverse categories. The identification of genomic alterations including mutations, copy number alterations, and gene fusions has enabled a genomic classification of AML. In a recent study of 1540 patients with AML who were tested by targeted next-generation sequencing, 11 nonoverlapping classes, each with a distinct clinical phenotype and outcome distinguished based on patterns of co-occurrence and mutual exclusivity of genetic changes, were described [30]. Such genomic classifications will enable targeted clinical trials and better definition of prognosis. Based on these genomic alterations, multiple clinical trials for protein kinase inhibitors, epigenetic modulators, immune checkpoint inhibitors, cellular immunotherapy, and drugs targeting the microenvironment are in development [31, 32].

References

1. Levine RL, Pardanani A, Tefferi A, et al. Role of JAK2 in the pathogenesis and therapy of myeloproliferative disorders. Nat Rev Cancer. 2007;7(9):673–83.
2. Kennedy JA, Ebert BL. Clinical implications of genetic mutations in myelodysplastic syndrome. J Clin Oncol. 2017;**35**(9):968–74.
3. Baxter EJ, et al. Acquired mutation of the tyrosine kinase JAK2 in human myeloproliferative disorders. Lancet. 2005;365(9464):1054–61.
4. James C, et al. A unique clonal JAK2 mutation leading to constitutive signalling causes polycythaemia vera. Nature. 2005;434(7037):1144–8.
5. Kralovics R, et al. A gain-of-function mutation of JAK2 in myeloproliferative disorders. N Engl J Med. 2005;352(17):1779–90.
6. Defour JP, et al. Oncogenic activation of MPL/thrombopoietin receptor by 17 mutations at W515: implications for myeloproliferative neoplasms. Leukemia. 2016;30(5):1214–6.
7. Papaemmanuil E, et al. Clinical and biological implications of driver mutations in myelodysplastic syndromes. Blood. 2013;122(22):3616–27; quiz 3699.
8. Yoshida K, et al. Frequent pathway mutations of splicing machinery in myelodysplasia. Nature. 2011;478(7367):64–9.
9. Delhommeau F, et al. Mutation in TET2 in myeloid cancers. N Engl J Med. 2009;360(22):2289–301.
10. Losman JA, et al. (R)-2-hydroxyglutarate is sufficient to promote leukemogenesis and its effects are reversible. Science. 2013;339(6127):1621–5.
11. Haferlach T, et al. Landscape of genetic lesions in 944 patients with myelodysplastic syndromes. Leukemia. 2014;28(2):241–7.
12. West AH, Godley LA, Churpek JE. Familial myelodysplastic syndrome/acute leukemia syndromes: a review and utility for translational investigations. Ann N Y Acad Sci. 2014;1310:111–8.
13. Ok CY, et al. TP53 mutation characteristics in therapy-related myelodysplastic syndromes and acute myeloid leukemia is similar to de novo diseases. J Hematol Oncol. 2015;8:45.
14. Aldoss I, et al. Favorable impact of allogeneic stem cell transplantation in patients with therapy-related myelodysplasia regardless of TP53 mutational status. Haematologica. 2017;102(12):2030–8.
15. Christiansen DH, Andersen MK, Pedersen-Bjergaard J. Mutations with loss of heterozygosity of p53 are common in therapy-related myelodysplasia and acute myeloid leukemia after exposure to alkylating agents and significantly associated with deletion or loss of 5q, a complex karyotype, and a poor prognosis. J Clin Oncol. 2001;19(5):1405–13.
16. Busque L, et al. Recurrent somatic TET2 mutations in normal elderly individuals with clonal hematopoiesis. Nat Genet. 2012;44(11):1179–81.
17. Link DC, Walter MJ. 'CHIP'ping away at clonal hematopoiesis. Leukemia. 2016;30(8):1633–5.
18. Bejar R. CHIP, ICUS, CCUS and other four-letter words. Leukemia. 2017;31(9):1869–71.
19. Steensma DP, et al. Clonal hematopoiesis of indeterminate potential and its distinction from myelodysplastic syndromes. Blood. 2015;126(1):9–16.
20. Cancer Genome Atlas Research, N, et al. Genomic and epigenomic landscapes of adult de novo acute myeloid leukemia. N Engl J Med. 2013;368(22):2059–74.
21. Patel JP, et al. Prognostic relevance of integrated genetic profiling in acute myeloid leukemia. N Engl J Med. 2012;366(12):1079–89.
22. Dufour A, et al. Acute myeloid leukemia with biallelic CEBPA gene mutations and normal karyotype represents a distinct genetic entity associated with a favorable clinical outcome. J Clin Oncol. 2010;28(4):570–7.
23. Thiede C, et al. Prevalence and prognostic impact of NPM1 mutations in 1485 adult patients with acute myeloid leukemia (AML). Blood. 2006;107(10):4011–20.
24. Tang JL, et al. AML1/RUNX1 mutations in 470 adult patients with de novo acute myeloid leukemia: prognostic implication and interaction with other gene alterations. Blood. 2009;114(26):5352–61.
25. Renneville A, et al. Wilms tumor 1 gene mutations are associated with a higher risk of recurrence in young adults with acute myeloid leukemia: a study from the Acute Leukemia French Association. Cancer. 2009;115(16):3719–27.
26. Gaidzik VI, et al. TET2 mutations in acute myeloid leukemia (AML): results from a comprehensive genetic and clinical analysis of the AML study group. J Clin Oncol. 2012;30(12):1350–7.
27. Paschka P, et al. ASXL1 mutations in younger adult patients with acute myeloid leukemia: a study by the German-Austrian Acute Myeloid Leukemia Study Group. Haematologica. 2015;100(3):324–30.
28. Boissel N, et al. Prognostic impact of isocitrate dehydrogenase enzyme isoforms 1 and 2 mutations in acute myeloid leukemia: a study by the Acute Leukemia French Association group. J Clin Oncol. 2010;28(23):3717–23.
29. Bowen D, et al. TP53 gene mutation is frequent in patients with acute myeloid leukemia and complex

karyotype, and is associated with very poor prognosis. Leukemia. 2009;23(1):203–6.
30. Papaemmanuil E, et al. Genomic classification and prognosis in acute myeloid leukemia. N Engl J Med. 2016;374(23):2209–21.
31. Dohner H, et al. Diagnosis and management of AML in adults: 2017 ELN recommendations from an international expert panel. Blood. 2017;129(4):424–47.
32. Stein EM, Tallman MS. Emerging therapeutic drugs for AML. Blood. 2016;127(1):71–8.

32 Predictive Biomarkers and Targeted Therapies for Lymphoid Malignancies

Raju K. Pillai, Bharat N. Nathwani, and Lixin Yang

Overview

Neoplasms derived from lymphoid cells include mature B-cell neoplasms, plasma cell myeloma, mature T- and NK-cell neoplasms, and Hodgkin lymphoma. Precursor lymphoid neoplasms (acute lymphoblastic leukemia/lymphoma) are considered together with lymphoid neoplasms in this chapter because of their common origin. In addition, many of the targeted therapies used in lymphoid neoplasms overlap with their use in precursor lymphoid neoplasms and plasma cell myeloma.

In 2017, non-Hodgkin lymphoma was the seventh commonest cancer in the United States with approximately 72,240 cases and 20,140 deaths. Diffuse large B-cell lymphoma and follicular lymphoma comprise 37% and 29% of B-cell lymphomas [1]. Peripheral T-cell lymphoma and angioimmunoblastic lymphoma constitute 26% and 19% of the mature T-cell lymphomas.

The morphology and immunophenotype of B-cell neoplasms correlate with various stages of normal B-cell differentiation and are currently used as a basis for their classification and nomenclature (Fig. 32.1). Many of the pathogenic alterations seen in B-cell lymphomas trace their origins to physiological processes that occur in B-cell development. B-cell differentiation begins with the lymphoblast in the bone marrow which undergoes immunoglobulin heavy chain and light chain gene rearrangement. The variable region of the heavy chain (Vh) is assembled from approximately 40 variable, 23 diversity, and 6 joining gene segments, and the variable region of the light chains is assembled from approximately 30–35 variable and 5 joining gene segments. These segments are rearranged by a process of double-stranded DNA breaks followed by recombination mediated by the RAG (recombination activating gene) enzymes. Aberrant recombination leads to chromosomal translocations, as exemplified by the RAG-mediated t(14;18) translocation in follicular lymphoma and t(11;14) translocation in mantle cell lymphoma.

Naïve B cells that are surface immunoglobulin positive circulate in the peripheral blood and enter the lymphoid follicles in secondary lymphoid organs. Recognition of antigen that fits the surface immunoglobulin B-cell receptor leads to proliferation and ultimately maturation into antibody secreting plasma cells and memory B cells. In the

Chapter includes adapted text extracts from: Pillai RK, Chan WC. Pathogenesis of lymphomas. In: Zain J, Kwak L. (eds) Management of Lymphomas: A Case-Based Approach. Adis, Cham; 2017: 11–31. With permission from Springer Nature.

R. K. Pillai (✉) · B. N. Nathwani · L. Yang
Department of Pathology, City of Hope National Medical Center, Duarte, CA, USA
e-mail: rpillai@coh.org

S. Badve, G. L. Kumar (eds.), *Predictive Biomarkers in Oncology*,
https://doi.org/10.1007/978-3-319-95228-4_32

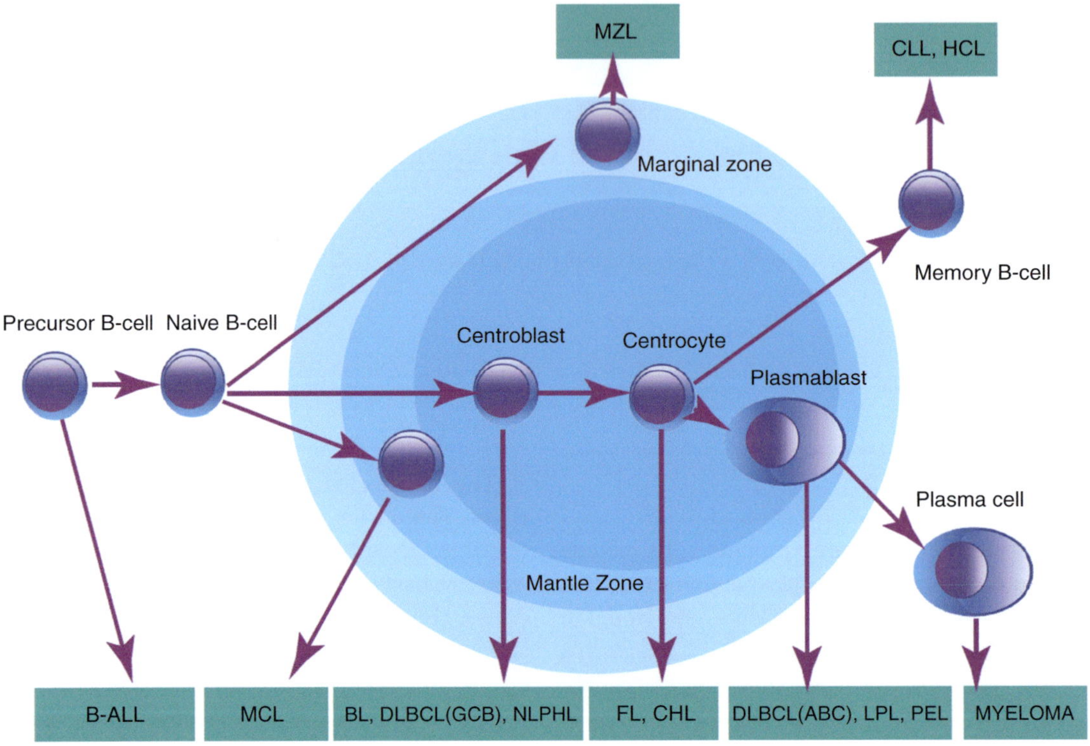

Fig. 32.1 Histogenetic correlation of various B-cell lymphomas with corresponding B-cell development stages. B-ALL B-cell acute lymphoblastic leukemia, BL Burkitt lymphoma, CHL classical Hodgkin lymphoma, CLL/SLL chronic lymphocytic leukemia (CLL)/small lymphocytic lymphoma (SLL), DLBCL diffuse large B-cell lymphoma, activated B-cell type (ABC) or germinal center type (GCB), FL follicular lymphoma, HCL hairy cell leukemia, LPL lymphoplasmacytic lymphoma, MCL mantle cell lymphoma, MZL marginal zone lymphoma, NLPHL nodular lymphocyte predominant Hodgkin lymphoma, PEL primary effusion lymphoma. (Adapted from Pillai and Chan [31]. With permission from Springer Nature)

germinal center, the affinity of the B-cell receptor is further improved by selection of cells that have undergone somatic hypermutation of the immunoglobulin variable region (IgV) genes mediated by the AID enzyme. Heavy chain class switching alters the functional properties of the antibody molecule. Both these processes also create double-stranded breaks and aberrant chromosomal translocations. Somatic mutations mediated by AID may also affect non-immunoglobulin genes in germinal center B-cells. Up to 10% of actively transcribed genes may be involved by aberrant somatic hypermutation.

The morphologic, immunophenotypic, and cytogenetic characteristics of the major subtypes of mature B-cell and T-cell lymphomas are shown in Tables 32.1 and 32.2.

Predictive Biomarkers in Lymphoid Neoplasms (Table 32.3)

Pathogenetic Mechanisms in Lymphoid Neoplasms

Alterations in the major physiologic signaling pathways in B cells play a major role in neoplastic transformation – examples of their alteration

Table 32.1 Morphologic, typical immunophenotypic and cytogenetic characteristics of major subtypes of mature B cell lymphomas [2]

	Morphology	Immunophenotype	Cytogenetics
CLL/SLL	Monomorphic small, round to slightly irregular B cells admixed with prolymphocytes and paraimmunoblasts	CD20+, CD5+, CD23+, LEF1+, cyclin D1−	
FL	Centrocytes and centroblasts with a follicular pattern	CD20+, CD5−, CD10+, BCL 6+, BCL-2+	t(14;18)(q32;q21) + (80%), loss of 1p, 6q, 10q, 17p, gains of chr 1, 6p, 7, 8, 12q, X, and 18q
DLBCL	Large B cells with nuclear size equal to histiocytes with a diffuse growth pattern	CD20+, CD79a+, CD30+−, CD5+−, CD10+−, BCL 6+-, MUM1+	t(14;18), gain 3q, 9p
BL	Monomorphic medium-sized transformed cells	CD20+, CD5−, CD10+, BCL2−, TdT−	t(8;14)(q24;q32)
MCL	Monomorphic small to medium-size cells with irregular nuclear contours	CD20+, CD5+, CD23−, cyclin D1+	t(11;14)(q13;q32), gain 3q26, 7p21, 8q24, trisomy12, loss of 1p13–p31, 6q23–q27, 9p21, 11q22–q23, 13q11–q13, 13q14–q34, 17p13
SMZL	Small lymphocytes replacing the splenic white pulp with a prominent marginal zone	CD20+, CD5−, CD10−, CD103−, cytoplasmic Ig−	Loss of 7q31, trisomy3q
MZL	Monocytoid B cells in the marginal and expanding to interfollicular areas	CD20+, IgM+, CD5−, CD10−, CD23+	Trisomies 3, 18, 7
LPL/WM	Small B cells, plasmacytoid lymphocytes and plasma cells	CD20+, IgM+, CD5−, CD10−, CD23+	
HCL	Small mature B cells with round nuclei and abundant cytoplasm with hairy projections	CD20+, CD22+, CD11c+, CD103+, CD25+, CD123+, Tbet+,AnnexinA1+,CD5−,CD10−	
HCL-V	Pro lymphocytes and hairy cells	CD20+,CD25−,CD123−,CD200−	
MM	Atypical plasma cells in interstitial clusters or diffuse sheets of plasma cells	CD20−, CD79a+, CD138+, CD38+, CD19−, CD56+, cytoplasmic Ig+	del17p13, t(11;14)(q13;q32), t(4;14)(p16;q32), t(14;16)(q32;q23), del13, amplification chr1

Adapted from Onaindia et al. [2]. With permission from Springer Nature
BL Burkitt lymphoma, *CLL/SLL* chronic lymphocytic leukemia (CLL)/small lymphocytic lymphoma (SLL), *DLBCL* diffuse large B-cell lymphoma, *FL* follicular lymphoma, *HCL* hairy cell leukemia, *LPL* lymphoplasmacytic lymphoma, *MCL* mantle cell lymphoma, *SMZL* splenic marginal zone lymphoma, *MZL* marginal zone lymphoma, *HCL* hairy cell leukemia, *HCL-V* hairy cell leukemia, variant, *MM* plasma cell myeloma

in lymphomas will be described in this section. An understanding of these pathways will enable rational therapeutic intervention.

B-Cell Receptor Pathway

The B-cell receptor (BCR) complex consists of the immunoglobulin molecule linked to a signaling subunit which is a heterodimer of Ig-alpha (*CD79A*) and Ig-beta (*CD79B*) proteins. Antigen binding to the Ig molecule leads to phosphorylation of ITAM motifs in CD79A and CD79B by src family kinases such as LYN followed by activation of SYK kinase, Bruton's tyrosine kinase (BTK), phospholipase C gamma (PLCg), and protein kinase C beta (PKCb) (Fig. 32.2). Activation of PKCb leads to CARD11 phosphorylation, recruitment of MALT1 and BCL10 into the CBM multi-protein complex, and nuclear factor-kappa B (NF-kB) pathway activation.

Table 32.2 Morphologic, typical immunophenotypic, and cytogenetic characteristics of major subtypes of mature T-cell lymphomas

	Morphology	Immunophenotype	Cytogenetics
AITL	Small to medium-sized, clear cytoplasm	CD3+, CD5+, CD4+, CD10+, CXCL13+, BCL6+, PD1+	
ALCL, ALK+	Variable morphology, characteristic cells with kidney shaped nuclei (hallmark cells)	CD30+, ALK+, EMA+, limited expression of pan T-cell antigens, most cases positive for cytotoxic markers TIA1/granzyme B/perforin	t(2;5)(p23;q35), t(1;2)(q25;p23)
EATL	Variable morphology, medium, large, anaplastic cells	CD3+, CD5−, CD7+, CD4−, CD8−, CD103+, cytotoxic markers+ (TIA1/granzyme B/perforin)	
HSTL	Medium-sized cells, loosely condensed chromatin, inconspicuous nucleoli	CD3+, TCR gamma-delta+, TCR alpha-beta-, CD56+/−, CD8 −/+, CD5-	
ATLL	Variable morphology, small, medium, large, anaplastic cells	CD3+, CD5+, CD4+, CD25+, CD7−	
ENKTL	Variable morphology, small, medium, large, anaplastic cells	CD2+, CD5−, CD56+, surface CD3−, cytoplasmic CD3+, EBV+	

AITL angioimmunoblastic T-cell lymphoma, *ALCL* ALK positive: anaplastic large cell lymphoma, *ALK positive* EATL: entropy-associated T-cell lymphoma, *HSTL* hepatosplenic T-cell lymphoma, *ATLL* adult T-cell leukemia/lymphoma, *ENKTL* extranodal NK/T-cell lymphoma

Table 32.3 Predictive biomarkers in lymphoid neoplasms

Gene/RNA/ protein biomarkers	Function/pathogenic process	Patient population	Clinical use and limitations	Approved drug(s) generic/trade name
CD19	Chimeric antigen receptor T-cells	DLBCL, PMBCL, HGBL	Relapsed disease	Axicabtagene ciloleucel (Yescarta™)
CD19	Chimeric antigen receptor T-cells	B-ALL	Refractory disease	Tisagenlecleucel (Kymriah®)
CD19	Bispecific antibody for targeted immunotherapy	B-ALL	Relapsed disease	Blinatumomab (Blincyto)
CD20	Targeted immunotherapy	B-cell non-Hodgkin lymphoma	Previously untreated, relapsed or refractory disease	Rituximab (Rituxan®)
CD20	Targeted radioimmunotherapy	B-cell non-Hodgkin lymphoma	Relapsed low-grade lymphoma, newly diagnosed follicular lymphoma	Ibritumomab tiuxetan (Zevalin®)
CD20	Targeted immunotherapy	B-cell non-Hodgkin lymphoma	Relapsed CLL and follicular lymphoma	Obinutuzumab (Gazyva®)
CD22	Antibody drug conjugate	B-cell ALL	Relapsed or refractory disease	Inotuzumab ozogamicin (Besponsa®)
CD25	Diphtheria toxin conjugate with IL-2	Cutaneous T-cell lymphoma	Persistent or recurrent disease	Denileukin viftitox (Ontak®)
CD30	Antibody drug conjugate	CHL, ALCL	Recurrent disease	Brentuximab vedotin (Adcetris®)
CD38	Targeted immunotherapy	Multiple myeloma	Relapsed or refractory disease	Daratumumab (Darzalex™)
CD52	Targeted immunotherapy	CLL	Previously untreated or relapsed disease	Alemtuzumab (Campath®)

Table 32.3 (continued)

Gene/RNA/ protein biomarkers	Function/pathogenic process	Patient population	Clinical use and limitations	Approved drug(s) generic/trade name
BCL2	BCL2 inhibitor, faciltates apoptosis	CLL	CLL with 17p deletion who have received at least one prior therapy	Venetoclax (Venclexta™)
BCR-ABL fusion	Tyrosine kinase inhibitor	B-cell ALL, Ph+	Previously untreated or relapsed disease	Imatinib mesylate (Gleevec®)
BCR-ABL fusion	Tyrosine kinase inhibitor	B-cell ALL, Ph+	Resistance or intolerance to prior therapy	Dasatinib (Sprycel®)
BCR-ABL fusion	Tyrosine kinase inhibitor	B-cell ALL, Ph+	T315I positive patients	Ponatinib hydrochloride (Iclusig®)
BTK	B-cell receptor signaling	B-cell lymphoma		Ibrutinib (Imbruvica®)
BTK	B-cell receptor signaling	B-cell lymphoma		Acalabrutinib (Calquence®)
Histone deacetylase	Histone deacetylase inhibitor	Cutaneous T-cell lymphoma	Recurrent or persistent disease	Vorinostat (Zolinza®)
Histone deacetylase	Histone deacetylase inhibitor	Cutaneous T-cell lymphoma	Recurrent or persistent disease	Romidepsin (Istodax®)
Histone deacetylase	Histone deacetylase inhibitor	Peripheral T-cell lymphoma	Recurrent or persistent disease	Belinostat (Beleodaq®)
Histone deacetylase	Histone deacetylase inhibitor	Multiple myeloma	Refractory disease	Panobinostat (Farydak®)
PD-1	Targeted immunotherapy	CHL		Nivolumab (Opdivo®)
PD-1	Targeted immunotherapy	CHL		Pembrolizumab (Keytruda®)
PI3K	B-cell receptor signaling	FL	Relapsed disease	Copanlisib hydrochloride (Aliqopa™)
PI3K	B-cell receptor signaling	CLL, FL	Relapsed disease	Idelalisib (Zydelig®)
Retinoid receptors	Retinoid X receptor activation	Cutaneous T-cell lymphoma	Refractory disease	Bexarotene (Targretin®)
26S proteasome	Proteasome inhibitor	Multiple myeloma, MCL	Combination therapy	Bortezomib (Velcade®)
20S proteasome	Proteasome inhibitor	Multiple myeloma	Relapsed or refractory disease	Carfilzomib (Kyprolis®)
20S proteasome	Proteasome inhibitor	Multiple myeloma	Relapsed or refractory disease	Ixazomib citrate (Ninlaro®)
SLAMF7	Targeted immunotherapy	Multiple myeloma	Relapsed or refractory disease	Elotuzumab (Empliciti™)

ALCL anaplastic large cell lymphoma, *B-ALL* B-cell acute lymphoblastic leukemia, Philadelphia positive (Ph+), *BL* Burkitt lymphoma, *BTK* Bruton's tyrosine kinase, *CHL* classical Hodgkin lymphoma, *CLL/SLL* chronic lymphocytic leukemia (CLL)/small lymphocytic lymphoma (SLL), *DLBCL* diffuse large B-cell lymphoma, *FL* follicular lymphoma, *HCL* hairy cell leukemia, *HGBL* high-grade B-cell lymphoma, *LPL* lymphoplasmacytic lymphoma, *MCL* mantle cell lymphoma, *PD-1* programmed cell death 1, *PI3K* phosphatidylinositol 3' kinase, *PMBCL* primary mediastinal large B-cell lymphoma, *SLAMF7* self-ligand receptor of the signaling lymphocytic activation molecule family member 7

A prominent feature of the activated B-cell type of DLBCL (ABC-DLBCL) is the constitutive activation of NF-kB signaling pathway and a "chronic active" form of BCR signaling. Chronic active BCR signaling is reminiscent of antigen-stimulated B cells and signals through the CBM complex to activate NF-kB. More than 20% of ABC-DLBCL patients harbor somatic mutations in *CD79B* and to a lesser extent in *CD79A* [3], and approximately 9% have oncogenic mutations in the *CARD11* gene [4]. Interruption of the BCR signaling pathway is specifically toxic to ABC-DLBCL cells in vitro [3].

The TCF3 transcription factor, in resting conditions, is maintained inactive through binding with ID3 (Fig. 32.2). Release of TCF3 from the ID3-TCF3 complex promotes BCR signaling by upregulating the expression of immunoglobulin heavy and light chain genes and repression of the BCR signaling inhibitor SHP1. Most BL (70%) harbor gain-of-function mutations affecting the TCF3 gene or mutations disrupting the TCF3 inhibitor ID3, which ultimately results in intensification of "tonic" BCR signaling [5].

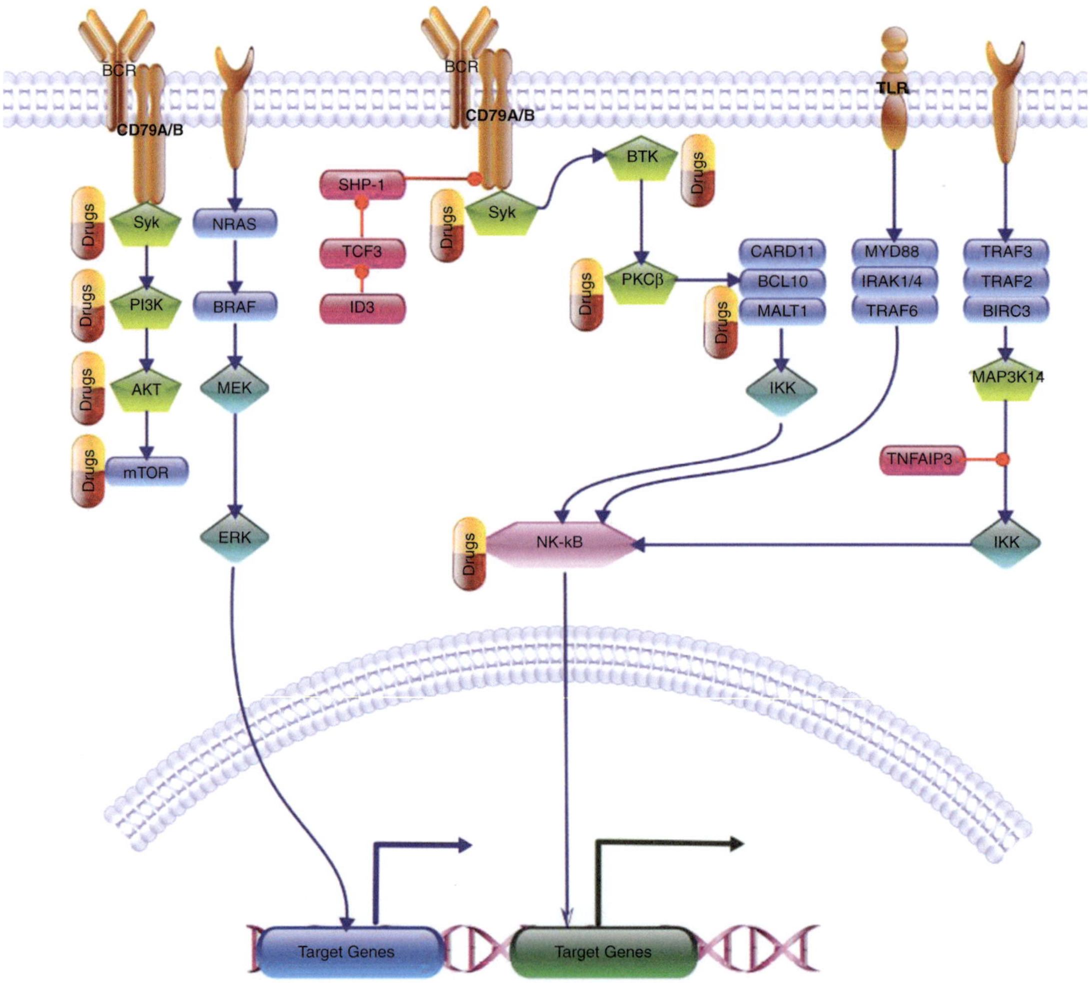

Fig. 32.2 Major signaling pathways involved in the development of B-cell lymphomas. (Adapted from Pillai and Chan [31]. With permission from Springer Nature)

Toll-Like Receptor Pathway

Toll-like receptors (TLRs) recognize a variety of pathogen associated molecular patterns derived from bacteria, viruses, and fungi in a BCR-independent manner. Ligand binding causes aggregation of TLRs and activation of cytoplasmic adapters such as MYD88 and triggering NF-kB pathway activation (Fig. 32.2). MYD88 mutations are seen in approximately 90% of all Waldenstrom's macroglobulinemia, 30% of ABC-DLBCL, and 10% of SMZL [6–9].

Notch Signaling

Notch receptor proteins are transmembrane receptors that function as ligand-activated transcription factors. Ligand binding causes proteolytic cleavage of the Notch intracellular domain which then translocates to the nucleus and induces transcription of target genes including MYC and NF-kB pathway components. The C-terminal PEST domain of the Notch protein is recognized by ubiquitin ligase leading to proteolytic degradation. Most mutations in lymphoma target the PEST domain resulting in truncation with impaired degradation and dysregulated Notch signaling. Notch pathway mutations have been identified in SMZL, CLL, and MCL. NOTCH mutations are seen in 5–10% of CLL and increase with relapse and transformation.

Nuclear Factor-Kappa B Pathway

NF-kB comprises a family of transcription factors including RELA/p65, RELB, c-Rel, NF-kB1, and NF-kB2 that function downstream of the BCR, TLR, and Notch signaling pathways. NF-kB proteins are bound in a complex with an inhibitor protein, IkB-alpha, which keeps them in the cytoplasm. Activation of IkB kinase (IKK) by external stimuli leads to phosphorylation of IkB-alpha and subsequent proteasomal degradation. Release of NF-kB transcription factors and translocation to the nucleus leads to activation of gene expression (Fig. 32.2).

Negative regulators of NF-kB signaling are inactivated by mutations/deletions in many B-cell lymphomas. TNFAIP3 is homozygously inactivated or deleted in 30% of ABC-DLBCL [10]. BIRC3, TRAF2, and TRAF3 are negative regulatory proteins that form a complex with the enzyme MAP3K14 (Fig. 32.2). When recruited to the active receptor, the complex is disrupted causing release and stabilization of MAP3K14, which in turn activates IKK. Loss of function mutations in *TRAF3* and *BIRC3* is detected in SMZL leading to activation of MAP3K14 and NF-kB signaling [11]. *BIRC3* gene is inactivated in CLL by mutations, and *BIRC3* mutations correlates with clinical course and chemo-refractoriness in CLL [12].

G-Protein-Coupled Receptors

G-protein-coupled receptors (GPCR) are the largest family of cell surface receptors involved in signal transduction with over 800 members. In conjunction with their cognate heterotrimeric G proteins, GPCRs activate multiple downstream targets that regulate cells survival, proliferation, and differentiation. Sequencing studies have revealed mutations or copy number alterations in members of the *S1PR2-GNA13* (encodes the Gα13 G-protein) – *RHOA* pathway in DLBCL as well as Burkitt lymphoma [5, 13].

Mitogen-Activated Protein Kinase Signaling Pathway

The mitogen-activated protein kinase pathway (MAPK) consists of receptor tyrosine kinases transmitting a variety of signals from the external environment to the nucleus. MAPK signaling in B cells is initiated at the BCR and propagated through RAS and RAF proteins. The *BRAF* V600E mutation is the pathognomonic genetic event for hairy cell leukemia (HCL). *BRAF* is also frequently mutated in Langerhans cell histiocytosis [14].

PI3K/AKT1/MTOR Signaling Pathway

The PI3K-AKT1-MTOR (phosphatidylinositol 3′-kinase/v-akt murine thymoma viral oncogene homolog 1/mechanistic target of rapamycin) pathway is a key component of tonic BCR signaling, a process that is required for B-cell survival. BCR signaling in Burkitt lymphoma activates the PI3K pathway, as demonstrated by phosphorylation of AKT1 and p70S6 kinase [15]. Activating mutations in *TCF3* as well as deleterious mutations in *TCF3* inhibitor, *ID3* potentiates BCR signaling by repression of SHP1 and enhancement of Ig production (Fig. 32.2). Seventy percent of Burkitt lymphoma cases have mutations in TCF3 or ID3 proteins, as described previously. Activation of the PI3K pathway due to loss of PTEN expression has been described in 37% of the DLBCL and 19% of MCL [16, 17]. Activating mutations in exons 9 and 20 of *PI3K* has also been reported [16].

Role of MYC in Lymphomagenesis

The MYC family includes C-MYC, N-MYC, L-MYC, and S-MYC which are involved in the control of cell growth, differentiation, and apoptosis. MYC-MAX heterodimers bind DNA sequences called E-box motifs to activate transcription of a large number of targets, estimated at 10–15% of all human genes. Enhanced MYC expression is seen in up to 70% of human malignancies and usually requires cooperation with other genetic lesions for oncogenic transformation.

Burkitt lymphoma is characterized by a translocation that places *MYC* gene under the control of the immunoglobulin enhancer. MYC deregulation is frequent in DLBCL – MYC translocations are seen in 5–14% of DLBCL, and MYC gains and amplification in 21–38% of DLBCL. Concurrent translocations of BCL2 or BCL6 gene with MYC, which are designated "double-hit lymphomas," impart a significantly worse prognosis. In addition, high MYC expression is also observed in 28–41% of DLBCL cases without MYC gene abnormalities, presumably induced by other upstream pathways. BCL2 protein is also concurrently overexpressed in about 60% of MYC-positive DLBCL cases, independent of the presence of gene rearrangement. These cases described as "double-expressor lymphoma" also behave more aggressively than cases with single protein overexpression. In addition, MYC alterations are also seen in plasmablastic lymphoma (PBL) and ALK-positive large B-cell lymphoma.

Role of p53 and DNA Repair Pathways

p53 is induced and activated by a number of stress signals, including DNA damage, oxidative stress, telomere shortening, hypoxia, and activated oncogenes leading to cell cycle arrest, cellular senescence, or apoptosis. Inactivation of p53 is seen in BL and DLBCL, including those derived from transformation of FL. Somatic and germline mutations have been identified in various DNA repair pathways, mainly in diffuse large B-cell lymphomas (DLBCLs) [18].

Regulation of Apoptosis

Apoptosis is a genetically controlled mechanism of cell death involved in the regulation of tissue homeostasis controlled by enzymes called caspases. The BCL2 family of proteins, which include pro-apoptotic (BAX, BAK, BIM, BID) and anti-apoptotic (BCL-2, BCL-XL, BCL-w, MCL-1, and BFL-1) members, regulates apoptosis by controlling mitochondrial outer membrane permeabilization (MOMP). Translocations affecting the BCL2 gene represent the genetic hallmark of FL and is seen in 80–90% cases, and it is also detected in about 30% of GCB-DLCBL.

Impairment of Differentiation to Plasma Cells

BCL6 is the master regulator of the germinal center reaction. BCL6 maintains the proliferative status of centroblasts and prevents terminal differentiation to plasma cells via suppression of the PRDM1/BLIMP1, the master regulator of plasma cell differentiation. The *PRDM1/BLIMP1* gene on 6q21 is biallelically inactivated in ~25% of ABC-DLBCL cases [19] and represents another mechanism of impairing differentiation with the tumor cells arrested at the plasmablastic stage.

Epigenetic Alterations

The epigenome is composed primarily of cytosine modifications, histone modifications, and expression of noncoding RNA molecules and is particularly important in the germinal center reaction [20, 21]. Histone octamers composed of H2A, H2B, H3, and H4 proteins form the scaffold upon which DNA is wound to form nucleosomes. Histones can be chemically modified by acetylation, methylation, phosphorylation, sumoylation, or ubiquitinylation. Trimethylation of H3 lysine 27 (H3K27me3) by the polycomb repressive complex 2 (PRC2) which includes EZH2, SUZ12, and EED is associated with transcriptional repression. EZH2 is upregulated in centroblasts and mediates H3K27 methylation and epigenetic silencing of genes required for memory and plasma cell differentiation. EZH2 gain-of-function mutations have been observed in >20% of FL and 23% of the GCB-DLBCL but are rare in ABC-DLBCL [13].

EP300 and CREBBP are histone acetyltransferases which are commonly mutated in FL and DL BCL [21] and may alter the enhancer profile of the tumor. Such loss of function mutations also lead to defective acetylation of other proteins including BCL6 and TP53 that may promote GCB cell lymphomagenesis [22]. Additional chromatin modifiers that are often mutated in lymphoma include components of the SNF/SWI complex, ARID1A and B, and SMARC family proteins.

Pathogenesis of T-Cell Lymphomas

Next-generation sequencing studies in angioimmunoblastic T-cell lymphoma (AITL) have shown frequent mutations of *TET2*, *DNMT3A*, *IDH2*, and *RHOA* with *IDH2* mutations exclusively involving R172 [23]. In contrast to the situation in AML, *IDH2*, and *TET2* mutations are not exclusive and frequently occur together. Other mutations are far less common, and some mutations such as *CD28* are quite unique to PTCL. *CD28* mutations appear to enhance its signaling either through increased binding affinity to its ligand CD80/86 on antigen presenting cells or its interaction with cytoplasmic adaptor molecules such as GRB2 eventually resulting in increased activation of the NF-kB pathway [24].

The genetic hallmark of ALK-positive ALCL is a chromosomal translocation involving the *ALK* gene. Translocations involving *DUSP2* and *TP63* have been described in some ALK-negative ALCL with good and poor prognosis, respectively [25].

Many cytokine signal through the JAK/STAT (Janus-associated kinase/signal transducer and activator of transcription) pathway includes four cytoplasmic JAK kinases and seven STAT proteins. Cytokine/receptor binding leads to activation of the associated JAK kinase, docking and phosphorylation of the cognate STAT protein, and translocation to the nucleus where it functions as a transcription factor. A systematic genomic analysis in ALK-ALCLs identified activating mutations of *JAK1* and/or *STAT3* genes in 18% of ALK-ALCLs and 5% of cutaneous ALCLs.

Two major subgroups have been described within the PTCL-NOS category by gene expression profiling. The first group is characterized by the high expression of *GATA3* and some of its target genes and is associated with poor prognosis. The second group demonstrates high expression of *TBX21* and some of its target genes–increased expression of the cytotoxic gene signature in this group also showed poor clinical outcome [26]. The mutational landscape of PTCL-NOS is not well defined, but some of the mutations seen in AITL and other lymphomas including epigenetic mediators (*MLL*, *TET2*, *DMNT3A*), signaling pathway proteins (*TNFAIP3*, *APC*, *CHD8*), and tumor suppressors (*TP53*, *FOXO1*, *ATM*) have been detected [27, 28].

Pathogenesis of Hodgkin Lymphoma

Classical Hodgkin lymphoma (cHL) arises from post-germinal center B cells, as evidenced by clonally rearranged and somatically mutated immunoglobulin genes. Surface Ig is not expressed indicating independence from antigen stimulation. Other important pathogenetic features include loss of expression of many B-cell-specific genes, activation of NF-kB pathway, EBV infection in a subset of cases, and genetic changes that modulate the tumor environment to favor HL cell survival.

Constitutive activation of NF-κB pathway occurs through multiple mechanisms in CHL, including *REL* amplification [29], increased expression of the positive NF-κB regulator *BCL3*, or inactivating mutations in negative regulators such as *NFKBIA* (20% of cases), *NFKBIE* (15%), and *TNFAIP3* (40%) [30]. Interestingly, TNFAIP3 loss and EBV positivity were mutually exclusive. Increased expression of PD-L1 (CD274), PD-L2 (CD273), and CIITA translocation has been shown to modulate the microenvironment in favor of immune escape.

Prognostic Biomarkers in Lymphoid Neoplasms

Data from next-generation sequencing-based testing has revealed the prognostic significance of mutations in the major oncogene's and tumor suppressor genes (Table 32.4).

Table 32.4 Representative biomarkers of prognostic significance in lymphoid neoplasms

Gene Mutations	Patient population	Prognostic significance [2]
ATM	CLL/SLL	May predict treatment failure
BIRC3	CLL/SLL	Chemorefractoriness and poor prognosis
TP53	CLL/SLL	Shorter OS, PFS, refractoriness to fludarabine regimens
NOTCH1	CLL/SLL	Shorter OS, PFS, and time to treatment
SF3B1	CLL/SLL	Shorter OS and time to treatment
TP53	FL	Higher FLIPI score, shorter PFS and OS
MYD88	LPL	Better response to ibrutinib
CXCR4	LPL	Associated with higher disease activity and resistance to ibrutinib
CCND1	PCM	Associated with shorter OS
TP53	PCM	Deletions of 17p and TP53 mutations associated with increased disease progression and treatment refractoriness

PFS progression-free survival, *OS* overall survival, *FLIPI* follicular lymphoma international prognostic index

Summary and Future Directions

Most lymphomas arise from a combination of somatic mutations and aberrantly regulated genes that lead to activation of biochemical and signaling pathways promoting tumor growth. Advances in understanding of the molecular pathogenesis of lymphomas in recent years especially using next-generation sequencing have enabled identification of novel diagnostic, prognostic, and predictive molecular biomarkers. Many of these biomarkers have already been included in the latest NCCN guidelines and the 2016 WHO classification of lymphoid neoplasms. Novel-targeted therapies have revolutionized the therapeutic landscape for relapsed and refractory diseases. Techniques for minimal residual detection especially with liquid biopsies will enable monitoring of treatment efficacy.

References

1. Swerdlow SH, Campo E, Harris NL, Jaffe ES, Pileri S, Stein H, et al. WHO classification of tumors of hematopoietic and lymphoid tissues. 4th ed. Lyon: IARC Press; 2017.
2. Onaindia A, Medeiros LJ, Patel KP. Clinical utility of recently identified diagnostic, prognostic, and predictive molecular biomarkers in mature B-cell neoplasms. Mod Pathol Off J U S Can Acad Pathol Inc. 2017;30(10):1338–66. PubMed PMID: 28664939.
3. Davis RE, Ngo VN, Lenz G, Tolar P, Young RM, Romesser PB, et al. Chronic active B-cell-receptor signalling in diffuse large B-cell lymphoma. Nature. 2010;463(7277):88–92. PubMed PMID: 20054396. Pubmed Central PMCID: 2845535.
4. Lenz G, Davis RE, Ngo VN, Lam L, George TC, Wright GW, et al. Oncogenic CARD11 mutations in human diffuse large B cell lymphoma. Science. 2008;319(5870):1676–9. PubMed PMID: 18323416.
5. Love C, Sun Z, Jima D, Li G, Zhang J, Miles R, et al. The genetic landscape of mutations in Burkitt lymphoma. Nat Genet. 2012;44(12):1321–5. PubMed PMID: 23143597. Pubmed Central PMCID: 3674561.
6. Ngo VN, Young RM, Schmitz R, Jhavar S, Xiao W, Lim KH, et al. Oncogenically active MYD88 mutations in human lymphoma. Nature. 2011;470(7332):115–9. PubMed PMID: 21179087.
7. Pasqualucci L, Trifonov V, Fabbri G, Ma J, Rossi D, Chiarenza A, et al. Analysis of the coding genome of diffuse large B-cell lymphoma. Nat Genet. 2011;43(9):830–7. PubMed PMID: 21804550. Pubmed Central PMCID: 3297422.
8. Rossi D, Trifonov V, Fangazio M, Bruscaggin A, Rasi S, Spina V, et al. The coding genome of splenic marginal zone lymphoma: activation of NOTCH2 and other pathways regulating marginal zone development. J Exp Med. 2012;209(9):1537–51. PubMed PMID: 22891273. Pubmed Central PMCID: 3428941. Epub 2012/08/15. eng.
9. Treon SP, Xu L, Yang G, Zhou Y, Liu X, Cao Y, et al. MYD88 L265P somatic mutation in Waldenstrom's macroglobulinemia. N Engl J Med. 2012;367(9):826–33.
10. Compagno M, Lim WK, Grunn A, Nandula SV, Brahmachary M, Shen Q, et al. Mutations of multiple genes cause deregulation of NF-kappaB in diffuse large B-cell lymphoma. Nature. 2009;459(7247):717–21. PubMed PMID: 19412164. Pubmed Central PMCID: 2973325.
11. Arcaini L, Rossi D. Nuclear factor-kappaB dysregulation in splenic marginal zone lymphoma: new therapeutic opportunities. Haematologica. 2012;97(5):638–40. PubMed PMID: 22556352. Pubmed Central PMCID: 3342963.
12. Rossi D, Fangazio M, Rasi S, Vaisitti T, Monti S, Cresta S, et al. Disruption of BIRC3 associates with fludarabine chemorefractoriness in TP53 wild-type chronic lymphocytic leukemia. Blood. 2012;119(12):2854–62.
13. Lohr JG, Stojanov P, Lawrence MS, Auclair D, Chapuy B, Sougnez C, et al. Discovery and prioritization of somatic mutations in diffuse large B-cell lymphoma (DLBCL) by whole-exome sequencing. Proc Natl Acad Sci U S A. 2012;109(10):3879–84. PubMed PMID: 22343534. Pubmed Central PMCID: 3309757.
14. Badalian-Very G, Vergilio JA, Degar BA, MacConaill LE, Brandner B, Calicchio ML, et al. Recurrent BRAF mutations in Langerhans cell histiocytosis. Blood. 2010;116(11):1919–23. PubMed PMID: 20519626. Pubmed Central PMCID: 3173987.
15. Schmitz R, Young RM, Ceribelli M, Jhavar S, Xiao W, Zhang M, et al. Burkitt lymphoma pathogenesis and therapeutic targets from structural and functional genomics. Nature. 2012;490(7418):116–20. PubMed PMID: 22885699. Pubmed Central PMCID: 3609867.
16. Abubaker J, Bavi PP, Al-Harbi S, Siraj AK, Al-Dayel F, Uddin S, et al. PIK3CA mutations are mutually exclusive with PTEN loss in diffuse large B-cell lymphoma. Leukemia. 2007;21(11):2368–70. PubMed PMID: 17657213.
17. Rudelius M, Pittaluga S, Nishizuka S, Pham TH, Fend F, Jaffe ES, et al. Constitutive activation of Akt contributes to the pathogenesis and survival of mantle cell lymphoma. Blood. 2006;108(5):1668–76. PubMed PMID: 16645163. Pubmed Central PMCID: 1895501.
18. de Miranda NF, Peng R, Georgiou K, Wu C, Falk Sorqvist E, Berglund M, et al. DNA repair genes are selectively mutated in diffuse large B cell lymphomas. J Exp Med. 2013;210(9):1729–42. PubMed PMID: 23960188. Pubmed Central PMCID: 3754869.
19. Mandelbaum J, Bhagat G, Tang H, Mo T, Brahmachary M, Shen Q, et al. BLIMP1 is a tumor suppressor gene frequently disrupted in activated B cell-like diffuse large B cell lymphoma. Cancer Cell. 2010;18(6):568–79. PubMed PMID: 21156281. Pubmed Central PMCID: 3030476.
20. Chi P, Allis CD, Wang GG. Covalent histone modifications – miswritten, misinterpreted and mis-erased in human cancers. Nat Rev Cancer. 2010;10(7):457–69. PubMed PMID: 20574448. Pubmed Central PMCID: 3262678.
21. Jiang Y, Melnick A. The epigenetic basis of diffuse large B-cell lymphoma. Semin Hematol. 2015;52(2):86–96. PubMed PMID: 25805588. Pubmed Central PMCID: 4374125.
22. Pasqualucci L, Dominguez-Sola D, Chiarenza A, Fabbri G, Grunn A, Trifonov V, et al. Inactivating mutations of acetyltransferase genes in B-cell lym-

phoma. Nature. 2011;471(7337):189–95. PubMed PMID: 21390126. Pubmed Central PMCID: 3271441.
23. Cairns RA, Iqbal J, Lemonnier F, Kucuk C, de Leval L, Jais JP, et al. IDH2 mutations are frequent in angioimmunoblastic T-cell lymphoma. Blood. 2012;119(8):1901–3. PubMed PMID: 22215888. Pubmed Central PMCID: 3293643.
24. Rohr J, Guo S, Huo J, Bouska A, Lachel C, Li Y, et al. Recurrent activating mutations of CD28 in peripheral T-cell lymphomas. Leukemia. 2015;5:1062–70.
25. Zeng Y, Feldman AL. Genetics of anaplastic large cell lymphoma. Leuk Lymphoma. 2016;57(1):21–7. PubMed PMID: 26104084.
26. Iqbal J, Wright G, Wang C, Rosenwald A, Gascoyne RD, Weisenburger DD, et al. Gene expression signatures delineate biological and prognostic subgroups in peripheral T-cell lymphoma. Blood. 2014;123(19):2915–23. PubMed PMID: 24632715. Pubmed Central PMCID: PMC4014836. Epub 2014/03/19. eng.
27. Palomero T, Couronne L, Khiabanian H, Kim MY, Ambesi-Impiombato A, Perez-Garcia A, et al. Recurrent mutations in epigenetic regulators, RHOA and FYN kinase in peripheral T cell lymphomas. Nat Genet. 2014;46(2):166–70. PubMed PMID: 24413734. Pubmed Central PMCID: 3963408.
28. Schatz JH, Horwitz SM, Teruya-Feldstein J, Lunning MA, Viale A, Huberman K, et al. Targeted mutational profiling of peripheral T-cell lymphoma not otherwise specified highlights new mechanisms in a heterogeneous pathogenesis. Leukemia. 2015;29(1):237–41. PubMed PMID: 25257991. Pubmed Central PMCID: 4286477.
29. Martin-Subero JI, Gesk S, Harder L, Sonoki T, Tucker PW, Schlegelberger B, et al. Recurrent involvement of the REL and BCL11A loci in classical Hodgkin lymphoma. Blood. 2002;99(4):1474–7. PubMed PMID: 11830502.
30. Schmitz R, Stanelle J, Hansmann ML, Kuppers R. Pathogenesis of classical and lymphocyte-predominant Hodgkin lymphoma. Annu Rev Pathol. 2009;4:151–74. PubMed PMID: 19400691.
31. Pillai RK, Chan WC. Pathogenesis of lymphomas. In: Zain J, Kwak L, editors. Management of lymphomas: a case-based approach. Adis, Cham: Springer International Publishing; 2017. p. 11–31.

Targeted Therapies for Pediatric Central Nervous System Tumors

33

Nicholas Shawn Whipple and Amar Gajjar

Abbreviations

CNS	Central nervous system
HGG	High-grade glioma
LGG	Low-grade glioma
MAPK	Mitogen-activated protein kinase
MEK	Mitogen-activated extracellular signal-regulated kinase
mTOR	Mammalian target of rapamycin
NF1	Neurofibromatosis type 1
PN	Plexiform neurofibroma
PTCH1	Patched 1
SEGA	Subependymal giant cell astrocytoma
SHH	Sonic hedgehog
SMO	Smoothened
SUFU	Suppressor of fused
TSC	Tuberous sclerosis complex
WHO	World Health Organization

Introduction

Central nervous system (CNS) tumors are the most common solid tumors in childhood and can be malignant or nonmalignant. Primary malignant CNS tumors represent approximately 20% of all childhood cancers yet account for 30% of all childhood cancer deaths in the United States, having superseded leukemia as the leading cause of death from childhood cancer [1, 2].

The prognosis for patients with CNS tumors is based on many factors, including the tumor type, its location and histologic grade, and the available treatment options. Historically, pediatric CNS tumors were diagnosed, classified, and treated based on their location and histologic criteria. Recent discoveries in pediatric neuro-oncology have greatly enhanced our understanding of the biology of these tumors, including their molecular and genetic characteristics. The molecular characterization of CNS tumors has led to improved diagnostic accuracy and risk stratification. In 2016, these advances resulted in a revised classification of CNS tumors by the World Health Organization (WHO), in which molecular parameters and histology define many tumor entities [3].

Based on our recent understanding of molecular markers, the use of targeted therapies has begun to transform our approach to treating many pediatric CNS tumors. Currently, a few targeted therapies are being used to treat subgroups of pediatric CNS tumors, mostly in the setting of clinical trials. The

N. S. Whipple (✉)
Division of Hematology/Oncology, Department of Pediatrics, University of Utah and Primary Children's Hospital, Salt Lake City, UT, USA
e-mail: nicholas.whipple@hsc.utah.edu

A. Gajjar
Department of Oncology, St. Jude Children's Research Hospital, Memphis, TN, USA

S. Badve, G. L. Kumar (eds.), *Predictive Biomarkers in Oncology*,
https://doi.org/10.1007/978-3-319-95228-4_33

Table 33.1 Pediatric central nervous system tumors with molecularly defined therapeutic targets

Tumor type	Molecular subgroup	Therapeutic target	Molecular pathway	Clinically tested agents
Medulloblastoma	Sonic Hedgehog	SMO	SHH	Sonidegib (Odomzo®), Vismodegib (Erivedge®)
Subependymal giant cell astrocytoma	–	mTOR	mTOR	Everolimus (Afinitor®)
Low-grade glioma	*BRAF* V600E	BRAF V600E	MAPK	Dabrafenib (Tafinlar®), vemurafenib (Zelboraf®)
High-grade glioma	*BRAF* V600E	BRAF V600E	MAPK	Dabrafenib (Tafinlar®), vemurafenib (Zelboraf®)
Pilocytic astrocytoma	*KIAA1549:BRAF* fusion	KIAA1549:BRAF fusion	MAPK	Selumetinib (AZD6244), trametinib (Mekinist®)
Plexiform neurofibroma	–	MEK 1/2	MAPK	Selumetinib (AZD6244)

SHH sonic hedgehog, *mTOR* mammalian target of rapamycin, *MAPK* mitogen-activated protein kinase, *MEK* mitogen-activated extracellular signal-regulated kinase, *SMO* smoothened

tumors being treated by this approach include sonic hedgehog (SHH) medulloblastoma, subependymal giant cell astrocytoma (SEGA), *BRAF* V600E-mutated low-grade and high-grade gliomas (LGG, HGG), *KIAA1549:BRAF* fusion-positive pilocytic astrocytoma, and plexiform neurofibroma (PN).

For many patients, including patients with recurrent or refractory disease, the use of targeted therapies for these tumor subtypes has resulted in significant tumor regression and improved survival. In this chapter we provide an overview of pediatric CNS tumors for which key driver mutations and targeted therapies have created a paradigm shift in the treatment approach (Table 33.1).

Medulloblastoma

Medulloblastoma is a heterogeneous disease consisting of four main molecularly defined subgroups: wingless (WNT; group 1), sonic hedgehog (SHH; group 2), group 3 (characterized by *MYC* amplification and *GFI* activation), and group 4 (characterized by *MYCN* and *CDK6* amplifications and alterations in *SNCAIP*) [4]. To date, targeted therapy is only applicable to the SHH subgroup, which represents approximately 25% of all medulloblastomas. Nodular desmoplastic histology is pathognomonic for the SHH subgroup, although SHH tumors also exhibit classic or large-cell/anaplastic histology. SHH medulloblastoma affects patients of all ages, but it primarily occurs in children younger than 5 years and in individuals older than 16 years. The 5-year overall survival for patients with tumors of this subgroup is 70%; however, the current treatment often results in significant morbidity, and the prognosis in recurrent or refractory disease is dismal [5].

SHH medulloblastoma most frequently arises from a cerebellar hemisphere, and cerebellar granule neuron precursors are its imputed cells of origin. This tumor subtype is characterized by aberrant activation of the SHH signaling pathway (Fig. 33.1). In a normal cell, binding of a SHH ligand to the patched 1 (PTCH1) receptor releases its inhibition of smoothened (SMO), the main upstream activator in the pathway. Activated SMO then releases suppressor of fused (SUFU) inhibition of GLI proteins, which can then translocate to the nucleus and activate transcription of SHH target genes (*GLI1*, *GLI2*, *PTCH1*, and *MYCN*). In SHH medulloblastoma, a ligand-independent pathway disruption occurs as a result of somatic or germline mutations involving one of several genes in the SHH pathway. This disruption leads to aberrant expression of SHH target genes, which allows cell proliferation and tumorigenesis [5].

The activating mutations most commonly found to occur in SHH medulloblastoma and to disrupt the SHH signaling pathway include mutations of *PTCH1*, *SMO*, *SUFU*, and *TP53*, as well as *GLI2* and *MYCN* amplification. Patients who harbor upstream SHH-pathway mutations (e.g.,

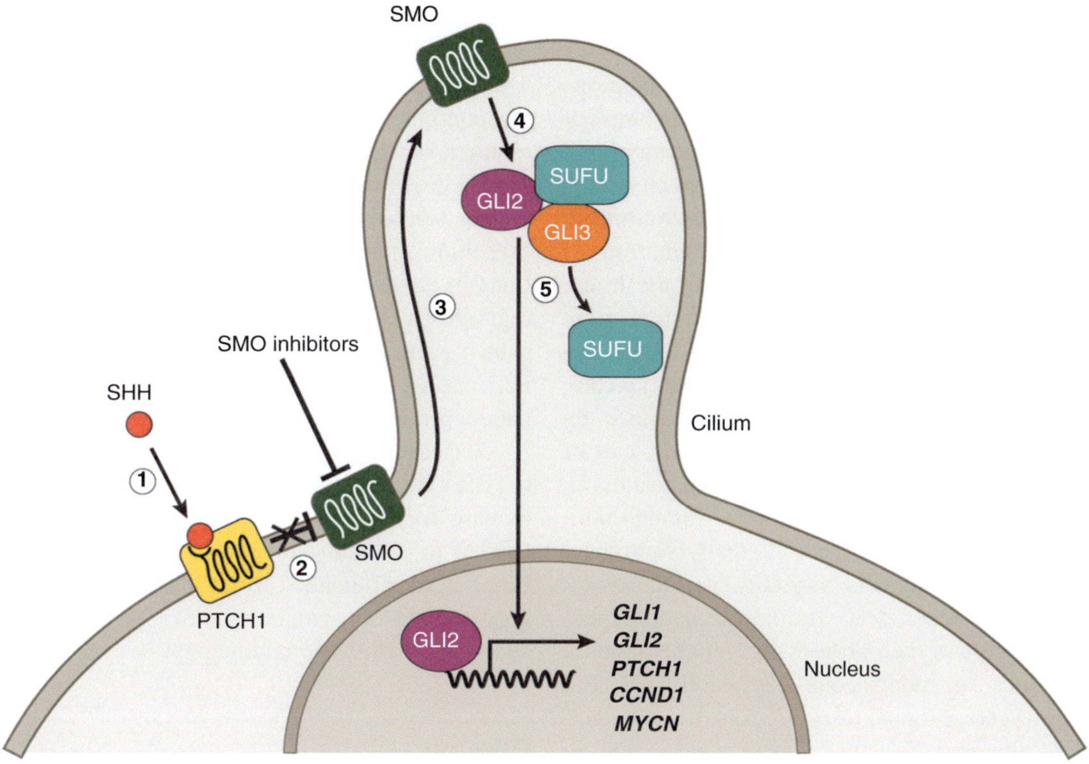

Fig. 33.1 Illustration of sonic hedgehog (SHH) signaling pathway. (1) SHH ligand binds to PTCH1 transmembrane protein. (2) Binding of SHH to PTCH1 relieves inhibition of smoothened (SMO). (3) Activated SMO localizes to cilium. (4) SMO releases suppressor of fused (SUFU) inhibition of GLI proteins. (5) Activated GLI proteins translocate to nucleus and activate transcription of SHH target genes (i.e., GLI1, GLI2, PTCH1, and MYCN). In SHH-subgroup medulloblastoma, disruptions to SHH pathway occur through mutation of PTCH1, SMO, or SUFU and/or amplification of GLI2 or MYCN. (Robinson et al. [5]. Reprinted with permission. © 2015 American Society of Clinical Oncology. All rights reserved]

PTCH1 and *SMO* mutations) have shown sensitivity to SMO inhibitors, which first emerged as a potentially effective targeted therapy for SHH medulloblastoma after an adult with relapsed metastatic disease was treated with vismodegib and experienced a profound initial response [6]. Agents in this class of targeted inhibitors act as competitive antagonists of the SMO receptor, inhibiting signaling downstream of SMO. Unfortunately, patients with SHH medulloblastoma who harbor downstream mutations (e.g., *SUFU* mutations or *GLI2* amplification) are resistant to these agents.

In several pediatric and adult clinical trials for recurrent SHH medulloblastoma, the SMO inhibitors vismodegib and sonidegib have been well tolerated and have demonstrated promising efficacy [5, 7]. Objective tumor responses were seen in as many as 33% of reported cases, with the responses in several patients being sustained for 4–8 months [5, 7]. Two clinical trials are currently investigating the effectiveness of vismodegib in treating this tumor subgroup [8, 9]. Children treated with SMO inhibitors should be closely monitored for premature physeal closure, as the SHH pathway plays a role in bone development [10, 11].

Subependymal Giant Cell Astrocytoma

Tuberous sclerosis complex (TSC) is an autosomal dominant genetic disorder with an estimated prevalence of 1 in 6000 live births. In more than

85% of affected individuals, TSC is caused by mutations in the tumor suppressor genes TSC1 (hamartin) or TSC2 (tuberin). These mutations cause hyperactivation of the mammalian target of rapamycin (mTOR) signaling pathway and upregulation of mTOR complex 1, which results in abnormal cellular growth and proliferation and the development of benign tumors (hamartomas) in multiple organ systems, including the brain, kidneys, lungs, and skin. Subependymal giant cell astrocytomas (SEGAs) are slow-growing low-grade glioneuronal tumors that typically arise near the foramen of Monro and occur in up to 20% of patients with TSC. Generally, half of all patients will become symptomatic, usually in adolescence or young adulthood. Although non-malignant, SEGAs carry a clinically significant risk of morbidity and mortality, including seizures and sudden death from acute hydrocephalus, because of their progressive volume increase and lack of spontaneous regression [12]. For many years, surgical resection was the only standard therapy available. Unfortunately, not all tumors are resectable because of their location, and numerous postoperative complications have been reported, including intraventricular hemorrhage, cognitive impairment, and eventual recurrence if gross total resection is not achieved.

Targeted inhibition of the mTOR pathway has significantly improved outcomes in patients with TSC. Everolimus was the first mTOR inhibitor to be approved for treating SEGA associated with TSC, after clinical trials demonstrated a rapid, marked reduction in tumor volume (over 50% in some cases) and improved quality of life after only a few months of therapy [12]. Everolimus functions by inhibiting mTOR complex 1, thereby correcting the molecular defect responsible for TSC and tumor development. There are numerous reports of cases in which treatment with everolimus led to the resolution of tumor-associated ventricular dilation, a reduction in seizure frequency, and a decrease in tumor size in organ systems other than the brain. Treatment with sirolimus (formerly called rapamycin) has demonstrated similar results [12, 13].

Long-term follow-up of patients being treated with everolimus continues to demonstrate the sustained efficacy of mTOR inhibition with respect to SEGA tumor reduction after more than 5 years of continuous therapy; no patients receiving continuous treatment with everolimus have required surgical intervention for tumor progression [13]. However, because SEGAs do not completely resolve with therapy, continuous use of everolimus may be necessary to maintain reductions in tumor volume and prevent lesions from regrowing. The studies performed to date have shown everolimus to be safe and effective, with no limiting toxicities and no adverse effect on patient growth or maturation.

Consensus guidelines recommend targeted mTOR inhibition with everolimus as the standard of care for treating symptomatic, unresectable SEGAs in patients with TSC. In addition, everolimus is recommended as an alternative to surgery in cases of asymptomatic SEGAs that show signs of growth on serial imaging.

Gliomas

Low-grade gliomas (WHO grade 1 and grade 2 tumors) are the most common CNS tumors in children. Based on their histology, they are categorized in three major classes: astrocytic tumors, oligodendroglial tumors, and neuronal and mixed neuroglial tumors. LGGs are characterized by slow growth and are often considered a chronic disease. They commonly arise in the cerebral hemispheres or posterior fossa and are frequently cured via gross total resection. Tumors arising from midline structures (e.g., the hypothalamus, basal ganglia, and brainstem) and the optic pathway are less amenable to resection and typically require alternative treatment approaches. Some cancer predisposition syndromes, such as tuberous sclerosis and neurofibromatosis type 1, are associated with an increased frequency of LGGs [14, 15].

Although they rarely undergo malignant transformation, LGGs can cause significant morbidity, including headaches, seizures, vision loss, endocrine dysfunction, and impaired cognition. Chemotherapy is the initial approach for treating unresectable or subtotally resected tumors.

Carboplatin, vincristine, temozolomide, vinblastine, thioguanine, procarbazine, lomustine, and Avastin are among the agents most commonly used, with variable response rates being reported in the medical literature. Radiation therapy is generally reserved for individuals who experience treatment failure after chemotherapy. However, these standard therapeutic approaches are not always successful [14, 15].

Recent genomic discoveries have altered the landscape of pediatric LGG therapy by identifying key driver mutations in the mitogen-activated protein kinase (MAPK) pathway that contribute to cellular proliferation and tumorigenesis. Duplication or mutation of the *BRAF* gene is the main molecular alteration in pediatric LGGs. The *BRAF* V600E mutation and *KIAA1549:BRAF* fusion (caused by duplication of the 7q34 region) are the two *BRAF* aberrations most frequently identified and result in constitutive activation of the MAPK pathway [14–16]. The prognostic implications of these genetic alterations in pediatric LGGs have not been determined [15, 16]. Recent use of therapies that target these alterations and inhibit the MAPK pathway has demonstrated that such approaches hold considerable promise for treating LGGs, including tumors that are refractory to conventional therapy. These personalized, selective approaches to therapy offer an alternative to the "one treatment fits all" strategy [14].

Approximately 90% of pilocytic astrocytomas (mostly extracerebellar tumors) harbor the *KIAA1549:BRAF* fusion; the prevalence of this fusion in other pediatric LGGs has not been determined [15]. Inhibitors of mitogen-activated extracellular signal-regulated kinase 1 (MEK1) and MEK2 activation and activity, of which selumetinib and trametinib are two examples, have been studied in *KIAA1549:BRAF* fusion-positive LGGs in preclinical and clinical settings, and sustained responses have been demonstrated, with a reduction in tumor size exceeding 60% in some cases [17, 18].

As many as 70% of pleomorphic xanthoastrocytomas, 20% of gangliogliomas, and 10% of pilocytic astrocytomas harbor the *BRAF* V600E mutation, and other LGG subtypes have also been found to carry the mutation [16]. Dabrafenib and vemurafenib are competitive small molecules that inhibit the ATP-binding domain of mutant BRAF V600E and have shown efficacy at slowing tumor growth and inducing tumor regression in a variety of *BRAF* V600E-mutated LGGs, including numerous tumors that were refractory to conventional therapy. BRAF inhibition has been reported to induce a reduction in tumor size of up to 70%, with the responses being sustained for up to 1 year. Those patients treated by this method, including infants as young as 2 months, have experienced clinical improvements in their neurodevelopment, ambulation, and vision in cases involving hypothalamic/chiasmatic tumors [19–21]. Retreatment with vemurafenib after tumor progression has been shown to induce tumor regression, which suggests that some patients will benefit from continuous therapy [19].

Agents that target BRAF can be easily administered orally, cause minimal myelosuppression, and are often less toxic than conventional agents. However, BRAF V600E inhibitors have been reported to induce proliferation of malignant cutaneous lesions [22]. Regular comprehensive assessments by a dermatologist should be part of the routine monitoring of pediatric patients with LGGs being treated with dabrafenib or vemurafenib, especially as the duration of therapy required in these cases has yet to be determined.

The *BRAF* V600E mutation and *KIAA1549:BRAF* fusion are valuable diagnostic markers and should be considered part of the standard workup in cases of pediatric LGG, especially when the tumors are refractory to conventional therapy. Clinical trials are currently being conducted to further investigate the efficacy of BRAF-targeting therapies for pediatric LGGs.

High-grade gliomas (WHO grade 3 and grade 4 tumors) are the least common malignant brain tumors in children, but as a group they remain the most lethal and difficult to treat, with an overall survival rate of less than 10% [15]. They arise most frequently from the brain or brainstem and are typically characterized by rapid growth. Diffuse intrinsic pontine gliomas are considered an incurable pediatric cancer type and significantly decrease the overall survival rate of pediatric HGGs.

Current treatment approaches for pediatric HGGs include surgical resection followed by radiation therapy and/or chemotherapy. As with most tumor types, the extent of resection is a strong clinical prognostic factor. To date, no chemotherapeutic regimen has proven highly effective in treating this class of tumors. For this reason, clinical trials continue to investigate ways to improve survival, including through the use of novel therapies. The biologic and molecular subgrouping of these tumors is expected to alter the treatment landscape by identifying actionable driver mutations.

Although less commonly seen than in LGGs, *BRAF* V600E mutations have been detected in pediatric HGGs, including glioblastoma multiforme, anaplastic astrocytoma, anaplastic pleomorphic xanthoastrocytoma, and anaplastic ganglioglioma. Vemurafenib has been reported to induce tumor regression in a few cases of recurrent or progressive *BRAF* V600E-mutated HGG [23–25]. Most notably, vemurafenib induced complete clinical regression of a recurrent glioblastoma multiforme in a 9-year-old patient within 4 months of treatment initiation. Clinical and radiographic response has now been maintained for more than 6 months, with the patient remaining on therapy [25]. Similarly, vemurafenib induced a partial response in a 2-year-old patient with anaplastic ganglioglioma; this patient was reported to have maintained significant clinical and neurological improvement at 20 months after treatment initiation [23]. BRAF inhibition has important therapeutic potential in pediatric HGGs; it may extend survival, improve quality of life, allow for a safer surgical resection, or increase the time to radiation treatment in order to preserve neurocognitive development. Diagnostic workup for the *BRAF* V600E mutation should be considered in cases of pediatric HGG, especially when the tumors are refractory to conventional therapy.

Plexiform Neurofibroma

Neurofibromatosis type 1 (NF1) is an autosomal dominant condition that affects 1 in 3000 live births. It is the most common human cancer predisposition syndrome and is characterized by multiple clinical manifestations, including tumors of the nervous system. Plexiform neurofibromas (PNs) are benign peripheral nerve sheath tumors characterized by differentiated Schwann cells. The tumors develop in 20–50% of individuals with NF1, generally during childhood, and are often characterized by rapid growth during this period. They can grow to be quite large and cause significant morbidity, including pain, disfigurement, functional impairment, and various other neurologic complications. Because of their location and tendency to extend through multiple layers of tissues and involve multiple nerves or nerve plexuses, complete surgical resection of PNs (historically, the most effective treatment) is often not feasible [26, 27]. An additional concern is that PNs can transform into malignant peripheral nerve sheath tumors, which affect approximately 10% of individuals with NF1 [27].

The development of PNs is a consequence of mutations in the NF1 gene, a tumor suppressor gene that encodes a protein called neurofibromin. Neurofibromin is a negative regulator of RAS activity that is nonfunctional in patients with NF1. The lack of functional neurofibromin leads to dysregulated RAS, tumorigenesis, and the development of PNs [26, 27]. It has been demonstrated in preclinical mouse models that targeted MEK inhibition can induce regression of PNs by suppressing the RAS/MAPK signaling pathway [26]. These data prompted a recent phase I clinical trial of selumetinib in NF1-related PNs, in which selective inhibition of MEK 1 and MEK2 produced profound, sustained tumor regression in children. Among the 24 patients enrolled on the trial, 71% experienced a partial response (a tumor volume decrease from baseline of at least 20%), and all patients experienced some decrease in tumor volume. The median decrease in tumor volume from baseline was 31%, with the largest decrease being 47%. Nearly all patients receiving long-term selumetinib therapy have experienced a sustained reduction in tumor volume and minimal toxic effects [26].

MEK inhibition has important therapeutic potential for pediatric PNs, particularly inoperable tumors. By reducing the tumor volume, treatment with selumetinib may improve the patient's quality of life by decreasing disfiguration and increasing motor function. Because PNs did not completely resolve with selumetinib therapy, continuous targeted MEK inhibition may be necessary to maintain the reductions in tumor volume. It is uncertain if MEK inhibition has the potential to decrease the incidence of malignant transformation of PNs.

Conclusions and Future Directions

Although the field of targeted therapy for pediatric CNS tumors is still in its infancy, it has the potential to revolutionize the care of children with primary CNS tumors by improving survival and limiting treatment-associated toxicity. The recently revised classification of CNS tumors by the WHO has enhanced our understanding of the underlying pathogenesis of pediatric brain tumors and highlighted the importance of molecular characterization in the identification, risk stratification, and treatment of CNS tumors.

Just as everolimus has become the standard of care for treating subgroups of patients in which SEGA has been diagnosed, targeted therapy based on molecular markers may become standard practice in subgroups of patients with other tumor types. Most notably, targeted therapy may play an increasingly significant role in treating patients with SHH medulloblastoma, *BRAF* V600E-mutated gliomas, *KIAA1549:BRAF* fusion-positive pilocytic astrocytoma, or plexiform neurofibroma (Table 33.1). Clinicians should actively pursue additional diagnostic testing in patients being treated for tumors that commonly harbor targetable molecular aberrations, especially when the tumors are recurrent or refractory to conventional therapy.

Laboratory-based research will continue to elucidate key driver mutations, expand our understanding of their associated molecular pathways, and lead to the use of additional targeted agents in other tumor subtypes, as well as to the development of novel agents. The challenge for clinicians and neuro-oncology consortiums will be to design the next generation of clinical trials to investigate the clinical potential of molecularly defined targeted therapies.

References

1. Ward E, et al. Childhood and adolescent cancer statistics, 2014. CA Cancer J Clin. 2014;64(2):83–103.
2. Curtin SC, Minino AM, Anderson RN. Declines in cancer death rates among children and adolescents in the United States, 1999–2014. NCHS Data Brief. 2016;257:1–8.
3. Louis DN, et al. The 2016 World Health Organization classification of tumors of the central nervous system: a summary. Acta Neuropathol. 2016;131(6):803–20.
4. Cavalli FMG, et al. Intertumoral heterogeneity within medulloblastoma subgroups. Cancer Cell. 2017;31(6):737–54. e6
5. Robinson GW, et al. Vismodegib exerts targeted efficacy against recurrent sonic hedgehog-subgroup medulloblastoma: results from phase II pediatric brain tumor consortium studies PBTC-025B and PBTC-032. J Clin Oncol. 2015;33(24):2646–54.
6. Rudin CM, et al. Treatment of medulloblastoma with hedgehog pathway inhibitor GDC-0449. N Engl J Med. 2009;361(12):1173–8.
7. Rodon J, et al. A phase I, multicenter, open-label, first-in-human, dose-escalation study of the oral smoothened inhibitor Sonidegib (LDE225) in patients with advanced solid tumors. Clin Cancer Res. 2014;20(7):1900–9.
8. Berard CL. Study of vismodegib in combination with temozolomide versus temozolomide alone in patients with medulloblastomas with an activation of the sonic hedgehog pathway. In: ClinicalTrials.gov [Internet]. Bethesda: National Library of Medicine (US). 2000 [cited 2017 Jan 1]. Available from: https://clinicaltrials.gov/ct2/show/NCT01601184: NCT01601184.
9. St. Jude Children's Research Hospital. A clinical and molecular risk-directed therapy for newly diagnosed medulloblastoma. In: ClinicalTrials.gov [Internet]. Bethesda: National Library of Medicine (US). 2000 [cited 2017 Jan 1]. Available from: https://clinicaltrials.gov/ct2/show/NCT01878617:NCT01878617.
10. Erivedge [package insert]. San Francisco: Genentech; 2012.
11. Lucas JT Jr, Wright KD. Vismodegib and physeal closure in a pediatric patient. Pediatr Blood Cancer. 2016;63(11):2058.
12. Krueger DA, et al. Everolimus for subependymal giant-cell astrocytomas in tuberous sclerosis. N Engl J Med. 2010;363(19):1801–11.
13. Franz DN, et al. Everolimus for subependymal giant cell astrocytoma: 5-year final analysis. Ann Neurol. 2015;78(6):929–38.
14. Gajjar A, et al. Pediatric brain tumors: innovative genomic information is transforming the diagnostic and clinical landscape. J Clin Oncol. 2015;33(27):2986–98.
15. Gajjar A, et al. Molecular insights into pediatric brain tumors have the potential to transform therapy. Clin Cancer Res. 2014;20(22):5630–40.

16. Schindler G, et al. Analysis of BRAF V600E mutation in 1,320 nervous system tumors reveals high mutation frequencies in pleomorphic xanthoastrocytoma, ganglioglioma and extra-cerebellar pilocytic astrocytoma. Acta Neuropathol. 2011;121(3):397–405.
17. Miller C, et al. Report of effective trametinib therapy in 2 children with progressive hypothalamic optic pathway pilocytic astrocytoma: documentation of volumetric response. J Neurosurg Pediatr. 2017;19(3):319–24.
18. Olow A, et al. BRAF status in personalizing treatment approaches for pediatric gliomas. Clin Cancer Res. 2016;22(21):5312–21.
19. Aguilera D, et al. Successful retreatment of a child with a refractory brainstem ganglioglioma with vemurafenib. Pediatr Blood Cancer. 2016;63(3):541–3.
20. Lassaletta A, et al. Profound clinical and radiological response to BRAF inhibition in a 2-month-old diencephalic child with hypothalamic/chiasmatic glioma. Pediatr Blood Cancer. 2016;63(11):2038–41.
21. Shih KC, et al. Successful treatment with dabrafenib (GSK2118436) in a patient with ganglioglioma. J Clin Oncol. 2014;32(29):e98–e100.
22. Su F, et al. RAS mutations in cutaneous squamous-cell carcinomas in patients treated with BRAF inhibitors. N Engl J Med. 2012;366(3):207–15.
23. Bautista F, et al. Vemurafenib in pediatric patients with BRAFV600E mutated high-grade gliomas. Pediatr Blood Cancer. 2014;61(6):1101–3.
24. Lee EQ, et al. Successful treatment of a progressive BRAF V600E-mutated anaplastic pleomorphic xanthoastrocytoma with vemurafenib monotherapy. J Clin Oncol. 2016;34(10):e87–9.
25. Robinson GW, Orr BA, Gajjar A. Complete clinical regression of a BRAF V600E-mutant pediatric glioblastoma multiforme after BRAF inhibitor therapy. BMC Cancer. 2014;14:258.
26. Dombi E, et al. Activity of selumetinib in neurofibromatosis type 1-related plexiform neurofibromas. N Engl J Med. 2016;375(26):2550–60.
27. Farid M, et al. Malignant peripheral nerve sheath tumors. Oncologist. 2014;19(2):193–201.

Predictive Biomarkers and Targeted Therapies in Adult Brain Cancers

34

Jose M. Bonnin

Introduction

In the last two decades, considerable advances have been made in our understanding of the biological and genetic features of primary central nervous system (CNS) tumors both in children and adults. The rapid identification and development of various types of biomarkers have the potential to significantly improve and speed up the development of new therapies to patients with central nervous system tumors [1]. They are the basis of a more precise approach to clinical and pathologic diagnosis (diagnostic biomarkers), predict the patients' response to particular therapies (predictive biomarkers), and determine the likelihood for recurrence or progression of primary and metastatic tumors to the nervous system (prognostic biomarkers).

The value of biomarkers is reflected in the current version of the World Health Organization (WHO) classification of the nervous system tumors which now integrates the classical histopathological features of tumors with some key molecular parameter [2, 3]. A comprehensive molecular profiling in neuro-oncology has led to a better understanding of the biological features and classification of numerous tumor groups, particularly the gliomas and the embryonal tumors [4–7]. In the case of primary brain tumors that most commonly affect adults, an emerging set of disease-defining biomarkers have allowed to a new understanding of previously unknown differences of tumor groups that are morphologically similar [5, 8]. Biomarkers now allow classification of tumors into more accurate and biologically relevant subgroups that often display significant differences in their growth rate, risk for recurrence, and response to therapy.

Substantial progress has been made in this regard for common primary brain tumors in both children and adults. Large-scale profiling efforts of diffuse gliomas in adults have led to the identification of highly prevalent molecular alterations that allow not only biologically based classifications as adjuncts to the traditional histopathologic diagnosis but also to the development of promising targeted therapies [4, 7, 8].

Despite years of basic research and clinical trials, diffuse gliomas and particularly glioblastoma (GBM) remains one of the deadliest primary brain tumors in adults. Standard treatments for GBM and other high-grade gliomas with a survival benefit in randomized studies include maximal safe surgical resection, radiation therapy, and chemotherapy with temozolomide (TMZ). In spite of these treatment advances, however, only ~15–20% of GBM patients survive to 5 years, and no therapies have

J. M. Bonnin
Department of Pathology and Laboratory Medicine,
Indiana University School of Medicine,
Indianapolis, IN, USA
e-mail: jbonnin@iupui.edu

S. Badve, G. L. Kumar (eds.), *Predictive Biomarkers in Oncology*,
https://doi.org/10.1007/978-3-319-95228-4_34

demonstrated consistent survival benefits in patients with recurrent tumors.

GBMs are part of a broader category of diffusely infiltrating gliomas consisting of grades II through IV tumors. Gliomas have historically been classified and treated according to the World Health Organization (WHO) criteria, which until recently has been based mainly on histopathological features which segregated them into focal and diffuse infiltrating tumors. The diffuse gliomas have been further classified as low-grade [WHO grade II: diffuse astrocytoma and oligodendroglioma] and high-grade gliomas [WHO grade III: anaplastic astrocytoma (AA), anaplastic oligodendroglioma (AO), and WHO grade IV: glioblastoma (GBM)]. Approximately 10–20% of GBMs are the result of the anaplastic transformation from a low-grade glioma (secondary GBMs) but most present de novo and are designated as primary GBMs [9]. Remarkable advances during the last decade had provided a better understanding of the molecular alterations in gliomas, and this has resulted in a reformulation of the classification criteria and the integration of specific molecular changes in core pathways that defined the various tumor categories and the traditional morphological classification system [2, 3].

These analyses highlighted the most common genetic alterations in GBM, including amplification of epidermal growth factor receptor (EGFR), platelet-derived growth factor receptor alpha (PDGFRA); mutation of TP53, PIK3CA, PTEN, isocitrate dehydrogenase 1 (IDH1), RB1, and TERT promoter; and deletions of PTEN, CDKN2A/B, and MGMT, as well as frequent alterations in chromatin remodeling genes (See Table 34.1). Previously, GBM clinical trials have included patients segregated by histological diagnoses, without molecular stratification. In retrospect, these trials had populations composed of morphologically similar tumors representing a mix of molecular and genetic subtypes. Understandably, studies of therapies for tumors with different molecular alterations, prognoses, and underlying biology may lead to meaningless results and variable outcomes.

The combined analyses of the comprehensive, multiplatform molecular studies have led to the realization that low-grade and high-grade gliomas may be subdivided into three major discrete molecular subtypes, with additional subdivisions within those larger groups. In the first group are tumors (predominantly grade II and III) that harbor IDH1/2 mutation and chromosome 1p/19q codeletion, which are also associated with DNA hypermethylation and CIC, FUBP1, TERT, and PIK3CA alterations and correspond to molecular oligodendrogliomas. These tumors have a very good prognosis with treatment, with a median survival of 14.7 years in one randomized study of grade III oligodendroglioma with 1p/19q loss treated with a combination of radiation and chemotherapy at diagnosis. Second group includes the astrocytomas which are IDH-mutant but lack 1p/19q codeletion but often express alterations in TP53 and ATRX as well as a hypermethylated phenotype. These tumors have a worse prognosis than oligodendrogliomas but better than IDH wild-type gliomas. Absence of hypermethylation may negatively impact the overall survival of patients with these tumors. Lastly, lower-grade IDH wild-type gliomas are clinically and molecularly similar to GBMs and have a relatively poor overall survival of 1.7 years. In this group of tumors, GBM-specific alterations in PTEN, EGFR, TERT, and CDKN2A, may occur. IDH wild-type GBMs may also include various subgroups depending on the presence or absence of other molecular alterations. The diagnostic and prognostic significance of these biomarkers suggests great biologic relevance. Their association with the activation of specific signaling pathways highlights the potential for the identification or development of therapies aimed at particular tumor groups or subgroups [5, 7, 10, 11].

Isocitrate Dehydrogenase (IDH1/2) Mutations

One of the most important biomarkers in gliomas is a mutation in the genes for IDH, an enzyme of the tricarboxylic acid cycle. IDH mutations are

Table 34.1 This table includes only a listing of the most common glial neoplasms in adults. Ongoing clinical trials are numerous, and possible therapies differ in recurrent tumors from those given at the time of initial diagnosis, MGMT methylation status, and/or IDH status (mutant/wild type)

Tumors	Genetic	Epigenetic	Chromosomal	Therapeutic target	Clinically tested agents	Candidate therapies[c]
Diffuse astrocytic gliomas						
Diffuse astrocytoma, IDH-mutant	*IDH1* or *IDH2*, *TP53*, *ATRX* mutation	G-CIMP	Trisomy 7 or 7q gain; LOH 17p			
Anaplastic astrocytoma, IDH-mutant	*IDH1* or *IDH2*, *TP53*, *ATRX* mutation	G-CIMP	Trisomy 7 or 7q gain; LOH 17p		[a]	
Glioblastoma, IDH-mutant	*IDH1* or *IDH2*, *TP53*, *ATRX* mutation; *CDKN2A* homozygous deletion	G-CIMP	Trisomy 7 or 7q gain; LOH 17p; 10q deletion		[a]	
Glioblastoma, IDH wild type	*TERT*, *PTEN*, *TP53*, *PIK3CA*, *PIK3R1*, *NF1*, *H3F3A*-G34 mutation; CDKN2A, *PTEN* homozygous deletion; *EGFR*, *PDGFRA*, *MET*, *CDK4*, *CDK6*, *MDM2*, *MDM4* amplification; *EGFRvIII* rearrangement	*MGMT* promoter methylation	Trisomy 7 or 7q gain; monosomy 10; double minute chromosomes	*TERT*, *PTEN*, *TP53*, *PIK3CA*, *PIK3R1*, *NF1*, CDKN2A, *PTEN*, *EGFR*, *PDGFRA*, *MET*, *CDK4/6*, *MDM2*, *MDM4*, *EGFRvIII*	[a]	Rindopepimut, Dasatinib, Cabozantinib, Voxtalisib, Buparlisib, Trametinib, Afanitib, Vemurafenib, Ribociclib, Onartuzumab, Bevacizumab, Erlotinib, Nivolumab, Ipilimumab, Pazopanib[d]
Other astrocytic gliomas						
Pilocytic astrocytoma	*BRAF*, *RAF1*, *NTRK2* gene fusions; *BRAF*-V600E, NF*1*, *KRAS*, F*GFR1*, *PTPN11* mutation	–	–	KIAA1549:BRAF fusion	Selumetinib Trametinib	
Pleomorphic xanthoastrocytoma	*BRAF*-V600E mutation, $CDKN2A/p14^{ARF}$ homozygous deletion	–	–	BRAF V600E	Dabrafenib Trametinib[b]	

(continued)

Table 34.1 (continued)

Tumors	Genetic	Epigenetic	Chromosomal	Therapeutic target	Clinically tested agents	Candidate therapies[c]
Oligodendrogliomas						
Oligodendroglioma, IDH-mutant and 1p/19q-codeleted	*IDH1* or *IDH2*, *TERT*, *CIC*, *FUBP1* mutation	G-CIMP	1p/19q codeletion			
Anaplastic oligodendroglioma, IDH-mutant and 1p/19q-codeleted	*IDH1* or *IDH2*, *TERT*, *CIC*, *FUBP1*, *TCF12* mutation; *CDKN2A* deletion	G-CIMP	1p/19q codeletion			
Ependymomas						
Supratentorial Ependymomas (ST)						
Ependymoma, *RELA*-fusion positive	*C11orf95–RELA* fusion	–	11q aberrations			
Ependymoma	*YAP1* gene fusions	–	11q aberrations			
Posterior Fossa Ependymomas (PF)						
Ependymoma PF-A	–	PF-A DNA-methylation profile with global hypermethylation	Stable genotype			
Ependymoma PF-B	–	PF-B DNA-methylation profile	Multiple copy-number imbalances (CIN)			
Spinal Ependymomas (SP)						
Ependymoma	*NF2* mutation	–	22q deletion			

[a]Treatment protocols for glioblastomas are at times applied to other high-grade diffuse astrocytic gliomas (anaplastic astrocytomas)
[b]Therapy in some cases of anaplastic pleomorphic xanthoastrocytomas
[c]Vaccines, antibodies, and drugs
[d]Partial listing; ongoing trials (recurrent glioblastomas)

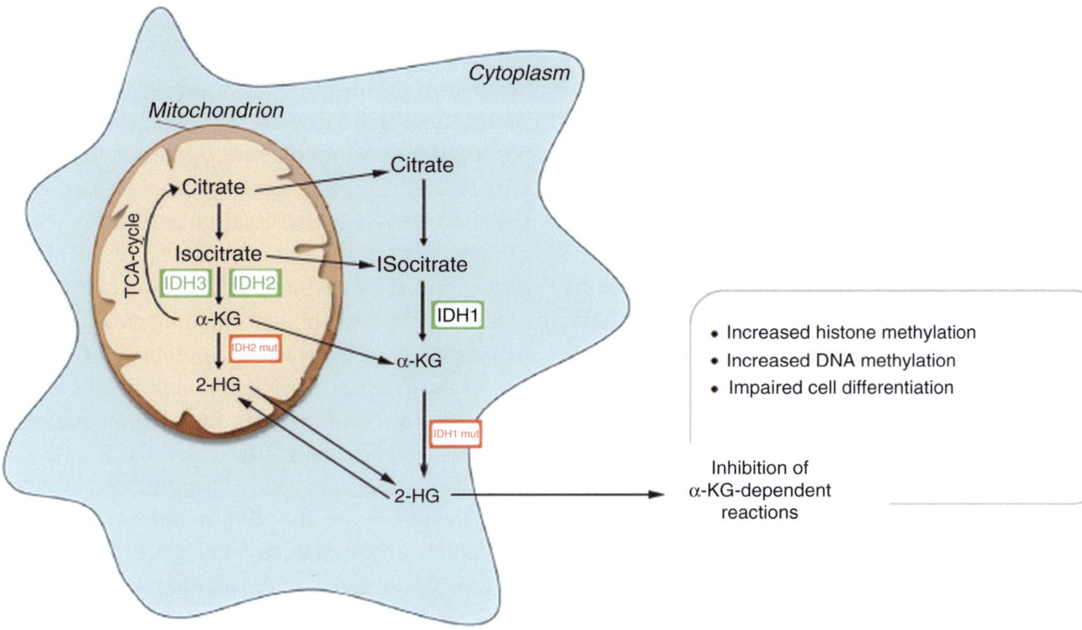

Fig. 34.1 An isolated missense mutation of IDH1/2 at arginine 132 (R132H) or the analogous residue 172 (R172) results in metabolic reprogramming with the ability to convert α-ketoglutarate (α-KG) to 2-hydroxyglutarate (2-HG). 2pHG alters the epigenetic factors that contribute to gliomagenesis, chromatin modification, and dysregulation of gene expression. (Modified from Kickingereder et al. [15]. With permission from Springer Nature)

associated with epigenetic reprogramming resulting in altered cancer cell differentiation [6–10]. Mutations in IDH1 and IDH2 result in a change of function that generates 2-hydroxyglutarate (2-HG) instead of the normal products of NADPH and α-ketoglutarate. Although primary GBMs most often lack IDH mutations, most secondary GBMs are IDH-mutant (Fig. 34.1).

Patients with tumors harboring IDH1/2 mutations had a survival benefit while those without a detectable mutation by immunohistochemical R132H profiling or IDH1/2 sequencing did not. The occurrence of IDH1/2 mutation precedes acquisition of other oncogenic alterations believed to drive malignant transformation in gliomas. Because mutation of the IDH gene alters the catalytic function of the enzyme, they are potentially targetable via drug inhibitors, and trials of direct inhibitors of IDH1 and IDH2-mutant proteins are ongoing.

One of the consequences of IDH mutation is elevation of 2-HG. This has been suggested to be a potential oncometabolite important in gliomagenesis, but the tumor-initiating capacity has not been confirmed. 2-HG has been shown to inhibit a range of α-ketoglutarate dioxygenases, including a hypoxia-inducible factor, histone demethylases, and 5-methylcytosine hydroxylases. Global hypermethylation has been shown to occur as a consequence of IDH1-mutant overproduction of 2-HG. Depletion of 2-HG via an IDH1 mutant-specific inhibitor, however, has been demonstrated to be insufficient to halt the formation or progression of IDH1-mutant gliomas. Preclinical studies have also shown efficacy for peptide vaccines directed against the IDH1 R132 mutation, and ongoing clinical trials are testing their efficacy in patients with high-grade gliomas.

IDH mutation results in a hypermethylation phenotype and broad epigenetic alterations in tumor cells. Hypermethylation can also alter chromosomal topology and the expression of multiple genes, including potential oncogenes. Other studies have suggested an epigenetic mechanistic link between IDH mutation and activation of PDGFRA. Hypomethylating agents have been ineffective in unselected high-grade gliomas, but trials in IDH-mutated gliomas may be warranted, although in many instances, capability of such agents to penetrate the blood-brain barrier remains questionable.

The potential application of epigenetic modulating agents in gliomas is relevant beyond alterations seen in IDH-mutant tumors. Whole-exome sequencing studies have uncovered specific mutations in histone proteins in pediatric GBM, suggesting a direct driver function of epigenetic mutations (see Chap. 36).

O6-Methylguanine-DNA Methyltransferase (MGMT) Promoter Hypermethylation in High-Grade Astrocytic Tumors

The recommended treatment for patients with glioblastoma after surgical resection or biopsy has been radiation therapy (RT) and chemotherapy with temozolamide (TMZ). An association of MGMT promoter methylation and response of high-grade astrocytomas to alkylating agents has been reported. In general, patients with methylated MGMT promoters had significantly longer survival with the addition of TMZ but not with RT, demonstrating that MGMT promoter methylation status predicts the response to TMZ. A study with hypofractionated RT ± TMZ for elderly patients with GBM showed similar results. While all patients have a survival benefit from the addition of TMZ, the magnitude of the benefit for unmethylated patients is less than that of patients with tumors harboring methylated MGMT promoters [4, 7–10].

Epidermal Growth Factor Receptor Gene (EGFR)

EGFR gene amplification is one of the activation mechanisms in GBMs. It is an apparently attractive target for receptor tyrosine kinase inhibitors (which have shown activity in other solid tumors), but EGFR has remained an inconsistent therapeutic target in GBM. Initial studies suggested activity of some EGFR inhibitors in molecular subsets of GBM, but subsequent larger studies could not replicate such promising results, perhaps because of the less than optimal central nervous system (CNS) penetration of many of these agents. A subset of GBMs with EGFR amplification also harbors a mutation known as EGFRvIII, identified in ~30% of newly diagnosed primary GBMs. EGFRvIII is characterized by in-frame deletion of exons 2–7, encoding the extracellular component of this cell membrane protein which is important for activation of the receptor. Such deletion promotes tumor cell migration and protects them from radiation and TMZ therapy [6–8, 10].

EGFR inhibitors and vaccines have been extensively tested in glial neoplasm, but the results have been discouraging. EGFR, however, remains an attractive molecular target and marker of distinct biologic glioma subtypes, and other approaches to targeted therapies for EGFR are ongoing [10].

Chromosomes 1p and 19q Codeletion

Oligodendroglial tumors are associated with better survival and are more chemosensitive than astrocytic neoplasm, and this has been demonstrated to be secondary to 1p/19q codeletion detected in oligodendrogliomas but not in other tumors of glial derivation. Two large randomized studies have suggested that the codeletion, as detected by fluorescent in situ hybridization (FISH), has some predictive capacity for PCV (procarbazine, lomustine, vincristine) therapy. Follow-up analyses of patients carrying the codeletion compared the treatment with radiation therapy (RT), and chemotherapy, with PCV versus RT alone, showed striking differences in median survival between patients who received combination therapy and RT alone. Median survival in patients treated with PCV followed by RT was twice as long as that of a patient treated with radiation alone. Currently, the presence of codeletion of 1p/19q (whole arm losses of 1p and 19q) as well as IDH1/2 mutation is required for the definitive diagnosis of oligodendrogliomas [2, 3].

BRAF V600E and H1AA1949-BRAF Fusion Mutations

BRAF is a member of the RAF kinase family of growth signal transduction protein kinases. The protein plays a role in regulating the MAP kinase/ERK-signaling pathway, which affects cell division and differentiation. Mutations of the BRAF gene have been identified in a variety of tumors, including some central and peripheral nervous system neoplasm. The BRAF V600E mutation has been documented in 60–70% of pleomorphic xanthoastrocytomas, approximately 50% of gangliogliomas and a small number of pilocytic astrocytomas. On the other hand, the KIAA1549-BRAF fusion mutation has been identified in approximately 80% of pilocytic astrocytomas. Both mutations have become diagnostic biomarkers for this group of tumors [12].

Targeted therapies for tumors, expressing the BRAF V600E mutation, appear to be relatively effective in inhibiting the growth of some types of tumors in the initial stages. In addition, some second-generation inhibitors of BRAF that do not result in paradoxical activation of cell proliferation have been reported to be helpful in the treatment of tumors expressing KIAA1549-BRAF fusion proteins (see Table 34.1). Although these mutations are more common in pediatric tumors, they are not specific for a particular diagnostic entity, and they have been observed in both low-grade and high-grade tumors in adults, including diffuse astrocytomas, glioblastomas, and gliosarcomas [6].

With the exception of pleomorphic xanthoastrocytomas, these tumors (pilocytic astrocytomas and gangliogliomas) are more prevalent in the pediatric population. Their importance as diagnostic biomarkers and the development of targeted therapies have been discussed in greater detail in Chap. 33.

Angiogenesis

Microvascular proliferation is a prominent feature of high-grade gliomas. It is promoted by high levels of vascular endothelial growth factor A (VEGF-A) within the tumor and the surrounding uninvolved parenchyma. The anti-angiogenic agent cediranib, a VEGF receptor tyrosine kinase inhibitor, and cilengitide, an integrin inhibitor, were both ineffective in altering the outcome in GBMs. Initially, bevacizumab, a humanized monoclonal antibody that targets VEGF-A, showed a high radiographic response rate prolonged progression-free survival (PFS). Bevacizumab also showed a reduction in glucocorticoid requirements, leading to the accelerated approval for treatment of patients with recurrent GBM. Unfortunately, no overall survival benefit of bevacizumab has been documented despite the apparent improvements in radiological imaging studies. Similarly, overall survival has not been impacted when treatment with bevacizumab and TMZ was given to patients with newly diagnosed GBM. Although no survival benefit of bevacizumab was observed in the overall population, a potential increase in median survival of ~4 months was observed when bevacizumab was added to TMZ in a patient with IDH1 wild-type GBMs [7–9].

Ependymal Tumors

Multiple genomic profiling studies of the tumors of the ependyma have resulted in a radical reevaluation of this seemingly uniform (morphologically) groups of tumors affecting the brain and spinal cord. Such studies have identified significant differences between the tumors arising in the supratentorial compartment, the posterior fossa, and the spinal cord [13, 14].

It has become evident that the grading of ependymomas according to morphological criteria must be revised, and this is partly reflected in the most recent version of the WHO classification of tumors of the nervous system [2]. It is well-known that myxopapillary ependymomas of the filum terminale and subependymomas are slow-growing. It has become evident that WHO grade I tumors such as the supratentorial and posterior fossa ependymomas are distinct diseases, and that the traditional grading approach into grade II or III tumors is controversial and lacks consistent association between tumor grades with patient outcomes. The listing of such groups in Table 34.1 summarizes the current views on ependymal neoplasms. Despite morphological similarities, posterior fossa ependymomas, group A (PF-A ependymomas) that occur mostly in infants and young children, are associated with high recurrence rate, while posterior fossa ependymomas, group B (PF-B ependymomas) that are diagnosed mostly in adolescents and young adults, have a more favorable prognosis. In the supratentorial compartment, more than 70% of ependymomas have fusions between C11ORF95 and the RELA gene and occur both in children and adults. The remaining supratentorial ependymomas express recurrent fusions of the YAP1 gene and are more frequent in the children.

Although these studies have provided a better understanding of the biology of the ependymal tumors as well as guidance in the design of more appropriate therapeutic protocols, they have not yet resulted in the identification of effective targeted therapies.

Conclusions

Remarkable advances have been made in the identification of the molecular pathways, genetic mutations, and alterations of the normal composition or functions of various proteins leading to the development of nervous system tumors. Despite such advances, the identification of diagnostic, predictive biomarkers and the development of targeted therapies are still in its early phases. To date, treatment of diffuse gliomas, particularly newly diagnosed glioblastomas in adults, still includes maximal safe surgical resection, radiotherapy, and chemotherapy. Recurrence rate within 2 years of initial diagnosis is common, and a variety of therapies in recurrent tumors have not been proven to result in prolonged survival in most cases.

As an example, genome-wide analysis has identified IDH mutations, MGMT promoter methylation status, and EGFRvIII as important prognostic and predictive biomarkers. IDH-mutant glioblastomas with MGMT promoter methylation have been found to have a slightly better prognosis, but this appears to be independent of better response to radiation therapy and chemotherapy. The value of MGMT methylation status remains somewhat controversial. MGMT promoter methylation has been found to be associated with benefit from TMZ therapy and prolonged survival in patients with glioblastoma, but this appears to be restricted to IDH wild-type tumors.

Although the biologic relevant genomic variants of high-grade gliomas are possible targets for drugs or vaccine therapies, none of the variants of glioblastoma have a prognostic or predictive utility. It is likely that the identified alterations in the various molecular pathways are not obligate cancer drivers in high-grade gliomas. In addition, there is a consensus that glioblastomas are remarkably heterogeneous, and this heterogeneity extends to various cell groups within the same tumor. It has become apparent that targeted therapies aimed at a single receptor or molecular pathway may not be effective in improving survival in these tumors which often harbor multiple genetic alterations. Future improvements in survival for patients with diffuse gliomas will likely require a combination of standard therapies together with one or more targeted therapies.

References

1. Cagney DN, Sul J, Huang RY, et al. The FDA NIH Biomarkers, EndpointS, and other Tools (BEST) resource in neuro-oncology. Neuro-Oncology. 2017; https://doi.org/10.1093/neuonc/nox242.
2. Louis DN, Perry A, Reifenberger G, et al. The 2016 World Health Organization classification of tumors of the central nervous system: a summary. Acta Neuropathol. 2016;131:803–20.
3. Velazquez Vega JE, Brat DJ. Incorporating advances in molecular pathology into brain tumor diagnostics. Adv Anat Pathol. 2018;25:143.
4. Staedtke V, a Dzaye OD, Holdhoff M. Actionable molecular biomarkers in primary brain tumors. Trends Cancer. 2016;2:338–49.
5. Reifenberger G, Wirsching HG, Knobbe-Thomsen CB, et al. Advances in the molecular genetics of gliomas – implications for classification and therapy. Nat Rev Clin Oncol. 2017;14:434–52.

6. Ballester LY, Fuller GN, Powell SZ, et al. Retrospective analysis of molecular and immunohistochemical characterization of 381 primary brain tumors. J Neuropathol Exp Neurol. 2017;76:179–88.
7. Khan IN, Ullah N, Hussein D, et al. Current and emerging biomarkers in tumors of the central nervous system: possible diagnostic, prognostic and therapeutic applications. Semin Cancer Biol. 2017;17:30114–1.
8. Chen R, Cohen AL, Colman H. Targeted therapeutics in patients with high-grade gliomas: past, present, and future. Curr Treat Options in Oncol. 2016;17:42.
9. Lieberman F. Glioblastoma update: molecular biology, diagnosis, treatment, response assessment, and translational clinical trials. F1000Res. 2017;6:1892.
10. Johanns TM, Dunn GP. Applied cancer immunogenomics: leveraging neoantigen discovery in glioblastoma. Cancer J. 2017;23:125–30.
11. Diamandis P, Aldape KD. Insights from molecular profiling of adult glioma. J Clin Oncol. 2017;35:2386–93.
12. Collins VP, Jones DT, Giannini C. Pilocytic astrocytoma: pathology, molecular mechanisms and markers. Acta Neuropathol. 2015;129:775–88.
13. Pajtler KW, Mack SC, Ramaswamy V, et al. The current consensus on the clinical management of intracranial ependymoma and its distinct molecular variants. Acta Neuropathol. 2017;133:5–12.
14. Mack SC, Pajtler KW, Chavez L, et al. Therapeutic targeting of ependymoma as informed by oncogenic enhancer profiling. Nature. 2018;553:101–5.
15. Kickingereder P, et al. IDH mutation status is associated with a distinct hypoxia angiogenesis transcriptome signature which is non-invasively predictable with rCBV imaging in human glioma. Sci Rep. 2015;5:16238.

35 Predictive Biomarkers and Targeted Therapies in Breast Cancer

Sunil Badve

Overview

One in eight women has a lifetime risk of developing breast cancer. This high prevalence, particularly when not associated with a strong environmental factor such as smoking, leaves all women vulnerable to this often-deadly cancer. Intense efforts have been focused on developing newer therapies and controlling this cancer for the last few decades. This has led to some very dramatic successes and made it a poster child for targeted therapies. However, as discussed in this chapter, progress has been slow albeit steady in converting this deadly disease into a chronic form of cancer.

Biology of Breast Cancer

Breast cancer can be classified in multiple ways. The commonly used classification system is based on the expression of hormone (estrogen and/or progesterone) receptors and human epidermal growth factor receptor 2 (HER2). This broadly divides breast cancer into those that express ER/PR and those that don't. Most ER+ tumors lack HER2+ expression; however tumors that are HER2+ can be ER+ or negative. Tumors that lack ER/PR and HER2 are commonly referred to as triple-negative tumors. Molecular classification of breast tumors has been performed using several techniques. The most commonly used molecular classification, intrinsic classification, recognizes two classes of ER+ tumors (luminal A and B) and three classes of ER- tumors (basal-like, HER2-enriched, and normal-like) [1]. This classification was originally devised using microarrays but has been adapted to qRT-PCR and is commercially available as the Prosigna assay® (NanoString Technologies). The integral classification recognizes ten categories of breast tumors, which have distinct molecular alterations and clinical outcomes [2]. Additional classification systems based on mutational patterns (21 subtypes) [3] and reverse phase protein arrays (8 subtypes) have been also described. Triple-negative tumors have been also further subclassified into six distinct types [4]; of these, luminal androgen receptor-positive (LAR) subtype has been documented to respond to antiandrogenic therapies.

Hormone Receptor

As discussed earlier (and in Chap. 13), steroid hormone receptors play a major role in the progression and management of breast cancer. Sir George Beatson, in the late 1800s, discovered

S. Badve
Department of Pathology and Lab Medicine,
Indiana University School of Medicine,
Indianapolis, IN, USA
e-mail: sbadve@iupui.edu

S. Badve, G. L. Kumar (eds.), *Predictive Biomarkers in Oncology*,
https://doi.org/10.1007/978-3-319-95228-4_35

that breast cancers can be slowed down by performing oophorectomies. This initial finding led workers to discover the importance of estrogen and develop therapies targeting estrogen synthesis (aromatase inhibitors), receptor blockers (e.g., tamoxifen, fulvestrant), and ovarian function suppressors (GnRH antagonists such as goserelin). These endocrine therapy agents still remain the mainstay for therapeutics in tumors that express the estrogen receptor (ER). Endocrine therapy has to be taken for a long period of time as recurrences from ER+ tumors can be observed decades after the original diagnosis. Recent data suggests that 10 years of endocrine therapy is better than 5 years of therapy. It has also been reported that as many as 20–30% of patients discontinue the prescribed therapy even when cost of drugs is not an issue.

A number of gene expression signatures such as Oncotype Dx (Genomic Health, Inc.) and MammaPrint ® (Agendia) have been developed to prognosticate breast cancer. Studies with these have shown that patients with "high" scores/risk are most likely to respond to chemotherapy. Dissection of the gene signatures has revealed that most, if not all, signatures for ER+ breast cancer are based on proliferation. This realization has led to an increased focus on the development of agents that affect the cell proliferation. Recent years have seen the successful adaptation of CDK4/CDK6 cell cycle inhibitors for the treatment of ER+ breast cancer. These are being increasingly used in many patients instead of chemotherapy. Metastases from ER+ tumors continue to be an important clinical problem. Numerous studies have analyzed pathways of resistance to these agents and identified novel targets for therapeutics. These have been used in combination with endocrine therapy with significant success. Mutations in ER were thought to be rare; however recent studies using NGS have documented a high incidence (~20%) in recurrent/metastatic tumors (Fig. 13.3, Chap. 13). In addition, NGS studies have identified a number of alterations in a large number of genes/pathways in luminal (ER+) breast cancers (Fig. 35.1).

Human Epidermal Growth Factor Receptor 2 (HER2)

HER2 was first described as a prognostic factor, expression being associated with poor prognosis. However, the development of humanized monoclonal antibody directed against it converted it into a predictive factor. Tumors that overexpress this receptor protein or exhibit DNA level amplification of the gene (see later for scoring guidelines) are candidates for anti-HER2 therapies. This receptor can be now targeted using multiple humanized antibodies including trastuzumab and pertuzumab. These can be used as single agents, in combination with each other or in combination with chemotherapy. Drugs that inhibit the tyrosine kinase activity of HER2 (e.g., lapatinib) as well as drug antibody conjugates (e.g., T-DM1) have also been approved by the FDA for treating HER2+ breast cancer. Biosimilars, which are very similar to the FDA-approved antibodies, have been developed and are being actively tested in clinical trials. The mechanisms of resistance to anti-HER2 agents are poorly understood, and numerous novel agents are being developed to treat HER2 resistance (Fig. 35.2).

The postulated role of other HER receptors (HER1, HER3, and HER4) has led to the development of pan-HER inhibitors such as neratinib. Additionally, the postulated role of PI3K/AKT, downstream of HER3, has resulted in the introduction of multiple agents to inhibit the function of this pathway. Targeting PI3K has proven to be difficult due to dose-limiting toxicity. The combination of mTOR inhibitors (e.g., everolimus) with anti-HER2 agents has been more successful (BOLERO-1 and BOLERO-3 clinical trials) [7].

Triple-Negative Tumors

Tumors that lack expression of ER/PR and HER2 receptors are called triple-negative breast cancers (TNBCs). These cancers are more difficult to treat and have a higher incidence in African-Americans. They often march rapidly in spite of chemotherapy

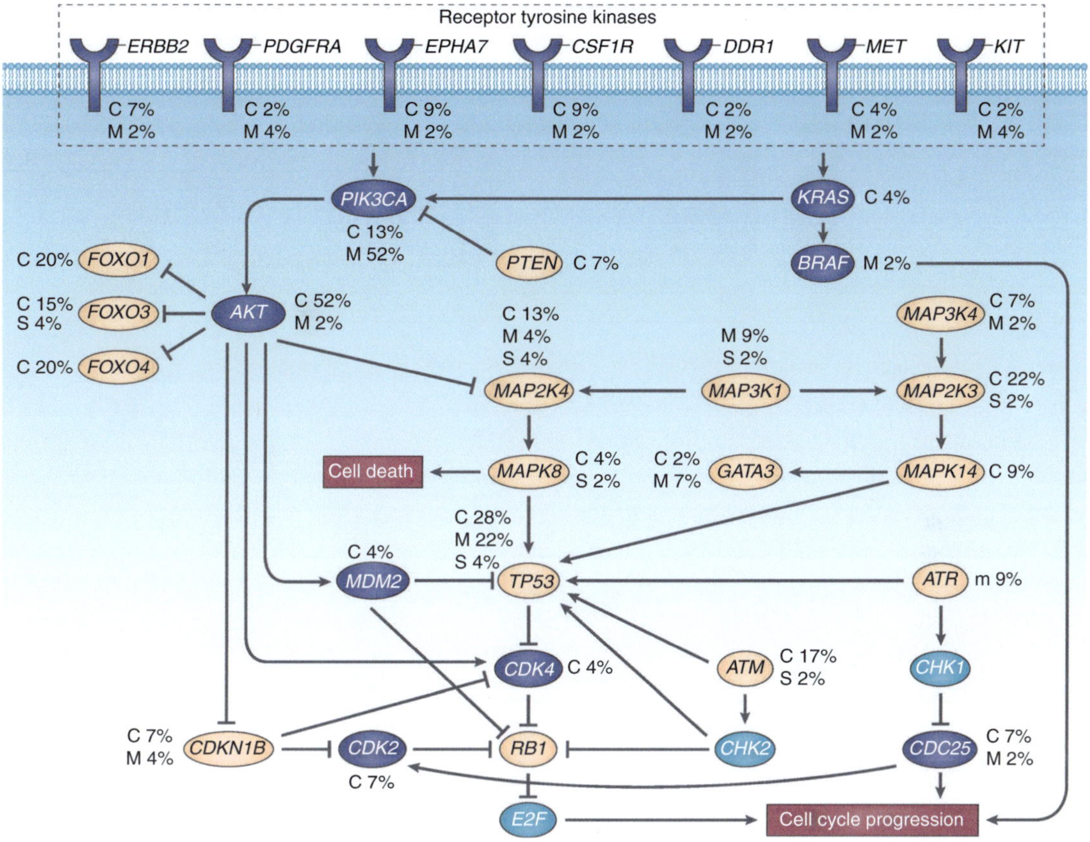

Fig. 35.1 Alterations in key cancer pathway components in luminal breast cancers. Frequencies of genetic alterations identified in 46 estrogen receptor-positive (ER+) breast cancers by whole-genome sequencing are shown. Key pathways affected include receptor tyrosine kinases, PI3K-AKT-mTOR, RAS-RAF-MAPK, and p53-RB, leading to cell cycle progression and resistance to cell death, which could potentially subject cells to estrogen-independent growth. Genes shown in blue are predicted to be functionally activated, and genes shown in yellow are predicted to be functionally inactivated. C refers to copy number alteration, M refers to mutation, and S refers to structural variation. ATM ataxia telangiectasia mutated, ATR ataxia telangiectasia and Rad3-related, CDC25 cell division cycle 25, CDK cyclin-dependent kinase, CDKN1B CDK inhibitor 1B, CHK checkpoint kinase, CSF1R colony-stimulating factor 1 receptor, DDR1 epithelial discoidin domain receptor 1, EPHA7 ephrin A7, FOXO forkhead box protein O, PDGFRA platelet-derived growth factor receptor-α, PIK3CA PI3K catalytic subunit-α. Ellis [5]. (Adapted from Ellis et al. [5]. With permission from Springer Nature)

and continue to give rise to metastases leading to death. Molecular classification of these tumors has shown that the subtypes might have different rates of relapse after neoadjuvant therapy (Fig. 35.3).

These have also identified novel therapeutic targets. The tumors in patients with BRCA1 or 2 mutations often have TNBC phenotype. Recognizing the importance of BRCA1/2 in DNA repair, it was postulated that inhibiting other DNA repair mechanisms could lead to cell death (often referred to as synthetic lethality) [11]. This concept, when put into practice using PARP inhibitors, showed that majority of patients with mutations responded very well. Further analysis of these agents in other TNBCs was being attempted. However, during these early studies, it was noted that PARPi (BSI-201) has significant off-target activities. Newer more specific PARPi are now in clinical trials for assessing utility in TNBC patients.

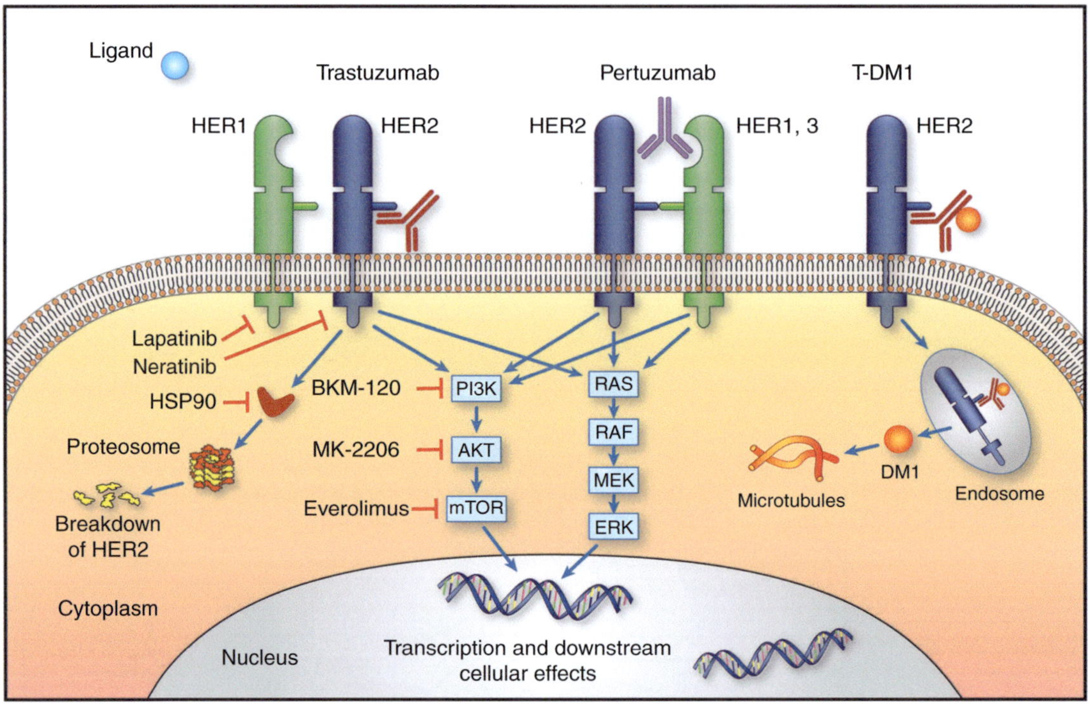

Fig. 35.2 Molecular approaches to HER2 targeted therapy depicting multiple approaches to target HER2 resistance [6]. (Reprinted from Singh et al. [6]. With permission from Springer Nature)

Many of these tumors are associated with prominent lymphocytic infiltrate, tumor-infiltrating lymphocytes (TILs) [12]. Recent exciting studies suggest that these tumors might be amenable to immune therapies. Another new exciting target in TNBC is Trop-2, a glycoprotein initially identified in a trophoblast cancer cell line. It is overexpressed in various solid cancers, including triple-negative breast cancer. Recent data from a Phase I/II trial suggests that anti-Trop-2 antibody-drug conjugate (sacituzumab govitecan) has significant activity in metastatic TNBCs (Bardia et al. ASCO 2017). Additional confirmatory studies are necessary.

Metastatic Breast Cancer

Metastases from breast cancers are unfortunately still common in spite of standard of care therapy. Next-generation sequencing (NGS) is often performed on these tumors in the hopes of identifying "actionable" mutations. Mutations in ER are observed in significant number of cases. Similarly HER2 mutations have been identified in invasive lobular carcinomas. Mutations in PI3K/AKT pathway are prevalent in these tumors. A number of therapies have been instituted based on the mutational status. The role of NGS-based therapeutics is discussed in detail in Chaps. 10 and 53 by Leyland-Jones's group.

Guidelines for Predictive Biomarkers in Breast Cancer

1. Estrogen receptor: The expression of ER was initially analyzed using ligand binding assay on fresh frozen tumor; however, this assay was problematic. There were additional issues regarding whether representative sample of the tumor was submitted for analysis. Immunohistochemical methods are now widely used for analysis; these enable detection of ER

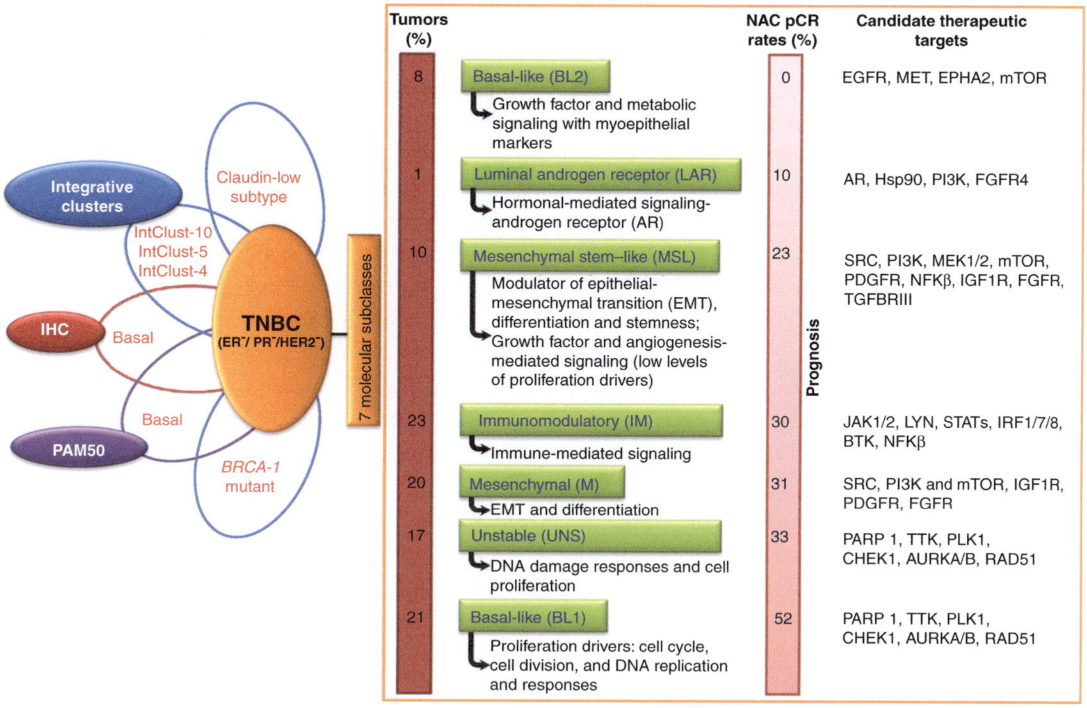

Fig. 35.3 Nomenclature and classification of triple-negative breast cancer (TNBC). TNBCs overlap predominantly (up to 70%) with basal-like breast cancers. Classically, TNBC was defined using immunohistochemistry (IHC) by evaluating three well-known predictive biological markers, namely, estrogen receptor (ER), progesterone receptor (PR), and human epidermal growth factor receptor 2 (HER2). Taking advantage of advances in the gene expression profiling, prediction analysis of microarray-50 (PAM50) classifier [8] (using a set of 50 genes) was introduced to classify different subtypes in breast cancers. More recently, integrative gene expression and copy number profiling (integrative clusters: IntClust) have delineated the diversity in breast cancers [2]. Both of these classifications include basal-like/TNBC tumors as a distinct subtype. A proportion of BRCA-dysfunction tumors and claudin-low tumors also overlap with TNBC. TNBC subtype in itself is molecularly heterogeneous and has been subdivided into six definable molecular subclasses, governed by distinct sets of genes and pathways according to the classification of Lehmann et al. [4]. Each of these subclasses show varying pathological complete response (pCR) rates following neoadjuvant chemotherapy (NAC) [9] and may be amenable to targeted therapies using different molecular targets [10]. (Reprinted from Kalimutho et al. [10]. With permission from Elsevier)

expression at the level of a single cell. RT-PCR methods are not recommended for detection of ER; however, when available they provide a good confirmation of the IHC data. The ASCO-CAP guidelines recommend the use of more than 1% of ER in tumor cells as a cutoff for determining positivity [13]. It is recognized that tumors with low ER expression may not respond well to endocrine therapy; however, these patients are candidates for therapeutic trial with these agents.

2. HER2: The analysis for HER2 can be performed using either in situ hybridization (ISH) or IHC. ISH detects the presence of DNA amplification of the HER2, while IHC detects the protein expression on the cell membrane. The proponents of ISH believe that it is a more stable assay and therefore more reliable, while the IHC proponents point to the fact that the protein on the cell surface is the therapeutic target. In any case, these assays are highly correlated giving concordant results in most cases. The IHC assay is reported as the percentage of cells exhibiting strong complete membranous staining. The ISH assay is reported as a ratio of the HER2 copies to that

of control CEP17 probe. There was initially a fear that many assays were falsely called positive; this concern resulted in the development of the ASCO-CAP guidelines. The current (2013) guidelines recommend the use of greater than 10% positivity by IHC and/or a ratio greater than 2 by FISH [14]. It is strongly recommended that FDA-approved kits and reagents be used for the analysis. The use of alternative probes for CEP17 is controversial and will be addressed in the as yet unpublished new ASCO-CAP guidelines. These new guidelines will also take into consideration data obtained from the NSABP B-47 clinical trial. This trial showed that patients who were HER2-normal (IHC <10%; ISH ratio <2, HER2 copies <4) do not have significant benefit from anti-HER2 therapies [15].

3. Ki67: Ki67 is expressed throughout the cell cycle and serves as a good marker for the assessment of proliferative activity. The St Gallen guidelines recommend consideration of chemotherapy treatment for patients with high (>20%) Ki67 expression [16]. However, there is some controversy in the assessment of Ki67 particularly regarding the methodology used and interobserver consistency. A group of international breast cancer researchers are actively working on these issues [17].

Future Directions

Breast cancer has been one of the first cancers to be treated as a group of diseases. This was based on the presence of ER with further modification after the identification of HER2. However, the classification of breast cancer has not advanced beyond that. The identification of novel therapeutic agents and their associated biomarkers will lead to further classification of breast cancer. Cyclin D, CDK4/CDK6, and Rb1 were thought to be specific markers for CDK4/CDK6 inhibitors, but the response to the agents has not correlated with expression of these markers. This is true for a number of other agents including PARPi. The lack of specific biomarkers has been a major limitation for progress. It is hoped that the Clinical Trials Sequencing Program, a new version of The Cancer Genome Atlas (TCGA), will identify biomarkers associated with therapeutic resistance. This will enable evidence-based identification of targets and novel agents and significantly improve the quality of care of patients with breast cancer.

The high incidence of breast cancer and the frequent failure of the current therapies have led to a significant amount of investment of energy and money in finding new therapies. Almost every pathway associated with cancer has been targeted in breast cancer, and the following is a brief synopsis of major research areas:

(a) p53 pathway: p53 is the most commonly mutated gene in breast cancer. However, targeting this pathway has been difficult. This has resulted in efforts to target downstream events or its regulatory factors. MDM2 and MDMX are potent negative regulators of p53. There are at least 20 drugs in clinical trials that target these molecules [18].

(b) EGFR pathway: EGFR pathway has been well established in other cancers as a therapeutic target. However, the role of this pathway in breast cancer remains unclear. One of the reasons for this could be that normal epithelium of the breast expresses EGFR and it is lost in most cancers apart from TNBCs. Most of the current efforts using drugs directed against this pathway are focused on TNBCs. The pathway has also been suggested as a cause of resistance to HER2-directed therapies. Multiple Pan-HER inhibitors are in clinical trials.

(c) AKT-PI3K-PTEN-mTOR pathway: PI3K is the second most commonly mutated gene in breast cancer and is particularly present in ER+ breast cancers. Targeting this pathway either by suppressing AKT-PI3K or increasing the PTEN activity is being attempted. Most of the currently available PI3K (specific or pan) inhibitors have been found to either have limited clinical activity or are too toxic for clinical use. AKT inhibitors are in clinical trials. The success of BOLERO trials, using downstream mTOR inhibitors, in ER+ and

HER2+ cancers has documented the importance of targeting this pathway [7, 19].

(d) FGFR: The fibroblast growth factor receptor has been targeted in ER+, HER2- breast cancer using small molecular inhibitors (e.g., AZD4547, dovitinib, and lucitanib). These trials are ongoing and the results are not currently available.

(e) Androgen receptor: Although it has been known for a long time that AR is expressed in breast cancer, the recognition of LAR subtype of TNBCs has led to a renewed interest in targeting this pathway. AR antagonists as well as cytochrome enzyme inhibitors are showing promise in treating AR+ tumors.

(f) Immune checkpoint therapies: Our group [12] contributed significantly to the recognition of the role of lymphocytes in TNBCs; this has important implications. It has established the hypothesis that immune therapies could have a role in the treatment of these aggressive cancers. Initial results suggest that therapies directed at T-cell activation using CTLA-4 or attacking the PD-1/PDL-1 pathway might play an important role in the treatment of TNBCs.

(g) Other pathways: As illustrated in Fig. 35.3, there are multiple targets that have been identified in triple-negative cancer subtypes; these are being actively investigated in multiple clinical trials.

Therapies for breast cancer continue to evolve with many agents showing promise in early Phase I/Phase II clinical trials. It however remains to be seen whether these promised will be realized. The community has been scarred by the experience with anti-VEGF (bevacizumab) therapy, where the Phase III trials failed to confirm the promise shown by earlier Phase I–II studies; this resulted in withdrawal of FDA approval for this drug (Table 35.1).

Table 35.1 Predictive markers and therapies in breast cancer

Marker	Method and sample type assay	Function/ pathogenic process	Approved drug (s)	Company	Clinical trials
Luminal (ER+) breast cancer					
ER	IHC	Receptor modulator	Tamoxifen citrate (Nolvadex)	AstraZeneca	FDA approved
			Raloxifene (Evasta®)	Eli Lilly	FDA approved
			Toremifene citrate (Fareston®)	Kyowa Kirin Inc.	Multiple ongoing trials
		Receptor downgrader	Fulvestrant (Faslodex®)	AstraZeneca	FDA approved
			Elacestrant	Radius Health, Inc.	FDA approved
		Aromatase inhibitors	Exemestane Aromasin®	Pfizer	FDA approved
			Anastrozole Arimidex®	AstraZeneca/ANI Pharmaceuticals, Inc.	FDA approved
			Letrozole Femara®	Novartis	FDA approved
		GnRH agonists	Goserelin acetate Zoladex®	TerSera Therapeutics LLC	FDA approved
mTOR	None	PI3K-AKT-mTOR	Everolimus (Afinitor®)	Novartis	FDA approved
PI3K	None	PI3K-AKT-mTOR	PI3K inhibitors	Multiple companies	Multiple ongoing trials

(continued)

Table 35.1 (continued)

Marker	Method and sample type assay	Function/ pathogenic process	Approved drug (s)	Company	Clinical trials
AKT	None	PI3K-AKT-mTOR	AKT inhibitors	Multiple companies	Multiple ongoing trials
CDK4/6	None	Cell cycle	Palbociclib (Ibrance®)	Pfizer	FDA approved
			Ribociclib (Kisqali®)	Novartis	FDA approved
			Abemaciclib (Verzenio™)	Eli Lilly	FDA approved
HER2+ breast cancer					
			Trastuzumab (Herceptin®)	Genentech/Roche	FDA approved
HER2	IHC 3+ or FISH+	Receptor tyrosine kinase	Pertuzumab (Perjeta®)	Genentech/Roche	FDA approved
			Ado-trastuzumab emtansine (Kadcyla®)	Genentech/Roche	FDA approved
			Lapatinib (Tykerb®)	Novartis	FDA approved
			Neratinib (Nerlynx™)	Puma Biotechnology Inc.	FDA approved for metastatic cancers
			Biosimilars	Multiple companies	Multiple ongoing trials
mTOR	None	PI3K-AKT-mTOR	Everolimus (Afinitor®)	Novartis	FDA approved
PI3K	None	PI3K-AKT-mTOR	PI3K inhibitors	Multiple companies	Multiple ongoing trials
AKT	None	PI3K-AKT-mTOR	AKT inhibitors	Multiple companies	Multiple ongoing trials
Triple-negative breast cancers (ER-/PR-/HER2-)					
BRCA1/2	Mutational analysis	DNA repair/ PARP inhibitors	Veliparib (ABT-888)	AbbVie	Multiple ongoing trials
			Olaparib (Lynparza™)	AstraZeneca	Multiple ongoing trials
			Rucaparib (Rubraca™)	Clovis	Multiple ongoing trials
			E7449 Talazoparib	Eisai Inc.	Multiple ongoing trials
				Pfizer	Multiple ongoing trials
Androgen receptor	IHC	AR antagonists	Bicalutamide Casodex®	AstraZeneca	Multiple ongoing trials
			Enzalutamide (Xtandi®)	Astellas/Medivation	Multiple ongoing trials
			Sacituzumab govitecan	Immunomedics	
Trop-2	None		Pembrolizumab (Keytruda®)	Merck/MSD	Multiple ongoing trials

Table 35.1 (continued)

Marker	Method and sample type assay	Function/ pathogenic process	Approved drug (s)	Company	Clinical trials
PD-1/ PDL-1	None (possible IHC, tumor mutational burden)	Immune checkpoint	Nivolumab (Opdivo®)	Bristol-Myers Squibb	Multiple ongoing trials
			Atezolizumab (Tecentriq™)	Genentech/Roche	Multiple ongoing trials
			Durvalumab (Imfinzi™)	MedImmune	Multiple ongoing trials
CTLA-4	None	Immune activation	Ipilimumab (Yervoy®)	Bristol-Myers Squibb	Multiple ongoing trials
VEGF	None	Angiogenesis	Bevacizumab (Avastin®)	Genentech/Roche	No longer FDA approved

Other compounds tested in breast cancer are atamestane (SH-489), also known as metandroden, an aromatase inhibitor, and leuprorelin, a GnRH agonist

References

1. Perou CM, Sorlie T, Eisen MB, van de Rijn M, Jeffrey SS, Rees CA, et al. Molecular portraits of human breast tumours. Nature. 2000;406(6797):747–52.
2. Curtis C, Shah SP, Chin SF, Turashvili G, Rueda OM, Dunning MJ, et al. The genomic and transcriptomic architecture of 2,000 breast tumours reveals novel subgroups. Nature. 2012;486(7403):346–52.
3. Alexandrov LB, Nik-Zainal S, Wedge DC, Aparicio SA, Behjati S, Biankin AV, et al. Signatures of mutational processes in human cancer. Nature. 2013;500(7463):415–21.
4. Lehmann BD, Bauer JA, Chen X, Sanders ME, Chakravarthy AB, Shyr Y, et al. Identification of human triple-negative breast cancer subtypes and preclinical models for selection of targeted therapies. J Clin Invest. 2011;121(7):2750–67.
5. Ellis MJ, Ding L, Shen D, Luo J, Suman VJ, Wallis JW, et al. Whole-genome analysis informs breast cancer response to aromatase inhibition. Nature. 2012;486(7403):353–60.
6. Singh JC, Jhaveri K, Esteva FJ. HER2-positive advanced breast cancer: optimizing patient outcomes and opportunities for drug development. Br J Cancer. 2014;111(10):1888–98.
7. Andre F, Hurvitz S, Fasolo A, Tseng LM, Jerusalem G, Wilks S, et al. Molecular alterations and everolimus efficacy in human epidermal growth factor receptor 2-overexpressing metastatic breast cancers: combined exploratory biomarker analysis from BOLERO-1 and BOLERO-3. J Clin Oncol. 2016;34(18):2115–24.
8. Parker JS, Mullins M, Cheang MC, Leung S, Voduc D, Vickery T, et al. Supervised risk predictor of breast cancer based on intrinsic subtypes. J Clin Oncol. 2009;27(8):1160–7.
9. Masuda H, Baggerly KA, Wang Y, Zhang Y, Gonzalez-Angulo AM, Meric-Bernstam F, et al. Differential response to neoadjuvant chemotherapy among 7 triple-negative breast cancer molecular subtypes. Clin Cancer Res. 2013;19(19):5533–40.
10. Kalimutho M, Parsons K, Mittal D, Lopez JA, Srihari S, Khanna KK. Targeted therapies for triple-negative breast cancer: combating a stubborn disease. Trends Pharmacol Sci. 2015;36(12):822–46.
11. Ashworth A. A synthetic lethal therapeutic approach: poly(ADP) ribose polymerase inhibitors for the treatment of cancers deficient in DNA double-strand break repair. J Clin Oncol. 2008;26(22):3785–90.
12. Adams S, Gray RJ, Demaria S, Goldstein L, Perez EA, Shulman LN, et al. Prognostic value of tumor-infiltrating lymphocytes in triple-negative breast cancers from two phase III randomized adjuvant breast cancer trials: ECOG 2197 and ECOG 1199. J Clin Oncol. 2014;32(27):2959–66.
13. Hammond ME, Hayes DF, Dowsett M, Allred DC, Hagerty KL, Badve S, et al. American Society of Clinical Oncology/College of American Pathologists guideline recommendations for immunohistochemical testing of estrogen and progesterone receptors in breast cancer. J Clin Oncol. 2010;28(16):2784–95.
14. Wolff AC, Hammond ME, Hicks DG, Dowsett M, McShane LM, Allison KH, et al. Recommendations for human epidermal growth factor receptor 2 testing in breast cancer: American Society of Clinical Oncology/ College of American Pathologists clinical practice guideline update. J Clin Oncol. 2013;31(31):3997–4013.
15. Fehrenbacher L, Cecchini RS, Geyer CE, Rastogi P, Costantino JP, Atkins JN, et al. NSABP B-47 (NRG

oncology): phase III randomized trial comparing adjuvant chemotherapy with adriamycin (A) and cyclophosphamide (C) -> weekly paclitaxel (WP), or docetaxel (T) and C with or without a year of trastuzumab (H) in women with node-positive or high-risk node-negative invasive breast cancer (IBC) expressing HER2 staining intensity of IHC 1+or 2+with negative FISH (HER2-Low IBC). Cancer Res. 2018;78(4).

16. Esposito A, Criscitiello C, Curigliano G. Highlights from the 14(th) St Gallen international breast cancer conference 2015 in Vienna: dealing with classification, prognostication, and prediction refinement to personalize the treatment of patients with early breast cancer. Ecancermedicalscience. 2015;9:518.
17. Leung SCY, Nielsen TO, Zabaglo L, Arun I, Badve SS, Bane AL, et al. Analytical validation of a standardized scoring protocol for Ki67: phase 3 of an international multicenter collaboration. NPJ Breast Cancer. 2016;2:16014.
18. Burgess A, Chia KM, Haupt S, Thomas D, Haupt Y, Lim E. Clinical overview of MDM2/X-targeted therapies. Front Oncol. 2016;6:7.
19. Hurvitz SA, Andre F, Jiang Z, Shao Z, Mano MS, Neciosup SP, et al. Combination of everolimus with trastuzumab plus paclitaxel as first-line treatment for patients with HER2-positive advanced breast cancer (BOLERO-1): a phase 3, randomised, double-blind, multicentre trial. Lancet Oncol. 2015;16(7):816–29.

36 Predictive Biomarkers in Lung Cancer

Reinhard Buettner

Introduction

Most lung cancers are diagnosed in advanced, surgically non-resectable stages, and hence, 85% of all lung cancer patients undergo systemic therapies after primary diagnosis. Based on histology, conventional platinum or pemetrexed regimes reached median overall survival rates of only 8–10 months. Thus, lung cancer is one of the most frequent and most deadly cancer diseases worldwide with very high medical need for more effective therapies. Small cell lung cancers (SCLC), which tend to spread systemically very early on, differ from NSCL in their initial high response rate to combined radio-/chemotherapies. However, after relapse within a very limited time to progression, there is practically no effective second-line chemotherapy.

Depending on geographical provenience, approximately 25–50% of non-small cell lung cancers are driven by tyrosine kinase coupled, dominant oncogenes (EGFR, ALK, ROS1, and others) which are typically constitutively activated by somatic mutations or translocations. Such tumors are frequently not related to cigarette smoking and typically harbor flat genomes (i.e., tumors without large number of mutations) with very few other genomic alterations. As these tumors critically depend in their growth on their activated tyrosine kinases, they harbor specific genomic vulnerability and benefit massively from targeted therapies with selective tyrosine kinase inhibitors (TKIs). However, treatment by TKIs inevitably elicits secondary, acquired resistance, and hence, constant monitoring is necessary to detect acquired resistance and to administer second- or third-line TKIs overcoming resistance [1].

Another 30% of cases, frequently smoking-associated, harbor highly mutated cancer genomes (high tumor mutational burden (TMB) and thereby present many tumor (neo)antigens eliciting strong immunological host responses. Such tumors depend on immune escape, and recently the PD1/PD-L1 axis has been identified as a potent signaling pathway rendering tolerance to cytotoxic T cell. These lung cancers massively benefit from treatment by immune checkpoint inhibitors, as recently shown by a first-line study comparing the monoclonal PD-1 antibody pembrolizumab against conventional histology-adapted first-line chemotherapy [2]. When these tumors emerge with acquired resistance, it is currently under evaluation whether intensified and combined therapies with immune checkpoint inhibitors or second-line chemotherapies should be administered [3] (see Fig. 36.1). It is possible that clinical studies being currently under way will reveal that combination of high TMB and

R. Buettner
Department of Pathology, University Hospital Cologne, Cologne, Germany
e-mail: reinhard.buettner@uk-koeln.du

S. Badve, G. L. Kumar (eds.), *Predictive Biomarkers in Oncology*,
https://doi.org/10.1007/978-3-319-95228-4_36

Biomarker - guided therapy of NSCLC				
1L	Targeted TKIs (EGFR / ALK / ROS)		PD1 mAB + CTX (PD-L1 < 50%)	CTX or CTX combined with Immune Therapy (IO)
2L	TKIs overcoming acquired resistance / TKI combinations	Targeted TKIs (BRAF/ RET / MET/ NTRKs)	Immune checkpoint inhibitors - Combinations (PD1 / CTLA4 mAB) (PD1 / CTX)	2L-CTX / PD1 or PD-L1 mABs

Fig. 36.1 Overview of biomarker-guided decisions for first (1L)- and second-line (2L) therapies of NSCLC. *Abbreviations*: NSCLC non-small cell lung cancer, TKI tyrosine kinase inhibitor, PD-1 programmed death receptor-1, PD-L1 programmed death receptor ligand-1, mAB monoclonal antibody, EGFR epidermal growth factor receptor-1, ALK anaplastic lymphoma kinase, ROS1 ROS proto-oncogene 1 receptor, CTX chemotherapy, NTRK neurotrophic receptor tyrosine kinase, IO immune oncology: PD1, PDL1, CTLR4 antibodies or combinations thereof

high PD-L1 expression will allow even better selection of patients for first-line PD-1/PD-L1 therapies than PD-L1 expression alone.

As a dramatic change in therapeutic approaches, approximately 50% of NSCLC are now eligible for first-line, biomarker-guided therapies other than standard, nonselective chemotherapy. Thus, the current challenges are tumors driven by dominant oncogenes which are not druggable, i.e., by mutated KRAS genes, and tumors without single dominant oncogenes lacking high expression of PD-L1 or low mutational burden (TMB). Current research is therefore aiming to clinically introduce a new generation of RAS inhibitors and to elicit antitumoral immune responses in immunologically "silent" tumors.

Targeting Oncogenic Driver Mutations

Approved First-Line Therapies

Kinase-activating mutations in EGFR or BRAF and kinase-activating gene fusions in ALK and ROS1 belong to the most frequent druggable oncogenic driver lesions in lung carcinomas [4]. There are significant geographic differences in prevalence from 12% to 15% of EGFR-mutated lung adenocarcinomas in patients from Western Europe to well over 50% in patients from Japan. Such tumors are highly responsive to treatment with orally available tyrosine kinase inhibitors, and thus, rapid determination of EGFR, ALK, ROS1, and BRAF status within ten working days is mandatory after diagnosis of lung adenocarcinomas, adenosquamous carcinomas, and squamous carcinomas in never smokers. Only rapid and highly sensitive technology allows delivering timely biomarker diagnostics for first-line therapies by erlotinib, gefitinib, and afatinib for EGFR mutations or crizotinib for ALK and ROS1 fusions. Clinical guidelines disfavor selection of patients by gender, geographic provenience, smoking status, or histology, although signet ring adenocarcinoma is frequently associated with ALK fusions and mucinous carcinomas frequently harboring K-RAS mutations (that are mutually exclusive with alterations in EGFR and ALK/ROS1) [5, 6].

Initially tumors driven by single dominant oncogenes respond very well to therapies with

TKIs but ultimately relapse with acquired resistance [7]. Typically, between 60% (in EGFR-mutated tumors) and approximately 30% (in ALK and ROS1 fusion-positive tumors) occur as gatekeeper mutations in the kinase domains and are highly sensitive to further treatment with second- or third-line TKIs. Gatekeeper mutations usually occur in the kinase domain and alter or limit the access of TKIs to the binding pocket. Thus, osimertinib has been approved for therapies of EGFR-mutated carcinomas with T790M resistance mutations [8] and ceritinib for second-line therapies of ALK-positive tumors with resistance to crizotinib [9], and further available TKIs are alectinib and lorlantinib for ALK and ROS1 resistance. Results from recent clinical trials may alter the sequence in TKIs, as first-line osimertinib treatment in EGFR-mutant lung cancer resulted in a very durable clinical benefit of almost 19 months progression-free survival in the FLAURA study [10], and first-line alectinib was superior to crizotinib especially in terms of CNS progression in the ALEX study [11]. Other bypass mechanisms of resistance include high-level HER2 or MET amplification, mutations in downstream AKT- or MAP-kinase pathways, or a change in histology to small cell or sarcomatoid cancers. Since all these acquired resistance mechanisms require different therapeutic modalities and may occur simultaneously in different tumor sites, consequent monitoring of these patients by re-biopsies and liquid biopsies is necessary.

Although many clinical trials are not designed to show a benefit in overall survival due to crossover between therapy arms, systematic comparison between patient cohorts receiving consequent all possible lines of TKIs beyond acquired resistance and patients receiving conventional chemotherapies has shown a massive prolongation in overall survival. For the first time, the German Network Genomic Medicine (NGM) showed a mean survival of 31 months after first-line TKIs in EGFR-mutated adenocarcinomas and a mean survival of 60 months after third-line TKIs [6 and further unpublished data].

Emerging Novel Gene Fusions and Oncogenic Driver Mutations

A list of novel gene fusions and further oncogenic drivers activated by mutations or gene amplifications is given in Tables 36.1, 36.2, and 36.3.

Significant responses have been observed in clinical trials for B-RAF V600E-mutated lung adenocarcinomas to the combination of dabrafinib (B-RAF-inhibitor) and trametinib (MEK-inhibitor) as well as for tumors with high-level MET amplification or exon 14 skipping mutations to MET-TKIs (reviewed in [13]). Approval of novel MET inhibitors is expected soon, and determination of these biomarkers may be required for first-line therapies. Novel lesions under current clinical investigation are kinase-activating fusions in NTRK1, NTRK2, and NTRK3 responding to treatment with the inhibitor entrectinib. Translocations in NRG1, causing activation of HER signaling, may be too rare to study systematically in a prospective clinical trial; however, treatment by panHER inhibitors in combination with HER2mABs is recommended.

Treatment of FGFR-dependent lung cancer has been described as a promising targeted approach in lung squamous cell carcinomas. However, only a subset of FGFR1-amplified lung cancers respond to FGFR-TKIs [14], and it seems that currently only patients with gene fusions (FGFR1,2,3) or kinase-activating mutations should be selected for therapies until better biomarkers are available to select patients with gene amplifications. Also results from clinical trials of patients with RET fusions and HER2 exon20 mutations or amplifications yielded mixed results, and it appears that better TKIs are needed for these patients. Specifically, the exon 20 insertions in EGFR and HER2 are not being inhibited at single-digit nanomolar concentrations by current TKIs.

Therefore, a number of strong oncogenic drivers need development of better and clinically applicable inhibitors, such as KRAS mutations, EGFR- and HER2 exon20 insertions, and RET fusions. Also other constitutively activated

Table 36.1 Predictive biomarkers and targeted therapies in NSCLC-approved first-line therapies in NSCLC

Predictive biomarker	Alteration and frequency (%)	Detection method	Approved drug(s)
EGFR	Mutation 12–15%	Preferred methods: next-generation parallel sequencing of multiplex PCR panels or hybrid capture panels	Erlotinib, gefitinib, afatinib, Osimertinib for acquired resistance with T790M
	Geographic differences in prevalence 0.10–50%	Exons 18,19,20 and 21	
ALK	Rearrangement 4%	IHC and/or FISH	Alectinib, crizotinib, ceritinib
		Hybrid generation sequencing of hybrid capture panels, RT-PCR, RNAseq, Nanostring assays	
		Ventana ALK (D5F3) CDx assay for the qualitative detection of ALK protein by IHC in FFPE specimens	
		Abbott Molecular Inc. Vysis ALK break apart FISH probe kit to qualitative detect rearrangements involving the ALK gene via FISH in FFPE specimens	
ROS	Rearrangement 1.5%	IHC and/or FISH	Crizotinib
		Parallel sequencing, RT-PCR, RNAseq, immunoassays	
		For details refer to Bubendorf L et al., *Virchows Arch* (2016):469:489–503	
BRAF	1–2%	Mutation, multiplex PCR panels or hybrid capture panels	Dabrafenib and trametinib
PD-L1	PD-L1 overexpression	PD-L1 clinical trial assays or lab-developed assays (reviewed in [12])	Pembrolizumab
	High tumor mutational burden	Hybrid capture panels, under development	PD-1 and PD-L1 antibodies (not yet approved)

Table 36.2 Predictive biomarkers and targeted therapies in NSCLC off-label approvals

Predictive biomarker	Alteration and frequency	Detection method	Off-label approved drug(s)
MET	Amplification, mutation, exon 14 skipping, gene fusion 2–4%	IHC, FISH, and RT-PCR hybrid capture panels	Crizotinib
HER2	Mutation 2%	IHC and/or FISH	Afatinib

oncogenic pathways need better selection by biomarkers and combined therapies as results from clinical trials with PIK3CA and PTEN mutations yielded disappointing results. It is expected that continuous improvements in TKIs, combinatorial treatment regimens, and also discovery of new targetable lesions will ultimately identify 25% of NSCLC patients with druggable lesions. Thus, comprehensive and rapid biomarker profiling is mandatory for every treatment decision in first- and second-line therapies [15].

Table 36.3 Predictive biomarkers and targeted therapies in NSCLC targets for drugs under clinical investigation

Predictive biomarker	Alteration and frequency	Off-label other cancers
RET	Rearrangement 1%	Cabozantinib, vandetanib, ponatinib
MEK1	Mutation 1%	Trametinib
NRG1	Rearrangement <1	Pan-HER inhibitors
NTRK1,2,3	Rearrangement 0.1%	Entrectinib
FGFR1	Amplification mutation Fusion 12%	Debrafinib + trametinib

Immune Checkpoint Inhibitors

Many lung cancers harbor highly mutated tumor genomes and therefore elicit intense responses to the adaptive immune system. In fact, NSCLC has been among the first tumor types to apply successfully inhibitors of PD-1/PD-L1 and CTLA4. Two monoclonal antibodies targeted against PD-1 (nivolumab and pembrolizumab) have received approval for second-line therapies of NSCLC both in the USA and Europe. A companion diagnostic is required for pembrolizumab (for tumors expressing PD-L1 in 1% or more of tumor cells), and complimentary testing is recommended but not mandatory for nivolumab or the PD-L1 mAB atezolizumab. In addition, pembrolizumab has received approval for first-line therapy in NSCLC with expression of PD-L1 in 50% or more of the tumor cells. However, clinical trials also showed that patients with EGFR mutations or ALK fusions did not reach better overall survival, and hence, patients with druggable oncogenic driver lesions should be treated by TKIs as long as possible. Thus, current clinical recommendations ask for comprehensive and reflex PD-L1 diagnostics for the diagnostics of all NSCLC.

Specifically, for first-line therapy decisions, PD-L1 status will be a clinically relevant biomarker. A clinical trial with pembrolizumab showed improved overall survival in patients with maximally positive tumors, i.e., equal or higher than 50% PD-L1 expression, and reached all positive endpoints. It is expected that further improvements will be reached by combinations of PD-1 mABs with ipilimumab, tremelimumab, or with chemotherapy and ultimately will significantly prolong overall survival of advanced-stage lung cancer patients. Results of ongoing clinical trials are expected to be reported in 2017 and 2018.

As a number of clinically validated assays and kits have been applied in studies with PD1 and PD-L1 mABs, the German Harmonization Trial and the US Blueprint project have compared performance of these assays. As a result, a 6-tired harmonized reporting system has been recommended by the German Trial, which allows for rapid reflex testing in a first-line setting by several different assays and retains this information throughout different therapies and second-line decisions [16]. Thus, we strongly recommend using this harmonized tumor cell proportion scoring system and add such biomarker information to every clinical pathology report [16].

Further emerging biomarkers are under development. Specifically, high mutational load, intrinsic DNA repair deficiencies, and gene expression signatures indicative of a strong antitumoral immune recognition are currently being clinically validated. In summary, comprehensive testing of genomic alterations and of markers indicative of escape from immune recognition is mandatory for all therapeutic decision in treatment of lung cancer. Areas in high need of improvement remain treatment of small cell lung cancer and treatment of NSCLC devoid of strong immune recognition and lacking druggable oncogenic driver lesions.

Novel Therapies for Small Cell Lung Cancer (SCLC)

While dramatic improvements in treatment of stage IV NSCLC were made for those tumors revealing druggable oncogenic driver mutations or high susceptibility for immune checkpoint inhibitors, treatment of SCLC has remained with unchanged fatal outcome for many years. More recently, two therapy regimes by rovalpituzumab tesirine [17], a DLL3-targeted antibody-drug conjugate, and by combined treatment with PD-1 and PD-L1 antibodies [18] were introduced and shown to provide significant progress after first-line therapy. As the majority of patients usually respond strongly to combined radio-/chemotherapies, it is unlikely that these therapies will replace current first-line therapies but rather prolong survival thereafter.

Delta-like ligand 3 (DLL3) is a negative regulator of Notch signaling and expressed at very high levels on the surface of most small cell lung cancer cells. As DLL3 is mostly expressed during development, there is a therapeutic window to deliver a very strong toxin conjugated to DLL3-targeted antibody selectively to SCLC. Current studies are designed for second-line therapy after recurrent disease or for extension therapy directly after finishing first-line therapy.

Interestingly, the concept of specifically targeting tumors with high TMB may also hold promises for SCLC. Very recently a clinical trial reported durable clinical benefit in patients with high TMB-SCLC undergoing combined treatment with nivolumab (PD1 antibody) and ipilimumab (CTLA4 antibody).

Thus, results of these clinical trials may show that the survival of NSCLC patients with relapses may improve for the first time after 20 years of therapeutic nihilism.

Summary and Future Directions

Currently, tyrosine kinase inhibitors and immune checkpoint inhibitors targeting PD1 or PD-L1 have reached clinical practice and improved survival of advanced stage lung cancer patients significantly. Hence, future challenges are to provide more durable therapies avoiding acquired resistance and better treatments for tumors that lack druggable mutations or high expression of immune checkpoints.

One strategy will be to combine tyrosine kinase inhibitors that limit the space for resistance. Current clinical trials therefore have applied third-generation TKIs as first-line therapies for EGFR-mutant or ALK fusion-positive tumors, i.e., osimertinib and alectinib. While these trials showed encouraging results, further trials will apply these inhibitors with combination of MEK, MET, or panHer inhibitors, representing the most frequent bypass track resistance mechanisms. It is expected that avoiding resistance by drug combinations may be more durable than adding sequentially single tyrosine kinase inhibitor therapies. This lesson has been learned from treating HIV infection.

Better selection of immune checkpoint inhibitors is expected to result from measurement of tumor mutational burden. It has been shown that the number of non-synonymous mutations in the cancer genome is one of several mechanisms triggering strong antitumoral responses by the adaptive immune system. Thus, rapid and clinically relevant measurement of tumor mutational burden is expected to add on information on tumors that benefit from reactivating antitumoral responses. In addition, monitoring precisely immune reactions in the tumor micromilieu by multiparametric immune fluorescence imaging platform is another path to better understand and select tumors for IO.

Finally, then, combining immune therapies with chemo- or radiation therapies is also in current clinical trials and has already generated encouraging results. However, tumors that are deficient in essential components of immune effector pathways, such as loss of ß2microglobulin or MHC class I expression, will remain a challenge. Here, artificially designed immune modalities including CART-T cells, BITE immunotherapies, or personalized vaccines will provide new opportunities for effective immunotherapies.

References

1. American Cancer Society. (2016). Cancer facts & figures 2016. http://www.cancer.org/research/cancerfactsstatistics/cancerfactsfigures2016/index. Accessed 13 Dec 2016.
2. Reck M, Rodríguez-Abreu D, Robinson AG et al. Pembrolizumab versus chemotherapy for PD-L1-positive non-small-cell lung cancer. N Engl J Med. 2016. https://doi.org/10.1056/NEJMoa1606774.
3. ClinicalTrials.gov Identifier: NCT02477826. www.clinicaltrials.gov. https://clinicaltrials.gov/ct2/show/NCT02477826.
4. Dearden S, Stevens J, Wu YL, Blowers D. Mutation incidence and coincidence in non small-cell lung cancer: meta-analyses by ethnicity and histology (mutMap). Ann Oncol. 2013;24:2371–6. https://doi.org/10.1093/annonc/mdt205.
5. Postmus PE, Kerr KM, Oudkerk M, Senan S, Waller DA, et al. Early and locally advanced non-small-cell lung cancer (NSCLC): ESMO Clinical Practice Guidelines for diagnosis, treatment and follow-up. Ann Oncol. 2017;28:iv1–21. https://doi.org/10.1093/annonc/mdx222.
6. Lindeman NI, Cagle PT, Aisner DL, Arcila ME, Beasley MB, Bernicker E, Colasacco C, Dacic S, Hirsch FR, Kerr K, Kwiatkowski DJ, Ladanyi M, Nowak JA, Sholl L, Temple-Smolkin R, Solomon B, Souter LH, Thunnissen E, Tsao MS, Ventura CB, Wynes MW, Yatabe Y. Updated molecular testing

guideline for the selection of lung cancer patients for treatment with targeted tyrosine kinase inhibitors: guideline from the College of American Pathologists, the International Association for the Study of Lung Cancer, and the Association for Molecular Pathology. Arch Pathol Lab Med. 2018, in press. 2018 Jan 22. https://doi.org/10.5858/arpa.2017-0388-CP. [Epub ahead of print].
7. Seidel D, et al. A genomics-based classification of human lung tumors. Sci Transl Med. 2013;5(209):209ra153. https://doi.org/10.1126/scitranslmed.3006802.
8. Pasi A, Jänne MD, Yang JC-H, Kim D-W, Planchard D, Ohe Y, et al. AZD9291 in EGFR inhibitor–resistant non–small-cell lung cancer. N Engl J Med. 2015;372:1689–99. https://doi.org/10.1056/NEJMoa1411817.
9. Crino L, Ahn MJ, De Marinis F, Groen HJM, Wakelee H, Hida T, et al. Multicenter phase II study of whole-body and intracranial activity with ceritinib in patients with ALK-rearranged non–small-cell lung cancer previously treated with chemotherapy and crizotinib: results from ASCEND-2. J Clin Oncol. 2016;34:2866–73. https://doi.org/10.1200/JCO.2015.65.5936.
10. Ramalingam S, Reungwetwattana T, Chewaskulyong B, et al. Osimertinib vs standard of care (SoC) EGFR-TKI as first-line therapy in patients (pts) with EGFRm advanced NSCLC: FLAURA. Presented at: ESMO Congress. Madrid. 2017. 9–12 September 2017. Abstract LBA2_PR.
11. Peters S, Camidge DR, Shaw AT, Gadgeel S, Ahn JS, Kim DW, et al. Alectinib versus crizotinib in untreated ALK-positive non–small-cell lung cancer. N Engl J Med. 2017;377:829–38. https://doi.org/10.1056/NEJMoa1704795.
12. Büttner R, Gosney JR, Skov BG, Adam J, Motoi N, Bloom KJ, Dietel M, Longshore JW, López-Ríos F, Penault-Llorca F, Viale G, Wotherspoon AC, Kerr KM, Tsao MS. Programmed death-ligand 1 immunohistochemistry testing: a review of analytical assays and clinical implementation in non-small-cell lung cancer. JCO. 2017. https://doi.org/10.1200/JCO.2017.74.7642. PMID: 29053400.
13. Schallenberg S, Merkelbach-Bruse S Buettner R. Lung cancer as a paradigm for precision oncology in lung cancer. Virchows Arch. 2017 (in press).
14. Nogova L, Sequist LV, Perez Garcia JM, Andre F, Delord JP, Hidalgo M, Schellens JH, Cassier PA, Camidge DR, Schuler M, Vaishampayan U, Burris H, Tian GG, Campone M, Wainberg ZA, Lim WT, LoRusso P, Shapiro GI, Parker K, Chen X, Choudhury S, Ringeisen F, Graus-Porta D, Porter D, Isaacs R, Buettner R, Wolf J. Evaluation of BGJ398, a fibroblast growth factor receptor 1-3 kinase inhibitor, in patients with advanced solid tumors harboring genetic alterations in fibroblast growth factor receptors: results of a global phase I, dose-escalation and dose-expansion study. J Clin Oncol: JCO. 2016. 2016672048. [Epub ahead of print]. PMID:27870574.
15. Meder L, König K, Fassunke J, Ozretić L, Wolf J, Merkelbach-Bruse S, et al. Implementing amplicon-based next generation sequencing in the diagnosis of small cell lung carcinoma metastases. Exp Mol Pathol. 2015;99(3):6826. https://doi.org/10.1016/j.yexmp.2015.11.002. PMID:26546837
16. Scheel AH, Dietel M, Heukamp LC, et al. Harmonized PD-L1 immunohistochemistry for pulmonary squamous-cell and adenocarcinomas. Mod Pathol. 2016;29(10):1165–72. https://doi.org/10.1038/modpathol.2016.117.
17. Rudin CM, Pietanza MC, Bauer TM, Ready N, Morgensztern D, Glisson BS, et al. Rovalpituzumab tesirine, a DLL3-targeted antibody-drug conjugate, in recurrent small-cell lung cancer: a first-in-human, first-in-class, open-label, phase 1 study.; SCRX16-001 investigators. Lancet Oncol. 2017;18(1):42–51. https://doi.org/10.1016/S1470-2045(16)30565-4. Epub 2016 Dec 5.
18. Sharma P, Callahan MK, Bono P, Kim J, Spiliopoulou P, Calvo E, et al. Nivolumab monotherapy in recurrent metastatic urothelial carcinoma (CheckMate 032): a multicentre, open-label, two-stage, multi-arm, phase 1/2 trial. Lancet Oncol. 2016;17(11):1590–8. https://doi.org/10.1016/S1470-2045(16)30496-X. Epub 2016 Oct 9.

Predictive Biomarkers and Targeted Therapies in Genitourinary Cancers

37

Li Yan Khor and Puay Hoon Tan

The genitourinary tract includes the adrenal glands, kidneys, ureters, bladder, and urethra and, specifically in men, the prostate, penis, and testicles.

Adrenal Cancer

Adrenocortical carcinoma (ACC) is rare, with an overall age-adjusted incidence of 7 million per year in the USA. There is a bimodal age distribution of presentation, with peaks at the first decade of life and the fifth or sixth decade. The cancer is also associated with well-defined genetic syndromes including Beckwith-Wiedemann, Li-Fraumeni, and Lynch syndromes, as well as multiple endocrine neoplasia type 1 (MEN1).

Although rare, ACC is an aggressive tumor which presents at an advanced stage. The median survival is less than 12 months even with complete surgical resection. Mitotane, an adrenolytic agent, is the standard FDA-approved adjuvant therapy for metastatic ACC, despite its toxicity and low response rate. RRM1 gene expression has been shown in limited studies to be functionally associated with mitotane sensitivity and may predict for response to adjuvant treatment. Further clinical studies are warranted. Reasonable objective response has been reported in the platinum-based chemotherapy combination of etoposide/doxorubicin/cisplatin/mitotane (EDP-M) (FIRM-ACT trial). Molecular predictors of resistance to these agents, such as TOPO2A, PGP, and ERCC1 protein expression by immunohistochemistry, have been described in a small study but have yet to be validated.

A range of targeted therapies has been tested in the clinical trial setting but have yielded disappointing results, including VEGF receptor blockers (bevacizumab and axitinib), EGFR blockers (erlotinib and gefitinib), multiple receptor tyrosine kinase inhibitor (sorafenib), IGF-1R blocker (linsitinib), and monoclonal antibody (cixutumumab). A clinical trial testing sunitinib, another multi-receptor tyrosine kinase inhibitor, in combination with mitotane, showed 5 of 35 (14.3%) patients had stable disease. However, it was postulated that mitotane could have suppressed the effect of sunitinib due to its ability to induce cytochrome p450-3A4. Future study of sunitinib alone is needed.

Phase I trial studies of a combination of mTOR inhibitor (temsirolimus) with an immunomodulatory agent, either lenalidomide or cixutumumab, showed some positive results with disease stabilization even in patients previously

L. Y. Khor (✉)
Department of Anatomical Pathology, Singapore General Hospital, Singapore, Singapore
e-mail: khor.li.yan@singhealth.com.sg

P. H. Tan
Division of Pathology, Singapore General Hospital, Singapore, Singapore
e-mail: tan.puay.hoon@singhealth.com.sg

S. Badve, G. L. Kumar (eds.), *Predictive Biomarkers in Oncology*,
https://doi.org/10.1007/978-3-319-95228-4_37

treated with IGF-1R therapy. Larger clinical studies are warranted.

Future Directions

Phase I trials are ongoing for FGFR and c-MET blockers (NCT01752920).

Small exploratory studies evaluating PD-L1 expression have been performed. Expression is reported in ACC tumor cells and tumor-infiltrating lymphocytes. However, no association with tumor stage, grade, functionality, or overall survival was seen. A prospective case-control preclinical immunotherapy study (NCT00457587) is ongoing.

Other potential targets under investigation include the Wnt/beta-catenin signaling pathway and SF1 transcription factor blockade [1, 2]. Figitumumab against the IGF type 1 receptor (IGF-IR) has been explored in a phase I trial [3].

Kidney Cancer

Kidney cancer is among the ten most common cancers in both men and women, with a higher risk in men (overall lifetime risk of 1.6%). The rate of new kidney cancers has been on the rise since the 1990s with a recent leveling off, and this rate of increase is partially attributed to increased availability of imaging resulting in the detection of incidental renal masses. Early-stage renal cell carcinoma (RCC) is treated by surgery, and 20–40% of patients relapse after initial surgery with curative intent. RCC with clear cell histology accounts for 90% of all malignant neoplasms of the kidney with approximately 20–30% of patients presenting with metastases.

FDA-approved drugs for metastatic RCC include anti-angiogenic agents, mammalian target of rapamycin therapies (mTOR), and immune therapies. Immunotherapy using anti-PD-1 antibody nivolumab is the only therapy to demonstrate statistically significant overall survival of more than 2 years in a phase III trial of previously treated advanced RCC patients. However, PD-L1 expression was not found to significantly impact the efficacy of nivolumab [4]. There are currently no reliable clinically approved predictive markers of response to these agents; many diverse potential candidates are being studied (see Table 37.1).

Two clinical trials have shown high levels of interleukin 6 (IL-6) to be a potentially favorable predictor of response to the anti-angiogenic agents, pazopanib (VEG105192 trial) and bevacizumab (CALGB90206 trial). IL-6, a pro-inflammatory cytokine, is elevated in RCC and involved in signaling via the JAK-STAT pathway.

VEGF receptors and ligands play key roles in angiogenesis and tumor growth. In particular, VEGFR-3 is primarily involved in lymphangiogenesis. In a study randomizing patients to interferon alpha (IFNα) or the tyrosine kinase inhibitor (TKI) sunitinib, low baseline levels of VEGFR-3 predicted increased benefit from sunitinib over IFNα.

A phase III trial comparing temsirolimus, a TORC-1 inhibitor (PI3K/Akt/TORC1/HIF pathway) to interferon alpha (IFNα), showed high serum lactate dehydrogenase (LDH) predicted treatment benefit in terms of overall survival (OS) [5].

Functional single nucleotide polymorphisms (SNP) in angiogenesis-related genes can provide information on a patient's inherent resistance to anti-angiogenic agents. A lower potential clinical benefit of sunitinib was reported in patients with IL-4 and *PTGS2* (prostaglandin-endoperoxide synthase 2) SNP variants. As such, it is speculated that patients with these genotypes are unlikely to benefit from sunitinib and should be offered alternatives [6]. An independent validation of these findings is however necessary.

MicroRNAs (miRNAs) are regulators of hypoxia and angiogenesis in renal cell carcinoma. Two miRNAs in particular, miR-1307 and miR-425-5p, identified by next-generation sequencing (NGS), are predicted for disease progression under TKI treatment, using both training and validation study sets. Thus, the 2-miRNA classifier has the potential to discriminate for TKI-refractory patients [7].

Table 37.1 Potential biomarkers under investigation

Gene/RNA protein biomarkers	Function/pathogenic process	Patient selection method and sample type	Observational studies/ successfully completed clinical trials and number of patients enrolled	Clinical use and limitations	Approved drug(s) Generic/trade name	Company
Prostate						
Androgen receptor splice variant-7 (AR-V7)	Loss of ligand- binding domain that binds to androgens	CTC RNA assay	Observational study,	Shorter OS	Enzalutamide/ Xtandi®	Medivation
		CTC IF protein assay	31 patients		Abiraterone/Zytiga®	Janssen
			161 patients	Superior OS	Taxanes versus Enzalutamide/ Xtandi®	
					Abiraterone/Zytiga®	Medivation
						Janssen
AR copy number or mutation	Increased AR drives resistance; AR mutation confers lower androgen affinity	Pretreatment presence of gain/ mutation in ctDNA	Observational study,	Worse OS	Abiraterone/Zytiga®	Medivation
			274 patients			
		Pretreatment presence of gain/ mutation in ctDNA	65 patients	Worse PFS	Enzalutamide/ Xtandi®	Janssen
Kidney						
PD-1/PD-L1	Human monoclonal antibody directed against PD-L1; blocks the interaction of PD-L1 with PD-1 and CD80 (B7.1)	PD-1/PD-L1 expression[a]	Checkmate 025, 406 patients	OS and ORR benefit	Nivolumab (Opdivo®)	Bristol-Myers Squibb
IL-6	Pro-inflammatory cytokine signals via mitogenic JAK/STAT pathway	Elevated plasma levels	VEG105192, 344 patients	PFS benefit	Pazopanib/Votrient®	GlaxoSmithKline
			CALGB9020,	Predict OS	Bevacizumab/ Avastin®	Genentech
			424 patients			
VEGFR-3	Mediates lymphangiogenesis	Low baseline plasma levels	Observational study,	PFS and ORR benefit	Sunitinib/Sutent®	Pfizer
			33 patients		versus IFN-α	

(continued)

Table 37.1 (continued)

Gene/RNA protein biomarkers	Function/pathogenic process	Patient selection method and sample type	Observational studies/ successfully completed clinical trials and number of patients enrolled	Clinical use and limitations	Approved drug(s) Generic/trade name	Company
Lactate dehydrogenase (LDH)	Associated with high cell turnover and stress	Serum levels	Phase III trial, 404 patients	Improved OS	Temsirolimus/ Torisel® versus IFN-α	Pfizer
IL-4 and Prostaglandin-endoperoxide synthase (PTGS2) SNP variants	Increased mRNA leads to elevated IL-4 and COX2	Peripheral blood genomic DNA	Observational study, 75 patients	Shorter CSS	Sunitinib/Sutent®	Pfizer
miRNA (miR-1307/ miR-425-5p)	Regulator of hypoxia and angiogenesis	NGS on FFPE tissue	Observational study, 76 patients	Shorter PFS and OS	Sunitinib/Sutent®, sorafenib/Nexavar® or pazopanib/Votrient	Pfizer Bayer/Onyx GlaxoSmithKline

[a]Clinical trial testing for PD-L1 expression by immunohistochemical staining on FFPE tissue did not show any predictive value

Guidelines

The European Society for Medical Oncology (ESMO) has published guidelines for diagnosis, treatment, and follow-up of RCC. These updated ESMO guidelines cover incidence and epidemiology, diagnosis and pathology/molecular biology, staging and risk assessment, management of local and metastatic disease, response evaluation and follow-up, long-term implications, and survivorship. For more information, refer to "Escudier et al. [8]."

Cancer Immunotherapy Guidelines

The Society of Immunotherapy of Cancer (SITC) has issued a consensus statement on immunotherapy for the treatment of RCC (Refer to "Rini et al. [9]"). The immunotherapy treatment algorithm for stage IV clear cell RCC is highlighted here. Nonsurgical candidates are stratified into those with good/intermediate and poor/sarcomatoid risk groups (per MSKCC/IMDC classification). Observation is recommended for good/intermediate risk group patients with small volume indolent metastasis. An immunotherapy-based clinical trial should thus be considered. High dose IL-2 should be considered with metastatic RCC patients with good performance status. For poor/sarcomatoid risk group patients, clinical trial enrollment should be considered. Anti-VEGF TKIs are recommended as a second-line option. In refractory RCC, nivolumab is an appropriate initial treatment in the absence of contraindications. TKIs, high-dose IL-2, and mTOR inhibitors can be considered based on patient performance, comorbidities, prior therapy, and preference.

Future Directions

Initial results from a phase II retrospective study show that the unsupervised hierarchical clustering approach identifies two patient groups as angiogenic or inflammatory subtypes by using a candidate plasma-based protein signature including VEGF ligands and soluble receptors. This could be useful in selecting patients for sorafenib treatment and improving progression-free survival. Preliminary study results found N-cadherin on circulating mesenchymal or stem cell CTC may be predictive for response to anti-VEGF therapies.

Circulating miRNAs in serum show promise as more accessible predictive markers of response to anti-angiogenic therapies.

Bladder Cancer

It is estimated that in 2017, approximately 79,000 Americans will be diagnosed with bladder cancer and almost 17,000 will die from this disease. Men are three to four times more likely to suffer from this cancer than women. In 2010, bladder cancer resulted in 170,000 deaths up from 114,000 in 1990.This is an increase of 19.4%, adjusted for increase in total world population. Muscle (muscularis propria)-invasive bladder cancer is the current focus of efforts in developing personalized therapeutic regimens. Neoadjuvant/adjuvant platinum-based chemotherapy confers a survival benefit, which is modest and short-lived in some patients. High expression levels of DNA nucleotide excision repair proteins ERCC1 and ERCC2 are associated with resistance to platinum-based chemotherapy. Thus, mutations or elevated wild-type expression of these proteins may be useful to predict responders and spare unsuitable patients the cytotoxic side effects of current standard chemotherapy [10].

Immune checkpoint inhibitors, such as atezolizumab (Tecentriq®), nivolumab (Opdivo®), durvalumab (IMFINZI™), pembrolizumab (Keytruda®), and avelumab (Bavencio®), are showing promise in the treatment of locally advanced or metastatic urothelial cancer. Atezolizumab and pembrolizumab have been approved for first-line use in patients with metastatic urothelial cancer who are not eligible for cisplatin chemotherapy. Programmed cell death protein 1 (PD-1) and programmed death-ligand 1 (PD-L1) expression levels on T-lymphocytes and tumor cells function as the predictive markers of

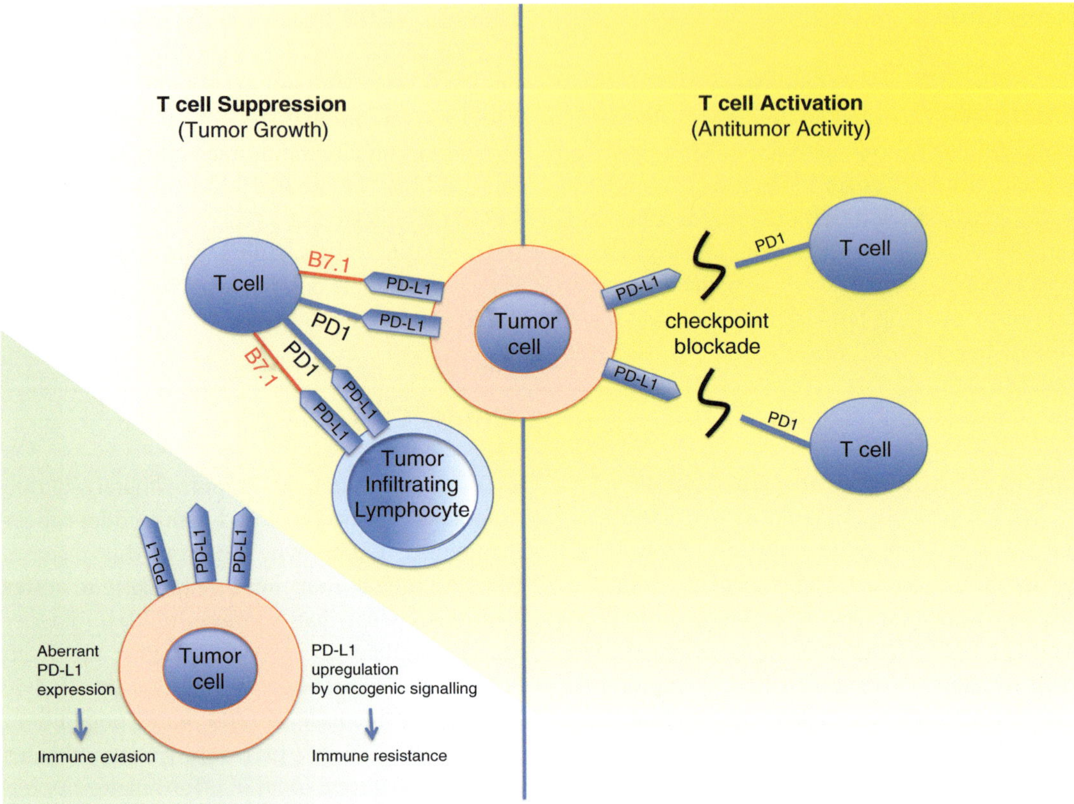

Fig. 37.1 PD-1/PD-L1 interaction in tumor and immune cells

response for these agents (See Fig. 37.1). Specific FDA-approved immunohistochemical stains and staining platforms form the basis of the companion diagnostic tests.

PD-L1 (programmed death-ligand 1) , which is expressed on tumor cells and tumor-infiltrating immune cells, interacts with PD-1 and B7.1, both found on the surface of T cells. This interaction suppresses T-cell function and restricts tumor cell killing. The overexpression of PD-L1 in tumor cells can avoid T-cell cytolysis and facilitate cancer growth. Inhibiting the interaction of PD-1 and its ligands can significantly enhance T-cell function, resulting in antitumor activity.

Targeted therapy still continues to be an emerging modality for the treatment of bladder cancer. A recent study describing complete response of bladder cancer to everolimus, an mTOR inhibitor, suggested that the presence of a *TSC1* mutation conferred excellent sensitivity to the targeted therapy. Successful treatment of metastatic bladder cancer in combination with the TKI, pazopanib, further suggests that activating mutations of the PI3K pathway may be useful predictive biomarkers of response to these agents.

A small series of studies reported excellent bladder cancer response to the pan-FGFR kinase inhibitor, BGJ398, where screening for the FGFR3 mutation proved useful as a predictive marker of response. In a phase I clinical trial, the presence of FGFR3-TACC3 translocation in patients with metastatic bladder cancer showed promise as a biomarker of response for JNJ-42756493, another pan-FGFR TKI.

Preclinical and phase I trials have shown ERBB2 to be a robust marker of response for urothelial cancer treated with HER2 receptor antagonists. Phase II trials are underway for trastuzumab and lapatinib (NCT01828736 and NCT00949455, respectively). HER2 has been associated with increased chemotherapy sensitivity in urothelial cancer and potentially predicts response to platinum-based therapies [10].

Guidelines

The European Society for Medical Oncology (ESMO) has published guidelines for diagnosis, treatment, and follow-up of bladder cancer. Refer to "Bellmunt et al. [11]" (http://www.esmo.org/Guidelines/Genitourinary-Cancers/Bladder-Cancer). Other useful resources for the diagnosis and pathologic staging and grading of tumors are the "Cancer Protocol Templates" from the College of American Pathologists (CAP).

PD-L1 Expression Assay Guidelines

Ventana PD-L1 SP142 Assay is the companion diagnostic test for atezolizumab (Tecentriq®).

PD-L1 expression on tumor-infiltrating immune cells in formalin-fixed paraffin-embedded tissue is scored.

Ventana PD-L1 SP263 Assay is the companion diagnostic test for durvalumab (IMFINZI™).

PD-L1 expression on both tumor cells and tumor-infiltrating immune cells in formalin-fixed paraffin-embedded tissue is scored.

Future Directions

Urothelial cancer studied by molecular analyses is unique as a heterogeneous and divergently clustered disease. Three distinct pan-cancer subtypes show molecular characteristics, which are either bladder-specific or overlapping with lung adenocarcinoma, or overlap with p53 mutation-associated head and neck squamous cancers. It is postulated that personalized therapies of these cancers (e.g., lung adenocarcinoma) could similarly be applied to the relevant urothelial cancer subtype.

Clinical trials matching targeted therapies based on patient's genomic and molecular alterations are underway [10].

Additionally, a SWOG trial (S1314; NCT02177695) involves applying predictive gene signatures of response to neoadjuvant chemotherapy agents derived from bladder and non-bladder cancer cell lines to patient samples. Intrinsic bladder subtypes (basal and luminal), which have been identified to predict differences in sensitivity to neoadjuvant chemotherapy, and specific mutations in predicting response will also be tested [12].

Prostate Cancer

Prostate cancer represents 26% of all newly diagnosed cancers in males and 4.4% of cancer related deaths. The overall 5-year survival rate for prostate cancer is high at 98.6% for the period of 2007–2013 (www.seer.cancer.gov). At the time of diagnosis, most men have localized or regional prostate cancer, but 4% will have distant metastases, and their 5-year survival rate is only 28.2%. Moreover, the median survival of men with metastatic castration-resistant prostate cancer ranges from 15 to 36 months. Therapies incorporating newer drugs available for such scenarios are still being optimized and necessitate the identification of suitable predictive biomarkers. Predictive biomarkers in prostate cancer are not as well-developed as for other solid tumors, despite androgen ablation being a long-established form of targeted therapy in urology. Several potential markers have been identified in observational studies but have yet to be validated. With surgical castration and prolonged conventional androgen ablation therapy, mutations and gene amplifications of the androgen receptor (AR) gene are believed to lead to the emergence of castration resistant prostate cancer (CRPC), with or without metastasis. Additionally, resistance mechanisms to novel antiandrogen agents, such as CYP17 (abiraterone acetate, Zytiga©) inhibitors that inhibit androgen synthesis or direct inhibitors of AR (enzalutamide, Xtandi©; ARN-509/apalutamide), are being elucidated. As such, CYP17 and AR expression are key candidates for predictive biomarkers in this scenario. For example, AR amplification can predict response to both enzalutamide and abiraterone; this can be detected in pretreatment plasma circulating tumor DNA [13]. Similarly, the presence of active, truncated AR splice variant, AR-V7, in circulating tumor cells (CTC) correlated with resistance to second-generation antiandrogens [14, 15].

Sipuleucel-T (Provenge®) is an FDA-approved cellular immunotherapy which utilizes the patient's peripheral blood mononuclear cells to enhance the activity of their autologous antigen-presenting cells and elicit a cytotoxic T-lymphocyte response against prostate cancer epithelial cells and shown to confer a survival advantage on patients with metastatic castration-resistant prostate cancer. However, this benefit is not accompanied by favorable effects on serum prostate-specific antigen (PSA), tumor regression, time to progression, or quality of life. Although there is reported increased T-cell proliferation and IFN-γ in peripheral blood, and increased infiltrating CD3+, CD4+, FOXP3−, and CD8+ T-cells in the tissues, their utility as predictive markers are not yet determined. Transient increases in serum eosinophil count observed 6 weeks posttreatment have been beneficial to the patient and are being studied.

Ipilimumab, a fully human monoclonal antibody targeting the cytotoxic T-lymphocyte antigen 4 (CTLA-4) checkpoint, failed in a phase III trial to improve the overall survival of men with chemotherapy-naive metastatic castration-resistant prostate cancer over a placebo.

Guidelines

Cancer Immunotherapy Guidelines

The Society of Immunotherapy of Cancer (SITC) has issued a consensus statement on immunotherapy for the treatment of prostate carcinoma. (Refer to "McNeel et al. [16]"). The treatment recommendations for metastatic prostate carcinoma are highlighted here. All patients undergo continuous testosterone suppression with or without denosumab or zoledronic acid. Additionally, patients with minimal/no symptoms may be offered sipuleucel-T, enzalutamide, abiraterone, clinical trial enrolment, or docetaxel. Symptomatic patients pre-docetaxel are offered similar options, excluding sipuleucel-T. Symptomatic patients post-docetaxel are treated with cabazitaxel, enzalumatide, abiraterone, and Ra-223 or offered clinical trial enrolment.

Future Directions

Several phase III immune-oncology studies are in progress. For more information, refer to the "National Cancer Institute" prostate cancer website (https://www.cancer.gov/types/prostate) or to the "Prostate Cancer Research Institute" (PCRI), a not-for-profit organization (http://pcri.org/clinicaltrials).

Testicular (Germ Cell) Tumors

Testis cancer represents 0.5% of all new cancer cases in the USA with about 410 estimated deaths in 2017. The overall 5-year survival rate for prostate cancer is high at 95.1% for the period of 2007–2013 (www.seer.cancer.gov). Germ cell tumors (GCT) are the most common malignancies among men between ages 15 and 35.

Cisplatin-based first-line treatment is effective in testicular GCT. However, the 10–15% of refractory GCT, especially in young patients, create the need for effective targeted therapy. Thus far, studies have been disappointing with only single-patient responses documented, and there is currently no defined predictive marker.

Future Directions

PD-L1 expression was shown to be about 64% in a tissue microarray study, suggesting potential for PD-1/PD-L1 immune checkpoint inhibitors. However, these agents have yet to be tested clinically. AKT1 and PIK3CA mutations of the PI3K-AKT-mTOR pathway have been described in cisplatin-resistant GCT and are potentially useful predictors requiring further investigation [17, 18]. For latest testicular cancer news, refer to the "National Cancer Institute" site. (https://www.cancer.org/cancer/testicular-cancer.html).

Penile Cancer

Penile cancer is rare, with only 1640 new cases diagnosed in the USA in 2014. They account for approximately 17% of cancers in the developing world. Ninety-five percent of penile cancers are squamous cell carcinoma (SCC). Conventional treatment is surgical resection with or without radiation/chemotherapy. Increased EGFR expression has been reported in some of these cancers, and a small number of case reports have attributed some success to immunotherapy agents targeting the gene in the setting of locally recurrent and metastatic penile SCC. Cetuximab and panitumumab are a few examples of such promising agents. However, additional studies are needed [19].

About one-third of penile SCC are human papillomavirus (HPV)-related, with a prevalence rate of between 20% and 80%. An ongoing phase I trial tests the combination of a T-cell receptor gene therapy targeting HPV-16 E7 in combination with pembrolizumab (NCT02858310).

Additional phase II immuno-oncology clinical trials are ongoing. (https://www.cancer.gov/types/penile) (Tables 37.1, 37.2, and 37.3).

Table 37.2 Predictive biomarkers in bladder cancer

Gene/RNA protein biomarkers	Function/pathogenic process	Patient selection method and sample type	Observational studies/successfully completed clinical trials and # of patients enrolled	Drug generic (and brand name)	Company[a]
Bladder					
PD-1	Human monoclonal antibody directed against PD-1; blocks PD-1 interaction with PD-L1 and PD-L2	Locally advanced unresectable or metastatic urothelial carcinoma following progression on a platinum-containing therapy;	Phase II CheckMate-275[a],	Nivolumab (Opdivo®)	Bristol-Myers Squibb
		FFPE tissue	270 patients		
PD-1		Locally advanced unresectable or metastatic urothelial carcinoma following progression on a platinum-containing therapy/first line therapy in cisplatin-ineligible patients;	Keynote-045, 270 patients	Pembrolizumab (Keytruda®)	Merck
		FFPE tissue	Keynote-052, 370 patients		
PD-L1	Human monoclonal antibody directed against PD-L1; blocks the interaction of PD-L1 with PD-1 and CD80 (B7.1)	Locally advanced unresectable or metastatic urothelial carcinoma following progression on a platinum-containing therapy; FFPE tissue	Phase I/II study 1108, 182 patients	Durvalumab (IMFINZI™)	AstraZeneca
		Locally advanced unresectable or metastatic urothelial carcinoma following progression on a platinum-containing therapy;	Phase I JAVELIN solid tumor trial[b], 242 patients	Avelumab (Bavencio®)	Pfizer
		N/A			

(continued)

Table 37.2 (continued)

Gene/RNA protein biomarkers	Function/pathogenic process	Patient selection method and sample type	Observational studies/ successfully completed clinical trials and # of patients enrolled	Drug generic (and brand name)	Company[a]
PD-L1		Locally advanced unresectable or metastatic urothelial carcinoma following progression on a platinum-containing therapy/first line therapy in cisplatin-ineligible patients;	Phase II IMvigor210[a], 310 patients	Atezolizumab (Tecentriq™)	Genentech/ Roche
		FFPE tissue	Phase II IMvigor210, 119 patients		

[a]Continued approval may be contingent upon verification and description of clinical benefit in confirmatory trials
[b]Patients included regardless of PDL-1 expression

Table 37.3 Approved drugs and their targets

Cancer	Gene/RNA protein targets	Function/pathogenic process	Drug generic (and brand name)	Company
Metastatic renal cell carcinoma	VEGF	Inhibits angiogenesis	Bevacizumab (Avastin®)	Genentech/ Roche
	VEGFR, PDGFR	Inhibits genetic transcription involving cell proliferation and angiogenesis	Sorafenib (Nexavar®)	Bayer/Onyx
	PDGFR, VEGFR, c-KIT (CD117)	Inhibits cellular signaling involving cell survival, cell proliferation, and angiogenesis	Sunitinib (Sutent®)	Pfizer
	VEGFR, PDGFR, c-KIT, FGFR	Inhibits cellular signaling involving cell survival, cell proliferation, and angiogenesis	Pazopanib (Votrient®)	Novartis
	Mammalian target of rapamycin (mTOR)	Inhibits cell division and tumor growth	Temsirolimus (Torisel®)	Pfizer
	mTOR		Everolimus (Afinitor®)	Novartis
	VEGFR1, VEGFR2, VEGFR3	Inhibits angiogenesis and tumor growth	Axitinib (Inlyta®)	Pfizer
	MET, VEGFR1, 2 and 3, AXL	Inhibits angiogenesis and cell proliferation	Cabozantinib (Cabometyx™)	Exelixis
	VEGFR 1, 2, and 3 FGFR 1, 2, 3, and 4 PDGFR alpha, c-KIT, and RET	Inhibits angiogenesis and tumor growth	Lenvatinib mesylate (Lenvima®)	Eisai

Table 37.3 (continued)

Cancer	Gene/RNA protein targets	Function/pathogenic process	Drug generic (and brand name)	Company
Metastatic castrate-resistant prostate cancer	Androgen receptor	Inhibits cancer cell proliferation	Enzalutamide (Xtandi®),	Astellas Medivation
Locally advanced/ metastatic urothelial carcinoma; Refractory renal cell carcinoma (nivolumab only)	PD-1	See Table 37.1	Nivolumab (Opdivo®)	Bristol-Myers Squibb
			Pembrolizumab (Keytruda®)	Merck
	PD-L1		Atezolizumab (Tecentriq™)	Genentech/ Roche
			Durvalumab (IMFINZI™)	AstraZeneca
			Avelumab (Bavencio®)	Pfizer
			Atezolizumab (Tecentriq™)	Genentech/ Roche

Summary and Future Directions

To date, PD-L1 is the only predictive marker with FDA-approved validated assays for a targeted treatment in genitourinary tumors, i.e., advanced urothelial carcinoma. Studies in other genitourinary cancers are underway for similar responses. Initial observational studies and clinical trials performed have also revealed other promising markers, but these have yet to be validated. Newer markers showing promise are those identified in individual patient's cancers by genotyping or next-generation sequencing using either tumor tissue or circulating tumor cells/DNA.

References

1. Konda B, Kirschner LS. Novel targeted therapies in adrenocortical carcinoma. Curr Opin Endocrinol Diabetes Obes. 2016;23(3):233–41. https://doi.org/10.1097/MED.0000000000000247.
2. Millis SZ, Ejadi S, Demeure M. Molecular profiling of refractory adrenocortical cancers and predictive biomarkers to therapy. J Biomark Cancer. 2015;7:69–76. https://doi.org/10.4137/BIC.S34292. eCollection 2015.
3. Haluska P, Worden F, Olmos D, et al. Safety, tolerability, and pharmacokinetics of the anti-IGF-1R monoclonal antibody figitumumab in patients with refractory adrenocortical carcinoma. Cancer Chemother Pharmacol. 2010;65(4):765–73.
4. Motzer RJ, Escudier B, McDermott DF, et al. Nivolumab versus everolimus in advanced renal-cell carcinoma. N Engl J Med. 2015;373(19):1803–13.
5. Zhang T, Zhu J, George DJ, Nixon AB. Metastatic clear cell renal cell carcinoma: circulating biomarkers to guide antiangiogenic and immune therapies. Urol Oncol. 2016;34(11):510–8. https://doi.org/10.1016/j.urolonc.2016.06.020.
6. Cebrián A, Gómez Del Pulgar T, Méndez-Vidal MJ, Gonzálvez ML, Lainez N, Castellano D, et al. Functional PTGS2 polymorphism-based models as novel predictive markers in metastatic renal cell carcinoma patients receiving first-line sunitinib. Sci Rep. 2017;7:41371. https://doi.org/10.1038/srep41371.
7. García-Donas J, Beuselinck B, Inglada-Pérez L, Graña O, Schöffski P, Wozniak A, et al. Deep sequencing reveals microRNAs predictive of antiangiogenic drug response. JCI Insight. 2016;1(10):e86051.
8. Escudier B, et al. Renal cell carcinoma: ESMO Clinical Practice Guidelines for diagnosis, treatment and follow-up. Ann Oncol. 2016;27(suppl 5):v58–68.
9. Rini BI, McDermott DF, Hammers H, et al. Society for Immunotherapy of Cancer consensus statement on

immunotherapy for the treatment of renal cell carcinoma. J Immunother Cancer. 2016;4:81.
10. Jones RT, Felsenstein KM, Theodorescu D. Pharmacogenomics: biomarker-directed therapy for bladder cancer. Urol Clin N Am. 2016;43(1):77–86. https://doi.org/10.1016/j.ucl.2015.08.007.
11. Bellmunt J, et al. Bladder cancer: ESMO Practice Guidelines for diagnosis, treatment and follow-up. Ann Oncol. 2014;25(suppl 3):iii40–8.
12. Contreras-Sanz A, Roberts ME, Seiler R, Black PC. Recent progress with next-generation biomarkers in muscle-invasive bladder cancer. Int J Urol. 2017;24(1):7–15. https://doi.org/10.1111/iju.13193.
13. Crawford ED, Higano CS, Shore ND, Hussain M, Petrylak DP. Treating patients with metastatic castration resistant prostate cancer: a comprehensive review of available therapies. J Urol. 2015;194(6):1537–47. https://doi.org/10.1016/j.juro.2015.06.106.
14. Barbieri CE, Chinnaiyan AM, Lerner SP, Swanton C, Rubin MA. The emergence of precision urologic oncology: a collaborative review on biomarker-driven therapeutics. Eur Urol. 2017;71(2):237–46. https://doi.org/10.1016/j.eururo.2016.08.024.
15. Miyamoto DT, Lee RJ. Cell-free and circulating tumor cell-based biomarkers in men with metastatic prostate cancer: tools for real-time precision medicine? Urol Oncol. 2016;34(11):490–501. https://doi.org/10.1016/j.urolonc.2016.09.001.
16. McNeel DG, Bander NH, Beer TM, et al. The Society for Immunotherapy of Cancer consensus statement on immunotherapy for the treatment of prostate carcinoma. J Immunother Cancer. 2016;4:92.
17. Fankhauser CD, Honecker F, Beyer J, Bode PK. Emerging therapeutic targets for male germ cell tumors. Curr Oncol Rep. 2015;17(12):54. https://doi.org/10.1007/s11912-015-0479-4.
18. Oing C, Kollmannsberger C, Oechsle K, Bokemeyer C. Investigational targeted therapies for the treatment of testicular germ cell tumors. Expert Opin Investig Drugs. 2016;25(9):1033–43. https://doi.org/10.1080/13543784.2016.1195808.
19. McDaniel AS, Hovelson DH, Cani AK, et al. Genomic profiling of penile squamous cell carcinoma reveals new opportunities for targeted therapy. Cancer Res. 2015;75(24):5219–27. https://doi.org/10.1158/0008-5472.CAN-15-1004.

Predictive Biomarkers and Targeted Therapies in Colorectal Cancer

38

Susan D. Richman and Bharat Jasani

Overview

Colorectal cancer (CRC) is one of the most common malignancies worldwide, with up to one million new diagnoses and 500,000 deaths, each year. Back in 1990, Fearon and Vogelstein [1] published their seminal model of cancer progression, starting with the inactivation of the APC gene, and evolving *thereafter* in a stepwise fashion, with the sequential activation of oncogenes *and/or* inactivation of tumour suppressor genes. This model has *certainly* stood the test of time and is still recognised as the most common mechanism; however, subsequent to this, several alternative pathways have been identified in smaller subsets of colorectal tumours, namely, chromosome instability (CIN) and tumours with a 'mutator phenotype'. The latter cohort includes tumours with aberrant DNA methylation and those with defective DNA mismatch repair pathways. A recent publication of gene expression data collected on over 4000 CRC patients, by an international consortium of experts, has suggested the categorisation of colorectal tumours into four 'consensus molecular subtypes': CMS1-4. The CMS1 tumours, accounting for 14% of the tumours assessed, are hypermutated, with microsatellite instability (MSI), and associated with a strong immune activation. The CMS2 tumours (37% of the tumours studied) are chromosomally unstable and tend to show WNT and MYC signalling activation. The CMS3 group, also referred to as 'metabolic tumours' and comprising 13%, show clear metabolic dysregulation. The CMS4 tumours representing 23% of the colorectal cancers have mesenchymal characteristics and display TGF-beta activation, stromal invasion and angiogenesis [2]. The remaining 13% show mixed features shared between different categories as well as tumour heterogeneity. This increase in our understanding of the molecular mechanisms driving these colorectal tumours, and the subsequent improved molecularly based classification system, should ultimately increase the ease with which patients are stratified for personalised therapies [3]. Even more recently, Dunne et al. have gone on to show that stromal-derived intra-tumoural heterogeneity (ITH) may actually confound the ability to correctly molecularly stratify patients and have developed a CRC intrinsic signature (CRIS), which can more accurately cluster samples into patient-of-origin, rather than region-of-origin [4].

Over the past 10–15 years, we have seen a marked increase in the amount of biomarker research, carried out in both clinical and research

S. D. Richman
Department of Pathology and Tumour Biology, Leeds Institute of Cancer and Pathology, Leeds, UK
e-mail: s.d.richman@leeds.ac.uk

B. Jasani (✉)
Department of Pathology, Targos Molecular Pathology GmbH, Kassel, Hessen, Germany
e-mail: Bharat.Jasani@targos-gmbh.de

S. Badve, G. L. Kumar (eds.), *Predictive Biomarkers in Oncology*,
https://doi.org/10.1007/978-3-319-95228-4_38

laboratories, resulting in a move towards personalised treatment, particularly for patients with advanced disease. However, such biomarkers tend to fall into the category of 'prognostic', where they may provide additional guidance as to a patient's likely outcome, independent of the treatment options available. In order to predict a patient's likely response to a therapy, 'predictive' biomarkers are keenly sought. These need to be able to stratify patients into those who will benefit from, and those who will gain no benefit from, a particular treatment. Furthermore, the biomarker has to be targetable, in that a therapy specifically acting upon the biomarker must be available. To date, there are very few predictive biomarkers which have accompanying targeted therapies, so this chapter describes those currently available to colorectal cancer patients and also those which, although not yet available in the clinic, have a promising future.

Predictive Biomarkers and Approved Targeted Therapy

RAS Mutation Status

As stated above, despite the increase in the volume of research, currently the only biomarker which is predictive of response to a targeted therapy is *RAS* mutation status. In 2011, Vaughan et al. determined the extended mutation status in a subset of colorectal tumours, which were wild-type for *KRAS* codons 12 and 13, and identified a further 27.3% of tumours with mutations in other *RAS* codons or *BRAF* [5]. Data from several clinical trials including OPUS, PRIME and CRYSTAL demonstrated that patients, whose tumour contained a mutation in exon 2 (codons 12 and 13) of *KRAS*, did not respond to anti-EGFR therapy. In addition, only a small proportion of patients, who were *KRAS-wt* at exon 2, actually benefitted from the monoclonal antibody therapy. This identified the need to further refine the population of metastatic CRC patients who were likely to benefit, and thus a retrospective analysis was carried out of both the PRIME and CRYSTAL clinical trials, which expanded the testing panel to include *KRAS* and *NRAS* exons 2, 3 and 4. This identified a further 17% and 14.7% of patients in these trials, respectively, whose tumours contained an activating mutation outside of *KRAS* exon 2 and provided evidence that extended *RAS* testing panels should be adopted [6, 7]. It is now compulsory that patients, who are being considered for anti-EGFR therapies such as cetuximab or panitumumab, are subjected to *RAS* mutation screening, to assess their eligibility to receive the agent.

Anti-VEGF Therapy

Vascular endothelial growth factor (VEGF) is an angiogenic growth factor, expressed by most cancer cells and stromal cells within the tumor. Signalling pathways involving VEGF and its isoforms are responsible for endothelial cell survival, proliferation, migration and vascular permeability, upon binding to VEGFR-1 or 2. The breakthrough trial back in 2004 by Hurwitz et al. demonstrated a benefit in mCRC upon the addition of bevacizumab to fluorouracil-based combination chemotherapy regimens [8]. Direct inhibition of VEGF-A, using the monoclonal antibody bevacizumab, has shown promising results in several other phase III clinical trials, where increased RR, PFS and OS were observed. Recent advances now include the recombinant fusion protein aflibercept (an angiogenic factor trap), which in addition to blocking VEGF-A binding, it also blocks VEGF-B and placental growth factor (PlGF) [9]. Other strategies for targeting the VEGF pathway include the use of monoclonal antibodies such as ramucirumab, which directly binds the extracellular domain of the VEGF receptor (VEGFR-2), to exert its anti-angiogenic influence [10], and tyrosine kinase inhibitors (TKIs) such as regorafenib, which targets all three VEGFRs [11].

PD-1/PD-L1 Inhibition

Over the past few years, the use of immune checkpoint inhibitors has emerged as a novel therapeutic area. The programmed death 1 (PD-1) negative feedback loop regulates cytotoxic T-cell responses. When PD-L1, on the surface of tumour cells, binds to the PD-1 receptor on T-cells, the resultant inhibitory signal blocks the antitumour immune response, thus allowing tumour proliferation. Blockade of PD-1 or PD-L1, with mAbs, has resulted in impressive clinical responses in melanomas, NSCLC, bladder cancer, renal cell carcinoma and ovarian cancer, to name but a few. A response rate of 60% to the anti-PD-L1 inhibitor pembrolizumab has been observed in mCRCs with dMMR status [12]. This effect was not seen in proficient mismatch repair (pMMR) tumours, suggesting that immune checkpoint inhibition may be a novel therapy to treat the small cohort of dMMR mCRC patients. As recently as May 2017, the US Food and Drug Administration (FDA) took the unprecedented step and approved the use of pembrolizumab in unresectable or metastatic, MSI-H or dMMR solid tumours, which had progressed following treatment. This is the first time that the FDA has approved a drug, for the treatment of cancer patients, which is not based upon tissue type. Furthermore, the approval was based on data from 149 patients who had been enrolled in 5 uncontrolled, single-arm clinical trials (Table 38.1).

Future Directions

Amphiegulin (AREG) and Epiregulin (EREG)

AREG and EREG are both EGFR ligands. Their RNA expression has recently been studied in relation to the efficacy of anti-EGFR therapy, and an evaluation of patients enrolled in the PICCOLO clinical trial demonstrated that high mRNA expression of either ligand was a predictive biomarker for PFS after anti-EGFR therapy in patients who were *RAS*-wt [19]. An assessment of AREG/EREG status may therefore prove beneficial in *RAS*-wt patients as a strategy for identifying the subset of patients who will actually gain benefit from treatment.

PIK3CA Mutation Status

Phosphatidylinositol-4,5-bisphosphate 3-kinase (PIK3CA) is the p110 catalytic subunit of phosphatidylinositol 3-kinase (PI3K) and is involved in the regulation of signalling downstream of the EGFR. Activating mutations within exons 9 and 20 of *PIK3CA* occur in approximately 10–20% colorectal tumours and result in constitutive activation of the PI3K/AKT pathway. These *PIK3CA* mutations can occur within tumours also harbouring *KRAS* or *BRAF* mutations. Data regarding the

Table 38.1 Targeted therapies currently used in clinical practice

Approved drug (generic/trade name)	Target	Reviews of phase III clinical trials leading to approval	Company
Panitumumab (Vectibix)	Wild-type KRAS and NRAS	Hocking et al. [13]	Amgen
Cetuximab (Erbitux)	Wild-type KRAS and NRAS	Yazdi et al. [14]	Bristol-Myers Squibb
Bevacizumab (Avastin)	Vascular endothelial growth factor (VEGF)	Ilic et al. [15]	Genentech
Ramucirumab (Cyramza)	$VEGFR_2$	Verdaguer et al. [16]	Eli Lilly
Regorafenib (Stivarga)	Dual targeted $VEGFR_2$-tyrosine kinase inhibition	Yoshino et al. [17]	Bayer
Ziv-Aflibercept (Zaltrap)	VEGF	Perkins et al. [18]	Sanofi-Genzyme
Pembrolizumab (Keytruda)	MSI-H or dMMR	n/a	Merck & Co.

predictive value of *PIK3CA* mutations is still contradictory, although there is a growing body of evidence to suggest that in the presence of WT *KRAS*, the presence of *PIK3CA* mutations is predictive of a lack of response to anti-EGFR mAbs. Epidemiological evidence encompassing stage I–IV CRC suggests that patients with *PIK3CA*-mut tumours benefit from aspirin [20]. This is currently being evaluated in the MRC FOCUS4 clinical trial [21] (discussed later in the chapter).

HER2

Overexpression of the tyrosine kinase receptor, HER2, is seen in approximately 20% of breast cancers and is associated with poor prognosis. Treatment of such patients with an anti-HER2 monoclonal antibody such as trastuzumab has seen improvements in patient survival. Moving into the CRC scenario, it has been postulated that one mechanism of anti-EGFR therapy resistance may be upregulation and subsequently increased signalling through other members of the HER family of receptors. Recently a phase II, proof-of-concept trial (Siena et al.) HERACLES demonstrated that, in chemotherapy-refractory HER2-positive mCRC patients, the combination of trastuzumab and the tyrosine kinase inhibitor, lapatinib, was an active drug combination [22]. The HER2 amplification is seen in approximately 3–5% of *KRAS* codon 12/13 wild-type patients, and identification of this sub-cohort of patients through routine laboratory screening may improve the response rates of chemorefractory patients.

Influence of Prognostic Factors

It is important that every predictive stratification of CRC takes into account the natural history or prognosis (i.e. an end outcome regardless of the treatment) associated with the morphological and/or molecular subtype of the tumour as well as the stage of the disease in the individual patient. Thus CRCs arising from a serrated adenoma pathway (30–35% of all CRC) [23] have been shown to have a poor prognosis regardless of the treatment [24]. A proportion of these type of tumours (~10% of CRC) have the activation of BRAF mutations as the initiating event instead that of the classical APC gene locus [25]. Interestingly, a recent study showed that stage III patients with BRAF or KRAS mutations with MMR-proficient tumours had poorer prognosis compared to those who lacked these mutations [26] (Fig. 38.1).

Guidelines

In terms of clinical guidelines for the prescribing of anti-EGFR therapies to CRC patients, the American Society of Clinical Oncology Provisional Opinion Update (ASCO PCO) was recently published, basing its recommendations on not only the original PCO but also a meta-analysis of more than 11 reviews, meta-analyses and retrospective studies. This vast volume of data provided the evidence to show that any mCRC patients, being considered for anti-EGFR therapy, should undergo extended *RAS* testing of their tumour, which should encompass *KRAS* exons 2 (codons 12 and 13), 3 (codons 59 and 61) and 4 (codons 117 and 146) and *NRAS* exons 2 (codons 12 and 13), 3 (codons 59 and 61) and 4 (codons 117 and 146) [27].

In 2013, the Food and Drug Administration (FDA) approved the prescribing of bevacizumab in combination with FP-irinotecan- or FP-oxaliplatin-based chemotherapy regimens, in mCRC patients, whose disease has progressed on a first-line treatment containing bevacizumab. They based this decision on the results of a randomised, open-label, multinational clinical trial (NCT00700102).

As none of the other biomarkers listed in Table 38.2 are licenced for clinical use, there are, as yet, no clinical prescribing guidelines.

Summary and Future Directions

We are entering an exciting era of personalised medicine, where patients are beginning to have their cancer treated in a very unique way. This is only possible through the identification of

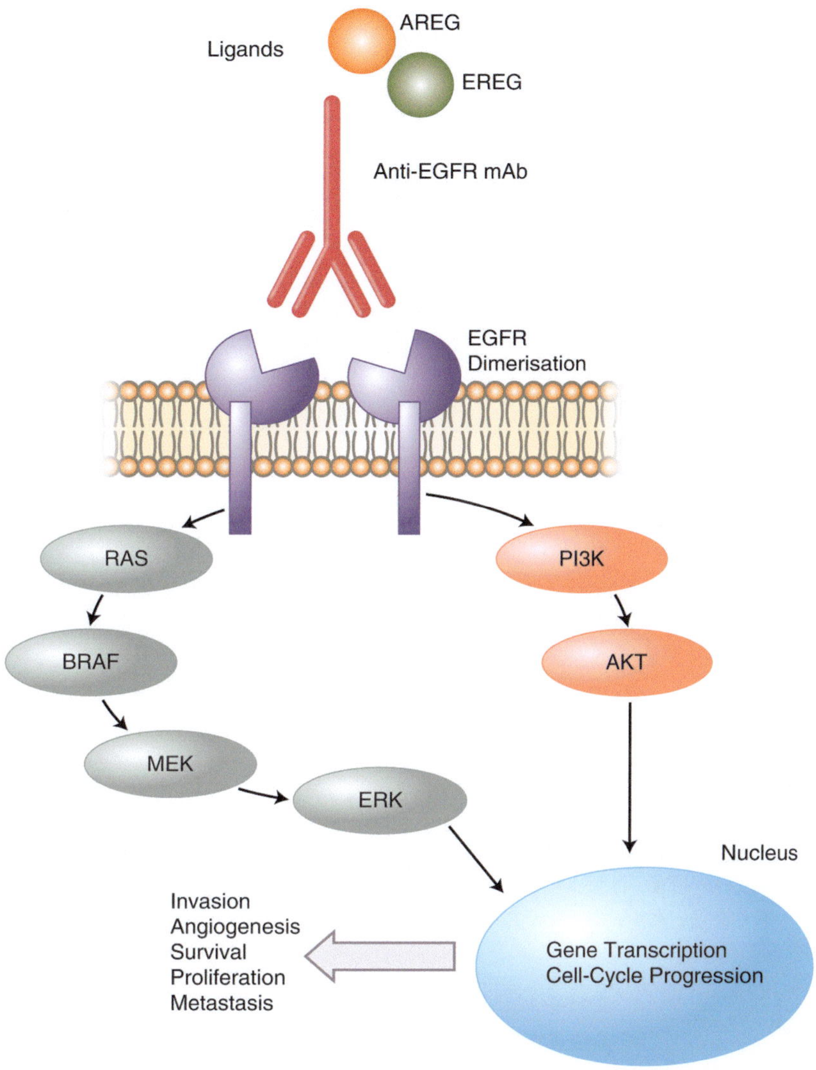

Fig. 38.1 A schematic representation of signalling through the epidermal growth factor receptor (EGFR). In a 'normal' cell, the regulation of gene transcription is tightly regulated. In a patient who is *RAS*-wt, the addition of an anti-EGFR monoclonal antibody (mAb) prevents receptor dimerisation and thus inhibits downstream signalling. In a *RAS*-mut patient, the *RAS* mutation causes constitutive activation of *RAS*, and subsequent downstream signalling is not regulated, resulting in increased gene transcription

predictive biomarkers, which categorise patients into those who will and will not gain benefit from a particular therapy or treatment combination. Drug companies are now developing novel, targeted agents, but as with any new agent, these need validating in a clinical setting.

In the UK, there is an adaptive molecularly stratified trial programme currently underway, called FOCUS4. This has been designed to allow phase II and III testing of new targeted agents in biomarker-selected cohorts. The multi-arm, multistage (MAMS) design is a totally new concept in clinical trial design and allows the assessment of new agents at predefined time points, to ensure safety and efficacy. The molecular characterisation of colorectal cancer is rapidly evolving, and FOCUS4 provides the ideal clinical scenario to keep up with these developments, by enriching patient cohorts for response to new agents. This trial programme is expected to run for 5 years, and it is hoped that sufficient data will be generated in this time to potentially identify drug treatments and predictive biomarkers that can be rolled out worldwide.

The majority of predictive biomarkers identified in solid tumours tend to predict a lack of response to a particular treatment, with the presence of *RAS* mutations being a perfect example

Table 38.2 Potential targeted therapies

DNA/RNA protein biomarkers	Function/pathogenic process	Patient selection: Method and sample type	Successfully completed clinical trials and number of patients enrolled	Clinical use and limitations	Approved drug(s): Generic or trade name	Company
Amphiregulin (AREG)	Ligands of EGFR	High mRNA expression as determined by RT-PCR	n/a	Potential treatment of patients with high expression of AREG/EREG	POTENTIALLY anti-EGFR monoclonal antibodies as above	Merck and Amgen
Epiregulin (EREG)						
mRNA expression						
PIK3CA	Involved in signalling pathways which result in cell survival and proliferation	Activating mutation present in either exon 9 or 20	FOCUS4 ongoing	Potential treatment of mCRC patients with PIK3CA mutation in exons 9/20	Aspirin	Bayer
			Add-aspirin ongoing			
HER2	Receptor tyrosine kinase	IHC for HER2 and/or FISH	HERACLES (914 patients)	Benefit in KRAS-WT, HER2-positive mCRC patients	POTENTIALLY Herceptin	

of one such biomarker, in relation to response to anti-EGFR therapy. It is encouraging to now see that positive predictive biomarkers, such as high expression of AREG and EREG or PD-L1, are being sought. There is a real need to move forward from preclinical data to robust data generated in the context of a randomised clinical trial, so that more of these markers make their way into clinical practice.

References

1. Fearon ER, Vogelstein B. A genetic model for colorectal tumorigenesis. Cell. 1990;61(5):759–67.
2. Guinney J, Dienstmann R, Wang X, de Reynies A, Schlicker A, Soneson C, et al. The consensus molecular subtypes of colorectal cancer. Nat Med. 2015;21(11):1350–6.
3. Dienstmann R, Vermeulen L, Guinney J, Kopetz S, Tejpar S, Tabernero J. Consensus molecular subtypes and the evolution of precision medicine in colorectal cancer. Nat Rev Cancer. 2017;17(2):79–92.
4. Dunne PD, Alderdice M, O'Reilly PG, Roddy AC, McCorry AMB, Richman S, et al. Cancer-cell intrinsic gene expression signatures overcome intratumoural heterogeneity bias in colorectal cancer patient classification. Nat Commun. 2017;8:15657.
5. Vaughn CP, Zobell SD, Furtado LV, Baker CL, Samowitz WS. Frequency of KRAS, BRAF, and NRAS mutations in colorectal cancer. Genes Chromosom Cancer. 2011;50(5):307–12.
6. Douillard JY, Oliner KS, Siena S, Tabernero J, Burkes R, Barugel M, et al. Panitumumab-FOLFOX4 treatment and RAS mutations in colorectal cancer. N Engl J Med. 2013;369(11):1023–34.
7. Van Cutsem E, Lenz HJ, Kohne CH, Heinemann V, Tejpar S, Melezinek I, et al. Fluorouracil, leucovorin, and irinotecan plus cetuximab treatment and RAS mutations in colorectal cancer. J Clin Oncol. 2015;33(7):692–700.
8. Hurwitz H, Fehrenbacher L, Novotny W, Cartwright T, Hainsworth J, Heim W, et al. Bevacizumab plus irinotecan, fluorouracil, and leucovorin for metastatic colorectal cancer. N Engl J Med. 2004;350(23):2335–42.
9. Van Cutsem E, Tabernero J, Lakomy R, Prenen H, Prausova J, Macarulla T, et al. Addition of aflibercept to fluorouracil, leucovorin, and irinotecan improves survival in a phase III randomized trial in patients with metastatic colorectal cancer previously treated with an oxaliplatin-based regimen. J Clin Oncol. 2012;30(28):3499–506.
10. Tabernero J, Takayuki Y, Cohn AL. Correction to Lancet Oncol 2015; 16: 499–508. Ramucirumab versus placebo in combination with second-line FOLFIRI in patients with metastatic colorectal carcinoma that progressed during or after first-line therapy with bevacizumab, oxaliplatin, and a fluoropyrimidine (RAISE): a randomised, double-blind, multicentre, phase 3 study. Lancet Oncol. 2015;16(6):e262.
11. Grothey A, Van Cutsem E, Sobrero A, Siena S, Falcone A, Ychou M, et al. Regorafenib monotherapy for previously treated metastatic colorectal cancer (CORRECT): an international, multicentre, randomised, placebo-controlled, phase 3 trial. Lancet. 2013;381(9863):303–12.
12. Le DT, Uram JN, Wang H, Bartlett BR, Kemberling H, Eyring AD, et al. PD-1 blockade in tumors with mismatch-repair deficiency. N Engl J Med. 2015;372(26):2509–20.
13. Hocking CM, Price TJ. Panitumumab in the management of patients with KRAS wild-type metastatic colorectal cancer. Ther Adv Gastroenterol. 2014;7(1):20–37.
14. Yazdi MH, Faramarzi MA, Nikfar S, Abdollahi M. A comprehensive review of clinical trials on EGFR inhibitors such as cetuximab and panitumumab as monotherapy and in combination for treatment of metastatic colorectal cancer. Avicenna J Med Biotechnol. 2015;7(4):134–44.
15. Ilic I, Jankovic S, Ilic M. Bevacizumab combined with chemotherapy improves survival for patients with metastatic colorectal cancer: evidence from meta analysis. PLoS One. 2016;11(8):e0161912.
16. Verdaguer H, Tabernero J, Macarulla T. Ramucirumab in metastatic colorectal cancer: evidence to date and place in therapy. Ther Adv Med Oncol. 2016;8(3):230–42.
17. Yoshino T, Komatsu Y, Yamada Y, Yamazaki K, Tsuji A, Ura T, et al. Randomized phase III trial of regorafenib in metastatic colorectal cancer: analysis of the CORRECT Japanese and non-Japanese subpopulations. Investig New Drugs. 2015;33(3):740–50.
18. Perkins SL, Cole SW. Ziv-aflibercept (Zaltrap) for the treatment of metastatic colorectal cancer. Ann Pharmacother. 2014;48(1):93–8.
19. Seligmann JF, Elliott F, Richman SD, Jacobs B, Hemmings G, Brown S, et al. Combined epiregulin and amphiregulin expression levels as a predictive biomarker for panitumumab therapy benefit or lack of benefit in patients with RAS wild-type advanced colorectal cancer. JAMA Oncol. 2016;2(5):633–42.
20. Liao X, Lochhead P, Nishihara R, Morikawa T, Kuchiba A, Yamauchi M, et al. Aspirin use, tumor PIK3CA mutation, and colorectal-cancer survival. N Engl J Med. 2012;367(17):1596–606.
21. Kaplan R, Maughan T, Crook A, Fisher D, Wilson R, Brown L, et al. Evaluating many treatments and biomarkers in oncology: a new design. J Clin Oncol. 2013;31(36):4562–8.
22. Sartore-Bianchi A, Trusolino L, Martino C, Bencardino K, Lonardi S, Bergamo F, et al. Dual-targeted therapy with trastuzumab and lapatinib in treatment-refractory, KRAS codon 12/13 wild-type, HER2-positive metastatic colorectal cancer (HERACLES): a proof-

of-concept, multicentre, open-label, phase 2 trial. Lancet Oncol. 2016;17(6):738–46.
23. Pereira L, Mariadason JM, Hannan RD, Dhillon AS. Implications of epithelial-mesenchymal plasticity for heterogeneity in colorectal cancer. Front Oncol. 2015;5:13.
24. De Sousa EMF, Wang X, Jansen M, Fessler E, Trinh A, de Rooij LP, et al. Poor-prognosis colon cancer is defined by a molecularly distinct subtype and develops from serrated precursor lesions. Nat Med. 2013;19(5):614–8.
25. Rad R, Cadinanos J, Rad L, Varela I, Strong A, Kriegl L, et al. A genetic progression model of Braf(V600E)-induced intestinal tumorigenesis reveals targets for therapeutic intervention. Cancer Cell. 2013;24(1):15–29.
26. Sinicrope FA, Shi Q, Smyrk TC, Thibodeau SN, Dienstmann R, Guinney J, et al. Molecular markers identify subtypes of stage III colon cancer associated with patient outcomes. Gastroenterology. 2015;148(1):88–99.
27. Allegra CJ, Rumble RB, Hamilton SR, Mangu PB, Roach N, Hantel A, et al. Extended RAS gene mutation testing in metastatic colorectal carcinoma to predict response to anti-epidermal growth factor receptor monoclonal antibody therapy: American Society of Clinical Oncology provisional clinical opinion update 2015. J Clin Oncol. 2016;34(2):179–85.

Predictive Markers and Targeted Therapies in Gastroesophageal Cancer (GEC)

39

Josef Rüschoff

Overview

Adenocarcinomas of the stomach and esophagus are the fifth and eighth most common cancers worldwide and comprise the second most cause of cancer-related death with the highest incidence in Asia (24.2/100,000) and lowest in Africa and North America (2.2–7.0/100,000). Within the last three decades, stomach cancer has steadily decreased in developed countries, but carcinomas at the gastroesophageal junction have, however, almost doubled [1]. At the time of diagnosis, most patients already suffer from advanced disease, and only about one third can be operated with a cure rate of less than 10%. The median survival of patients with advanced inoperable or metastatic disease is less than 1 year (Fig. 39.1a). Up to now, four targeted drugs have been approved for the treatment of advanced gastric and esophageal adenocarcinomas (GEC) (See Table 39.1).

Therapies Targeting Transmembrane Receptors

Several combination therapies have been used in the treatment of advanced GEC raising the median survival rate from about 3 to 4 months (best supportive care) to about 8–11 months. This changed in 2010 to nearly 16 months after the publication of phase III randomized controlled trial called ToGA (**T**rastuzumab f**o**r **G**astric C**a**ncer) wherein 584 advanced HER2-positive adenocarcinomas of stomach and gastroesophageal junction treated with trastuzumab significantly prolonged the overall survival as compared to chemotherapy alone [2] (Fig. 39.1a).

HER2 is the second of the four-member epidermal growth factor receptor family (ERBB/HER1-4) involved in the regulation of cell proliferation and apoptosis. In contrast to the other members, the HER2 receptor has no specific ligand and is activated by homo- or heterodimerization leading to phosphorylation at intracellular tyrosine kinase sites and thus signal transduction. Currently, three different drugs have been tested in HER2-positive GEC that specifically inhibit the HER2 receptor. Two are monoclonal antibodies – trastuzumab and pertuzumab directed against the extracellular domain of the HER2 receptor – and the third one is lapatinib, a small molecule that passes through the cell membrane and binds to the part of HER2 found inside the cell. So far only trastuzumab in combination with chemotherapy has shown some benefit and therefore been approved for first-line palliative therapy. Lapatinib has not demonstrated any benefit [3], and pertuzumab did not prolong overall survival in HER2+ metastatic gastric and gastroesophageal cancer [4]. Another trial using a chemotherapeutic

J. Rüschoff
Targos Molecular Pathology GmbH,
Kassel, Germany
e-mail: rueschoff@pn-nordhessen.de

S. Badve, G. L. Kumar (eds.), *Predictive Biomarkers in Oncology*,
https://doi.org/10.1007/978-3-319-95228-4_39

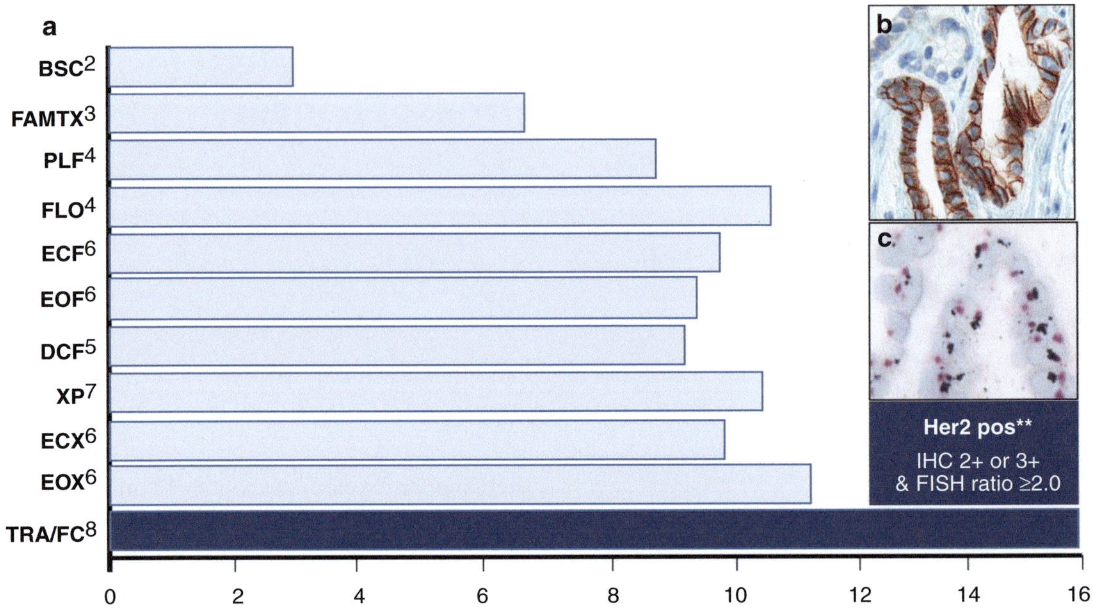

Fig. 39.1 (**a**) Median overall survival in months for various chemotherapies compared to trastuzumab in advanced GEC [2]; (**b**, **c**) HER2 overexpression (IHC 3+) and HER2 gene amplification (ratio ≥ 2.0) shown by immunohistochemistry (IHC) and chromogenic in situ hybridization (CISH). Abbreviations: BSC = best supportive care; FAMTX = 5-fluorouracil, doxorubicin, methotrexate; PLF = cisplatin, leucovorin, 5-fluorouracil; FLO = 5-fluorouracil, leucovorin, oxaliplatin; ECF = epirubicin, cisplatin, 5-fluorouracil; EOF = epirubicin, oxaliplatin, 5-fluorouracil; DCF = docetaxel, cisplatin, 5-fluorouracil; XP = capecitabin, cisplatin; ECX = epirubicin, cisplatin, capecitabin; EOX = epirubicin, oxaliplatin, capecitabin; FLOT = 5-fluorouracil, leucovorin, oxaliplatin, docetaxel; TRA/FC = trastuzumab, 5-fluorouracil, cisplatin

Table 39.1 Predictive biomarkers in gastroesophageal adenocarcinoma

Gene/protein	Function/pathogenic process	Patient selection: Method and sample type	Successfully completed clinical trials and number of patients enrolled	Clinical use and limitations	Approved drug(s): Generic/trade name	Company
HER2	Proliferation	E.g., HER2 IHC and ISH	E.g., ToGA 584 patients	First-line Advanced or metastatic GEC	Trastuzumab/herceptin	Roche
VEGFR2	Neovascularization	Not applicable	REGARD, RAINBOW	Second-line Advanced or metastatic GEC	Ramucirumab	Elli Lilly
MSI/MMRd	Mismatch repair	IHC or PCR	KEYNOTE-059	Advanced GEC	Pembolizumab	Merck/MSD
PD-L1	Immune checkpoint regulation	PD-L1 IHC	KEYNOTE-059		Pembrolizumab	Merck/MSD

compound ado-trastuzumab emtansine bound to trastuzumab (Kadcyla) was recently stopped due to lack of survival benefit [5].

Prerequisite of trastuzumab therapy in GEC is the demonstration of HER2 overexpression (Fig. 39.1b). Overall 22.1% of 3807 patients tested in the ToGA trial showed HER2 positivity defined by either HER2 overexpression (IHC 3+) or amplification (Ratio ≥2.0). Within the group of HER2-amplified cancers, only those with overexpression (at IHC2+ and IHC3+ level) showed a significant improvement of survival of 16.0 months as compared to 11.8 months in the chemotherapy alone arm (Fig. 39.1a). Thus the European Medicines Agency (EMA) was the first to approve trastuzumab for GEC with immunostaining as the primary choice for testing. Our group introduced the modified scoring system that was validated in the ToGA trial. This system has been adopted by EMA and FDA and has recently been recommended by ASCO/CAP as the *"Rüschoff/Hofmann method"* [6].

The major challenge of HER2 testing in GEC is false negative scoring which may affect up to 27% of cases, particularly in intestinal type carcinomas [7]. Besides heterogeneity, we regard intensity assessment of staining as one of the major pitfalls in HER2 evaluation of GEC. We, therefore, introduced a reliable and reproducible semiquantitative approach by taking the microscope magnification into account [8]. Membrane staining unequivocally visible at low magnification (2.5–5×) corresponds to IHC3+. However, cases that need higher magnification (10–20×) for the demonstration of specific HER2 membrane staining should be confirmed by ISH (equivocal, IHC2+). If a higher magnification (40×) is needed to demonstrate any staining, the case is scored negative for HER2 (IHC1+ or IHC 0).

Based on our current knowledge of molecular subtypes and signal transduction pathways in gastric cancer [9], a number of other targeted drugs have been investigated in GEC (Fig. 39.2).

So far only one drug (Ramucirumab (Cyramza®), directed against VEGFR2, has been approved in 2014 for second-line therapy in GEC. However, no biomarker has been established. The clinical benefit is limited to overall survival improvement from 3.8 months in the placebo to 5.2 months in the therapy arm (p = 0.0.0473, REGARD trial). In combination with paclitaxel (9.6 month) versus paclitaxel only (7.4 month), ramucirumab performed better, and in a third study, survival was prolonged to about 5.2 month in comparison with standard treatment (RAINBOW trial). Main indication is the HER2-negative GEC [10].

Studies are ongoing for antibodies directed against hepatocyte growth factor (HGF),

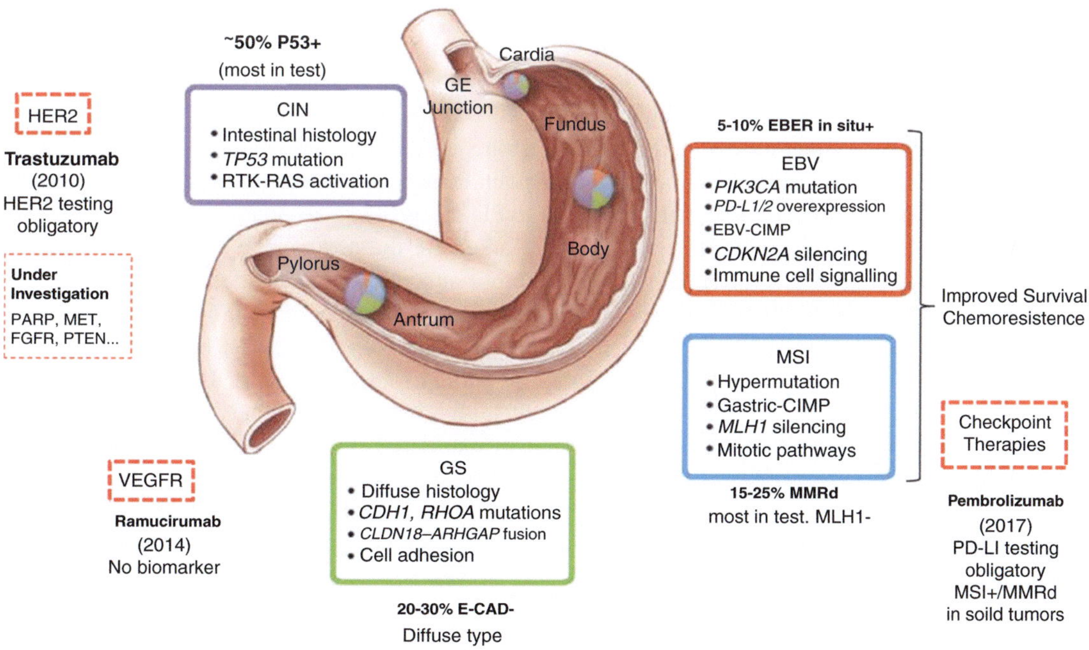

Fig. 39.2 Targets and targeted therapies in gastric cancer (acc. to [9, 17]). (Adapted from Cancer Genome Atlas Research Network [9]. With permission from Springer Nature)

e.g., rilotumumab and ficlatuzumab and for small MET kinase inhibitory molecules such as cabozantinib, crizotinib, and tivantinib. Patient selection is based on c-MET overexpression and/or amplification in HER2-negative GEC. However, the phase III trial using onartuzumab as a monoclonal antibody against c-MET has recently been stopped showing no benefit in the c-MET overexpression group [11].

Although overexpression of EGF receptor protein has been shown to be a negative prognostic marker [12], it is of no prognostic value [13]. In a number trials using drugs directed against EGFR (ErbB-1/HER1), no clinical benefit could be demonstrated either for antibodies (cetuximab, panitumumab) or small molecules (erlotinib, gefitinib, lapatinib) [10]. Studies targeting FGFR-2 are ongoing.

Therapies Targeting Signal Transduction Pathways

Other therapeutic interventions are aimed at signal transduction pathways such as PI3K-Akt-mTOR that regulates apoptosis and cell growth. Amplification or overexpression (PIK3CA, AKT1) or activating mutations (PIK3CA) or loss of PTEN is frequently observed in GEC. However, in a trial using everolimus as an mTOR inhibitor, the primary endpoint, improved overall survival, was not reached. Studies of ipatasertib, an AKT inhibitor, in GEC with loss of PTEN expression are still ongoing. Interventions targeting the RAS-RAF-MEK-ERK pathway are in early phase II trials currently without decisive data [10].

Therapies Using Checkpoint Inhibitors

A somewhat new approach in tumor therapy has gained much interest during the last few years that involves therapeutic suppression of inhibitory signaling pathways involved in T-cell homeostasis and autoimmunity. It turned out that tumor cells protect themselves against cytotoxic T-cell attack, e.g., by expressing the programmed cell death ligand 1 (PD-L1) which blocks the programmed cell death receptor (PD1) on CD8-positive cytotoxic T cells. Different therapeutic antibodies directed either against the PD1 receptor (pembrolizumab, nivolumab) or the PD-L1 ligand (atezolizumab) have been developed. Inhibition of the PD1/PD-L1 checkpoint leads to reactivation of the tumor suppressing the immune response and thus to tumor regression. First study data for pembrolizumab (Keytruda®) are encouraging for PD-L1-positive advanced GEC and has led to the FDA approval of this drug just recently [14]. In phase III trial by nivolumab, endpoint (overall survival) was met in advanced GEC irrespective of PD-L1 expression [15].

According to data from Lynch syndrome-related metastatic cancer, microsatellite instability (MSI) is another marker predicting response to checkpoint inhibitors. Accordingly, the overall response rate to pembrolizumab in MSI-high advanced gastric cancer was 57.1% versus 9.0% to stable microsatellite tumors [16]. Thereby loss of MLH1 seems to be an easy to assess biomarker for this new promising kind of targeted therapy in GEC [17].

In summary (Table 39.1), since the introduction of the first biomarker-based targeted therapy, i.e., trastuzumab (Herceptin®) in GEC, a new group of therapeutics, namely, immune checkpoint treatments, have been approved. In addition to PD-L1 as the corresponding target-related biomarker, for the first time, a molecular test that is tumor agnostic (i.e., whose tumors are microsatellite instability-high “MSI-H” or mismatch repair deficiency “MMRd”) has been approved for patient selection. For MSI analysis two types of studies have been validated: first PCR by using a primer panel that was originally introduced and validated by our group and later adopted by NCI [18] and second by immunohistochemistry demonstrating loss of mismatch repair gene expression (MMRd). More promising data is expected to come from ongoing clinical trials that may be of predictive value for diagnosing and treating gastric cancer.

References

1. World Cancer Research Fund International. Stomach cancer statistics. Retrieved from: http://www.wcrf.org/int/cancer-facts-figures/data-specific-cancers/stomach-cancer-statistics.
2. Bang YJ, Van Cutsem E, Feyereislova A, et al. Trastuzumab in combination with chemotherapy versus chemotherapy alone for treatment of HER2-positive advanced gastric or gastro-oesophageal junction cancer (ToGA): a phase 3, open-label, randomised controlled trial. Lancet. 2010;376:687–97.
3. Lorenzen S, Riera Knorrenschild J, Haag GM, et al. Lapatinib versus lapatinib plus capecitabine as second-line treatment in human epidermal growth factor receptor 2-amplified metastatic gastro-oesophageal cancer: a randomised phase II trial of the Arbeitsgemeinschaft Internistische Onkologie. Eur J Cancer. 2015;51:569–76.
4. Tabernero J, Hoff P, Shen L, et al. Pertuzumab (P) + trastuzumab (H) + chemotherapy (CT) for HER2-positive metastatic gastric or gastro-oesophageal junction cancer (mGC/GEJC): final analysis of a phase III study (JACOB). Ann Oncol. 2017;28(suppl_5):v209–68.
5. Thuss-Patience PC, Shah MA, Ohtsu A, et al. Trastuzumab emtansine versus taxane use for previously treated HER2-positive locally advanced or metastatic gastric or gastro-oesophageal junction adenocarcinoma (GATSBY): an international randomised, open-label, adaptive, phase 2/3 study. Lancet Oncol. 2017;18:640–53.
6. Bartley AN, Washington MK, Colasacco C, et al. HER2 testing and clinical decision making in gastroesophageal adenocarcinoma: guideline from the College of American Pathologists, American Society for Clinical Pathology, and the American Society of Clinical Oncology. J Clin Oncol. 2017;35:446–64.
7. Doucet L, Terris B, Chenard MP, et al. Concordance of HER2 status between local and Central Review in Gastric (GC) and gastroesophageal Junction Cancers (GEJC): a French observational study of 394 specimens: HERable study. Mod Pathol. 2016;29:170A. (Abstract 671).
8. Rüschoff J, Dietel M, Baretton G, et al. HER2 diagnostics in gastric cancer-guideline validation and development of standardized immunohistochemical testing. Virchows Arch. 2010;457:299–307.
9. Cancer Genome Atlas Research Network. Comprehensive molecular characterization of gastric adenocarcinoma. Nature. 2014;513:202–9.
10. Cetin B, Gumusay O, Cengiz M, Ozet A. Advances of molecular targeted therapy in gastric cancer. J Gastrointest Cancer. 2016;47:125–34.
11. Shah MA, Cho JY, Tan IB, et al. A randomized phase II study of FOLFOX with or without the MET inhibitor onartuzumab in advanced adenocarcinoma of the stomach and gastroesophageal junction. Oncologist. 2016;21:1085–90.
12. Marx AH, Zielinski M, Kowitz CM, et al. Homogeneous EGFR amplification defines a subset of aggressive Barrett's adenocarcinomas with poor prognosis. Histopathology. 2010;57:418–26.
13. Han SW, Oh DY, Im SA, et al. Phase II study and biomarker analysis of cetuximab combined with modified FOLFOX6 in advanced gastric cancer. Br J Cancer. 2009;100:298–304.
14. Fuchs CS, Doi T, Woo-Jun Jang R, et al. KEYNOTE-059 cohort 1: efficacy and safety of pembrolizumab (pembro) monotherapy in patients with previously treated advanced gastric cancer. JCO. 2017;35(suppl):4003. (Abstract 4003).
15. Kang YK, Boku N, Satoh T, et al. Nivolumab in patients with advanced gastric or gastro-oesophageal junction cancer refractory to, or intolerant of, at least two previous chemotherapy regimens (ONO-4538-12, ATTRACTION-2): a randomised, double-blind, placebo-controlled, phase 3 trial. Lancet. 2017;390:2461–71.
16. Le DT, Uram JN, Wang H, et al. PD-1 blockade in tumors with mismatch-repair deficiency. N Engl J Med. 2015;372:2509–20.
17. Setia N, Agoston AT, Han HS, et al. A protein and mRNA expression-based classification of gastric cancer. Mod Pathol. 2016;29:772–84.
18. Umar A, Boland CR, Terdiman JP, et al. Revised Bethesda Guidelines for hereditary nonpolyposis colorectal cancer (Lynch syndrome) and microsatellite instability. J Natl Cancer Inst. 2004;96:261–8.

Predictive Biomarkers and Targeted Therapies in Hepatic, Pancreatic, and Biliary Cancers

40

Steven Alexander Mann and Romil Saxena

Overview

Hepatocellular carcinoma is among the top five most common malignancies in the world and is the second most common cause of cancer-related deaths. HCC in developed countries most commonly develops in the setting of hepatic cirrhosis due to alcohol use, hepatitis C virus infection, or other causes of chronic inflammation in the liver [1]. Pancreatic cancer has become increasingly more common in the past decade [2] and is currently the fourth most common cause of cancer-related death in the United States [3]. According to the American Cancer Society, the risk factors for the development of pancreatic cancer include obesity, smoking, and chronic pancreatitis. Cholangiocarcinoma is a relatively uncommon malignancy in the western world; however, it is a top ten cause of cancer-related mortality in some Asian countries where liver fluke infection is more common [3, 4]. Other proposed risk factors for cholangiocarcinoma include smoking, primary sclerosing cholangitis, and chronic liver disease.

Hepatic and pancreatic and biliary cancers often present at advanced stages when systemic therapies are usually indicated. Despite the relatively poor prognoses of these malignancies, currently, there are very few US Food and Drug Administration (FDA)-approved therapeutic options. This chapter focuses on potential predictive biomarkers for these current therapies.

S. A. Mann · R. Saxena (✉)
Department of Pathology and Laboratory Medicine, Indiana University School of Medicine, Indianapolis, IN, USA
e-mail: rsaxena@iupui.edu

Hepatocellular Carcinoma

Partial hepatectomy and orthotopic liver transplantation are the treatments of choice for liver-confined cases of HCC. Locoregional therapies such as ablation and chemoembolization are also available. For unresectable or metastatic HCC, conventional chemotherapy has not shown an objective benefit, and the multikinase inhibitor sorafenib is currently the only FDA-approved first-line therapy. Though the benefits of sorafenib in advanced HCC have been shown to be statistically significant, the difference in overall survival compared to placebo is only a few months [5]. This limitation, coupled with common side effects of this drug, has left a need for alternative treatment options as well as methods to predict which patients are most likely to respond to therapy.

In a Phase III trial of sorafenib for advanced HCC, plasma biomarkers related to the mechanism of action of sorafenib were evaluated for their use as predictors of therapeutic response [5]. Soluble c-KIT and hepatocyte growth factor (HGF) showed a trend towards predicting a

S. Badve, G. L. Kumar (eds.), *Predictive Biomarkers in Oncology*,
https://doi.org/10.1007/978-3-319-95228-4_40

survival benefit with sorafenib, but this was not statistically significant. Other targets of sorafenib, including VEGF, were not associated with therapeutic response.

Though alpha-fetoprotein (AFP) is a relatively insensitive and nonspecific marker for HCC, changes in serum levels may correlate with postsurgical and therapeutic reductions in tumor burden. As a predictive marker, there is evidence that AFP may be useful in predicting responses to therapy. In a Phase III trial of ramucirumab, a VEGFR-2 inhibitor, for advanced HCC after sorafenib failure, the overall survival was similar for the treatment and placebo arms. However, there was a significant improvement in overall survival for the group of patients with elevated AFP who received ramucirumab [6].

Glypican-3 (GPC3) is expressed in a variety of tumors and is commonly expressed in HCC, conferring a poor prognosis. Because this antigen is absent in normal hepatocytes, GPC3 represents a potential therapeutic target. Sawada et al. [7] demonstrated in a Phase II trial of postsurgery GPC3 peptide vaccines that GPC3-positive tumors have lower recurrence rates compared to GPC3-negative tumors. Though the data was inconclusive, it was suggested that the frequency of GPC3-specific cytotoxic T cells measured by an interferon-gamma assay after vaccination could predict an increase in overall survival.

The cellular mesenchymal-epithelial transcription factor (c-MET) receptor tyrosine kinase inhibitor tivantinib has shown benefits in a Phase II trial of patients with advanced HCC who have previously received sorafenib. When patients were subcategorized according to tumor c-MET expression, high c-MET cases had significantly improved overall survival on tivantinib. No difference was found between low c-MET cases on tivantinib and the placebo treatment group [8].

Since the above example roles of predictive markers were not associated with FDA-approved therapy, additional studies are necessary before recommendations outside of clinical trials can be made for management of HCC. Fortunately, research efforts across the medical community continue to exponentially increase our understanding of these and many other new potential biomarkers. By utilizing next-generation sequencing and other powerful molecular testing techniques, a wealth of information has been gathered that could lead to clinically significant breakthrough therapies and predictive markers for HCC [1, 9, 10].

Pancreatic Ductal Adenocarcinoma

Pancreatic ductal adenocarcinoma carries a dismal prognosis with the median survival for the lowest stage tumors after resection being approximately 2 years. Adjuvant chemotherapy with either gemcitabine or 5-fluorouracil is recommended. For advanced pancreatic cancer, gemcitabine-based combination chemotherapy or FOLFIRINOX is used. The epidermal growth factor receptor (EGFR) inhibitor erlotinib is approved for first-line therapy in combination with gemcitabine in unresectable pancreatic adenocarcinoma. However, there are currently no large trials that stratify patients by EGFR amplification status or according to downstream Ras protein mutations. Therefore, the use of EGFR testing as a predictive biomarker of therapeutic response to this inhibitor is uncertain.

Though it is not sensitive or specific enough to be recommended for screening purposes, serum CA 19-9 measurement is a part of the initial assessment of biopsy-proven pancreatic cancer. Elevated levels in the absence of jaundice are an indication for neoadjuvant chemotherapy in potentially resectable cases. Changes in CA 19-9 can correlate with therapeutic response and are used in combination with imaging for follow-up surveillance [2]. A study of adjuvant chemotherapy for pancreatic cancer reported that cases with low postoperative CA 19-9 had significantly improved survival with adjuvant chemotherapy. Survival in cases with elevated postoperative CA 19-9 was similar regardless of whether adjuvant therapy was received [11]. An important consideration for CA 19-9 testing is that patients with Lewis antigen-negative blood types are unable to produce CA 19-9.

Human equilibrative nucleoside transporter 1 (hENT1) has been identified as the major mechanism of gemcitabine entry into tumor cells. High expression of hENT1 has been shown to predict significantly improved overall survival in both resected and unresectable cases of pancreatic cancer after gemcitabine therapy [12, 13]. Cases with low hENT1 expression may be candidates for non-gemcitabine-based first-line chemotherapy options. A standardized method of hENT expression testing is still needed as mixed results have been reported based on immunohistochemistry [14].

SPARC has potential prognostic value in many types of malignancy. There has been some evidence that SPARC expression can predict therapeutic responses in pancreatic cancer. However, the results of studies in pancreatic cancer have been mixed, depending on plasma levels and tumor cell or stromal expression of SPARC. The overall trend is that high stromal SPARC expression correlates with poorer responses to gemcitabine and nanoparticle albumin-bound paclitaxel [12, 15]. Two Phase II trials for pancreatic cancer that include SPARC measurement are currently recruiting patients.

RRM1 is part of a larger human enzyme that is involved in producing the deoxyribonucleotide triphosphates necessary for synthesizing DNA. A small number of studies have reported that high levels of RRM1 predict decreased benefit from gemcitabine in pancreatobiliary cancers [4]. However, larger studies are needed to confirm RRM1 as a predictive marker in this setting.

Poly ADP-ribose polymerase (PARP) is an intracellular protein that is essential to numerous cellular functions. PARP inhibitors have been proposed to have a selective toxicity toward cell lines with deficiencies in DNA repair. The high mutational burden observed in pancreatic ductal carcinomas, as well as the occurrence of BRCA1 and BRCA2 mutations in these tumors, offers another potential mechanism for therapy. Ongoing trials of pancreatic tumor patients receiving combination chemotherapy plus PARP inhibitors may yield further information in the future [16].

Pancreatic Neuroendocrine Tumors

Neuroendocrine tumors represent about 1% of all pancreatic tumors. The work-up, management, and prognosis of neuroendocrine tumors of the pancreas differ significantly from that of pancreatic ductal adenocarcinoma. There is no consensus on the use of adjuvant systemic therapy for pancreatic neuroendocrine tumors, and cases are often curable with surgery alone. In unresectable and metastatic cases, the optimum choice for cytotoxic chemotherapy is unknown. Conventional somatostatin analogues may be used in symptomatic cases of functional, hormone-secreting tumors.

Both everolimus and sunitinib have been approved for unresectable and metastatic pancreatic neuroendocrine tumors. Neither drug indication incorporates molecular testing as a prerequisite for treatment. There is no predictive biomarker currently available for mammalian target of rapamycin (mTOR) inhibitor everolimus [17]. Sunitinib is a drug that inhibits multiple receptor tyrosine kinases, preventing many downstream mechanisms of oncogenesis including angiogenesis. None of the described targets of sunitinib have been shown in a large clinical trial to have a predictive role thus far.

Cholangiocarcinoma

Though cholangiocarcinoma is relatively uncommon compared to HCC and pancreatic cancer, the prognosis is similarly dismal with many cases presenting at advanced stages. Surgical resection when possible is the primary treatment. Data from clinical trials on cholangiocarcinoma is limited, but either fluoropyrimidine-based or gemcitabine-based chemotherapy is currently used in adjuvant or unresectable settings. As such, the evaluation of other therapeutic options and their predictive biomarkers are at preliminary stages.

Regarding hENT1, results similar to those seen in pancreatic cancer are reported in advanced cholangiocarcinoma cases treated with adjuvant gemcitabine. The high hENT1 expression is correlated with significantly improved survival in patients who receive adjuvant gemcitabine. Worse outcomes are reported in cases with low hENT1 expression, with similar overall survival in this group regardless of whether gemcitabine was received [4]. Similar to pancreatic cancers, cholangiocarcinomas with low hENT1 expression may benefit more from current non-gemcitabine therapeutic options. However, additional clinical trials are necessary for confirmation.

Numerous common mutations have been identified in cholangiocarcinoma cases, including KRAS mutations. The presence of these mutations may become significant decision points for future therapies such as EGFR inhibitors. Furthermore, a minority of cholangiocarcinoma expresses HER2, the target of trastuzumab in breast and gastric cancers. Additionally, miRNAs have shown promise for future roles in as both predictive markers of response and as potential therapeutic targets for hepatic and pancreatobiliary tumors. Just as with HCC, next-generation sequencing and other molecular methods are providing valuable information about potential predictive biomarkers for future use in the management of pancreatic and biliary cancers [12] (Table 40.1).

Signaling Pathways

There are multiple types of receptor tyrosine kinases and other receptor types implicated in the oncogenesis and progression of HCC (Fig. 40.1). Many of these kinases are inhibited by sorafenib [5]. Another receptor tyrosine kinase, c-MET, is inhibited by the novel drug tivantinib. The main ligand of c-MET is HGF [8]. GPC3 performs several roles in the promotion of HCC. When bound to tumor cell surfaces, GPC3 stabilizes growth factors such as HGF and Wnt, allowing for improved interaction with their respective receptors. The downstream effects of the activation of these receptors lead to angiogenesis, tumor cell proliferation, invasion, and metastasis. Immunotherapeuctic modalities under development exploit the specificity of GPC3 for tumor cells in HCC by inducing GPC3-specific cytotoxic T cells against these tumors [7].

Most of the current potential predictive biomarkers of therapeutic response for pancreatobiliary cancers involve gemcitabine, the major component of most approved chemotherapy regimens for these malignancies. hENT1 is the transporter responsible for getting gemcitabine inside of tumor cells where the drug can inhibit deoxyribonucleic acid (DNA) synthesis. Specifically for pancreatic cancer, stromal secreted protein acidic and rich in cysteine (SPARC) is thought to prohibit gemcitabine and perhaps nanoparticle albumin-bound paclitaxel from entering tumor cells [15]. Inside the cell, gemcitabine must become phosphylated by deoxycytidine kinase before the drug can disrupt the function of DNA synthetizing enzymes [12, 13]. One hypothesis is that increased amounts of ribonucleotide reductase subunit M1 (RRM1) contributes to gemcitabine resistance by competitive inhibition with increased deoxyribonucleotide triphosphate production [4] (Fig. 40.2).

Guidelines

Currently, there are no ASCO guidelines for hepatic, biliary, or non-exocrine pancreatic tumors. The most recent guidelines released for pancreatic cancer recommend the use of CA 19-9 in combination with other clinical, laboratory, and radiologic information for the initial work-up, postoperative, adjuvant therapy, and surveillance phases of disease management [2]. In general, additional clinical trials are necessary before consensus statements can be made regarding the use of predictive biomarkers in the treatment of hepatic, biliary, and pancreatic cancers.

Table 40.1 Potential predictive biomarkers in hepatocellular and pancreatobiliary cancers

Biomarker	Relevant cancer type	Biomarker function/process	Sample type: evaluation method	Highest level study, number of patients	Predictive use	Limitations	Approved drug (generic/trade name) company
AFP	HCC	Nonspecific antigen secretion	Serum: various assays	Phase III trial (REACH), 565 patients	Increased AFP levels may predict better responses to VEGFR inhibitors	More data from clinical trials are needed	Ramucirumab (Cyramza®)
				ClinicalTrials.gov identifier: NCT01140347			Eli Lilly
GPC3	HCC	Nonspecific antigen expression	Tissue: GPC3 IHC	Phase II trial, 41 patients	GPC3-positive tumors are potential candidates for GPC3-specific immunotherapy	More data from clinical trials are needed	GPC3 peptide vaccine
			Blood: IFN-gamma ELISPOT assay	Sawada Y, Yoshikawa T, Ofuji K, et al. Phase II study of the GPC3-derived peptide vaccine as an adjuvant therapy for hepatocellular carcinoma patients. *Oncoimmunology*. 2016;5(5):e1129483			American Peptide Company, Sunnyvale, CA
c-MET	HCC	Receptor tyrosine kinase with downstream oncogenic processes	Tissue: IHC, mRNA analyses	2 Phase III trials, 383 and 386 patients, respectively	High c-MET tumors are potential candidates for c-MET-targeted therapy	No standardized method of marker quantification	Tivantinib. Small molecule inhibitor of c-met
				ClinicalTrials.gov identifiers: NCT01755767 NCT02029157		More data from clinical trials are needed	ClinicalTrials.gov identifier: NCT02150733 Daiichi Sankyo, Inc.
CA 19-9	PDAC	Nonspecific antigen secretion	Serum: various assays	Retrospective cohort study, 260 patients	Low CA 19-9 levels may predict better responses to adjuvant chemotherapy	Jaundice	Adjuvant chemotherapy
				Humphris JL, Chang DK, Johns AL, et al. The prognostic and predictive value of serum CA19.9 in pancreatic cancer. *Ann Oncol*. 2012;23(7):1713–22		Lewis antigen negative,	
						More data from clinical trials are needed	
hENT1	PDAC, CCA	Gemcitabine transport across cell membrane	Tissue: IHC, RT-PCR	Phase III trial (ESPAC-3), 380 pancreatic cancer patients	High hENT1 tumor expression may predict better responses to gemcitabine	No standardized method of marker quantification	Gemzar (gemcitabine)
				The Australasian Gastro-Intestinal Trials Group (AGITG) https://gicancer.org.au/about-us/		More data from clinical trials are needed	

AFP alpha-fetoprotein, *CCA* cholangiocarcinoma, *c-MET* cellular mesenchymal-epithelial transcription factor, *ELISPOT* enzyme-linked immunospot, *GPC3* glypican-3, *HCC* hepatocellular carcinoma, *hENT1* human equilibrative nucleoside transporter 1, *IHC* immunohistochemistry, *IFN* interferon, *PDAC* pancreatic ductal adenocarcinoma, *RT-PCR* reverse transcription polymerase chain reaction

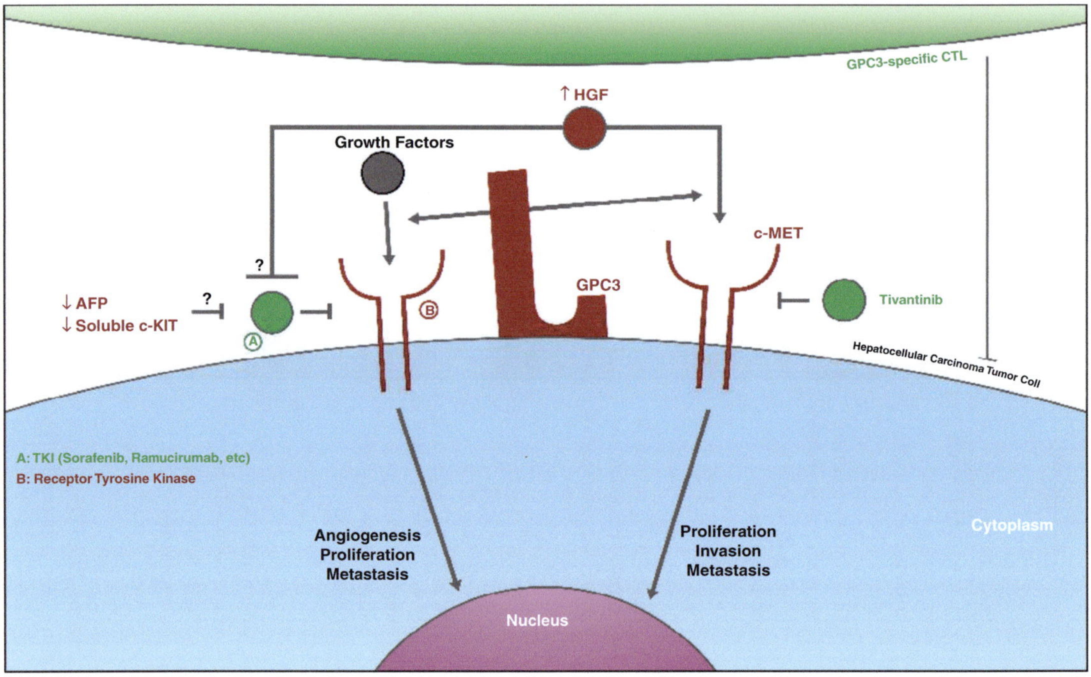

Fig. 40.1 A summary of the signaling pathways relevant to potential predictive markers of hepatocellular carcinoma therapy. Green indicates a therapy. Red indicates a potential predictive biomarker. Lines with arrowheads indicate stimulatory effects. Lines with perpendicular heads indicate inhibitory effects. AFP alpha-fetoprotein, C-MET cellular mesenchymal-epithelial transcription factor, CTL cytotoxic T cell, GPC3 glypican-3, HGF hepatocyte growth factor, TKI tyrosine kinase inhibitor

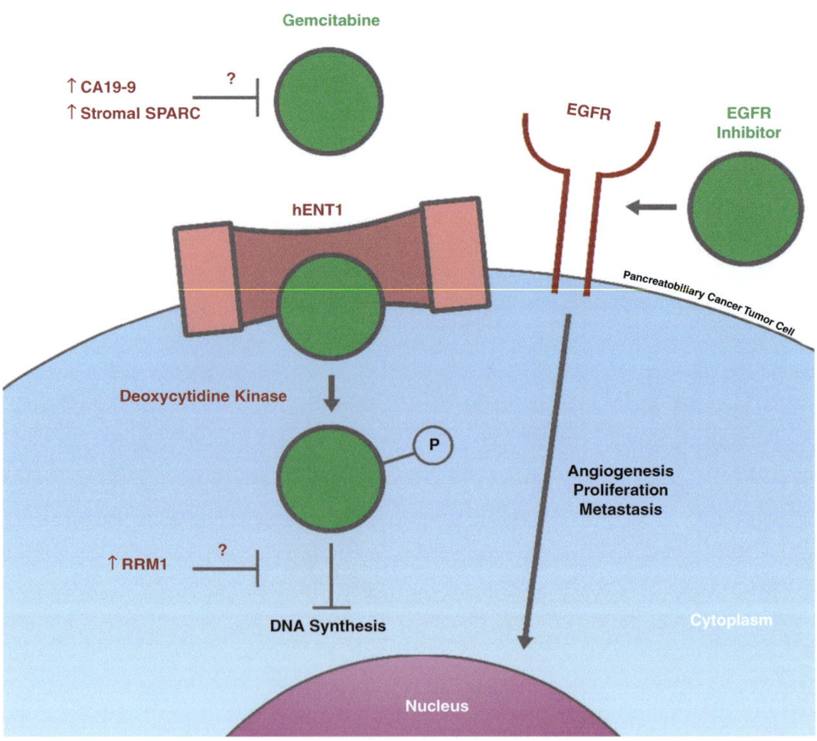

Fig. 40.2 A summary of the signaling pathways relevant to potential markers predictive of pancreatobiliary cancer therapy. Green indicates a therapy. Red indicates a potential predictive biomarker. Lines with arrowheads indicate stimulatory effects. Lines with perpendicular heads indicate inhibitory effects. DNA deoxyribonucleic acid, EGFR epidermal growth factor receptor, hENT1 human equilibrative nucleoside transporter 1, RRM1 ribonucleotide reductase subunit M1, SPARC secreted protein acidic and rich in cysteine

Summary and Future Directions

There are several ongoing clinical trials designed to confirm and expand the roles of predictive biomarkers for therapy in these cancers. GPC3 vaccine trials in patients with GPC3-positive HCC patients are pending [7]. There are two ongoing Phase III trials of c-MET inhibitors in advanced stage HCC expressing high c-MET. A Phase III trial of ramucirumab in advanced HCC with elevated AFP is currently recruiting participants.

For pancreatic cancer, a trial of adjuvant gemcitabine versus 5-fluorouracil based on tumor hENT1 immunohistochemistry status is currently recruiting participants. Standardized methods of grading hENT1 expression in pancreatobiliary tumors are needed. Also, additional markers such as deoxycytidine kinase, which phosphylates gemcitabine inside tumor cells, may add further predictive value to hENT1 [12]. The significance of EGFR amplification status in pancreatic cancer patients receiving inhibitor therapy is unknown. The future results of ongoing trails may shed light on other potential predictive markers in pancreatic cancer such as DNA repair function (for PARP inhibitors) and SPARC and RRM1 expression (for gemcitabine response). Finally, more information about the predictive response of hENT1 expression for gemcitabine therapy in both pancreatic cancer and cholangiocarcinoma is warranted.

Recent FDA approvals for immune checkpoint inhibitors have made these revolutionary therapies available for use against hepatocellular and pancreatobiliary malignancies. Pembrolizumab (Keytruda®), a PD-1 inhibitor, is approved for soft tissue tumors with no other viable alternative therapy if they are mismatch repair deficient or have high microsatellite instability. Nivolumab (Opdivo®), another PD-1 inhibitor, is approved for HCC after sorafineb [18]. At this point, positivity for the ligand to PD-1 (PD-L1) is a predictive biomarker only for use in pembrolizumab therapy for non-small cell lung cancer. Given the ongoing success of immune checkpoint inhibitors in clinical trials, generating numerous breakthrough therapy designations over the past 3 years, it is likely that predictive biomarker use in oncology will continue to evolve rapidly.

The most common hepatic, pancreatic, and biliary cancers, HCC, pancreatic ductal adenocarcinoma, and cholangiocarcinoma, respectively, collectively represent a significant portion of worldwide cancer mortality. Each may present aggressively with most cancers being diagnosed at advanced stages. Limited and controversial screening and diagnostic methods no doubt contribute to the overall poor prognosis of these malignancies. Despite the fact that there are currently only a few FDA-approved therapies for hepatic and pancreatobiliary cancers, there is building evidence for new treatment options and predictive biomarkers to support them. The incorporation of recent advances in molecular analysis and immunotherapy into clinical trials promises new hope for these dismal diagnoses.

References

1. Tsuchiya N, et al. Biomarkers for the early diagnosis of hepatocellular carcinoma. World J Gastroenterol. 2015;21(37):10573–83.
2. Khorana AA, et al. Potentially curable pancreatic cancer: American Society of Clinical Oncology clinical practice guideline. J Clin Oncol. 2016;34(21):2541–56.
3. Siegel RL, Miller KD, Jemal A. Cancer statistics, 2018. CA Cancer J Clin. 2018;68(1):7–30.
4. Sasaki H, et al. Concurrent analysis of human equilibrative nucleoside transporter 1 and ribonucleotide reductase subunit 1 expression increases predictive value for prognosis in cholangiocarcinoma patients treated with adjuvant gemcitabine-based chemotherapy. Br J Cancer. 2014;111(7):1275–84.
5. Llovet JM, et al. Plasma biomarkers as predictors of outcome in patients with advanced hepatocellular carcinoma. Clin Cancer Res. 2012;18(8):2290–300.
6. Trojan J, Waidmann O. Role of regorafenib as second-line therapy and landscape of investigational treatment options in advanced hepatocellular carcinoma. J Hepatocell Carcinoma. 2016;3:31–6.
7. Sawada Y, et al. Phase II study of the GPC3-derived peptide vaccine as an adjuvant therapy for hepatocellular carcinoma patients. Oncoimmunology. 2016;5(5):e1129483.
8. Pievsky D, Pyrsopoulos N. Profile of tivantinib and its potential in the treatment of hepatocellular carcinoma: the evidence to date. J Hepatocell Carcinoma. 2016;3:69–76.

9. Scaggiante B, et al. Novel hepatocellular carcinoma molecules with prognostic and therapeutic potentials. World J Gastroenterol. 2014;20(5):1268–88.
10. Chauhan R, Lahiri N. Tissue- and serum-associated biomarkers of hepatocellular carcinoma. Biomark Cancer. 2016;8(Suppl 1):37–55.
11. Humphris JL, et al. The prognostic and predictive value of serum CA19. 9 in pancreatic cancer. Ann Oncol. 2012;23(7):1713–22.
12. Viterbo D, Gausman V, Gonda T. Diagnostic and therapeutic biomarkers in pancreaticobiliary malignancy. World J Gastrointest Endosc. 2016;8(3):128–42.
13. Nordh S, Ansari D, Andersson R. hENT1 expression is predictive of gemcitabine outcome in pancreatic cancer: a systematic review. World J Gastroenterol. 2014;20(26):8482–90.
14. Sinn M, et al. Human equilibrative nucleoside transporter 1 expression analysed by the clone SP 120 rabbit antibody is not predictive in patients with pancreatic cancer treated with adjuvant gemcitabine – results from the CONKO-001 trial. Eur J Cancer. 2015;51(12):1546–54.
15. Han W, et al. Prognostic value of SPARC in patients with pancreatic cancer: a systematic review and meta-analysis. PLoS One. 2016;11(1):e0145803.
16. Steffen JD, et al. Structural implications for selective targeting of PARPs. Front Oncol. 2013;3:301.
17. Chan DL, Segelov E, Singh S. Everolimus in the management of metastatic neuroendocrine tumours. Ther Adv Gastroenterol. 2017;10(1):132–41.
18. El-Khoueiry AB, et al. Nivolumab in patients with advanced hepatocellular carcinoma (CheckMate 040): an open-label, non-comparative, phase 1/2 dose escalation and expansion trial. Lancet. 2017;389(10088):2492–502.

41 Predictive Biomarkers and Targeted Therapies in Gynecological Cancers

Louise De Brot and Fernando Augusto Soares

Overview

Gynecologic cancers form a heterogeneous group of tumors, affecting the tissue and organs of the female reproductive system, and include ovarian, cervical, endometrial, vaginal, and vulvar cancers. They are a very common malignancy, with about 105,890 new diagnoses and 30,890 deaths, per year in the United States. In underdeveloped nations data regarding incidence and prevalence is scarcer, but undoubtedly it remains as an important health issue with a great necessity of advances in pathologic comprehension, development of biomarkers, and improvements in diagnosis and therapy [1].

The most common malignant gynecologic tumor is ovarian cancer, with around 239,000 new cases in women and over 152,000 deaths yearly. Incidence and mortality numbers demonstrate the severity of the disease, making ovarian cancer the most lethal gynecologic neoplasm [1, 2]. The current standard of care is based on optimal surgery (cytoreduction) and platinum-based chemotherapy. Important advances in genomic tumor assessment and targeted therapy with new drugs such as pazopanib (an angiogenesis inhibitor) and olaparib (a PARP inhibitor) have shown some promising results. Despite the improvement in the biological understanding of the disease, recurrences are inevitable with poor survival rates and a dismal prognosis in the advanced stages of disease [3].

Cervical cancer is more frequent in low-resource countries, figuring as the second cause for mortality due to malignancies in women, only after breast cancer. In 2012, 528,000 women were affected, and 266,000 died of their disease [1]. Human papillomavirus (HPV) is the major etiological agent of cervical cancer [4], although infection alone is not enough to induce the malignant process [5]. Other individual genetic alterations are important in the development of cervical cancer, especially due to the provision of a phenotype less likely to promote HPV clearance. These alterations include genes involved with immunity, cytokine production, and tumor suppressor pathways, among others. A lot has been done for the primary prevention of cervical cancer, especially with HPV vaccination and screening by HPV testing and subtyping [6, 7]. Preventive measures, however, are mostly accessible in countries with higher incomes, relegating the countries with higher prevalence of the disease to late-stage or metastatic disease.

Endometrial cancer has a more homogeneous distribution [8]. In 2012, there were an estimated

L. De Brot
Department of Anatomic Pathology, A.C. Camargo Cancer Center, São Paulo, SP, Brazil

F. A. Soares (✉)
Department of Pathology, Rede D'Or Hospital Network, São Paulo, SP, Brazil
e-mail: Fernando.asoares@rededor.com.br

S. Badve, G. L. Kumar (eds.), *Predictive Biomarkers in Oncology*,
https://doi.org/10.1007/978-3-319-95228-4_41

320,000 new cases and 76,000 deaths related to the disease. It is the second most fatal gynecological cancer in developed countries. Possible reasons to the development of endometrial cancer are related to hormonal imbalance, especially high estrogen and low progesterone levels. It has been reported that low expression of the progesterone receptor is associated with recurrent endometrial cancer. Epigenetic changes, such as aberrant DNA methylation, atypical histone modification, and deregulation of miRNAs, resulting in altering gene expression patterns, are also frequent [9]. Treatment also had limited advance in the latter years, with the mainstay of therapy still based on radical surgery followed by chemotherapy and/or radiotherapy [10]. Early stage and locally advanced disease is amenable to cure with a relative good prognosis. On the other hand, therapeutic options are limited for patients with metastatic or persistent/recurrent disease after platinum-based chemoradiotherapy [11].

Vaginal and vulvar cancers are rarer. For vaginal cancer in 2017, the American Cancer Society (ASCO) estimates 4810 new diagnosis. Risk factors include the following: aged 60 or older, exposure to diethylstilbestrol (DES) before birth, human papillomavirus (HPV) infection, and a history of abnormal cells in the cervix or uterus. Squamous cell carcinoma (SCC) is the most common type of vaginal cancer (70%), followed by adenocarcinoma (15%), melanomas, and sarcomas [12].

Cancer of the vulva accounts for 3% of all female genital cancers and occurs mainly in women older than 60 years. According to ASCO, in 2014, there were about 4900 new cases of vulvar cancer and 1030 deaths from the disease. SCC accounts for 90% of vulvar cancers, followed by melanoma (8–10%), basal cell carcinoma (1–2%), and Bartholin gland carcinoma [13].

Recent years are marked by increasing knowledge in understanding genetic alterations and biologic pathways of neoplasia, leading to the development of new drugs. The marked advancement in the understanding of targets and treatment, however, was not encompassed by improvement in survival, demanding the understanding of predictive biomarkers for therapy [14, 15]. This chapter describes those biomarkers that are available for each type of gynecological cancer.

Ovarian Cancer

Antiangiogenic Therapies

Angiogenesis is a complex and multistep process which is controlled by several pathways such as the vascular endothelial growth factor (VEGF) pathway, the platelet-derived growth factor (PDGF) pathway, the fibroblast growth factor (FGF) pathway, and the angiopoietin-Tie2 receptor pathway [16]. The natural angiogenic process is based on a balance between pro- and antiangiogenic signaling pathways, and its regulation is fundamental for cell survival and tissue homeostasis [17]. In neoplasia, angiogenesis has been implicated in a myriad of tumors. Figures 41.1 and 41.2 demonstrate some physiologic functions of VEGF that provide the rationale for targeting it. For ovarian cancer, it has a definite role in the early and later pathogenesis, survival, drug resistance, and progression.

Development and sustainment of blood flow are essential for the neoplastic process. When a tumor exceeds 1 mm in diameter, it needs new blood vessel formation to support further growth. Tumor cells induce an angiogenic switch in response to hypoxia and genetic alterations to produce angiogenic growth factors that promote proangiogenic signaling, such as the VEGF pathway [18]. VEGF signaling can be blocked by several molecules. Bevacizumab, a monoclonal antibody that acts by direct inhibition of VEGF-A, has shown promising results in diverse tumors in several phase III clinical trials. Its role in ovarian cancer was studied for "adjuvant" treatment in metastatic disease, especially in platinum-sensitive and platinum-resistant disease [20].

Results, although positive, are relatively discordant between different trials. Extended progression-free survival (PFS) was demonstrated in numerous trials with bevacizumab. Overall survival (OS), nonetheless, is still a barrier to be broken, with benefits remaining

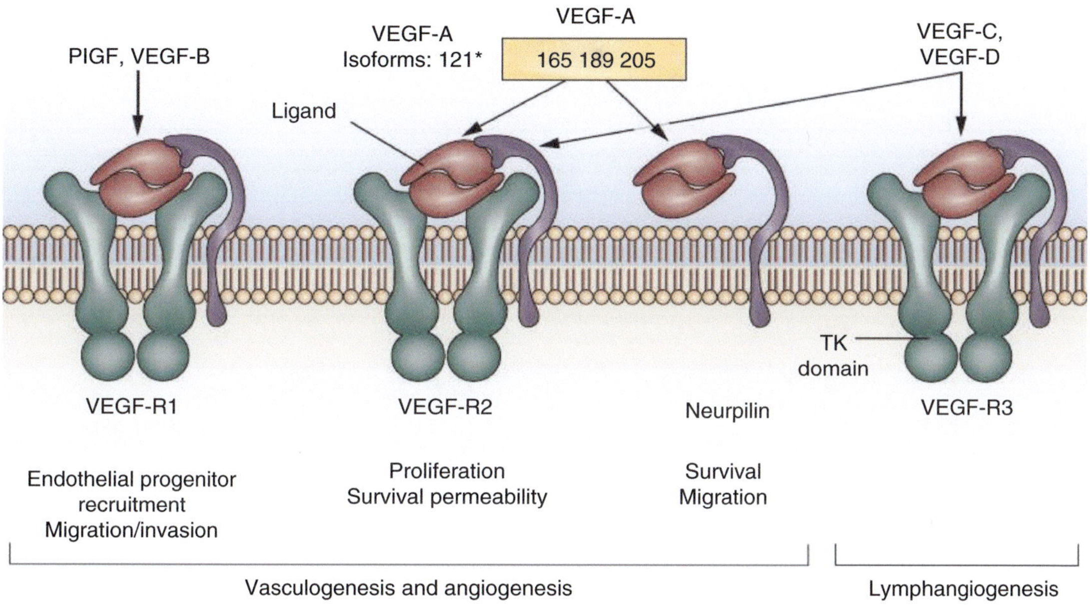

Fig. 41.1 VEGF receptors and their specific ligands. The VEGF family consists of various ligands (VEGF-A, VEGF-B, VEGF-C, VEGF-D, and PlGF) with VEGF-A being the dominant mediator of proangiogenic signaling in human cancers. VEGF-A has different isoforms that are generated by alternate splicing, and these ligands bind to the main receptors, VEGF-R2 and neuropilin, which can regulate the signaling activity mediated by the VEGF-A and VEGF-R2 interaction. Abbreviations: PlGF, placental growth factor; TK, tyrosine kinase. (Reprinted from Grothey and Galanis [18]. With permission from Springer Nature)

controversial in recent phase III trials. In advanced and metastatic tumors, bevacizumab in the post-cytoreduction scenario was evaluated by two clinical trials for first-line therapy: GOG218 [21] and ICON7 [22]. Both trials studied the role of bevacizumab in combination with chemotherapy (carboplatin + paclitaxel) in patients at high risk for relapse, leading to a significant increase in PFS and response rates, with little benefit demonstrated for OS. For high-risk and worse prognosis patients, there was a trend for a better OS with the addition of bevacizumab, and the magnitude of benefit was greater across all endpoints.

Addition of bevacizumab after relapse was also studied. For platinum-sensitive disease (i.e., those with a relapse after 6 months of completion of first-line chemotherapy), the OCEANS [23] trial demonstrated an increment of 4 months in PFS and response rates when bevacizumab was added to carboplatin + gemcitabine (plus optional bevacizumab maintenance). For platinum-resistant patients (relapse prior to 6 months of completion of first-line chemotherapy), the AURELIA [24] trial showed meaningful benefit for the combination of bevacizumab with single-agent chemotherapy, with doubling on PFS and reduction of about 52% in the risk of disease progression compared with chemotherapy alone.

The increments in survival and response rates with bevacizumab, however, were not accompanied by the characterization of a robust biomarker for the selection of antiangiogenic agents. Some gene polymorphisms, such as IL-18, and the levels of circulating VEGF were independent prognostic factors but could not be characterized as predictive. Other trials with oral tyrosine kinase inhibitors were also described (trebananib, pazopanib, nintedanib), with modest results and without the description of a useful predictive biomarker [3].

PARP Inhibitors

Enzymes such as poly(ADP-ribose) polymerase (PARP) play a crucial role in cellular DNA repair.

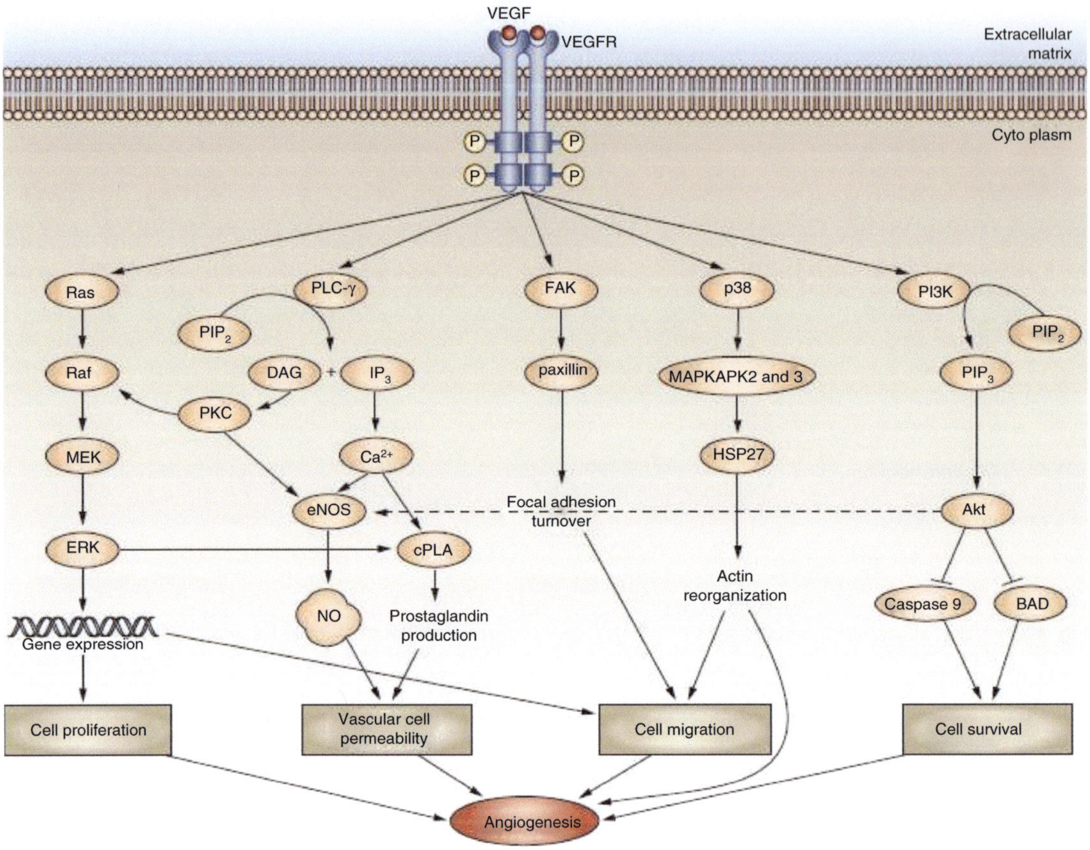

Fig. 41.2 The role of VEGFR signaling in tumor angiogenesis. Activation of VEGFR mediates proliferation, vascular permeability, cell migration, and cell survival, leading to angiogenesis. VEGFR signaling via the Ras/MEK/ERK pathway induces gene expression and results in cell proliferation. VEGFR can also signal through PLC-γ, which activates PKC by the generation of diacylglycerol and increases the concentration of intracellular calcium via inositol triphosphate. Increased intracellular calcium results in induction of vascular permeability via activation of endothelial nitric oxide synthase and generation of nitric oxide, but also via activation of cytosolic phospholipase A and prostaglandin production. Signaling through the focal adhesion kinase pathway leads to focal adhesion turnover and cell migration. VEGFR signaling through p38 results in heat shock protein 27 induction, which leads to actin reorganization and cell migration. Activation of the PI3K pathway maintains cell survival. PI3K catalyzes the conversion of PIP_2 to PIP_3. In turn PIP_3 recruits Akt to the cell membrane where Akt is activated via phosphorylation. Akt can activate endothelial nitric oxide synthase, which leads to vascular cell permeability, but the kinase also inhibits the proapoptotic proteins caspase 9 and BAD, which enhances cell survival. Small-molecule VEGFR inhibitors, such as the VEGFR2 inhibitors, block these pathways and thus inhibit angiogenesis [19]. (Reprinted from Ivy et al. [19]. With permission from Springer Nature)

Inhibition of PARP results in an excess of single-strand breaks, which causes double-strand breaks during replication [25]. Such problems in replication are repaired by homologous recombination, a process that requires intact BRCA proteins. Neoplasms with defective homologous recombination, including *BRCA1/2*, display sensitivity to PARP inhibition [26]. In the most common form of malignant epithelial ovarian cancer, and high-grade serous carcinoma (HGSC), defects in homologous recombination occur in up to 50% of cases, including germline or somatic loss-of-function mutations of BRCA1 or BRCA2, epigenetic silencing of BRCA1, and defects in other genes in this class including RAD51D, ATM, PALB2, RAD51C, and BRIP1. Patients with homologous recombination-deficient ovarian cancer typically demonstrate a

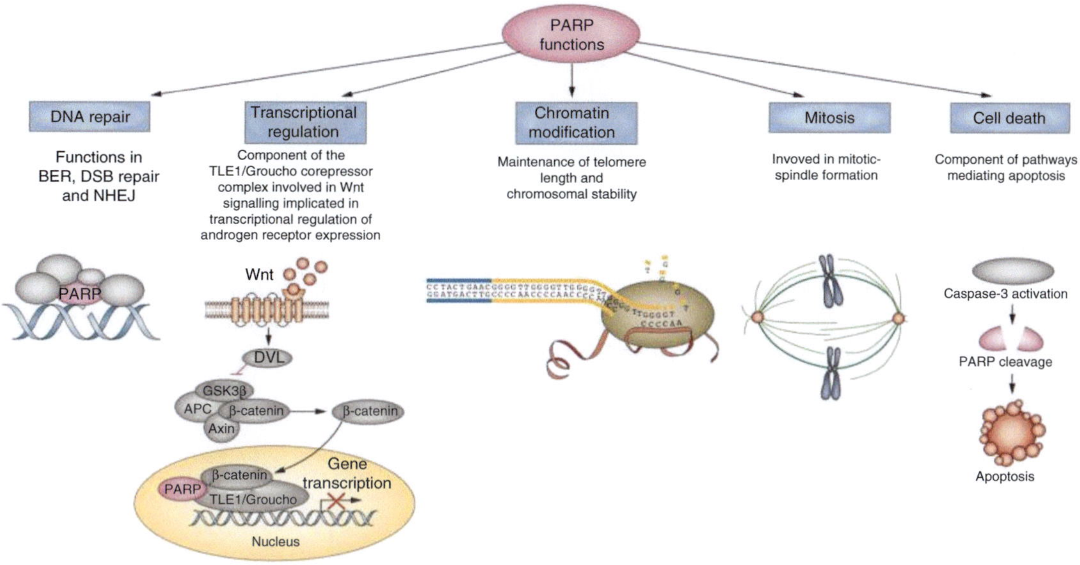

Fig. 41.3 The different biological PARP functions relevant to cancer. In addition to the classic activity of PARP in BER, the PARP family members have diverse functions in other biological processes, including transcriptional regulation, chromatin modification, mitosis (mitotic-spindle formation), and apoptosis, as well as intracellular trafficking, and energy metabolism (not shown). PARP1 and, to a lesser extent, PARP2 are important in maintaining telomere length and chromosomal stability. PARP1 also forms part of the Groucho/TLE1 corepressor complex and has been implicated as a transcriptional regulator of androgen receptor expression. Other PARPs function in the repair of DSBs and in progression of mitosis (PARP3), and some have potential roles in Wnt signaling and telomere maintenance (PARP5 and PARP6). PARP1 is also a regulator of NHEJ, a mechanism of DSB repair. *Abbreviations*: APC adenomatous polyposis coli protein, BER base excision repair, DSB double-strand break, DVL dishevelled homologue, GSK-3β glycogen synthase kinase-3β, NHEJ nonhomologous end joining, PARP poly(ADP-ribose) polymerase, TLE1 transducin-like enhancer protein 1 (Groucho homologue) [34]. (Reprinted from Sonnenblick et al. [34]. With permission from Springer Nature)

"BRCAness" phenotype that is similar to that of patients with BRCA-mutated tumors, who generally exhibit better outcomes compared with patients with sporadic ovarian cancer, including improved platinum sensitivity and overall survival [27]. Figure 41.3 depicts the major functions related to the intrinsic relationship displayed between PARP and BRCA.

Differently, from the antiangiogenic setting, the presence of a BRCA mutation is important as a predictive biomarker for PARP inhibition. Current studies showed that drugs such as olaparib and niraparib demonstrated a significant increase in PFS as maintenance therapy for platinum-sensitive relapse disease independently of BRCA status. Despite the independent benefit, patients with BRCA mutations derived longer PFS. For olaparib patients that progressed after first-line treatment and had platinum-sensitive disease, maintenance showed a doubling in PFS in a phase II trial (Study 19) independent of BRCA status [28], leading to the FDA approval of the drug. A phase III trial (SOLO2) demonstrated greater benefit in PFS for BRCA-mutated patients, with a gain of 14 months until disease relapse [29]. Two more drugs niraparib and rucaparib showed similar results in large randomized clinical trials. Niraparib showed benefit in PFS for maintenance therapy in platinum-sensitive relapsed disease regardless of BRCA status in a phase III trial [30]. Rucaparib showed benefit in maintenance and as monotherapy, in maintenance therapy, differently from the niraparib and olaparib trials. Rucaparib allowed patients with residual disease to enter a phase III trial, with improvement of PFS unrelatedly to BRCA status [31].

For PARP inhibition, deleterious *BRCA1* and *BRCA2* mutations are thought to have predictive value for overall treatment of ovarian cancer, with greater response to platinum chemotherapy. BRCA-mutated patients have better PFS and OS. What is especially true for BRCA2-mutated patients is that the survival curves are up to three to five times greater when compared to BRCA-proficient patients [32]. Studies are ongoing for further validation of these results, but PARP inhibition is already a reality, and BRCA testing for all high-grade serous carcinoma (HGSC) patients should be encouraged.

Other trials are also ongoing for germline BRCA-mutated (gBRCAm) advanced ovarian cancer. Drugs such as veliparib and talazoparib are currently being tested alone or in multiple combinations, not only for ovarian cancer but also for other cancers with BRCA mutations, such as breast, prostate, and lung cancers [33] (Fig. 41.4).

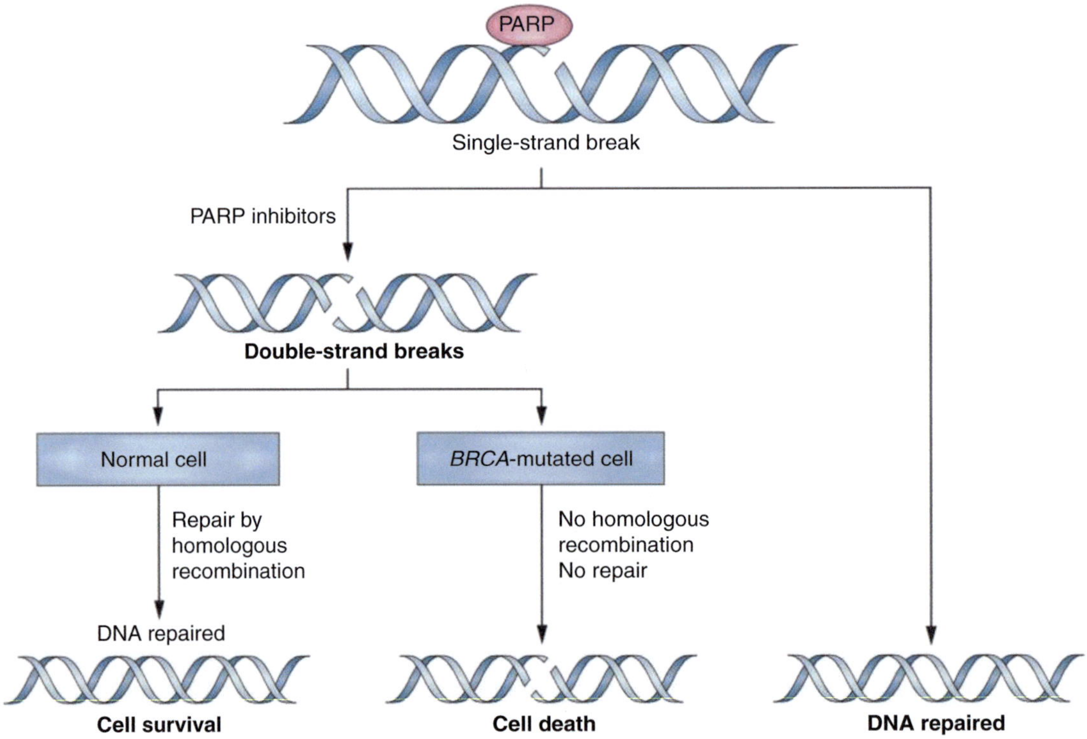

Fig. 41.4 The role of PARP inhibitors in synthetic lethality. PARP enzymes play a key part in the repair of DNA damage. In particular, PARP1 binds to single-strand breaks in DNA and recruits other enzymes to repair the DNA damage. Failure to repair single-strand breaks can result in double-strand breaks during DNA replication; thus, PARP inhibition can induce further DNA damage. However, DNA damage, which is a frequent occurrence during each cell cycle, can also be repaired through homologous recombination mechanisms. The *BRCA1* and *BRCA2* genes encode key components of these homologous recombination repair pathways, and, therefore, *BRCA*-mutant tumors are inherently deficient in DNA repair. This vulnerability forms the foundation for selective targeted synthetic lethal therapy with PARP inhibitors in patients with *BRCA*-mutant breast cancer. The DNA damage that occurs after inhibition of PARP activity cannot be adequately repaired in these cancers and eventually results in chromosomal instability, cell cycle arrest, and subsequent apoptotic cell death. As DNA repair processes remain intact in noncancerous cells, which generally retain at least one functional copy of both *BRCA1* and *BRCA2*, PARP inhibition is hypothesized to selectively kill cancer cells, sparing normal tissue [34]. *Abbreviation*: PARP poly(ADP-ribose) polymerase. (Reprinted from Sonnenblick et al. [34]. With permission from Springer Nature)

The HER Family

The ERBB/HER family are a group of tyrosine kinase receptors that play a key role in the pathogenesis of numerous malignancies, specially breast, lung, and colorectal cancer. EGFR (epidermal growth factor receptor) is expressed in 25–50% of ovarian cancer, but the prognostic implications are still controversial. Clinical trials of EGFR inhibition in ovarian cancer have been disappointing [15].

Immunotherapy

The increasing interest in immunotherapy for the treatment of cancer was struck with disappointment when the first studies of anti-PD-1 and anti-PD-L1 drugs in ovarian cancer came out [35]. Ovarian cancer has typically a low mutational burden, with the exception of some patients with pathogenic germline BRCA mutations [36]. Currently, there is some evidence for response for the PD-1 inhibitor nivolumab in a phase I trial of genetically unselected platinum-resistant ovarian cancer patients [37]. Pembrolizumab, another PD-1 inhibitor, was tested in PD-L1-positive platinum-resistant patients, and preliminary results seem to be more promising than in the unselected population [38]. Anti-PD-L1 therapy with avelumab and atezolizumab was also studied, but results are still immature or pending.

As in lung cancer, there has been a great effort to characterize biomarkers predictive for immunotherapy response. High tumor mutational burden, PD-L1 expression in the neoplastic tissue, and inflammatory infiltrate were prognostic in some trials but not sufficient enough to select patients. A basket trial for patients with microsatellite instability was demonstrated by immunohistochemistry. Patients derived great benefit from immunotherapy with pembrolizumab, and the drug was granted approval for these patients independent of the type of cancer encountered [39].

Most recently, further evidence that an immunogenic tumor microenvironment may be associated with better outcomes was observed in gene expression analysis of ovarian cancer. A subset of genes related to immune response was associated with one molecular type designated as an immunoreactive subtype (C2). These tumor types are more responsive to immune checkpoint blockade [35]. Other immune-based approaches to ovarian cancer include vaccines (anti-CA 125) and immune cellular therapy (genetically engineered CAR-T cells). These latter approaches, however, are still very incipient, and additional work must be performed before this becomes a reality.

Folate Antagonists

Folate receptors (FR) are highly expressed in non-mucinous ovarian cancer. These receptors are involved in tumor folate transportation and to chemoresistance and poor survival outcomes. In recent years, it had been investigated as a therapeutic target. A current phase III trial is investigating a FR-monoclonal antibody in combination with carboplatin/paclitaxel in the first-line setting [3, 15]. Phase I and II trials for folate receptor inhibitors were disappointing until the present day, not only for ovarian cancer but also for other tumor subtypes.

Insulin Signaling Inhibition

The insulin-like growth factor (IGF) pathway consists of three ligands (IGF-1, IGF-2, and insulin) and their cell surface receptors (IGF-R1, IGF-R2, IR). The most important interactions are with Ras-Raf-MAPK and PI3K-Akt pathways, which facilitate tumorigenesis by promoting cellular proliferation, angiogenesis, invasion, and metastatic potential. The dual IR/IGFR inhibitor has been investigated in patients with platinum-resistant disease. The results of this study are pending at the moment. Studies with a combination of anti-IGFR/PI3K/mTOR inhibitors are also currently underway [15]. Toxicity, nevertheless, is a major barrier for these approaches.

Cervical Cancer

Antiangiogenic Therapies

VEGF promotes angiogenesis, a key role in the progression of cervical cancer. As in ovarian cancer, bevacizumab was demonstrated to have activity in heavily pretreated patients, inspiring a phase III randomized controlled trial: GOG 240 [40]. This trial investigated the incorporation of bevacizumab with and without platinum combination chemotherapy in the treatment of advanced cervical cancer. For the first time, a monoclonal antibody showed promising results for cervical cancer. They concluded that the regimens that included bevacizumab were associated with reduced hazard of death, with an improvement of almost 4 months in OS. Based on this study, FDA and EMA approved bevacizumab in combination with paclitaxel plus either cisplatin or topotecan as a treatment for patients with persistent, recurrent, or metastatic cervical cancer. Despite better results, no predictive biomarker was shown to be valid in the study.

Immunotherapy

It is very important to mention that the novel approaches arising in the last few years did not put cervical cancer in the spotlight. Being a disease more prevalent in poor countries, the economic appeal is lacking [5]. In theory, immunotherapy would be a great approach for this disease, which has a high mutational burden [36] and a virus intrinsically related to its pathogenesis and can be treated with radiotherapy [41].

Recent evidence supports a potential role for immune checkpoint inhibitors as a therapeutic strategy in cervical cancer, but data is still limited. The cervical cancer cohort of the KEYNOTE-028 trial, a phase Ib trial including multiple tumors to be treated with pembrolizumab, demonstrated interesting results regarding PFS in heavily pretreated patients with PD-L1-positive tumors [42]. As per the author's knowledge, there is now published phase II data available.

In the light of a disease with restricted therapeutic approaches, new immunotherapy approaches are currently being developed, such as vaccines and molecules linked with viruses and bacteria. One molecule already studied in phase I is called ADXS11-001 (axalimogene filolisbac) [35]. FDA has granted axalimogene filolisbac orphan drug designation as well as a Special Protocol Assessment for the phase III AIM2CERV trial in cervical cancer and Fast Track designation. Besides from monotherapy approaches, collaborations between companies are now trying to focus efforts in clinical trials directed to cervical cancer.

Currently, there are no predictive biomarkers for cervical cancer. Molecular markers that have been studied for cervical cancer include EGFR, Bcl2, and VEGF. The precise role of these markers requires further elucidation, and special attention to immunotherapy approaches should be prompted [7]. Recently, FDA approved immunotherapy as second-line therapy for advanced or recurred cervical cancer. A combined score (immune plus cancer cells) should be used and positive cases are considered that above score 1 using immunohistochemistry. (No auhtors. Cancer Discov. 2018 Aug;8(8):904; Saglam et al. [43]).

Endometrial Cancer

No targeted therapy is currently available for endometrial carcinomas. There are also no biomarkers that are precise and definite to determine derived benefit from any specific therapy. Despite the fact that most endometrial carcinomas are diagnosed at an early stage, 15–20% [8] still recur after surgery, radiation therapy, cytotoxic chemotherapy, or hormonal therapy. These patients are orphan to specific treatments, and chemotherapy remains the ultimate approach.

The attempts in bringing biologics into endometrial cancer care have shown limited success in early trials. A phase II trial with bevacizumab in recurrent and metastatic endometrial cancer demonstrated that almost 40% of patients had a progression-free survival of at least 6 months [44]. This was not translated into phase III trials, and evidence for the usage of antiangiogenics in endometrial cancer is lacking. Despite strong evidence that the PI3k/AKT/mTOR pathway is often deregulated in endometrial cancer, targeted therapy with temsirolimus was studied, and evidence was limited to less than 20 patients. Various other molecules such as fibroblast growth factor receptor (FGFR) 2 have been investigated but without much success [5].

Currently, much attention is being drawn for immunotherapy for some special subtypes of endometrial cancer. The tumor mutational burden is normally high throughout different subtypes [36], especially for patients bearing Lynch syndrome, BRCA mutations, and microsatellite instability [9]. These patients may benefit from immunotherapy, and patients with high microsatellite instability are strong candidates for pembrolizumab therapy in advanced lines of treatment [39]. The KEYNOTE-028 cohort for endometrial cancer showed meaningful activity for PD-L1-positive advanced refractory endometrial cancer, not selected according to mutational status [45].

Although some advances were shown in endometrial cancer therapy, there is still a great need for a definition of biomarkers that could acknowledge research for novel drugs and approaches. Currently, endometrial cancer is treated based on anatomic and histological features, with the potential to hormonal status (making patients amenable to receiving anti-hormonal therapy) and potential biomarkers for immunotherapy response [11].

Vaginal Cancer

Vaginal cancer carries the same characteristics as cervical cancer, including the potential for immunotherapy. Squamous cell carcinoma of the vagina does not have any approved drug or any drug in development that could target specific biomarkers in order to obtain response [12]. It is important to notice that melanomas arising in the vagina are rare, and therefore there is minimal data specific to this malignancy. However, it is advocated that BRAF mutations encountered in these lesions could be of potential for targeted therapy. Also, novel immunotherapeutic uses in melanoma are considered in cases of vulvovaginal melanoma [46]. Patients with vulvovaginal melanoma are encouraged to be tested for at least *c-KIT* and *BRAF V600E* mutations.

Vulvar Cancer

Vulvar cancer comprises 5% of gynecologic malignancies, mostly squamous in histology, and approximately 40–70% are HPV-induced. The most frequently overexpressed proteins are EGFR (95%), MRP1/TOP2A (76%), and PD1+ tumor-infiltrating lymphocytes (72%) [15]. Genomic analysis shows that the most frequently mutated genes are EGFR-amplified in 6% and TP53 in 33% followed by PIK3CA (8%), BRCA2 (7%), and HRAS (6%). Treatment options for vulvar cancers are limited after first-line therapy. Human papillomavirus (HPV) infection may elicit an immune reaction in vulvar cancer, and an open-label, multi-cohort, phase I/II study of nivolumab in patients with virus-associated tumors could be useful not only for vulvar but also for vaginal and cervical SCC. Currently, the knowledge about biomarkers in SCC of the vagina and vulva is lacking.

Summary and Future Directions

Gynecologic cancers are still a problem to women who are diagnosed with them. These cancers threaten their current health status and their quality of life. Significant progress has been made in reducing the incidence of cervical carcinoma with widespread screening and HPV vaccination. However, the rates of the other gynecologic cancers have remained steady in recent decades. In addition, the efforts in preventing cervical cancer are not widespread worldwide. The small numbers of patients affected, relative to other malignancies, and the heterogeneity of these cancers have slowed progress in large clinical trials, hampering efforts in developing new treatments [14, 15]. Table 41.1 tries to summarize the main indications for therapy based on biomarker approach.

Table 41.1 Major targets and possible biomarkers for gynecologic cancers

Potential biomarker	Function	Drug in study/ company	Diagnosis	Population and positive clinical trials	Comments
VEGF (vascular endothelial growth factor)	Signaling protein for promoting and sustaining angiogenesis	Bevacizumab (Avastin®) Genentech /Roche	Ovarian cancer	Platinum-sensitive ovarian cancer in first line (ICON7 [22]/GOG218 [21])	High circulating VEGF levels are correlated with worst prognosis for recurrence and death, without demonstrated predictive value
				Platinum-sensitive ovarian cancer after first recurrence (OCEANS [23])	
				Platinum-resistant ovarian cancer (AURELIA [24])	
			Cervical cancer	Metastatic setting first line (GOG240 [40])	No correlation between results and biomarker status
		Cediranib (AZD2171)	Ovarian cancer	Metastatic setting first line in combination with platinum chemotherapy (ICON 6 [47])	No predictive biomarker. Drug still in study by AstraZeneca. Benefit still dubious
PARP (BRCA1/2)	Control of DNA repair, cell death, and genomic stability	Olaparib (Lynparza™) AstraZeneca	Ovarian cancer	Maintenance therapy, platinum-sensitive disease (SOLO2)	Increase in survival independent of BRCA status. Greater benefit for BRCA-mutated patients
			Ovarian cancer	Maintenance therapy, platinum-sensitive disease (NOVA) [30]	
		Niraparib Tesaro®	Ovarian cancer	Maintenance therapy, platinum-sensitive disease (ARIEL3) [31]	
		Rucaparib (Rubraca™) Clovis Oncology			

Despite the fact that survival rates for ovarian cancer have improved since the 1990s, the disease continues to pose a significant mortality threat to patients. Improved understanding about tumor microenvironment has led to the development of potent antiangiogenic agents that are effective in various other tumors, and these drugs have become a current new option in advanced and recurrent ovarian and cervical cancer. Though questions remain regarding the benefit of antiangiogenic therapy, emerging biomarkers that can stratify patient populations may soon allow clinicians to administer this new treatment modality to those who are most likely to benefit.

In addition to antiangiogenic drugs, promising results from trials with PARP inhibitors, immunotherapies, and repurposed drugs offer the potential for improved treatment on several fronts for patients. The main objective of those target strategies is to improve survival rates with fewer toxic side effects.

The recent revolution in DNA sequencing and RNA detection technologies has produced an extensive list of new candidate biomarkers and the possibility of routinely using profiles of multiple biomarkers that have the potential to improve prognostic power and allow clinicians to match patients to therapies that are most likely to produce positive results. The convergence of new diagnostic tools and new therapeutic agents has raised the prospect that a major improvement in the diagnosis and care of gynecologic malignancies is imminent.

References

1. Torre LA, Bray F, Siegel RL, Ferlay J, Lortet-Tieulent J, Jemal A. Global cancer statistics, 2012. CA Cancer J Clin. 2015;65(2):87–108.
2. Jayson GC, Kohn EC, Kitchener HC, Ledermann JA. Ovarian cancer. Lancet. 2014;384(9951):1376–88.
3. Coleman RL, Monk BJ, Sood AK, Herzog TJ. Latest research and treatment of advanced-stage epithelial ovarian cancer. Nat Rev Clin Oncol. 2013;10(4):211–24.
4. Schiffman M, Castle PE, Jeronimo J, Rodriguez AC, Wacholder S. Human papillomavirus and cervical cancer. Lancet. 2007;370(9590):890–907.
5. Monk BJ, Tewari KS. Evidence-based therapy for recurrent cervical cancer. J Clin Oncol. 2014;32(25):2687–90.
6. Arrossi S, Temin S, Garland S, Eckert LON, Bhatla N, Castellsagué X, et al. Primary prevention of cervical cancer: American Society of Clinical Oncology resource-stratified guideline. J Glob Oncol. 2017;3(5):611–34.
7. Bailey HH, Chuang LT, NC DP, Eng C, Foxhall LE, Merrill JK, et al. American Society of Clinical Oncology statement: human papillomavirus vaccination for cancer prevention. J Clin Oncol. 2016;34(15):1803–12.
8. Morice P, Leary A, Creutzberg C, Abu-Rustum N, Darai E. Endometrial cancer. Lancet. 2016;387(10023):1094–108.
9. Kandoth C, Schultz N, Cherniack AD, Akbani R, Liu Y, Shen H, et al. Integrated genomic characterization of endometrial carcinoma. Nature. 2013;497(7447):67–73.
10. Colombo N, Creutzberg C, Amant F, Bosse T, González-Martín A, Ledermann J, et al. ESMO-ESGO-ESTRO consensus conference on endometrial cancer: diagnosis, treatment and follow-up. Ann Oncol. 2016;27(1):16–41.
11. Burke WM, Orr J, Leitao M, Salom E, Gehrig P, Olawaiye AB, et al. Endometrial cancer: a review and current management strategies: part I. Gynecol Oncol. 2014;134(2):385–92.
12. Gadducci A, Fabrini MG, Lanfredini N, Sergiampietri C. Squamous cell carcinoma of the vagina: natural history, treatment modalities and prognostic factors. Crit Rev Oncol Hematol. 2015;93(3):211–24.
13. Tyring SK. Vulvar squamous cell carcinoma: guidelines for early diagnosis and treatment. Am J Obst Gynecol. 2003;189(3 Suppl):S17–23.
14. Huang J, Hu W, Sood AK. Prognostic biomarkers in ovarian cancer. Cancer Biomark. 2010;8(0):231–51.
15. Tarala DN. Biomarkers for cancers in women: present status and future perspectives. Curr Women Health Rev. 2012;8(4):297–311.
16. Ferrara N, Adamis AP. Ten years of anti-vascular endothelial growth factor therapy. Nat Rev Drug Discov. 2016;15(6):385–403.
17. Carmeliet P, De Smet F, Loges S, Mazzone M. Branching morphogenesis and antiangiogenesis candidates: tip cells lead the way. Nat Rev Clin Oncol. 2009;6(6):315–26.
18. Grothey A, Galanis E. Targeting angiogenesis: progress with anti-VEGF treatment with large molecules. Nat Rev Clin Oncol. 2009;6(9):507–18.
19. Ivy SP, Wick JY, Kaufman BM. An overview of small-molecule inhibitors of VEGFR signaling. Nat Rev Clin Oncol. 2009;6(10):569–79.
20. Chase DM, Chaplin DJ, Monk BJ. The development and use of vascular targeted therapy in ovarian cancer. Gynecol Oncol. 2017;145(2):393–406.
21. Burger RA, Brady MF, Bookman MA, Fleming GF, Monk BJ, Huang H, et al. Incorporation of bevacizumab in the primary treatment of ovarian cancer. N Engl J Med. 2011;365(26):2473–83.
22. Perren TJ, Swart AM, Pfisterer J, Ledermann JA, Pujade-Lauraine E, Kristensen G, et al. A phase 3 trial of bevacizumab in ovarian cancer. N Engl J Med. 2011;365(26):2484–96.
23. Aghajanian C, Blank SV, Goff BA, Judson PL, Teneriello MG, Husain A, et al. OCEANS: a randomized, double-blind, placebo-controlled phase III trial of chemotherapy with or without bevacizumab in patients with platinum-sensitive recurrent epithelial ovarian, primary peritoneal, or fallopian tube cancer. J Clin Oncol. 2012;30(17):2039–45.
24. Pujade-Lauraine E, Hilpert F, Weber B, Reuss A, Poveda A, Kristensen G, et al. Bevacizumab combined with chemotherapy for platinum-resistant recurrent ovarian cancer: the AURELIA open-label randomized phase III trial. J Clin Oncol. 2014;32(13):1302–8.
25. Holthausen JT, Wyman C, Kanaar R. Regulation of DNA strand exchange in homologous recombination. DNA Repair. 2010;9(12):1264–72.
26. Renkawitz J, Lademann CA, Jentsch S. Mechanisms and principles of homology search during recombination. Nat Rev Mol Cell Biol. 2014;15(6):369–83.
27. Muggia F, Safra T. 'BRCAness' and its implications for platinum action in gynecologic cancer. Anticancer Res. 2014;34(2):551–6.
28. Ledermann J, Harter P, Gourley C, Friedlander M, Vergote I, Rustin G, et al. Olaparib maintenance therapy in platinum-sensitive relapsed ovarian cancer. N Engl J Med. 2012;366(15):1382–92.
29. Pujade-Lauraine E, Ledermann JA, Selle F, Gebski V, Penson RT, Oza AM, et al. Olaparib tablets as maintenance therapy in patients with platinum-sensitive, relapsed ovarian cancer and a *BRCA1/2* mutation (SOLO2/ENGOT-Ov21): a double-blind, randomised, placebo-controlled, phase 3 trial. Lancet Oncol. 2017;18(9):1274–84.
30. Mirza MR, Monk BJ, Herrstedt J, Oza AM, Mahner S, Redondo A, et al. Niraparib maintenance therapy in platinum-sensitive, recurrent ovarian cancer. N Eng J Med. 2016;375(22):2154–64.
31. Coleman RL, Oza AM, Lorusso D, Aghajanian C, Oaknin A, Dean A, et al. Rucaparib maintenance

treatment for recurrent ovarian carcinoma after response to platinum therapy (ARIEL3): a randomised, double-blind, placebo-controlled, phase 3 trial. Lancet. 2017;390(10106):1949–61.
32. Scott CL, Swisher EM, Kaufmann SH. Poly (ADP-ribose) polymerase inhibitors: recent advances and future development. J Clin Oncol. 2015;33(12):1397–406.
33. Mullard A. PARP inhibitors plough on. Nat Rev Drug Discov. 2017;16(4):229.
34. Sonnenblick A, de Azambuja E, Azim HA Jr, Piccart M. An update on PARP inhibitors – moving to the adjuvant setting. Nat Rev Clin Oncol. 2015;12(1):27–41.
35. Heong V, Ngoi N, Tan DSP. Update on immune checkpoint inhibitors in gynecological cancers. J Gynecol Oncol. 2017;28(2):e20.
36. Alexandrov LB, Nik-Zainal S, Wedge DC, Aparicio SA, Behjati S, Biankin AV, et al. Signatures of mutational processes in human cancer. Nature. 2013;500(7463):415–21.
37. Hamanishi J, Mandai M, Ikeda T, Minami M, Kawaguchi A, Murayama T, et al. Safety and antitumor activity of anti–PD-1 antibody, nivolumab, in patients with platinum-resistant ovarian cancer. J Clin Oncol. 2015;33(34):4015–22.
38. Varga A, Piha-Paul SA, Ott PA, Mehnert JM, Berton-Rigaud D, Morosky A, et al. Pembrolizumab in patients (pts) with PD-L1–positive (PD-L1+) advanced ovarian cancer: updated analysis of KEYNOTE-028. J Clin Oncol. 2017;35(15_suppl):5513.
39. Le DT, Uram JN, Wang H, Bartlett BR, Kemberling H, Eyring AD, et al. PD-1 blockade in tumors with mismatch-repair deficiency. N Engl J Med. 2015;372(26):2509–20.
40. Tewari KS, Sill MW, Long HJI, Penson RT, Huang H, Ramondetta LM, et al. Improved survival with bevacizumab in advanced cervical cancer. N Engl J Med. 2014;370(8):734–43.
41. Cancer Genome Atlas Research Network, Albert Einstein College of Medicine, Analytical Biological Services, et al. Integrated genomic and molecular characterization of cervical cancer. Nature. 2017;543(7645):378–84.
42. Frenel JS, Le Tourneau C, O'Neil B, Ott PA, Piha-Paul SA, Gomez-Roca C, et al. Safety and efficacy of pembrolizumab in advanced, programmed death ligand 1-positive cervical cancer: results from the phase Ib KEYNOTE-028 trial. J Clin Oncol. 2017;35(36):4035–41.
43. Saglam O, Conejo-Garcia J. PD-1/PD-L1 immune checkpoint inhibitors in advanced cervical cancer. Integr Cancer Sci Ther. 2018;5(2). https://doi.org/10.15761/ICST.1000272.
44. Aghajanian C, Sill MW, Darcy KM, Greer B, McMeekin DS, Rose PG, et al. Phase II trial of bevacizumab in recurrent or persistent endometrial cancer: a gynecologic oncology group study. J Clin Oncol. 2011;29(16):2259–65.
45. Ott PA, Bang Y-J, Berton-Rigaud D, Elez E, Pishvaian MJ, Rugo HS, et al. Safety and antitumor activity of pembrolizumab in advanced programmed death ligand 1–positive endometrial cancer: results from the KEYNOTE-028 study. J Clin Oncol. 2017;35(22):2535–41.
46. Kaufman HL, Kirkwood JM, Hodi FS, Agarwala S, Amatruda T, Bines SD, et al. The Society for Immunotherapy of Cancer consensus statement on tumour immunotherapy for the treatment of cutaneous melanoma. Nat Rev Clin Oncol. 2013;10(10):588–98.
47. Ledermann JA, Embleton AC, Raja F, Perren TJ, Jayson GC, Rustin GJS, et al. Cediranib in patients with relapsed platinum-sensitive ovarian cancer (ICON6): a randomised, double-blind, placebo-controlled phase 3 trial. Lancet. 2016;387(10023):1066–74.

42

Predictive Biomarkers and Targeted Therapies in Head and Neck Cancer

Felipe D'Almeida Costa
and Fernando Augusto Soares

Overview

Head and neck cancer (HNC) can occur in a variety of sites in the head and neck region, such as the upper aerodigestive system, thyroid, and the salivary glands [1]. Head and neck squamous cell carcinoma (SCC), the most common histologic type, represents the sixth most frequent human cancer, with an incidence of 560,000 cases and over 350,000 deaths annually. The estimated incidence for oral cavity and pharyngeal malignancies together in 2013 was 11.37 per 100,000 people [2]. Smoking and excessive consumption of alcohol are the classic leading causes of SCC. In the last decade, it was discovered that the human papillomavirus (HPV) infection plays a significant role as a risk factor for HNC, especially in the oropharynx [3]. Due to tumor type and site heterogeneity, there are many therapeutic alternatives for patients with HNC, such as surgery alone, surgery plus neoadjuvant or postoperative radiation therapy, and/or chemotherapy and targeted therapies. Choosing the appropriate treatment for each patient is a challenge that has to be overcome, raising the need for the identification of biological markers capable of recognizing cancers with more aggressive or indolent behavior. This allows the attending physician to implement more specific effective therapies upfront and spare patients from being unnecessarily exposed to toxic chemotherapeutic drugs, higher radiation doses, and mutilating surgeries.

Biomarkers are defined as any biological finding that can be measured and evaluated as indicators of normal biological and pathological states or pharmacological responses to a therapeutic strategy [4]. In HNC, as in many other oncology areas, a vast number of biomarkers are being recently studied, leading head and neck oncology to move toward personalized cancer care. This chapter will address the most relevant predictive biomarkers for head and neck tumors, emphasizing their role in patient management [5].

Human Papillomavirus (HPV)

The association between HPV and HNC arose from observations of an increasing incidence of tumors in patients under the age of 50, most of them without a history of tobacco use or alcohol intake. These malignancies are generally located in the oropharynx (base of the tongue and tonsils), with HPV 16 being detected in over 90% of these cancers [6]. The carcinogenic effect of HPV is related to its proteins E6 and E7, which degrade and inactivate p53 and

F. D. Costa
Department of Anatomic Pathology, A.C. Camargo Cancer Center, São Paulo, São Paulo, Brazil

F. A. Soares (✉)
Department of Pathology, Rede D'Or Hospital Network, São Paulo, SP, Brazil
e-mail: Fernando.asoares@rededor.com.br

S. Badve, G. L. Kumar (eds.), *Predictive Biomarkers in Oncology*,
https://doi.org/10.1007/978-3-319-95228-4_42

retinoblastoma (Rb) gene products, respectively. These molecules are key regulators of cell cycle. Disruption of Rb function leads to an increase in expression of p16^{INK4A}, allowing the latter to be used as a surrogate marker for HPV-related oropharyngeal squamous cell carcinomas. Several clinical trials have shown better outcomes for patients with HPV-related SCC, both for progression-free and overall survival rates. For these patients, it is feasible to reduce treatment intensity, achieving lower toxicities without negative impact on survival [1]. The identification of an HPV-related tumor begins at the morphological level, as most of these lesions are "non-keratinizing" SCC in hematoxylin and eosin stained slides. Other morphologic patterns associated with HPV are lymphoepithelial, undifferentiated, sarcomatoid, and papillary. The virus can be directly detected through PCR or in situ hybridization-based assays, but the 2017 World Health Organization classification of head and neck tumors permits the use of p16 immunohistochemistry (IHC) for indirect detection, a low cost and faster method, as a reliable HPV surrogate marker. Together with correct tumor site (oropharynx) and morphology (non-keratinizing and others), diffuse p16 expression (over 70% of tumor cells with nuclear and cytoplasmic p16 strong staining) has a sensitivity approaching 100% for detecting transcriptionally active HPV [7].

Epidermal Growth Factor Receptor (EGFR)

EGFR is a transmembrane tyrosine kinase receptor member of the ErbB family. Its activation leads to downstream phosphorylation and signaling of several pathways, such as PI3K/PTEN/AKT, MAPK, and JAK/STAT, promoting cell proliferation, stromal invasion, increased angiogenesis, and metastatic spread. Hyperactivation of EGFR signaling in HNC can occur by gene amplification and protein overexpression, gene mutations, and collateral activation by other receptor tyrosine kinases. EGFR is highly expressed in virtually all head and neck (H&N) cancers, and its expression is inversely associated with prognosis. Its inhibition through monoclonal targeted antibodies such as cetuximab in curative and palliative settings has been used in combination with radiotherapy for patients with advanced H&N cancers. EGFR detection is through protein expression or activation, gene copy number, polymorphisms, mutations, and the expression of the mutated form EGFR vIII [8].

In patients eligible for exclusive radiation therapy, high EGFR protein expression measured by IHC and quantitative assays seems to be predictive of poor locoregional control. For patients who underwent surgery combined with radiation therapy and chemotherapy and patients eligible for EGFR inhibition, the results regarding EGFR protein expression are discordant or are not able to predict response to therapy [9]. Neither gene copy number, mutation, activation, nor polymorphisms have shown to be predictive factors in head and neck SCC [8]. Despite the lack of a predictive marker for response, cetuximab is recommended to most patients with oral, oropharyngeal, hypopharyngeal, and laryngeal SCC in metastatic, recurrent, or unresectable cases [10]. This raises the need for continued efforts to develop or discover a biomarker able to better select patients who would really benefit from multimodal or single-agent chemotherapeutic regimens which include anti-EGFR targeted drugs.

Programmed Death 1 and Its Ligand (PD-1 and PD-L1)

HNC can evade the immune system by several mechanisms. Tumors can downregulate antigen presentation molecules, such as MHC 1, and induct the secretion of immunosuppressive cytokines by the inflammatory milieu or by upregulating and expressing co-inhibitory molecules, particularly CTLA-4 and PD-L1, to decrease the immune response. Targeting the immune checkpoint pathway of the latter is being extensively studied in many solid tumors, either with drugs directed against PD-1 or PD-L1. HNC has raised attention in utilizing immune checkpoint inhibitors because up to 70% of HPV-related SCC express PD-L1 in tumor cells, which seems to predict response to anti-PD-1/PD-L1 therapies [3]. Recent trials showed a positive correlation between PD-L1 expression, evaluated through immunohistochemistry, and response to nivolumab and pembrolizumab in recurrent or metastatic settings (CHECKMATE-141 (ClinicalTrials.gov Identifier, NCT02105636), KEYNOTE-012 (ClinicalTrials.gov Identifier, NCT01848834), and KEYNOTE-055 (ClinicalTrials.gov Identifier, NCT02255097)), regardless of HPV status. The cutoffs and interpretation methods vary among studies. For trials studying nivolumab, PD-L1 positivity (≥1%) in tumors was associated with better response to anti-PD-L1 therapy [11]. In those who studied pembrolizumab, a combined positive score (CPS) was used, considering positivity in tumor cells and inflammatory mononuclear cells, both within tumor nests and adjacent stroma. Positive cases were defined as CPS ≥1%, which showed higher overall response rates in comparison to negative cases. In addition, for both drugs, the higher the expression observed, regardless of the method applied, the better the response achieved. It is important to address that even for patients with PD-L1 staining considered negative, some degree of response was observed [12]. These findings suggest a possible role for other molecules in this inhibitory pathway, leaving an open gap for searching additional predictive biomarkers in this field. Despite not reaching its prespecified difference for statistical significance in overall survival (OS) (primary endpoint), a recent trial (KEYNOTE-040, ClinicalTrials.gov Identifier, NCT02252042) showed better median OS for patients in the pembrolizumab arm versus investigator's choice standard of care (SOC) (8.4 versus 7.1 months, hazard ratio [HR] 0.81; 95% CI 0.66–0.99, $P = 0.0204$). The benefit was even higher for patients with CPS >50% (median OS 11.6 versus 7.9 months HR 0.54; 95% CI 0.35–0.82, $P = 0.0017$). The OS analysis may have been confounded by using immunotherapy in the SOC arm while the study was ongoing. Additionally, patients in the treatment arm experienced lower rates of adverse effects, which may represent a good option for patients with elevated levels of PD-L1 expression [13, 14].

Epstein-Barr Virus (EBV)

EBV is strongly associated with non-keratinizing squamous cell nasopharyngeal carcinoma (NPC), usually undifferentiated or poorly differentiated. It is associated with numerous tumor-infiltrating lymphocytes, which may obscure the epithelial component and lead to the former designation of lymphoepithelioma. These patients usually present with advanced locoregional disease, and the treatment options are generally not surgical. Exclusive radiation therapy and radiation combined with platinum-based chemotherapy are commonly used therapeutic regimens. Although not a predictive marker, high serum EBV circulating DNA is associated with recurrent disease and worse survival. It will be incorporated in an upcoming trial as a biomarker to individualize treatment for patients with NPC [2].

HER-2/neu

In HNC, up to 40% of salivary duct carcinomas (SDC) show HER-2/neu gene amplification and HER-2/neu protein overexpression. The identification of this group of tumors allows clinical use of anti-HER-2/neu drugs, in an extrapolation of breast cancer experience. There are only case reports and small case series of objective responses in patients submitted to specific therapy, but the complete response is rare in this subset of patients, probably because SDC have an additional molecular alteration in genes such as TP53, HRAS, and PTEN. Due to its rare occurrence, there are no randomized trials to date that standardize treatment choices and outcomes with HER-2/neu antagonists in SDC [12].

Androgen Receptor (AR)

Salivary duct carcinomas (SDC) with apocrine phenotype uniformly express AR. This can be used as an important diagnostic tool in the differential diagnosis of high-grade salivary gland carcinomas and as a potential therapeutic target. Some studies have shown tumor reduction in androgen deprivation therapy, alone or in combination with radiation therapy. For the same reasons described in the topic above, there is no correlation between AR expression intensity and response, as there is no consensus threshold for considering a tumor as AR positive [12] (Table 42.1).

Table 42.1 Table of predictive biomarkers in head and neck cancer

Gene/protein	Function/pathogenic process	Patient selection method and sample type	Successfully completed clinical trials and # of patients enrolled	Clinical use and limitations	Approved drug(s) generic/trade name	Company
EGFR (HER1/ERBB1)	Proliferation	IHC	ClinicalTrials.gov Identifier: CT00003809 117 patients [9]	Metastatic or recurrent squamous cell carcinoma of the head and neck	Cetuximab (Erbitux®)	Eli Lilly
PD-L1	Immune checkpoint regulation	IHC	ClinicalTrials.gov Identifier: NCT02105636 361 patients [11]	Recurrent squamous-cell carcinoma of the head and neck	Nivolumab (Opdivo®)	Bristol-Myers Squibb
PD-L1	Immune checkpoint regulation	IHC or PCR	ClinicalTrials.gov Identifier: NCT02252042 495 patients [12]	Recurrent or metastatic head and neck cancer	Pembrolizumab (Keytruda®)	Merck/MSD

Other Markers and Novel Therapeutics

Several other biomarkers have been studied in HNC, most of them proto-oncogenes and tumor suppressor genes. Activating mutations can occur in genes such as FGFR gene family, CCND1, which encodes cyclin D1 c-MET (hepatocyte growth factor receptor) and PIK3CA, involved in the mTOR signaling pathway. Tumor suppressor genes usually mutated in HNC are TP53 and NOTCH. Although most of them are associated with prognosis and have a potential to be targetable, additional evidence is still lacking for its use in clinical practice. New target drugs are also being developed and evaluated every year. For example, VEGF inhibitors, such as bevacizumab, and newer generation versions axitinib and pazopanib have been currently investigated in recurrent or metastatic H&N cancer patients (ClinicalTrials.gov Identifier, NCT00588770). Small molecule pan receptor tyrosine kinase inhibitors sorafenib and sunitinib have been tested in metastatic HNSCC with modest activity [15, 16]. Another exciting target is NTRK gene alterations in cases of mammary analogue secretory carcinoma of the salivary glands. This rare and unique carcinoma frequently harbors the translocation ETV6-NTRK3. Usually, it has an indolent course, but it is more aggressive than other similar tumors as acinic cell carcinoma and can evolve with lymph node and distant metastasis. Several clinical trials have shown the efficacy of TRK fusion kinase receptor inhibitors and open the possibility of treatment of advanced cases with larotrectinib and entrectinib. The gold standard to identify the fusions is gene sequencing, but immunohistochemistry can be used as a useful screening method [17].

Guidelines

The molecular knowledge and translational medicine are still evolving in HNC; thus despite a large number of biomarkers under investigation, there are no guidelines for use in clinical practice.

Summary and Future Directions

We are experiencing the era of personalized medicine. In many solid tumors, the role of targeted therapies and biological markers is well established and is already incorporated in treatment decisions. The head and neck oncology field is evolving quickly, and probably we will be able to see in the next few years the emergence of new possibilities and strategies for most of the markers discussed in this chapter. The newer molecules studied such as PD-1 and PD-L1 are the best candidates to be added to our daily routines. This is due to the promising results reported in clinical trials, with advanced diseases that failed treatment with conventional therapies. Additionally, there are already many drugs which target known molecules and biomarkers that have shown some level of evidence in terms of predicting response. Thus, as stated above, this is still an open field of research, and more investigating studies are needed in head and neck predictive biomarkers. Gene therapy approaches utilizing replicating adenovirus [18] and a specially modified virus called ONYX-015 that may be able to kill tumor cells while leaving normal cells undamaged have been tried (ClinicalTrials.gov Identifier, NCT00006106) in advanced head and neck cancer patients. However, these studies have been stopped due to unknown reasons. In summary, systemic, targeted, and immunotherapy are an integral part of head and neck cancer treatment, and preliminary results with immunotherapy hold promise in the near future.

References

1. Kang H, Kiess A, Chung CH. Emerging biomarkers in head and neck cancer in the era of genomics. Nat Rev Clin Oncol. 2015;12(1):11–26. https://doi.org/10.1038/nrclinonc.2014.192.2.
2. SEER Cancer Statistics Review [Internet]. Available from: https://seer.cancer.gov/csr/1975_2013/browse_csr.php?sectionSEL=20&pageSEL=sect_20_table.06.html. Accessed 31 Mar 2017.
3. Giefing M, Wierzbicka M, Szyfter K, Brenner JC, Braakhuis BJ, Brakenhoff RH, et al. Moving towards personalised therapy in head and neck squamous

cell carcinoma through analysis of next generation sequencing data. Eur J Cancer. 2016;55:147–57. https://doi.org/10.1016/j.ejca.2015.10.070.
4. Biomarkers Definitions Working Group. Biomarkers and surrogate endpoints: preferred definitions and conceptual framework. Clin Pharmacol Ther. 2001;69:89–95.
5. Cosway B, Paleri V, Wilson J. Biomarkers predicting chemotherapy response in head and neck squamous cell carcinoma: a review. J Laryngol Otol. 2015;129(11):1046–52. https://doi.org/10.1017/S0022215115002479.
6. Bishop JA, Lewis JS Jr, Rocco JW, Faquin WC. HPV-related squamous cell carcinoma of the head and neck: an update on testing in routine pathology practice. Semin Diagn Pathol. 2015;32(5):344–51. https://doi.org/10.1053/j.semdp.2015.02.013.
7. Westra WH, Lewis JS Jr. Update from the 4th edition of the World Health Organization classification of head and neck tumors: oropharynx. Head Neck Pathol. 2017;11(1):41–7. https://doi.org/10.1007/s12105-017-0793-2.
8. Bossi P, Resteghini C, Paielli N, Licitra L, Pilotti S, Perrone F. Prognostic and predictive value of EGFR in head and neck squamous cell carcinoma. Oncotarget. 2016;7(45):74362–79. https://doi.org/10.18632/oncotarget.11413.
9. Burtness B, Goldwasser MA, Flood W, Mattar B, Forastiere AA, Phase III. Randomized trial of cisplatin plus placebo compared with cisplatin plus cetuximab in metastatic/recurrent head and neck cancer: an eastern cooperative oncology group study. J Clin Oncol. 2005;23:8646–54.
10. National Comprehensive Cancer Network. Head and neck cancers (Version 1.2017). Available from: https://www.nccn.org/professionals/physician_gls/PDF/head-and-neck.pdf. Accessed 31 Mar 2017.
11. Ferris RL, Blumenschein G Jr, Fayette J, Guigay J, Colevas AD, Licitra L, et al. Nivolumab for recurrent squamous-cell carcinoma of the head and neck. N Engl J Med. 2016;375(19):1856–67. Epub 2016 Oct 8.
12. Seethala RR, Griffith CC. Molecular pathology: predictive, prognostic, and diagnostic markers in salivary gland tumors. Surg Pathol Clin. 2016;9(3):339–52. https://doi.org/10.1016/j.path.2016.04.002.
13. Cohen EE, Harrington KJ, Le Tourneau C, Dinis J, Licitra L, Ahn M-J, Soria A, Machiels J-P, Mach N, Mehra R, Burtness B, Wang Y, Tuozzo AJ, Swaby R, Soulieres D. Abstract LBA45_PR 'Pembrolizumab (pembro) vs standard of care (SOC) for recurrent or metastatic head and neck squamous cell carcinoma (R/M HNSCC): phase 3 KEYNOTE-040 trial' will be presented by Dr. Cohen during proffered paper session 'Head and neck cancer' on Monday, 11 September 2017, 15:00 to 16:20 (CEST) in Granada Auditorium.
14. Bauml J, Seiwert TY, Pfister DG, Worden F, Liu SV, Gilbert J, et al. Pembrolizumab for platinum- and cetuximab-refractory head and neck cancer: results from a single-arm, phase II study. J Clin Oncol. 2017;22:JCO2016701524. https://doi.org/10.1200/JCO.2016.70.1524.
15. Elser C, Siu LL, Winquist E, et al. Phase II trial of sorafenib in patients with recurrent or metastatic squamous cell carcinoma of the head and neck or nasopharyngeal carcinoma. J Clin Oncol. 2007;25:3766–73.
16. Machiels JP, Henry S, Zanetta S, et al. Phase II study of sunitinib in recurrent or metastatic squamous cell carcinoma of the head and neck: GORTEC 2006-01. J Clin Oncol. 2010;28:21–8.
17. Hechtman JF, Benayed R, Hyman DM, Drilon A, Zehir A, Frosina D, Arcila ME, Dogan S, Klimstra DS, Ladanyi M, Jungbluth AA. Pan-Trk immunohistochemistry is an efficient and reliable screen for the detection of NTRK fusions. Am J Surg Pathol. 2017;41(11):1547–1551.
18. McNally LR, Rosenthal EL, Zhang W, Buchsbaum DJ. Therapy of head and neck squamous cell carcinoma with replicative adenovirus expressing tissue inhibitor of metalloproteinase-2 and chemoradiation. Cancer Gene Ther. 2009;16(3):246–55.

Predictive Biomarkers and Targeted Therapies in the Skin

43

Aaron Phelan and Simon J. P. Warren

Melanoma

Traditional therapies for melanoma previously centered on excision with limited options for metastatic disease. Dacarbazine has been used since its approval in 1976 as first-line treatment for metastatic melanoma. However, clinical trials have shown a response rate of 7–12% and median overall survival of 5.6–7.8 months after initiation of treatment. Only recently with the approval of interferon in 1996 and high-dose IL2 in 1998 have we seen the development of targeted therapies that have begun to replace more traditional therapies. The following sections will present the more significant recent advances in targeted therapy and predictive biomarkers for melanoma (see Table 43.1).

A. Phelan
Department of Pathology and Laboratory Medicine, Indiana University School of Medicine, Indianapolis, IN, USA

S. J. P. Warren (✉)
Departments of Pathology and Dermatology, Indiana University, Indianapolis, IN, USA
e-mail: simwarre@iupui.edu

Predictive Biomarkers for Melanoma

BRAF Inhibitors in Melanoma

Approximately 80% of nevi and 50% of melanomas [1] (as well as 7–8% of all cancers) have an activating BRAF mutation. This mutation is the only mutation present in many nevi. A small percentage of nevi evolve into melanoma, and this typically requires one or several additional mutations or amplifications [2] such as p16, CCND1, CDK4/6, PTEN, BRAF amplification, or RAS amplification. However, the original mutation in BRAF is retained as a driver mutation, making it an exciting prospect for targeted therapies in melanoma [3].

BRAF protein is a component of the mitogen-activated protein kinase signaling pathway (MAPK; see Fig. 43.1). The pathway begins with binding of extracellular signaling molecules to receptor tyrosine kinases on the cell surface. This binding results in the activation of RAS and downstream BRAF. In non-mutated cells, the BRAF protein dimerizes with other RAF enzymes (ARAF or CRAF) resulting in phosphorylation of MEK kinases and activation of ERK. ERK activation regulates growth and survival of the cell. Constitutively active BRAF mutations can have a 400-fold increase in activity and result in uncontrolled cell growth.

S. Badve, G. L. Kumar (eds.), *Predictive Biomarkers in Oncology*,
https://doi.org/10.1007/978-3-319-95228-4_43

Table 43.1 Predictive biomarkers and targeted therapies in the skin

Gene/RNA protein biomarkers	Function/pathogenic process	Patient selection method and sample type	Successfully completed clinical trials and # of patients enrolled	Clinical use and limitations	Approved drug (generic/trade name)	Company
BRAF	Key protein in MAPK signaling transduction pathway	BRAF mutation testing on primary or metastatic melanoma, using formalin-fixed paraffin-embedded (FFPE) tissue	BRIM 3, Chapman PB, et al. *N Engl J Med.* 2011 Jun 30;364(26):2507–16 675 patients	Metastatic or unresectable melanoma withV600E or V600K mutation	Vemurafenib (Zelboraf®)	Genentech/Roche
			Larkin J, et al. *N Engl J Med.* 2014;371:1867–1876 495 patients		Vemurafenib and Cobimetinib	
			Hauschild A, et al. *Lancet.* 2012 Jul 28;380(9839):358–65. 250 patients		Dabrafenib/Tafinlar	Novartis
			MEK115306, ClinicalTrials.gov Identifier: NCT01584648 340 patients		Dabrafenib (Tafinlar®)	
MEK	Key protein in MAPK signaling transduction pathway	BRAF mutation testing on primary or metastatic melanoma	METRIC, Flaherty KT, et al. *N Engl J Med.* 2012;367:107–114 322 patients	Metastatic or unresectable melanoma with V600E mutation, no prior BRAF inhibitor therapy	Trametinib (Mekinist®)	GlaxoSmithKline
		FFPE tissue	See above	Only approved as combination	Cobimetinib (Cotellic™)	Genentech/Roche
c-KIT	C-KIT receptor tyrosine kinase	c-KIT mutation testing in mucosal and acral melanoma FFPE tissue	Hodi FS, et al. *J Clin Oncol.* 2013;31(26):3182–3190 24 patients	Metastatic melanoma with C-kit mutation or amplification	Imatinib mesylate (Gleevec®)	Novartis
CTLA-4	Immune suppression/evasion	Histology FFPE tissue	EORTC 18071, 951 patients ClinicalTrials.gov Identifier: NCT00636168	Melanoma in regional lymph nodes of more than 1 mm, after resection	Ipilimumab (Yervoy®)	Bristol-Myers Squibb
			Postow M 2015, *N Engl J Med.* 2015, 21;372(21):2006–17 142 patients	BRAF wild-type with metastatic or unresectable disease	Ipilimumab (Yervoy®) Nivolumab (Opdivo®)	

PD-L1	Immune suppression/evasion	PD-L1 immunostain available for FFPE tissue	Robert C, et al., *N Engl J Med.* 2015. 372(26):2521–2532, 834 patients KEYNOTE-002, 540 patients	Metastatic melanoma. Progression after ipilimumab and BRAF inhibitor failure	Pembrolizumab (Keytruda®)	Merck
PD-L1	Immune suppression/evasion	PD-L1 immunostain available for FFPE tissue	CheckMate-037 Larkin J, et. al. *Lancet Oncol.* 2015, 6(4):375–84 370 patients	Disease progression after ipilimumab and BRAF inhibitor failure	Nivolumab (Opdivo®)	Bristol-Myers Squibb
PTCH-1 and SMO	Sonic hedgehog signaling pathway	Histology FFPE tissue	Sekulic A, et al. *N Engl J Med.* 2012;366:2171–9 104 patients	Basal cell carcinoma: Locally advanced, recurrent after surgery, or not candidate for surgery or radiation	Vismodegib (Erivedge®)	Genentech/Roche
			BOLT, Dummer R, et al., 2016 annual meeting of the American Society of Clinical Oncology (ASCO), Chicago, IL, June 4, 2016, poster #9538. 230 patients	Basal cell carcinoma: locally advanced, recurrent after surgery or radiation, or not candidate	Sonidegib (Odomzo®)	Novartis
PDGFB-like protein	COL1A1-PDGFB translocation resulting in PDGF-like fusion protein	FISH for 17: 22 translocation on FFPE	McArthur GA 2005, *J Clin Oncol.* 2005 Feb 1;23(4):866–73. 10 patients	Unresectable, recurrent or metastatic DFSP	Imatinib mesylate (Gleevec®)	Novartis
PD-L1	Immune suppression/evasion	PD-L1 immunostain on FFPE, patients eligible irrespective of PD-L1 status	JAVELIN ClinicalTrials.gov Identifier: NCT02155647 88 patients	Metastatic Merkel cell carcinoma	Avelumab (Bavencio®)	Pfizer/Merck KGaA

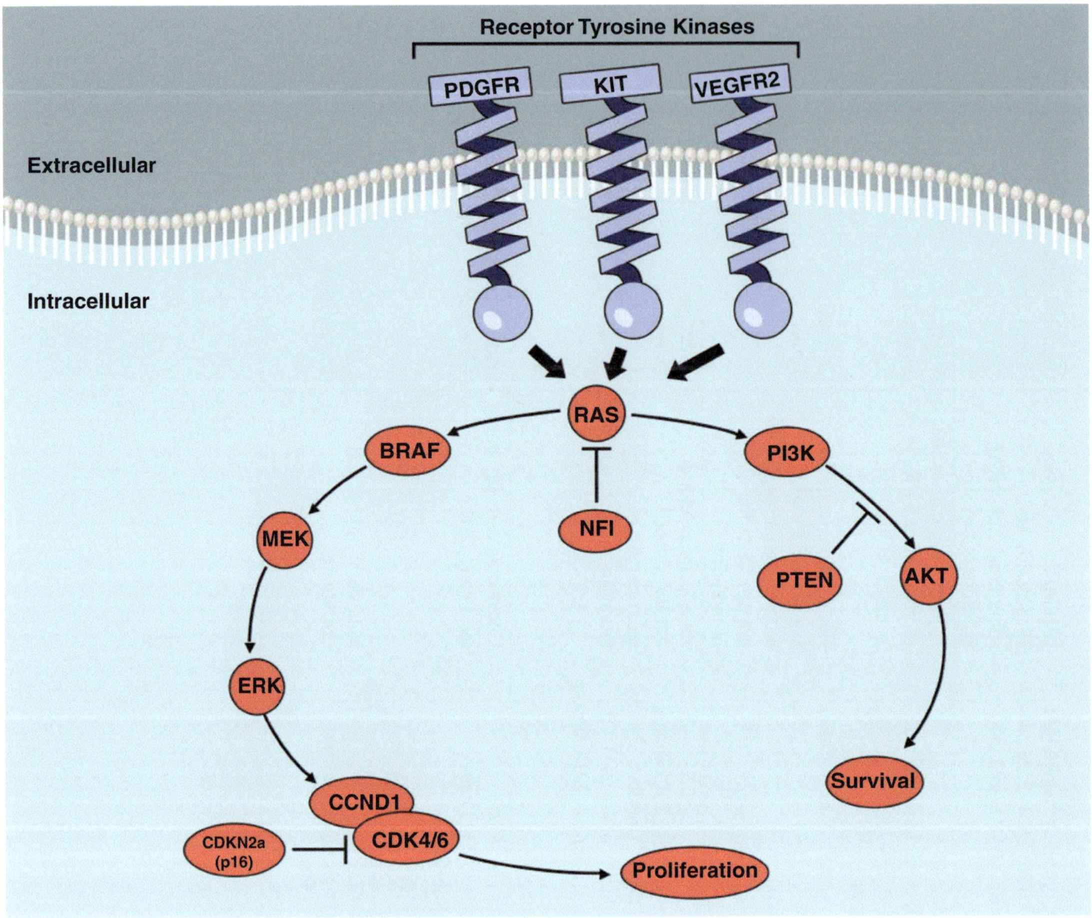

Fig. 43.1 Mitogen-activated protein kinase signaling pathway (See text for details)

Vemurafenib is a drug developed to bind to and inhibit a specific mutated form of BRAF (V600E) in which valine is replaced by glutamic acid at position 600. V600E represents 75% of mutations at this site; other common mutations are V600K (17% of cases) and V600R (2.6% of cases). The drug has some activity against V600K and V600R. Of note, only V600E is reliably detected by the FDA approved cobas® 4800 V600 test, with approximately 70% of V600K detected by this technique.

Vemurafenib treatment results in response rates of 48% in metastatic melanoma with median progression-free survival of 5.3 months. Since the approval of vemurafenib, another selective BRAF inhibitor dabrafenib has been added to the treatment of this disease.

Sequencing of the tumor to identify the BRAF mutation is important for two reasons: (1) targeted therapy with BRAF inhibitors and MEK inhibitors have no activity against the wild-type (non-mutated) BRAF, and (2) these agents might accelerate the growth of some melanomas with wild-type BRAF.

MEK Inhibitors in Melanoma

A major issue encountered with BRAF inhibitors is the development of resistance to the drug, in approximately half of patients by 5 months after starting treatment [4]. Resistance mechanisms may involve reactivation of the MAPK pathway, through either acquisition of additional mutations in RAS or MEK, amplification of BRAFV600E, or the loss of tumor suppressor genes such as P16/CDKN2a (see Fig. 43.1). Resistance mechanisms such as an activating mutation in MEK can be present in a small subpopulation of the original tumor, so that when BRAF inhibitor therapy is initiated, this subpopulation could have a survival advantage.

MEK is a kinase downstream of BRAF in the MAPK pathway (see Fig. 43.1). When used as monotherapy, MEK inhibitors have less activity in melanoma than BRAF inhibitors. However, trials of combining MEK and BRAF inhibitors have offered an improved response over BRAF inhibitors alone. Specifically, combination therapy with vemurafenib plus the MEK inhibitor cobimetinib offers an improved objective response of 68% (vs. 45% for monotherapy with vemurafenib), with complete responses in 10% (vs. 4% in the control group).

Squamous cell carcinomas have been reported during BRAF monotherapy in 19–26% of patients, and some other secondary malignancies have been reported in case reports including chronic myeloid leukemia, pancreatic adenocarcinoma, and colonic adenocarcinoma. The postulated mechanism is paradoxical activation of wild-type BRAF by vemurafenib in other cell types that already are in a premalignant state due to RAS mutation. Interestingly, the use of combination therapy has been reported to reduce the incidence of secondary cutaneous squamous cell carcinomas (3% of patients vs. 11% with BRAF monotherapy). Presumably, use of a MEK inhibitor compensates for the paradoxical stimulation of BRAF by BRAF inhibitors in keratinocytes with wild-type BRAF.

c-Kit Inhibitors in Melanoma

c-Kit (also known as CD117 or mast/stem cell growth factor receptor (SCFR)) is a receptor tyrosine kinase that mediates cell proliferation (see Fig. 43.1). Its activity is inhibited by imatinib (which also inhibits other tyrosine kinases such as the PDGF receptor tyrosine kinase and the Abl kinase). Previous phase II studies using imatinib in *unselected* melanoma patients or patients selected by immunohistochemistry for imatinib targets failed to demonstrate clinical activity [5]. Activating mutations or amplifications in c-Kit have been identified in approximately 25% of acral and mucosal melanomas, and these patients typically do not have BRAF mutations. In this subgroup, c-Kit mutation status is useful as a predictive biomarker for imatinib response [5]. Mutations in the juxtamembrane region (exons 11, 13) appear to predict responsiveness to imatinib. In a phase II trial of imatinib in 25 patients with melanoma in mucosal, acral, and chronically damaged skin with mutations or amplifications in c-Kit, there was a disease control rate of 50%. This was predominantly in patients with mutated rather than amplified c-Kit [5].

CTLA-4 Blockade

Melanoma is a tumor in which the immune system may sometimes be effective in combating primary and metastatic disease. Ipilimumab gained FDA approval in 2010. This drug augments the immune response to melanoma. Ipilimumab is a monoclonal antibody to cytotoxic T-lymphocyte antigen-4 (CTLA-4), a cell surface protein primarily present on activated T lymphocytes [6]. CTLA-4 is expressed approximately 2 days after T-cell activation and acts to prevent T-cell overactivation (see Fig. 43.2). CTLA-4 competes with CD28 (another protein expressed on T cells) to bind with B7 receptors on antigen-presenting cells. The binding of CD28 to B7 acts as a co-stimulatory signal for T-cell activation. In contrast, CTLA-4 binding to B7 downregulates T-cell activity. Blockade of CTLA-4 by ipilimumab, therefore, results in unopposed co-stimulation by CD28 and potentiates the antitumor T-cell response [6].

Ipilimumab is effective in some, but not in all melanoma patients, suggesting that a predictive biomarker would be helpful in selecting patients that would benefit from treatment. A high mutational load (as seen in melanomas from severely sun damaged skin) correlates with prolonged benefit, presumably because a high mutational load results in the synthesis of immunogenic

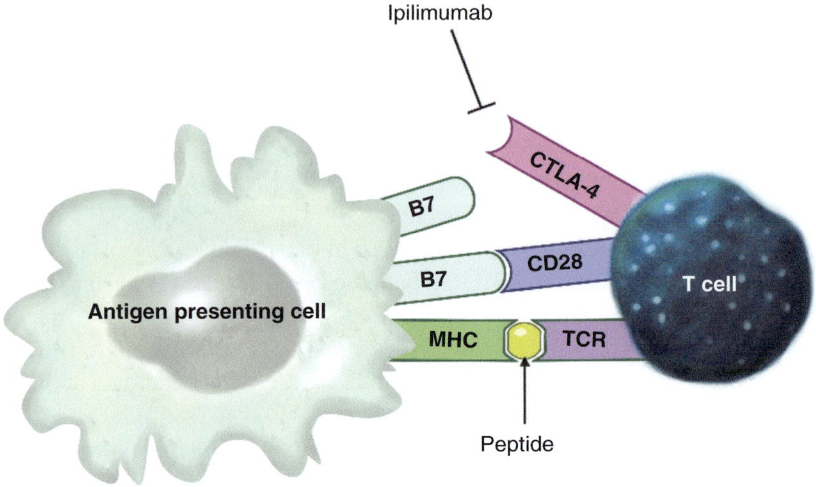

Fig. 43.2 CTLA-4 and the mechanism of action of ipilimumab. Binding of B7 molecule from the antigen-presenting cell onto the CD28 receptors on the T cell results in T-cell activation. Binding of ipilimumab to CTLA-4 results in potentiation of T-cell proliferation. MHC major histocompatibility complex, TCR T-cell receptor, CTLA-4 cytotoxic T-lymphocyte antigen-4

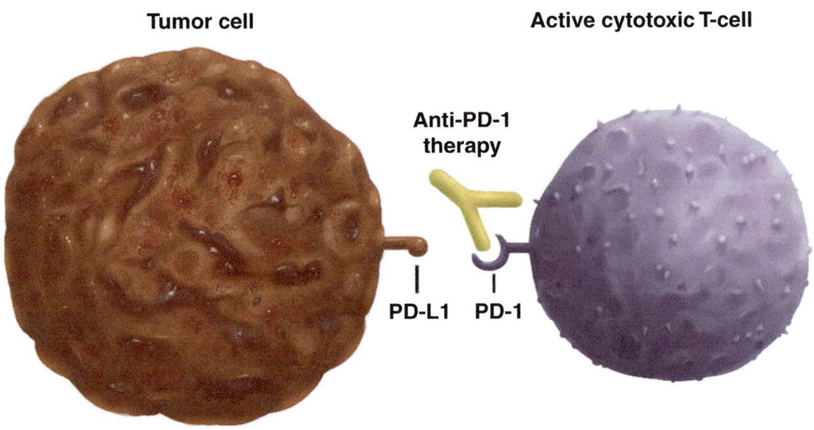

Fig. 43.3 PD-1 pathway in melanoma. Anti-PD-1 antibodies such as nivolumab and pembrolizumab that impair the ability of PD-L1 to interact with PD-1 may prevent the ability of PD-L1 to suppress T-cell activation, helping to promote the antitumor response

neoantigens by tumor cells. Specific tumor neoantigens have been identified that are shared among patients with a prolonged benefit but not with patients with a minimal benefit [7]; however, subsequent studies have failed to reproduce these findings [8]. Whole transcriptome studies designed to identify features of the host immune response to the tumor have identified transcription of CTLA-4 itself, as well as perforin and granzyme B as significant predictive biomarkers for good response to CTLA-4 blockade [8].

PD-1 Inhibitors

Another pathway that has been successfully targeted in melanoma therapy is the interaction of programmed cell death ligand 1 (PD-L1) and the programmed cell death receptor PD-1 (see Fig. 43.3). PD-L1 can be expressed by melanoma cells, as well as other cancer cells [9]. PD-L1 on melanoma cells binds to PD-1 receptors on the surface of T lymphocytes and suppresses their activity. In this way, the tumor can evade the host immune defenses. There are currently at least two monoclonal antibodies (nivolumab and pembrolizumab) against the PD-1 receptor that are FDA approved for the treatment of metastatic or unresectable melanoma [6]. These monoclonal antibodies bind to PD-1 on the surface of lymphocytes and block inactivation of the lymphocyte by PD-L1.

About a third of patients with melanoma respond to PD-1 targeted treatment. Higher response rates have been reported with combination of anti-CTLA-4 and anti-PD1 therapies (with an overall response rate of 57.6%) but are associated with more severe (grade 3 or grade 4) adverse effects. In patients whose melanomas express PD-L1 (defined as any degree of expression on >5% of cells, totaling approximately 23% of patients), no additional benefit appears to be derived from combination therapy over monotherapy with the PD-1 inhibitor nivolumab alone. In contrast, the PD-L1 negative cohort did better with combination therapy. This suggests that combination therapy might be reserved for this subgroup of patients, and for them, the risk of more severe adverse events might be more acceptable.

Although detection of PD-L1 on the surface of melanoma cells would in theory help to predict response, several studies have shown a response to PD-1 inhibitors despite tumor cells testing negative for PD-L1 [10]. The presence of CD8+ cells at the periphery of the melanoma in pretreatment biopsies, however, has been linked to a good response to PD-1 inhibition. Also, an increase in CD8 positive cells in biopsies following treatment has been associated with a good response [6, 10]. Further, PD-L1 gene amplifications have been identified in a subset of BRAF-mutated melanomas and may prove to be a useful predictive biomarker for good response to PD-1 inhibitors.

Use of Talimogene Laherparepvec (T-VEC) Viral Oncolytic Immunotherapy in Melanoma

Talimogene laherparepvec (T-VEC) is a modified herpes virus that is capable of infecting tumor cells in melanoma after direct injection into metastatic lesions [11]. T-VEC was approved for metastatic melanoma to the skin or lymph nodes by the FDA in 2015. Cutaneous metastases are common in melanoma so that direct injection of the virus is often feasible (see Fig. 43.4). The T-VEC modified herpes virus lacks the *ICP34.5* gene. In healthy cells, this gene allows replication of the virus by counteracting the interferon-induced block to virus replication. Healthy cells are therefore resistant to infection by the modified virus. However, tumor cells often have a disabled antiviral response pathway and therefore are still susceptible to infection (Fig. 43.4a).

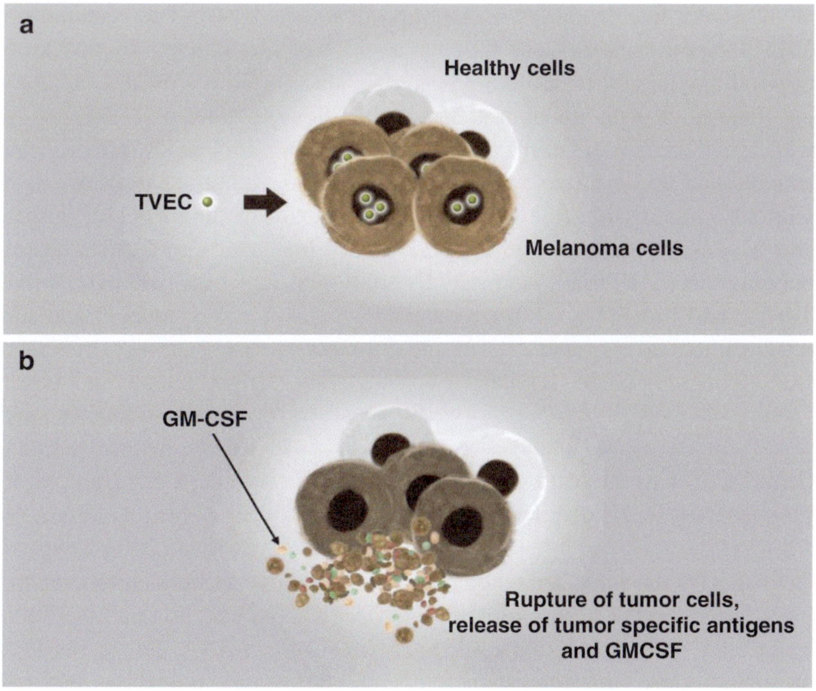

Fig. 43.4 Mechanism of action of T-VEC. (**a**) Inside a healthy cell, the virus is unable to replicate. This leaves the healthy cell unharmed. (**b**). Inside a melanoma cell, the virus replicates resulting in cell lysis, releasing more virus, GM-CSF, and tumor-specific antigens. (**c**) GM-CSF attracts dendritic cells to the site, which process and present the antigens to the T cells. The T cells are now "programmed" to identify and destroy cancer cells throughout the body including those that are not directly injected with the virus

Fig. 43.4 (continued)

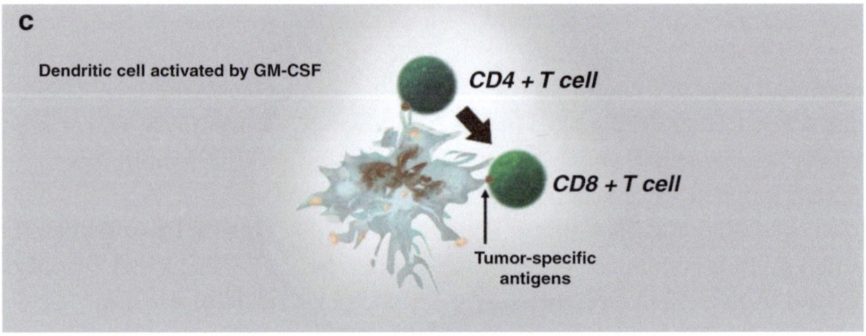

The modified virus also lacks *ICP47*, which normally blocks antigen processing in infected cells, resulting in increased antigen processing and an improvement in the immune stimulating properties of the virus. Thirdly, the modified virus expresses granulocyte-macrophage colony-stimulating factor (GM-CSF) which, after lysis of the infected cell, promotes T-cell activation and differentiation of progenitor cells into dendritic cells for antigen presentation in the tumor microenvironment (Fig. 43.4b, c). Tumor cell lysis consequently generates a patient-specific in situ tumor cell vaccine with enhanced presentation of tumor neoantigens. This could explain tumor regression that can occur at non-injected sites. A phase III trial demonstrated a 26% overall response rate leading to FDA approval in 2015. Responses seem to be durable, as 93% of responding patients were alive 1 year after starting therapy. A subsequent trial in combination with the CTLA-4 inhibitor ipilimumab in 19 patients had an objective response rate of 50%, with 50% progression-free survival at 18 months and a complete response in 22%. There is an ongoing phase II trial. A phase Ib trial with T-VEC and the PD-1 inhibitor pembrolizumab demonstrated a response rate of close to 60% with an ongoing phase III trial.

Basal Cell Carcinoma and Sonic Hedgehog Pathway Inhibitors

Activation of the sonic hedgehog pathway (SHH) has been identified in the initiation of basal cell carcinoma. It also has a role in the later stages of growth of some cutaneous squamous cell carcinomas [12]. Binding of the extracellular sonic hedgehog protein (SHH) to the membrane protein "Patched" (PTCH-1) results in the release of the smoothened receptor (SMO) and downstream activation of GLI-1 (Fig. 43.5, left panel). SMO is a G protein-coupled receptor, which activates the GLI-1 transcription factor. GLI-1 then travels to the nucleus where it promotes transcription of genes that increase cell growth and survival. This pathway is active in embryogenesis but inactive in most adult tissues. Mutations that inactivate PTCH-1 and allow it to release SMO or mutations that activate SMO can result in tumor formation (Fig. 43.5, middle panel). In sporadic basal cell carcinoma, inactivating mutations in PTCH-1 are responsible for the disease in 80% of cases. The remainder are caused by activating mutations in SMO, the molecule downstream from PTCH-1. In nevoid basal cell carcinoma syndrome (NBCCS), germ line deletion of one copy of the PTCH-1 gene results in the development of dozens to thousands of basal cell carcinomas [12], although sun exposure and fair skin also seem to be cofactors in the development of basal cell carcinomas in NBBCS.

Vismodegib and sonidegib inhibit the hedgehog pathway by inactivating SMO (Fig. 43.5, right panel) and blocking subsequent cell survival/proliferation signals. Vismodegib and sonidegib received FDA approval for the treatment of recurrent, locally advanced or metastatic basal cell carcinoma in 2012 and 2015, respectively. Locally advanced and metastatic basal cell carcinomas generally retain overactivity in the hedgehog pathway as measured by raised GLI-1 mRNA. The response rate in metastatic basal cell carcinoma is approximately 33% [12], with

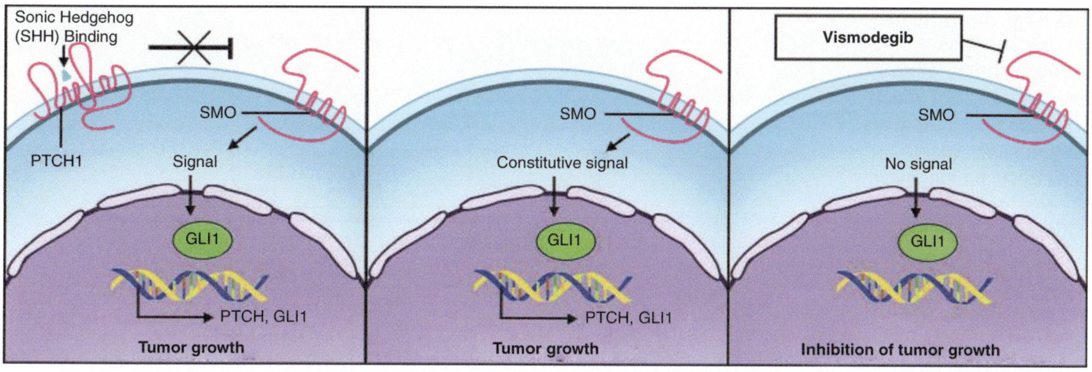

Fig. 43.5 Hedgehog signaling pathway and mechanism of action of vismodegib. Binding of the extracellular sonic hedgehog protein (SHH) to the membrane protein "Patched" (PTCH-1) results in the release of the smoothened receptor (SMO) and downstream activation of GLI-1 (left panel). GLI-1 travels to the nucleus where it promotes transcription of genes that increase cell growth and survival. Mutations that inactivate PTCH-1 and allow it to release SMO or mutations that activate SMO can result in tumor formation (middle panel). Vismodegib inhibits the hedgehog signaling pathway by inactivating SMO (right panel) and blocking subsequent cell survival/proliferation signals

induction of new mutations in SMO that leads to loss of vismodegib binding apparently responsible for acquired resistance. The response rate in locally advanced basal cell carcinoma is higher at 47%. CD56 expression by immunoperoxidase has been reported as a predictive biomarker that weakly predicts lack of response to vismodegib in advanced basal cell carcinoma. GLI-1 levels in metastatic tumor have so far not been predictive of response to vismodegib. Multiple primary basal cell carcinomas in NBCCS have also been successfully treated with these drugs [12] with two new surgically eligible basal cell carcinomas/year compared to 34 new basal cell carcinomas/year in controls. However, a significant number of patients discontinue therapy due to adverse events and existing tumors recur after discontinuation of therapy.

Dermatofibrosarcoma Protuberans and PDGF Pathway

Imatinib has also been FDA approved [13] for targeted treatment of dermatofibrosarcoma protuberans (DFSP). Most DFSPs (more than 90%) exhibit a translocation between chromosome 17 and 22 [t(17, 22)] which results in platelet-derived growth factor B (PDGFB) coming under the control of an active collagen 1A1 (COL1A1) promoter (Fig. 43.6). PDGFB protein is a ligand for the platelet-derived growth factor receptor (PDGFR) present on the surface of tumor cells which regulates cell growth and survival. Imatinib is a tyrosine kinase inhibitor and has activity against PDGFR (see also Fig. 43.1). There are no extensive head-to-head clinical trials evaluating the effectiveness of imatinib in translocation-positive versus translocation-negative DFSP. However, in smaller studies, translocation negative tumors show no response to imatinib with translocation positive tumors showing response rates of approximately 45–80% so that testing for the presence of the t(17–22) translocation is advisable before therapy. In isolated case reports, this also appears to be true in cases of fibrosarcoma and undifferentiated pleomorphic sarcoma arising from DFSP, where retention of the t(17–22) translocation is predictive of a response to imatinib. As DFSP typically infiltrates locally along fascial planes and requires wide margins to adequately excise, preoperative therapy with imatinib has also been advocated to enable resection of previously unresectable disease.

Merkel Cell Carcinoma

Merkel cell carcinoma (MCC) is an aggressive skin cancer that had been thought to arise from

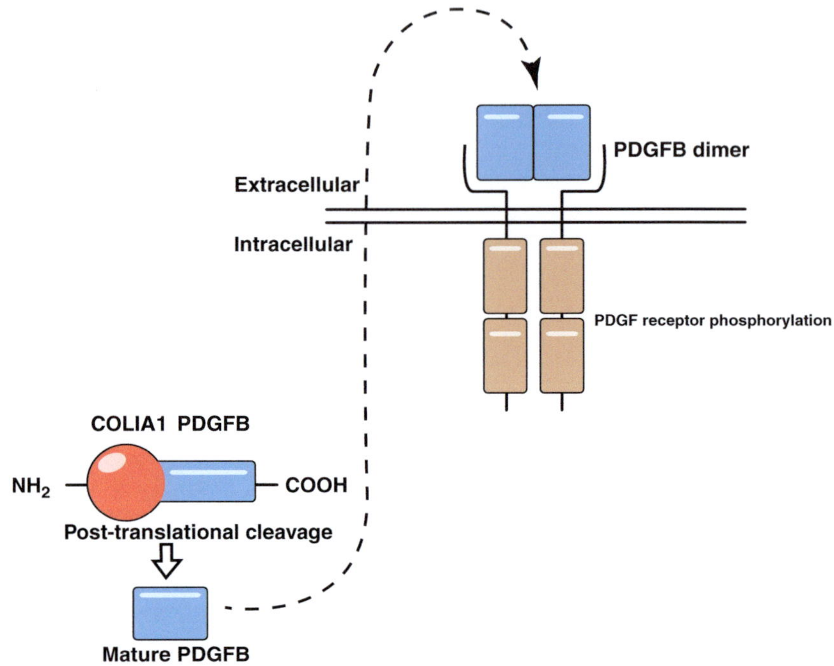

Fig. 43.6 In DFSP there is a translocation between chromosome 17 and 22 [t(17, 22)] which results in platelet-derived growth factor B (PDGFB) coming under the control of the collagen 1A1 (COL1A1) promoter. The COL1A1-PDGFB fusion protein undergoes posttranslational cleavage, and then PDGFB protein binds platelet-derived growth factor receptor present on the surface of tumor cells to stimulate tumor growth

Merkel cells in the epidermis. However, recent research has raised the possibility of origin from pluripotent dermal stem cells. A link between Merkel cell carcinoma and the Merkel cell polyomavirus is identified in approximately 80–97% of patients. The virus is clonally integrated into the MCC genome and unable to replicate. Merkel cell polyomavirus is oncogenic in MCC via the activity of its large T (LT) and small T (sT) proteins. LT protein targets cellular pocket proteins such as pRB, p107, and p130 as well as p53. Immunosuppression including AIDS and solid organ transplant are also important cofactors for MCC and are present in 10% of MCC patients. UV light is another cofactor. Polyomavirus-negative tumors have a 100-fold higher mutational load than polyomavirus-positive tumors [14], reflecting the importance of UV light in this subset of patients.

There are case reports of MCC regressing after improvement in immune function, and the presence of intratumoral CD8-positive cells detected by immunohistochemistry has been associated with a favorable prognosis. Specifically, in a subgroup of MCC patients (26 of a total of 146 patients) with more than 60 CD8-positive cells per high-power field, there was 100% disease-specific survival at 5 years [15], compared with 60% survival among the remaining 120 patients with lesser degrees of intratumoral CD8 cells. For these reasons, treatments that improve immune response may have potential in MCC.

Recently, encouraging results have been reported with avelumab, an antibody targeting PD-L1 (Fig. 43.3) that has led to FDA approval [16]. Of the 88 patients in this phase II study with chemotherapy-refractory metastatic Merkel cell carcinoma, 28 (32%) achieved an objective response including eight complete responses. Responses occurred irrespective of PD-L1 status, defined as more than 1% of Merkel cells expressing PD-L1 by immunostain at any intensity, although a higher percentage of PD-L1-positive patients benefited compared to PD-L1-negative patients (36% vs. 18%). Responses also occurred irrespective of Merkel cell polyomavirus status. Pembrolizumab, which targets PD-1 (Fig. 43.3), has also recently been evaluated in Merkel cell carcinoma [14]. In a group of 25 patients who had not previously received chemotherapy, there was an objective response in 14 (56%), with 12 of the 14 responses ongoing at last follow-up. In this paper, there was no correlation between response and PD-L1 status or Merkel cell polyomavirus status.

Guidelines/Consensus Statements

No ASCO or CAP guidelines exist for the testing of biomarkers in skin cancers. The FDA approvals for the BRAF and MEK inhibitors require testing for BRAF V600 mutations using an FDA-approved test prior to treatment.

Future Directions

N-RAS Inhibitors in Melanoma

N-RAS is mutated in 15–20% of melanomas and is a driver mutation of similar significance to BRAF mutation [4, 17]. N-RAS mutations are only rarely present in benign nevi, in contrast to BRAF. RAS proteins are a component of the MAP kinase pathway (Fig. 43.1). Activating mutations are present in N-RAS codon 61 in more than 80% of cases, while mutations in codons 12 and 13 are less common. Direct targeting of N-RAS mutants has so far been unsuccessful. Although BRAF is downstream of RAS, BRAF inhibitors cannot be used as they may paradoxically activate wild-type BRAF in RAS-mutated malignancies and consequently promote proliferation. Therefore, targets further downstream such as MEK have been proposed. Binimetinib is a MEK1/2 inhibitor; in a phase II trial and an ongoing phase III trial, a progression-free survival of 3.7 months has been reported in patients with N-RAS Q61 mutant melanoma. A phase 1/2 trial using binimetinib plus ribociclib (a CDK4/6 inhibitor, also see Fig. 43.1) has reported an overall response rate of 33%. In addition, as N-RAS mutants are capable of activating the P13K/AKT pathway (see Fig. 43.1), several trials are examining the efficacy of combinations of MEK inhibitors and P13K/AKT inhibitors.

Atypical BRAF Mutations in Melanoma

The more common BRAF mutations (i.e., BRAF V600E/K/R) have been discussed above. A large study [18] of 1112 cases of melanoma sequenced at BRAF exons 11 and 15 showed that the V600E mutation represents only 75% of mutations analyzed, with V600K the second most common at 17% and V600R at 2.6%. There are also case reports of V600M and a complex mutation V600D K601del being treated with both clinical and radiological response. In addition, there are mutations at other positions such as BRAF L597, K601, G466, and BRAF fusion genes that together comprise up to 5% of all melanomas. These mutations may not be detected if the only sequencing method used is the FDA-approved COBAS 4800 test for V600E but are important as they may be sensitive to MEK inhibitors.

Summary

Targeted therapies in skin cancer have become increasingly important as they offer improved survival for patients due to specific features of the tumor or a specific subgroup of that tumor. Predictive biomarkers will be increasingly important to guide therapy as possibilities for therapy increase, and all members of the skin cancer diagnosis and treatment teams will need to be aware of these exciting advances.

References

1. Foth M, Wouters J, de Chaumont C, et al. Prognostic and predictive biomarkers in melanoma: an update. Expert Rev Mol Diagn. 2016;16:223–37.
2. Shain AH, Bastian BC. From melanocytes to melanomas. Nat Rev Cancer. 2016;16:345–58.
3. Rajkumar S, Watson IR. Molecular characterisation of cutaneous melanoma: creating a framework for targeted and immune therapies. Br J Cancer. 2016;115:145–55.
4. Fedorenko IV, Gibney GT, Sondak VK, et al. Beyond BRAF: where next for melanoma therapy? Br J Cancer. 2015;112:217–26.

5. Hodi F, Corless C, Giobbie Hurder A, et al. Imatinib for melanomas harboring mutational activated or amplified Kit arising on mucosal, acral or chronically sun damaged skin. J Clin Oncol. 2013;31:3182–90.
6. Zhu Z, Liu W, Gotlieb V. The rapidly evolving therapies for advanced melanoma – towards immunotherapy, molecular targeted therapy, and beyond. Crit Rev Oncol Hematol. 2016;99:91–9.
7. Snyder A, Makarov V, Merghoub T, et al. Genetic basis for clinical response to CTLA4 blockade in melanoma. New Eng J Med. 2014;371(23):2189–99.
8. Allen V, Miao D, Schilling B, et al. Genomic correlates of response to CTLA4 blockade in metastatic melanoma. Science. 2015;350(6257):207–11.
9. Melero I, Hervas-Stubbs S, Glennie M, et al. Immunostimulatory monoclonal antibodies for cancer therapy. Nat Rev Cancer. 2007;7:95–106.
10. Tumeh PC, Harview CL, Yearley JH, et al. PD1 blockade induces responses by inhibiting adaptive immune resistance. Nature. 2014;515:568–71.
11. Ott PA, Hodi FS. Talimogene Laherparepvec for the treatment of advanced melanoma. Clin Cancer Res. 2016;22(13):3127–31.
12. Athar M, et al. Sonic hedgehog signaling in Basal cell nevus syndrome. Cancer Res. 2014;74(18):4967–75.
13. Rutkowski P, Van Glabbeke M, Rankin CJ, et al. Imatinib Mesylate in advanced dermatfibrosarcoma protuberans: pooled analysis of two phase II clinical trials. J Clin Oncol. 2010;28:1772–9.
14. Nghiem PT, Bhatia S, Lipson EJ, et al. PD-1 blockade with Pembrolizumab in advanced Merkel cell carcinoma. N Engl J Med. 2016;374:2542–52.
15. Paulson GP, et al. Transcriptome wide studies of Merkel cell carcinoma and validation of Intratumoral CD8+ lymphocyte invasion as independent risk factor of survival. J Clin Oncol. 2011;29:1539–46.
16. Kaufman HL, Russell J, Hamid O, et al. Avelumab in patients with chemotherapy-refractory metastatic Merkel cell carcinoma: a multicentre single-group, open-label phase 2 trial. Lancet Oncol. 2016;17:1374–85.
17. Johnson DB, Puzanov I. Treatment of NRAS-mutant melanoma. Curr Treat Options in Oncol. 2015;16(15):1–12.
18. Greaves WO, Verma S, Patel KP, et al. Frequency and spectrum of BRAF mutations in a retrospective, single institution study of 1112 cases of melanoma. J Mol Diagn. 2013;15(2):220–6.

Predictive Biomarkers and Targeted Therapies in Sarcomas

44

Hans-Ulrich Schildhaus and Sebastian Bauer

Introduction

There are more than 50 malignant soft tissue sarcomas, most of which are extremely rare on its own. Diagnosis and treatment require multidisciplinary expert teams including reference pathologists as well as highly specialized surgeons and medical oncologists. There is considerable overlap in morphologic appearance between entities, and multiple diagnostic modalities are usually applied to come to the correct diagnosis. These modalities also include molecular tests. Mutational analyses as well as methods to detect gene fusions and amplifications are employed. Technologies comprise various DNA- and RNA-based sequencing approaches as well as in situ hybridization and immunohistochemistry.

Soft tissue tumors frequently arise in unfavorable locations that preclude radical surgery or are associated with significant morbidity. This may explain the high number of local recurrences, e.g., in the context of retroperitoneal primaries. Radiotherapy greatly decreases the risk of local relapses but does not affect the survival of sarcomas. More than half of all sarcomas represent high-grade tumors that are associated with an aggregate risk of metastatic spread of approximately 50%. Systemic treatments that improve this rate are therefore direly needed. While chemotherapy is unequivocal standard in high-grade bone sarcomas, the clinical use in localized soft tissue sarcoma is still disputed – despite strong evidence of a moderate improvement of survival. Nonetheless, progress will only be made if novel treatments capitalize on biological insights about driver mutations or histologic features that can be addressed therapeutically. Most of these approaches are directly related to predictive biomarkers.

In the past, most sarcoma subtypes have been treated within trials as one disease allowing very little generalization – except that patients with locally advanced or metastatic disease are faced with a mostly dismal prognosis with a median survival that in the past decades has only marginally been improved in spite of rapid progress in other cancers. The growing understanding of the underlying biology, in particular of genetic aberrations which initiate and propagate tumor growth, has led to the development of targeted treatment options for sarcoma patients. Inhibition of receptor tyrosine kinases, blocking of activated downstream signaling molecules, and interaction with transcription factors are

H.-U. Schildhaus (✉)
Institute of Pathology, Universitätsmedizin Göttingen, Göttingen, Germany
e-mail: hans-ulrich.schildhaus@med.uni-goettingen.de

S. Bauer
Department of Medical Oncology, Sarcoma Center, West German Cancer Center, University Duisburg-Essen, Medical School, Essen, Germany

German Cancer Consortium (DKTK), Heidelberg, Germany

S. Badve, G. L. Kumar (eds.), *Predictive Biomarkers in Oncology*,
https://doi.org/10.1007/978-3-319-95228-4_44

fundamental pharmacologic mechanisms applied to sarcoma treatment. These therapies are also highly dependent on predictive biomarker assays. Many novel and innovative targeted treatments are currently being explored in clinical trials with GIST and sarcoma patients.

Gastrointestinal Stromal Tumors

Treatment of gastrointestinal stromal tumors (GISTs) is the most paradigmatic example of success for targeted treatment in solid cancer. GISTs are predominantly driven by activating mutations in *KIT* and *PDGFRA (platelet-derived growth factor receptor alpha)*, two genes which encode closely related tyrosine kinases. There is a strong genotype–phenotype correlation in this disease which extends not only to morphologic appearance and clinical presentation but also to response to certain drugs. Mutational analysis in GIST can reliably predict the effects of tyrosine kinase inhibitors, and genotyping is considered standard of care in patients who receive systemic treatment. GIST patients may experience a relapse in the disease after initial response to drugs due to secondary mutations which confer resistance. In the past years, further genotyping of patients failing treatment has revealed highly specific and also predictive resistance mutations, which may soon be relevant for treatment decisions as well. Thus, GIST can be regarded as a prototype of a mesenchymal malignancy where predictive biomarkers play a decisive role in treatment decisions. The first part of this chapter is therefore dedicated to this entity.

Gastrointestinal stromal tumors represent the most frequent mesenchymal tumors of the gastrointestinal tract; it is most likely the most common sarcoma. These tumors occur throughout the entire gastrointestinal tract from the esophagus to the rectum. Most common sites are the stomach and the small bowel. Tumors are less frequent in the rectum and relatively rare in the esophagus. They are almost never located in the colon. Sometimes GISTs are found in the abdominal cavity, especially in the upper parts without any evidence for an anatomic connection to the tubular GI tract. These lesions are referred to as extra-gastrointestinal GIST (E-GIST). Patients with GIST present frequently with gastrointestinal bleeding from ulcerated mucosa due to tumor growth. Sometimes, however, GISTs are simply accidental findings in asymptomatic patients. Small bowel or gastric tumors can grow up to an enormous size without provoking any symptoms. In contrast, minute GIST are assumed to occur very frequently – especially in the proximal stomach – but very rarely grow into a clinically relevant tumor.

Histologically, GISTs derive from or have some similarities with the interstitial cell of Cajal, which are pacemaker cells in the outer muscle wall of the GI tract. Thus, GISTs start growing between the inner and outer muscle layer where the myenteric nerve plexus is located. Most GISTs have a spindled morphology; however, also epithelioid subtypes occur. The latter morphologic variant and spindled-epithelioid mixed types are more or less confined to gastric location. Plump spindle cell morphology with a more ovoid appearance can also occur in the small bowel. This phenotype is, however, associated with more malignant behavior in these tumors (in contrast to the true epithelioid forms which are associated with less aggressive biology).

The clinical course of GIST is dependent on several factors: (i) tumor size, (ii) mitotic activity, and (iii) tumor location. These parameters have been included in a scoring system, which allows the estimation of a patient's risk for a clinical progression of the disease. Basically, the larger the tumor and the higher the mitotic count, the higher is the risk for a metastatic progression. While size, as with other tumors, is an important factor, recent studies have underscored that particularly the mitotic count is most predictive of relapse. Overall, GISTs of the stomach behave less aggressive than tumors which originate from the jejunum or ileum. Additional parameters have been identified to be associated with clinical outcome: tumor rupture or incomplete surgical removal is indicative of a significantly increased risk for an aggressive clinical behavior. Furthermore, certain molecular findings, i.e., specific mutations (see below), are prognosticators for a worse outcome.

Local recurrence is extremely rare in GIST with the exception of duodenal or rectal GIST, where radical surgery is limited by the associated morbidity. Metastases occur to the liver or to the peritoneal cavity – or very rarely outside the abdomen. Surgery is the cornerstone in localized GIST, and treatment with the KIT inhibitor imatinib is offered in patients with large GIST before surgery and as adjuvant treatment in patients with high risk of relapse. No chemotherapy has yet been found to be clinically meaningful. GISTs are considered radiosensitive, but radiotherapeutic strategies can only in rare exceptions (e.g., palliative radiation of bone metastases, symptomatic, non-resectable bulky tumors) be used due to the intra-abdominal localization.

For patients with metastatic disease, three drugs have been approved based on randomized trials: imatinib, which is the standard first-line treatment, as well as sunitinib and regorafenib.

Pathologists apply several biomarker assays to a GIST sample. Diagnostic biomarkers include mainly immunohistochemistry to establish the diagnosis. The most important IHC marker is DOG-1 (detected on GIST-1, synonymous with anoctamin 1), a calcium-dependent chloride channel which is specifically expressed in nearly all gastrointestinal stromal tumors irrespective of morphologic or genetic subtypes. Also, KIT (CD117) is expressed in the vast majority of GIST. However, there are also KIT-negative tumors (especially among gastric tumors with *PDGFRA* mutations) which express no or only very low levels of KIT (CD117). Additional (ancillary) diagnostic markers include CD34 and bcl-2. Furthermore, markers of a smooth muscle differentiation may be expressed, such as actin or h-caldesmon. Prognostic biomarkers are mainly tumor size, location, and mitotic count (with the cutoff of five mitoses in 5 mm^2 tissue). The third group of biomarkers comprises assays which allow prediction of the effects of certain systemic treatments. Some of these predictive markers, however, have also a prognostic significance.

Molecular Subtypes and Predictive Biomarkers in GIST

Predictive biomarkers in gastrointestinal stromal tumors are directly linked to tumor genetics. The molecular hallmark of GIST is an activated signaling of receptor tyrosine kinases (RTK) which cause an increase in proliferation together with blocking of apoptosis. This effect is predominantly achieved by activating mutations in genes encoding RTK or rarely by activating gene fusions. In a minority of cases activating mutations in downstream signaling molecules such as BRAF play a causative role.

Most GIST harbor activating mutations in the *KIT* gene, which encodes a class III RTK. Less frequently comparable mutations occur in the *PDGFRA* gene, which encodes another kinase (platelet-derived growth factor receptor alpha). These mutations do not appear randomly in the genes but affect either regulatory domains or the catalytically active kinase domains (Fig. 44.1). Since all these different mutational variants are activating, it is not surprising that frameshift mutations or changes which induce stop codons are not observed. Mutational subtypes consist of single nucleotide substitutions (point mutations), small deletions, combined deletion-insertion (delins) mutations, and insertion or duplications. The latter variants affect all multiples of three nucleotides to keep the mutation in frame.

Primary *KIT* mutations can affect the extracellular domains (which represent the immunoglobulin-like structures at the extracellular side), i.e., exon 8 and 9, the juxtamembrane domain (exon 11), or the tyrosine kinase domains (exons 13 and 17). In *PDGFRA* exon 12 (juxtamembrane domain), 14 or 18 (kinase domains) can be affected. All these changes lead to a ligand-independent homodimerization of receptors with subsequent autophosphorylation followed by activated downstream signaling.

Common KIT/PDGFRA Mutations

KIT exon 11 mutations form the largest molecular subtype (60% of all patients). Patients with

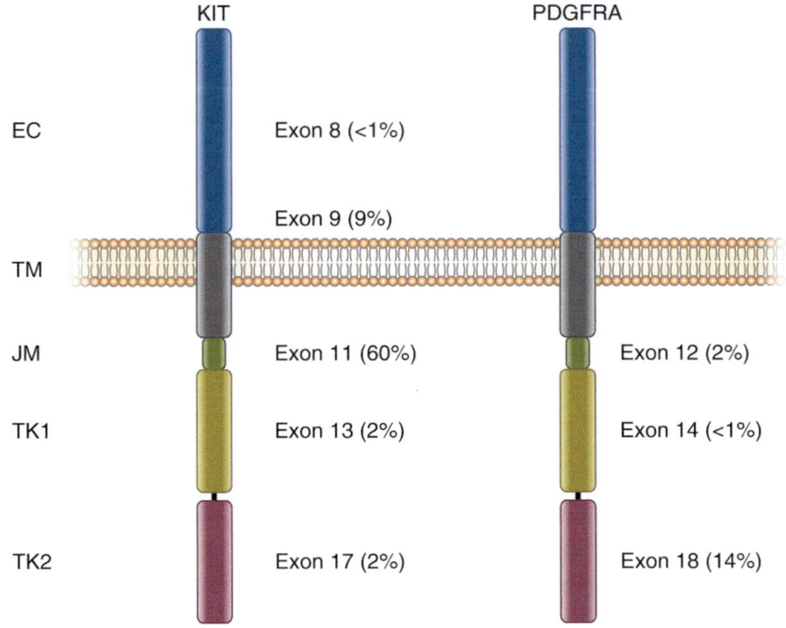

Fig. 44.1 Local distribution and relative frequencies of *KIT/PDGFRA* mutations in gastrointestinal stromal tumors. EC: extracellular domains, TM: transmembrane domain, JM: juxtamembrane domain, TK1 and TK2: kinase domains

these alterations have the best response rates under imatinib treatment. *KIT* exon 11-mutated GIST can occur anywhere in the GI tract. There are, however, some differences in the clinical impact of certain variants. In exon 11, insertions, point mutations, and deletions/delins mutations occur (Fig. 44.2). It has been shown that deletions/delins mutations affecting codons 557 and 558 are associated with a higher rate of metastases and a worse outcome (Table 44.1). They are, however, as sensitive to imatinib as to the other *KIT* exon 11 mutations [1].

The second largest group (14%) consists of GIST with *PDGFRA* exon 18 mutations which are almost restricted to gastric tumors. Some extragastrointestinal stromal tumors (E-GIST) also harbor these mutations. Tumors of this molecular subtype show nearly an exclusive epithelioid- or mixed-type morphology. In addition, multinuclear giant cells can frequently be seen and are indicative of this particular morphologic and genetic subtype. *PDGFRA* exon 18-mutant GIST behave mostly in a benign fashion. However, in multivariate analyses it seems not to be an independent prognosticator; more likely, the gastric location itself is associated with better outcome. Tumors of the stomach tend to have low mitotic counts (and gastric tumors harbor more frequently *PDGFRA* mutations). However, those tumors that exhibit typical risk factors do have a high risk of metastatic spread. In terms of predictive significance, this *PDGFRA* exon 18 mutations mostly confer primary resistance to imatinib as well as sunitinib. Novel inhibitors are under clinical development, which are effective in these genetic variants (Table 44.1). By far the most frequent variant, the D842V mutation, can be regarded as the prototype of a GIST mutation, which is associated with primary resistance to imatinib. Tumors with this genetic change do not benefit from this drug, neither if they are metastatic nor in an adjuvant or neoadjuvant setting. There is, however, roughly a total of 60 mutations which have been described in this exon, whose clinical predictive impact is highly variable. For example, the single nucleotide exchange D846V confers resistance as well – however, another mutation in the same codon (leading to D846Y mutation) is responsive. Additionally, mutations affecting codon D842 such as D842_M844del might be sensitive to imatinib. An overview of the complete list has been published [2]. Basically, *PDGFRA* exon 18 mutations can be assigned to 4 subgroups (Table 44.1): (i) D842V being the most common one, (ii) non-D842V mutations which confer primary resistance to imatinib as

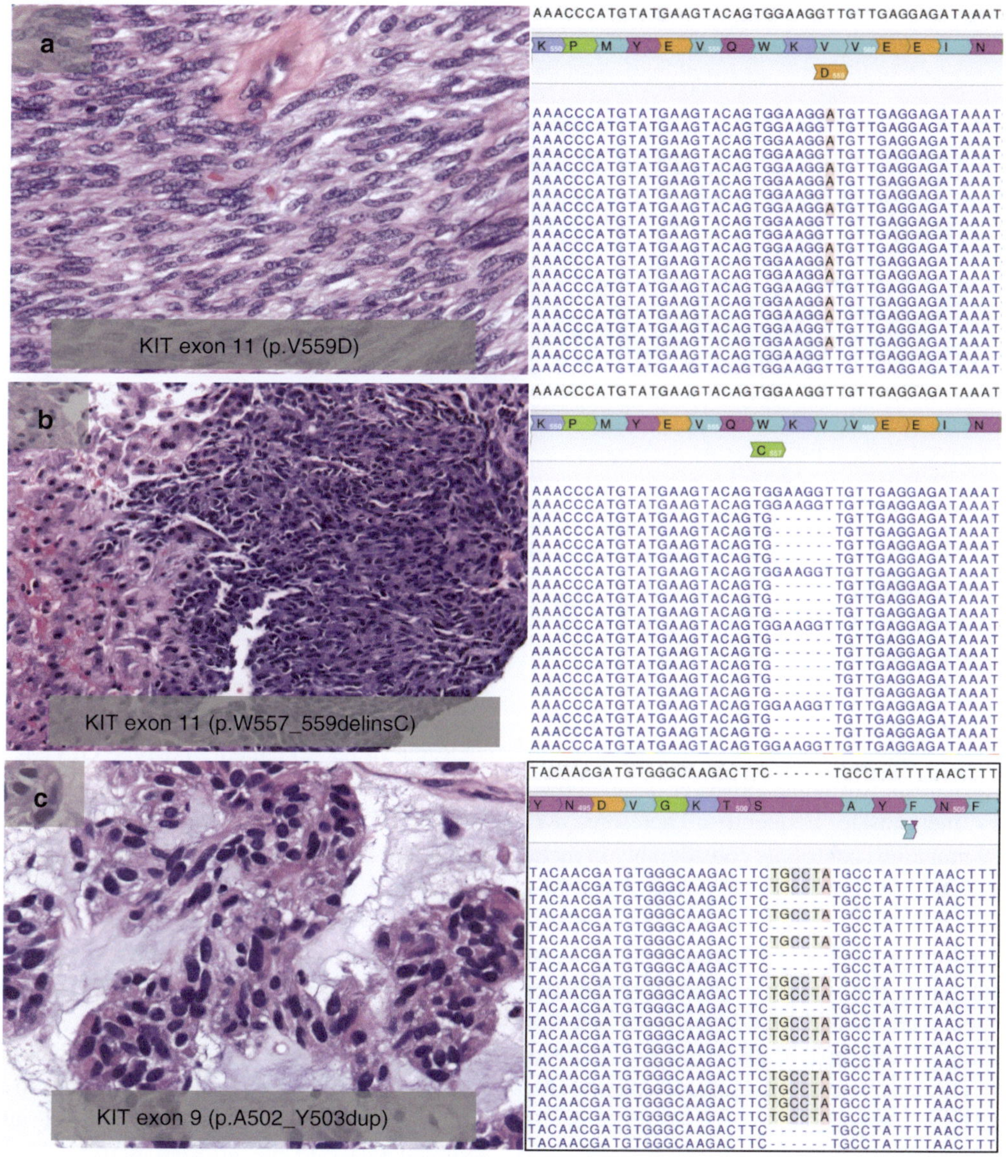

Fig. 44.2 Common *KIT/PDGFRA* mutations in gastrointestinal stromal tumors. (**a**, **b**) *KIT* exon 11 mutations represent by far the most common molecular subtype. (**a**) Exon 11 mutations are strongly associated with spindled morphology. This gastric tumor shows low mitotic activity, has a low risk for aggressive clinical behavior, and harbors a point mutation. (**b**) Liver metastasis of a gastrointestinal stromal tumor with spindled morphology. Deletions and delins mutations including codons 557 and 558 are associated with higher risk of tumor progression. This mutational subtype is, however, sensitive to imatinib. (**c**) Peritoneal metastasis of a GIST originating from the small bowel. The tumor harbors the common mutation in exon 9, a 6-base pair duplication. This example shows only slight regressive changes to prior imatinib treatment. GISTs with *KIT* exon 9 mutations respond better to an increased dose of this drug. (**d**) Epithelioid subtype of GIST. These tumors occur nearly exclusively in the stomach and are associated with mutations in the *PDGFRA* gene. This particular mutational subtype of a 12-base pair deletion confers sensitivity to imatinib (in contrast to the more common p.D842V mutation in *PDGFRA* exon 18). All mutations are designated according to the Human Genome Variation Society (HGVS)standard nomenclature. Frequencies of mutated alleles are high, in the range of 40–60%, correlating with a high tumor cell content (>90%) of all samples and in the presence of heterozygous mutations

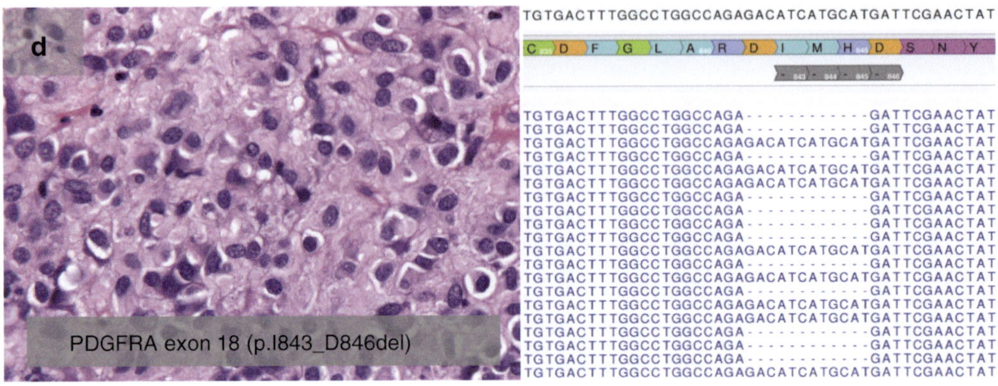

Fig. 44.2 (continued)

well, (iii) non-D842V mutations which confer sensitivity to imatinib, and (iv) mutations with unknown predictive value.

KIT exon 9 mutations account for 9% of all GIST. Tumors with these changes are predominantly located in the small bowel. Less frequently, they originate from the stomach or the rectum. Exon 9 mutations themselves do not have a specific prognostic meaning; however, small bowel GISTs metastasize frequently. Therefore, *KIT* exon 9 mutations contribute considerably in metastatic GIST. By far the predominant mutational subtype in these tumors is a six base pair duplication leading to the A502_Y503dup change at the protein level. The important predictive meaning of this change is that higher doses of imatinib (800 mg daily) are more effective, whereas patients do not or only incompletely respond to the standard dose (400 mg) (Fig. 44.2c). It is currently unknown whether other less frequent molecular subtypes of *KIT* exon 9 mutations respond to the regular or increased imatinib dose (overview published in [3]).

Rare Mutations and Wild-Type GIST

KIT exon 11, *PDGFRA* exon 18, and *KIT* exon 9 mutations account for approximately 85% of all pathogenic mutations in gastrointestinal stromal tumors. Roughly 5% of the tumors harbor rare mutations in other parts of *KIT* and *PDGFRA*. Many of them are associated with imatinib sensitivity; some are, however, the source of primary resistance such as *KIT* exon 17 mutations (Table 44.1). The remaining 10% of GIST are referred to as "wild-type GIST" since they do not harbor known mutations in *KIT* and *PDGFRA*. These tumors have been recently characterized, and various genetic subgroups have already been established.

Some of the so-called wild-type GIST harbor activating *BRAF* mutations in exon 15, namely, the V600E mutation which is also known from malignant melanomas, colorectal cancer, and many other malignancies. So far, a limited number of cases from this rare subgroup have been treated with BRAF inhibitors. Similarly, mutations in RAS genes have been associated with gastrointestinal stromal tumors.

GIST can also originate against a hereditary background. Several *KIT* mutations mostly in exon 11 have been found as germline mutations causing multiple GIST in the affected family members. Penetrance of this autosomal dominant familial GIST syndrome is quite high (nearly 100%). Clinically aggressive cases with metastatic spread are, however, less frequent. Another more common hereditary disorder is neurofibromatosis type 1 (NF1). Affected individuals may develop multiple GIST, predominantly in the small bowel. Sometimes several hundred GISTs of variable size occur in NF1 patients, which may mimic a clinically diffuse metastatic spread in the peritoneal cavity. These lesions represent merely multiple synchronous tumors rather than metastases and are genetically defined by bi-allelic *NF1* muta-

Table 44.1 Genetic subgroups of gastrointestinal stromal tumors and their prognostic and predictive impact

Gene	Exon	Mutational subtype	Prognostic significance	Personalized treatment options
KIT	**Exon 8**	D419del T417_D419delinsY	Probably higher risk for metastatic disease	Probably imatinib-sensitive
	Exon 9	(1) A502_Y503dup (most frequent subtype of *KIT*-exon 9 mutants)	No prognostic significance. Peritoneal metastases more frequent (if progressive)	Higher dose of imatinib (800 mg) associated with better progression-free survival (PFS) and overall response rate (ORR); trend for better survival in the metastatic setting; also in the adjuvant setting (high-risk patients) preferred by some centers
		(2) Non-A502_Y503dup mutations (currently 16 activating subtypes described; rare)		Rare mutations. Predictive impact unknown
	Exon 11	(1) Point mutations	No specific prognostic value	Mutations affecting exon 11 of *KIT* are highly predictive of therapeutic response to imatinib in general
		(2) Deletions/delins mutations	Worse prognosis if codon 557/558 is affected	
		(3) Insertions	No specific prognostic significance	
	Exon 13	K642E	Probably worse prognosis	(Probably) sensitive to imatinib and sunitinib. Same mutation as secondary mutation confers resistance to imatinib
	Exon 17	N822K	No specific prognostic significance	Confers resistance to imatinib. Patients respond to BLU-285 (personal communication; SB)
	Secondary mutations	V654A V654E	Occurrence of secondary mutations after first-line treatment is always associated with tumor progress and worse outcome	All secondary mutations confer resistance to imatinib
	Exon 13			Sensitive to sunitinib, considered insensitive to regorafenib
	Secondary mutations **Exon 14**	T670I T670E		Sensitive to sunitinib
	Secondary mutations **Exon 17**	C809G		Rare mutation. Predictive impact unknown
		D816E/G/H/V		Resistant to sunitinib and sorafenib, sensitive to ponatinib and avapritinib
		D820A/E/G/Y N822H/K/Y Y823D A829P		Sensitive to regorafenib/ sorafenib
	Secondary mutations	Additional rare mutations described in exons 15 (D716N) and 16 (L783V)		Rare mutations. Predictive impact unknown

(continued)

Table 44.1 (continued)

Gene	Exon	Mutational subtype	Prognostic significance	Personalized treatment options
PDGFRA	**Exon 12**	(1) V561D (2) Non-V561D mutations (at least six additional mutations have been described, among them deletions, delins mutations, and insertions; rare)	Associated with better outcome	Sensitive to imatinib and sunitinib
	Exon 14	N659K N659Y		Sensitive to imatinib and sunitinib
	Exon 18	Four groups of mutations, ca. 60 mutational subtypes: (1) D842V (most frequent variant)		Resistant to imatinib and sunitinib. Crenolanib is effective
		(2) Non-D842V mutations which provide resistance to imatinib		Resistant to imatinib
		(3) Non-D842V mutations which provide sensitivity to imatinib (ca. 10 subtypes including point mutations and deletions, e.g., I843del, I843_D846del, D846Y)		Sensitive to imatinib
		4. Non-D842V mutations with unknown clinical significance		Rare mutations. Predictive impact unknown
BRAF	**Exon 15**	V600E		Dabrafenib + trametinib, vemurafenib (off-label use)
NF1		Various mutations (inherited germline mutation plus somatic mutation)	Clinically less aggressive. Occurrence of multiple tumors indicates synchronous primary tumors rather than metastatic disease	Resistant to imatinib, patients may benefit from sunitinib and regorafenib
SDH genes		Succinate dehydrogenase deficiency related to: (1) Mutations in the genes encoding the SDH subunits *SDHA, SDHB, SDHC*, or *SDHD* (routinely detected by IHC showing loss of SDHB expression) (2) Methylation of one of these genes	Clinically less aggressive	Resistant to imatinib, patients may derive benefit from sunitinib and regorafenib, presumably due to VEGFR-mediated effects
HRAS/ NRAS/ KRAS		Rare activating mutations		Resistant to imatinib, sunitinib, and regorafenib
FGFR1		Activation of FGFR1 by mutations or *FGFR1* gene fusions (*FGFR1-TACC1* or other translocation partners)		FGFR1 inhibitor (hypothetically; various drugs with anti-FGFR activity in clinical development for a number of solid cancers)
NTRK3		*NTRK3-ETV6* fusion (rare)		Entrectinib (hypothetically; current phase II trial ongoing, ClinicalTrials.gov Identifier: NCT02568267. Also sarcoma patients with NTRK fusions can be included)

tions. Neurofibromatosis type 1-associated GISTs do not harbor activating *KIT* or *PDGFRA* mutations and do not respond to imatinib. That said, NF1 patients may also suffer from sporadic GIST with common *KIT/PDGFRA* mutations which, contrarily, show the common clinical appearance.

More recently, recurrent gene fusions have been identified, which involve receptor tyrosine kinases, e.g., *fibroblast growth factor receptor 1* (FGFR1) and *neurotrophic tyrosine kinase inhibitor 3* (*NTRK3*). Whether these tumors still should be termed as GIST remains to be seen, but nonetheless, these changes may represent targets for treatment with potent kinase inhibitors, which are currently under development.

Secondary Resistance

It has been mentioned that some of the mutations confer primary resistance to imatinib or to comparable drugs. In addition, secondary resistance has also been clinically recognized. This phenomenon is basically defined by a tumor progression after an initial response to the treatment. Secondary resistance in metastatic GIST is not uncommon. Secondary, additionally acquired mutations provide the genetic basis in most cases. These secondary mutations usually affect exons of the *KIT* gene which encode parts of the kinase domains (Table 44.1). These genetic alterations lead to a change in the sterical conformation of the KIT protein, which prevents imatinib from fitting to its binding pocket in the molecule. Some tumors develop a polyclonal evolution under imatinib treatment with subsequent occurrence of different simultaneous secondary mutations in different tumor nodules. Very rarely, secondary mutations can be found in *PDGFRA*. Secondary mutations can be diagnosed if secondarily progressive tumors are re-biopsied. Results from such re-biopsies can help to choose an appropriate second-line treatment in metastatic GIST since secondary mutations confer variable sensitivity to available drugs depending on the exact genetic subgroup. Novel drugs are under clinical development, which will hopefully cover secondary resistance mutations more effectively.

Molecular Subtypes and Targeted Treatment of Soft Tissue Sarcomas

Soft tissue sarcomas represent a heterogeneous group of neoplasms with more than 50 sub-entities. The number of subtypes that are exceedingly rare greatly increases the challenge for diagnosis- and evidence-based treatment making treatment in high-volume centers or even international networks self-evident. Many of these subtypes are associated with specific molecular changes which may provide a rationale for targeted treatment.

Soft tissue sarcomas are historically classified on the basis of their differentiation and resemblance to non-neoplastic components of the connective and soft tissue. The current WHO classification describes, for example, tumors with adipocytic (liposarcomas), fibroblastic/myofibroblastic, smooth and skeletal muscle (leiomyosarcomas, rhabdomyosarcomas), or vascular differentiation (angiosarcomas). There are, however, many sarcoma subtypes which do not fit to the described categories. Thus many sarcomas of uncertain differentiation are recognized.

Diagnosis of an individual case is based on many parameters:

(i) *Clinical information such as location and patient's age*. The clinical spectrum of sarcomas is highly related to these parameters incl. patients' age. There are entities which are more or less restricted to children. For example, malignant rhabdoid tumor, a highly aggressive sarcoma subtype, occurs almost exclusively in newborn babies or during the first few years of life. On the other hand, dedifferentiated liposarcoma is one of the commonest sarcomas in adults where patients are usually in their 60s. In contrast, dedifferentiated liposarcomas are not known among pediatric patients. Moreover, some tumors occur typically at certain locations within the human body, e.g. dedifferentiated liposarcomas in the retroperitoneum or myxofibrosarcomas at the proximal parts of the lower extremities.

(ii) *General morphologic appearance of tumor cells*. Some examples include shape (such as spindled, round cell, or epithelioid) and growth pattern of tumor cells such as fascicular, storiform, etc.

(iii) *Analysis of the tumor matrix*. Examples of extracellular matrix specifically produced by sarcoma cells include collagenous, myxoid, or mucinous stroma. Some tumors do not produce a specific matrix. Also, this constellation may be diagnostic.

(iv) *Immunohistochemistry*. Frequently, immunohistochemistry helps to identify the lineage of differentiation of a given tumor. For example, a small blue round cell sarcoma can be recognized as an (alveolar) rhabdomyosarcoma based on protein expression and IHC detection of myogenin or MyoD1, two transcription factors which play a role in skeletal muscle differentiation.

(v) *Molecular tests*. Molecular tests help to identify individual entities by detecting (or excluding) tumor-specific recurrent genetic abnormalities.

Genetically, soft tissue tumors can be subdivided into four groups:

(i) *Sarcomas with recurrent translocations (translocation-positive tumors)*

Translocations can be found in a minor but rapidly growing subgroup of mesenchymal tumors. Many sarcomas are more or less defined by entity-specific translocations such as synovial sarcoma with its specific t(X;18) rearrangement. This particular fusion has never been described in other tumors than synovial sarcomas, and nearly all synovial sarcomas harbor this aberration. The term "X;18" sarcoma would be in this context much more appropriate than "synovial" sarcoma as there is neither a strong association with synovia nor a resemblance.

Other translocations can be shared by a (small) number of entities where the final diagnosis is based on the integration of molecular, morphologic, and clinical information. For example, the *EWSR1-ATF1* fusion has been described in angiomatoid fibrous histiocytoma (a low-grade malignant tumor, which occurs in the extremities of children and adolescents) as well as in clear cell sarcoma-like tumors that originate from the gastrointestinal tract.

Sarcoma-specific translocations are mostly pathogenetic and directly linked to tumor formation. Frequently, certain transcription factors are involved, which become overactive through linkage to an active promoter by this rearrangement. However, very few fusion molecules, e.g., receptor tyrosine kinases, yet serve as therapeutically addressable targets.

(ii) *Tumors with recurrent entity-specific mutations*

Rarely, sarcomas are defined by specific mutations. One common example is desmoid fibromatosis where activating mutations in the gene *CTNNB1* occur which encodes ß-catenin. These mutations stabilize ß-catenin and cause an activated wnt-signaling.

Some sarcomas can even develop additional activating mutations on an individual basis. These changes can sometimes represent actionable mutations, such as *BRAF*-V600 mutations. They are, however, mostly independent from entity-defining mutations.

(iii) *Sarcomas and soft tissue tumors which occur as part of an underlying genetic tumor syndrome*

Very rarely, sarcomas can be part of hereditary tumor syndromes such as Li-Fraumeni, Werner, or (anecdotally) Lynch syndrome. Tuberous sclerosis is an autosomal dominant hereditary disorder which is based on mutations in *TSC1* on chromosome 9q34 or *TSC2* on chromosome 16p13.3. Affected individuals develop angiomyolipomas or other tumors which belong to the PEComa group: soft tissue tumors with differentiation of perivascular epithelioid cells. Neurofibromas and plexiform neurofibromas are syndrome-defining mesenchymal tumors in neurofibromatosis type 1 (NF1, formerly known as Recklinghausen disease). Affected individuals carry a germline mutation in *NF1*, a gene located on chromosome 17 which encodes neurofibromin. This protein is a negative regulator of RAS signaling and acts, therefore, as a tumor suppressor. Because RAS signaling is involved in many biologic processes, neurofibromatosis type 1 is considered as a complex multi-organ disease regarded a "RASopathy" – a developmental syndrome caused by germline mutations. A subset of NF1 patients – mostly adolescents – develop malignant peripheral nerve sheath tumors (MPNST), a highly malignant sarcoma genetically characterized by bi-allelic mutations or inactivations of *NF1*. Germline *NF1* mutations are frequently microdeletions anywhere in this huge gene, which extends over 60 exons. Less frequently patients with NF1 develop gastrointestinal stromal tumors. Those – in the vast majority of patients – do not harbor *KIT* or *PDGFRA* mutations but mutations that cause a conjoined activation of signaling pathways that are typically activated by KIT. These are the PI3K pathway (usually by *PI3K* mutations or mutations/deletions of PTEN) in combination with the activation of RAS/RAF by inactivated NF1.

Familial adenomatous polyposis (FAP) coli represents another tumor syndrome where mesenchymal tumors may occur. FAP is related to *APC* mutations, and affected individuals suffer from multiple colonic adenomas, colorectal, as well as duodenal carcinomas. Since *APC* mutations cause activation of wnt-signaling, it is not surprising that FAP patients can also develop desmoid fibromatoses. In contrast to sporadic cases, FAP-associated desmoids (with underlying *APC* mutations) do not harbor *CTNNB1* mutations.

(iv) *Sarcomas that do not harbor recurrent genetic alterations*

A major group of sarcomas do not show any of the abovementioned types of genetic alterations. Frequently, these tumors are characterized by an enormous magnitude of multiple changes leading to a rather "chaotic" genotype. Genomic aberrations include multiple nonrecurrent chromosomal translocations, amplifications, and major deletions of chromosomal material as well as non-synonymous mutations. As far as we know, chromosomal instability seems to exceed mutational load in the majority of such sarcomas. Leiomyosarcoma represents a good example of tumors with a chaotic genotype. There are, however, other sarcomas, which are thought to be associated with a high mutational burden, among them cutaneous angiosarcomas, which are related to UV light exposure or irradiation. The latter tumors may be candidates for immune checkpoint inhibitor treatment, which is explored in currently ongoing clinical trials.

Treatment Modalities in Soft Tissue Sarcomas

Treatment of sarcomas is based on three modalities with surgery representing the fundamental basis. Whenever possible, complete removal of sarcomas with clear margins should be achieved. Radiation therapy is a standard perioperative treatment for patients with high-grade and deep-seated sarcomas and substantially improves local control, particularly in patients with close margins. The role of chemotherapy in localized sarcomas is not yet considered to be a standard, as randomized trials have resulted in conflicting results. This is mainly thought to be a systematic problem, as until recently adjuvant chemotherapy trials have subsumed all sarcoma subtypes as a single disease. Based on the observations from patients with metastatic disease, soft tissue sarcomas exhibit vastly different sensitivity to cytotoxic chemotherapy. For subtypes that are considered chemosensitive, such as undifferentiated sarcomas, subgroup analyses have also strongly implicated a survival benefit in the adjuvant setting. Unfortunately, predictive molecular markers have yet not been identified that would help to improve the selection of patients. Nonetheless, the exact histological subtype may trigger highly specific treatment algorithms, regardless of specific targets. Specific neoadjuvant therapies such as isolated limb perfusion with chemotherapy may increase local operability and may contribute to a better outcome. Furthermore, recent scientific progress has led to the introduction of several targeted treatments for sarcoma treatment with improved potency with reduced side effects; and many more therapeutics are in preclinical and clinical research settings.

Apart from empiric treatments, recent scientific progress led to the introduction of several targeted treatments in the context of personalized oncology which combine improved potency with reduced side effects. Targeted therapeutics mostly require predictive biomarker tests to demonstrate that the target of the specific medication is apparent in an individual tumor. First targeted drugs have already been approved for sarcoma treatment, and many more therapeutics are still subject to clinical and preclinical research.

Approved Targeted Therapeutics

Dermatofibrosarcoma protuberans (DFSP) is a spindle cell sarcoma of the skin with fibroblastic differentiation. Tumors tend to be locally aggressive and to recur after surgical removal. DFSP metastasize only rarely, but progression to overtly biologically aggressive fibrosarcomas occurs. In some patients, relapsed DFSP cannot be surgically excised due to size or location of the tumor. On the genetic level, DFSP is characterized by a gene fusion, which includes *PDGFb*, the gene which encodes the platelet-derived growth factor beta (Fig. 44.3d-f). If fused to the *COL1A1* gene, *PDGFb* is highly expressed since the *COL1A1* promoter is highly active in fibroblastic cells (*COL1A1* encodes a collagen chain). DFSP cells also express the receptor for PDGFb, which provides the biologic basis for an autocrine stimulation and tumor growth. The tyrosine kinase inhibitor, imatinib, blocks potently the platelet-derived growth factor receptors, and its clinical effect in DFSP patients has been demonstrated. Therefore, imatinib has received approval for the treatment of locally advanced or metastatic dermatofibrosarcoma protuberans (Table 44.2) [4].

As mentioned above, tumors of the PEComa group can occur against the genetic background of tuberous sclerosis. Fortunately, most tumors in tuberous sclerosis patients are benign, e.g., angiomyolipomas of the kidneys. Sporadic subtypes of PEComas, however, occur at various locations throughout the body. Sporadic PEComas can be clinically more aggressive with local recurrences, local invasive growth, and metastatic spread. The tumors, like their hereditary counterparts, are biologically characterized by activating *TSC1/2* mutations or comparable ways to activate mTOR (mechanistic target of rapamycin) signaling. Everolimus blocks mTOR, and its efficacy in PEComas has been demonstrated. Thus, everolimus has been approved for the treatment of advanced or metastatic PEComas [5].

Trabectedin, another anticancer drug with efficacy in sarcomas, received approval for systemic therapy of myxoid liposarcomas (after the failure of conventional chemotherapy) [6]. The drug has been originally discovered in sea squirts but is now pharmaceutically synthetized. The mode of action is not fully understood, but it has been shown that the substance binds DNA and blocks the effects of the chimeric *FUS-DDIT3* transcription factor which is the consequence of the recurrent gene fusion in myxoid liposarcomas. Since that particular gene fusion and the subsequent action of the chimeric protein as a transcription factor directly initiate and propagate formation of myxoid liposarcomas, the drug is thought to have the capacity to differentiate tumor cells and to inhibit tumor progress. Furthermore, trabectedin seems to alkylate DNA and to induce DNA strand breaks. Trabectedin is also approved for the treatment of leiomyosarcomas without a specific predictive biomarker.

Targeted therapies are also brought into action in the mode of off-label use. Drugs, which are approved for anticancer treatment in any tumor entity, might be used for certain mesenchymal tumors if there is a convincing rationale. This is an area of oncology where predictive biomarkers play a decisive role (see Table 44.3 for an overview). One prominent and frequently applied example is the detection of *ALK* gene fusions in inflammatory myofibroblastic tumors (IMFT). This soft tissue tumor frequently affects adolescents and young adults. IMFT can basically arise anywhere in the human body with some more pronounced locations such as thoracic cavity (e.g., in the chest wall, lungs, mediastinum), orbit, and abdomen (such as mesentery, soft tissues around the urinary bladder, etc.). Many tumors can be cured by complete local excision. There are, however, cases of these tumors which

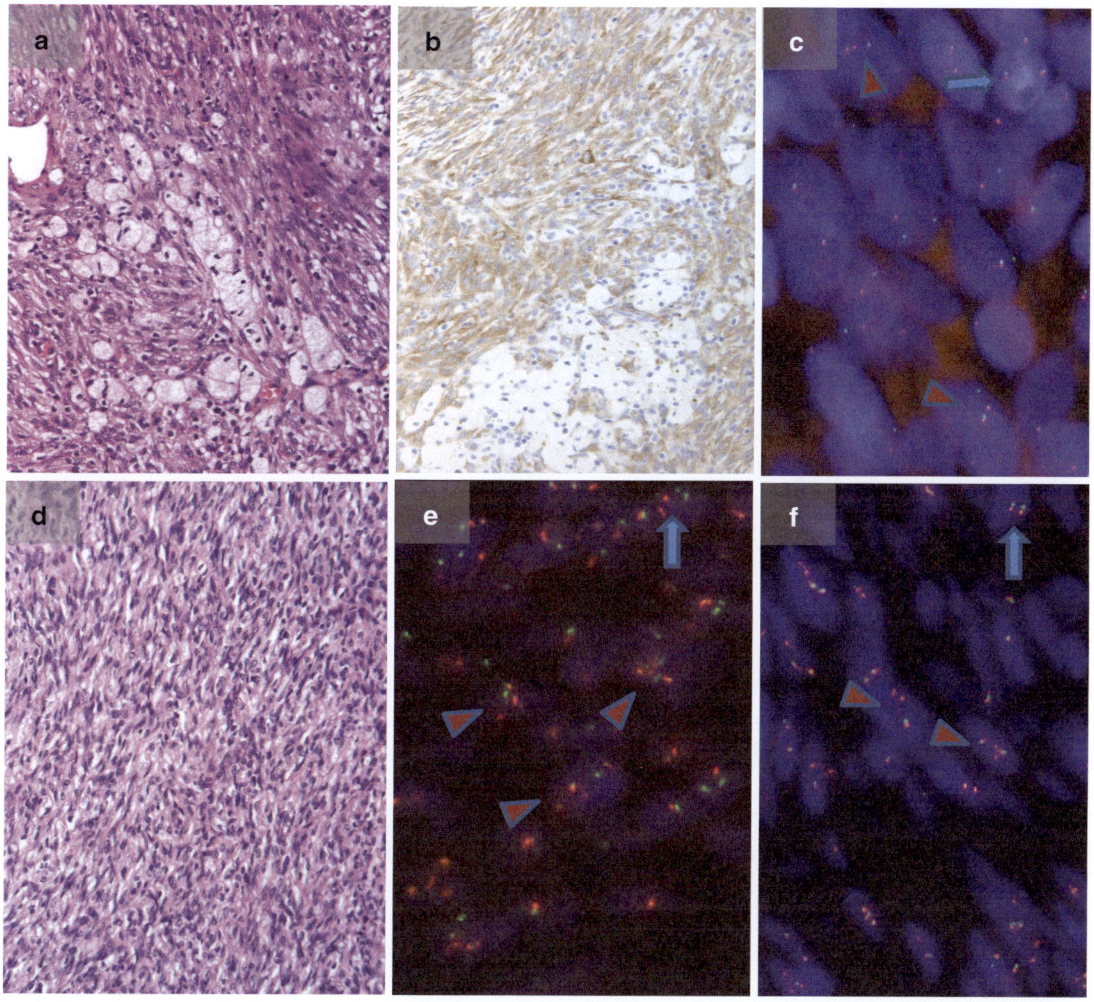

Fig. 44.3 Predictive biomarker assays in soft tissue tumors. (**a–c**) Inflammatory myofibroblastic tumor (IMFT) showing an *ALK* rearrangement. Tumors with these changes have been proven to respond to tyrosine kinase inhibitor treatment, such as crizotinib. Comparable alterations can also affect the *ROS1* gene. (**a**) Typical morphology of IMFT – spindled myofibroblastic cells against a background of inflammatory cell, namely, lymphocytes, macrophages, and foam cells (H&E). (**b**) ALK immunohistochemistry showing weak to moderate ALK expression (antibody clone 1A4). (**c**) Fluorescence in situ hybridization (FISH) demonstrating an *ALK* rearrangement. Nonneoplastic cells such as lymphocytes harbor regular orange/green fused signals (blue arrow). Tumor cells show break apart signals (arrowheads) which are indicative of an *ALK* rearrangement. (**d–f**) *PDGFb-COL1A1* fusion in dermatofibrosarcoma protuberans (DFSP). Imatinib treatment is approved for this constellation. (**d**) DFSP represent fibroblastic spindled neoplasms of the skin (H&E). (**e–f**) FISH demonstrating *PDGFb-COL1A1* fusion. This genomic alteration can be detected either by dual-color dual-fusion probes (**e**) or by *PDGFb* break apart probes (**f**). (**e**) Normal cell with two sets of orange and green signals (blue arrow). Rearranged tumor cells are characterized by occurrence of two fused signals (arrowheads). (**f**) In the break apart approach, nonneoplastic cells display normal fused signals which flank the break point region within the *PDGFb* gene (blue arrow). Rearranged cells are characterized by break apart signals (arrowheads). Gene fusion can also be detected by RNA- or DNA-based sequencing approaches

Table 44.2 Predictive biomarkers in sarcomas. Selected examples of approved treatments

Biomarker	Function/pathogenic process	Patient selection method	Clinical use	Approved drug (generic/trade name)	Company
DDIT3	Chromosomal translocation/*DDIT3* fusion	Histology and biomarker studies, e.g., FISH or sequencing	Myxoid liposarcoma (with DDIT3 fusion) and leiomyosarcoma (without specific molecular biomarker)	Trabectedin (Yondelis®)	Jansen Oncology, Pharma Mar S.A
PDGFb	Chromosomal translocation/*PDGFb* fusion	FISH or sequencing	Dermatofibrosarcoma protuberans	Imatinib mesylate (Gleevec®)	Novartis
TSC1/2	mTOR activation by *TSC1/2* mutations	Histology; sequencing (in the context of genetic counseling of patients with tuberous sclerosis)	Approved for the treatment of angiomyolipomas in the context of tuberous sclerosis. Angiomyolipomas represent a subtype of PEComas	Everolimus (Afinitor®)	Novartis

behave malignantly with frequent local recurrences and distant metastases. A significant number of these cases harbor various subtypes of *ALK* gene fusions (which are separated by different translocation partners). A common effect is activation of ALK signaling and subsequent tumor growth. It is known from *ALK*-positive pulmonary adenocarcinomas that multitude of tyrosine kinase inhibitors (e.g., alectinib, crizotinib, ceritinib, lorlatinib, brigatinib, entrectinib) are effective in these tumors. Alectinib, crizotinib, and ceritinib are already approved for the treatment of *ALK*-fused lung cancer. These drugs are also used for the therapy of locally advanced or metastatic inflammatory myofibroblastic tumors ("off label") if an *ALK* rearrangement can be demonstrated (Fig. 44.3a-c). Based on the clinical experience, this treatment is highly potent. As also known from lung cancer, IMFT can become resistant to ALK inhibitors (e.g., crizotinib) under treatment. This is mainly due to secondary mutations, which evolve under therapy as source of secondary resistance. There are, however, second- (ceritinib) and third-generation ALK inhibitors (e.g., alectinib, lorlatinib) which may overcome this situation.

Another example of off-label use represents clear cell sarcomas with BRAF-V600 mutations. This genomic variant is common in malignant melanomas where vemurafenib as a BRAF inhibitor is approved. Furthermore, a subset of pulmonary adenocarcinomas harbors BRAF-V600 mutations. A dual BRAF/MEK inhibition with dabrafenib and trametinib is approved for BRAF-mutated lung cancer. Advanced or metastatic clear cell sarcomas may be treated with either vemurafenib or dabrafenib/trametinib if a BRAF mutation in exon 15 at codon 600 can be detected.

Molecular Targets Which Provide Rationale for Emerging Treatments

Many novel sarcoma therapeutics are currently being investigated in clinical trials, the majority of them are accompanied by predictive biomarkers. Table 44.3 summarizes recent examples. Most biomarker approaches address basic biologic processes, which are driven by genetic changes. Some current attempts are directed against receptor tyrosine kinases – e.g., *NTRK* (*neurotrophic tyrosine kinase inhibitors*) fusions – which may occur in various (rare) sarcoma subtypes. Another example of tyrosine kinase activation is tenosynovial giant cell tumor where a recurrent *CSF1-COL6A3* gene fusion causes overexpression of CSF1 (colony-stimulating factor 1). CSF1 expression subsequently activates its

Table 44.3 Biomarkers in sarcomas. Selected examples of ongoing clinical trials and off-label use of targeted drugs

Entity	Aberration	Method of detection	Treatment	Level of evidence (approved, off-label use, clinical trial, other)
PEComas	mTOR activation, *TSC1/2* mutations	Sequencing (in the context of genetic counseling of patients with tuberous sclerosis)	Everolimus (Afinitor®)	Approved for treatment of angiomyolipomas in the context of tuberous sclerosis. Angiomyolipomas represent a subtype of PEComas
Inflammatory myofibroblastic tumor	*ALK* fusion	FISH, sequencing, IHC	Crizotinib (Xalkori®) Alectinib (Alecensa®) Ceritinib (Zykadia™)	Off-label use
Dedifferentiated liposarcoma	*MDM2* amplification	FISH	Mdm2 inhibitor hdm201	Phase I trial ClinicalTrials.gov Identifier: NCT02343172
Dedifferentiated liposarcoma	*CDK4* amplification, RB expression	FISH IHC	Palbociclib (Ibrance®) (Loss of RB as negative predictor)	Phase II trial ClinicalTrials.gov Identifier: NCT01209598
Synovial sarcoma	NY-Eso-1 expression	IHC	Vaccines, genetically engineered T cells	Phase II trial for vaccination, CAR-T cell against NY-ESO ClinicalTrials.gov Identifier: NCT01967823
Low-grade endometrial stromal sarcoma	ER/PgR expression *JAZF1* fusion	IHC FISH, sequencing	Aromatase inhibitors	Off-label use
Tenosynovial giant cell tumor/ pigmented villonodular synovitis	*CSF1* fusion	FISH, sequencing	Imatinib mesylate (Gleevec®)	Off-label use
Clear cell sarcoma	*BRAF*-V600E	Sequencing, mutation-specific IHC	Vemurafenib (Zelboraf®) Dabrafenib (Tafinlar®) and Trametinib (Mekinist®)	Off-label use
Various sarcomas (rare)	*NTRK* fusions	FISH, sequencing (IHC)	Entrectinib (RXDX-101)	Phase II trial ClinicalTrials.gov Identifier: NCT02568267
Chondrosarcomas	*IDH1/2* mutations	Sequencing	IDH inhibitors	Phase II trial ClinicalTrials.gov Identifier: NCT02073994
Alveolar soft part sarcoma	*TFE3* fusion	FISH, sequencing, IHC	Cediranib (AZD2171) Sunitinib (Sutent®)	Phase II trial (approval pending) ClinicalTrials.gov Identifier: NCT01337401

FISH fluorescence in situ hybridization, *IHC* immunohistochemistry receptor CSF1R in other cellular components of the tumor, which finally lead to tumor growth. Imatinib and other, even more specific CSFR inhibitors (e.g., pexidartinib) are currently explored as a targeted treatment in tenosynovial giant cell tumors. Same alterations can also be found in the diffuse variant of this tumor, which is referred to as pigmented villonodular synovitis.

Molecular-based treatments in dedifferentiated liposarcomas are currently under clinical investigation. On a molecular level, dedifferentiated liposarcomas are characterized by formation of supernumerary giant or marker chromosomes, which contain highly amplified chromosomal material from 12q13-15, including *MDM2* among many other genes such as *CDK4*. On the other hand, dedifferentiated liposarcomas do not harbor inactivating TP53 mutations. MDM2, as an oncogene, binds and inactivates p53 by ubiquitination and proteasomal degradation. Therefore, MDM2 inhibition restores and stabilizes wild-type p53 activity, thus leading to a p53-dependent growth inhibition and induction of apoptosis. Recent studies have provided proof-of-concept evidence that MDM2 inhibitors have clinical effects in patients – even though a monotherapy appears to have only moderate effects. Given the high-level co-amplification of CDK4, which occurs in the majority of dedifferentiated liposarcoma cases, this change may provide another target for a specific treatment. Palbociclib, an oral inhibitor of cyclin-dependent kinases cdk4 and cdk6, is an already approved, well-tolerated agent (approved for treatment of advanced breast cancer). As with MDM2 inhibitors, some patients have had responses, but inhibiting this pathway may not be enough for long-term tumor control.

Technologies to Detect Molecular Targets in Mesenchymal Tumors

As explained in previous parts of this chapter, predictive biomarkers in gastrointestinal stromal tumors and sarcomas are very closely related to tumor genetics. Therefore, immunohistochemistry (IHC) plays a minor role in this field. There are, however, a few exceptions. IHC-based assays may serve as surrogate markers, which display protein (over)expression instead of the underlying genomic alteration. Thus, ALK protein expression can be utilized as a surrogate marker for *ALK* gene fusions in inflammatory myofibroblastic tumors (Fig. 44.3). Additional examples are ROS1 expression in *ROS1* fused IMFT, BRAF-V600E mutation-specific antibody staining in clear cell sarcomas, or mdm2 expression in *MDM2* amplified dedifferentiated liposarcomas. It is noteworthy in this context that KIT expression in GIST is independent of *KIT* mutations. Wild-type GISTs show usually significant levels of KIT expression, which are indistinguishable from *KIT* mutants.

IHC may receive a more pronounced role in association with the emerging field of immuno-oncology treatments. In currently ongoing clinical trials, the NY-Eso-1 protein expression in synovial sarcomas serves as a target to immunological approaches, i.e., for vaccines or CAR-T cells. NY-Eso-1 expression can be detected immunohistochemically. The role of PD-L1 (programmed death ligand 1) as a predictive biomarker for immune checkpoint inhibitors, i.e., anti-PD1 and anti-PD-L1 antibodies, is currently being explored in clinical trials.

Genetic aberrations in mesenchymal tumors comprise basically four groups of changes: (i) gene fusions (translocations, rearrangements), (ii) amplifications, (iii) (large) deletions, and (iv) mutations (consisting of nucleotide substitutions, i.e., point mutations, small deletions, delins mutations, insertions, and duplications). All these changes can serve as predictive biomarkers.

The detection of gene fusions can be carried out by in situ hybridization (mostly FISH) or sequencing approaches. FISH as a method is frequently established in departments with expertise in sarcoma diagnosis since many diagnostic markers are based on this technology. FISH can also be easily applied to determine predictive biomarkers, which are based on gene fusions (Fig. 44.3). Next-generation sequencing (NGS) approaches are another useful alternative method. There are basically two ways to detect gene fusions by NGS – one based on DNA and another based on RNA. Hybrid capture DNA-based NGS has the capacity to cover all genomic changes, i.e., gene mutations, amplifications/deletions, and gene fusion, in one single assay. On the other hand, RNA-based assays have the advantage to analyze directly transcribed RNA. Rearrangements in neoplasms do not always result in the transcription of the affected fused genes with subsequent expression of a chimeric protein. Therefore, direct measurement on the RNA level may be useful. RNA-based NGS assays are, however, more labor intensive since RNA extraction from paraffin blocks and transcription into cDNA by reverse transcriptase-PCR require additional steps. For both NGS approaches, commercial kits and platforms are readily available. Overall, sequencing approaches are superior to FISH by their capacity to detect unexpected or unknown gene fusions and by their ability to test for multiple changes simultaneously in a multiplex approach and/or by providing comprehensive genomics-based information. In contrast, FISH represents a relatively easy and fast method. FISH can produce reliable results of predictive biomarker assays within only 1 or 2 working days.

There are some more RNA-based technologies such as RT-PCR (reverse transcriptase-PCR), which can detect gene fusion. However, RT-PCR is highly dependent on the RNA quality and requires relatively high amounts of intact RNA. Another disadvantage is that specific assays for every subtype of rearrangements are needed. This might be challenging if several fusion partners and/or various breakpoints within the affected genes occur. Recently, novel methods, such as NanoString technology and nuclease protection assays, have been introduced which can capture RNA fragments and analyze them directly from tissue lysates without specific RNA extraction and purification. Moreover, these techniques have no or only limited PCR-based amplification steps which make them faster and more robust. These methods have the capacity to detect gene fusions also in the context of predictive biomarker assays.

Gene amplifications or deletions are relatively infrequent biomarker applications in predictive diagnostics of soft tissue tumors. These changes can be detected either by FISH assays or by DNA-based NGS approaches.

Mutational analyses of GIST and sarcomas are by far the most frequent applications. In contrast to other tumor entities, sensitivity of the assays is not so much of an issue. Mostly, resection specimens are used which usually contain a large amount of tumor cells. Even core needle biopsies or open surgical biopsies are well suited. However, sequencing may be technically challenging if tumor samples show a higher proportion of necrotic tissue or if samples are decalcified or if small biopsies (e.g., from the gastrointestinal tract) are used. Basically, the spectrum of applicable technologies is wide and ranges from Sanger sequencing over pyrosequencing, high-resolution melting assays, and mass spectrometry to next-generation sequencing assays. Since *KIT*, *PDGFRA*, *CTNNB1*, *BRAF*, and comparable genes are usually included in nearly all commercially available NGS kits for tumor diagnostics, most high-volume centers will use these technologies.

In terms of reporting mutations, it is noteworthy that laboratories should strictly stick to a common standard for naming and reporting of DNA variants. Not infrequently, there are different potential ways to describe deletions, delins mutations, insertions, or duplications. Even bioinformatic NGS platforms do not always name these changes correctly. Therefore, rules and recommendations provided by the Human Genome Variation Society (www.hgvs.org) should be applied to allow interlaboratory comparability of sequencing results.

Summary and Future Directions

It has been shown that the correct treatment of GIST and many sarcomas are highly dependent on predictive biomarker analyses. Pathologists should be aware of fundamental therapeutic principles and should have knowledge on biomarkers, which are related to certain therapies. Predictive biomarkers in GIST and sarcomas are mostly based on molecular findings. Activating mutations in genes encoding receptor tyrosine kinases and recurrent gene fusions are the most common genetic aberrations, which can be detected by sequencing approaches or in situ hybridizations. Many innovative treatments are currently being investigated in clinical trials, and improved sequencing technologies will be able to discover additional therapeutic targets in mesenchymal tumors. Therefore, we can expect an increase in the number of available treatment options together with rising requests for biomarker tests. Patients should receive access to related diagnostic procedures.

References

1. Pogorzelski M, Falkenhorst J, Bauer S. Molecular subtypes of gastrointestinal stromal tumour requiring specific treatments. Curr Opin Oncol. 2016;28:331–7.
2. Künstlinger H, Binot E, Merkelbach-Bruse S, Huss S, Wardelmann E, Buettner R, et al. High-resolution melting analysis is a sensitive diagnostic tool to detect imatinib-resistant and imatinib-sensitive PDGFRA exon 18 mutations in gastrointestinal stromal tumors. Hum Pathol. 2014;45:573–82.
3. Künstlinger H, Huss S, Merkelbach-Bruse S, Binot E, Kleine MA, Loeser H, et al. Gastrointestinal stromal tumors with KIT exon 9 mutations: update on genotype-phenotype correlation and validation of a high-resolution melting assay for mutational testing. Am J Surg Pathol. 2013;37:1648–59.
4. Rubin BP, et al. Molecular targeting of platelet-derived growth factor B by imatinib mesylate in a patient with metastatic dermatofibrosarcoma protuberans. J Clin Oncol. 2002;20(17):3586–91. 8.
5. Bissler JJ, et al. Everolimus for angiomyolipoma associated with tuberous sclerosis complex or sporadic lymphangioleiomyomatosis (EXIST-2): a multicentre, randomised, double-blind, placebo-controlled trial. Lancet. 2013;381(9869):817–24.
6. Demetri GD, et al. Efficacy and safety of trabectedin or dacarbazine for metastatic liposarcoma or leiomyosarcoma after failure of conventional chemotherapy: results of a phase III randomized multicenter clinical trial. J Clin Oncol. 2016;34(8):786–93.

Predictive Markers and Targeted Therapies in Thyroid Cancer and Selected Endocrine Tumors

45

Juan C. Hernandez-Prera and Bruce M. Wenig

Overview

Endocrine tumors are ubiquitous and can arise in many organs such as the pituitary, the pancreas, the ovaries, the testicles, the thyroid, the parathyroid, and the adrenal glands. Many of these tumors retain the ability to secrete hormones which can be the source of their symptomatology. In this chapter, we will mainly cover biomarkers associated with thyroid cancer – the most common malignancy of the endocrine system – and briefly touch upon other tumors such as the parathyroid carcinoma and adrenal gland neoplasia.

Thyroid Cancer

Approximately 56,870 adults (14,400 men and 42,470 women) in the USA are diagnosed with thyroid cancer. This is the most common malignancy of the endocrine system, and the majority of them are derived from follicular cells and are further classified as either papillary (PTC) or follicular thyroid carcinoma (FTC). Due to their histological appearance in addition to their ability to uptake iodine and synthetize thyroglobulin under the stimulus of thyrotropin-stimulating hormone (TSH), PTC and FTC are generically referred to as differentiated thyroid cancer (DTC). The treatment of DTC relies mainly on surgery followed by radioactive iodine (RAI) to eliminate suspected metastases, residual or recurrent disease. This therapeutic strategy achieves excellent survival rates in most cases; however, some patients with DTC do not respond or become refractory to RAI. A similar management challenge is faced in patients with less differentiated follicular-derived tumors, such as poorly differentiated (PDTC) and anaplastic thyroid cancer (ATC).

A more complicated scenario is encountered in patients with medullary thyroid carcinoma, a neoplasm derived from parafollicular C cells that do not benefit from RAI therapy. Our growing understanding of the molecular mechanisms of thyroid cancer has revealed specific alterations in signaling molecules and receptor tyrosine kinases. Given the oncogenic role of these molecular events, inhibitors are currently available to target RAI-refractory DTC and advanced MTC (Table 45.1).

Guidelines

The 2015 American Thyroid Association (ATA) management guidelines for DTC in adult patients states that the use of a Food and Drug Administration (FDA)-approved kinase inhibitor should be considered in RAI-refractory DTC

J. C. Hernandez-Prera (✉) · B. M. Wenig
Department of Anatomic Pathology,
Moffitt Cancer Center, University of South Florida,
Tampa, FL, USA
e-mail: juan.hernandez-prera@moffitt.org

S. Badve, G. L. Kumar (eds.), *Predictive Biomarkers in Oncology*,
https://doi.org/10.1007/978-3-319-95228-4_45

Table 45.1 Biomarkers in thyroid cancer

Gene/RNA or protein biomarkers	Patient selection method and sample type	Successfully completed clinical trials and # of patients enrolled	Clinical use and limitations	Approved drugs Generic/trade name	Company
VEGFR-1, VEGFR-2, and VEGFR-3, RET, RAF (including BRAF V600E), platelet-derived growth factor receptor beta	Radioactive iodine for selecting refractory locally advanced or metastatic, progressive, differentiated thyroid cancer	DECISION Trial (NCT00984282) Total patients: 417 Sorafenib group: 207 Placebo group: 210	Improved median progression-free survival compared with placebo (10·8 vs. 5·8 months, respectively)	Sorafenib (Nexavar®)	Bayer
VEGFR-1, VEGFR-2, and VEGFR-3, FGFR-1, FGFR-2, FGFR-3, and FGFR-4, PDGFRα, RET, and KIT	Progressive differentiated thyroid cancer refractory to iodine-131	SELECT trial (NCT01321554)	Improved median progression-free survival compared with placebo (18.3 vs. 3.6 months, respectively)	Lenvatinib (Lenvima®)	Eisai Inc.
		Total patients:392 Lenvatinib group: 261 Placebo group: 131	Improved response rate compared with placebo (64.8% in vs. 1.5%)		
RET, VEGFR, and EGFR		ZETA trial (NCT00410761) Total patients: 331 Vandetanib group: 231 Placebo group: 100	Improved median progression-free survival compared with placebo (30.5–19.3 months)	Vandetanib (Caprelsa®)	Sanofi Genzyme
MET, VEGFR-2, and RET	Unresectable, locally advanced, or metastatic hereditary or sporadic medullary thyroid cancer	EXAM trial (NCT00704730)	Improved median progression-free survival compared with placebo (11.2–4 months)	Cabozantinib (Cometriq®)	Exelixis
		Total patients: 330 Vandetanib group: 219 Placebo group: 111	Improved response rate compared with placebo (28% in vs 0%)	(EXAM trial, NCT00704730)	

with metastatic, rapidly progressive, symptomatic, and/or imminently threatening disease not otherwise amenable to local control using other approaches (recommendation 96). So far, based on phase III randomized placebo-controlled clinical trials, the FDA has approved the use of sorafenib and lenvatinib. These two multikinase inhibitors are considered as first-line targeted therapy; however additional clinical trials comparing the efficacy of these drugs with each other are pending. In addition, the ATA recommends that in patients who have disease progression while on sorafenib or lenvatinib, a second-line kinase inhibitor therapy could be considered within the context of therapeutic clinical trials (recommendation 97). Sorafenib may also play an important role in the management of advance RAI-refractory DTC in children. However, the experience of kinase inhibitor therapy in pediatric patients is limited, and further studies are required. For patients affected by MTC with significant tumor burden and symptomatic or progressive metastatic disease, the 2015 ATA management guidelines for MTC recommend the systemic therapy with vandetanib or cabozantinib (recommendation 65). FDA approval of these two multikinase tyrosine kinase inhibitors has been granted after successful completion of phase III clinical trials.

Pathway(s) Where the Biomarker Is Involved

Regardless of histological type, most follicular cell and parafollicular C-cell-derived carcinomas have genetic alterations involving components of the mitogen-activated protein kinase (MAPK) signaling pathway including transmembrane receptor tyrosine kinases and downstream signaling molecules (Fig. 45.1). For instance, the most common alteration in PTC includes point mutations of the BRAF and RAS genes and RET/PTC gene rearrangements. Meanwhile, the most frequent genetic alterations in FTC are RAS mutations. RET gene point mutations are molecular cornerstones in medullary thyroid carcinomas. Of note, all these molecular events are mutually exclusive.

The most common mutations in PTC involve the BRAF gene, and with a few exceptions, all point mutations result in an amino acid substitution at position 600 from valine (V) to glutamic acid (E). The resulting BRAF V600E mutation is detected in approximately 60% of cases and is typically observed in PTC with classical or tall cell morphology. In addition, BRAF V600E mutations are found in PDTC and ATC arising in the background of PTC. PTCs harboring BRAF V600E mutation are more refractory to RAI therapy. The latter phenomenon seems to be secondary to suppression of genes enquired for iodide incorporation by the BRAF V600E oncoprotein. The degree of constitutive activation of the MAPK signaling driven by BRAF V600E is the highest compared to other abnormalities present in PTC.

BRAF is a serine-threonine kinase that belongs to the family of RAF proteins. After activation by RAS, BRAF phosphorylate and activate MEK leading to activation of ERK. Activated ERK migrates to the nucleus and regulates transcription of the genes involved in cell differentiation, proliferation, and survival. Alternative mechanisms for BRAF activation is K601E mutation, which results in an amino acid substitution at position 601 from a lysine (K) to glutamic acid (E). BRAF K601E mutant tumors fall within the histological spectrum of follicular variant of PTC. Gene fusion involving BRAF can also lead to its constitutive activation, and different gene partners have been described. For instance, a paracentric inversion of chromosome 7q results in chimeric gene fusion between AKAP9 and BRAF. The latter chromosomal abnormality has been reported in some radiation-induced PTC.

After BRAF abnormalities, mutations in the RAS gene family are the second most frequent pathogenic mutations in thyroid cancer, and they are identified in 10–20% of PTC, 40–50% of FTC, and 20–40% of PDT and ATC. Of note, characteristically most RAS-mutated PTCs are follicular variants. The RAS gene family includes three highly homologous genes: HRAS, KRAS, and NRAS. The RAS genes encode highly related proteins with GTPases whose activity is located at the inner surface of the cell membrane. RAS proteins cycle between inactive GDP-bound and active GTP-bound forms and play a critical role as intracellular mediators downstream of cell membrane growth factor receptor signaling. In their active state, RAS proteins subsequently activate several downstream effectors, including the MAPK pathway and the PI3K-AKT-mTOR pathway. Secondary to their intrinsic GTPase property, activated RAS proteins become quickly inactive; however, point mutations at specific hot spots (codons 12, 13, and 61) result in permanently active GTP-bound forms and activation of downstream signaling pathways. The most common mutations involve NRAS codon 61 and HRAS codon 61. The ability to uptake and metabolized iodide appears to be preserved in RAS-mutated tumors.

RET proto-oncogene codes for a cell membrane receptor tyrosine kinase and its constitutive activation is an important oncogenic

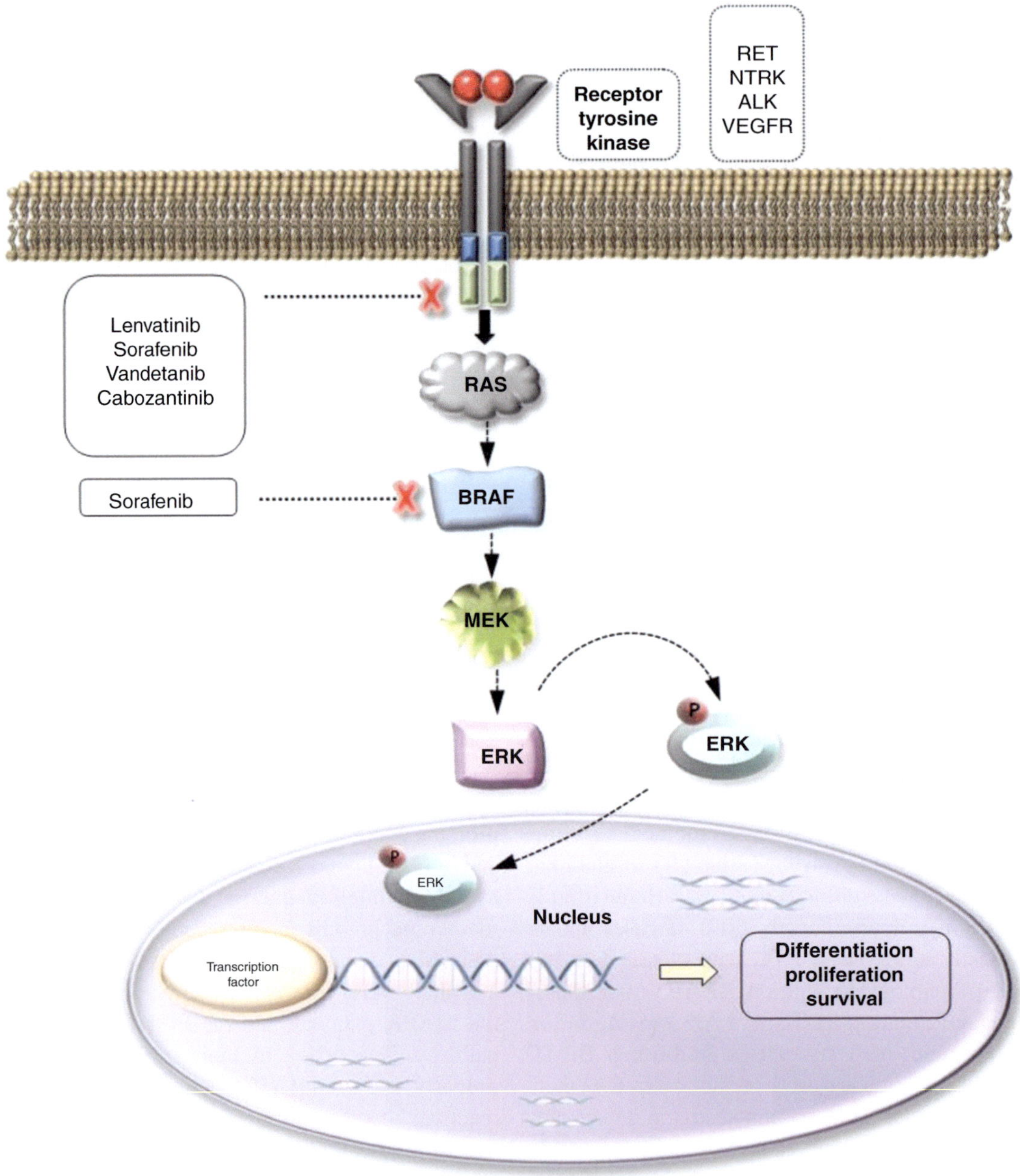

Fig. 45.1 Genetic alterations in thyroid cancer involve components of the mitogen-activated protein kinase (MAPK) signaling pathway. Receptor tyrosine kinase (RET, NTRK, ALK, and VEGFR) activation is initiated by a ligand that induces receptor dimerization and kinase activation leading to autophosphorylation. Downstream effector molecules (RAS, BRAF, MEK, and ERK) with kinase activity subsequently propagate the signaling leading to translocation of phosphorylated ERK to the nucleus and regulation of transcription factors involved in cell differentiation, proliferation, and survival. Multikinase tyrosine kinase inhibitors (e.g., sorafenib and lenvatinib for RAI-refractory DTC and vandetanib and cabozantinib for advanced MTC) showing improved progression-free survival have been approved by FDA. (Courtesy of Dr. Janis de la Iglesia)

event in PTC and MTC. Similar to other receptor tyrosine kinase (RTK), RET receptor protein is composed of an extracellular ligand-binding domain, a transmembrane domain, and a cytoplasmic tyrosine kinase domain. Receptor activation is initiated by a ligand that induces receptor dimerization and kinase activation leading to autophosphorylation. Subsequently, activated RET phosphorylates its substrates resulting in downstream signal transduction of different pathways, including the MAPK pathway.

In PTC, chromosomal rearrangements are the primary mechanisms for RET proto-oncogene activation. These rearrangements lead to fusion of the intracellular tyrosine kinase-encoding domain of the RET gene to the 5′ portion of various unrelated genes characterized by the presence of a coiled–coiled domain. Multiple gene fusions have been reported to date with RET/PTC1 (CCDC6-RET), RET/PTC2 (PRKAR1A-RET), and RET/PTC3 (NCOA4-RET) which constitute the most common ones. Initial studies reported that 20% of PTCs were characterized by RET/PTC gene fusion. However, most recent data estimates a prevalence of 6.7%. In general, RET/PTC-positive tumors occur more commonly in patients with the history of radiation exposure in children. The gene partners appear to influence the morphology tumors. For instance, RET/PTC3 tumors tend to exhibit solid morphology, while RET/PTC1 show classic papillary histology.

In contrast to PTC, germline or somatic point mutations are the primary mechanisms for RET proto-oncogene activation in MTC. Technically all patients with multiple endocrine neoplasia type 2A (MEN2A), MEN type 2B (MEN2B), and familial MTC (FMTC) have RET germline mutations, while somatic mutations are noted in approximately 50% of cases of sporadic MTCs.

The mutations can either affect the extracellular or the intracellular domain of RET receptor protein. In patients with MEN2A and FMTC mutations, they typically affect the cysteine-rich extracellular domain. In up to 90% of MEN2A cases, the mutations result from amino acid substitutions at position 634, from cysteine to either tryptophan (W), arginine (R), or tyrosine (Y); however other cysteine codons can be affected (C609, C611, C618, C620, and C630). Alteration in the cysteine residues result in ligand-independent dimerization through disulfide bridges and constitutive activation of receptor kinase. In the majority of MEN 2B and sporadic MTCs, the RET mutation involve codon 918 of the intracellular tyrosine kinase domain resulting in an amino acid substitution at position 918, from methionine to threonine. This mutation activates RET independently of dimer formation by altering the substrate specificity of RET receptor kinase.

In addition to RET, alterations in other genes coding for a receptor tyrosine kinase have been noted in minority of DTC and dedifferentiated thyroid cancer. These alterations include rearrangements of ALK, NTRK1, and NTKR3. Not all gene fusions have been functionally characterized. However, the resulting chimeric proteins retain the kinase domain. Radiation-induced PTC and tumors occurring in the pediatric population have a high prevalence of fusion oncogenes.

Angiogenesis is a key event in maintaining tumor growth. The signaling pathways that lead to new blood formation are initiated by the interaction of members of the vascular endothelial growth factor (VEGF) family and their receptor (VEGFR) primarily located on the cell surface of endothelial cells. VEGFR is also a tyrosine kinase receptor, and it exerts its effects via the MAPK pathway, the PI3K-Akt pathway, and the Src-eNOS pathway. Activation of the aforementioned pathways results in increased endothelial cell proliferation and migration, endothelial cell survival, and vascular permeability. In DTC and MTC, VEGF receptors (VEGFR-1 and VEGFR-2) are often overexpressed. A higher level of VEGF expression has been particularly noted in BRAF V600E-mutated PTC and C634 RET-mutated MTC.

Parathyroid Carcinoma

Parathyroid carcinoma is a very rare malignant neoplasm accounting for 0.005% or about 1 case in every 200 patients according to the US National Cancer Data Base – a jointly run program by the American College of Surgeons and the American Cancer Society. The Surveillance, Epidemiology, and End Results (SEER) Program and SEER data indicate that its annual incidence in the USA is less than one case per million people. Men and women are equally affected with a mean age of diagnosis of 56 years. The primary clinical feature of parathyroid carcinoma is associated

Table 45.2 Biomarkers in adrenocortical carcinoma

Gene/RNA or protein biomarkers	Patient selection method and sample type	Successfully completed clinical trials and # of patients enrolled	Clinical use and limitations	Drug(s) Generic/ trade name	Company
IGF-1 receptor	Locally advanced or metastatic adrenocortical carcinoma	GALACCTIC trial (NCT00924989) Total patients: 139 Linsitinib group: 90 Placebo group: 49	Failure to improve progressive-free survival or overall survival	Linsitinib (OSI-906)	Astellas Pharma

with severe hyperparathyroidism and high calcium levels often higher than 14 mg/dL. This is also the most common cause of mortality in recurrent or metastatic cases.

Most cases of parathyroid carcinoma are sporadic but may occur in hyperparathyroidism-jaw tumor syndrome (HJTS) – an inherited condition that causes overactivity of the parathyroid glands. HJTS has an autosomal dominant pattern of inheritance and is associated with germline mutations in the tumor suppressor gene CDC73 (also known as HRPT2). Moreover, inactivation of CDC73 is also the major molecular alteration in sporadic cases. Due to the rarity of parathyroid carcinoma, very few studies have addressed the role of predictive or prognostic biomarkers in this tumor.

Surgery, including en bloc resection of the parathyroid gland with ipsilateral hemithyroidectomy and central neck dissection, is the treatment of choice for parathyroid carcinoma. Treatment options for unresectable, recurrent, or metastatic cases are limited and primarily directed to control the hypercalcemia. Therefore, there is a need for effective targeted treatment modalities in the management of complicated parathyroid carcinoma.

Adrenocortical Carcinoma

Adrenocortical carcinoma (ACC) is a rare malignant neoplasm with an estimated incidence of 0.5–2 cases per million. Most cases are diagnosed during the fifth or sixth decade of life, but tumors also occur in the pediatric population. Currently, radical adrenalectomy is the treatment of choice, and the adrenolytic agent mitotane is the only FDA-approved drug used as adjuvant therapy or for the treatment of locally advanced or metastatic disease. Different studies have contributed to the comprehensive molecular characterization of ACC, and these discoveries promise to advance the application of targeted therapies in this tumor (Table 45.2).

Pathway(s) Where the Biomarker Is Involved

ACC is a molecularly heterogeneous disease and has genetic alterations mainly involving PI3K/Akt/mTOR and the RAS-MAPK signaling pathways. Most studies have collectively demonstrated overexpression of IGF-2 in approximately 90% of cases, and it has been shown that IGF-2 promotes cancer cell growth via activation of IGF-1 receptor. Another molecular hallmark in these tumors is the high frequency of TP53 mutations, especially in pediatric and aggressive cases. In addition, mutations in PRKAR1A gene that encodes the enzyme cAMP-dependent protein kinase type I-alpha regulatory subunit and alteration in Wnt/β-catenin signaling pathway via activation of the β-catenin gene (*CTTNB1*) have been identified as important molecular events that contribute to the development of these tumors. Moreover, the abovementioned molecular events involved in the pathogenesis of adrenocortical carcinoma explain the genetic susceptibility of Li-Fraumeni syndrome (TP53), Beckwith-Wiedemann syndrome (IGF-2), and Carney complex (PRKAR1A) for development of this neoplasm.

The response to mitotane is variable among patients with ACC, and currently, there are no biomarkers that predict a response to this toxic drug. However, the utility of potential biomarkers, such as CYP2W1 (a member of the cytochrome P450 superfamily of enzymes) and RRM1 (an enzyme essential for the conversion of ribonucleotides into deoxyribonucleotides), requires further clinical investigation.

Pheochromocytoma and Paraganglioma

Pheochromocytomas (PCC) and paragangliomas (PG) arise from catecholamine-producing chromaffin cells of the adrenal medulla and extra-adrenal sympathetic and/or parasympathetic paraganglion cells, respectively. Historically, the definition of malignancy in PCC/PG was the presence of metastatic disease. However, the current perspective is that all tumors have some metastatic potential, and consequently the use of the categories benign and malignant is no longer recommended. Moreover, PCC and sympathetic PG show high morbidity due to excessive catecholamine production.

Approximately 30% of PCC/PG are known to be hereditary, and germline mutations in at least 19 susceptible genes have been so far reported. Overall, these tumors are the most hereditary neoplasm in the human body, and an autosomal dominant pattern of inheritance is most commonly seen. Mutations in succinate dehydrogenase (SDH) genes are the most common one implicated in the pathogenesis of hereditary cases. SDH is a mitochondrial enzyme complex with important roles in oxidative phosphorylation and intracellular oxygen sensing and signaling. Mutations involve different subunits of the SDH complex (SDHA, SDHB, SDHC, SDHD, SDHAF2). The identification of these causative mutations allows for detection of individuals at risk for these tumors. In addition, the SDHB-mutated tumors have higher metastatic potential. For instance, this molecular abnormality accounts for 55% of metastatic cases. Loss of SDHB expression by immunohistochemical staining is a valuable ancillary tool that helps to identify patients with germline mutations. Other genes associated with hereditary PCC/PG include VHL, NF1, and RET.

Our current understanding of the molecular alterations involved in the pathogenesis of PCC/PG may identify potential targeted therapies for locally aggressive and metastatic disease. Currently, there are prospective phase I and II clinical trials, and it is expected that these trials would identify effective medications for the treatment of PCC/PG.

Future Directions

In the past few years, our understanding of the molecular mechanisms of thyroid cancer has rapidly grown. This has happened in parallel with the expansion of new therapeutic options primarily targeting tyrosine kinases. As of today, four successfully completed, randomized, placebo-controlled phase III trials showing improved progression-free survival have led to the approval of sorafenib and lenvatinib for RAI-refractory DTC and vandetanib and cabozantinib for advanced MTC. Despite these encouraging advances, there are many unsolved questions. For instance, the main mechanism of action of these drugs remains unknown, primarily because they have the ability to inhibit multiple targets. The lack of specificity could be responsible for the toxic effects of these drugs that in a significant percentage of cases lead to dose reduction, interruption, or withdrawal. It is also unclear if the molecular signature of the tumor predicts the treatment response since the progression-free survival benefit associated with these drugs appears independent of mutation status. All clinical trials have shown improvements in the progression-free survival. However, the extent to which tyrosine kinases inhibitors may prolong overall survival is not available at the moment. Currently, multiple clinical trials are active and under development, and without a doubt these investigations of new therapeutic target agent will solve many of these questions.

In contrast to thyroid cancer, the clinical utility of targeted therapy for the management of parathyroid carcinoma and adrenal gland neoplasia is under evaluation. Promising clinical trials are being conducted, and, hopefully, their results will improve the outcome of patients with advance disease.

Acknowledgment We would like to thank Dr. Janis de la Iglesia from the Department of Head and Neck/Endocrine Cancer at Moffitt Cancer Center for creating Fig. 45.1 of this chapter.

Suggested Readings

1. Angelousi A, et al. Molecular targeted therapies in adrenal, pituitary and parathyroid malignancies. Endocr Relat Cancer. 2017;24(6):R239–59.
2. Brose MS, et al. Sorafenib in radioactive iodine-refractory, locally advanced or metastatic differentiated thyroid cancer: a randomised, double-blind, phase 3 trial. Lancet. 2014;384(9940):319–28.
3. Cancer Genome Atlas Research Network. Integrated genomic characterization of papillary thyroid carcinoma. Cell. 2014;159(3):676–90.
4. Elisei R, et al. Cabozantinib in progressive medullary thyroid cancer. J Clin Oncol. 2013;31(29):3639–46.
5. Erovic B, et al. Biomarkers of parathyroid carcinoma. Endocr Pathol. 2012;23:221.
6. Fagin JA, Wells SA Jr. Biologic and clinical perspectives on thyroid cancer. N Engl J Med. 2016;375(11):1054–67.
7. Fassnacht M, et al. Linsitinib (OSI-906) versus placebo for patients with locally advanced or metastatic adrenocortical carcinoma: a double-blind, randomised, phase 3 study. Lancet Oncol. 2015;16:426–35.
8. Fishbein L, et al. Comprehensive molecular characterization of pheochromocytoma and paraganglioma. Cancer Cell. 2017;31(2):181–93.
9. Haugen BR, et al. 2015 American Thyroid Association management guidelines for adult patients with thyroid nodules and differentiated thyroid cancer: the American Thyroid Association guidelines task force on thyroid nodules and differentiated thyroid cancer. Thyroid. 2016;26(1):1–133.
10. Mete O, et al. Immunohistochemical biomarkers of adrenal cortical neoplasms. Endocr Pathol. 2018; https://doi.org/10.1007/s12022-018-9525-8. [Epub ahead of print].
11. Nikiforov YE. Molecular diagnostics of thyroid tumors. Arch Pathol Lab Med. 2011;135(5):569–77.
12. Schlumberger M, et al. Lenvatinib versus placebo in radioiodine-refractory thyroid cancer. N Engl J Med. 2015;372(7):621–30.
13. Viola D, et al. Treatment of advanced thyroid cancer with targeted therapies: ten years of experience. Endocr Relat Cancer. 2016;23(4):R185–205.
14. Wells SA Jr, et al. Revised American Thyroid Association guidelines for the management of medullary thyroid carcinoma: the American Thyroid Association Guidelines Task Force on medullary thyroid carcinoma. Thyroid. 2015;25(6):567–610.
15. Wells SA Jr, et al. Vandetanib in patients with locally advanced or metastatic medullary thyroid cancer: a randomized, double-blind phase III trial. J Clin Oncol. 2011;30(2):134–41.
16. Zheng S, et al. Comprehensive pan-genomic characterization of adrenocortical carcinoma. Cancer Cell. 2016;30(2):363.

The Response Evaluation Criteria in Solid Tumors (RECIST)

46

Kate Lathrop and Virginia Kaklamani

Introduction

While cancer has always been a major health concern, the proportion of people affected by cancer has grown considerably as major advances in medicine have enabled people to live longer. In parallel fashion, the treatment and evaluation of cancer and cancer therapies are becoming more and more complex and varied as the number of cancer therapeutics increases at an exponential rate. It is therefore critical to have standardized and accurate means to assess the effectiveness of new cancer therapies. Radiologic imaging is the primary means of assessing the effects of new antineoplastic therapies in clinical trials, as well as standard of care therapies. Antineoplastic therapies vary in their mechanisms of actions, and this needs to be considered when evaluating response. It is thus critical to have an accurate, efficient, and reproducible means of measuring response to treatment.

The fundamental concept supporting radiologic evaluation of tumors is that the shrinkage or stability in the tumor size is a surrogate for improved survival [1]. Thus, a decrease in tumor size in an early phase clinical trial, such as a phase II trial, would be supportive for further evaluation of that therapy in a larger phase III clinical trial. Objective radiologic response is the most commonly used endpoint in clinical trials for assessment of efficacy of new therapeutics [2]. This endpoint is based on anatomical measurement of tumor size and disease burden. A shared system for enumerating and measuring tumor burden at baseline, and assessing changes after exposure to antineoplastic therapy, is essential for combining data from different clinical sites and for comparison across different clinical trials. It also provides a familiar framework to evaluate response to standard of care therapies for the patient being treated outside of clinical trials.

The first widely adopted standardized approach to assessing tumor response by imaging was the World Health Organization (WHO) criteria which were published in 1981 and primarily utilized in a clinical trial where tumor response rate was the primary endpoint. The WHO system was the first guideline to use the sum of two-dimensional tumor products to measure response to therapy. This system developed different categories of response based on the change in tumor size. In 2000, the first version of the Response Evaluation Criteria in Solid Tumor (RECIST) was published to provide a more standardized approach for assessing tumor response for patients enrolled in clinical trials [1]. Major

K. Lathrop
Department of Medical Oncology and Hematology, University of Texas Health Science Center San Antonio, San Antonio, TX, USA

V. Kaklamani (✉)
Department of Medicine, University of Texas Health Science Center San Antonio, San Antonio, TX, USA
e-mail: kaklamani@uthscsa.edu

S. Badve, G. L. Kumar (eds.), *Predictive Biomarkers in Oncology*,
https://doi.org/10.1007/978-3-319-95228-4_46

modifications from the WHO guideline included specifications on the number of lesions to measure total and per organ site, change to single dimension measurements, and guidelines for measuring overall tumor burden.

RECIST criteria were widely adopted by the research community and created a common system for evaluating tumor response across different clinical trials. The guideline was then revised in 2008 to address questions and add additional guidance and was published as RECIST 1.1, which replaced the initial 1.0 version [2]. The modifications of RECIST 1.0–1.1 are outlined in Table 46.1.

Major changes include the number of lesions to be assessed, method of assessing pathologic lymph nodes, confirmation of response in clinical trials with certain endpoints, defining disease progression, and an appendix on imaging guidance. The RECIST working group was also established to ensure that modifications to the guideline are continually evaluated and validated and that updates keep pace with advancing imaging and therapeutic modalities [3].

Table 46.1 Comparison of tumor response criteria methods

RECIST	RECIST 1.1	iRECIST	PERCIST 1.0	RANO-BM
Measurable (target) lesions at baseline				
Unidimensional LD only with 20 mm for conventional techniques and 10 mm with spiral CT	Unidimensional LD only with 20 mm on conventional techniques and 10 mm on spiral CTs	Unidimensional LD only with 20 mm on conventional techniques and 10 mm on spiral CTs	The SUL is determined for up to 5 tumors (up to 2 per organ) with most intense 18F-FDG uptake	Bidimensional ≥10 mm, up to five lesions
	≥15 mm for nodes lesions	≥15 mm for nodes lesions		
	Max 5 lesions, 2 per organ	Max 5 lesions, 2 per organ		
Nonmeasurable (nontarget) lesions at baseline				
All other lesions, including small lesions	All other lesions, including small lesions (LD ≤10 mm or pathologic lymph nodes 10–15 mm short axis)	All other lesions, including small lesions (LD ≤10 mm or pathologic lymph nodes 10–15 mm short axis)	All other lesions, including small lesions	Lesions less than 10 mm
New measurable lesions				
Must be a minimum of 5 mm	No specific criteria – new lesions must be unequivocal	Assessed as per RECIST 1.1 but recorded separately on case form and incorporated into total tumor burden (>5 mm)	Clinical judgment should be used in identifying new lesions	Must be a minimum of 5 mm
	When unequivocal = PD			
New nonmeasurable disease				
Always PD (<5 mm)	An increase in overall disease burden based on the change in nonmeasurable disease is comparable in magnitude to the increase that would be required to declare PD for measurable disease	Not defining of progression but precludes irCR (<5 mm)	Not specifically defining of progressive disease	Not defining of progressive disease

Table 46.1 (continued)

RECIST	RECIST 1.1	iRECIST	PERCIST 1.0	RANO-BM
Complete response				
Disappearance of all lesions in two consecutive assessments at least 4 weeks apart	Disappearance of all nontarget lesions and normalization of tumor marker level. All lymph nodes must be non-pathological in size (<10 mm short axis)	Disappearance of all lesions in two consecutive assessments at least 4 weeks apart	Complete metabolic response (CMR): disappearance of all evidence of disease. Complete resolution of uptake within measurable target lesions and all other lesions so that it is less than mean liver activity and indistinguishable from surrounding background blood-pool level. No new lesions. If PD based on standard RECIST, must verify with follow-up scan	Complete disappearance of all lesions sustained for at least 4 weeks. No new lesions. No ongoing steroid use and either clinically stable or improved
Partial response				
≥50% decrease in SPD of all index lesions compared to baseline in two consecutive assessments at least 4 weeks apart. Absence of new lesions or unequivocal progression of non-index lesions	At least a 30% decrease in the sum of diameters of target lesions, taking as reference the baseline sum diameters	≥50% decrease in tumor burden compared with baseline in two assessments at least 4 weeks apart	Partial metabolic response (PMR): reduction of minimum of 30% in target measurable tumor SUL peak. Absolute drop in SUL must be at least 0.8 SUL units. Measurement is commonly in the same lesion as baseline but can be a different lesion if that lesion was present at baseline and if now the most after lesion after treatment. No increase ≥30% in SUL or size of any lesion (target or nontarget). If PD by RECIST, must verify with follow-up. Reduction in extend of lesion 18F-FDG uptake is not required. No new lesions	≥30% decrease in the SLD of CNS target lesions compared to baseline, sustained for at least 4 weeks, no new lesions Stable/less steroid requirement Clinically stable or improved
Stable disease				
50% decrease in SPD compared with baseline cannot be established nor a 25% increase compared to the nadir. Absence of new lesions	Neither sufficient shrinkage to qualify for PR nor sufficient increase to qualify for PD, taking as reference the smallest sum diameters while on study	50% decrease in tumor burden compared with baseline cannot be established nor a 25% increase compared to baseline nadir	Stable metabolic disease (SMD): does not meet criteria for CMR, PMR, or PMD	Does not meet criteria for CR, PR, PD

(continued)

Table 46.1 (continued)

RECIST	RECIST 1.1	iRECIST	PERCIST 1.0	RANO-BM
Progressive disease				
At least 25% increase in SPD compared with nadir and/or unequivocal progression or non-index lesions and/or appearance of new lesions	At least a 20% increase in the sum of diameters of target lesions, taking as reference the smallest sum on study. Sum must also demonstrate an absolute increase of at least 5 mm. Appearance of one or more new lesions is also considered progression	At least 25% increase in tumor burden compared with nadir in two consecutive observations at least 4 weeks apart	Progressive metabolic disease: ≥30% increase in 18F-FDG SUL peak with ≥0.8 SUL unit increase in tumor SUV peak from baseline and not typical pattern for infection or treatment effect. Visible increase in extent of 18F-FDG uptake – 75% of the total lesion glycolysis (TLG) volume and no decline in SUV. New 18F-FDG-avid lesions that are typical of cancer	≥20% increase in SLD compared to nadir, with minimum absolute increase of 5 mm in at least one lesion Significant T2 signal increase Clinical status worsening Unequivocal progression of existing enhancing nontarget CNS lesions In case of immune therapy, new lesions alone may not constitute progressive disease

Based on data from Refs. [2, 11, 15, 17, 18]
SLD sum of the longest diameter, *SUL* SUV correct for mean body mass

Response Evaluation Criteria in Solid Tumor: RECIST 1.1

Assessment Modalities

The first step in the clinical and radiologic evaluation of tumors is to ensure the proper modality is being utilized. Imaging should be completed as close to the start of the therapeutic invention as possible and not more than 4 weeks prior to the start of treatment. All measurements should be recorded in metric notation and calipers should be used for clinical measurements. There should be consistency in the imaging modality in the baseline and comparison studies. The preferred method of evaluation is by imaging, and clinical measurements should be utilized only if the lesion is not well visualized on imaging and more accurately measured by clinical exam. For skin-based lesions, color photography with a ruler in the field is suggested for documentation of changes.

Overall, computed tomography (CT) is the preferred imaging method. Chest X-ray can be used in the setting of a well-aerated lung, but the minimal lesion size increases to 20 mm. Ultrasound is not recommended because the exam cannot be completely reproduced for independent review, and there is operator variation in measurement technique. If a new lesion is identified via ultrasound, confirmation is required by CT scan or magnetic resonance imaging (MRI). Likewise, endoscopy and laparoscopy for objective tumor response are not recommended in these guidelines. These methods can be utilized for biopsies in determining recurrent disease in protocols where disease-free survival is an endpoint. Changes in tumor markers alone are not sufficient to determine response or progression. If tumor markers had been measured and were above normal prior to the start of treatment, normalization of this marker is required for a patient to be considered having a complete response.

Measurable Versus Nonmeasurable Disease

In order to assess response to therapy, the overall tumor burden at baseline needs to be evaluated and recorded as the comparator to subsequent imaging. Once baseline imaging is completed, lesions are categorized as measurable vs. nonmeasurable. At baseline, measurable tumor lesions include those that can be accurately measured in at least one dimension with the longest diameter recorded and measuring 20 mm or more with conventional techniques or at least 10 mm with spiral CT scans when the CT scan slice thickness is no greater than 5 mm. If calipers are used for clinically assessable disease, the lesion must measure at least 10 mm. In the case of a chest X-ray as the imaging modality, lesions must measure 20 mm or more to be considered measurable. Lymph nodes can be considered measurable target lesions but must have a baseline measurement of at least 15 mm in the short axis when assessed by CT scan. The short axis of the lymph node should be followed on subsequent assessments.

The determination of nonmeasurable disease can, at times, be more complicated than determining measurable lesions. All other lesions measuring less than 10 mm at the longest diameter or pathologic lymph nodes that measure more than 10 mm but less than 15 mm on the short axis are considered nonmeasurable lesions. Nonmeasurable disease also includes leptomeningeal disease, ascites, pleural or pericardial effusions, cystic lesions, organomegaly, lymphangitic spread in the skin or lung, and inflammatory breast cancers.

There are several caveats in regard to bone lesions in the updated RECIST 1.1 guidelines. These guidelines do not endorse bone scans, positron emission tomography (PET) scans, or plain films as adequate imaging modalities for evaluation of the size of bone lesions. For lytic bone lesions with a soft tissue component that meets the measurability criteria for measurable disease, the lesion can be considered measurable if evaluated by cross-sectional CT scan or MRI. Bone lesions with a blastic component, however, are considered nonmeasurable regardless of size or imaging modality.

Cystic disease requires some clinical judgment as to the nature of the cystic lesions. Cystic lesions that meet the radiographic definition of a simple cyst should not be considered as measurable or nonmeasurable as these are benign. For malignancies that form cystic lesions, these cystic-appearing lesions can be considered measurable and follow the previously described criteria for measurability. Non-cystic lesions should preferentially be identified as the target lesions over cystic lesions if available.

Lesions in the area of previous local therapy, such as radiation, are generally not considered target lesions. However, the guideline allows for some flexibility with this criteria and permits the determination of a target lesion if there has been a significant progression of the lesion after regional therapy. A particular protocol may also specify that lesions located in the area of previous local therapy are permitted to be followed as a target lesion.

Baseline Documentation of Disease

The baseline documentation of lesions is important as the sum of the greatest diameter of these lesions will be compared to subsequent measurements. A maximum of five lesions can be defined as target lesions. All other lesions will be recorded as nontarget lesions. Target lesions include no more than two lesions per organ and should include representative lesions from all involved organs. Target lesions should be selected based on size with lesions with the longest diameter taking priority. However, if it is anticipated that a large lesion cannot be measured in a reproducible manner, it is reasonable to exclude and use smaller lesions as the target lesions.

In the initial RECIST criteria, all nontarget lesions or sites of disease were recorded at baseline. Measurements of these lesions were not required, but the presence or absence of the lesions was required to be noted at each tumor assessment. RECIST 1.1 allows for documentation of nontarget lesions at baseline but does not require specific assessment of each nontarget lesion. For example, a clinician can note multiple liver metastases or multiple enlarged mediastinal lymph nodes without detailing the exact number or measurement of each lesion.

Definitions of Response

There are four general categories of response: complete response (CR), partial response (PR), stable disease (SD), and progressive disease (PD). In RECIST version 1.1, a CR is defined as the disappearance of all target lesions including any pathologically enlarged lymph nodes, regardless of whether they were designated as target or nontarget lesions, and the normalization of tumor marker levels. In the original version, it was not defined how to address the change in lymph node size in regard to a complete response. To be considered a complete response, lymph nodes need to measure 10 mm or less on the short axis. Of note, the sum may not be "0" if lymph nodes are included as target lesions. Persistence of one or more nontarget lesions or the maintenance of tumor marker level above normal limits excludes classification of complete response. A PR is defined as at least a 30% decrease in the sum of diameters of target lesions, taking as reference the baseline sum diameters. PD is defined as at least a 20% increase in the sum of diameters of target lesions, taking as reference the smallest sum on study. The 20% increase of the sum must also demonstrate an absolute increase of at least 5 mm as this is above the error of CT scanning. The appearance of one or more new lesions is also considered progression. SD is neither sufficient shrinkage to qualify for PR nor sufficient increase to qualify for PD, taking as reference the smallest sum diameters while on study (Figs. 46.1 and 46.2).

The original RECIST criteria also did not provide for exact or absolute increase in tumor sizes to be considered PD. Version 1.1 defines target lesions that increase by more than 20% in the sum of the longest diameters (SLD) as PD when compared to the smallest SLD recorded since the initiation of treatment. This lowest SLD is often referred to as the nadir. The increase must also be at least 5 mm absolute increase over the nadir to be considered PD as changes less than 5 mm may be contributable to measurement error.

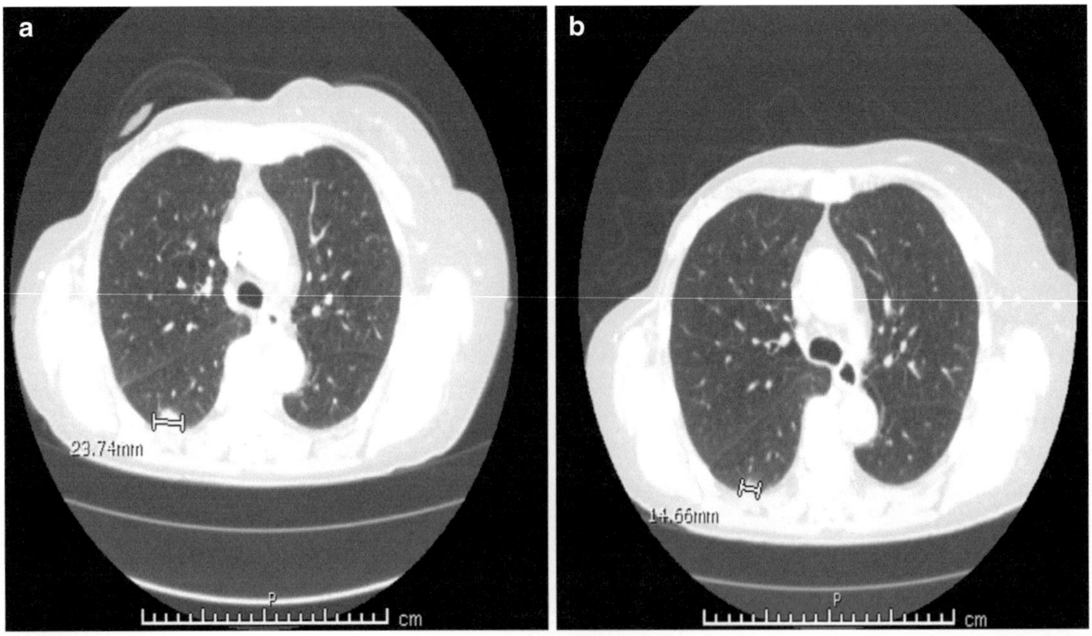

Fig. 46.1 Target lesion is identified in the right lung and measures 24 mm on the longest dimension (**a**). Follow-up imaging after initiation of treatment has the same lesion measuring 15 mm when measured in a similar fashion (**b**). This is about a 38% decrease which is consistent with partial response if this was the only target lesion

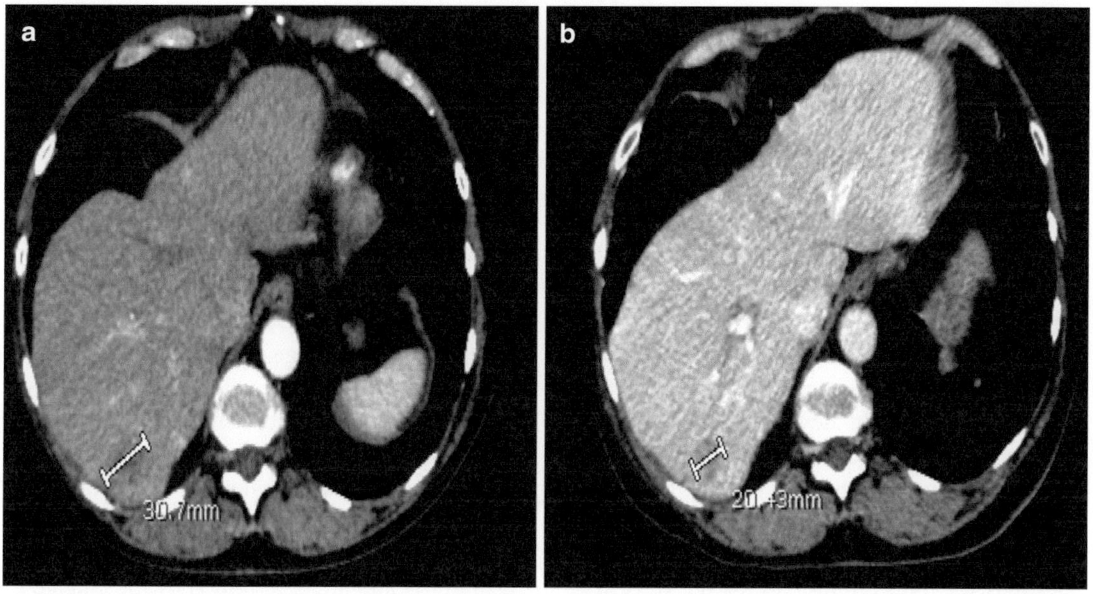

Fig. 46.2 Target lesion is identified in the liver and measures 31 mm on the longest dimension (**a**). Follow-up imaging after initiation of treatment has the same lesion measuring 20 mm when measured in a similar fashion (**b**). This is about a 35% decrease which is consistent with partial response if this was the only target lesion

There is also the concept of unequivocal progression to consider. Initially, this was included to allow for the determination of progression when there was a concern for clinically significant progression of nontarget lesions. RECIST 1.1 defines unequivocal progression as an overall level of substantial worsening in the nontarget disease, even in the setting of SD or PR in the target lesions. This would include overall tumor burden that has increased sufficiently to merit discontinuation of therapy. In patients that do not have measurable disease, unequivocal progression is the change in nonmeasurable disease to the size criteria that would be required to determine progressive disease for measurable lesions. Examples of this significant change in nonmeasurable disease would be increase in effusions or ascites from small to large or progression in lesions in previously radiated fields.

In general, the appearance of new lesions is consistent with PD. New lesions should not be attributable to other causes such as difference in imaging techniques or when the new lesion could just as likely represent something other than the malignancy being evaluated. Some particular examples include change in bone appearance or liver lesions becoming necrotic and change in appearance or size. If it is uncertain to the investigator if the new lesion is truly PD, it is reasonable to continue treatment and reevaluate at the next imaging interval.

Determination of Best Overall Response

The best overall response is defined as the best response since the start of treatment until the end of the study period. Some therapies have continued response after the completion of treatment, and thus the planned assessments should capture the full anticipated effect of the treatment. The best overall response includes evaluation of target lesions, nontarget lesions, and new lesions. Certain protocols will require confirmatory assessments with repeat tumor measurements after the best overall response is defined to determine the duration of that response. An example of confirmatory testing is in clinical trials where response is an endpoint. In this situation, a repeat assessment is needed to determine if a PR or CR is the best overall response. The interval for repeat imaging will be defined by the treatment protocol.

Immune Response Evaluation Criteria in Solid Tumor: iRECIST

In recent years, the development of immune therapies has provided significantly improved overall survival for some tumor types with few therapeutic options including metastatic melanoma [4] and metastatic non-small cell lung cancer [5]. These novel therapies are also currently being tested across diverse tumor types in early phase and phase III clinical trials [6]. It is now well established that tumors respond differently to immune therapies than they do to cytotoxic therapies. Immune therapies in general work through activation of the T-cells, and this activation can lead to a radiologic response that resembles a tumor flare. Alternatively, effective immune therapies may not induce tumor shrinkage yet still be effective in decreasing the rate of progression. If this deference in tumor response to immune therapies is not anticipated, then potentially effective therapies may be discontinued prematurely [4]. A new shared guideline was needed to help standardized the approach to accessing response to immune therapies as many trials that were using RECIST 1.1 had variability in tumor assessments leading to difficulty in comparing trials and pooling data [7].

RECIST criteria were subsequently modified in 2009 to account for this difference in response pattern, and the immune-related response criteria (irRC) were published [8]. Phase II clinical trial data from ipilimumab in metastatic melanoma was the most comprehensive dataset at the time and was used to develop assessment criteria. Using this dataset, patients fell into four distinct response groups: shrinkage in baseline lesions without new lesions, durable stable disease with a slow steady decline in total tumor burden in some patients, response after an initial increase in tumor burden, and response in the presence of new lesion. Importantly, all of these patterns were associated with a more favorable overall survival despite the last two representing PD based on RECIST 1.1 criteria.

Thus the iRECIST criteria evaluate the total measurable tumor burden when determining response. For iRECIST, only index and measurable new lesions are taken into account which are in contrast to standard RECIST criteria which do not require measurement or inclusion of new lesions in the overall tumor burden assessment. Measureable lesions for iRECIST are equal to or greater than 5 mm × 5 mm. Tumor burden is based on the calculation of the sum of the products of the two largest perpendicular diameters (SPD) of all index lesions. Tumor burden = SPD $_{\text{index lesions}}$ + SPD $_{\text{new, measurable lesions.}}$ Index lesions are limited to five per organ, a total of ten visceral lesions, and five cutaneous lesions. At each subsequent tumor measurement, the SPD of the index lesions, as well as any new measurable lesions, are utilized to calculate the total tumor burden.

Responses are still characterized as CR, PR, SD, and PD and are determined by the change in the overall tumor burden. CR is the disappearance of all lesions in two consecutive observations not less than 4 weeks apart. PR is at least a 50% decrease in tumor burden compared with baseline in two observations at least 4 weeks apart. To be considered PD, the tumor burden must increase at least 25% compared to the nadir at any single time point in two consecutive observations measured at least 2 weeks apart. This is an important difference in comparison to RECIST criteria that considers PD as an increase of 25% compared to the nadir for a single time point or the appearance of any new lesions. The tumor assessment is considered SD when there is neither a 50% decrease in tumor burden nor at least a 25% increase compared to nadir.

Therefore, patients are considered to have irPR or irSD even if new lesions appear as long as the overall threshold for tumor burden falls within the established percentage change from the nadir. For patients with a rapid increase in tumor burden, or with rapid clinical deterioration, repeat assessment 4 weeks after the first assessment showing an increase in tumor burden is not needed to establish progressive disease. In the phase II, a clinical trial program of a patient with malignant melanoma of which the iRECIST criteria were first applied, 9.7% (22 of 227) of the treated patients who were initially designated as PD by standard RECIST criteria was redefined as having response to ipilimumab therapy. In this patient group, overall survival was similar between patients defined as having responsive disease by conventional criteria and patients shifted from progressive disease to responsive disease by iRECIST criteria. Therefore, iRECIST provides a framework for oncologists evaluating immune therapies base assessment on clinically relevant criteria and time points.

Positron Emission Tomography Response Criteria in Solid Tumors: PERCIST

In recent years, 18F-FDG PET imaging has become more widely used to assess tumor response, especially for certain tumor types like breast, colorectal, esophageal, head and neck, and non-small cell lung cancer and melanoma [9]. There is extensive literature to support an 18F-FDG PET in the assessment of early treatment response as well its limitations of anatomical imaging as compared to standard CT imaging [10]. As PET scanning became part of standard disease assessments, a standardized framework for PET tumor assessment was needed. Subsequently, a guideline for assessing response with PET imaging was first published in 2009 by Wahl et al. as the PET Response Criteria in Solid Tumors (PERCIST) [11]. The authors of this guideline based the framework on the premise that tumor response as assessed by PET is a continuous and time-dependent variable. Tumors should be evaluated at multiple time points during treatment, and assessments may vary on metabolic activity and glucose levels. For example, it is recommended to wait at least 10 days after the last administration of chemotherapy before PET imaging as this allows for bypassing of an immediate chemotherapy flare effect in the tumor [12].

The main tracer used in PET CT scans, 18F-FDG, has an established correlation between uptake concentrations and the number of cancer cells in a region of interest (ROI). The driving principle is that reductions in tumor 18F-FDG are seen following the loss of viable cancer cells, and the inverse, increase in tumor uptake of 18F-FDG, is seen with cancer cell proliferation [13]. The percent decrease in standard uptake value max (SUVmax) between baseline and end of treatment scans in patients with non-small cell lung cancer has been shown to correlate with likihood of achieving a complete pathologic response [14]. However, it is also important to note that a completely negative PET scan after completion of treatment, while being associated with a better prognosis than imaging consistent with persistent uptake, cannot differentiate between minimal tumor burden and no tumor burden.

In designing a guideline for PET response, the authors needed to determine the main measurement of response. PET scans for cancer staging and treatment use a qualitative measurement of the distribution and intensity of 18F-FDG uptake is a ROI and compare that to what is deemed normal tissue. This normal tissue can be normal structures such as the blood pool, muscle, brain, or liver. The standard uptake value (SUV) is the most widely used metric for assessing tissue concentration of tracers such as 18F-FDG and can be normalized for patient factors such as body mass, lean body mass (SUL), and body surface area. SUV is also influenced by consistent patient preparation for the scan and scan quality. This is particularly important when using sequential PET scans to determine tumor response. The NCI has published recommendations for the use of 18F-FDG PET as an indicator of therapeutic response with attention to the variables that can influence consistency in scanning results [15]. Compliance with these standards is assumed in the PERICT recommendations. Patients should fast at least 4–6 h prior to scanning, the measured serum glucose level must be less than 200 mg/dl, and PET scan should be obtained at 50–70 min after injection of the tracer. It is recommended that the same tracer dose and same scanner be used for sequential imaging to improve reproducibility.

The primary measurement for PERICT is SUL, which is SUV lean (SUV corrected for lean body mass). The SUL is determined for up to five tumors with the greatest 18F-FDG uptake, maximum of two per organ site. These are usually the same lesions that would be identified as target lesions by RECIST1.1 criteria. Tumor size should also be noted, and ideally, these lesions should measure at least 2 cm for accurate PET avidity measurements. However, lesions smaller than 2 cm with appropriately elevated SUL levels can be recorded as target lesions. On baseline imaging, tumor SUL peak must be 1.5 × mean liver SUL + 2 standard deviations of the mean liver SUL to be considered as the minimal metabolically measurable tumor activity for target lesions. In the setting of liver dysfunction, the SUL should be based on 2.0 × blood pool 18F-FDG activity + 2 standard deviations in the mediastinum for determination of the minimum metabolic activity.

In PERCIST, response to therapy is expressed as a percentage change in the SUL peak between scans. A complete metabolic response is defined as the complete visual resolution of all metabolically active tumor. Partial response is defined as at least 30% and a 0.8 unit decline in the peak SUL between the most metabolically active lesions at baseline and the most intense lesions after treatment. Importantly, the comparison does not have to occur between the same lesion, and this difference can be calculated from two different anatomical lesions. Inversely, more than a 30% and an increase of 0.8 units in the peak SUL, or new lesions, is considered a progressive disease.

Evaluating Brain Metastasis

Evaluating response to treatment for patients with primary central nervous system tumors or metastatic cancer to the brain poses a set of unique challenges. First, the treatment of malignant disease in the brain often differs from the non-CNS tumor site treatment, as many antineoplastic therapies do not pass the blood-brain barrier, and thus surgery and radiation therapy are often utilized. Furthermore, many clinical trials for solid tumors exclude patients with brain metastases, or the metastatic lesions must be treated and stable prior to enrollment. Finally, the criteria used to evaluate the response to treatment of brain metastases have differed between clinical trials when these patients are included [1].

The original system for assessing response in CNS tumors was the McDonald criteria for the assessment of high-grade gliomas [16], and more recently the response assessment in neuro-oncology (RANO) group published updates to the McDonald criteria [17]. Both primary brain malignancies such as high-grade gliomas and brain metastases present challenges in assessing response through imaging. Included in these challenges are criteria for the minimal size of a measurable lesion, determining response to therapy versus treatment effect and incorporating the clinical condition of the patient. Because of the need to standardize the different criteria, the Response Assessment in Neuro-oncology (RANO) and RECIST 1.1 were used to form a new guideline for assessment named the RANO-BM criteria [18]. One important addition to the RANO-BM criteria is that it recognizes potential differences between primary CNS tumors and metastatic disease to the brain. This is needed as CNS malignancies can respond differently than non-CNS malignancies to therapeutic interventions. Thus, the RANO-BM criteria suggest separate assessment for the CNS with RANO-BM criteria used for the CNS disease and RECIST 1.1 used for the non-CNS disease. The RANO-BM criteria use unidimensional measures with a minimum lesion size of 10 mm, allow for up to five lesions to be followed as target lesions, and include clinical status and steroid administration when determining response.

RANO-BM defines a complete response as complete disappearance of all lesions without recent steroid use and with a clinically stable or improved status. Partial response is defined as a greater than or equal to 30% decrease in the sum of the longest diameter compared to baseline scan with stable or decreased steroid administration and clinically stable or clinically improved status. Progressive disease is defined as a greater than or equal to 20% increase in the sum of the longest diameters compared to the nadir measures. One important caveat is that an absolute minimum increase of 5 mm in one of the lesions is required for progressive disease. Also qualifying for progressive disease is a significant increase in T2 signal or a worsening clinical status.

Immune therapies are increasingly important in malignancies with high rates of CNS metastasis such as malignant melanoma, breast cancer, and non-small cell lung cancers. Based on the RANO-BM criteria, just like with iRECIST, the appearance of new lesions alone does not constitute progressive disease. When immunotherapies are being evaluated, new lesions are measured and included in the sum of the longest diameters when determining if there is an equal to or greater than 20% increase.

Summary and Future Directions

The development of RECIST criteria was an important advancement in oncology research with the implementation of a single system for evaluating response to therapies which can be universally applied to clinical research and standard of care therapy. It has made possible the accurate aggregation of data across clinical trials. Just like oncology therapies themselves, the criteria for assessing tumor response are dynamic and will need continued review and modification. All of these tumor response criteria rely on skilled clinicians to select appropriate sites of disease and to perform appropriate measurements. There remains the need for proper clinical judgment in the selection of target site and the incorporation of patient symptoms into the global response analysis.

References

1. Therasse P, Arbuck SG, Eisenhauer EA, et al. New guidelines to evaluate the response to treatment in solid tumors (RECIST guidelines). J Natl Cancer Inst. 2000;92:205–16.
2. Eisenhauer E, Therasse P, Bogaerts l, et al. New responses evaluation criteria in solid tumours: revised RECIST guidelines (version 1.1). Eur J Cancer. 2009;45:228–47.
3. Schwartz LH, Litiere S, de Vries E, et al. RECIST 1.1- update and clarification: from the RECIST committee. Eur J Cancer. 2016;62:132–7.
4. Hodi FS, O'Day SJ, McDermott DF, et al. Improved survival with ipilimumab in patients with metastatic melanoma. N Engl J Med. 2010;363:711–23.
5. Topalian SL, Drake C, Pardoll D. Immune checkpoint blockade: a common denominator approach to cancer therapy. Cancer Cell. 2015;27:450–61.
6. Pardoll D. The blockade of immune checkpoints in cancer immunotherapy. Nat Rev Cancer. 2012;12:252–64.
7. Seymour L, Bogaerts J, Perrone A, et al. iRecist: guidelines for response criteria for use in trials testing immunotherapeutics. Lancet Oncol. 2017;18:143–52.
8. Wolchok JD, Hoos A, O'Day S, et al. Guideline for the evaluation of immune therapy activity in solid tumors: immune-related response criteria. Cancer Therapy. 2009;15:7412–20.
9. Juweid ME, Cheson BD. Positron-emission tomography and assessment of cancer therapy. N Engl J Med. 2006;354:496–507.
10. ZinZani P, Tani M, Alinari S, et al. Early positron emission tomography (PET) restaging: a predictive final response in Hodgkin's disease patients. Ann Oncol. 2006;17:1296–300.
11. Wahl R, Jacene H, Kasamon Y, et al. From RECIST to PERCIST: evolving considerations for PET response criteria in solid tumors. J Nucl Med. 2009;50:122–50.
12. Dehdashti F, Flanagan FL, Mortimer JE, Katzenellenbogen JA, Welch MJ, Siegel BA. Positron emission tomographic assessment of "metabolic flare" to predict response of metastatic breast cancer to antiestrogen therapy. Eur J Nucl Med. 1999;26:51–6.
13. Bos R, van Der Hoeven JJ, van der WE, et al. Biologic correlates of (18)fluorodeoxyglucose uptake in human breast cancer measured by positron emission tomography. J Clin Oncol. 2002;20:379–87.
14. Cerfolio R, Bryant A, Winokur T, et al. Repeat FDG-PET after neoadjuvant therapy is a predictor of pathologic response in patients with non-small cell lung cancer. Ann Thorac Surg. 2004;78:1903–9.
15. Shankar LK, Hoffman JM, Bacharach S, et al. Consensus recommendations for the use of 18F-FDG PET as an indicator of therapeutic response in patients in National Cancer Institute Trials. J Nucl Med. 2006;47:1059–66.
16. MacDonald DR, Cascino TL, Scold SC, et al. Response criteria for phase II studies of supratentorial malignant glioma. J Clin Oncol. 1990;8:1277–80.
17. Wen P, Macdonald D, Reardon D, et al. Update response assessment criteria for hight-grade gliomas: response assessment in neuro-oncology working group. J Clin Oncol. 2010;28(11):1963–72.
18. Lin N, Aoyama H, Barani I, et al. Response assessment criteria for brain metastases: proposal from the RANO group. Lancet Oncol. 2015;16:e270–8.

Part IV

Regulatory Processes, Quality and Policy Issues, Companion Diagnostics, and Role of Central Laboratories

IVDs and FDA Marketing Authorizations: A General Overview of FDA Approval Process of an IVD Companion Diagnostic Device in Oncology

47

Shyam Kalavar and Reena Philip

Introduction

Scientific advances leading to increased knowledge and understanding of the underlying molecular mechanisms in cancer have given rise to the development of targeted therapies in the field of oncology. In vitro diagnostics (IVDs) play a key role in the success of these targeted therapies. IVDs play an important part in biomarker-based oncology trials by identifying patients with specific biomarkers for treatment with targeted drugs or function as a stratification tool in order to eventually be able to assess patient response to these drugs. Since the performance of the IVD is critical in determining the success of these targeted therapies, the importance of well-designed, adequately validated, and high-performing diagnostics cannot be understated. The recognition of the key role that IVDs play in assessing patients for treatment with targeted therapies has ushered in the era of co-development of the IVD companion diagnostic and the associated therapeutic agent. Co-development involves the preclinical and clinical development of the IVD and the therapeutic agent concurrently and testing them in clinical trials in order to gather supporting data for regulatory approval of these two products. Along with this development, the regulatory review and marketing authorization model for IVDs has seen the needed adaptations that enable concurrent review and approval of these products.

IVDs: General Regulatory Concepts

IVDs are defined as "reagents, instruments, and systems intended for use in diagnosis of disease or other conditions, including a determination of the state of health, in order to cure, mitigate, treat, or prevent disease or its sequelae. Such products are intended for use in the collection, preparation, and examination of specimens taken from the human body" [1]. The US Food and Drug Administration (FDA) review of device regulatory applications is to ensure the safety [2] and effectiveness [3] of the device. IVDs are classified based on the risk posed to the patient based on the intended use (IU) of the device and any mitigating measures that might affect the level of risk. In assessing the risk of an IVD to patients, a major consideration is the consequences of incorrect results from the IVD. Incorrect results may be either false positive or false negative results,

Note: This book chapter reflects the views of the authors and should not be construed to represent FDA's views or policies.

S. Kalavar (✉) · R. Philip
Center for Devices and Radiological Health, Office of In Vitro Diagnostics and Radiological Health, Division of Molecular Genetics and Pathology, US Food and Drug Administration, Silver Spring, MD, USA
e-mail: Shyam.kalavar@fda.hhs.gov

S. Badve, G. L. Kumar (eds.), *Predictive Biomarkers in Oncology*,
https://doi.org/10.1007/978-3-319-95228-4_47

and each of these may have its own specific effect. For example, false positive results may erroneously trigger inappropriate treatment such as an invasive procedure or unnecessary treatment with therapeutic product which may have adverse side effects. A false negative result may result in a delay or withholding of treatment which can adversely impact the medical management of disease. Thus, incorrect results may, especially in the case of companion diagnostic tests, result in substantial harm to the patient. Each IVD is assigned to one of three risk-based regulatory classes, Class I, Class II, or Class III, based on the level of regulatory control necessary to provide reasonable assurance of its safety and effectiveness. The regulatory controls increase as the device regulatory class increases from Class I to Class III.

The risk-based classification of IVDs and regulatory controls is listed in Table 47.1.

General controls are the basic provisions of the May 28, 1976, Medical Device Amendments to the Food, Drug, and Cosmetic Act that provide the FDA with the means of regulating devices to ensure their safety and effectiveness. General controls encompass a variety of requirements including device registration and listing; good manufacturing practices; requirements related to adulteration; misbranding; banned devices; notification requirements, including repair, replacement, or refund; maintenance of records and reports; etc. [4]. Generally Class II devices require a 510(k) submission for marketing authorization. These devices need performance data to support their substantial equivalence to a previously cleared (predicate) device and, depending on the intended use of the device, may be required to comply with special controls. The term "substantially equivalent" or "substantial equivalence" means, with respect to a device being compared to a predicate device, that the device has the same intended use as the predicate device and similar performance characteristics. The de novo pathway is another regulatory process for Class I or Class II devices. The de novo process provides a means for a new device, without a valid predicate, to be classified into Class I or II if it meets a certain criteria [5]. For Class III devices, a premarket approval application (PMA) is required. PMA applications involve a more in-depth and comprehensive review of safety and effectiveness of the device. The 510(k) review standard is comparative, whereas the PMA standard relies on an independent demonstration of safety and effectiveness. Additional information is available on the FDA website at the following link: http://www.fda.gov/MedicalDevices/DeviceRegulationandGuidance/Overview/ClassifyYourDevice/.

Table 47.1 Risk-based classification of IVDs

Class	Risk level	Regulatory control
I	Low	General controls
II	Moderate	General controls and special controls
III	High	General controls and premarket approval

The FDAs pre-submission program is a mechanism by which device manufacturers can engage in an earlier communication with FDA prior to the marketing application. The pre-submission program assists applicants such as device manufacturers (sponsors) or others by allowing requests for feedback from the FDA regarding any planned device regulatory application such as an investigational device exemption (IDE) application, 510(k) notification, PMA, and other applications. FDA's pre-submission guidance [6] provides information related to this program and specifically provides detailed instruction about how to request meetings with the FDA and request feedback related to device development and performance assessment, etc.

Use of IVDs in Oncology Clinical Trials

IVDs are being increasingly relied upon to guide therapy in the field of oncology. IVDs, by identifying specific biomarkers in patients, enable targeted therapy which helps achieve better clinical outcomes for biomarker-positive patients. A biomarker is defined as a "characteristic that is measured as an indicator of normal biological processes, pathogenic processes, or responses to an exposure or intervention, including therapeutic interventions" [7]. Biomarkers are often detected or measured by IVDs. In clinical trials biomarkers may serve different purposes such as functioning as diagnostic biomarkers in order to select patients for treatment with specific therapies; as prognostic biomarkers to assess prognostic value, i.e. disease progression; or as predictive biomarkers to predict treatment responses [8]. In the field of oncology, IVDs have played an important role in identifying markers belonging to specific categories such as molecular (e.g., EGFR mutation testing in non-small cell lung cancer (NSCLC), BRAF V600E mutation testing in melanoma) and histologic (e.g., tissue-based testing such as HER2 IHC and FISH testing in breast cancer, PD-L1 IHC testing in NSCLC).

It is very useful to be able to understand the role of the biomarker in response or safety in earlier stages of therapeutic product development. Adequate analytical and preclinical testing of the prototype IVD in the early stages of development can help generate a strong hypothesis toward this end. Based on this understanding, appropriate strategies can be developed to streamline the development, assessment, and a coordinated regulatory process for the diagnostic device as well as the therapeutic product. It is important to ensure that the companion diagnostic IVD test for the biomarker is ready for a regulatory submission at the same time as the therapeutic product submission by ensuring that the companion diagnostic device has adequate performance data to be approved contemporaneously with the therapeutic product.

Investigational Device Exemption (IDE)

When conducting a clinical trial with an investigational device (device being studied in the trial), a key issue to consider is whether an IDE application should be filed with the FDA. If a diagnostic device is used to identify a biomarker-based patient population in a therapeutic product trial and if the diagnostic device is not already FDA cleared or approved for that indication, then one needs to assess whether use of the device will pose a significant or nonsignificant risk to the patient. Sponsors can determine if an IDE is needed for their diagnostic device by obtaining feedback from the FDA via a "Study Risk Determination" pre-submission. Risk determination submissions are for the purpose of determining whether an IDE is needed to allow the use of an investigational device in a clinical trial.

Bridging Study

Clinical trials may use prototype or nonfinal versions of the proposed companion diagnostic IVD, colloquially call clinical trial assays (CTA) to assess biomarker status of trial subjects. These CTAs may differ in important ways from the version of the IVD that will be the subject of the PMA application [also called market ready assay (MRA)]. In this case, a bridging study will likely be needed to demonstrate the concordance between the CTA and the MRA in order to "bridge" the clinical data (e.g., overall survival) from CTA to MRA and to evaluate the therapeutic product efficacy in IVD companion diagnostic intended use population [9]. The goal of the bridging study is to demonstrate that the therapeutic product efficacy is maintained when the MRA is used to test the clinical trial patient samples. In order to perform an adequate bridging study, a robust plan for banking of all marker positive samples and a random subset of market negative cases along with the associated clinical and demographic information is very important.

Contacting CDRH through the pre-submission process is recommended when the need for a bridging study is anticipated.

IVD Companion Diagnostic Devices

An IVD companion diagnostic device is defined as "an in vitro diagnostic device that provides information that is essential for the safe and effective use of a corresponding therapeutic product" [10]. Companion diagnostic devices are only required when essential for safe and effective use of the therapeutic product, and in such a case, the use of a companion diagnostic device is stipulated in the associated therapeutic product labeling. Similarly, the therapeutic product is specified in the companion diagnostic device labeling. FDA guidance "In Vitro Companion Diagnostic Devices, Guidance for Industry and Food and Drug Administration Staff, February 18, 2014," specifies that "An IVD Companion diagnostic device could be essential for the safe and effective use of a corresponding therapeutic product to:

- Identify patients who are most likely to benefit from a particular therapeutic product;
- Identify patients likely to be at increased risk for serious side effects as a result of treatment with a particular therapeutic product; or
- Monitor response to treatment with a particular therapeutic product for the purpose of adjusting treatment to achieve improved safety or effectiveness."

Regulatory requirements for companion diagnostics stem from the Federal Food, Drug, and Cosmetic Act (FD&C Act) and relevant medical device and therapeutic product (i.e., drug products) regulations, Section 505 of the FD&C Act (i.e., drug products) or Section 351 of the Public Health Service Act (i.e., biological products). Similar to the risk-based classification of devices, the regulatory pathway applicable to IVD companion diagnostic devices is also based on the level of risk to patients posed by these devices as determined by the process discussed previously. IVD companion diagnostic devices authorized by FDA to date in the field of oncology have been classified as Class III devices, because the risk to the patient of an incorrect result is high, and no mitigations of the risks have been identified that would be adequate to justify classification in Class II or I. If an IVD companion diagnostic is essential for assuring safety or effectiveness of the therapeutic product, FDA generally will not approve the therapeutic product if the IVD companion diagnostic will not receive contemporaneous marketing authorization for use with that therapeutic product for that indication. However, as stated in the "In Vitro Companion Diagnostic Devices" FDA guidance, the FDA may approve a therapeutic product without the prior or contemporaneous marketing authorization of an IVD companion diagnostic in certain circumstances (i.e., when a therapeutic product is intended to treat a serious or life-threatening condition for which no satisfactory available therapy exists or when the labeling of an approved therapeutic product needs to be revised to address a serious safety issue) [11].

The FDA approved the first IVD companion diagnostic in 1998. The companion diagnostic device is an immunohistochemistry (IHC) device HercepTest™ (Dako), and the drug is trastuzumab (Herceptin®, Roche/Genentech). The companion diagnostic device is intended to be used to identify HER2 overexpression in patients with advanced breast cancer. The approval of the HercepTest™ was followed by other companion diagnostic device approvals. A complete list of companion diagnostic device approvals is available on the FDA website [12].

An IVD companion diagnostic device may be developed as a new diagnostic test for a new therapeutic product (e.g., HER2/neu IHC testing for treatment with Herceptin), as a new diagnostic test for an old therapeutic product (e.g., KRAS testing for treatment with cetuximab), or may be submitted for review as an existing diagnostic test for a new therapeutic product (e.g., EGFR IHC testing for treatment with panitumumab) [13].

Follow-On Companion Diagnostic Devices

A second-generation companion diagnostic device or a follow-on companion diagnostic device is an IVD companion diagnostic device that is intended for the same indication as the original FDA-approved IVD companion diagnostic device. More specifically, a follow-on companion diagnostic device is intended to be used with the same therapeutic product to identify the same indicated patient population, as specified in the labeling of the original FDA-approved companion diagnostic device. Companion diagnostic devices are used to direct patient therapy, so it is important that follow-on companion diagnostic devices correctly identify the same intended use patient population. The benefit and risk profiles associated with the use of a follow-on companion diagnostic device and the original FDA-approved companion diagnostic device would be comparable. Relying on a simple method comparison study between the original approved companion diagnostic device and its follow-on companion diagnostic device to assess comparability between these two devices is generally not acceptable for approval, because it is unknown how different levels of analytical comparability between the two companion diagnostic devices would translate into clinical performance of the follow-on companion diagnostic device. Therefore, the regulatory review of the follow-on companion diagnostic device generally would also include some type of assessment of clinical performance to ensure that use of the follow-on companion diagnostic device would not alter the established therapeutic efficacy and safety profile. The VENTANA ALK (D5F3) CDx Assay and cobas® KRAS Mutation Test are two follow-on companion diagnostic devices intended for treatment of patients with NSCLC and colon cancer, respectively. The performance of the ALK (D5F3) CDx Assay was evaluated in a prospective trial, where patient samples for which associated clinical outcome was known were tested with both the follow-on companion diagnostic device and the original approved companion diagnostic device [14] The performance of the cobas® KRAS Mutation Test was evaluated using an innovative statistical approach that allowed FDA to determine that the safety and effectiveness of the approved companion diagnostic are preserved, even though there were no patient samples with associated therapeutic clinical outcome known [15]. Additional details may be obtained by referring to the device Summary of Safety and Effectiveness Data at the following link: http://www.accessdata.fda.gov/cdrh_docs/pdf14/P140023b.pdf.

FDA has also approved IVDs together with a therapeutic product approval where the IVD is not considered to be a companion diagnostic device. These IVDs are used to assess biomarker positivity in patients (e.g., PD-L1 in NSCLC, etc.) to determine if patients may have enhanced response to a particular therapeutic product based on the level of biomarker expression, relative to the biomarker negative patient population who may have an acceptable response to the same therapeutic product. An example of such a diagnostic device is the PD-L1 IHC 28-8 pharmDx where "PD-L1 expression as detected by PD-L1 IHC 28-8 pharmDx in non-squamous NSCLC may be associated with enhanced survival from OPDIVO® (nivolumab)" [16]. In the clinical trial to assess the activity of nivolumab in patients with metastatic NSCLC, subjects were treated with nivolumab regardless of PD-L1 expression as measured by the PD-L1 IHC 28-8 pharmDx. In the overall study population, patients treated with nivolumab had higher overall survival (OS) when compared to those treated with docetaxel. These results were statistically significant and independent of PD-L1 expression in the tumor biopsies. However, in this trial, a pre-specified retrospective analysis of the efficacy of nivolumab in patients based on specific levels of PD-L1 expression (1%, 5% and 10%) revealed that the higher the level of PD-L1 expression, the greater the benefit in terms of OS from nivolumab in comparison to docetaxel. This information can be used in treatment planning and is included in the therapeutic product label.

FDA Regulatory Review Process for IVD Companion Diagnostic Devices

FDA review of IVD companion diagnostic device regulatory applications is conducted concurrent to and within the context of the review of the associated therapeutic product, through a collaborative interaction between the reviewing offices. FDA review of IVD companion diagnostic device regulatory applications includes the assessment of the device analytical and clinical performance to determine the safety and effectiveness of the IVD companion diagnostic when used with the corresponding therapeutic product. Review of the clinical trial data to assess the clinical activity of the therapeutic product, usually in the population identified by the companion diagnostic device, and the determination that an IVD companion diagnostic is essential for the safe and effective use of the therapeutic product are made by the FDA therapeutic product review center [Center for Drug Evaluation and Research (CDER) or Center for Biologics Evaluation and Research (CBER)]. The device review center (CDRH) performs a detailed review of device information such as analytical performance, clinical performance in the intended use population, manufacturing information, device software, and other applicable information (Fig. 47.1).

The companion diagnostic device is reviewed based on the analytical and clinical performance data provided by the device sponsor. Analytical performance is the ability of the test to accurately and reliably measure the analyte of interest and is supported by studies that evaluate device sensitivity, specificity, accuracy, precision, reproducibility, etc. In a regulatory submission, analytical performance evaluation of an IVD companion diagnostic should be performed with the final version of the device, i.e., the device that is intended to be marketed (market ready assay). Validation studies are generally performed on specimens from the intended use population with specific attention to characterizing performance at the clinical decision point. For example, if the IVD companion diagnostic device is an IHC assay, then performance of the device around the scoring cutoff should be carefully assessed. Clinical performance is assessed by a determination of how well the test results are correlated with the specified clinical action and outcome in the therapeutic product trial(s). A list of general analytical and clinical performance studies that are reviewed as part of the regulatory application for an IVD companion diagnostic device is shown in Table 47.2.

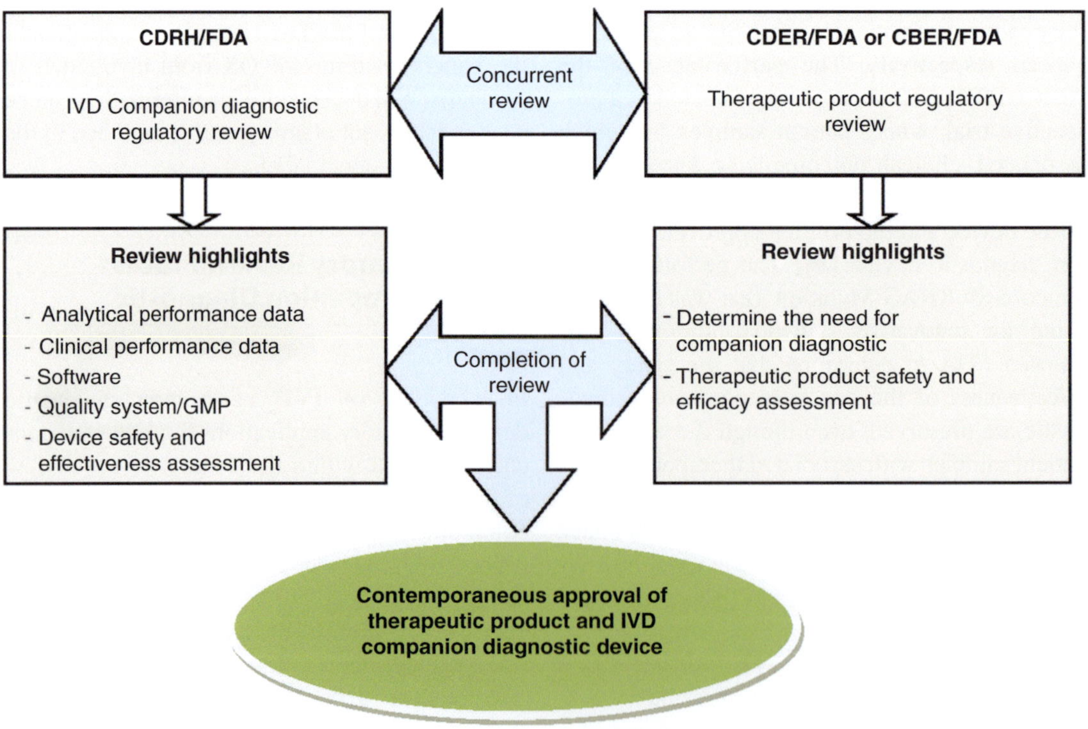

Fig. 47.1 FDA regulatory review process for IVD companion diagnostic devices

Table 47.2 Overview of analytical and clinical performance data[a]

Analytical performance studies	Clinical performance studies
Pre-analytical variables	Performance of device assessed in a clinical trial or study
Sensitivity	
Specificity	Comparison to a reference method, if applicable
Precision: repeatability and reproducibility, reader precision/reproducibility	Bridging study, if applicable
Controls	
Robustness	
Limit of blank, limit of detection	
Linearity, as applicable	
Carry-over and cross-contamination	
Interference	
Matrix equivalence, as applicable	
Sample and product stability	

[a]Additional performance studies may be needed as applicable

Regulatory Approval of Companion Diagnostic Device: An Example

Although 21CFR 814.20 specifies that all information for a traditional PMA be submitted at the same time, avenues exist for alternate methods, i.e., the modular PMA process. The modular PMA approach accommodates challenges present in the IVD companion diagnostic development environment such as late identification for the need for a companion diagnostic in a therapeutic product clinical trial which leaves insufficient time for preparing the requisite information for a PMA application. The modular PMA process allows for specific portions of the PMA application to be submitted as modules at specified intervals which allows the sponsor to adequately prepare for the PMA application. Additionally, it allows planning and scheduling of any FDA-required inspections of manufacturing facilities, clinical laboratory testing sites, etc. Typically, four modules are submitted each with the following specific information: device software/hardware, GMP/Quality Systems, analytical performance data, and clinical performance data which is usually the final module. Software/hardware PMA module, GMP/Quality Systems PMA module, and analytical performance data module are submitted in succession even as the therapeutic product clinical trial is in progress and the device clinical performance data is being gathered.

The recent PMA application for the PD-L1 IHC 22C3 pharmDx device followed the modular submission [17] approach where the performance data/information supporting the device safety and effectiveness are submitted to the FDA in "modules" at successive time points. The companion diagnostic is intended for use in the detection of PD-L1 in formalin-fixed, paraffin-embedded (FFPE) non-small cell lung carcinoma (NSCLC) tissue as an aid in identifying NSCLC patients for treatment with KEYTRUDA® (pembrolizumab). The IVD companion diagnostic was approved for use in the detection of PD-L1protein in formalin-fixed, paraffin-embedded (FFPE) NSCLC tissue [18].

The drug was granted breakthrough therapy designation, and the therapeutic product application was reviewed under the provision of accelerated approval by CDER. The drug is indicated for the treatment of patients with advanced NSCLC whose tumors express PD-L1 as determined by an FDA-approved test and who have disease progression on or after platinum-containing chemotherapy. Patients with EGFR or ALK genomic tumor aberrations should have disease progression on FDA-approved therapy for these aberrations prior to receiving KEYTRUDA [19].

In this and many other IVD companion diagnostic device PMA applications, the modular PMA approach was used and was helpful in assuring that the therapeutic product approval and the diagnostic product approval were contemporaneous. This is especially true when therapeutic product review times are short, as it allows the review of the device to begin prior to therapeutic product review.

The highlights of the analytical and clinical performance data that was reviewed are shown in Fig. 47.2.

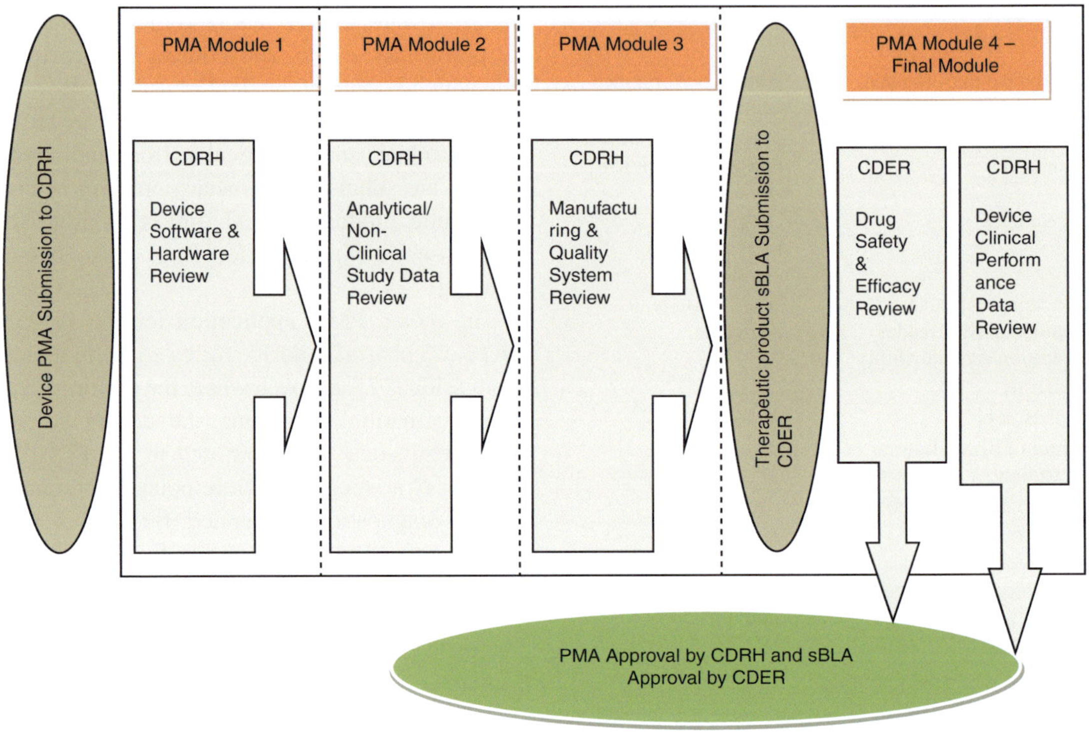

Fig. 47.2 Timeline of the regulatory review

Future Directions

The growth of personalized medicine has underscored the importance of IVD companion diagnostics in identifying appropriate patient populations for the treatment of cancer with specific therapeutic products. IVD companion diagnostic devices will continue to facilitate the approval of therapeutic product by identifying specific biomarker-defined population groups. In keeping pace with these developments, the FDA has also adopted efficient strategies to perform regulatory reviews of marketing applications. These strategies have enabled concurrent reviews for IVD companion diagnostic devices and the associated therapeutic product between the appropriate FDA review centers and to synchronize the review timelines to accommodate the concurrent accelerated or regular approval of the therapeutic product-diagnostic pair. From the industry perspective, early collaboration between the therapeutic product and the device manufacturers is very helpful to be able to gather performance data and successfully submit simultaneous regulatory applications for their respective products. Advanced and complex diagnostic technologies such as next-generation sequencing (NGS) are expected to play a more significant role in the future of personalized medicine. Tests relying on these technologies present new challenges in the volume and complexity of the data generated, assessing the accuracy of such data and their clinical implications, and the subsequent use of this data (test results) in the clinical setting in order to treat patients.

References

1. US FDA: Code of Federal Regulations Title 21 §809.3(a).
2. US FDA: Code of Federal Regulations Title 21 §860.7(d)(1).
3. US FDA: Code of Federal Regulations Title 21 §860.7(e)(1).

4. US FDA. General controls for medical devices. http://www.fda.gov/MedicalDevices/DeviceRegulationandGuidance/Overview/GeneralandSpecialControls/ucm055910.htm.
5. US FDA. Evaluation of automatic class III designation (De Novo) Summaries. http://www.fda.gov/AboutFDA/CentersOffices/OfficeofMedicalProductsandTobacco/CDRH/CDRHTransparency/ucm232269.htm.
6. Requests for feedback on medical device submissions: the pre-submission program and meetings with food and drug administration staff. Guidance for industry and food and drug administration staff. Document issued on: February 18, 2014.
7. BEST (Biomarkers, Endpoints, and other Tools) resource. Food and Drug Administration (US) Silver Spring (MD) and National Institutes of Health (US) Bethesda (MD).
8. IOM (Institute of Medicine). Evaluation of biomarkers and surrogate endpoints in chronic disease. Washington, DC: The National Academies Press; 2010.
9. Li M. Statistical consideration and challenges in bridging study of personalized medicine. J Biopharm Stat. 2015;25:397–407.
10. In Vitro Companion Diagnostic Devices. Guidance for industry and food and drug administration staff. Document issued on: August 6, 2014.
11. In Vitro Companion Diagnostic Devices. Guidance for industry and food and drug administration staff. August 6, 2014.
12. US FDA. List of cleared or approved companion diagnostic devices (In vitro and imaging tools) http://www.fda.gov/MedicalDevices/ProductsandMedicalProcedures/InVitroDiagnostics/ucm301431.htm.
13. Reena P, Mansfield E, Carrington L, et al. In vitro companion diagnostic and anti-cancer drug co-development: regulatory perspectives. In: Tan D, Lynch HT, editors. Principles of molecular diagnostics and personalized Cancer medicine. Philadelphia: Lippincott Williams & Wilkins; 2013. p. 301–6.
14. PMA P140025: FDA summary of safety and effectiveness data. VENTANA ALK (D5F3) CDx Assay. http://www.accessdata.fda.gov/cdrh_docs/pdf14/P140025b.pdf.
15. PMA P140023: FDA summary of safety and effectiveness data. cobas® KRAS Mutation Test. http://www.accessdata.fda.gov/cdrh_docs/pdf14/P140023b.pdf.
16. PMA P150025: FDA summary of safety and effectiveness data. PD-L1 IHC 28-8 pharmDx (P150025). http://www.accessdata.fda.gov/cdrh_docs/pdf15/p150025b.pdf.
17. US FDA. Public workshop–next generation sequencing-based oncology panels, February 25, 2016. http://www.fda.gov/MedicalDevices/NewsEvents/WorkshopsConferences/ucm480046.htm.
18. PMA P150013: FDA summary of safety and effectiveness data. PD-L1 IHC 22C3 pharmDx http://www.accessdata.fda.gov/cdrh_docs/pdf15/P150013B.pdf.
19. US FDA. Approved drugs. Pembrolizumab (KEYTRUDA) Checkpoint inhibitor. http://www.fda.gov/Drugs/InformationOnDrugs/ApprovedDrugs/ucm465650.htm.

48 Quality Control of Immunohistochemical and In Situ Hybridization Predictive Biomarkers for Patient Treatment: Experience from International Guidelines and International Quality Control Schemes

Petra Heinmöller, Gudrun Bänfer, Marius Grzelinski, Katya Victoria Alexander, Kathrina A. Alexander, and Bharat Jasani

Abbreviations

BC	Breast cancer
CAP	College of American Pathologist
CDx	Companion diagnostics
EQA	External quality assurance
GC	Gastric cancer
HER2	Human epidermal growth factor receptor 2
IHC	Immunohistochemistry
IQC	Internal quality control
MMR	Mismatch repair (MLH1, PMS2, MSH2, MSH6)
PT	Proficiency testing

P. Heinmöller
Department of Quality Management, Targos Molecular Pathology GmbH, Kassel, Hessen, Germany

G. Bänfer
Department of Advance -Training and Consulting, Targos Molecular Pathology GmbH, Kassel, Germany

M. Grzelinski · K. V. Alexander
Laboratory for Study Analytics 2, Targos Molecular Pathology GmbH, Kassel, Germany

K. A. Alexander · B. Jasani (✉)
Department of Pathology, Targos Molecular Pathology GmbH, Kassel, Hessen, Germany
e-mail: Bharat.Jasani@targos-gmbh.de

Overview

A predictive biomarker is a test of a measurable variable associated with a disease state that is able to predict patient response to a specific treatment [1, 2]. Forming the basis for so-called "companion diagnostic" [3, 4] tests, predictive biomarkers are designed and developed with the capacity to prospectively and reliably classify patients – through pretreatment analyses of their tumors – into those with a positive status in whom a specific treatment is predicted to be effective and those with a negative result in whom a specific treatment is likely to fail. The biomarker-specific treatment is therefore selected to be given only to the former subset. Because of their crucial role in determining the treatment choice, these tests are designated as Class III devices by the FDA and are deemed to

S. Badve, G. L. Kumar (eds.), *Predictive Biomarkers in Oncology*,
https://doi.org/10.1007/978-3-319-95228-4_48

require the highest standards of quality assurance (QA) and quality control (QC) for their safe and effective application [5].

Multiple steps and stages are involved in the assay procedure including a wide variety of reagents and assay platforms used in different laboratories across the world for conducting a given predictive test. Because of this, an effective QA/QC of an assay is both difficult and a complex process as illustrated in Figs. 48.1 and 48.2, unless approved standardized and validated procedures employing optimized, ready-to-use reagents on fully automatic assay platforms are used. The latter are designed to minimize avoidable variability and to maximize precision, reproducibility, as well as accuracy. Day-to-day quality of the analysis is further assured and controlled by the inclusion of batched and/or individual on-slide controls selected to provide effective monitoring of assay sensitivity and specificity. Besides monitoring of assay quality, the outcome of every test reader's evaluation is monitored on an ongoing quality assurance program and annual competency assessment. This process is collectively referred to as a set of internal quality control (IQC) checks as exemplified in Figs. 48.1 and 48.2.

In addition, a system of externally operated and imposed QA/QC checks is necessary for managing variability in sample quality and minimizing any subjective and systematic bias in the interpretation/scoring of results. Quality monitoring schemes, such as those provided by UKNEQAS and NordicQC, include both in-house and externally circulated control samples to comparatively evaluate the quality of the total performances of assay runs in different laboratories. Such external comparisons are performed at regular (3–6 months) intervals. Internal and external audits are also required to allow

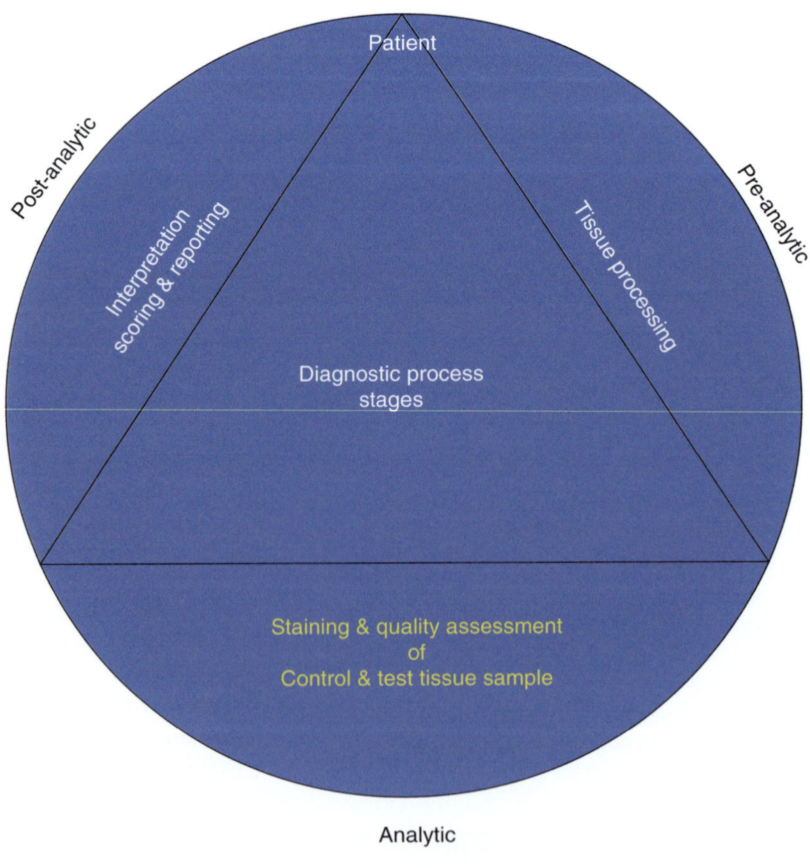

Fig. 48.1 Predictive biomarker quality assurance cycle

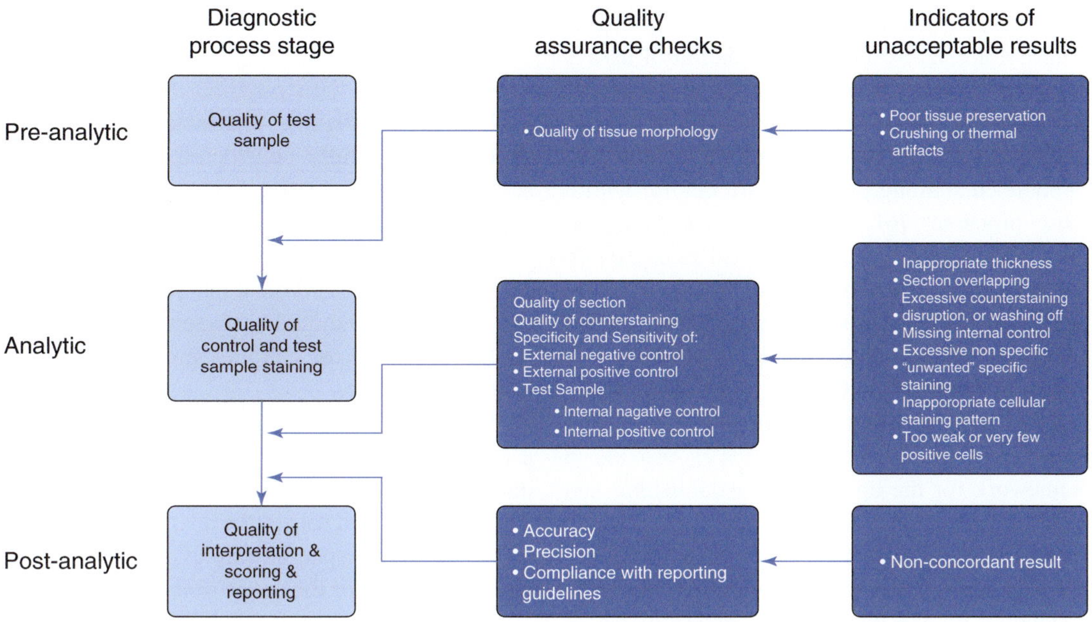

Fig. 48.2 Assay quality assurance

comparison of the mean and the range of positive and negative results recorded from year to year within a laboratory and between laboratories at the regional, national, and international levels. This chapter will be devoted to an examination of the principles of assay validation, IQC, and IQA, as applied collectively to IHC and in situ hybridization (ISH) assays.

The above issues will be presented under three different sections for the sake of simplicity and clarity:

Section 1: To provide an overview of the principles and the methodology involved in the validation of predictive biomarker assays

Section 2: To describe internal quality control and assurance measures required to assure reproducibility of results, accuracy between labs over time, and general reliability

Section 3: To give an account of the operation of the various external quality assurance (EQA) and audit schemes which function to minimize variability in assay quality due to local/demographic factors

Section 1

Validation of Predictive Biomarker Assays

Tissue-based predictive cancer biomarker assays involve three stages of analysis referred to as pre-analytical, analytical, and post-analytical phases (Fig. 48.1), respectively. Steps involved in each of these phases are subject to stringent validation and internal quality assurance procedures (Fig. 48.2).

The analysis phase of a routinely applied tissue-based IHC assay consists of dewaxing of the sections, antigen retrieval, and application of the primary antibody to the sections followed by the detection system reagent development of signal, counterstaining, and coverslip mounting of the stained sections.

An accurate and reproducible assay performance is crucial at every step and for every testing algorithm. Hence, elements influencing assay variation should be diminished and if possible eliminated. For laboratories working in a clinical environment, various guidelines exist, serving as a basis for establishing and validation of assays for usage in personalized medicine. Professional scientific organizations offering participation in proficiency testing schemes and publishing recommended staining protocols are helpful within these processes [6]. In addition, for particular labs, institutions certifying clinical labs provide the necessary checklists or publications [7] to ensure that the appropriate procedures and/or protocols and methods exist.

The ICH guideline Q2(R1) on method validation [8] and the guideline from the College of American Pathologists (CAP) [9] provide a framework for the approach in establishing and validating a clinical assay. Since it is virtually impossible to deliver an exact plan for every assay, a case-by-case decision must be made depending on the purpose and requirements of the developed assay. In case of protocol modifications or other changes of critical components, a revalidation may be necessary.

Intended Use of Assays

- Research-use-only (RUO) tests: The term RUO refers to devices that are in the laboratory phase of development. These are usually not used for providing patient health-care management decisions.
- Investigational use only (IUO): The term IUO refers to assays and or devices that are in the product testing phase of development.
- Laboratory-developed tests (LDTs): LDTs are tests that hospitals, academic, and clinical laboratories develop as testing services according to their own procedures. These tests are often created in response to unmet clinical needs and are commonly used in Europe and other countries for early and precise diagnosis, monitoring, and guiding patient treatment. Validations are done according to "the clinical laboratory improvement amendments (CLIA)/CAP guidelines".
- Commercial, unmodified In Vitro Diagnostic (IVD) tests: Use for patient health-care management decisions based on tests performed at a lab.

Typical validation characteristics which should be generally considered for assays are accuracy, precision, specificity, linearity, range, and robustness. The analytical procedure should describe in detail the steps necessary to perform each analytical test. Typical items are the sample, the reference standards, reagent preparations, handling of devices, calibration curves, calculation, formula, etc.

The initial step of assay validation is based on a profound search of any scientific and clinical data available for a biomarker. It helps to elucidate the biological role of the analyzed biomarker and to develop a specific way to go for each biomarker in terms of the scope of the assay.

It is advantageous if predefined protocols or even manufacturer's manuals can be used for this approach. Nevertheless, even for FDA-approved assays, a good clinical practice (GCP) compliant validation should be done in order to demonstrate an assay is functioning in a particular laboratory using the available equipment and performed by certain operators. If applicable, certain items can be omitted if tested elaborately by the original laboratory. For fully validated assays, a well-planned assay transfer including major aspects of the ICH guidelines paralleled by a risk analysis (e.g., FMEA – failure mode and effects analysis) may be sufficient.

An assay validation project based on the ICH guidelines requires a project plan where all steps are described in detail and a subsequent analysis of the tested parameters is feasible. All raw data generated during the testing must be collected and well documented. After completion of the measurements and testing, a comprehensible project report must be written, which reflects all items planned in advance. Only if all major preconditions are fulfilled, an assay release can take place. Otherwise further modifications of the protocol and additional validation work are needed.

Identification of Appropriate Controls (Expression Level, Ideally Based on a Clinical Threshold)

The choice of optimal clinical material for IHC assay validation represents a crucial step in the process. Appropriate samples allow to receive representative results and to establish stable and robust protocols which do not involve subsequent modifications on a patient specimen. The type of tissue and its pre-analytical preparation should reflect the application area of the assay (FFPE or frozen tissue, serum, and its origin, i.e., human, cell culture, xenograft, etc.). In addition, the prevalence and expression level of the biomarker in analyzed tissue play a major role in the choice of suitable tissue specimens. Sources of validation material range from leftover material from routine pathology labs to commercial tissue providers, occasionally offering even pre-screened samples. Regardless of the origin, the tissue must feature an ethics committee approval for research usage or for scientific purposes.

Control material used within the validation should be ideally utilized in the initial phase of clinical projects utilizing the assay. Further provision of controls must be exactly planned for every clinical trial to avoid sudden shortage or unnecessary changes during the project.

Devices and Reagents

Prerequisites desirable in an academic scientific environment like very innovative, or cheap, assays and mainly manual assays are often contrary to requirements in a clinical GCP setting. Stable and robust detection of an analyte in clinical trials lasting 5 years or more favors automated platforms and expensive "ready-to-use" kits eliminating operator's variability. The devices involved in the validation must be monitored, and parameters like software version and system setup cannot be modified during the project. Subsequent upgrades or changes must be revalidated.

Environmental Factors

The environmental factors such as temperature, humidity, or light exposure must be monitored as well and should not differ during the subsequent usage of the assay of interest. Many of the laboratory devices are tempered to avoid the influence of such factors.

The above criteria and technical issues involved in the validation of a predictive biomarker assay should be implemented to run and monitor the validity of a routinely applied assay to ensure reproducibility and accuracy of the analysis performed on a day-to-day basis.

The Role of Assay Operator

The role of an assay operator (e.g., technical assistant or a biomedical scientist) in running automated assays should not be undervalued. Highly computerized systems and their complex software applications in running automated assays require a high standard of training. Modification of controlled documents related to predictive biomarker assays should be avoided to deviate from standard protocols.

Initial Training and Competency Assessment

During the process of introducing a new biomarker in the laboratory, all technical personnel – from laboratory technicians to biomedical scientists and the responsible pathologist and/or residents – must be trained in the appropriate performance and interpretation of the assay. Initial training should be documented and followed by a competency assessment [10, 11]. For technical staff, competency is demonstrated by the production of slides or samples which are deemed acceptable and free of unreasonable artifacts by a qualified pathologist or an advanced practitioner scientist. This should include the generation of appropriate controls which are of acceptable staining intensity within predicted and measurable ranges. Of course, the design will necessarily vary depending on the assay.

For pathologists, training of predictive biomarker interpretation should ideally follow official recommendations such as the American Society of Clinical Oncology (ASCO)-CAP guidelines [12–14] or peer-reviewed publications (e.g., 15) which are specific to a biomarker. In the case of a new biomarker, or in the absence of other specific guidance, it is generally advisable to parallel the recommendations set forth for an already established biomarker. A CAP-accredited institute must always follow the official ASCO-CAP guidelines [12–14] if available for the specific biomarker. In any case, initial training should ideally include assessments of both inter- and intra-observer concordance rates to guarantee a high level of reliable results for the specific predictive biomarker. A goal of a high concordance rate usually of at least 90% should be achieved for measures of both intra- and inter-observer variability.

For HER2 and ER/PgR in breast cancer, a concordance rate of at least a 95% is recommended for annual result comparison [1]. This includes proficiency testing [16] as well as competency assessment when compared to either the results of an experienced reader or to a consensus score. In the case of ER/PgR, the concordance rates should be of similar magnitude, and it is best to adopt an internationally agreed rate for both positive and negative predictions.

Using PD-L1 as an example, initial training should include guidance specific to the actual antibody utilized and should include discussion of the features unique to the interpretation of that particular antibody and biomarker. For example, in some assays, immune cell staining is disregarded during evaluation, while the stained immune cell percentage or area in others is considered within the evaluation. Furthermore, different clinically relevant cutoffs have been established for different PD-L1 antibodies. Depending on the assay, the threshold of stained tumor cells considered clinically relevant is highly variable, ranging currently from a cutoff of 1–5%, 25%, and 50%. Finally, the interpretation scheme employed varies significantly depending on the tumor type on which the assay is performed. In the lung, a pathologist evaluating the 22C3 antibody will produce a "Tumor Proportion Score (TPS)" [17] including tumor cell staining only, while a "Combined Proportion Score (CPS)" would be generated in gastric cancer including both tumor and immune cell staining [18]. In assessing competency, pathologists should be tested to achieve the designated concordance rate at each cutoff or category which has been established as relevant for the particular assay.

In order to ensure enduring quality, regular and continuous competency assessments [12, 14] are required. Both pathologists and technical personnel (such as technicians who read slides for in situ hybridization) should undergo competency assessment on a recurring basis. CAP guidelines require that competency assessments be performed at least semiannually in the first year and then annually in subsequent years. Some elements of the ongoing competency assessment can be covered using internal or external proficiency testing. Like most biomarkers, the QC plan must be appropriate to the individual test, whether it is an already established biomarker or a newly introduced one. First, a risk-based approach is needed to determine the proper steps for quality assurance. This must include the general experience with a specific biomarker and/or the phase of introducing a new biomarker. In general, it seems prudent to adopt a paradigm parallel to that recommended for regular competency assessment with initial training followed by regular competency assessments twice in the first year and regular quality monitoring to avoid drift.

To comply with the recommendation for regular (annual/biannual) competency assessment and to avoid a drift/shift in the quality of the reported results, a lab needs to regularly monitor both intra- and interobservor variability especially of a newly implemented biomarker.

Section 2

Internal Quality Control (IQC) and Internal Quality Assurance (IQA)

A proper validation method and training of staff members is the first step in establishing high-quality laboratory testing. To detect changes in analytic performance and avoid evaluation errors, assays must be regularly monitored. This is usually done by daily quality control, periodic proficiency testing, and comparing positivity rates for selected markers (e.g., hormonal receptors, or HER2/neu) with expected positivity rates and result analogies for histopathology tests as part of the quality assurance. Ongoing monitoring of assay and reader performance is as important as initial assay validation.

Whenever possible, same-slide controls with defined expression levels should be used to detect false results and assay drift, which may be caused by the use of inadequate reagents or improperly functioning devices. For IHC testing, controls should include sections with no staining, strong staining, and staining that is equivocal or close to cutoffs. Controls should be prepared using similar fixation, processing, and paraffin embedding techniques as specimens, whenever possible.

Once controls are selected and established, laboratories should regularly audit their internal test results to identify any discrepancies and/or drift in the results (trending).

The following quality control steps should regularly be implemented, not only for central lab testing but also for labs involved in any clinical predictive biomarker assays.

A. *Tracking trends over time and location*: The prevalence of positive and negative results generated by a pathologist, by site, and by a period should be tracked and monitored to detect positive or negative trends and drift. If a significant change is noted in the positivity rate, an investigation should be undertaken to determine whether the trend is the result of an explained phenomenon (i.e., alterations in the patient base to include only individuals preselected as positive) or the result of a drift in assay performance or in the pathologist's interpretation.
B. *Random review*: A small percentage of cases (3–10% in our facility) should be selected at random and independently reviewed by a second pathologist.
C. *Regular inspection of failure or rejection rate for individual pathologists*: Tracking trends over time can serve as a harbinger of impending assay failure or a drift in its quality.

Central laboratories involved in the analysis of samples from clinical trials face additional challenges regarding their use of controls. In this setting, it is not uncommon for different controls to be utilized for the same test in different clinical trials, as imposed by the specific study protocol. This introduces an element of variability and rigorous quality control with a "four eyes principle" which must be applied at all critical points to ensure reliability and reproducibility.

As the nature of a routine lab differs from that of the central laboratory, e.g., monitoring of HER2 positivity [19], the routine labs role cannot be fulfilled by a central laboratory because the analyzed samples are often preselected.

Section 3

External Quality Assurance (EQA) and Audit Schemes

All laboratories measuring biomarkers for patient management should use analytically and clinically validated assays [20, 21], participate in external quality assurance programs [12, 14, 22], established assay acceptance and rejection criteria, and perform regular audits and be accredited by an appropriate organization.

The following section gives a brief description of some programs and providers. An overview of the offered proficiency testing schemes can be found in Table 48.1. Recently, some of the most important EQA organizations worldwide are incorporated in an international multi-stakeholder expert group called the "International Quality Network for Pathology" or commonly referred to as "IQN PATH" focused on improving quality of clinical biomarker testing in pathology with the aim of delivering high-quality patient care. IQN PATH is also involved in promoting EQA, exchanging expertise, and coordinating interactions among key stakeholders such as the European Society of Pathology (ESP), Gen&Tiss of France, NordiQC, UK NEQAS, AIOM

Table 48.1 Overview of proficiency testing schemes for predictive biomarkers

Country	PT provider	Biomarker assessments	Comment
Germany	QUIP – Qualitätssicherungs-Initiative Pathologie GmbH https://quip.eu/	HER2 IHC, HER2 ISH, ER, PgR, HER2 IHC GC, PD-L1, BRAF mutation analysis, RAS mutation analysis, EGFR mutation analysis	1 survey per year; separate PTs for laboratory part (staining) and reading (selected biomarker); provides detailed feedback
Belgium	ESP – European Society for Pathology https://www.esp-pathology.org/ http://lung.eqascheme.org/	ALK FISH, ALK IHC, NSCLC scheme (EGFR, KRAS, BRAF), ROS-1 FISH, ROS-1 IHC, PD-L1 (pilot)	Open to other countries, offers digital samples for EQA
Scandinavia	NordiQC http://www.nordiqc.org/	HER2 IHC BC, HER2 ISH BC, ER, PgR, PD-L1 IHC, ALK IHC, EGFR	IHC and ISH only; slides need to be returned; assessment team (assessors) provides feedback on staining protocol; open to other countries; offers assessment of a number of routine IHC markers
UK	UK NEQAS https://ukneqas.org.uk	HER2 IHC BC, HER2 ISH BC, ER, PgR, HER2 IHC GC, PD-L1 IHC, ALK IHC, MYC FISH, BRAF mutation analysis	International assessment team (assessors) evaluates the results
USA	CAP – College of the American Pathologists www.cap.org	HER2 IHC BC, HER2 ISH BC, ER, PgR, HER2 IHC GC, PD-L1, ALK IHC, BRAF, KRAS, EGFR	Focus on final result; no assessment team. A result needs to show 80% consensus by all participants to be included in the final evaluation
Canada	cIQc – Canadian Immunohistochemistry Quality control http://cpqa.ca/	ER PgR HER2 IHC BC, HER2 ISH BC, HER2 IHC GC, HER2 ISH GC, ALK IHC, ALK ISH, PD-L1, BRAF V600E IHC, MMR IHC panel, ATRX IHC, IDH1 R132H IHC	TMAs with up to 40 cores are distributed; stained slides need to be returned; collects protocol details; assessment team provides comments; open to international stakeholders
Australia	RCPAQAP – The Royal College of Pathologists of Australasia Quality Assurance Program https://www.rcpaqap.com.au/	IHC breast marker, HER2 BRISH BC, HER2 BRISH GC, Lung carcinoma IHC	6 cases, 2 surveys per year

French and Italian EQA schemes are not included because they are currently set up only to cater for EQA of molecular testing and not for IHC

SIAPEC from Italy, and the German Society of Pathology (DGP) [23].

Scandinavia

The Nordic Immunohistochemical Quality Control (NordiQC) is an international proficiency testing program established in 2003. Primarily aimed at assessing the analytical phases of the laboratory IHC quality [24, 25], NordiQC offers a general module which includes tests for the most common epitopes in pathology.

Besides a general module that includes tests for the most common epitopes demonstrated in surgical and clinical pathology to identify and subclassify neoplasms performed in three runs per year, NordiQC offers semiannual runs for HER2, ER/PgR IHC, and other markers relevant in breast cancer pathology, as well as HER2 ISH in breast cancer. Proficiency testing schemes for PD-L1 were introduced only recently. NordiQC recommends four different protocols for PD-L1 staining using the clones 22C3, 28-8, or SP263 on the Dako Autostainer Link, the Ventana BenchMark ULTRA, or the Ventana BenchMark XT, respectively. These three antibodies have shown a good concordance, while SP142 has shown lower scores when assessing NSCLC.

In general, TMA slides from standard processed formalin-fixed, paraffin-embedded (FFPE) material are used for all tests. For the breast cancer module and HER2 ISH module, the tissues have been fixed and processed according to the recommendations of the American Society of Clinical Oncology and College of American Pathologists (ASCO-CAP) guidelines [12–14].

Stained slides sent to NordiQC are assessed by a team of pathologists that provide individually tailored recommendations for protocol optimization as far as applicable, as well as a detailed summary of the results for an assessment. Results of past assessments can be found on the NordiQC homepage (http://www.nordiqc.org/).

About 700 laboratories from 80 countries are currently participating in the NordiQC PT programs [24].

United Kingdom

The UK National External Quality Assessment Service (UK NEQAS) (http://www.ukneqasiccish.org/) is an internationally active independent charitable consortium of external quality assessment providers that handles the assessment of over 5000 slides every quarter and over 20,000 slides per annum from over 50 countries worldwide, with the aim of ensuring optimal quality in testing for the benefit of patients. The External Quality Assessment Scheme for Immunocytochemistry was founded in 1985, and in 1988 the service was recognized by the UK Department of Health [26]. From that time, it became known as the UK National External Quality Assessment Service for Immunocytochemistry (UKNEQAS ICC). UKNEQAS has more than 20 modules that cover different aspects of pathology, cytology, and cytogenetics.

The combined immunohistochemistry and in situ hybridization module offer runs for breast cancer markers, as well as HER2 in gastric cancer, PD-L1, ALK IHC, and MYC FISH. Slides sent to the module are evaluated by a designated set of regularly competence assessed international team of assessors.

Germany

The Qualitätssicherungs-Initiative Pathologie GmbH (QuIP) (https://quip.eu/en_GB/) is a German initiative which cooperates with the Referenzinstitut für Bioanalytik (RfB) and offers separate lab and evaluation tests [27]. This is especially relevant for laboratories where more than one person is evaluating patient samples. QuIP offers the possibility to request a separate evaluation test for each pathologist, who receives an individual certificate, which then can be used for competency assessment.

Similar to NordiQC, QuIP also requests the return of stained slides and provides feedback on the staining quality. With the certificate, they also provide a detailed statistical analysis (weighted kappa test). It is also worthwhile to mention that the runs usually contain test and training samples, which are statistically analyzed separately. The test part reflects the requirements profile of everyday practice in a routine histopathology laboratory. In the training part, specially selected equivocal cases are included which place particularly high demands on the sensitivity and discriminative capacities of the analyses carried out and thus are sensitive indicators of the quality of the detection methods. In the case of HER2, this would mean that the test part contains relatively few equivocal cases, in contrast to the training part. However, only the test part is used for the assessment of successful participation in the proficiency test.

Belgium

The European Society of Pathology established an EQA program for testing biomarker mutations in colorectal cancer and non-small cell lung carcinoma (NSCLC) [28] (https://www.esp-pathology.org/esp-foundation.html). This program aims to ensure optimal accuracy and proficiency in colorectal cancer (Colon Scheme) and lung cancer (Lung Scheme) biomarker testing across all countries. For most of the biomarkers, three runs per year are provided, except for ROS-1 (two runs). A pilot run for c-Met will be offered in 2018. The proficiency testing samples are either provided in the form of glass slides with TMAs or digital scans. The molecular EQA scheme includes a mandatory EGFR analysis, whereas BRAF and KRAS testing are optional. ALK-, ROS1-, and PD-L1 IHC-stained slides can also be sent to EQA for a technical assessment. A special feature of the ESP Lung EQA scheme is that for some cases, written reports are sent that describe the results and the methods used.

USA

In the USA, the College of American Pathologists (CAP) (http://www.cap.org/) serves to foster and advocate excellence in the practice of pathology and laboratory medicine worldwide. CAP is served by a number of board-certified pathologists. Laboratories accredited by the (CAP) are required to participate in all survey programs concerning the test menu which they provide [29]. The CAP proficiency testing schemes are also open to other laboratories, including those based internationally. Samples for predictive biomarkers are shipped twice a year, and results can be entered online after logging into the *e-LAB Solutions Suite*™. IHC slides, however, are not returned for review, and thus there is no feedback on the staining quality or the protocol used. The CAP evaluation only provides the intended result and the participants' grade. The overall concordance rate (%) achieved by the participants in any given run [%] can be retrieved via the Analyte Scorecard in the *e-LAB Solutions Suite*™.

CAP provides a detailed result analysis within peer groups. In contrast to other EQA schemes (NordiQC and UKNEQAS), no assessor team reviews the results. A decision on whether a lab has passed or failed is based on the consensus of the participants. The consensus score needs to reach at least 80%; otherwise, the significance of an individual result is considered not to be evaluable in terms of participant's performance.

Canada

In Canada, proficiency testing schemes are offered by the Canadian Immunohistochemistry Quality Control (cIQc) (http://cpqa.ca/main/) in collaboration with the Canadian Partnership Against Cancer. cIQc is committed to systematically monitoring and improving the proficiency of IHC testing nationwide, and the organization distributes IHC challenges to pathology laboratories, assesses participant staining, and provides an anonymized summary of laboratory performance [30]. The program is also open to international stakeholders.

For the Breast and Mismatch Repair (MMR) IHC Module, as well as that for ALK IHC, two runs per year are organized. For all remaining biomarkers, only one run is organized. Participating laboratories enter their results via the online system TMA Scorer, enabling immediate comparison of submitted results against a reference laboratory and other pathology laboratories.

For each program, the assessment team publishes a detailed report on the website, which deals with the internal evaluation of cases by the assessment team as well as a detailed statistical analysis of the proficiency testing results. In the event that participant-specific feedback has been provided, the individual results of these laboratories are listed, along with all the relevant details of the staining protocols used. Laboratories with sub-optimal staining results are given the opportunity to request additional sections for a test repeat, optionally after protocol optimization.

Besides the proficiency testing schemes, cIQc offers an annual symposium or workshop.

Asia Pacific

The Royal College of Pathologists of Australasia Quality Assurance Programs (RCPAQAP) (https://rcpaqap.com.au/) was founded in 1988 and has evolved to become one of the world's leading external quality assurance (EQA) providers to pathology laboratories in Australia and many parts of the world [31]. RCPA Quality Assurance Programs are provided for all disciplines of pathology. The RCPAQAP has agents in over 60 countries supporting the enrollment in their region. The international enrollments account for about 40%.

Besides the typical breast markers ER, PgR, and HER2, *the Immunohistochemistry Breast Markers Module* includes two additional antibodies, D240 (alternate CD34) and CK5/CK6 (alternate CK14). The module is provided once per year with six exercises.

The HER2 BRISH (Brightfield ISH) Breast Diagnostic Module is based on an evaluation of digital slides, whereas the *HER2 BRISH Gastric Module* is a combined technical and diagnostic module.

RCPAQAP facilitates the enrollment of individual pathologists.

In addition to the PT schemes, the RCPAQAP would also offer an IHC breast marker audit and various diagnostic modules, e.g., for the breast, dermatopathology, gastrointestinal tract, hematopathology, gynecology, neuropathology, and electron microscopy.

Brazil

The External Quality Assessment (EQA) in Brazil is performed by the National Health Ministry for diseases that are under supervision of the Public Health Department. In addition to the government program, the Brazilian Society of Clinical Analysis and the Brazilian Society of Medical Pathology are allowed to provide their programs under the Supervision of National Agency for Sanitary Surveillance (ANVISA) that regulates laboratories to perform EQA programs [32]. The programs offered so far are aimed only at clinical and cytological laboratories. At the moment, there are no interlaboratory tests available in the field of routine pathology methods. For that reason, and since South American countries usually follow the EMEA guidelines, pathology labs in Brazil usually participate in the proficiency testing schemes of UKNEQAS or NordiQC.

China and India

In China, the "first-class hospital laboratories" are required to participate in a proficiency testing program by the Ministry of Health of the People's Republic of China.

For India (and China) in August 2013, the College of American Pathologists (CAP) and BD Diagnostics [33] formed a strategic alliance through which both organizations agreed to provide education and integrated quality improvement solutions to laboratories in mainland China and India [34]. The aim of the alliance is to improve access in both countries to external quality assurance and proficiency testing. BD Diagnostics was assigned to distribute to these nations CAP surveys, proficiency testing and interlaboratory comparison programs, and Q-Probes, Q-Tracks, and Quality Monitors.

The College of American Pathologists (CAP) proficiency testing programs have indeed become the preferred choice of laboratories due to the well-prepared specimens, comprehensive programs, scientific evaluation, and useful educational opportunities involved [35].

Summary and Conclusion

The field of developing predictive biomarkers has become increasingly complex. Not only are new predictive biomarkers being developed, but the manner in which existing biomarkers are interpreted continues to evolve with new and different cutoffs and diagnostic algorithms being continually introduced. For example, several different drugs have been developed to target the PD-L1 pathway, each having been developed independently in conjunction with its own independently developed biomarker test. This has resulted in the generation of a very complex and complicated testing arena in which multiple antibodies are used for PD-L1 testing, each utilizing a different interpretation scheme and diagnostic guideline to achieve clinical relevance. Furthermore, the interpretation guidelines vary between antibodies, generating a high risk for confusion and misapplication. Thus, extraordinary care must be exercised to ensure that appropriate methods and interpretation guidelines are employed for each case. Harmonization efforts are ongoing, but until a cohesive paradigm is achieved, additional caution must be exercised. The multiplicity of reagents and methods studies by external QC organizations, such as NordiQC and UKNEQAS [20], has shown an astonishing range of variation among different laboratories with respect to choice of reagents, retrieval methods, and staining protocols (automated or manual). For example, in one UKNEQAS survey encompassing 365 laboratories performing an "IHC stain" for keratin, 26 different primary antibodies were employed, with more than 20 different detection systems from 13 vendors, using 17 different auto-stainers or manual methods. This enormous diversity represents a great problem for the reproducibility of IHC in general and becomes a critical issue for "companion diagnostics," where the question asked of the test is much more rigorous, looking beyond the question of positivity but also evaluating the presence of protein on a comparative basis from case to case. Just as the question is more rigorous, so are the demands of test performance. Some of these problems which stem from diversity of reagents and protocols are addressed using "kits" (such as IVDs) approved by the FDA or by comparable agencies in other countries, and by automation, which has the side benefit of imposing some standardization by restriction in practical choice of reagents and protocol. The use of an FDA-approved test (Class III, IVD) restricts the performing laboratory to a specified reagent set and a closed protocol and to an internal validation process that provides some assurance of run-to-run stability. In such cases, any deviation from reagents or protocol negates validation. Such deviations include differences in antigen (or epitope) retrieval protocols (or solutions and heating methods), which have been shown to have profound effects upon the ability to demonstrate proteins by otherwise standard protocols. In the UKNEQAS survey cited above, 17 different retrieval solutions were employed.

With the pending flood of new companion diagnostics, some of which are IHC based, the problems encountered with the HER2 test will be continuously revisited. The outcome is not difficult to predict by reference to the history of the introduction of ELISA into the clinical laboratory. Much greater attention is already being given to all aspects of tissue sample preparation for IHC, with attempts to document and standardize across institutions. However, the logistical and cost issues are formidable, and it is not likely that the level of standardization achieved with blood or fluid samples for ELISA can ever be achieved for tissues and IHC. It appears probable, therefore, that some method of "qualifying" a tissue sample as suitable for performance of a particular IHC companion assay by use of an internal control will become necessary. Furthermore, by analogy with ELISA in the clinical laboratory, automation of IHC for companion diagnostics is inevitable, with reagents and protocols subject to increasingly rigorous approval processes, using closed assay systems, that cannot be "tweaked" to get more intense staining in order to compensate for a deficient FFPE process. In the face of poorly controlled sample preparation, the mantra – "don't tweak the protocol; fix the fixation"– comes to mind.

References

1. Duffy MJ, Harbeck N, Nap M, Molina R, Nicolini A, Senkus E, et al. Clinical use of biomarkers in breast cancer: updated guidelines from the European Group on Tumor Markers (EGTM). Eur J Cancer. 2017;75:284–98.
2. Taylor CR. Predictive biomarkers and companion diagnostics. The future of immunohistochemistry – 'in situ proteomics', or just a 'stain'? Appl Immunohistochem Mol Morphol. 2014;22(8):555–61.
3. Jørgensen JT. Companion and complementary diagnostics: clinical and regulatory perspectives. Trends Cancer. 2016;2:706–12.
4. Scheerens H, Malong A, Bassett K, Boyd Z, GuptaV, Harris J, et al. Current status of companion and complementary diagnostics: strategic considerations for development and launch. Clin Transl Sci. 2017;10:84–92.
5. Siegel J. Department of health & human services. 1998. https://www.accessdata.fda.gov/drugsatfda_docs/appletter/1998/trasgen092598L.pdf.
6. Nordi QC. Recommended protocols. 2018. http://www.nordiqc.org/recommended.php
7. Fitzgibbons PL, Murphy DA, Hammond ME, et al. Recommendations for validating estrogen and progesterone receptor immunohistochemistry assays. Arch Pathol Lab Med. 2010;134:930–5.
8. ICH Harmonised Tripartite Guideline – validation of analytical procedures: text and methodology Q2(R1). 2005.
9. Fitzgibbons PL, Bradley LA, Fatheree LA, Alsabeh R, Fulton RS, Goldsmith JD, et al. Principles of analytic validation of immunohistochemical assays – guideline from the College of American Pathologists Pathology and Laboratory Quality Center. Arch Pathol Lab Med. 2014;138:1432–43.
10. Grzybicki DM, et al. The usefulness of pathologists' assistants. Am J Clin Pathol. 1999;112:619–26.

11. Galvis CO, et al. Pathologists' assistants practice. A measurement of performance. Am J Clin Pathol. 2001;116:816–22.
12. Wolff AC, Hammond ME, Hicks DG, Dowsett M, McShane LM, Allison KH, et al. Recommendations for human epidermal growth factor receptor 2 testing in breast cancer: American Society of Clinical Oncology/College of American Pathologists clinical practice guideline update. Arch Pathol Lab Med. 2014;138(2):241–56.
13. Bartley AN, Washington MK, Ventura CB, Ismaila N, Colasacco C, Benson IIIAB, et al. HER2 testing and clinical decision making in gastroesophageal adenocarcinoma guideline from the College of American Pathologists, American Society for Clinical Pathology, and American Society of Clinical Oncology. Arch Pathol Lab Med. 2016;140:1345–63.
14. Hammond ME, Hayes DF, Wolff AC, Mangu PB, Temin S. American Society of Clinical Oncology/College of American Pathologists guideline recommendations for immunohistochemical testing of estrogen and progesterone receptors in breast cancer. J Oncol Pract. 2010;6(4):195–19.
15. Cree IA, Booton R, Cane P, Gosney J, Ibrahim M, Kerr K, et al. PD-L1 testing for lung cancer in the UK: recognizing the challenges for implementation, Cree et al. Histopathology. 2016;69:177–86.
16. Krusche CA, von Wasielewski R, Rüschoff J, Fisseler-Eckhoff R, Kreipe HH. Ringversuche zum Nachweis von therapeutischen Zielmolekülen beim Mammakarzinom in Deutschland. Pathologe. 2008;29:315–20.
17. Garon EB, Rizvi NA, Hui R, Leighl N, Balmanoukian AS, Eder JP, et al. Pembrolizumab for the treatment of non-small-cell lung cancer. N Engl J Med. 2015;372(21):2018–28.
18. Zhu J, Armstrong AJ, Friedlander TW, Kim W, Pal SK, George DJ, et al. Biomarkers of immunotherapy in urothelial and renal cell carcinoma: PD-L1, tumor mutational burden, and beyond. J Immunother Cancer. 2018;6:4.
19. Rüschoff J, Lebeau A, Kreipe H, et al. Assessing HER2 testing quality in breast cancer: variables that influence HER2 positivity rate from a large, multicenter, observational study in Germany. Mod Pathol. 2017;30:217–26.
20. Ibrahim M, Parry S, Wilkinson D, et al. ALK immunohistochemistry in NSCLC: discordant staining can impact patient treatment regimen. J Thorac Oncol. 2016;11(12):2241–7.
21. Scheel AH, Baenfer G, Baretton G, Dietel M, Diezko R, Henkel T, et al. Interlaboratory concordance of PD-L1 immunohistochemistry for non-small-cell lung cancer. Histopathology. 2018;72(3):449–59.
22. Cheung CC, Garratt J, Won J, et al. Developing ALK immunohistochemistry and in situ hybridization proficiency testing for non-small cell lung cancer in Canada: Canadian immunohistochemistry quality control challenges and successes. Appl Immunohistochem Mol Morphol. 2015;23(10):677–81.
23. International Quality Network for Pathology. Home page: http://www.iqnpath.org/.
24. Vyberg M, Nielsen S. Proficiency testing in immunohistochemistry – experiences from Nordic Immunohistochemical Quality Control (NordiQC). Virchows Arch. 2016;468:19–29.
25. NordiQC. NordiQC assessment scheme 2018. http://www.nordiqc.org/modules.php.
26. UK NEQAS. International quality expertise. UK NEQAS ICC & ISH modules. http://www.ukneqasiccish.org/modules/.
27. Quip. Qualitätssicherungs-initiative pathologie QuIP GmbH. Home page: https://quip.eu/en_GB/zerpa/trials/.
28. European Society of Pathology. EQA schemes; 2018: https://www.esp-pathology.org/esp-foundation/eqa-schemes.html.
29. College of American Pathologists. Home page: www.cap.org.
30. Canadian Partnership Against Cancer. Home page: http://cpqa.ca/.
31. The Royal College of Pathologists of Australasia: Quality Assurance Programs. Home page: https://www.rcpaqap.com.au/.
32. Kneip Fleury M, Menezes ME, Abol Correa J. Implementation of the external quality assessment program in Brazil. Biochem Med (Zagreb). 2017;27(1):93–6.
33. Becton, Dickinson and Company. Home page: http://www.bd.com/.
34. CAP Today. CAP partners with BD in China and India, August 1st, 2013. http://www.captodayonline.com/cap-partners-with-bd-in-china-and-india/.
35. Xiaojuan L, Qingkai D, Yongmei J. Proficiency testing experience with College of American Pathologists' Programs at a University Hospital in China from 2007 to 2011. Arch Pathol Lab Med. 2014;138(1):114–20.

49

Use of Companion Diagnostics (CDx) and Predictive Biomarkers for Cancer Targeted Therapy: Clinical Applications in Precision Medicine

Rosanne Welcher

Introduction

The idea of using a diagnostic assay to determine treatment course and patient management is not a new concept in oncological clinical practice. In the 1960s, measurement of estrogen levels in a tumor was frequently used to determine if a breast cancer patient had a likelihood to respond to anti-hormonal therapy. Recognizing the importance of these specific tests that personalize the medical treatment to a specific patient, the FDA and others formalized the category of these tests as "companion diagnostics" (CDx). Although pharmaceutical companies have used biomarkers to research and study the clinical effectiveness of oncology drugs, the use of a CDx model in which the drug and biomarker assay are co-developed notably started with the introduction of HercepTest in 1998. HercepTest is an immunohistochemistry (IHC) assay that detects HER2 overexpression and aids in the selection of breast cancer patients eligible for trastuzumab (Herceptin®) treatment. In the original clinical trials, patients with HER2 protein overexpression responded better to Herceptin than patients with no or low levels of HER2 expression on their tumor cells [1]. It also became known that HER2 overexpression was associated with poorer outcomes in patients with breast cancer [2] presenting a good rationale for testing a targeted anti-HER2 therapeutic such as trastuzumab.

In the late 1990s, two phase III trials were submitted in support of trastuzumab (Herceptin®) approval: the H0648g pivotal study (first-line treatment in 469 women) and the H0649g study (assessing second-line treatment in 222 women). Study H0648g was a randomized, controlled, open-label trial to evaluate the efficacy and safety of adding trastuzumab to chemotherapy in patients with metastatic breast cancer and HER2-overexpressing tumors (judged as 2+ and 3+ by IHC testing using a CTA). Addition of trastuzumab to chemotherapy increased TTP (time to tumor progression, the primary trial endpoint) from 4.6 to 7.4 months (61% increase) compared to chemotherapy alone [3]. Study H0649g evaluated the response to trastuzumab as a single agent in patients with metastatic breast cancer overexpressing HER2 (2+ or 3+ membrane staining in >10% of tumor cells, by IHC), who had relapsed after two cytotoxic chemotherapy regimens. The primary endpoint, overall response rate (ORR), was reached at 15% (95% confidence interval, 11–21%), with one of the secondary objectives, duration of response, indicating a durable median response of 9.1 months [4]. Because of the correlation between HER2 overexpression as determined by a clinical trial assay for HER2 and patient response, the FDA required that a fully analytically validated "companion diagnostics"

R. Welcher
Companion Diagnostics, Agilent Technologies, Carpinteria, CA, USA
e-mail: rosanne.welcher@agilent.com

S. Badve, G. L. Kumar (eds.), *Predictive Biomarkers in Oncology*,
https://doi.org/10.1007/978-3-319-95228-4_49

be made available to test patients to assess their eligibility for Herceptin.

Since the introduction of HercepTest in 1998, several other HER2 companion diagnostic assays have been approved, including additional IHC tests as well as fluorescent in situ hybridization assays (FISH) and colorimetric in situ hybridization tests (CISH) that determine gene amplification. In addition, HercepTest has received FDA approval for expanded claims to include gastric cancer patients, as well as the addition of other treatment options, such as trastuzumab emtansine (Kadcyla®) and pertuzumab (Perjeta®).

Companion Diagnostics (CDx)

Companion diagnostics are specialized in vitro diagnostic (IVD) assays that are critical adjuncts in the field of personalized medicine. Simply put, CDx help identify subsets of patients who may be more likely to respond to a targeted therapy than otherwise. According to the US FDA definition[1], a companion diagnostics is an:

- Assay or imaging tool that provides information essential for the safe and effective use of the drug.
- Drug label requires patient testing.
- Selective, predictive, exclusion, or monitoring indication for use.

Most CDx assays are considered in the USA as high risk, class III, requiring a pre-market approval (PMA) application that contains sufficient scientific evidence to assure that the laboratory test is safe and effective for its intended use. FDA regulations provide 180 days to review the submission and make a determination. However, close collaboration with FDA is required due to the various clinical trial designs, challenges, accelerated approvals, and circumstances to consider in a co-development program.

How Is a Companion Diagnostic (CDx) Assay Developed?

The first step in any CDx development is to identify the appropriate biomarker(s) and technology that produce the best “fit-for-purpose” assay to be used in pharma clinical trials [5]. Often, a biomarker assay may be first utilized in exploratory studies to test the hypothesis. Once established as a potential CDx, the assay is then used in prospective or retrospective clinical trials to establish clinical utility. In a traditional co-development paradigm, as illustrated in Fig. 49.1, CDx development aligns sequentially with the traditional pharmaceutical clinical trial phases. In this paradigm, drug-diagnostic co-development is fully aligned to enable timely and cost-effective commercialization of emerging cancer therapies.

Early engagement with pharma is required to enable a stepwise development approach with the potential CDx assay. A prototype assay may be used in early drug development clinical trials to

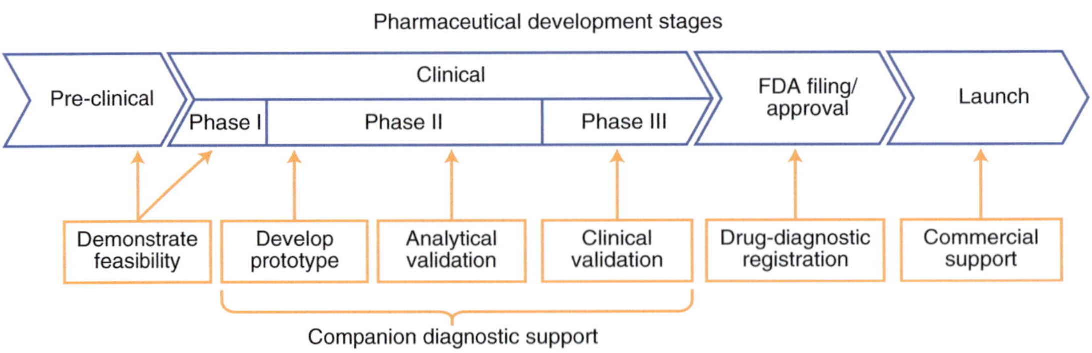

Fig. 49.1 Traditional paradigm for Rx/Dx co-development and co-approval

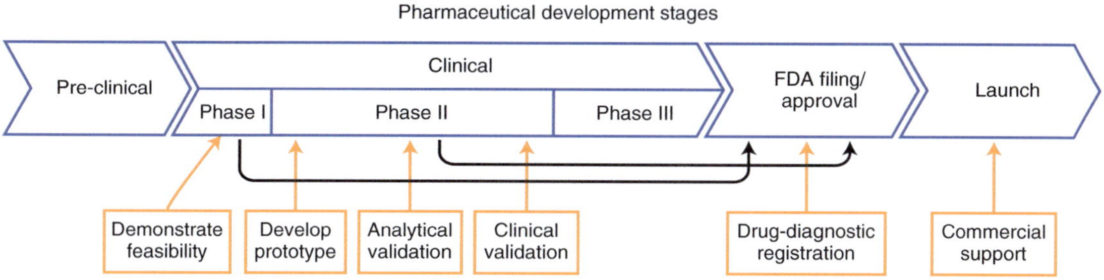

Fig. 49.2 Accelerated paradigm for Rx/Dx co-development and co-approval

test a biomarker hypothesis or for exploratory studies. As the drug moves through a typical phase I/II/III study design, ideally, an analytically validated assay is available to deploy into pivotal phase II or phase III clinical trials, of which test results are then used to support clinical utility and CDx regulatory submissions [6].

However, the reality is that CDx developers must be prepared to provide assays to accelerated drug programs, leading to insufficient time to validate an assay for use in registrational clinical trials (Fig. 49.2). In these cases, both pharma and CDx provider must seek guidance from the FDA and be prepared to retest clinical samples when the analytically validated assay is ready and bridge the clinical outcome data back to the retest results. Such a design is challenging due to many factors, including lack of clinical specimens to retest and complicated statistical analysis plans necessary for the bridging studies. To avoid these pitfalls, it is important that pharma and CDx providers engage early in the process, allowing adequate time to develop an analytically validated assay in time for pivotal trial.

FDA Concept and Guidance Document on CDx: The FDA Model

The FDA first published a concept paper in 2005 for discussion purposes only. The concept paper introduced many topics for consideration when embarking on the co-development of Rx and CDx that still hold true today. The FDA followed this concept paper with a draft guidance on In Vitro Companion Diagnostic (IVD CDx) Devices in 2011, with a final guidance issued on August 6, 2014.[1,2]

The 2014 guidance on IVD CDx described "what" but not "how" to plan and execute Rx/CDx co-development program. Although very useful, the guidance still lacked specific details about how IVD and pharma should work in concert in the co-development and co-approval process. Recognizing this, the FDA issued a draft guidance entitled "Principles for Co-development of an In Vitro Companion Diagnostic Device with a Therapeutic Product," issued on July 15, 2016.[3] The publication of this draft guidance is notably an important *practical* guide for both CDx manufacturers and pharma. Public comments have been positive, with the draft guidance providing more comprehensive descriptions of the real-life scenarios and challenges which can be encountered in co-development programs and how to address these. This guidance represents a culmination of experiences and enables partners and the FDA to have a same reference point.

[1] Guidance for Industry and Food and Drug Administration Staff - In Vitro Companion Diagnostic Devices. July 14, 2011. https://www.document-center.com/standards/show/FDA-IN-VITRO-COMPAN/history/2011%20DRAFT.

[2] In Vitro Companion Diagnostic Devices. Guidance for Industry and Food and Drug Administration Staff. August 6, 2014. https://www.fda.gov/downloads/MedicalDevices/DeviceRegulationandGuidance/GuidanceDocuments/UCM262327.pdf.

[3] Principles for Co-development of an In Vitro Companion Diagnostic Device with a Therapeutic Product: Draft Guidance for Industry and Food and Drug Administration Staff Document issued on July 15, 2016: https://www.fda.gov/downloads/MedicalDevices/DeviceRegulationandGuidance/GuidanceDocuments/UCM510824.pdf.

In reality, drug development and CDx development may not be on the same timeline. Some challenges faced in co-development programs include:

1. Expedited and/or accelerated oncology drug pathways, which are shorter than CDx development time
2. Late introduction of the CDx to the Rx clinical program
3. Uncertainty in the biomarker cutoff and technology
4. Sample availability or stability

The best path to success to ensure a smooth co-development and co-approval pathway for Rx and Dx is to have a dialogue with the FDA early and often. Through communication and strong collaboration, many scenarios in which co-approval seems difficult can be made less complicated by having a strategy, discussing with the FDA, and executing well on the plan.

Complementary Diagnostic Assays

Missing from the 2016 draft co-development guidance document from the FDA was a detailed description of "complementary diagnostics," a category of diagnostic assays that was an immediate consequence of recent "all-comers" clinical trial designs. Biomarker negative or diagnostic negative patients may respond to the drug, although to a lesser extent than high expressers. The term complementary diagnostics was used to describe these assays which are not required for treatment but provide valuable information to oncologists in determining treatment options [7]. Most notably, several programmed death/programmed death ligand (PD-1 and PD-L1) assays were approved in 2016 with complementary diagnostic claims [8]. The introduction of these types of assays, at times, has caused some confusion for clinicians and reimbursement agencies. Also, industry or laboratories developing assays with complementary claims must fulfill all requirements for high-risk assays, i.e., Class III, PMAs, in the USA, but the return on investment may not be worthwhile since the assay is not required for the prescription. Future development of complementary assays will require additional interaction with pharma and FDA centers to understand the risk associated with these assays and to determine if there are alternative regulatory pathways.

Companion Diagnostic Laboratory-Developed Tests (LDT)

In addition to the manufacturer-developed CDx IVD kits which are subject to FDA pre-market review and approval, there are companion diagnostic laboratory-developed tests available in the market.

The FDA defines laboratory-developed test (LDT) as an in vitro diagnostic test that is manufactured by and used within a single laboratory. While the LDT is regulated under the Clinical Laboratory Improvement Amendment (CLIA) program which regulates laboratories to ensure accurate and reliable test results when performing testing on patient specimens, the FDA regulates manufacturers and devices under the Federal Food, Drug, and Cosmetic Act (FDCA) to ensure that CDx IVD devices are safe and effective by meeting regulatory requirements including premarket review, quality system, and establishment registration and listing. Establishing performance characteristics relating to analytical validity for the intended use of the test is required under both CLIA program and FDCA; however, the FDA review under FDCA includes clinical validation, whereas CLIA does not.

LDTs are similar to other in vitro diagnostic tests and are considered as medical devices under FDCA. However, the FDA has exercised enforcement discretion, not to enforce the requirements under the FDCA. Consequently, LDTs are subject to regulatory oversight by the FDA. In 2014, the FDA issued the LDT draft guidance and proposed a comprehensive policy. However, after 2 years of collecting feedback from a broad range of stakeholders, the FDA announced in December 2016 that a final guidance on the oversight of LDT concerning the balance of patient protection with continued access and innovation would not be issued. Instead, the FDA released the discussion paper on January 13, 2017, to provide a possible approach for further public discussion and collaboration with other government agencies to assess existing quality controls are adequate as well as whether additional oversight is required to address LDT development activities.

It is important to keep in mind that an FDA-approved assay does not necessarily mean that a particular brand of test is superior to similar tests, including LDTs. However, an LDT must be analytically and clinically validated to ensure the performance characteristics are met and the scoring and interpretation are set appropriately to provide patient safety and treatment guidance. Many labs choose to use an FDA-approved assay if available, for routine use or to use as a comparator to their own LDT.

When choosing an approved assay or LDT, it is also wise to engage in the proficiency programs established by CAP, NordiQC, or UKNEQAS, to name a few. LDTs typically perform less optimally than FDA-approved assays, for several reasons, including the lack of proper analytical and clinical validation (scoring and interpretation) which can lead to greater variability and uncertainty of the test results. Each lab must choose the most appropriate assay for their infrastructure, capabilities, case volume, and turnaround time.

Current CDx Portfolio in Oncology

As of this publication date, 35 CDx IVDs tied to cancer therapeutics have been approved or cleared by the FDA including two NGS assays using solid tumors. The number of CDx utilizing NGS assays is expected to increase as the technology gains broader use in routine diagnostic laboratories, including in liquid biopsies. Regarding CDX approvals and its relationship with tumor types, lung cancer is one of the leading tumor types, addressing unmet clinical needs and a high prevalence rate of tumors, where the level of expression may provide information to physicians regarding treatment decisions.

For a list of current FDA-approved companion diagnostics, refer to the FDA link below: https://www.fda.gov/MedicalDevices/ProductsandMedicalProcedures/InVitroDiagnostics/ucm301431.htm (Tables 49.1 and 49.2).

CDx: Frequently Asked Questions

1. *Why are there multiple tests for the same drug?*

   When partnering with a diagnostic test provider, many pharmaceutical companies are concerned about access to the test in a worldwide market. By partnering with more than one diagnostic provider to offer the companion diagnostic test, pharmaceutical companies facilitate broader access to the test by enabling various diagnostic platforms. In addition, there may be more than one technology with proven clinical utility. In cases where either protein overexpression or gene amplification may be relevant, it is possible to see both a protein-based test (e.g., IHC) and gene-based test (e.g., FISH) as options (HER2, ALK).
2. *Are FDA-approved tests better than laboratory-developed tests (LDTs) that are not FDA-approved?*

   Not necessarily, however, LDTs may not be analytically or clinically validated to the same rigor as FDA-approved tests. To provide some distinction, LDTs or "home brew" tests are IVDs that are manufactured by and used within a single laboratory. The Clinical Laboratory Improvement Amendment (CLIA) program regulates laboratories performing testing on patient specimens, however, as LDTs are "devices," and they are subject to regulatory oversight by the FDA. Although all IVDs, including LDTs, must comply with regulatory requirements governing safety and efficacy, the FDA has not enforced these requirements for LDTs, including a pre-market review of analytical and clinical validation.

   Under the CLIA regulations, laboratories must establish certain performance specifications relating to analytical validation (42 CRR 493.1253(b)(2), but the CLIA program does not address the clinical validity of the test. Clinical utility of an FDA-approved test is proven in prospective or retrospective clinical trials, with results correlated to patient outcome. There are exceptions in which clinical specimens are not available in which case FDA has provided guidance in clinical study design [9]. In addition to development, diagnostic manufacturers must follow 21 CFR 820 (Quality System Regulations and Good Manufacturing Practices) and are subject to FDA inspections prior to approval and *post-approval.*

Not all laboratories developing their own tests have robust validation and QC methods, which can lead to variability within and between labs offering a similar LDT. There is increasing interest from pharmaceutical companies to partner with IVD providers for even commonly used biomarker tests, *where* unacceptably high variability in biomarker positive prevalence is observed with LDTs. Also, when reviewing quality assurance schemes, such as CAP, *UKNEQAS*, and NordiQC, FDA-approved tests typically outperform LDTs.

3. *What are the liabilities associated with the use of "home brew" tests?*
Many LDTs are not robustly and analytically validated. Those that claim "full validation" suffer from statistically insignificant sample sizes and may not have been clinically validated by association with patient outcome. Because of these issues, it is critical that the interpretation of a positive test provides the same result as a test that was used in the clinical trials to establish clinical utility. Some labs will optimize their LDT using an FDA test as the "reference," which may lead to the higher likelihood the LDT sensitivity and interpretation is appropriately set. Tracking prevalence data for the LDT is critical to ensure drift is not occurring. When LDTs are used, strict adherence to CLIA regulations and guidelines on assay validation and corrective action is critical to reducing liabilities associated with "home brew" tests.

Future Directions

It is clear that the paradigm of one assay/one drug is no longer representative of the future direction of companion and complementary diagnostics. With the emergence of *multiple*-analyte gene-based tests, validation of various analytes correlated to patient outcome has increased the complexity of determining clinical utility as well as the regulatory pathway for approval of such tests. In addition, many assays, such as those used to select patients for immunotherapies (e.g., PD1/PD-L1), will emerge with multiple drugs on one assay label, as well as multiple assays detecting the same analyte. As we learn more about the clinical relevance of biomarkers, it is important to keep the laboratory in mind in reducing complexity and confusion about which test to use. Patient safety and drug efficacy should always be in mind when determining the right test for the right drug.

FDA and industry experts are discussing how to harmonize scoring and interpretation of multiple assays for the same analyte, which will de-risk improper use and interpretation of tests that have not been cross-validated. Studies, such as the Blueprint for PD-L1 IHC testing [10], are a culmination of pharma, IVD providers, and Key Opinion Leaders working together to understand the interchangeability of multiple assays for the same biomarker. These important studies will help pave the way to inform clinicians and regulators on the validity of assay interchangeability in immunotherapy treatment options for patients.

Other important areas of consideration are adding multiple tumor indications onto FDA-approved assays. Presently, the FDA is working with IVD manufacturers to understand how studies can be leveraged across tumor types while still meeting standards for safety and effectiveness. The FDA is also considering ways to further improve the co-approval process, such as sharing data across the Rx and Dx centers and providing additional guidance documents for new technologies and the co-approval process [11, 12].

Table 49.1 List of FDA-approved companion diagnostic assays in oncology

Cancer	Drug	CDx assay	Supplier	Assay technology	Application
Melanoma	Dabrafenib (Tafinlar®) Trametinib (Mekinist®)	THXID® BRAF test	bioMérieux, Inc.	Real-time PCR	Detection of the BRAF V600E and V600K mutations in DNA samples extracted from formalin-fixed paraffin-embedded (FFPE) human melanoma tissue
	Vemurafenib (Zelboraf®)	cobas® 4800 BRAF V600 Mutation Test	Roche Molecular Systems, Inc.	Real-time PCR	Detection of the BRAF V600E mutation in DNA extracted from FFPE human melanoma tissue
Colorectal cancer	Olaparib (Lynparza™)	BRACAnalysis CDx™	Myriad Genetic Laboratories, Inc.	PCR and Sanger sequencing	FFPE human colorectal cancer (CRC) tumor tissue
	Panitumumab (Vectibix®)	Praxis® Extended RAS Panel	Illumina, Inc.	NGS	Detection of 56 specific mutations in RAS genes in DNA extracted from (FFPE) colorectal cancer (CRC) tissue samples. Indicated to aid in the identification of patients with colorectal cancer for treatment with Vectibix® (panitumumab) based on a no mutation detected test result
		The cobas® KRAS Mutation Test	Roche Molecular Systems, Inc.	Real-time PCR	Detection of seven somatic mutations in codons 12 and 13 of the KRAS gene in DNA derived from FFPE human colorectal cancer (CRC) tumor tissue
		therascreen KRAS RGQ PCR Kit	Qiagen Manchester, Ltd.	Real-time PCR	Detection of seven somatic mutations in the human KRAS oncogene, using DNA extracted from (FFPE), colorectal cancer (CRC) tissue
		DAKO EGFR pharmDx™ Kit	Dako North America, Inc.	IHC	Identifies epidermal growth factor receptor (EGFR) expression in normal and neoplastic tissues routinely fixed for histological evaluation. EGFR pharmDx specifically detects the EGFR (HER1) protein in EGFR-expressing cells

(continued)

Table 49.1 (continued)

Cancer	Drug	CDx assay	Supplier	Assay technology	Application
Breast	Trastuzumab (Herceptin®)	HercepTest™	Dako Denmark A/S	IHC	Determine HER2 protein overexpression in breast cancer tissues routinely processed for histological evaluation
		PATHWAY anti-HER2/neu (4B5) Rabbit Monoclonal Primary Antibody	Ventana Medical Systems, Inc.		Detection of c-erbB-2 antigen in sections of FFPE normal and neoplastic tissue on a Ventana automated immunohistochemistry slide staining device. It is indicated as an aid in the assessment of breast cancer patients for whom Herceptin treatment is being considered
		InSite HER2/neu Kit	Biogenex Laboratories, Inc.		For localizing the overexpression of HER2/neu in FFPE normal and neoplastic tissue sections. InSite HER2/neu is indicated as an aid in the assessment of breast cancer patients for whom Herceptin (trastuzumab) therapy is being considered
		Bond Oracle HER2 IHC System	Leica Biosystems, Inc.		Determine HER2 oncoprotein status in FFPE breast cancer tissue
		INFORM HER2	Ventana Medical Systems, Inc.	Fluorescence in situ hybridization (FISH)	Determines the qualitative presence of HER2/Neu gene amplification on FFPE human breast tissue as an aid to stratify breast cancer patients according to risk for recurrence or disease-related death
		INFORM HER2 Dual ISH DNA Probe Cocktail			Determines HER2 gene status by enumeration of the ratio of the HER2 gene to chromosome 17. The HER2 and chromosome 17 probes are detected using two-color chromogenic in situ hybridization (ISH) in FFPE human breast cancer tissue specimens
		PathVysion® HER2-DNA Probe Kit	Abbott Molecular, Inc.		Detects amplification of the HER2/neu gene via fluorescence in situ hybridization (FISH) in formalin-fixed, paraffin-embedded human breast cancer tissue specimens
		HER2 FISH pharmDx™ Kit	Dako Denmark A/S		Determine HER2 gene amplification in FFPE breast cancer tissue specimens
		SPOT-Light® HER2 CISH Kit	Thermo Fisher Scientific	Colorimetric in situ hybridization (CISH)	Determine HER2 gene amplification in FFPE breast carcinoma tissue sections using chromogenic in situ hybridization (CISH) and brightfield microscopy
		HER2 CISH pharmDx™ Kit	Dako Denmark A/S		Determine HER2 gene status in FFPE breast cancer tissue specimens
	Pertuzumab (Perjeta®)	HercepTest™	Dako Denmark A/S	IHC	Determine HER2 protein overexpression in breast cancer tissues routinely processed for histological evaluation
	Ado-trastuzumab emtansine (Kadcyla®)	HER2 FISH pharmDx™		FISH	Determine HER2 gene amplification in FFPE breast cancer tissue specimens

Gastric cancer	Trastuzumab (Herceptin®)	HercepTest™	Dako Denmark A/S	IHC	Determine HER2 protein overexpression in FFPE cancer tissue from patients with metastatic gastric or gastroesophageal junction adenocarcinoma
	Pertuzumab (Perjeta®)				
	Ado-trastuzumab emtansine (Kadcyla®)	HER2 FISH pharmDx™ Kit	Dako Denmark A/S	FISH	Determine HER2 gene amplification in FFPE breast cancer tissue specimens and FFPE specimens from patients with metastatic gastric or gastroesophageal junction adenocarcinoma
	Pembrolizumab (Keytruda®)	PD-L1 IHC 22C3 pharmDx™	Dako North America, Inc.	IHC	Detection of PD-L1 protein in FFPE in gastric or GEJ adenocarcinoma tissue. Indicated as an aid in identifying gastric or GEJ adenocarcinoma patients for treatment with Keytruda® (pembrolizumab)
Non-small cell lung cancer	Pembrolizumab (Keytruda®)	PD-L1 IHC 22C3 pharmDx™	Dako North America, Inc.	IHC	Detection of PD-L1 protein in FFPE non-small cell lung cancer (NSCLC) tissue. Indicated as an aid in identifying NSCLC patients for treatment with Keytruda® (pembrolizumab)
	Osimertinib (Tagrisso™)	cobas® EGFR Mutation Test v2	Roche Molecular Systems, Inc.	Real-time PCR	Detection of defined EGFR mutations non-small cell lung cancer (NSCLC) patients
	Erlotinib (Tarceva®)				
	Gefitinib (Iressa®)	therascreen® EGFR RGQ PCR Kit	Qiagen Manchester, Ltd.	Real-time PCR	Detection of exon 19 deletions and exon 21 (L858R) substitution mutations in EGFR gene derived from FFPE NSCLC tumor tissue
	Crizotinib (Xalkori®)	Vysis ALK Break Apart FISH Probe Kit	Abbott Molecular Inc.	FISH	Detection of rearrangements involving the ALK gene via FISH in FFPE NSCLC tissue specimens
		Ventana ALK (D5F3) CDx Assay	Ventana Medical Systems	IHC	Detection of the anaplastic lymphoma kinase (ALK) protein in FFPE NSCLC tissue
	Erlotinib (Tarceva®)	cobas® EGFR Mutation Test	Roche Diagnostics	Real-time PCR	Detection of exon 19 deletions and exon 21 (L858R) substitution mutations in EGFR gene derived from FFPE human NSCLC tumor tissue
	Dabrafenib (Tafinlar®), trametinib (Mekinist®), crizotinib (Xalkori®), or gefitinib (Iressa®)	The Oncomine Dx Target Test	Thermo Fisher Scientific	NGS	Simultaneously evaluates 23 genes clinically associated with NSCLC. Results from analysis of three of these genes can now be used to identify patients who may be eligible for treatment with one of the following: the combined therapy of dabrafenib (Tafinlar®), trametinib (Mekinist®), crizotinib (Xalkori®), or gefitinib (Iressa®)
	Afatinib dimaleate (Gilotrif®)	therascreen® EGFR RGQ PCR Kit	Qiagen	Real-time PCR	Detection of exon 19 deletions and exon 21 (L858R) substitution mutations in EGFR gene-derived FFPE NSCLC tumor tissue

(continued)

Table 49.1 (continued)

Cancer	Drug	CDx assay	Supplier	Assay technology	Application
Ovarian	Rucaparib (Rubraca®)	FoundationFocus CDx *BRCA*	Foundation Medicine, Inc.	NGS	Detection of sequence alterations in *BRCA1* and *BRCA2* (*BRCA1/2*) genes
	Olaparib (Lynparza®)	BRACAnalysis CDx™	Myriad Genetic Laboratories, Inc.	PCR and Sanger sequencing	Detection and classification of variants in the protein coding regions and intron/exon boundaries of the BRCA1 and BRCA2 genes using genomic DNA obtained from whole blood specimens collected in EDTA
B-cell chronic lymphocytic leukemia (CLL)	Venetoclax (Venclexta®)	Vysis CLL FISH Probe Kit	Abbott Molecular, Inc.	Fluorescence in situ hybridization	Detects deletion of the LSI TP53 probe target via fluorescence in situ hybridization (FISH) in peripheral blood specimens from patients with B-cell chronic lymphocytic leukemia (CLL)
Mastocytosis	Imatinib mesylate (Gleevec®)	*KIT* D816V mutation	ARUP Laboratories, Inc.	PCR	Detection of KIT D816V mutational status from fresh bone marrow samples of patients with aggressive systemic mastocytosis
Myelodysplastic syndrome/ myeloproliferative disease (MDS/ MPD)	Imatinib mesylate (Gleevec®)	*PDGFRB* FISH	ARUP Laboratories, Inc.	FISH	Detection of PDGFRB gene rearrangement from fresh bone marrow samples of patients with MDS/MPD with a high index of suspicion based on karyotyping showing a 5q31~33 anomaly
Gastrointestinal stroma tumors (GIST)	Imatinib mesylate (Gleevec®)	DAKO C-KIT pharmDx™	Dako North America, Inc.	IHC	Identification of c-kit protein/CD 117 antigen (c-kit protein) expression in normal and neoplastic FFPE tissues for histological evaluation. The c-Kit pharmDX rabbit polyclonal antibodies specifically detect the c-kit protein in CD 117 antigen-expressing cells
Acute myelogenous leukemia (AML)	Midostaurin (Rydapt®)	LeukoStrat® CDx FLT3 Mutation Assay	The Laboratory for Personalized Molecular Medicine, a subsidiary of Invivoscribe Technologies, Inc.	PCR	Detects internal tandem duplication (ITD) mutations and the tyrosine kinase domain mutations D835 and I836 in the FLT3 gene in genomic DNA extracted from mononuclear cells obtained from peripheral blood or bone marrow aspirates of patients diagnosed with acute myelogenous leukemia (AML)
	Enasidenib (Idhifa®)	RealTi*me* IDH2	Abbott Molecular, Inc.	PCR	In vitro (PCR) assay for the qualitative detection of single nucleotide variants (SNVs) coding nine IDH2 mutations in DNA extracted from human blood (EDTA) or bone marrow (EDTA)
					AML patients with an (IDH2) mutation eligible for treatment with enasidenib Idhifa®

CML	Nilotinib (Tasigna®)	MolecularMD MRDx® BCR-ABL Test	MolecularMD Corporation	Real-time PCR	Detection of BCR-ABL1 transcripts (e13a2/b2a2 and/or e14a2/b3a2) and the ABL1 endogenous control mRNA in peripheral blood specimens from patients previously diagnosed with t(9:22) positive chronic myeloid leukemia (CML)
NSCLC, melanoma, breast cancer, CRC, ovarian cancer	Dabrafenib (Tafinlar®), trametinib (Mekinist®), crizotinib (Xalkori®), gefitinib (Iressa®), rucaparib (Rubraca®), osimertinib (Tagrisso®), erlotinib (Tarceva®), cetuximab (Erbitux®), panitumumab (Vectibix®), afatinib (Gilotrif®), trastuzumab (Herceptin®), pertuzumab (Perjeta®), ado-trastuzumab emtansine (Kadcyla®), vemurafenib (Zelboraf), alectinib (Alecensa®), cobimetinib in combination with vemurafenib (Cotellic®/Zelboraf®)	FoundationOne CDx	Foundation Medicine, Inc.	Next-generation sequencing	Detection of substitutions, insertion and deletion alterations (indels), and copy number alterations (CNAs) in 324 genes and select gene rearrangements, as well as genomic signatures including microsatellite instability (MSI) and tumor mutational burden (TMB) using DNA isolated from formalin-fixed paraffin-embedded (FFPE) tumor tissue specimens

Table 49.2 List of FDA-approved complementary diagnostic assays in immuno-oncology

Cancer	Drug	CDx assay	Supplier	Assay technology	Application
Non-squamous NSCLC	Nivolumab (Opdivo®)	PD-L1 IHC 28-8 pharmDx™	Dako North America, Inc.	IHC	Provides PD-L1 status in FFPE human non-squamous non-small cell lung cancer tissue
Melanoma	Nivolumab (Opdivo®) in combination with ipilimumab (Yervoy®)	PD-L1 IHC 28-8 pharmDx™	Dako North America, Inc.	IHC	Provides PD-L1 status in FFPE human melanoma tissue
Squamous cell carcinoma of the head and neck (SCCHN)	Nivolumab (Opdivo®)	PD-L1 IHC 28-8 pharmDx	Dako North America, Inc.	IHC	Provides PD-L1 status in FFPE human SCC head and neck
Human urothelial cancer	Nivolumab (Opdivo®)	PD-L1 IHC 28-8 pharmDx	Dako North America, Inc.	IHC	Provides PD-L1 status in FFPE human urothelial cancer
NSCLC	Atezolizumab (Tecentriq™)	Ventana PD-L1 (SP142) Assay	Ventana Medical Systems	IHC	Provides PD-L1 status in FFPE human non-small cell cancer tissue
Metastatic urothelial cancer	Atezolizumab (Tecentriq™)	Ventana PD-L1 (SP142) Assay	Ventana Medical Systems	IHC	Provides PD-L1 status in FFPE human UC cancer tissue
	Durvalumab (IMFINZI™)	Ventana PD-L1 (SP263) Assay	Ventana Medical Systems	IHC	Provides PD-L1 status in FFPE human UC cancer tissue

New Technologies and Tools

NGS, Liquid Biopsies, Need for Quantitative IHC, Image Analysis, Multiplexing, RNA ISH

As technologies evolve and the need for gene expression and quantification of biomarkers becomes a reality, pharma will increasingly be using these tools and technologies to select and stratify patient populations most likely to respond to targeted therapies. We are entering a phase of more information, which for IVD manufacturers and FDA alike, present new sets of challenges in analytical and clinical validation. Clinical studies will adapt to the use of complex testing algorithms, with NGS and gene expression assays leading the way. FDA has built a community platform for NGS assay evaluation and regulatory science exploration called precisionFDA, (https://precision.fda.gov/). Through efforts such as precisionFDA, precision medicine will be advanced through multiple channels, including researchers, clinicians, industry, and regulatory bodies.

References

1. Romond EH, et al. Trastuzumab plus adjuvant chemotherapy for operable HER2-positive breast cancer. N Engl J Med. 2005;353:1673–84.
2. Ménard, et al. HER2 as a prognostic factor in breast cancer. Oncology. 2001;61(Suppl 2):67–72.
3. Vogel CL, Cobleigh MA, Tripathy D, Gutheil JC, Harris LN, Fehrenbacher L, et al. First-line Herceptin monotherapy in metastatic breast cancer. Oncology. 2001;61(Suppl 2):37–42.
4. Baselga J. Clinical trials of Herceptin(trastuzumab). Eur J Cancer. 2001;37(Suppl 1):S18–2.
5. Lee JW, Devanarayan V, Barrett YC, et al. Fit-for-purpose method development and validation for successful biomarker measurement. Pharm Res. 2006;23(2):312–28.
6. Olsen D, Jorgensen JT. Companion diagnostics for targeted cancer drugs – clinical and regulatory aspects. Front Oncol. 2014;4:105.
7. Audette J. FDA introduces a new test concept: "complementary diagnostics". Amplion; 2015.
8. Novotny JF Jr, Cogswell J, Inzunza H, et al. Establishing a complementary diagnostic for anti-PD-1 immune checkpoint inhibitor therapy. Ann Oncol. 2016;27(10):1966–9.
9. Meijuan L. Statistical methods for clinical validation of follow-on companion diagnostic devices via an external concordance study. Stat Biopharma Res. 2016;8(3):355–63.
10. Hirsch FR, et al. PD-L1 immunohistochemistry assays for lung cancer: results from phase 1 of the blueprint PD-L1 IHC Assay Comparison Project. J Thorac Oncol. 2016;12(2):208–22.
11. U.S. Department of Health and Human Services, Food and Drug Administration, Center for Devices and Radiological Health, Center for Biologics Evaluation and Research, Center for Drug Evaluation and Research. In vitro companion diagnostic devices: guidance for industry and food and drug administration staff. 2014.
12. U.S. Department of Health and Human Services, Food and Drug Administration, Center for Devices and Radiological Health, Center for Drug Evaluation and Research, Center for Biologics Evaluation and Research. Principles for codevelopment of an in vitro companion diagnostic device with a therapeutic product: draft guidance for industry and food and drug administration staff. 2016.

Policy Issues in the Clinical Development and Use of Predictive Biomarkers for Molecular Targeted Therapies

50

V. M. Pratt

Introduction

Policy issues in the clinical development and use of predictive biomarkers for molecularly targeted therapies must be viewed within the larger context of the current system for biomarker test regulation as well as the predominantly fee-for-service reimbursement system in the United States. The National Academy of Medicine issued a report in 2016, Biomarker Tests for Molecularly Targeted Therapies: Key to Unlocking Precision Medicine [1]. In this report, the authors emphasize that having appropriate and effective regulatory and reimbursement frameworks is critical to ensuring that clinicians and their patients have access to—and the ability to benefit from—the potential of biomarker tests for molecularly targeted therapies to optimize patient care and therapies. Regulatory and reimbursement policy directly impacts on how medical product industries evolve and grow. Payers require clarity about the types of information required to establish clinical utility or the test's usefulness in terms of its impact on clinical outcomes—or the actual effect on patients. Thus, policy challenges involve balancing the competing demands of the patient's need and desire for access to tests against the need for sufficient evidence to assess the potential risks and benefits of the tests.

Policy Issue Recommendations or Goals

Common clinical utility evidentiary standards that are applied for initial and ongoing coordinated regulatory, coverage, and reimbursement decisions for biomarker tests for molecularly targeted therapies are needed.

Implementation and Challenges

The current regulatory structure for biomarker tests for molecularly targeted therapies in the United States features two federal agencies: the US Food and Drug Administration (FDA) and the Centers for Medicare and Medicaid Services (CMS). Numerous state regulatory bodies and professional and accreditation organizations are also involved and provide complementary oversight of diagnostic tests and laboratory operations.

V. M. Pratt
Department of Medical and Molecular Genetics, Indiana University School of Medicine, Indianapolis, IN, USA
e-mail: vpratt@iu.edu

S. Badve, G. L. Kumar (eds.), *Predictive Biomarkers in Oncology*, https://doi.org/10.1007/978-3-319-95228-4_50

Food and Drug Administration (FDA)

The FDA has oversight of molecularly targeted therapies by both the therapy itself (e.g., drugs) and the kits and instruments used to perform clinical testing. The FDA's jurisdiction does not normally cover laboratory facilities or functions (see CLIA, below); rather it focuses on safety and efficacy. The FDA has inferred this to mean analytical validity and clinical validity of laboratory tests. The regulatory pathways for tests and drugs are not the same. Laboratory tests such as biomarker tests for molecularly targeted therapies are introduced into standard clinical practice in either of the two ways: by seeking FDA approval or clearance using the premarket approval or 510(k) process [2] or by developing a test for exclusive use within a specific laboratory, commonly referred to as a laboratory-developed test (LDT) or procedure (LDP), respectively. Manufacturers of test kits that are sold to laboratories claim an unequal playing field as clinical laboratories can develop a similar LDP without the expense of submitting to the FDA (estimated to be in the millions of dollars). While kit manufacturers have significant initial costs, these are usually recouped by passing along those costs to clinical laboratories which buy the test kits and to the patients who have the tests.

The FDA views laboratory-developed procedures (LDPs) to be a class of in vitro diagnostics (IVDs) and considers laboratories to be manufacturers which are subject to their oversight authority. The FDA defines "device" to include any "in vitro reagent, or other similar or related article, including any component" "(2) intended for use in the diagnosis of disease or other conditions, or in the cure, mitigation, treatment, or prevention of disease, in man or other animals" [3]. Historically, FDA requirements have applied only to medical device manufacturers that develop and distribute test kits in interstate commerce to laboratories. Because new innovative molecular test services where evidence may be constantly evolving do not fit neatly within the FDA's medical device framework, which was established in the 1970s to oversee the safety and effectiveness of such devices as implants and surgical tools, the FDA created two high-level draft guidance documents in 2014 that describe how it intends to regulate LDPs [4]. The FDA Commissioner explained that because of the advances in test complexity, particularly genomic sequencing, they believe it is necessary to establish a risk-based regulatory framework for the FDA to ensure the clinical validity of LDPs. The FDA announced in late 2016 that they would not issue a final guidance on the oversight of LDPs at the request of various stakeholders to allow for further public discussion on an appropriate oversight approach and to give congressional authorizing committees the opportunity to develop a legislative solution [5].

LDPs in the United States are already highly regulated under a three-part framework that consists of federal laws (i.e., CLIA), state laws, and accreditation by authorities, such as the College of American Pathologists (CAP). That framework requires extensive validation of the quality of diagnostic services yet also allows laboratories the flexibility to develop and validate laboratory tests quickly and, thus, adopt new scientific knowledge.

Companion Diagnostics

Biomarker tests that will be used to identify patients that are likely to benefit from a specific investigational targeted therapy may be co-developed with the drug; the biomarker and drug are both tested simultaneously in clinical trials, and the safety and efficacy of the test and the drug are evaluated in the same trial. Biomarker tests that are co-developed with a drug and co-approved by FDA are known as companion in vitro diagnostics [6]. For a biomarker test that is co-developed with a drug (e.g., ERBB2 (HER2)/neu for trastuzumab), the regulatory pathway enables concurrent approval of the test and the drug [7]. Although establishment of clinical utility for the drug-diagnostic combination would be expected to ensure reimbursement of the test, this is not always the case; one study found limited and variable reimbursement of drug-diagnostic combinations, stating that "even in cases of co-developed combinations, drug reimbursement does not necessarily imply diagnostic reimbursement" [8]. Given the emergence of new technologies such as massively parallel sequencing, the companion diagnostic model of single test, single drug may not be sustainable in the new era of multianalyte tests with various assay methodologies performed across multiple platforms [9, 10].

CMS Laboratory Oversight Through CLIA

Any clinical laboratory that reports tests for clinical management of patients falls under the purview of CMS' Clinical Laboratory Improvement Amendments of 1988 (a CLIA-certified laboratory) that provides a baseline level of oversight with respect to test development (i.e., analytical validity and clinical validity) and the quality of laboratory operations. Clinical laboratories have a professional service component for the interpretation and, as such, have additional opportunities to promote patient safety due to the professional judgment used when interpreting and reporting test results. The healthcare professional (e.g., molecular pathologist) is involved in designing and validating the test, purchasing manufactured products/instruments, determining appropriateness given the clinical presentation, and interpreting the results in the context of other medical information.

In addition to CLIA, several entities have deemed status. This means that these entities meet or exceed CLIA standards. One is the New York (NY) State Department of Health (NYSDOH)'s Clinical Laboratory Evaluation Program (CLEP) which certifies laboratories that serve NY residents. Many larger, national laboratories have NYSDOH accreditation. § 58–1.10(g) of Part 58 of Title 10 (Health) of the Official Compilation of Codes, Rules, and Regulations of the State of New York states that all technical procedures employed in a laboratory shall be of proven reliability and generally accepted by leading authorities in the specialties of laboratory medicine and/or approved by the Department. For NYSDOH, laboratories performing any non-FDA approved or LDPs must submit materials to have these assays reviewed and approved for use in New York State (or for NY State residents) (§ 58–1.10(g) of Part 58 of Title 10 (Health) of the Official Compilation of Codes, Rules, and Regulations of the State of New York). Formerly, there were significant delays in bringing new and/or improved testing services to NY residents (Victoria Pratt, personal experience). Due to the backlog, this approval process was taking greater than 2 years for approval in most cases. How these delays impacted NY residents is unknown. Responding to concerns expressed by various organizations, NYSDOH improved their process in which they are granting conditional approvals to tests in NY State licensed laboratories, resulting in advances in molecular diagnostic testing to be put into practice for providing high-quality healthcare to NY residents.

Coverage and Reimbursement Decisions

The US government, through programs like CMS, pays for approximately half of the country's healthcare. While diagnostic testing comprises less than 5% of hospital costs and about 1.6% of all Medicare costs, their findings influence as much as 60–70% of healthcare decision-making [11].

Current procedural terminology (CPT) codes are used to report procedures (e.g., medical, surgical, and diagnostic) and services to entities such as physicians, health insurance companies, and accreditation organizations. In the 2012 Medicare Physician Fee Schedule (MPFS) Proposed Rule [12], CMS articulated its concern with improving the accuracy of coding and payment for molecular pathology services. Prior to 2013, laboratories billed for molecular pathology services using unique combinations of CPT "stacking" codes that describe each step of the procedure required to perform the test. Because of concerns that payers could not determine the specific tests performed when billed under the stacking codes, the

American Medical Association's CPT® Editorial Panel adopted an entire new subsection of the Pathology Section of CPT to describe molecular pathology procedures. With the introduction of these new codes, the stacking codes were retired effective January 1, 2013, and laboratories were required to report molecular pathology tests using the new CPT codes that are analyte (gene)-specific and method agnostic.

In the absence of evidence of clinical utility, or consensus regarding evidentiary standards, payers rely on a variety of information sources to develop their coverage policies [13, 14]. In addition to peer-reviewed studies published in medical journals, payers consider:

- Reviews of published studies on a particular topic, such as those conducted by the Agency for Healthcare Research and Quality (AHRQ), the Blue Cross/Blue Shield Technology Evaluation Center, or the Duke Evidence-based Practice Center
- Evidence-based consensus statements or guidelines from professional societies or other nationally recognized healthcare organizations, such as the American Society of Clinical Oncology (ASCO) or National Comprehensive Cancer Network (NCCN)
- Guidance documents developed by multi-stakeholder groups such as the Center for Medical Technology Policy [1]

It is generally believed that private payers often follow Medicare's coverage determinations. However, a recent study found that the coverage decisions for medical devices by 16 private payers aligned with Medicare decisions only half the time [15].

Many, including CMS, have noted that DNA sequencing costs have decreased. While this is true with respect to reagent costs (i.e., with respect to the original cost to sequence the human genome), it does not account for the significant costs of equipment, informational technology (IT) infrastructure, personnel to analyze and interpret the results, and regulatory oversight (e.g., licensure, proficiency testing). Also, most of the FDA-cleared assays, especially molecularly based, have not decreased in price. In fact, many prices have increased annually to account for inflation. CMS, in the Physician Fee Schedule Final Rule [16], planned to reset rates on the Clinical Laboratory Fee Schedule (CLFS) based upon such technological changes. Many in the clinical laboratory industry feared that rates on the CLFS could be reduced by at least 30% or more for older established tests. The rule provided a great deal of latitude to CMS with no guardrails to prevent drastic cuts, and it was not clear if stakeholders would have the opportunity to request reconsideration. On April 1, 2014, President Obama signed into law H.R. 4302, Protecting Access to Medicare Act of 2014 (PAMA). This act prevented CMS from revaluing the CLFS based upon technological changes. In addition, PAMA has a provision that the Secretary of the Department of Health and Human Services (HHS) now has the authority to designate up to four Medicare Administrative Contractors (MACs) to establish coverage policies and/or process claims for payment for laboratory tests for the entire Medicare program. Another requirement is that laboratories must report the market data that CMS will use to determine CLFS prices. Failure to report this data can result in a penalty to the clinical laboratory of as much as $10,000 per day per unreported test. If the weighted median price is significantly less, it is possible that CMS will cut the price of a laboratory test by 55% over a 6-year period, compared to the base year. It is known that CMS wants to reset prices for the 20 highest-volume tests (note, none of the top 20 tests are molecular-based) (Office of the Inspector General Report. Comparing Lab Test Payment Rates: Medicare Could Achieve Substantial Savings (OEI-07-11-00010), 2013) that represent more than half of what is spent annually on Medicare Part B CLFS.

MolDX [17] was a program originally developed by one MAC, Palmetto GBA, to review the quality of the validation (analytical validity) to determine clinical utility of molecular tests and thus determine coverage and pricing. Inherently, CPT codes do not differentiate between IVDs and LDPs. Palmetto-designed MolDX, which includes McKesson-owned Z-Code Identifiers [18], is a unique five-character alphanumeric identifier code associated with a specific advanced diagnostic test and is assigned based on the uniqueness of each laboratory's test or manufacturer's product being registered. For example, a test for the same analyte but performed by a different laboratory methodology would be considered unique and thus would merit a unique Z-Code Identifier. This system is designed to complement the current CPT codes. IVDs can be distinguished from LDPs and differential pricing implemented. Each laboratory in Palmetto's jurisdiction as well as other MACs who have adopted the MolDX program (in 2017, more than 50% of the US states are in the MolDx program) and that would like to obtain coverage for a molecular test must meet the requirements of the MolDX program. The laboratory must obtain a Z-code, and if the test is a LDP, the laboratory must also submit a detailed technical assessment of published test data for coverage determination.

The FDA and CMS are well positioned to work together to support timely review, coverage, and reimbursement of new advanced tests; a key example of such collaboration is the FDA-CMS parallel review program [19]. Historically, the FDA and CMS have worked independently, with separate staff focused on different points of a product's development life cycle and with different evidentiary expectations from FDA's focus on safe and effective to CMS's focus on reasonable and necessary. The parallel review program was developed in response to the fact that attaining FDA approval of a product based on safety and efficacy does not necessarily result in a timely determination by Medicare that the product is medically necessary and therefore should be covered. The stated goal of the program was to reduce the time between FDA approval and CMS national coverage determinations, which are important for products to be integrated broadly into clinical practice [20].

Summary

Policies involved in biomarker tests for molecularly targeted therapies are diverse including patients, clinicians, academia, industry, government agencies, and payers, each with their own perspective on evidentiary bases for clinical utility. As Ginsburg and Kuderer note: "More dialogue and coordination among stakeholders is needed to facilitate the development of the necessary evidence base. It is equally apparent that test development and reimbursement need to focus on clinical utility of the test and the net benefit to patients" [21].

Glossary

Medicare Administrative Contractors (MACs) Contracted private healthcare insurers that process medical claims for Medicare beneficiaries

References

1. Graig LA, Phillips JK, Moses HL, editors. Biomarker tests for molecularly targeted therapies: key to unlocking precision medicine. Washington, DC: The National Academy of Science; 2017. http://www.nap.edu/catalog/21860/biomarker-tests-for-molecularly-targeted-therapies-key-to-unlocking-precision. Accessed 4/25/2017.
2. U.S. Food & Drug Administration (FDA). 510(k) clearances. 2017. http://www.fda.gov/MedicalDevices/ProductsandMedicalProcedures/DeviceApprovalsandClearances/510kClearances/. Accessed 4/25/2017.
3. U.S. Food & Drug Administration (FDA). Is the product a medical device? 2014. https://www.fda.gov/MedicalDevices/DeviceRegulationandGuidance/Overview/ClassifyYourDevice/ucm051512.htm. Accessed 4/25/2017.
4. U.S. Food & Drug Administration (FDA). Draft guidance for industry, food and drug administration staff, and clinical laboratories: framework for regulatory oversight of Laboratory Developed Tests (LDTs). 2014. https://www.fda.gov/downloads/MedicalDevices/DeviceRegulationandGuidance/GuidanceDocuments/UCM416685.pdf. Accessed 4/25/2017.
5. U.S. Food & Drug Administration (FDA). Discussion paper on Laboratory Developed Tests (LDTs). 2017. https://www.fda.gov/downloads/MedicalDevices/ProductsandMedicalProcedures/InVitroDiagnostics/LaboratoryDevelopedTests/UCM536965.pdf. Accessed 4/25/2017.
6. U.S. Food & Drug Administration (FDA). Companion diagnostics. 2016. https://www.fda.gov/MedicalDevices/ProductsandMedicalProcedures/InVitroDiagnostics/ucm407297.htm. Accessed 4/25/2017.

7. Frueh FW. Regulation, reimbursement, and the long road of implementation of personalized medicine – a perspective from the United States. Value Health. 2013;16(6 Suppl):S27–31.
8. Cohen J, Felix A. Personalized medicine's bottleneck: diagnostic test evidence and reimbursement. J Personalized Med. 2014;4(2):163–75.
9. Mansfield EA. FDA perspective on companion diagnostics: an evolving paradigm. Clin Cancer Res. 2014;20:1453–7.
10. IOM. Policy issues in the development and adoption of biomarkers for molecularly targeted cancer therapies: workshop summary. Washington, DC: National Academies Press; 2015.
11. UK Department of Health Pathology Modernisation Team. Modernising pathology services. 60–70% of NHS patients' diagnoses depend on laboratory tests, 2004. p. 7. Available at www.dh.gov.uk.
12. Centers for Medicare & Medicaid Services (CMS). Details for title: CMS-1524-FC. 2012. https://www.cms.gov/Medicare/Medicare-Fee-for-Service-Payment/PhysicianFeeSched/PFS-Federal-Regulation-Notices-Items/CMS1253669.html. Accessed 4/25/2017.
13. Graf M, Needham D, Teed N, Brown T. Genetic testing insurance coverage trends: a review of publicly available policies from the largest US payers. Pers Med. 2013;10(3):235.
14. Trosman JR, Van Bebber SL, Phillips KA. Health technology assessment and private payers' coverage of personalized medicine. J Oncol Pract. 2011;7(3 Suppl):18s–24s.
15. Chambers JD, Chenoweth M, Thorat T, Neumann PJ. Private payers disagree with medicare over medical device coverage about half the time. Health Aff. 2015;34(8):1376–82.
16. Federal Register: The Daily Journal of the United States Government. Medicare Program; Revisions to Payment Policies Under the Physician Fee Schedule and Other Revisions to Part B for CY 2017; Medicare Advantage Bid Pricing Data Release; Medicare Advantage and Part D Medical Loss Ratio Data Release; Medicare Advantage Provider Network Requirements; Expansion of Medicare Diabetes Prevention Program Model; Medicare Shared Savings Program Requirements. 2016. https://www.federalregister.gov/documents/2016/11/15/2016-26668/medicare-program-revisions-to-payment-policies-under-the-physician-fee-schedule-and-other-revisions. Accessed 4/25/2017.
17. Palmetto GBA. MolDX ®. 2017. http://www.palmettogba.com/MolDx. Accessed 4/25/2017.
18. McKesson Diagnostics Exchange. McKesson Z-Code™ Identifiers. http://mckessondex.com/z-codes. Accessed 4/25/2017.
19. Federal Register: The Daily Journal of the United States Government. Program for Parallel Review of Medical Devices. 2016. https://www.federalregister.gov/documents/2016/10/24/2016-25659/program-for-parallel-review-of-medical-devices. Accessed 4/25/2017.
20. Messner DA, Tunis SR. Current and future state of FDA-CMS parallel reviews. Clin Pharmacol Ther. 2012;91(3):383–5.
21. Ginsburg GS, Kuderer NM. Comparative effectiveness research, genomics-enabled personalized medicine, and rapid learning health care: a common bond. J Clin Oncol. 2012;30(34):4233–42.

Role of Central Laboratories in Research, Validation, and Application of Predictive Biomarkers

51

Oliver Stoss and Thomas Henkel

Introduction

Patient selection for targeted therapies using molecular biomarkers has become mandatory in clinical development and practice in the last 10 years. Consequently, pharmaceutical industry research and development (R&D) needed a reliable partner who is able to validate biomarker assays and apply them reproducibly over many years in a highly standardized manner within a regulatory environment, suited to fulfill the requirements for drug and diagnostic approval. Central labs fulfill this service through the application of a new biomarker test in a highly standardized manner, keeping the variation in sample preparation, test platform, material and devices, test performance, and test interpretation at a minimum. The role of a central lab starts with the choice of the most promising test platform and the development and analytical validation of an assay according to regional and international guidelines. Analysis of a biomarker within a clinical trial and cancer drug development requires a well-organized workflow of sample shipments to the central lab and back to the investigator as well as instructions of local sites on sample preparation in order to standardize pre-analytical factors as far as possible. In particular, these processes primarily rely on tissue biomarkers for response prediction requiring a high degree of expert knowledge and standardization due to challenges in sample preparation and readouts with inter-observer variability.

Clinical trials also require the application of different complex test algorithms that depend on multiple variables such as the type of informed consent, the sample type or visit code, or site-specific individual requirements. Especially the development of companion diagnostics adds a high degree of complexity to biomarker-driven clinical trials where central labs help to streamline the workflow by integrating processes such as biomarker validation, monitoring of test results, reporting to investigators, and studying monitors, sponsors, and regulatory authorities. Finally, a central lab may provide training activities for a successful introduction of a new biomarker into clinical routine.

Workflow of a Central Lab

The major components of a central lab workflow are shown in Fig. 51.1. As a first step, project teams of the pharmaceutical company, dedicated contact research organizations (CROs), and the central lab determine the scope of work. In case

O. Stoss · T. Henkel (✉)
Targos Molecular Pathology GmbH,
Kassel, Hessen, Germany
e-mail: thomas.henkel@targos-gmbh.de

S. Badve, G. L. Kumar (eds.), *Predictive Biomarkers in Oncology*,
https://doi.org/10.1007/978-3-319-95228-4_51

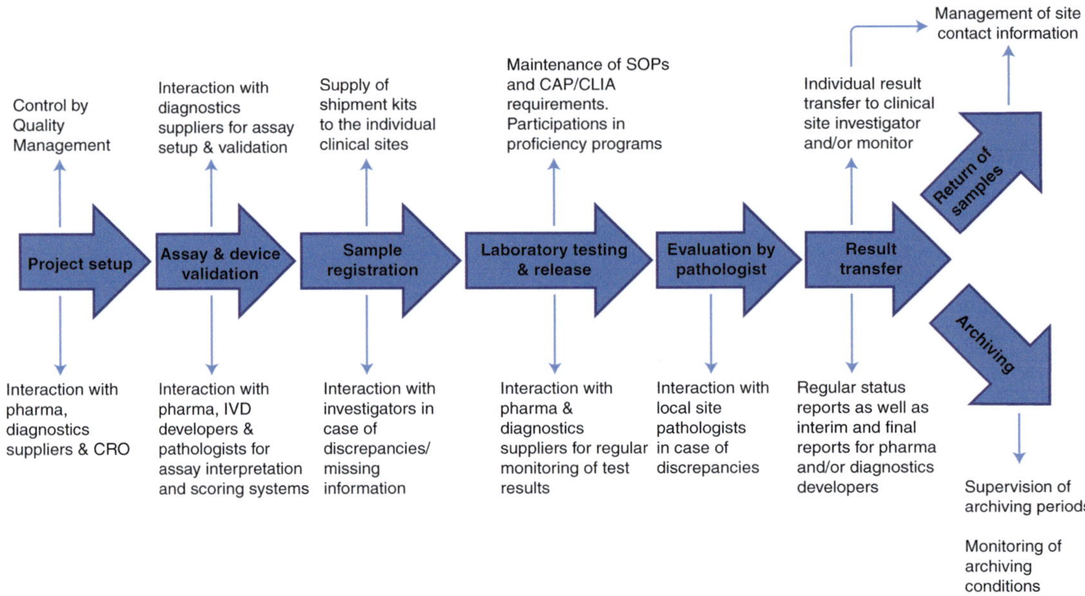

Fig. 51.1 Major components of a central lab workflow

of new assays to be implemented or in case of an IVD development, the manufacturer of the diagnostic assay is an essential party in the project setup as well. New markers typically require new assays that need to be validated prior to the conduction of the clinical project. Pharma, diagnostics industry, and the central lab closely work together in order to determine assay interpretation guidelines and assay protocols. The final product of this phase is a technical validation report. Once the assay is validated, the clinical testing of patient samples can be initiated. Samples are typically directly sent by the clinical centers to the central lab using dedicated shipment kits and specifications delivered by the central laboratory or a CRO. Sample registration at the central lab includes the collection of sample-related data delivered on a sample requisition form by the local site. In case inconsistencies or missing information is detected, a query is sent to the monitors or investigators by the central lab. Once essential issues have been sorted out, the sample undergoes testing according to a study-specific algorithm, which is defined in the clinical trial protocol and specified in a project plan and standard operating procedures (SOP). Trials that are testing a companion diagnostics require additional monitoring steps of assay and device results and quality parameters that include a close interaction with the diagnostics supplier. In case FDA-approved in vitro diagnostics (IVD) or laboratory-developed tests (LDTs) are applied, the central lab has to participate in regular proficiency testing programs for quality assurance. Histopathology tests such as immunohistochemistry or in situ hybridization require subsequent evaluation by a qualified pathologist. Predictive marker results categorized as patient selection criteria are directly communicated to the investigator for treatment decision. In case of discrepancies or critical cases, a direct contact between the central lab and the investigator or the local pathologist helps to find the best treatment option for a patient. The project manager of the central lab is in charge of providing regular status updates to the pharma or diagnostics partner. Depending on the study protocol, samples are either archived at the central lab or need to be returned to the investigator.

The growing trend to outsource as much services as possible and the desire of pharmaceutical companies and clinical trial organizers for a maximum of service under one roof led to expansion of the activities of central labs and contact research organizations (CROs). As a consequence, this has led to a growing importance of central labs in clinical trial testing and a growing service portfolio. These include:

1. Sample logistics including customs clearance, sample reception, sample registration, tracking, collection, and storage
2. Regulatory expertise for the validation of computerized systems, assay validation and testing, and the approval of tests and/or drugs
3. Data management, data flow, and data monitoring
4. IT expertise maintaining and configuring CFR21 part 11 compliant information management systems
5. Assay development responsible for assay establishment and validation
6. Sample archiving and data archiving
7. Biostatistical evaluation of clinical trial data
8. Training of pathologists and technicians for novel biomarkers

The constant management of a highly regulated and standardized but yet flexible workflow is one of the biggest challenges for any central lab.

Role of a Central Lab in a Predictive Biomarker-Driven Research and Clinical Trials

The integration of all aspects of clinical trial testing and predictive biomarker research within one or more CROs and other contractual partners has to be streamlined in a process from biomarker research up to the approval of a drug or companion diagnostics. In the following sections, the major roles of central labs in predictive biomarker research will be discussed.

The Quality of Test Results and Scientific Input

Central labs are associated both with retrospective testing with the aim to characterize novel biomarkers and to generate new potential markers to be investigated in more detail in future clinical trials and prospective clinical biomarker testing with the aim to prove the clinical benefit of a given marker or panel of markers and/or to deliver high-quality test results for drug approval. Obtaining reliable test results for a new biomarker using a new test in combination with a new instrument and a new drug is a challenge. Successful biomarker testing starts with the choice of the right test method and platform. Central labs typically run multiple competing test platforms in parallel, such as sequencing, PCR-based methods, nucleic acid hybridization-based methods, ELISA, or methods using histopathology such as immunohistochemistry or in situ hybridization. New proteomics- and metabolomics-based tests are rapidly joining the portfolio. Experience with most, if not all, these options is certainly of value when choosing the correct methodology. High-quality test results require the application of rigorous quality control procedures during assay validation and the routine application of a released test as described below.

For histology-based assays, an additional variable to be controlled is the interpretation of a tissue slide that is stained for a given biomarker. The biomarker research community is faced with three major challenges. The first one is to find a scoring system that allows for the optimal separation of different treatment groups. The example of PD-L1 or PD1 immunohistochemistry shows how complex this situation can get, with several different antibody clones as companion diagnostics for several different drugs having different scoring systems in different tissue types [1]. Second, the inter-observer variation between pathologists needs to be controlled. Although digital image analysis algorithms have become more and more robust and successful today, manual reading by pathologists is still state of the art for IVD approval trials. This is especially true for heterogeneous diseases and staining patterns where variation between different qualified readers can increase to over 30%. A central lab offers monitoring tools in order to align qualified readers prior to and during a clinical trial. Third, the scoring system has to be as simple as possible in order to easily penetrate the market of routine clinical application. The more complex the scoring system, the higher the inter-observer variation and the lower the test acceptance rate. Therefore, development of final scoring guidelines that may be part of the label of an in vitro diagnostics requires a tight interaction between laboratory and regulatory experts, pathologists, clinicians, and the clinical trial sponsor. The biomarker HER2 in solid tumors is a good example of how important it is to carefully design and improve the interpretation guidelines in order to lower the inter-observer variation. The introduction of the so-called magnification rule for HER2 IHC assessment in gastric cancer made test interpretation much easier and more standardized [2].

Switching a test from one tissue type to another one is not trivial, even if the same assay protocol is applied. The reference range and the scoring system may need to be adapted due to the different biologies of the respective tissues. For example, a pre-study in gastric cancer showed that the HER2 test scoring system as applied for breast cancer required modification in order to be suitable for HER2 assessment in gastric cancer [3]. This adoption then required prospective testing in a separate phase III trial, which led to the successful approval of the modified HER2 test in gastric cancer [4, 5].

In this context, it is important for the biomarker community that central labs actively publish scientific results collected during biomarker testing as soon as permission is granted by the sponsoring organization and the terms of the confidentiality and intellectual property agreements are met.

A lot of biomarker tests, such as mutation or RNA expression analyses, are dependent on computerized systems that perform the assay and – for a growing number of assays – also deliver final interpretation of the test results. The quality of data is therefore also dependent on the validation of the test environment. The validation of computerized systems needs to follow defined guidelines such as 21 CFR part 11 or EU-GMP 2003/94/EC Annex 11 and Annex 15. Thus, from an organizational point of view, it has become convenient for a biomarker-driven clinical trial sponsor to receive assay validation and training documents from one contractual partner either in order to discuss and implement changes that became evident after FDA pre-submission or for final review and approval steps. Centralized testing and assay development help to streamline this entire process.

A typical advantage of central lab testing is that the test is typically conducted at high volumes which offer the chance to reduce the price per test. The additional advantage is that at high test volumes, the detection of rare cases previously not detected during assay validation is possible. These cases can immediately be discussed and included into final test recommendations. The awareness of critical cases has a direct impact on the quality of the data reported.

In contrast to centralized testing, decentralized testing has a major disadvantage in terms of high interlaboratory variance that may interfere with the final interpretation of the test results or efficacy of companion diagnostics. Even if several sites are using the same method such as immunohistochemistry, this leaves the individual center with plenty of variables such as the clone, pretreatment conditions, detection systems, or interpretation guidelines. The biomarker HER2 is a well-studied example on how central and local testing varies even if all participants use FDA- or CE-approved tests [6]. Although the concordance between different testing sites is approaching an acceptable range, this example shows how long this process of alignment has taken.

Sample Preparation and Shipment

Sample preparation and shipment conditions are essential for high-quality testing. A central lab provides a perfect interface between the courier company and the clinical center in order to reduce the shipment time and the interlaboratory variation to a minimum. Ideally, the central lab is taken into consideration at the time of clinical protocol development in order to ensure that the tissue is delivered in a most optimal fashion. It is essential to analyze pre-analytic factors such as the exact location of the sample in the respective organ, the type of tissue collection, time of ischemia, date of tissue collection, fixative quality, and time of fixation or the date of tissue sectioning to get optimal results. Central lab testing can lead to a standardization of sample preparation. Based on the correlation of test results with preanalytic data obtained by the center, a central lab can give recommendations to individual sites in order to reduce test failure rates. In this manner, a central lab gets aware of a number of test robustness factors that can be incorporated in test recommendations and labels.

The shipment of patient samples from the investigator site or the pharmaceutical company requires an in-depth knowledge of national/international customs declarations, import and export licenses, and approved packaging material for sample transportation. In addition, the local sites need to be instructed regarding the preparation of tissues and the procedures for shipment. For example, in China, regulatory and legal hurdles hardly allow patient tissue samples to leave the country. Other countries such as Japan only allow sending out tissue slides instead of the complete tissue blocks which adds another complexity to testing regarding the quality of sample preparation, the use of the correct glass slides, and the stability of the biomarker to be measured under these conditions. In addition, for certain sample types such as fresh or frozen tissue, the transportation chain needs to be monitored for temperature or extensive mechanical stress. Finally, patient samples belonging to the patients need to be sent back to the local site after testing. This requires a significant amount of trust by the clinician in a central lab.

Complexity of Clinical Trials

Highly regulated complex clinical trials require efficient central labs. Several factors accelerate the complexity of biomarker-driven clinical trials:

(a) Approving an in vitro diagnostics (IVD) as a companion diagnostic (CDx) test at the same time as the CDx associated drug candidate based on the same clinical trial data leads to the fact that different study teams require different types of results in different formats from the same database at the central lab. While for drug approval the final result of a biomarker assay may be sufficient, IVD approval requires additional data such as raw data or detailed explanations for assay failures. In some cases, different scoring systems are applied on the same set of raw data. In other cases, the same samples have to be investigated by different qualified readers in a blinded fashion. Several datasets may exist for the same sample but have to be submitted in various ways to different clinical databases and/or other branches of the regulatory authorities. Due to the varied focus of drug and test development study teams, several monitoring strategies are required resulting in regular interim reports and quality control procedures for the same clinical study. These types of work can only be performed by a highly organized central lab and not by decentralized labs.

(b) Different regulatory requirements in the USA, Europe, and Asia may lead to different sample analysis workflows within the same clinical trial. This means that upon registration, analysis algorithms are dependent on the country and sometimes even dependent on the local site requirements. Biomarker test kits may differ in their composition or in the respective manufactures' guidelines between the USA, Europe, and Asia, which has to be taken into account in a global clinical trial. The PD-L1 IHC test (Clone 22C3, DAKO Agilent) used as a predictive marker for treatment of lung cancer patients with pembrolizumab (Keytruda, MSD) is an example where different cutoffs exist for the same assay in the USA and in Europe.

(c) There is a trend toward clinical trials with fewer number of patients but with more biomarkers to be tested per patient. For biomarker testing labs, this is a challenge since the diversity of the biomarker portfolio increases, but the individual test is only applied to a few patients. For a phase I clinical trial, this may have the consequence that the effort for test establishment and validation is much higher than tests on patient samples for the respective study. Increasing diversity of tests will likely put an additional pressure on the costs due to smaller test batches. In addition, the amount of quality control activities as defined by CAP and/or CLIA or other quality control organizations have to be maintained for each new test. This includes the maintenance of respective devices and their backups, regular staff qualification, and maintenance of SOPs or regular proficiency testing. Analysis algorithms are also dependent on multiple mandatory and optional patient-informed consents. One patient may give his consent to HER2 IHC testing but may not give his consent to an exome sequencing sub-analysis in the same study. Informed consents can change or can be withdrawn during or after the study, and adequate mechanisms need to exist to handle such situations at the central lab.

(d) Due to the trend in molecular oncology that a diagnosis is not necessarily dependent on a specific organ or where the cancer originated but rather on specific mutations in different cancers, clinical trials employing "basket studies" for cancer drugs with multiple different tissue types are increasingly becoming common. This means, interpretation of the same assay on different tissue types may differ, as it is the case for HER2 in breast or gastric cancer or PD1 in malignant melanoma or lung cancer. This requires tissue-dependent analysis algorithms to be applied on a given sample.

    In case a biomarker has to be investigated over time, multiple samples with different visits will have to be analyzed for the same patient. In this situation, it may be efficient not to apply all tests on all visits but to have visit code-dependent analysis algorithms.

(e) Frequently pharmaceutical companies or investigators ask central labs to apply different biomarkers on the same sample. For obvious reasons, patient samples are a limited resource, and sample sizes are increasingly smaller due to improvements in early disease detection and prevention. Remarkable advances have been seen in the development of tests that analyze several different markers with less sample input, as in the case of lung cancer DNA sequencing, copy number variation, and translocation analysis combined in one and the same assay. Other examples of combining different methods in one assay are the detection of multiple markers on the same tissue slide using IHC or FISH or the combination of IHC and bright field in situ hybridization (Ventana) or the combination of protein and RNA quantification (Nanostring). Nevertheless, the limitation of tissue availability still requires the definition of a clear priority list of different tests to be applied sequentially. For instance, each biomarker test requires its own sample preparation procedure. For example, IHC requires 2–4 μm tissue sections to be placed on specific positively charged glass slides, whereas for DNA- or RNA-based methods, section thickness can ideally be much thicker and not necessarily be prepared on tissue slides; however, both require sample preparation under nuclease-free conditions. The fact that certain sites cannot provide tissue blocks but only send tissue slides complicates the process. In addition, one nucleic acid isolation procedure may be suitable for one assay but might lead to an unacceptable failure rate for another assay. Regulatory agencies often prefer closed systems where the DNA or RNA isolation kit is part of the testing procedure. However, this might interfere with the clinical requirements to have different tests applied on the same DNA or RNA isolate.

    The different levels of complexity within a biomarker-driven clinical trial as described above show that biomarker testing requires a huge effort of sample logistics, staff training, and project and data management activities that is increasingly difficult to handle in a decentralized manner.

Regulatory

Regulatory requirements for assay establishment and validation are also constantly increasing. A central lab offers the possibility to provide an assay for a global clinical biomarker trial that satisfies regulatory requirements of the major regulatory authorities worldwide. In case of a laboratory-developed test that is used for treatment decisions, various aspects of an assay have to be investigated such as specificity, sensitivity, linearity, dynamic and reference range, precision, accuracy, reproducibility, or robustness. In addition, the respective samples for assay validation require an increasing regulatory and ethical standard and are a major driver of costs in case the prevalence of the respective biomarker to develop is low. These samples are not only used during assay establishment and validation but also serve as qualified controls for the subsequent clinical trial where the correlation of the given biomarker with clinical outcome is investigated. The analytical validation is followed by a clinical validation once the biomarker shows potential for discriminating between clinically meaningful patient cohorts. The biggest challenge is to identify the most appropriate clinical cutoff and scoring system. Typically, information from preclinical models as well as phase I or phase II trials are combined in order to define interpretation criteria that will have the highest chance of success to optimally stratify responders from non-responders during prospective testing in an approval trial.

In addition to the country-/region-specific quality criteria such as CAP or CLIA [7], FDA guidance for industry for "Bioanalytical method validation" [8], or the "Guidance on analytical method validation" by EMA [9], general assay validation quality criteria as defined by GCP, GLP, and the ICH Q2R1 guidelines [10] should be reflected in the procedures of the biomarker lab. A lot of CLSI guidelines for test validation are derived from tests applied on liquids and cannot be directly transferred to tests on solid tissue. Biomarkers cannot be titrated when testing them on solid tissue. In addition, it is often not possible to obtain the same size of replicates out of the same tissue block. Finally, tissue heterogeneity has to be considered. This requires different approaches of the validation of tests on solid tissues.

In biomarker translational research, the drug and IVD developers have the choice of either using a research-use-only (RUO) assay in the early phase of clinical development or to go for an extensive validation of the biomarker assay already in the very early stage of clinical research. The first strategy has the consequence that if the biomarker shows promising results, the biomarker has to be further validated in order to be applied as a selection criterion within a drug or test approval trial. It also harbors the risk that during validation, it becomes evident that the protocol applied so far is not sufficiently reproducible, robust, or specific and that a completely new test has to be established. On the other hand, a full validation in a very early phase has the problem of relatively high costs for assay development and validation in relation to only a small number of patients to be tested in an early setting.

The required regulatory efforts of both customer and the testing lab have significantly increased with the advent of companion diagnostic development which requires special know-how, such as how to document trainings, reagents, and results for tests during investigation-use-only (IUO) clinical trials according to the FDA or EMEA requirements. IUO tests need special processes for assay supply chain management since they are not commercially available. Precise accountability logs as well as special regular audits have to be conducted. In some cases, devices require special control, calibration, or preventive maintenance steps for the IUO that are not required for other tests running on the same machine. This complexity often requires dedicated devices for a given test under investigation including additional device validation steps as discussed above.

Driven by the development of companion diagnostics, consulting and lab activities based on ISO13485 are often requested which further extends the field of action for laboratories involved in biomarker-driven research. Although the test laboratory is typically not directly involved in the manufacturing of the in vitro diagnostics, it often gets involved in the analytical and clinical validation of the respective test.

Specialists are required for the setup of strategies and clinical trial design in order to test for the stratification of patients for a yet non-approved drug using a non-approved assay. Test development for in vitro diagnostics approval requires many additional tasks on its way to a fully validated assay that go far beyond the CLIA lab requirements for laboratory-developed tests. Typically, the test volume further increases due to extensions in precision, specificity, robustness, stability, or accuracy for analytical test validation. In addition, knowledge about the interaction and registration at national agencies, notified bodies, or ethics approval committees are required. In addition to GCP and GLP, the development of in vitro diagnostics also requires know-how about validated procedures for GMP activities in case the lab is involved in the production of the test kit. Procedures for the labeling of test kits, risk management, and process validation are additional features that a test lab needs to acquire when entering in the field of in vitro diagnostics development.

Data Management

The increasing amount of biomarker data with clinical trials means that a lab needs to be able to handle big data. Adequate laboratory information management systems, database programmers, system administrators, and of course efficient data storage and archiving solutions need to be continuously developed. Even in highly industrialized countries, big differences in the speed of internet connection can be observed. Nevertheless, efficient up- and download procedures are becoming mandatory in the era of digital imaging. Biomarker recognition technologies based on tissue scans and whole organ scans have quickly emerged. The same holds true for genomics, proteomics, and metabolomics approaches that are increasingly applied within clinical trials.

Big data also means that the lab requires the capacity for bioinformatics analysis in order to interpret raw data correctly. Next-generation sequencing, SNP chips, or expression-based tests such as Affymetrix or Nanostring require well-trained bioinformaticians.

Efficient automation of the complex biomarker algorithms among different clinical trials is mandatory in order to prevent the application of the wrong tests at the wrong time-point on a given sample. Since this aspect requires programming or configuration steps in a data management system as well as testing of algorithms according to defined specifications, the study setup at a biomarker-driven laboratory is getting less flexible. This is not so much a problem for late stage clinical trials since the clinical study protocol has to be changed accordingly. However, for translational research studies, the test algorithms are more variable. They are often not yet defined when the first samples come in, and they are prone to changes within the study based on most recent scientific advances published elsewhere.

Efficient data security mechanisms have to be in place at the central lab. Data security require multilayer procedures, starting with efficient mechanisms in order to prevent non-authorized access by externals, security concepts for physical as well as software access for internal members that needs to be released by the system owner, efficient training of lab personnel prior to the use of any equipment, and efficient backup as well as disaster recovery procedures in case of system failure.

Biorepository

Biosample repositories are of high value for biomarker research. In many cases, biomarkers to be assessed are not yet known at the time of a clinical trial and may evolve at a later time. This may include markers for de novo resistance or additional markers to enhance the predictive value using a combination of tests. Part of translational research programs is therefore the collection of a part of the tissue material prior to sending the remaining tissue back to the investigator. Many large laboratories do offer the capacity to store barcoded samples either in liquid nitrogen, −70 °C, −20 °C freezers, fridges, or at ambient temperature under controlled environmental conditions. It is convenient then to continue further biomarker tests and to combine already assessed data with newly incoming data. Experience on how to store samples in the most efficient way is required, including the question if DNA or RNA should be isolated from a given tissue prior to storage or to prepare tissue sections and store them. A central item when preparing tissue for long-term storage is an extensive histological characterization of the solid tissue via H&E staining and evaluation by a pathologist including the amount and percentage of tumor cells, necrotic cells, stroma cells, immune cells, or normal epithelial cells in a given tissue section. This makes it easy at a later stage to select for those samples that are suitable for a given test.

Launch of a New CDx Test

Once a CDx test is successfully established and analytically and clinically validated, it is important to launch the new assay efficiently in the biomarker community and the clinical diagnostic routine. Especially for tests that require manual test interpretation, this requires training for pathologists or lab personnel. Training courses need to be organized and harmonization studies to be initiated. The results of ring studies and harmonization efforts have to be published in recommendation papers and introduced into the national and international testing guidelines (e.g., ASCO CAP). The direct knowledge transfer from the central lab to the biomarker community is of high value in order to minimize inter-observer variation as far as possible.

Summary and Conclusions

In summary, central laboratories contribute to predictive biomarker research by defining and maintaining high test quality and by standardizing test methods. Central labs play an essential role in order to streamline complex clinical trials at all stages of a clinical trial, including the protocol and study setup, provision of analytical and clinical test validation documentation, implementation of novel tests in routine testing, as well as result submission. Finally, the know-how gained by a central lab should be used in order to share knowledge by training pathologists or lab personnel and to contribute to a successful introduction of a new test into the clinical routine.

References

1. Scheel AH, Dietel M, Heukamp LC, et al. Harmonized PD-L1 immunohistochemistry for pulmonary squamous-cell and adenocarcinomas. Mod Pathol. 2016;29:1165–72.
2. Rüschoff J, Hanna W, Bilous M, et al. HER2 testing in gastric cancer: a practical approach. Mod Pathol. 2012;25:637–50.
3. Hofmann M, Stoss O, Shi D, et al. Assessment of a HER2 scoring system for gastric cancer: results from a validation study. Histopathology. 2008;52:797–805.
4. Bang YJ, Van Cutsem E, Feyereislova A, et al. Trastuzumab in combination with chemotherapy versus chemotherapy alone for treatment of HER2-positive advanced gastric or gastro-oesophageal junction cancer (ToGA): a phase 3, open-label, randomized controlled trial. Lancet. 2010;376:687–97.
5. Van Cutsem E, Bang YJ, Feng-Yi F, et al. HER2 screening data from ToGA: targeting HER2 in gastric and gastroesophageal junction cancer. Gastric Cancer. 2015;18:476–84.
6. Rüschoff J, Lebeau A., Kreipe H. et al. Assessing HER2 testing quality in breast cancer: variables that influence HER2 positivity rate from a large, multicenter, observational study in Germany. Mod Pathol. 2017;30(2):217–26.
7. U.S. Food & Drug Administration (FDA). Clinical Laboratory Improvement Amendments (CLIA). 2014. http://www.fda.gov/MedicalDevices/DeviceRegulationandGuidance/IVDRegulatoryAssistance/ucm124105.htm
8. U.S. Department of Health and Human Services, Food and Drug Administration, Center for Drug Evaluation and Research (CDER), Center for Veterinary Medicine (CVM). Guidance for industry: bioanalytical method validation. May 2001. http://www.fda.gov/downloads/Drugs/Guidance/ucm070107.pdf
9. European Medicines Agency (Science Medicines Health). Guideline on bioanalytical method validation. July 21, 2011. http://www.ema.europa.eu/docs/en_GB/document_library/Scientific_guideline/2011/08/WC500109686.pdf
10. International Conference on Harmonisation of Technical Requirements for Registration of Pharmaceuticals for Human Use. ICH harmonised tripartite guideline. Validation of analytical procedures: text and methodology. 2005. http://www.ich.org/fileadmin/Public_Web_Site/ICH_Products/Guidelines/Quality/Q2_R1/Step4/Q2_R1__Guideline.pdf

Part V

Precision Medicine Clinical Trials and FDA-Approved Targeted Therapies

Prominent Precision Medicine Clinical Trials in Oncology Around the World

52

George Louis Kumar

In 2015, President Obama described "precision medicine (PM)" as "… health care tailored to you." [1] Later in the year, the White House further elaborated precision medicine as "…an innovative approach… that enables health care providers to tailor treatment and prevention strategies to people's unique characteristics; including their genome sequence, microbiome composition, health history, lifestyle, and diet." A more technical definition from Jameson and Longo in the *New England Journal of Medicine* describes precision medicine as "…treatments targeted to the needs of individual patients on the basis of genetic, biomarker, phenotypic, or psychosocial characteristics that distinguish a given patient from other patients with similar clinical presentations. Inherent in this definition is the goal of improving clinical outcomes for individual patients and minimizing unnecessary side effects for those less likely to have a response to a particular treatment." [2]

The term "precision medicine" (instead of the old phrase "personalized medicine") was used and recommended by Stephen J. Galli, MD, then Chair of Pathology at the Stanford University School of Medicine, Stanford, CA, USA, in a 2011 report of the US National Research Council (NRC) meeting – "Toward Precision Medicine" [3, 4]. Today, the terms precision medicine, individualized medicine, P4 (predictive, preventive, participatory, and personalized) medicine, precision health, precision oncology, precision immunotherapy, precision immunomedicine, precision immunotherapeutics, genomics medicine, predictive medicine, molecular medicine, and tailored therapy are often used interchangeably to describe the same meaning. All of these encompass an attempt to understand the disease at a molecular (i.e., genomic/epigenomic or immunological) level in order to develop "targeted therapy" with a technology-driven and participant-centered approach in addition to using traditional clinical information such as symptoms, personal history, patient history, and histology of the sample in order to tailor medical care to individual patients while sparing the patient the unnecessary treatments and costs that are unlikely to prove beneficial [5–7].

Precision medicine reaches beyond genomics and into the fields of gene therapy and immunology. The new frontier of immune targeting of cancer cells in the context of precision medicine ideology has spawned a promising new field called "Precision Immunomedicine (PIM)" [7]. Two excellent examples of precision immunomedicine are the new FDA-approved "made-to-order personalized drugs": sipuleucel-T (PROVENGE®) – a first-in-class therapeutic autologous vaccine approved for the treatment of men with asymptomatic or minimally

G. L. Kumar
Targos Inc., Issaquah, WA, USA
e-mail: george.kumar@targos-inc.com

S. Badve, G. L. Kumar (eds.), *Predictive Biomarkers in Oncology*,
https://doi.org/10.1007/978-3-319-95228-4_52

symptomatic castration-resistant metastatic prostate cancer – and tisagenlecleucel (Kymriah®), the first genetically modified autologous T-cell immunotherapy drug used in the treatment of acute lymphoblastic leukemia. Tisagenlecleucel (Kymriah®) based on gene therapy on immune cells is dubbed as a "living drug" because these genetically modified cells can persist in the body of patients for years, seeking out and attacking cancer cells as they arise. The approval of these drugs represents a paradigm shift in cancer therapeutics, especially with tisagenlecleucel because of their high remission rate of 83% within 3 months in some patients [8]. Currently, there are hundreds of these precision medicine immunotherapy clinical trials underway to evaluate new pharmaceuticals and biologics. (See Table 52.1 for some prominent trials and visit the ClinicalTrials.gov for more information.)

To advance the field of precision medicine, many countries have now launched government-driven projects focusing on this topic, including the Precision Medicine Initiative from the US White House, the US National Cancer Institute's "Cancer Moonshot program^{SM}," and HORIZON 2020 Work Programme from the European Commission.

Precision Medicine Clinical Trials

The development of a new drug, whether conventional or biologic, is lengthy, expensive, and risky. Currently, to bring a new drug to market, it costs between $ 1 and 2.5 billion (US), takes around 12–15 years, and involves the recruiting of a few hundred to thousands of patient volunteers. Despite this high cost and time needed to recruit new patients, 60–70% of phase III oncology trials either have negative results or fail to complete their course, resulting in raising costs of the drug and delaying or canceling the introduction of new and more effective therapies. One reason for this failure is the poor identification of the right patients at the right time, combined with the right drug and dosage that are most likely to respond to a therapeutic agent. With the advent and significant advances in tumor biology, multiplex genomics, imaging, bioinformatics, systems biology, immunotherapy, gene therapy, and related disciplines, pathologists and oncologists are now able to stratify subsets of cancer patients (Fig. 52.1) and make informed therapeutic decisions using precision medicine. Because of this "-omics" revolution, and the advent of the use of biomarkers and companion diagnostics in stratifying patients, the old paradigm of "reactive medical care" (i.e., diagnosis first, followed by iterative switching of drugs during therapy) is being replaced by targeted therapy and/or immunotherapy, thus leading to a decrease in adverse drug reactions, increase in drug safety, and the design of smaller, quicker, and smarter clinical trials [10, 11].

According to one report published in 2015, there were about 9094 adult interventional cancer trials registered on ClinicalTrials.gov – a publicly available clinical trial database developed and maintained by the United States National Library of Medicine [12]. Of the 9094 adult interventional cancer trials identified, only 795 (8.7%) were classified as precision cancer medicine trials, and the remaining 8299 (91.3%) were classified as non-precision cancer medicine trials. The precision medicine trials included both genetic alterations (684 out 8299 (7.5%) and protein alterations (111 out of 8299 (1.2%). However, the study did not include non-US trials, trials that involved gene expression profiling, or immunohistochemistry. It also included only one database that may have missed out many genetic alterations. In addition, the report excluded immunotherapy studies, "made-to-order personalized drugs" such as CAR-T therapy, ADAPT Trial of Autologous Dendritic Cell Immunotherapy, and sipuleucel-T (PROVENGE®).

A broad overview of some prominent, past and present, precision medicine clinical trials conducted around the world using the "-omics" technologies, immunohistochemistry, predictive biomarkers, cytogenetics, fluorescence in situ hybridization, and immuno- and gene therapy is provided in Table 52.1. This table is not intended to be an all-inclusive list, but helps to

Table 52.1 Prominent precision medicine clinical trials around the world (present and past)

Study name	Goal Tumor type	Location	Reference
American Society of Clinical Oncology (ASCO) study			
TAPUR The Targeted Agent and Profiling Utilization Registry	Testing the use of FDA-approved drugs that target a specific abnormality in a tumor gene in people with advanced stage cancer	ASCO, Alexandria, VA, USA	American Society of Clinical Oncology (ASCO)
	Lymphoma (non-Hodgkin), multiple myeloma, advanced solid tumors	In collaboration with: AstraZeneca Bayer Bristol-Myers Squibb Eli Lilly Genentech Merck Pfizer	http://www.tapur.org/ ClinicalTrials.gov Identifier NCT02693535 For details of the genes involved in the trial, refer to https://beta.mycancergenome.org/content/clinical_trials/NCT02693535/
Asian studies			
VIKTORY The Screening Protocol for The Targeted Agent Evaluation in Gastric Cancer Basket Korea Study: SMC-AZ GC Basket Trial Screening Protocol	To screen metastatic gastric cancer (GC) patients with RPTOR independent companion of MTOR, complex 2 (RICTOR) amplification who failed or progressed on first-line chemotherapy. This is a biomarker-based umbrella trial in GC	Samsung Medical Center, Seoul, South Korea	Jeeyun L et.al. *J Clin Oncol* 35, 2017 (suppl; abstr 4024) ClinicalTrials.gov Identifier NCT02299648
LC-SCRUM Lung Cancer Genomic Screening Project for Individualized Medicine in Japan	A nationwide lung cancer genomic screening project in Japan to identify SCLC patients harboring targetable genomic alterations such as TP53/RB1, MYC/MYCL1/MYCN, EGFR, KRAS, PI3K/AKT/mTOR, PTEN, and TSC2 for the development of novel targeted therapies	Shizuoka Cancer Center, Shizuoka, Japan	Haruyasu M et.al. *J Clin Oncol* 35, 2017 (suppl; abstr 8518)
ToGa Trastuzumab for Gastric Cancer	A Phase Ib/II Study of First-Line Pembrolizumab in Combination with Trastuzumab, Capecitabine, and Cisplatin in HER2-Positive Gastric Cancer	Yonsei University, South Korea	ClinicalTrials.gov Identifier: NCT02901301
Australian studies			
IMPaCT *I*ndividualized *M*olecular *P*ancreatic *C*ancer *T*herapy	The IMPaCT trial screened patients with advanced pancreas cancer for three genetic phenotypes matched to precision treatment: Her2 amplification, KRAS wild type, and DNA damage repair pathway defects. The study closed in December 2015, having delivered personalized treatment to only one patient	Brisbane, Australia	Clinical Trial Registry Number ACTRN12612000777897 Lorraine AC et.al. *J Clin Oncol* 35, 2017 (suppl 4S; abstract 314)

(continued)

Table 52.1 (continued)

Study name	Goal / Tumor type	Location	Reference
Canadian studies			
IMPACT (Integrated Molecular Profiling in Advanced Cancer Trial) and COMPACT (Community Oncology Molecular Profiling in Advanced Cancer Trial)	To provide molecular profiling data to the treating physician for patients with advanced breast, NSCLC, colorectal, genitourinary, pancreatobiliary gastrointestinal, upper aerodigestive tract, gynecological, melanoma, unknown primary, and rare carcinomas. Genotyping assays including AKT1, HRAS, AKT2, JAK2, AKT3, KIT, BRAF, KRAS, CDK, MEK1, CTNNB1, MET, EGFR, NOTCH1, ERBB2, NRAS, FGFR1, PDGFRA, FGFR2, PIK3CA, FGFR3, RET, FGFR4, SMO, and STK11	University Health Network, Toronto, Canada The Princess Margaret Cancer Centre, Toronto, Canada	Stockley et al. *Genome Medicine* (2016) 8:109 ClinicalTrials.gov Identifier: NCT01505400
CAPTUR Canadian Profiling and Targeted Agent Utilization Trial	Non-Hodgkin lymphoma Multiple myeloma Advanced solid tumors The Canadian Profiling and Targeted Agent Utilization Trial (CAPTUR) will test the activity of a list of commercially available targeted agents in patients who have undergone tumor profiling by one of the abovementioned programs and have "druggable" changes identified in their cancers	Various sites in Canada in collaboration with AstraZeneca Boehringer Ingelheim Bristol-Myers Squibb Hoffmann-La Roche Pfizer	ClinicalTrials.gov Identifier:NCT03297606 Implementing Personalized Cancer Genomics in Clinical Trials. Lillian L Sui, Special Sessions. Educational Track. 2016 ASCO Annual Meeting For a list of genes required for profiling, refer to ClinicalTrials.gov Identifier: NCT03297606
IND206 A Phase II Study of Sunitinib and Temsirolimus in Patients with Rare Tumours	Exploratory genetic cohort for sunitinib: rare tumors with somatic or germline mutations in sunitinib targets such as VEGFR, PDGFR, KIT, and RET	Canadian Cancer Trials Group	ClinicalTrials.gov Identifier:NCT01396408
	Exploratory genetic cohort for temsirolimus: rare tumors arising from known or suspected germline mutations in mTOR pathway such as PTEN, TS1/2, LKB1, and NF1/2 or somatic mutations in the mTOR pathway such as mutation or amplification of P13K or AKT	In collaboration with Pfizer	Dancey J, et al., *J Clin Oncol* 33, 2015 (suppl; abstr 2594)
European Union studies			
CREATE Cross-tumoral Phase II with Crizotinib	Cross-tumoral phase II clinical trial exploring crizotinib in patients with advanced tumors induced by causal alterations of ALK and/or MET Anaplastic large cell lymphoma Inflammatory myofibroblastic tumor Papillary renal cell carcinoma type 1 Alveolar soft part sarcoma Clear cell sarcoma Alveolar rhabdomyosarcoma	European Organisation for Research and Treatment of Cancer – EORTC A Pan-European organization connecting leading clinical centers across Europe	ClinicalTrials.gov Identifier:NCT01524926

WINTHER Worldwide Innovative Network Therapeutics	WINTHER study will explore matched tumoral and normal tissue biopsies for predicting efficacy of drugs. The aim is to provide a rational personalized therapeutic choice to metastatic patients of lung, breast, colon, head and neck, kidney, liver, and rhabdomyosarcomas harboring oncogenic events such as mutations, translocations, amplifications, etc.	Gustave Roussy, Cancer Campus, Grand Paris, France	ClinicalTrials.gov Identifier:NCT01856296
		Spain, Israel, Canada, and the USA	Jean-Charles Soria et al. *J Clin Oncol* 35, 2017 (suppl; abstr TPS11625)
			Rodon J, et al. *Ann Oncol* (2015), 26*(8)*, 1791–1798
AURORA	Aiming to understand the molecular aberrations in metastatic breast cancer by performing high coverage targeted gene and RNA sequencing	Breast European Adjuvant Studies Team, Brussels, Belgium	ClinicalTrials.gov Identifier: NCT02102165
Aiming to Understand the Molecular Aberrations in Metastatic Breast Cancer			
SPECTAColor	Pan-European Biomarker Screening Platform for KRAS, NRAS, BRAF, PI3K, and MSI in advanced colorectal cancer (CRC)	EORTC	Folprecht G, et al. *J Clin Oncol* 33, 2015 (suppl 3; abstr 575)
Screening Patients for Efficient Clinical Trial Access in advanced colorectal cancer			ClinicalTrials.gov Identifier: NCT01723969
SPECTALung	Screening patients with thoracic (lung, pleural mesothelioma, and thymic) tumors for efficient clinical trials access	EORTC and European Thoracic Oncology Platform (ETOP)	Besse B, et al. *Ann Oncol* (2015) 26 (suppl_1): i44
Screening Patients with Thoracic Tumors for Efficient Clinical Trials Access			ClinicalTrials.gov Identifier: NCT02214134
The Drug Rediscovery Protocol (DRUP trial)	This Dutch national study aims to simplify patient access to approved targeted therapies collaborating with pharmaceutical companies and to perform next-generation sequencing on tumor biopsies for biomarker analyses	Sponsor: The Netherlands Cancer Institute	ClinicalTrials.gov Identifier: NCT02925234
	Patients from the Netherlands and the USA will be included in two independent protocols (DRUP and TAPUR), allowing data exchange and empowering of both trials	Collaborators: Amgen AstraZeneca Bayer Bristol-Myers Squibb Novartis Hoffmann-La Roche Pharma AG	Van Der Velden DP, et al. *J Clin Oncol* 35, 2017 (suppl; abstr 2547)

(continued)

Table 52.1 (continued)

Study name	Goal / Tumor type	Location	Reference
French studies			
SHIVA A randomized proof-of-concept phase II trial comparing therapy based on tumor molecular profiling versus conventional therapy in patients with refractory cancer	Any type of cancer that is refractory to standard of care	Institut Curie, Paris, France	Le Tourneau C, et al. *Lancet Oncol.* 2015;16(13):1324–1334 ClinicalTrials.gov Identifier: NCT01771458
MOST My Own Special Treatment	A two-period, multicenter, randomized, open-label, phase II study evaluating the clinical benefit of a maintenance treatment targeting tumor molecular alterations in patients with progressive locally advanced or solid metastatic tumors	Centre Leon Berard, Lyon, France	ClinicalTrials.gov Identifier: NCT02029001 Toulmonde M, et al. *J Clin Oncol* 33, 2015 (suppl; abstr TPS2622)
SAFIR 01 High-throughput technologies (HTT) to drive breast cancer patients to specific phase I/II trials of targeted agents	HTT is used to identify deregulated genes. This study led to the conclusions that HTT is feasible (80%) and robust and identify "targetable" genomic alterations in around 40% of samples allowing for a more tailored targeted regimen	UNICANCER Gustave Roussy, Cancer Campus, Grand Paris	ClinicalTrials.gov Identifier: NCT01414933
SAFIR 02 Breast Breast – evaluation of the efficacy of high-throughput genome analysis as a therapeutic decision tool for patients with metastatic breast cancer	Open-label multicentric phase II randomized trial to evaluate whether treatment with targeted agents guided by HTT improves progression-free survival as compared to standard maintenance therapy in patients with metastatic breast cancer	UNICANCER Gustave Roussy, Cancer Campus, Grand Paris	ClinicalTrials.gov Identifier: NCT02299999
SAFIR 02 Lung Evaluation of the efficacy of high-throughput genome analysis as a therapeutic decision tool for patients with metastatic non-small cell lung cancer	To evaluate whether treatment with targeted agents guided by HTT improves progression-free survival as compared to standard maintenance therapy in patients with metastatic NSCLC	UNICANCER Gustave Roussy, Cancer Campus, Grand Paris	ClinicalTrials.gov Identifier:NCT02117167
MOSCATO (01,02) Molecular Screening for Cancer Treatment Optimization	To use high-throughput molecular analysis (CGH array and sequencing) to treat patients with metastatic cancer with targeted therapeutics to improve the progression-free survival compared to the previous treatment line Metastatic solid tumors (any localization)	Gustave Roussy, Cancer Campus, Grand Paris	ClinicalTrials.gov Identifier:NCT01566019

MEGA Met or EGFR Inhibition in Gastroesophageal Adenocarcinoma	FOLFOX alone or in combination with AMG 102 or panitumumab as first-line treatment in patients with advanced gastroesophageal adenocarcinoma: FNCLCC-FFCD-AGEO-GERCOR PRODIGE 17-ACCORD 20 randomized phase II trial To identify candidate predictive and prognostic biomarkers among functional of molecular alterations of the EGFR/RAS/RAF and HGF/c-Met pathways	Sponsor: UNICANCER	ClinicalTrials.gov Identifier:NCT01443065 Malka D, et al., *J Clin Oncol* 29: 2011 (suppl; abstr TPS178)
United Kingdom (UK)			
Lung MATRIX Multi-drug, Genetic Marker-directed, Non-comparative, Multi-centre, Multi-arm Phase II Trial in Non-small Cell Lung Cancer	Non-small cell lung cancer carcinoma, squamous cell adenocarcinoma	University of Birmingham	ClinicalTrials.gov Identifier: NCT02664935 Middleton G, et al. *Ann Oncol* 2015;00:1–6.
FOCUS 4 Molecularly stratified randomized controlled trial program for patients with advanced colorectal	Advanced colorectal cancer FOCUS4-A: Mutated BRAF FOCUS4-B: Mutated PIK3CA FOCUS4-C: Mutated KRAS or NRAS genes FOCUS4-D: No mutation sub-type in the BRAF, PIK3CA, KRAS, and NRAS genes FOCUS4-N: Nonclassified, the tests fail to work, so it is not possible to classify the tumor as any one of the specific sub-types above	The Medical Research Council (MRC), London, UK	Richman SD, et al. *J Clin Pathol* 2015;0:1–7. doi:https://doi.org/10.1136/jclinpath-2015-203097
CRUK SMP Cancer Research UK Stratified Medicine Programme	Screening of non-small cell lung (NSCLC) cancer patients by next-generation sequencing (NGS) for clinically actionable gene changes	Cancer Research UK, London	Cancer Research UK http://www.cancerresearchuk.org/about-cancer/find-a-clinical-trial/a-study-looking-how-to-test-genes-in-lung-cancer-cells-smp2 Poster http://www.winsymposium.org/wp-content/uploads/2015/07/P7.06-Rebecca-Cummings.pdf
REAL-III Randomized Trial of EOC ± Panitumumab for Advanced and Locally Advanced Esophagogastric Cancer	To determine whether adding panitumumab, an antibody against the epidermal growth factor receptor (EGFR), to standard chemotherapy with epirubicin, oxaliplatin, and capecitabine (EOX), improves the duration of survival of patients with advanced stomach and esophageal cancer	Royal Marsden NHS Foundation Trust, London, UK	ClinicalTrials.gov Identifier:NCT00824785

(continued)

Table 52.1 (continued)

Study name	Goal Tumor type	Location	Reference
OCTOPUS Ovarian Cancer Trials of Weekly Paclitaxel	A trial of AZD2014 and paclitaxel for women with ovarian cancer that has come back	Cancer Research UK Clinical Trials Unit	UK Clinical Trials Gateway ISRCTN16426935 http://www.isrctn.com/ISRCTN16426935
	AZD2014 is a potent (IC50 2.81 nM), selective (inactive against 220 other kinases) inhibitor of mTOR kinase	The Beatson West of Scotland Cancer Centre	EudraCT number: 2014-005221-12
	Immunohistochemistry staining in eight of ten evaluable paired tumor biopsies indicating mTORC1 activity	1053 Great Western Road, Glasgow, G12 0YN, UK	
TOPARP	The trial aims to evaluate the antitumor activity of olaparib in metastatic castration-resistant prostate cancer, identify molecular signatures of tumor cells in responding and nonresponding patients, and identify predictive biomarkers of olaparib response	The Institute of Cancer Research, London, UK	Clinical Trial Registry Number: NCT01682772
Trial of Olaparib in Patients with Advanced Castration-Resistant Prostate Cancer		AstraZeneca, Cambridge, UK	Mateo J, et al. *J Clin Oncol* 35, 2017 (suppl 6S; abstract 141)
National Lung MATRIX trial	The trial consists of a series of parallel multicenter single arm phase II trial arms, each testing an experimental targeted drug in a population stratified by multiple pre-specified actionable target putative biomarkers	Sponsor: University of Birmingham	ClinicalTrials.gov Identifier:NCT02664935
	Drug: AZD4547 FGFR inhibitor	Collaborators: Cancer Research UK AstraZeneca Pfizer Experimental Cancer Medicine Centre Network	For other active clinical trial programs in solid tumors, please refer to the following link: http://ciic.org.uk/clinical-themes/solid-tumours/solid-tumoursclinical-trial-programme/
	Drug: AZD2014 MTORC1/2 inhibitor		
	Drug: Palbociclib CDK4/6 inhibitor		
	Drug: Crizotinib ALK/MET/ROS1 inhibitor		
	Drug: Selumetinib MEK inhibitor		
	Drug: AZD5363 AKT inhibitor		
	Drug: AZD9291 EGFRm+ T790M+ inhibitor		
	Drug: MEDI4736 Anti-PDL1 Other Name: Durvalumab		

United States of America (USA)			
NCI-MATCH National Cancer Institute-Molecular Analysis for Therapy Choice	Genetic testing in treating patients with advanced solid tumors, lymphomas, or myeloma once they have progressed on standard treatment for their cancer or if they have rare cancer for which there is no standard treatment	National Cancer Institute, Rockville, MD, USA	ClinicalTrials.gov Identifier: NCT02465060
	Address molecular abnormalities in EGFR, ERBB2, PIK3CA, PTEN, ALK, ROS, BRAF, NRAS, NF2, KIT, NF1, SMO, PTCH1, MET, FGFR, AKT, NTRK, cell cycle, and PI3K pathway as well as other less common molecular abnormalities	Participating labs include: Foundation Medicine, Inc. Caris Life Sciences MD Anderson Cancer Center Memorial Sloan-Kettering Cancer Center	Conley BA, et.al. *J Clin Oncol* 34, 2016 (suppl; abstr TPS2606)
ALCHEMIST Adjuvant Lung Cancer Enrichment Marker Identification and Sequencing Trials	Randomized clinical trials for patients with early-stage non-small cell lung cancer (NSCLC) whose tumors have been completely removed by surgery	National Cancer Institute (NCI) and Center for Cancer Genomics (CCG). Rockville, MD, USA	ClinicalTrials.gov Identifier: NCT02194738
ALCHEMIST currently consists of three integrated protocols: ALCHEMIST Screening (A151216; NCT02194738), ALCHEMIST-EGFR (A081105; NCT02193282), and ALCHEMIST-ALK (E4512; NCT02201992)	Tested biomarkers include EGFR, ALK, PD1, PD-L1, and CTLA-4		Gerber DE, et al. *J Clin Oncol* 33, 2015 (suppl; abstr TPS7583)
Lung-MAP Lung Cancer Master Protocol	Master "umbrella" protocol in partnership between government, academia, patient advocacy organizations, and industry, designed to simultaneously and independently test multiple biomarker-driven (PI3KCA, CDK4, CCND1–3, FGFR 1–3, MET) therapies for patients (pts) with chemo-refractory squamous cell lung cancer (SCCA)	Southwest Oncology Group (SWOG) and NCTN, USA	Clinical Trial Registry Number: NCT02154490
			Papadimitrakopoulou V, et al. *J Clin Oncol* 34, 2016 (suppl; abstr 9088)
			Abrams J., Conley B., Mooney M, et al. National Cancer Institute's Precision Medicine Initiatives for the New National Clinical Trials Network. *2014 Am Soc Clin Oncol Educ Book*. 2014:71–76

(continued)

Table 52.1 (continued)

Study name	Goal Tumor type	Location	Reference
NCI-MPACT	Analytical validation study of a targeted next-generation sequencing mutation-detection assay used for patient selection in advanced malignant solid neoplasm and recurrent malignant solid neoplasm. MPACT is designed to assess whether response rate (CR+PR) and progression-free survival (PFS) are improved following treatment with WEE1 inhibitor MK-1775, everolimus, trametinib, temozolomide, and veliparib based on the presence of specific actionable mutations of interest in 20 genes belonging to 3 pathways: DNA repair, PI3K, and RAS/RAF	National Cancer Institute, Rockville, MD, USA	ClinicalTrials.gov Identifier: NCT01827384
National Cancer Institute-Molecular Profiling-Based Assignment of Cancer Therapy for Patients with Advanced Solid Tumors			Lih CJ, et.al. *J Mol Diagn*. 2016 Jan;18(1):51–67
NCI-COG Pediatric MATCH	This is a nationwide clinical trial for children and adolescents ages 1–21 who have solid tumors, including non-Hodgkin lymphomas, brain tumors, and histiocytoses, that no longer respond to standard treatment or have recurred after treatment. Pediatric MATCH will use a single sequencing test to screen for alterations in more than 160 genes associated with cancer at once	Biopsy specimens from all patients will be sent to the COG Biopathology Center at Nationwide Children's Hospital in Columbus, Ohio, for DNA and RNA processing. Sequencing analysis will be done at the MD Anderson Cancer Center and the Frederick National Laboratory for Cancer Research	ClinicalTrials.gov Identifier: NCT03155620
(Molecular Analysis for Therapy Choice)			News in Brief. Pediatric MATCH Trial Opens Enrollment. *Cancer Discov.* 2017 Aug 9. doi: https://doi.org/10.1158/2159-8290.CD-NB2017-113
			https://www.cancer.gov/about-cancer/treatment/clinical-trials/nci-supported/pediatric-match
Exceptional responders study	A study to investigate the molecular factors of tumors associated with exceptional treatment responses (i.e., unexpected and prolonged remissions following standard therapy) of cancer patients to drug therapies	National Cancer Institute (NCI). Rockville, MD, USA	ClinicalTrials.gov Identifier: NCT02243592
Molecular profiling of tumors from cancer patients who are exceptional responders		Isolation of DNA and RNA from tissue samples at the Biospecimen Core Resource at Nationwide Children's Hospital, Columbus, Ohio	https://www.nih.gov/news-events/news-releases/nih-exceptional-responders-cancer-therapy-study-launched
		DNA sequencing and analysis center at Baylor College of Medicine	

The Molecular Profiling Protocol trial (SCRI-CA-001, also known as Bisgrove trial)	All tumors	Sponsor: Scottsdale Healthcare, Scottsdale, Arizona, USA	ClinicalTrials.gov Identifier: NCT00530192
	A prospective clinical trial whose purpose is to validate this approach by comparing time on therapy (TOT) using a treatment regimen selected by molecular profiling (using IHC, FISH, and gene expression microarray) with TOT for the most recent regimen on which the patient has just progressed	Collaborator: Translational Genomics Research Institute, Phoenix, AZ, USA	Biankin AV, et al. *Nature* 526,361–370. (15 October 2015). doi:https://doi.org/10.1038/nature15819
IMPACT Initiative for Molecular Profiling in Advanced Cancer Therapy	Molecular profiling of patients with advanced cancer, with the goals of (1) providing a comprehensive characterization of the molecular profiles of individual patients and (2) correlating molecular profile with response to phase I therapies	Sponsor: M.D. Anderson Cancer Center, Houston, TX, USA	ClinicalTrials.gov Identifier: NCT00851032
		Collaborator: National Institute of Health's National Center for Research Resources (NCRR), Bethesda, MD, USA	Tsimberidou AM, et al. *J Clin Oncol* 35, 2017 (suppl; abstr 2513)
BATTLE Biomarker-Integrated Approaches of Targeted Therapy for Lung Cancer Elimination	This study is the "umbrella" study in a group of five studies known as the BATTLE	Sponsor: M.D. Anderson Cancer Center, Houston, TX, USA	ClinicalTrials.gov Identifier: NCT00409968 NCT00411671 NCT00411632 NCT00410059 NCT00410189
	The BATTLE-2 trial is a study that is designed to understand how tissue biomarkers (EGFR, KRAS/BRAF, VEGF/VEGFR2, RXR/CyclinD1) predict who will respond to targeted therapy or targeted therapy combinations in NSCLC	Collaborator: United States Department of Defense	Kim ES, et al. *Cancer Discov* 2011;1:44
I-SPY (1–3) Investigation of Serial Studies to Predict Your Therapeutic Response with Imaging and Molecular Analysis	Breast cancer. Tumor assay profile includes ER and HER2 status by IHC, FISH, and MammaPrint 70-gene signature test (Agendia)	Sponsor: QuantumLeap Healthcare Collaborative	ClinicalTrials.gov identifier: NCT01042379
Lung Cancer Mutation Consortium Protocol	The primary objective of this protocol is to determine the frequency of oncogenic mutations in patients with advanced adenocarcinoma of the lung	University of Colorado, Denver	ClinicalTrials.gov Identifier: NCT01014286
	The primary endpoint of this protocol is the mutation rate		Kris MG, et al. *JAMA*. 2014 May 21;311(19):1998–2006
	Testing for oncogenic drivers, KRAS, EGFR, ALK, ERBB2, BRAF, PIK3CA, MET, NRAS, MEK, and AKT		

(continued)

Table 52.1 (continued)

Study name	Goal Tumor type	Location	Reference
I-PREDICT	An open-label navigational investigation of molecular profile-related evidence determining individualized cancer therapy for patients with incurable malignancies and poor prognosis	Sponsor: University of California, San Diego	ClinicalTrials.gov Identifier: NCT02534675
Study of Molecular Profile-Related Evidence to Determine Individualized Therapy for Advanced or Poor Prognosis Cancers	The primary objective of this protocol is to determine the frequency of oncogenic mutations in patients with advanced adenocarcinoma of the lung	Collaborator:	
	The primary endpoint of this protocol is the mutation rate	Foundation Medicine	
WISDOM	This University of California (UC) Health study aims to uncover whether annual mammograms really are the best way to screen for breast cancer or whether a more personalized approach—driven by the data attached to each woman's genetic makeup, family history, and risk factors—could deliver better results	Sponsor: Athena Breast Health Network University of California, San Francisco	ClinicalTrials.gov Identifier: NCT02620852
Women Informed to Screen Depending on Measures of Risk			https://wisdom.secure.force.com/portal/
		Collaborators: Patient-Centered Outcomes Research Institute Robert Wood Johnson Foundation Color Genomics, Inc. Salesforce	The ASCO Post. September 25, 2017. p. 96
The **T**rial **A**ssigning **I**ndividua**L**ized	TAILORx seeks to incorporate a molecular profiling into clinical decision-making and thus spare women unnecessary treatment if chemotherapy is not likely to be of substantial benefit Breast adenocarcinoma Estrogen receptor and/or progesterone receptor positive HER2/Neu negative	Sponsor: National Cancer Institute (NCI)	ClinicalTrials.gov Identifier: NCT00310180
Options for Treatment (**Rx**), or TAILORx, Breast Cancer Trial			Sparano JA, et al. *N Engl J Med* 2015;373:2005–14.
Program for the Assessment of Clinical Cancer Tests (PACCT-1): Trial Assigning Individualized Options for Treatment – The TAILORx Trial		Collaborators: American College of Surgeons Cancer and Leukemia Group B NSABP Foundation Inc	
TAILORx is one of the first trials to examine a methodology for personalizing cancer treatment		NCIC Clinical Trials Group	
		North Central Cancer Treatment Group	
		Southwest Oncology Group	

MINDACT Microarray In Node-Negative and 1–3 Positive Lymph Node Disease May Avoid Chemotherapy	A prospective, randomized study comparing the 70-gene signature with the common clinical-pathological criteria in selecting patients for adjuvant chemotherapy in breast cancer with 0–3 positive nodes	The Netherlands Cancer Institute, Antoni van Leeuwenhoek Hospital, Amsterdam, Netherlands Sponsor: European Organisation for Research and Treatment of Cancer – EORTC Collaborators: Agendia Breast International Group (BIG) Roche Pharma AG Novartis Sanofi	ClinicalTrials.gov Identifier: NCT00433589 Cardoso F, et al. *N Engl J Med* 2016; 375:717–729
BEAUTY Breast Cancer Genome Guided Therapy Study	The purpose of this research study is to better understand the reasons why or why not breast cancers are destroyed by standard chemotherapy. This information will be used to develop new and better cancer therapies Identification of novel somatic changes within gene and gene pathways Determination of the frequency of known tumor mutations for which current drug therapies already exist (e.g., BRAF, C-KIT, EGFR mutation, KRAS, PTEN, PI3K)	Sponsor: Mayo Clinic	ClinicalTrials.gov Identifier: NCT02022202
NSABP B-47	Chemotherapy with or without trastuzumab after surgery in treating women with invasive breast cancer. Evaluation of HER2 status by IHC and FISH	Sponsor: National Cancer Institute	ClinicalTrials.gov Identifier: NCT01275677
Representative pharmaceutical industry-sponsored trials			
Novartis Signature Trial Program	US-based tissue-agnostic pilot study recruiting patients using a targeted screening approach to precisely match patients with investigational treatments based on their distinct tumor profiles	Sponsor: Novartis	The Novartis Signature Trial Program http://www.trials.novartis.com/en/clinical-trials/us-oncology/oncology/signature/about/

(continued)

Table 52.1 (continued)

Study name	Goal Tumor type	Location	Reference
My Pathway	An open-label phase IIa study evaluating trastuzumab/pertuzumab (HER2), erlotinib (EGFR), vemurafenib/cobimetinib (BRAF V600), vismodegib (SMO/PTCH-1), alectinib (ALK), and atezolizumab (PD-L1, MSI-H) in patients who have advanced solid tumors with mutations or gene expression abnormalities predictive of response to one of these agents	Sponsor: Genentech	ClinicalTrials.gov Identifier: NCT02091141
VE BASKET	An open-label, multicenter study to assess the efficacy and safety of Zelboraf (vemurafenib) in patients with BRAF V600 mutation-positive cancers (solid tumors and multiple myeloma, except melanoma and papillary thyroid cancer) and for whom Zelboraf is deemed the best treatment option in the opinion of the investigator	Hoffmann-La Roche	ClinicalTrials.gov Identifier: NCT01524978 Hyman DM, et al. *J Clin Oncol* 35, 2017 (suppl; abstr 2004)
ToGa (Trastuzumab for Gastric Cancer)	A Study of Herceptin (Trastuzumab) in Combination with Chemotherapy Compared with Chemotherapy Alone in Patients with HER2-Positive Advanced Gastric Cancer	Chugai Pharmaceutical Hoffmann-La Roche	ClinicalTrials.gov Identifier: NCT01041404
HERA	Herceptin (Trastuzumab) in Treating Women with Human Epidermal Growth Factor Receptor (HER) 2-Positive Primary Breast Cancer	Sponsor: Hoffmann-La Roche Collaborators: Breast International Group European Organisation for Research and Treatment of Cancer – EORTC NCIC Clinical Trials Group International Breast Cancer Study Group	ClinicalTrials.gov Identifier: NCT00045032
LOGiC (Lapatinib Optimization Study in the HER2-Positive Gastric Cancer)	Lapatinib Optimization Study in ErbB2 (HER2) Positive Gastric Cancer: A Phase III Global, Blinded Study Designed to Evaluate Clinical Endpoints and Safety of Chemotherapy Plus Lapatinib	Novartis Pharmaceuticals, USA	ClinicalTrials.gov Identifier: NCT00680901
TyTAN (Tykerb with Taxol in Asian HER2-Positive Gastric Cancer)	Lapatinib in Combination with Weekly Paclitaxel in Patients with ErbB2 (Her2) Amplified Advanced Gastric Cancer	Novartis Pharmaceuticals, USA China Japan S. Korea Taiwan	Clinical Trial Registry Number: NCT00486954 Yung-Jue Bang et al. *J Clin Oncol* 31, 2013 (suppl 4; abstr11) Satoh et. Al., *J Clin Oncol* 32, no. 19 (July 2014) 2039–2049

EXPAND	Clinical outcome according to tumor HER2 status and EGFR expression in advanced gastric cancer patients	Merck KGaA, Darmstadt, Germany	Lordick F, et al. *J Clin Oncol* 31, 2013 (suppl; abstr 4021)
		University Cancer Center Leipzig, Leipzig, Germany	
AVAGAST (Avastin in Gastric Cancer)	A Study of Bevacizumab in Combination with Capecitabine and Cisplatin as First-Line Therapy in Patients with Advanced Gastric Cancer (AVAGAST)	Responsible Party Genentech, Inc. USA	ClinicalTrials.gov Identifier: NCT00548548
	Previous platinum or anti-angiogenic therapy (i.e., anti-vascular endothelial growth factor [VEGF] or VEGF receptor tyrosine kinase inhibitor, etc.)	Collaborators: Hoffmann-La Roche, Basel, SwitzerlandChugai Pharmaceutical, Japan	Shah MA, et al. *J Clin Oncol* 30, 2012 (suppl 4; abstr 5)
		Chugai Pharmaceutical, Japan	
RILOMET-1 rilotumumab for MET-positive gastric or gastroesophageal junction cancer	A Phase III, Multicenter, Randomized, Double-Blind, Placebo-Controlled Study of Rilotumumab (AMG102) with Epirubicin, Cisplatin, and Capecitabine (ECX) as First-Line Therapy in Advanced MET-Positive Gastric or Gastroesophageal Junction Adenocarcinoma	Amgen, Thousand Oaks, CA, USA	ClinicalTrials.gov Identifier: NCT01697072
			Cunningham D, et al. *J Clin Oncol* 33, 2015. (suppl; abstr 4000)
METGastric	METGastric: A phase III study of onartuzumab plus mFOLFOX6 in patients with metastatic HER2-negative (HER2-) and MET-positive (MET+) adenocarcinoma of the stomach or gastroesophageal junction (GEC)	Hoffmann-La Roche, USA	Clinical Trial Registry Number: NCT01662869
			Shah MA, et al. *J Clin Oncol* 33, 2015 (suppl; abstr 4012)
GATSBY	A Study of Trastuzumab Emtansine Versus Taxane in Participants with Human Epidermal Growth Factor Receptor 2 (HER2)-Positive Advanced Gastric Cancer	Hoffmann-La Roche Multicenter	ClinicalTrials.gov Identifier: NCT01641939
			Yoon-Koo Kang et al. *J Clin Oncol* 34, 2016 (suppl 4S; abstr 5)
SHINE	Efficacy and Safety of AZD4547 Versus Paclitaxel in Patients with Advanced Gastric or Gastro-oesophageal Cancer	AstraZeneca	ClinicalTrials.gov Identifier: NCT01457846
	FGFR2 polysomy or gene amplification		Yung-Jue Bang et al. *J Clin Oncol* 33, 2015 (suppl; abstr 4014)
RAINBOW	Ramucirumab plus paclitaxel versus placebo plus paclitaxel in patients with previously treated advanced gastric or gastroesophageal junction adenocarcinoma	Eli Lilly and Company	Wilke H, et al. *Lancet Oncol.* 2014 Oct;15(11):1224–35
A double-blind, randomized phase III trial for ramucirumab in combination with paclitaxel	VEGFR-2		

(continued)

Table 52.1 (continued)

Study name	Goal Tumor type	Location	Reference
REGARD	Ramucirumab monotherapy for previously treated advanced gastric or gastroesophageal junction adenocarcinoma	Eli Lilly and Company (ImClone Systems)	Fuchs CS, et al. *Lancet*. 2014 Jan 4;383(9911):31–39
An international, randomized, multicenter, placebo-controlled, phase III trial for ramucirumab monotherapy	VEGFR-2		
PANORAMA	This is an observational cohort study of patients with locally advanced or metastatic NSCLC (non-small cell lung cancer)	Patients will be recruited from participating sites in Europe, Asia, and Canada	ClinicalTrials.gov Identifier: NCT03053297
Observational Study of Patients with Locally Advanced or Metastatic NSCLC (Non-small Cell Lung Cancer)	The primary cohort will include patients with EGFR mutation-positive locally advanced or metastatic NSCLC who have progressed while on or after receiving frontline EGFR-TKI (tyrosine kinase inhibitors) therapy (e.g., gefitinib, erlotinib, afatinib, or icotinib)		
B2225	A basket trial to examine (a) the effect(s) of imatinib mesylate (Gleevec) treatment on life-threatening rare diseases with known associations to one or more imatinib mesylate-sensitive tyrosine kinases and (b) to identify the contribution of specific protein tyrosine kinases of that specific disease	Novartis Pharmaceuticals and Oregon Health and Science University	ClinicalTrials.gov Identifier: NCT00154388
			Chugh H, et al. *J Clin Oncol*. 2009 Jul 1;27(19):3148–53
			Erratum in: *J Clin Oncol*. 2009 Sep 20;27(27):4630. Myers, Paul A [corrected to Meyers, Paul A]
BRAF V600	Histology-independent phase II "basket" study of vemurafenib in BRAF V600 mutation-positive nonmelanoma cancers	Hoffmann-La Roche	ClinicalTrials.gov Identifier: NCT01524978
			Hyman DM, et al. *N Engl J Med*. 2015 Aug 20;373(8):726–36
IMagyn050	A study of atezolizumab versus placebo in combination with paclitaxel, carboplatin, and bevacizumab in participants with newly diagnosed stage III or stage IV ovarian, fallopian tube, or primary peritoneal cancer	Hoffmann-La Roche	ClinicalTrials.gov Identifier: NCT03038100
Neo ALTTO (Neoadjuvant Lapatinib and/or Trastuzumab Treatment Optimisation) Study	Phase III study comparing the efficacy of neoadjuvant lapatinib plus paclitaxel, versus trastuzumab plus paclitaxel, versus concomitant lapatinib and trastuzumab plus paclitaxel given as neoadjuvant treatment in HER2/ErbB2 overexpressing and/or amplified primary breast cancer	Novartis	ClinicalTrials.gov Identifier: NCT00553358
			Baselga J, et al. *Lancet*. 2012; S140–6736(11)
MONALEESA-7	Study of efficacy and safety in premenopausal women with hormone receptor-positive, HER2-negative advanced breast cancer	Novartis	ClinicalTrials.gov Identifier: NCT02278120

New precision immunotherapy studies			
ADAPT	An international phase III randomized personalized immunotherapy trial of (AGS-003) for the treatment of mRCC. In this trial, dendritic cells are extracted from mRCC patients. They are cultured, amplified, and electroporated with total tumor RNA plus synthetic CD40L RNA. The mature electroporated dendritic cells called ROCAPULDENCEL-T (or AGS-003) are reintroduced into the patient. The theory is that the mature electroporated dendritic cells, when injected back into the patient, will present an antigen to the subject's T cells that will then attack the tumor. Combined with sunitinib which also prevents RCC-induced accumulation of myeloid-derived suppressor cells (MDSC) and normalizes type 1 T-cell function, the treatment is effective in yielding a median overall survival (OS) of over 30 months in newly diagnosed mRCC patients	Argos Therapeutics, Durham, NC, USA	ClinicalTrials.gov Identifier: NCT01582672
Trial of Autologous Dendritic Cell Immunotherapy Plus Standard Treatment (with receptor tyrosine kinase inhibitor (RTKI) sunitinib) of Metastatic Renal Cell Carcinoma (mRCC)		Europe and Israel	ESMO 2017 http://www.esmo.org/Conferences/ESMO-2017-Congress/News-Articles/Rocapuldencel-T-Shows-Potential-Benefit-in-Metastatic-Renal-Cell-Carcinoma
			Figlin RA, et al. *J Clin Oncol* 33, 2015 (suppl; abstr TPS4582)
JULIET	Global CAR-T trial	Novartis Pharmaceuticals	ClinicalTrials.gov Identifier: NCT02445248
Study of Efficacy and Safety of CTL019 in Adult DLBCL Patients	A Phase II, Single Arm, Multicenter Trial to Determine the Efficacy and Safety of CTL019 in Adult Patients with Relapsed or Refractory Diffuse Large B-Cell Lymphoma (DLBCL)	Austria, Australia, Belgium, Canada, France, Germany, Italy, Japan, the Netherlands, and the USA	
There are over 227 clinical CAR-T trials			
ELIANA	Global CAR-T trial	Novartis Pharmaceuticals	ClinicalTrials.gov Identifier: NCT02435849
Determine Efficacy and Safety of CTL019 in Pediatric Patients with Relapsed and Refractory B-cell ALL	A Phase II, Single Arm, Multicenter Trial to Determine the Efficacy and Safety of CTL019 in Pediatric Patients with Relapsed and Refractory B-cell Acute Lymphoblastic Leukemia	Austria, Australia, Belgium, Canada, France, Germany, Italy, Japan, Norway, Spain, and the USA	
A Randomized Phase II Trial of Combining Sipuleucel-T with Immediate vs. Delayed CTLA-4 Blockade for Prostate Cancer	The purpose of this study is to find out what effects taking ipilimumab, as an immediate or delayed treatment, following completion of sipuleucel-T (SipT) treatment, has on patients and their prostate cancer	Sponsor: University of California, San Francisco	ClinicalTrials.gov Identifier: NCT01804465
		Collaborators: M.D. Anderson Cancer Center Bristol-Myers Squibb Dendreon	

(continued)

Table 52.1 (continued)

Study name	Goal Tumor type	Location	Reference
Clinical Study of Atezolizumab (Anti-PD-L1) and Sipuleucel-T in Patients Who Have Asymptomatic or Minimally Symptomatic Metastatic Castrate Resistant Prostate Cancer	The purpose of the study is to compare the safety and tolerability of sequential atezolizumab followed by sipuleucel-T (Arm 1) vs. sipuleucel-T followed by atezolizumab (Arm 2) in patients who have asymptomatic or minimally symptomatic metastatic CRPC, not previously treated with docetaxel or cabazitaxel	Sponsor: University of Hawaii Collaborators: Genentech, Inc. Dendreon	ClinicalTrials.gov Identifier: NCT03024216
KEYNOTE-001	Phase I Study of Single Agent Pembrolizumab (MK-3475) in Patients with Progressive Locally Advanced or Metastatic Carcinoma, Melanoma, and Non-Small Cell Lung Carcinoma	Sponsor: Merck Sharp & Dohme Corp	ClinicalTrials.gov Identifier: NCT01295827
Currently, there are more than 112 studies on different indications using the search word KEYNOTE on ClinicalTrials.gov	Programmed cell death 1 ligand (PD-L1) immunohistochemistry and gene expression		
CheckMate 067 Currently, there are more than 44 studies on different indications using the search word CheckMate on ClinicalTrials.gov	Phase III Study of Nivolumab or Nivolumab Plus Ipilimumab Versus Ipilimumab Alone in Previously Untreated Advanced Melanoma	The purpose of this study is to show that nivolumab and/or nivolumab in combination with ipilimumab will extend progression-free survival and overall survival compared to ipilimumab alone	ClinicalTrials.gov Identifier: NCT01844505

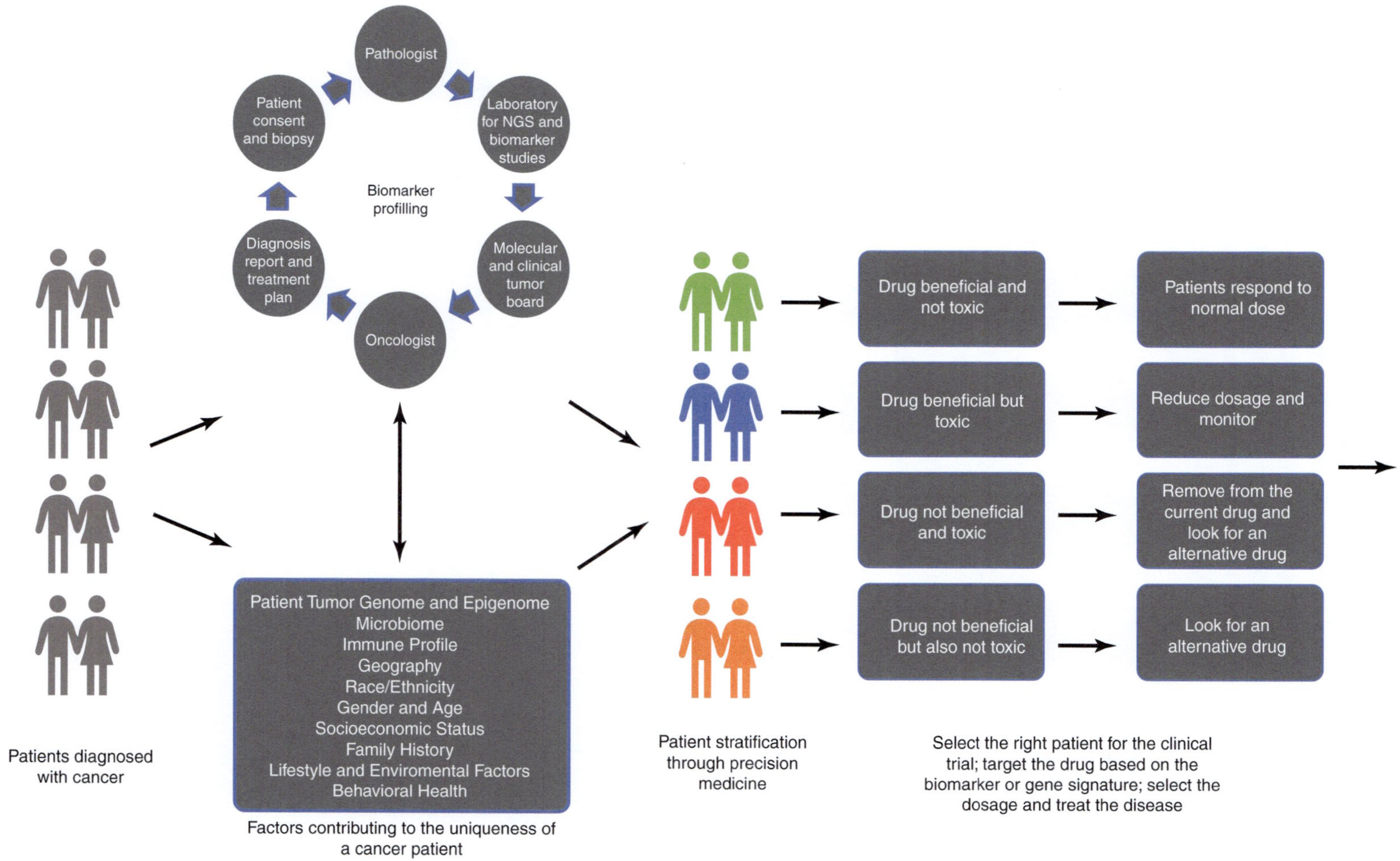

Fig. 52.1 The goal of precision medicine is to provide the *right patient* with the *right* drug at the *right* dose and at the *right time*. The figure above depicts key steps in a precision medicine clinical trial from patient consent to biopsy retrieval to biomarker profiling to patient stratification and treatment plan generation. Understanding the basis of variability in patients due to genome, epigenome, and microbiome differences, as well as other factors such as immune profiles, family history, or lifestyle patterns, is critical to our understanding of patients' response to medications and the potential for life-threatening drug toxicities. Adverse drug reactions are one of the leading causes of morbidity and mortality in hospitalized patients. The incidence of severe-/fatal-grade adverse reactions ranges anywhere from 1% to 40% depending on the nature of the drug therapy [9]. If clinicians could better predict *efficacy*, dosing, and drug-related toxicities, while also identifying those most likely to benefit, the overall care of patients could be improved considerably

illustrate major trials targeting different indications. Some clinical trials such as the HERA trial have already succeeded in bringing blockbuster drugs into the market (e.g., trastuzumab), whereas others are just beginning to reveal their potential (e.g., CAR-T therapy:tisagenlecleucel and axicabtagene ciloleucel due to their high remission rates in childhood leukemia or large B-cell lymphoma, respectively). There are many trials whose outcomes are not yet known (e.g., ADAPT trial for autologous dendritic cell immunotherapy), while other trials have hundreds of studies in various indications (e.g., Keynote trial with 112 studies as of this writing and CheckMate with 44 studies). For simplicity, the studies conducted are arranged by the organization (e.g., ASCO), continent/country of origin (e.g., Asia, Europe, USA), pharmaceutical industry-sponsored trials, and some precision immunotherapy studies. For those who are interested in the technical details of study protocols or classification of trials, the reader is referred to the National Cancer Institute website (https://www.cancer.gov/about-cancer/treatment/clinical-trials/nci-supported/nci-match), or reference [14], or the chapters in this book by Mark Abramovitz et. al or Karla Ballman on clinical trials, clinical trial designs, and statistical terminology used for predictive biomarker research and validation. The *New England Journal of Medicine* also includes a series of articles written by those performing clinical trials for that same audience [13]. This is an excellent source of material in the exploration of the changing face of clinical trials. What is common to all these studies is the complexity, cost, and risk involved in the conducting of these large multicenter trials. Another complication is the involvement of many entities such as large academic centers, central labs, clinical research organizations, pharma industry, patient groups, regulatory bodies, insurance companies, diagnostic companies, and many other stakeholders (see Fig. 52.2). It also takes many years to complete a study before the successful approval of a drug by the FDA (see also Chapter on FDA-approved drugs in oncology). In general, all the clinical trials mentioned in the table involve obtaining biopsy tissue or blood samples from the patient, patient profiling, biomarker-enabled reporting, molecular/clinical tumor board discussion, and a treatment plan forged in collaboration with an academic center or a pharmaceutical company.

In summary, precision medicine clinical trials focusses on an individual's genetic make-up rather than average responses to therapy. These trials are there to stay because therapeutically targeting specific genetic mutations has been shown to be successful clinically (or in patients). In the future, precision medicine clinical trials may also take into consideration lifestyle and environmental factors. Additionally, smaller clinical trials including one-person trials known as "N-of-1" trials may become a crucial part of the precision medicine trials. Will these developments lead to better results and faster drug approvals? Only time will tell. As President Abraham Lincoln once said, "The best way to predict your future is to create it." Precision medicine, as it stands today, is exactly in this situation.

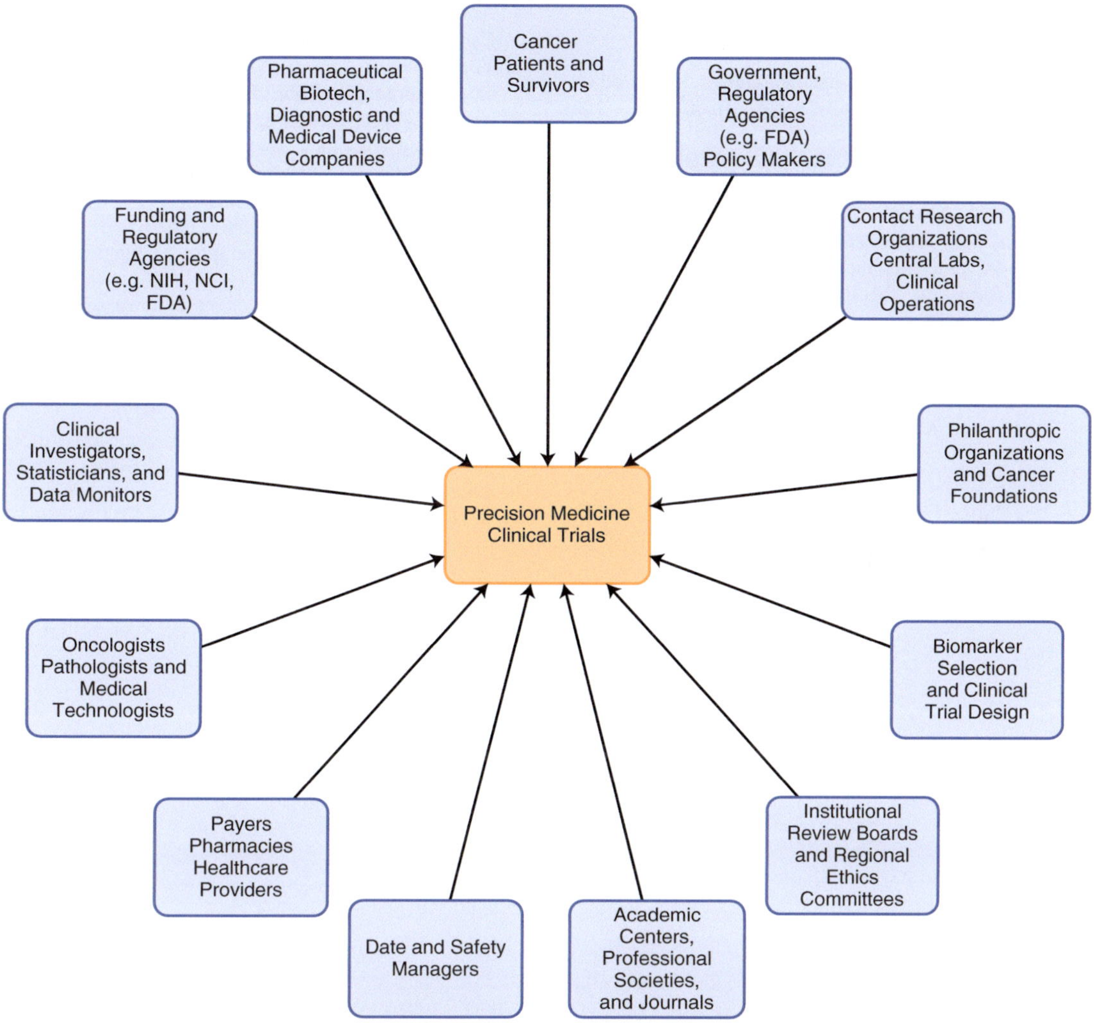

Fig. 52.2 Stakeholders in precision medicine clinical trials. Clinical trials are necessary for the independent determination of clinical utility and toxicity profile of the drugs. They can be conducted in all locations where patient-physician encounters occur. All trials result from a close collaboration between pharma (academics and community), clinicians, and governmental agencies such as the FDA. The research team includes statisticians, physicians, and health informatics people to provide methodology and analytical support for statistical design, determination of sample size, inclusion and exclusion criteria, and statistical methodology for data analysis. Protocols for analysis and data interpretation are predefined and funding procured from pharma, federal, state, or philanthropic organizations. Patient advocacy organizations and institutional review boards (IRBs) ensure protection of the patients' rights. The Data and Safety and Monitoring Board (DSMB) advises study investigators on drug and participant safety and, when appropriate, on the futility or efficacy of the trial. They also make recommendations concerning the continuation, modification, or termination of a trial [14]

References

1. The Precision Medicine Initiative. https://obamawhitehouse.archives.gov/precision-medicine. Last Accessed November 22, 2017.
2. Jameson JL, Longo DL. Precision medicine – personalized, problematic, and promising. N Engl J Med. 2015;372:2229–34.
3. Olson MV. Historical note: a behind-the-scenes story of precision medicine. Genomics Proteomics Bioinformatics. 2017;15:3–10. National Research Council. 2011.
4. Toward precision medicine: building a knowledge network for biomedical research and a new taxonomy of disease. Washington, DC: The National Academies Press. https://doi.org/10.17226/13284.
5. Ashley EA. Towards precision medicine. Nat Rev Genet. 2016;17(9):507–22.
6. Vargas AJ, Harris CC. Biomarker development in the precision medicine era: lung cancer as a case study. Nat Rev Cancer. 2016;16(8):525–37.
7. American Association of Cancer Research (AACR) Cancer Progress Report. 2017.
8. Ying T, Wen Y, Dimitrov D. Precision immunomedicine. Emerging Microbes Infect. 2017;6:e25.
9. Kroschinsky F, Stölzel F, von Bonin S, et al. New drugs, new toxicities: severe side effects of modern targeted and immunotherapy of cancer and their management. Crit Care. 2017;21:89.
10. Rosenbaum L. Tragedy, perseverance, and chance—the story of CAR-T therapy. N Engl J Med. 2017;377:1313–5.
11. Schacter B. The new medicine$. How drugs are created, approved, marketed, and sold. Westport: Praeger Publishers; 2006.
12. Hays P. Advancing healthcare through personalized medicine. Boco Raton: CRC Press; 2017.
13. Roper N, et al. The landscape of precision cancer medicine clinical trials in the United States. Cancer Treat Rev. 2015;41(5):385–90.
14. The Changing Face of Clinical Trials. NEJM. Retrieved from: http://www.nejm.org/page/clinical-trials-series. Last Accessed 22 Nov 2017.

Precision Medicine Clinical Trials: Successes and Disappointments, Challenges and Opportunities – Lessons Learnt

53

Mark Abramovitz, Casey Williams, Pradip K. De, Nandini Dey, Scooter Willis, Brandon Young, Eleni Andreopoulou, W. Fraser Symmans, Jason K. Sicklick, Richard L. Schilsky, Vladimir Lazar, Catherine Bresson, John Mendelsohn, Razelle Kurzrock, and Brian Leyland-Jones

M. Abramovitz
Avera Cancer Institute, Sioux Falls, SD, USA

C. Williams
Center for Precision Oncology, Avera Cancer Institute, Sioux Falls, SD, USA

P. K. De
Department of Molecular and Experimental Medicine, Avera Cancer Institute, Sioux Falls, SD, USA

N. Dey · B. Leyland-Jones (✉)
Department of Molecular and Experimental Medicine, Center for Precision Oncology, Avera Cancer Institute, Sioux Falls, SD, USA
e-mail: brian.leylandjones@avera.org

S. Willis · B. Young
Center for Precision Oncology, Avera Cancer Institute, Sioux Falls, SD, USA

E. Andreopoulou
Department of Medicine, Division of Hematology & Medical Oncology, Weill Cornell Medicine/New York Presbyterian Hospital, New York, NY, USA

W. F. Symmans
MD Anderson Cancer Center, Houston, TX, USA

J. K. Sicklick
Division of Surgical Oncology, General Surgery Residency, Biorepository and Tissue Technology Shared Resource, Moores Cancer Center, University of California San Diego (UCSD), School of Medicine, San Diego, CA, USA

R. L. Schilsky
American Society of Clinical Oncology, Alexandria, VA, USA

V. Lazar · C. Bresson
Win Consortium, Villejuif, Val de Marne, France

J. Mendelsohn
Khalifa Institute for Personalized Cancer Therapy, The University of Texas MD Anderson Cancer Center, Houston, TX, USA

R. Kurzrock
Moores Cancer Center, UC San Diego Moores Cancer Center, La Jolla, CA, USA

S. Badve, G. L. Kumar (eds.), *Predictive Biomarkers in Oncology*,
https://doi.org/10.1007/978-3-319-95228-4_53

Introduction

Clinical trials in cancer have been transformed from being empiric (one size fits all) into precision medicine (PM) (individualized treatment) that rely on our ability to detect molecular and genetic alterations in each patient and use that information to direct the administration of targeted therapy, ushering in the era of personalized therapy. Being able to uncover genomic alterations in a tumor in "real time" using next-generation sequencing (NGS) and liquid biopsies will undoubtedly expand the boundaries of PM. There is now a growing body of evidence that suggests that the use of therapy targeted at the molecular characteristics of a tumor can have substantial benefical effects compared with standard of care, positively affecting patient survival outcomes. What is becoming increasingly evident is that relying on histology and tumor stage as eligibility criteria for most clinical trials that use drugs that directly (target the mutated gene/protein directly) or indirectly (target the signaling pathway related to the mutated gene/protein) target mutations is woefully inadequate.

By establishing a patient's tumor molecular profile, guided therapy can now be based on the molecular alterations that have been identified. The goal of PM clinical trials, which can be designed to specifically deal with the issue of tumor molecular heterogeneity [1], is to evaluate whether tumor molecular profiling-guided treatment is superior to unselected treatment in maximizing survival outcomes (Fig. 53.1).

Algorithms are used in some of these trials to ascribe patients to a specific targeted treatment or treatments based on their tumor molecular alterations. As the design of these "Next-Generation Clinical Trials" evolve, and more data is accumulated over time, increasingly optimized algorithms will be implemented using rules that can be standardized and guarantee reproducibility across platforms and trials [1].

The use of high-throughput molecular omics technologies is being put to the test in more and more PM studies, and molecular profiles of patient tumors are being used to guide therapy. Initial results are encouraging as described in three recent meta-analysis studies [2–4], although not all trials have shown positive results and opinions vary regarding benefits derived from PM clinical trials [5–7]. For example, in the SHIVA trial (randomized phase II trial for patients with refractory cancer) [8], progression-free survival (PFS) did not improve with the use of molecularly targeted agents outside their indications compared with physicians' treatment choice in heavily pretreated cancer patients; however, much was learned from this PM trial that can help with future ones [9]. More recently, in the Molecular Screening for Cancer Treatment Optimization (MOSCATO) trial, a subset of patients with hard-to-treat cancers were shown to benefit from high-throughput genomics-guided therapy [10].

Currently, there are ongoing PM clinical trials using NGS-based molecular screening programs, including WINther (Worldwide Innovative Network (WIN) Consortium trial) [11], NCI-Molecular Analysis for Therapy Choice (MATCH) trial [12], American Society of Clinical Oncology (ASCO) Targeted Agent and Profiling Utilization Registry (TAPUR) study, as well as others [1]. Implementation of PM clinical trials has not been without challenges regarding several aspects including choice of molecular assay, collecting and interpreting gene variant data, tumor heterogeneity, uncovering resistance mechanisms, the need to screen a large number of patients, infrastructure, the need for investigators and industry to collaborate, and governmental policies and regulations.

We anticipate that the gathering of information and deeper understanding of tumor biology gained over time, along with the increasing precision and decreasing cost of NGS, and the increasing availability of agents that target "driver" mutations and/or critical pathways, will pay dividends in how we use these new tools and new knowledge to improve cancer treatment. There are still many hurdles that need to be overcome before PM will become the standard of care, but there is no doubt that this will happen with the appropriate clinical research. In this chapter, PM-based clinical trials are presented and discussed in terms of their advantages, drawbacks, and what has been learnt thus far.

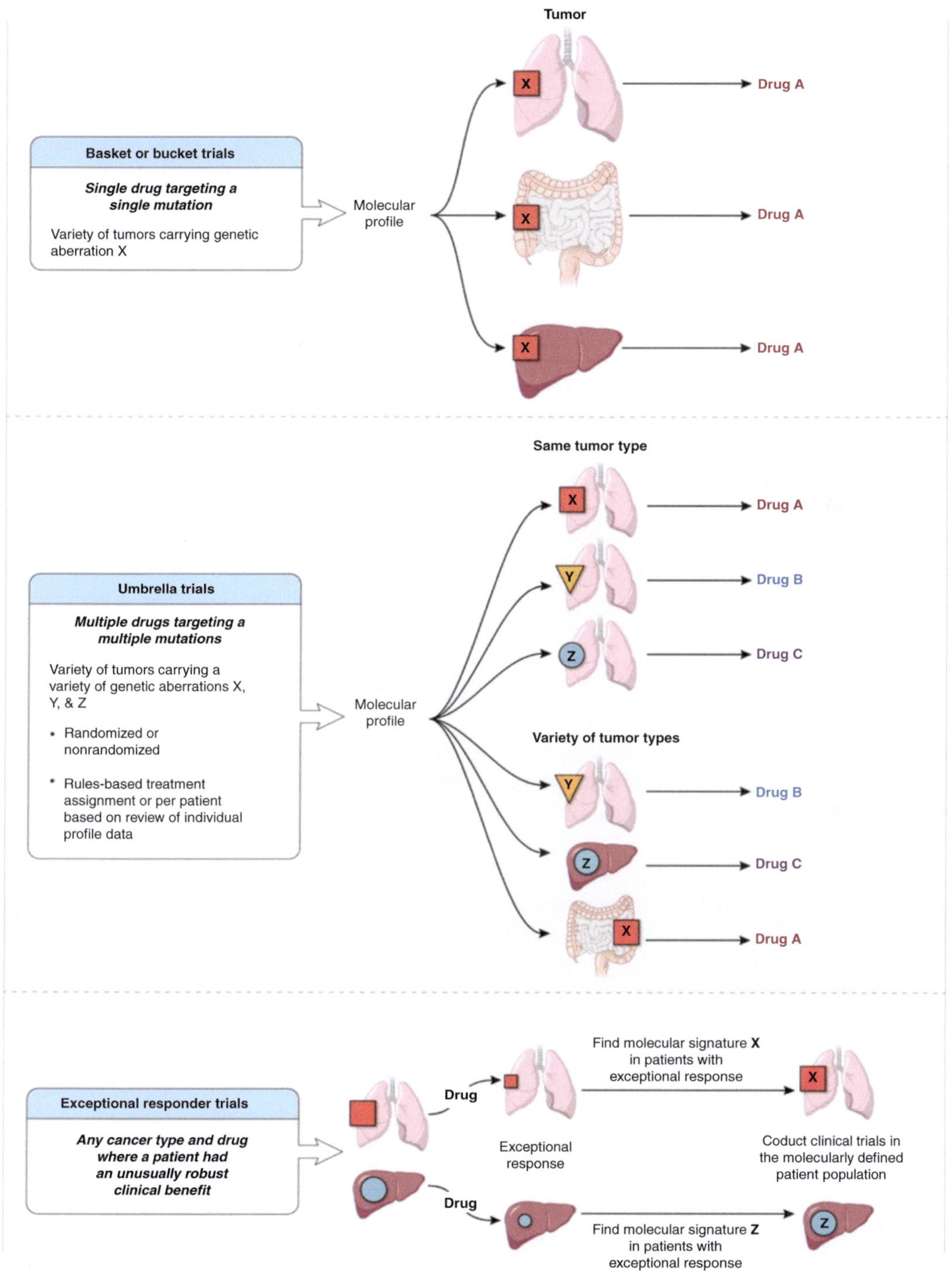

Fig. 53.1 NGS-guided precision medicine cancer clinical trials: basket versus umbrella trials. As a first step, NGS and/or other technologies are used to molecularly characterize tumor tissue (either a freshly obtained biopsy or an archived specimen). The profiling data are analyzed and a report of the results is generated. A predefined algorithm in conjunction with tumor board discussions can be used to interpret the results. It can then be recommended that the patient enroll in a specific genomic-based clinical trial: umbrella, basket, or hybrid – a mix of "umbrella" and "basket" trial designs within one protocol. For exceptional responder trials, the underlying aberration that explains the unusual degree or duration of clinical benefit derived from an otherwise relatively ineffective treatment is then tested in a patient population with that particular molecular aberration. (Adapted from Kummar et al. [15]. With permission from Oxford University Press]

Targeted Treatment and Precision Medicine Clinical Trials

Numerous clinical trials have been conducted that have used a targeted treatment approach to test whether potential actionable targets, for which there are either approved or experimental drugs, would provide greater survival benefit on a selected patient population compared with standard of care in the treatment of cancer patients. Very few trials, however, to date have implemented a PM approach employing molecular profiling to direct personalized treatment strategies (as shown in Fig. 53.1).

Targeted Clinical Trials

Targeted clinical trials, phase I, II, and III, have been subjected to analysis in three meta-analyses that have been published recently in an effort to assess the efficacy of targeted therapy in cancer treatment. Overall, these meta-analyses support the notion that targeted treatment is more effective than non-targeted or standard of care treatment.

A meta-analysis of phase I trials that included 346 studies for a total of 13,203 patients, in which single agents were tested within a 3-year period (January 1, 2011–December 31, 2013), determined that using a personalized strategy was the variable that correlated best with significantly improved outcomes, both in terms of response rate (RR) (30.6% [95% CI, 25.0–36.9%] vs. 4.9% [95% CI, 4.2–5.7%], $P < 0.001$) and PFS in multivariable analysis [2]. OS was not evaluated because it was assessed in only a small number of included studies. Thus, the use of targeted agents resulted in better outcomes compared with non-targeted agents.

Another recent meta-analysis of phase II single-agent trials (570 studies; 32,149 patients) published between January 1, 2010, and December 31, 2012, showed that across malignancies a personalized strategy was an independent predictor of better outcomes and fewer toxic deaths [3]. Using multivariable analysis, a personalized strategy compared with a nonpersonalized one was the only factor determined to be independently and strongly correlated with a higher median RR (31 vs. 10.5%, respectively, $P < 0.001$) and a prolonged median PFS (5.9 vs. 2.7 months, respectively, $P < 0.001$) and OS (13.7 vs. 8.9 months, respectively, $P < 0.001$). A personalized approach was also an independent predictor of fewer toxic deaths, with a mortality rate of 1.52% compared with 2.26% for nonpersonalized treatments.

A comprehensive review of phase III clinical trials across many tumor types that resulted in the FDA approval of cancer treatments between September 1998 and June 2013 was undertaken by Jardim and colleagues [4]. Meta-analysis of 112 registration trials (randomized and nonrandomized) showed that patient treatment using a personalized approach was both statistically significant and independently associated with a higher RR ($P < 0.001$) and a longer PFS ($P = 0.002$) and OS ($P = 0.04$). Thus, this study found that in registration trials in which the drugs tested became FDA approved, biomarker-based targeted treatment was associated with increased clinical benefit for patients with different cancers and mutations.

PM Clinical Trials

The Biomarker-integrated Approaches of Targeted Therapy for Lung Cancer Elimination (BATTLE) study was an important prospective randomized clinical trial that integrated PM rational into the trial strategy and used mandatory biopsies, in which 11 pre-specified biomarkers were evaluated in real time to match the patient to the right therapy [10] (see Chap. 10 for more details). Just prior to this, von Hoff and colleagues conducted the first nonrandomized trailblazing pilot study that examined the strategy of matching patients with various tumor types to treatments (including targeted, off-label, chemo-, and hormone therapy as single or combined drugs) based on molecular alterations that were evaluated using three different techniques: gene expression arrays, fluorescent in situ hybridization, and immunohistochemistry [11]. Another key innovation in the BATTLE trial was the use of each patient as his/her own control in evaluating how well a patient responded to the molecular profiling-selected treatment. Therefore, PFS observed when the patient was prescribed the matched treatment was compared with PFS observed when the patient was on the last prior therapy. Usually the PFS ratio would be <1 because PFS would be expected to decrease with each subsequent treatment. Therefore, a ratio (current matched treatment PFS/last prior treatment PFS) of >1.3 was taken to signify superior efficacy of the molecularly selected treatment, which was achieved in 27% of patients.

The SHIVA trial was notable in that it was the first randomized multicenter phase II trial to evaluate the efficacy of molecularly targeted treatment based on tumor molecular profiling versus conventional treatment in metastatic cancer patients (with any solid tumor) refractory to standard of care [8]. In this study, high-throughput NGS in conjunction with estimations of gene copy number and hormone receptor expression were used to obtain a molecular profile of each patient's biopsy. Of the 741 enrolled patients (between October 4, 2012, and July 11, 2014), a molecular alteration that matched one of the 11 approved targeted agents, as determined using a predefined algorithm, was identified in 293 (40%) patients. Of these, 195 patients were randomly assigned to receive either targeted (used off-label) or conventional (physician's choice) therapy. The median follow-up period was 11.3 months. The results of the trial were somewhat disappointing in that PFS was similar between the two treatment groups, 2.3 months for patients receiving targeted therapy vs. 2.0 months in heavily pretreated refractory cancer patients receiving therapy based on the physician's choice. It is important to note, however, that approximately 80% of patients in SHIVA received either single-agent everolimus or hormone modulator monotherapy. Therefore, the correct conclusion from SHIVA is that matched single-agent mTOR inhibitors or hormone modulators are not effective in the heavily pretreated, refractory cancer setting. Extrapolating from these observations to a judgment call on all of precision medicine, as seems to have happened in several post-SHIVA press releases, has no rational, data-based justification [9]. Even though the overall results of the trial were negative, PFS was improved in a subgroup of patients who received treatment targeting the RAF/MEK signaling pathway, as has been reported in other studies [11]. The SHIVA trial was important in that it was the first randomized multi-institutional study that attempted to match a targeted drug with a patient's tumor genomic profile. It also underscored the feasibility of performing randomized PM cancer clinical trials.

The recently described MOSCATO trial investigated the use of high-throughput genomics to match targeted treatment to patients with advanced "hard-to-treat" cancers [12]. From the 843 patients for whom a molecular portrait was obtained, 411 patients (49%) had an actionable target. In the end, a matched therapy was administered to 199 patients (~25%). A variety of molecular alterations in these patients were identified in 53 genes including 98 amplifications and 23 deletions or loss, 103 mutations, and 18 translocations, and 8 were based on IHC. Their genomic-based therapy matching strategy resulted in improved outcomes in 33% of patients with advanced cancers who obtained a PFS ratio above the predefined threshold.

This trial was positive compared with the SHIVA trial for a couple of reasons. First, many more targeted therapies were available that were also from the most recent generation of drugs (i.e., vemurafenib versus sorafenib), in which patients were treated in phase I/II trials. Second, given the number of previous precision medicine trials, lessons learned from those were applied in the MOSCATO trial. Limitations included the fact that a number of patients had high genomic instability, making it less likely that benefit could be derived from the targeted therapies, and furthermore, there were patients with multiple driver alterations that might have benefited from combination therapy. Furthermore, only 25% of patients could be enrolled.

Like the WINther trial, described below, larger multicenter PM cancer trials based on NGS tests are in the patient recruitment phase including NCI-MATCH, part of the NCI PM Initiative, and TAPUR, an ASCO-affiliated study. The NCI-MATCH trial involves clinical sites across the

United States that participate in NCI's National Clinical Trials Network. There will be a minimum of 24 treatment arms with a goal of enrolling 35 patients per arm. At least 6,000 patients will be screened using NGS to detect 4,000 variants across 143 genes regardless of tumor origin. The objective response rate (ORR) will be the primary endpoint with both PFS and time to progression (TTP) as secondary endpoints. If any of the treatments show potential, then they can be more definitively evaluated in larger clinical trials. If a match is not found for patients with rare tumors, they can choose to move to the DART (*D*ual **A**nti-CTLA-4 and Anti-PD-1 Blockade in *R*are *T*umors) study, the first national study of immunotherapy in rare tumors where patients would receive a combination of ipilimumab (CTLA-4 inhibitor) and nivolumab (PD1 inhibitor).

TAPUR is ASCO's first clinical research study that will allow patients access to targeted therapies prescribed off-label, and all the outcome data will be collected and analyzed. The trial is open to patients with advanced solid tumors, multiple myeloma, or B-cell non-Hodgkin lymphoma. The main goal will be to assess the efficacy and safety of the prescribed drugs. To date, 19 drugs from five pharmaceutical companies are accessible to patients at no cost, with the possibility of adding more drugs as they become available. The primary endpoint is objective tumor response or stable disease at 16 weeks following initiation of treatment, with PFS and OS as secondary endpoints, as well as duration of treatment on study. Various commercially available NGS tests used by participating oncologists will be recorded. Eligibility criteria used are purposely broad, unlike most clinical trials, allowing for those recruited patients to reflect the real-world situation (Table 53.1).

Table 53.1 Successful and disappointing PM clinical trials (see text for trial details)

Successful PM trials	Disappointing PM trails
BATTLE Achieved PFS >1.3 in 27% of patients	SHIVA First randomized PM trial Limited treatment options for patients PFS unchanged between treatment groups
MOSCATO 411 patients (49%) had actionable target Matched therapy administered to 25% of patients Greater PFS achieved in 33% of treated patients	
Ongoing PM trials	
WINther First PM trial to use both genomic and transcriptomic profiling	
NCI-MATCH US-wide trial Tumors of any origin 24 treatment arms	
TAPUR Advanced solid tumors, multiple myeloma, or B-cell non-Hodgkin lymphoma ≥19 drugs will be used	

WINther Trial: The First Precision Medicine Trial to Include Genomics and Transcriptomics

WINther, a Worldwide Innovative Network (WIN) Consortium trial, is the first study to assign patients to a therapy matched to their structural genomic profile (using NGS) or to their functional genomic profile (mRNA gene expression – transcriptome) [13]. Patients recruited to the trial are stage IV cancer patients with solid tumors that include malignancies of the lung, breast, colon, head and neck, kidney, and liver, as well as rhabdomyosarcomas. DNA aberrations (such as mutations, amplifications, and deletions) are assessed, and if no match with a targeted therapy is obtained, information on gene expression (transcriptome) obtained from the patient's tumor is used to select a potentially active targeted therapy. In this way, targeted therapies can be assigned to more patients. The WINther trial design is shown in Fig. 53.2.

Once the omics data are generated, an algorithm is used to assess the differential levels of expression of mRNA from the metastasis (tumor) and the non-tumoral (histologically matched normal tissue) samples and provides information

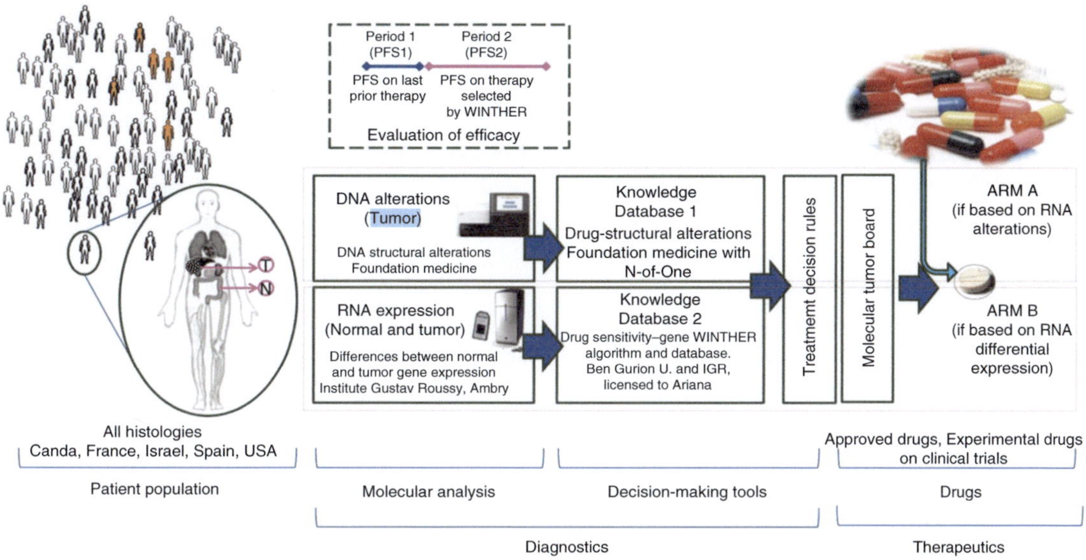

Fig. 53.2 The WINther trial design [13]. A biopsy of the tumor (or metastasis) and normal tissue from the organ of origin of the tumor was performed on each patient, and molecular profiling of both DNA and RNA was done. Matching of actionable targets to the choice of therapy was rationally guided based on mutations, amplifications, or gene rearrangements (arm A) or on differential gene expression between tumor and normal tissues (arm B). WINther, launched by the WIN Consortium, involved six international centers, and tumor and normal tissue samples were analyzed by two different laboratories. Treatment decisions were determined using two different platforms and knowledge bases as well as a clinical management committee, which met by teleconference. Both FDA-approved and experimental drugs were potential alternatives for patients. T tumor biopsy, N normal tissue biopsy, PFS progression-free survival, TTP time to progression. (Adapted from Rodon et al. [13]. With permission Oxford University Press)

that is used to identify potential driver genes. The trial is testing whether information about gene expression can identify driver genes that are likely to respond to targeted therapies and whether this approach is as effective as assigning therapy based on genetic aberrations.

Tumors can harbor many genetic aberrations that are potential drivers of the malignancy, but for which a targeted drug or combination of drugs is not available. However, with time this will improve as more and more targeted agents become available and a better understanding of driver genes and relevant signaling pathways are discovered. It will become more commonplace to incorporate omics testing that includes RNA, proteins, and the immunome. Algorithms used to assign targeted treatment will improve as more data are fed into the "system," and treatment efficacy should improve with time. Of special interest, the patients once believed to be the most difficult to treat, i.e., those whose tumors have chaotic genomes with multiple abnormalities, are now the patients whose prognosis may be the best. These tumors respond to immunotherapy, perhaps because the immune system can be reactivated by drugs, such as checkpoint inhibitors, to recognize the neo-antigens produced by the mutanome. Hence, the more mutations, the better the chance that the reactivated immune system will recognize and eradicate the malignant cells.

There are still several funding and regulatory approval hurdles that will need to be overcome to ensure that PM clinical trials can run more smoothly and can enroll patients in a timelier manner. Obtaining the necessary medication from numerous pharmaceutical companies for screening efficacy on a broad range of tumors and for testing targeted therapies in combinations is still a challenging task that requires much effort. We are seeing this starting to happen, as in the TAPUR and NCI-MATCH trials, in which several pharmaceutical companies have agreed to provide the necessary targeted agents. Thus, obtaining permission to use a wide range of approved and/or experimental drugs is paramount to overcome this major rate-limiting step, if patients are to acquire the requisite drugs from different sponsors.

WIN Consortium Experience

WIN is the Worldwide Innovative Network for Personalized Cancer Therapy. It is a unique organization whose goal is to accelerate global implementation of precision cancer medicine. Its membership includes >20 of the best-known academic centers on four continents, payers, large pharmaceutical companies, patient advocacy groups, contact research organizations, and molecular diagnostic companies. In other words, diverse stakeholders are brought together under the WIN umbrella. WIN's first international trial is WINther (described above – the first precision medicine trial that includes genomics and transcriptomics). This trial is now fully accrued and is undergoing analysis. The challenges faced by this international precision medicine trial have been previously published [13]. Interestingly, the most serious challenge is related to regulatory delays and lack of regulatory harmonization across nations. Building on the lessons of WINther, WIN is set to launch several additional international precision medicine lung cancer trials within the next year.

Precision Medicine Clinical Trial Challenges

There are many challenges associated with PM clinical trials. These are presented below along with possible opportunities for improvement (Table 53.2).

Table 53.2 PM clinical trials: challenges and opportunities

Challenge/pitfall	Opportunities for improvement
Study design	
Implementing the right PM trial design	New types of PM trials (see Fig. 53.1) are being evaluated
Selecting the right endpoint	
Evaluating patient outcomes	Choosing the most appropriate prospective trial design; nonrandomized, nonrandomized with PFS ratio as endpoint (BATTLE trial), or randomized
Conducting a PM trial	
Ethics approval and regulations in a national or international setting	Government agencies must deal with changing and streamlining regulations to give PM trials the best opportunities to be conducted
Acquiring drugs from different companies that are to be administered together	
Synchronization of numerous stakeholders including health authorities, pharmaceutical companies, physicians, pathologists, radiologists, bioinformaticians, and biostatisticians	Pharmaceutical companies are realizing the need for collaborative PM cancer clinical trials but are not fully committed
	Coordination between health authorities and IRBs is required
Obtaining regulatory approval for complex trials that use both single and combination treatments, including approved (for both on- and off-label use) and experimental agents	Development of an ad hoc fast-track review system by health authorities for each proposed PM trial
Tumor analysis	
Tissue sampling and tumor heterogeneity	As more and more commercial and academic groups are involved in performing molecular profiling-based PM clinical trials, implementation and standardization of tissue sampling, assays, and platforms should occur
Quality control aspects and biobanking	
Obtaining the necessary fresh-frozen biopsies and performing real-time molecular profiling	
Standardization of real-time NGS and other molecular analysis assays across centers and across countries	Experience gained will also contribute to the optimization of all steps involved in collecting and biobanking tumor specimens

Table 53.2 (continued)

Challenge/pitfall	Opportunities for improvement
Bioinformatics hurdles	
Optimizing algorithms used in selecting the most appropriate targeted drug or combination of drugs	As universal NGS testing becomes more widespread along with processing of all PM clinical trial data, algorithms will improve, and matching an optimized treatment to each cancer patient will increase
Data storage and security (hacking prevention)	Data storage and security are paramount issues under evaluation
Treatment decisions	
NGS testing is not useful if a therapy cannot be selected and may thus benefit only a subset of patients. Increasing the identification of patients that are likely to respond to targeted treatment is needed	NGS panels are identifying potentially actionable genomic alteration in ~40% of patients with diverse cancers [14], although many cases involve the off-label use of an FDA-approved drug. In addition, panels are growing in terms of actionable genes, in which there is a need for expansion of the number and type of biomarkers and incorporation of additional data from circulating tumor DNA in blood samples (liquid biopsies)
Significant benefit from NGS-guided treatment compared with standard of care must be shown for a broad range of patients in PM trials	
Conducting an algorithm-based trial	
Single versus combination treatment	Meta-analysis of phase I [2], II [3], and III [4] clinical trials with a combined >85,000 patients treated showed that personalized targeted therapy selected on the basis of genomic biomarkers was associated with higher RRs and longer survival (PFS in all trials and in OS in phase II and III), as well as fewer deaths related to toxic adverse events across a number of cancers
Availability of adequate treatment/drug after the trial ends	
Cost considerations	
The cost of implementing PM clinical trials must be justified given the expense related to NGS	NGS costs have decreased steeply and should continue to decrease over the next several years
The cost of cancer treatment, which includes prescribing chemotherapy, targeted therapy, and/or immunotherapy, which often are not effective in individual patients, is producing costs that are not sustainable	Data analysis is also becoming more manageable and less costly
Justifying universal PM for every patient as a standard policy will require demonstrating both financial feasibility and improved clinical outcomes	A complete molecular diagnosis afforded by NGS testing and development of other biomarkers may help to streamline costs through selection of optimized therapy and avoiding therapy that is ineffective
Additional benefits of PM trials	
NGS and, in the future, other biomarkers can also be used to determine genetic aberrations that can be used to predict prognosis and, thus, aid in patient care	
Information gained from NGS tumor analysis can also be used to predict contraindicated drugs (e.g., epidermal growth factor receptor (EGFR) inhibitors are contraindicated in *KRAS*-mutant colorectal cancer)	
Given that new driver genes are being identified more quickly, fueling targeted drug discovery efforts (although it is not possible to determine who will benefit in advance), new targeted drug choices may become available to a patient at a later time after the cancer's initial presentation	
Instead of continuing with standard chemotherapies, many of which have toxic and life-threatening adverse effects, embracing PM is the future of cancer treatment	
With time, universal NGS testing along with processing of all PM clinical trial data and the availability of more targeted agents will ensure that matching an optimized treatment regimen to each cancer patient is far more achievable	

Conclusions and Future Directions

The protracted era of grouping patients together based primarily on a histologic diagnosis to evaluate a cancer treatment in a clinical trial is rapidly coming to an end. Incremental improvements in survival over the years have been achieved in metastatic cancer patients; however, the bar must be raised if more meaningful gains in efficacy are to be achieved. From a fundamental scientific point of view, the omics revolution has shed light on the complexity of cancer, which has resulted in a deeper understanding of the complicated genomic landscape, and the increasing number of genomic drivers and pathways originating in both the same and different tissues that do not segregate by histology.

Targeting these diverse drivers and pathways that can benefit small subsets of patients has resulted in a paradigm shift to designing "next-generation" clinical trials, which is gradually being implemented. Therefore, to take full advantage of this omics revolution that has led to the explosive growth and availability of potent genomic and immunologic targeted agents, PM clinical trials are moving from the standard drug-centric model to a patient-centric one. As genomic, transcriptomic, and proteomic testing become more monetarily feasible and mainstream, they will be routinely performed on all tumors at the outset, enabling an accurate diagnosis and selection of the most appropriate treatment early on, which will inevitably help prevent metastasis and/or relapse from occurring.

However, for this bright future in cancer diagnosis, treatment, and prevention to materialize, there must be sufficient data and new practices that also consider costs to the healthcare system and society. PM clinical trials rigorously performed in broad populations subjected to universal molecular profiling of tumors will contribute the evidence needed to validate the biomarkers and the vast array of targeted drugs that will make this a reality. Finally, the creation of massive databases that will incorporate both omics and clinical information on thousands of PM trial patients will eventually lead to new discoveries that will advance us toward the goal of the successful treatment of many more patients suffering from cancer.

Acknowledgments We would like to acknowledge the Avera Cancer Institute, Sioux Falls, SD, for providing support for our research.

Conflicts of Interest The authors have no conflicts of interest to declare.

References

1. Catenacci DV. Next-generation clinical trials: novel strategies to address the challenge of tumor molecular heterogeneity. Mol Oncol. 2015;9(5):967–96.
2. Schwaederle M, Zhao M, Lee JJ, et al. Association of biomarker-based treatment strategies with response rates and progression-free survival in refractory malignant neoplasms: a meta-analysis. JAMA Oncol. 2016; https://doi.org/10.1001/jamaoncol.2016.2129.
3. Schwaederle M, Zhao M, Lee JJ, et al. Impact of precision medicine in diverse cancers: a meta-analysis of phase II clinical trials. J Clin Oncol. 2015;33(32):3817–25.
4. Jardim DL, Schwaederle M, Wei C, et al. Impact of a biomarker-based strategy on oncology drug development: a meta-analysis of clinical trials leading to FDA approval. J Natl Cancer Inst. 2015;107(11).
5. West HJ. No solid evidence, only hollow argument for universal tumor sequencing: show me the data. JAMA Oncol. 2016;2(6):717–8.
6. Subbiah V, Kurzrock R. Universal genomic testing needed to win the war against cancer: genomics IS the diagnosis. JAMA Oncol. 2016;2(6):719–20.
7. Prasad V. Perspective: the precision-oncology illusion. Nature. 2016;537(7619):S63.
8. Le Tourneau C, Delord JP, Goncalves A, et al. Molecularly targeted therapy based on tumour molecular profiling versus conventional therapy for advanced cancer (SHIVA): a multicentre, open-label, proof-of-concept, randomised, controlled phase 2 trial. Lancet Oncol. 2015;16(13):1324–34.
9. Le Tourneau C, Kurzrock R. Targeted therapies: what have we learned from SHIVA? Nat Rev Clin Oncol. 2016; https://doi.org/10.1038/nrclinonc.2016.164.
10. Kim ES, Herbst RS, Wistuba II, et al. The BATTLE trial: personalizing therapy for lung cancer. Cancer Discov. 2011;1(1):44–53.
11. Von Hoff DD, Stephenson JJ Jr, Rosen P, et al. Pilot study using molecular profiling of patients' tumors to find potential targets and select treatments for their refractory cancers. J Clin Oncol. 2010;28(33):4877–83.
12. Massard C, Michiels S, Ferte C, et al. High-throughput genomics and clinical outcome in hard-to-treat advanced cancers: results of the MOSCATO 01 trial.

Cancer Discov. 2017; https://doi.org/10.1158/2159-8290.CD-16-1396.
13. Rodon J, Soria JC, Berger R, et al. Challenges in initiating and conducting personalized cancer therapy trials: perspectives from WINTHER, a worldwide innovative network (WIN) consortium trial. Ann Oncol. 2015;26(8):1791–8.
14. Wheler JJ, Janku F, Naing A, et al. Cancer therapy directed by comprehensive genomic profiling: a single center study. Cancer Res. 2016;76(13):3690–701.
15. Kummar S, Mickey WP, Lih C-J, et al. Application of molecular profiling in clinical trials for advanced metastatic cancers. J Nat Cancer Inst. 2015;107(4):djv003.

FDA-Approved Targeted Therapies in Oncology

54

George Louis Kumar

Governments regulate the testing, manufacture, approval, sale, marketing, and post-marketing of modern pharmaceutical products. For example, in the United States, the "Food and Drug Administration" commonly referred to as the FDA regulates over-the-counter and prescription drugs to protect the public health and safety of its citizens. FDA's Office of International Programs (OIP) in Europe (Brussels, Belgium), India (New Delhi), China (Beijing), and Latin America (San Jose, Costa Rica) helps to assure that medical products exported to the United States from these foreign countries meet US standards. For an alphabetical list of countries covered by FDA's OIP, see Ref. [1]. Like the FDA, the European Medicines Agency (EMA), China Food and Drug Administration (CFDA), Japanese Ministry of Health and Welfare, Indian Central Drugs Standard Control Organization, National Health Surveillance Agency or ANVISA (Agência Nacional de Vigilância Sanitária) in Brazil, and Health Canada are, respectively, responsible for safety monitoring of medicines developed by pharmaceutical companies in their countries. These countries collaborate with the FDA to seek guidance for approval of drugs in their own countries.

The following section gives a 30,000-foot view of FDA-approved anticancer drugs that are specifically designed to "target" and kill cancer cells that have a "molecular lesion" – such as an aberrant protein (e.g., Her2) involved in cancer progression. The use of these "targeted agents" for cancer treatment is referred to as "targeted therapy." The idea of targeting molecules in cancer cells started in the 1970s with Imperial Chemical Industries' (ICI) failed postcoital contraceptive compound ICI 46,474 – now called tamoxifen – for targeting estrogen receptor-positive tumors in breast cancer patients [2]. This discovery was followed by research in the discovery of imatinib (Gleevec) to target *BCL-ABL* fusion oncogene created by the Philadelphia chromosome translocation in chronic myelogenous leukemia (Fig. 54.1) and development of targeted drugs against epidermal growth factor receptors (EGFR) such as gefitinib (Iressa) and erlotinib (Tarceva) for non-small cell lung cancer. The first FDA-approved targeted drug for cancer was rituximab (Rituxan) (Fig. 54.2) in 1997 to treat patients with B-cell non-Hodgkin's lymphoma followed by trastuzumab (Herceptin) – an anti-breast cancer drug in 1998. These discoveries and many other approvals of targeted drugs for cancer have now become a cornerstone of precision medicine [3].

Targeted drugs can be unconjugated humanized monoclonal antibodies (mAb) such as trastuzumab (Fig. 54.3), small-molecule inhibitors

G. L. Kumar
Targos Inc., Issaquah, WA, USA
e-mail: george.kumar@targos-inc.com

S. Badve, G. L. Kumar (eds.), *Predictive Biomarkers in Oncology*,
https://doi.org/10.1007/978-3-319-95228-4_54

such as imatinib and erlotinib, or an engineered antineoplastic protein like denileukin diftitox (Ontak) – a combination of interleukin-2 and diphtheria toxin. Some of these molecules can be used for treating more than one type of cancer. For example, trastuzumab that targets the Her2 protein is used for treating both breast and gastric cancer. Similarly, erlotinib that targets the EGFR receptor is used for treating both metastatic pancreatic and non-small cell lung cancer.

Unconjugated monoclonal antibodies such as trastuzumab work by themselves. They are clones of a single parent cell. They seek out cell surface transmembrane receptors or sometimes free-floating protein fragments in the blood (e.g., the extracellular domain of Her2 receptor in the blood circulation) to inhibit growth. In some cases, mAbs are "weaponized" by conjugating to a toxin (e.g., brentuximab vedotin (Adcetris) (Fig. 54.3c)), to a chemical substance

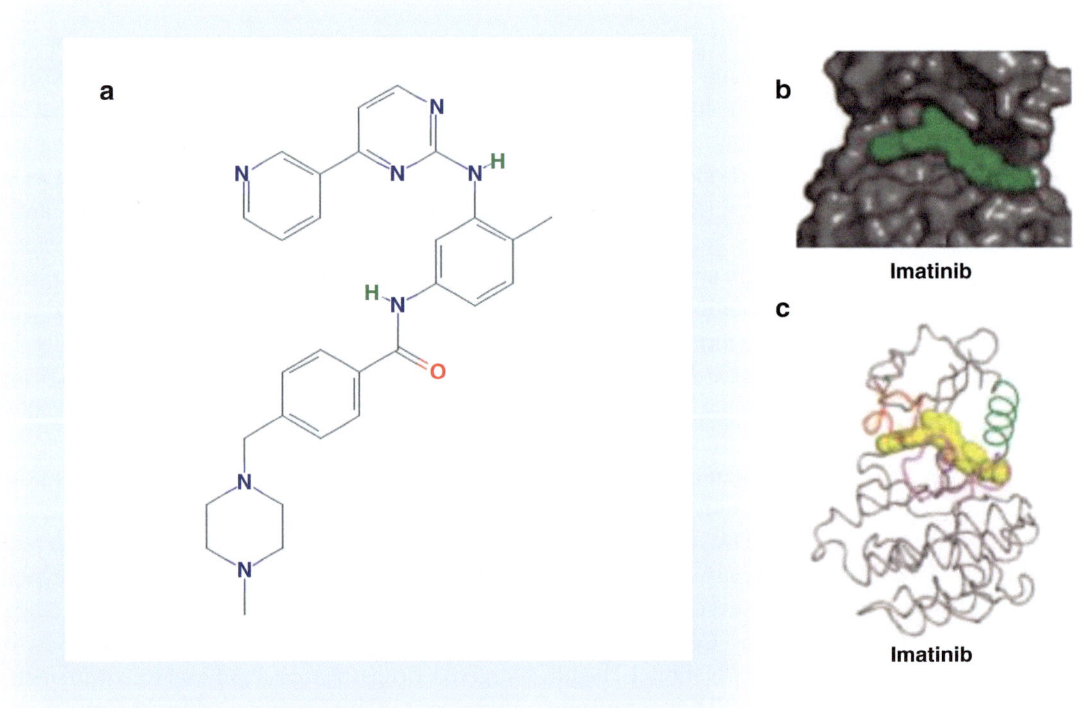

Fig. 54.1 (**a**) Small-molecule inhibitor imatinib mesylate (Gleevec®) 2D structure. The molecular weight of imatinib is 493.615 g/mol, and the topological polar surface area is 86.3 Å2. The molecular weight of imatinib is close to the molecular weight of adenosine triphosphate (ATP), which is 507.181 g/mol. However, the topological polar surface area of ATP is 279 Å2, which is three times larger than that of imatinib. This smaller size of imatinib gives the molecule the advantage of slipping into the ATP-binding cleft of the tyrosine kinase domain of Bcr-Abl and thereby preventing Bcr-Abl from transferring a phosphate group from ATP onto a tyrosine residue in a substrate protein. (Courtesy of the National Center for Biotechnology Information. PubChem Compound Database. Retrieved from: https://pubchem.ncbi.nlm.nih.gov/compound/5291). (**b**) Surface representation of crystal structure of ABL kinase in complex with imatinib (green). Residues from the nucleotide-binding loop (P-loop) and activation loop (A-loop) are omitted from the surface calculation for clarity. (Reprinted from Weisber et al. [10]. With permission from Nature Publishing Group.) (**c**) | Comparison of the different binding modes of imatinib. The positions of the nucleotide-binding loop (P-loop) (red) and activation loop (A-loop) (magenta) vary according to whether the kinase is in an active conformation or an inactive conformation. The green helix is helix C, which often moves between the active and inactive states of kinases. (Reprinted from Weisber et al. [10]. With permission from Nature Publishing Group.) (**d**) Blocking of Bcr-Abl protein activity by tyrosine kinase inhibitor imatinib (Gleevec®). Imatinib binds to the ATP-binding cleft of the kinase domain of Bcr-Abl protein and stabilizes the protein in its closed, inactive conformation. This action results in the inhibition of the kinase activity of Bcr-Abl. (Adapted from Alberts et al. [11]. With permission from Taylor & Francis Group LLC)

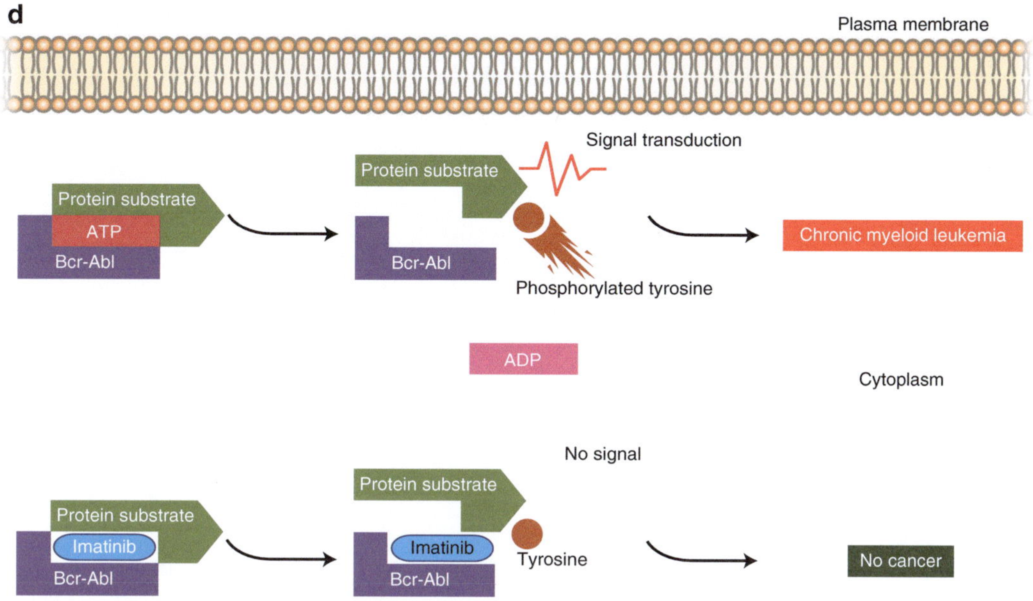

Fig. 54.1 (continued)

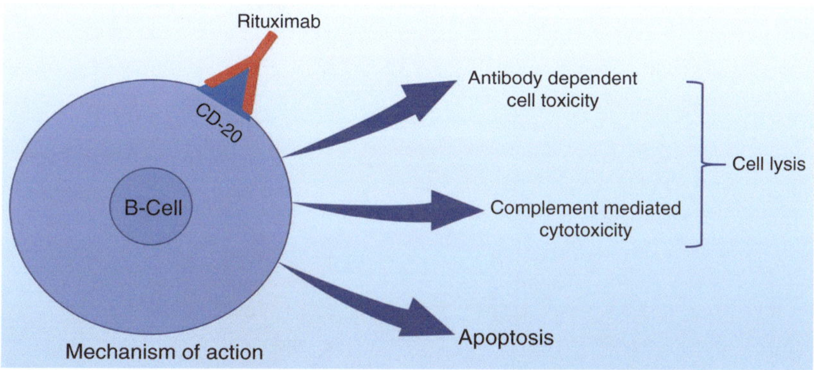

Fig. 54.2 Rituximab: Mechanism of action: Upon binding to CD-20 on the plasma membrane of B cells, rituximab can mediate antitumor responses by three mechanisms. (**a**) Antibody-dependent cell toxicity via macrophages and natural killer cells (not shown in the diagram), (**b**) complement-mediated B-cell lysis, and (**c**) apoptosis through selected chemo-and immune-sensitization signaling pathways. (Adapted from Seyfizadeh et al. [12]. With permission from Elsevier)

(e.g., trastuzumab emtansine (Kadcyla) (Figs. 54.3b and 54.4)), or to a radioactive particle (e.g., ibritumomab tiuxetan (Zevalin) (Fig. 54.5d)). Chemolabeled antibodies such as trastuzumab emtansine are also known as "antibody-drug conjugates (ADCs) (Fig. 54.3 for structures of selected first-, second-, and third-generation ADCs). Sometimes two different mAbs are used to attach two different proteins in two different cells at the same time. An example is blinatumomab (Blincyto) used in the treatment of acute lymphocytic leukemia (ALL) (Fig. 54.5f). Here, one part of blinatumomab gets attached to the CD19 protein of malignant B cells and another to CD3 part of the T-cell receptor. Blinatumomab works by linking these two cell types and activating the endogenous T cell to exert cytotoxic activity on the target cell.

Large molecules such as monoclonal antibodies whose molecular weight is greater than 185,000 daltons cannot pass through the plasma membrane. They are usually used to target cell surface receptors (Figs. 54.2 and 54.3). On the

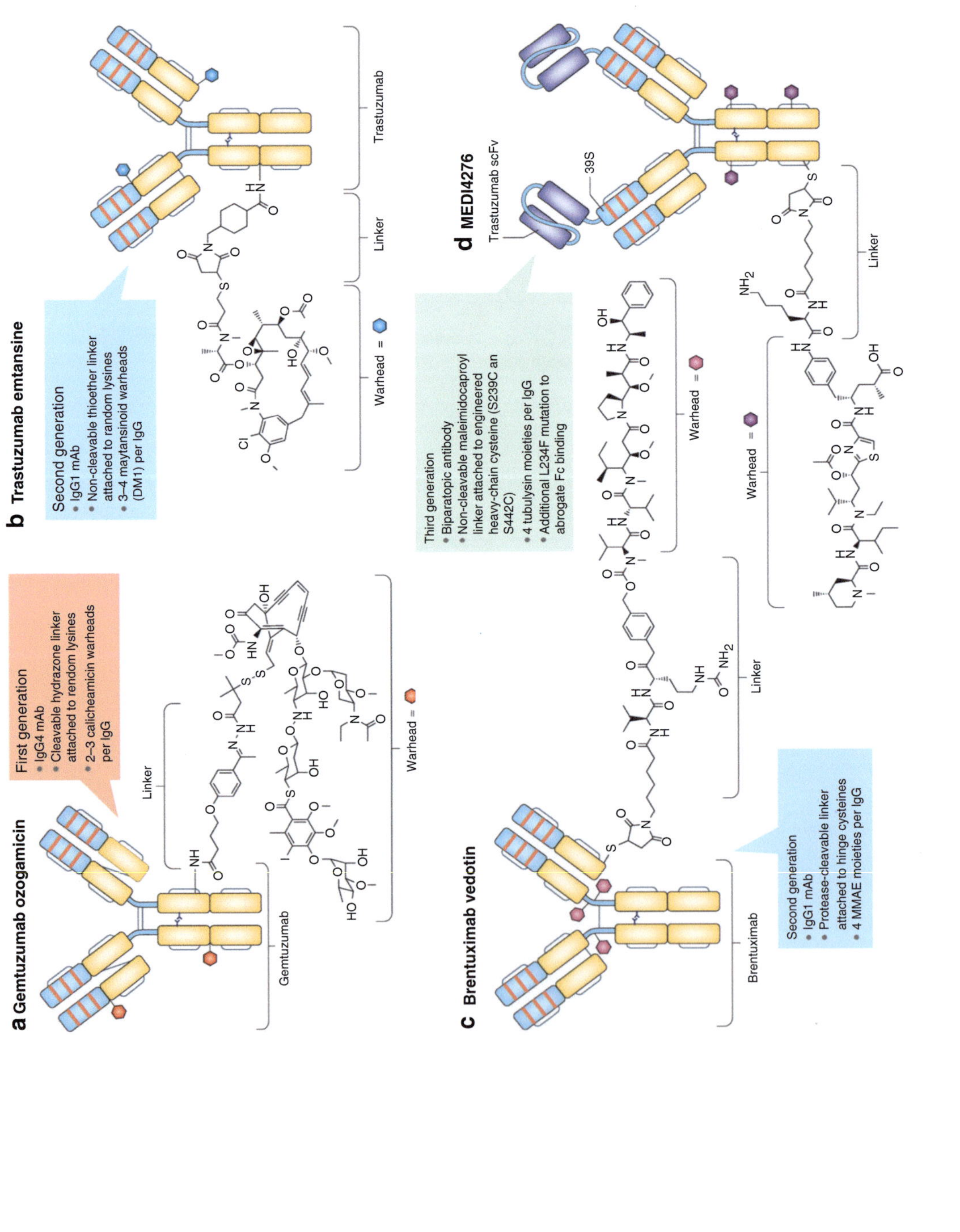
a Gemtuzumab ozogamicin
First generation
IgG4 mAb
Cleavable hydrazone linker attached to rendom lysines
2–3 calicheamicin warheads per IgG
Gemtuzumab
Linker
Warhead =
b Trastuzumab emtansine
Second generation
IgG1 mAb
Non-cleavable thioether linker attached to random lysines
3–4 maytansinoid warheads (DM1) per IgG
Warhead =
Linker
Trastuzumab
c Brentuximab vedotin
Second generation
IgG1 mAb
Protease-cleavable linker attached to hinge cysteines
4 MMAE moieties per IgG
Brentuximab
Linker
Warhead =
d MEDI4276
Third generation
Biparatopic antibody
Non-cleavable maleimidocaproyl linker attached to engineered heavy-chain cysteine (S239C an S442C)
4 tubulysin moieties per IgG
Additional L234F mutation to abrogate Fc binding
Trastuzumab scFv
39S
Warhead =
Linker

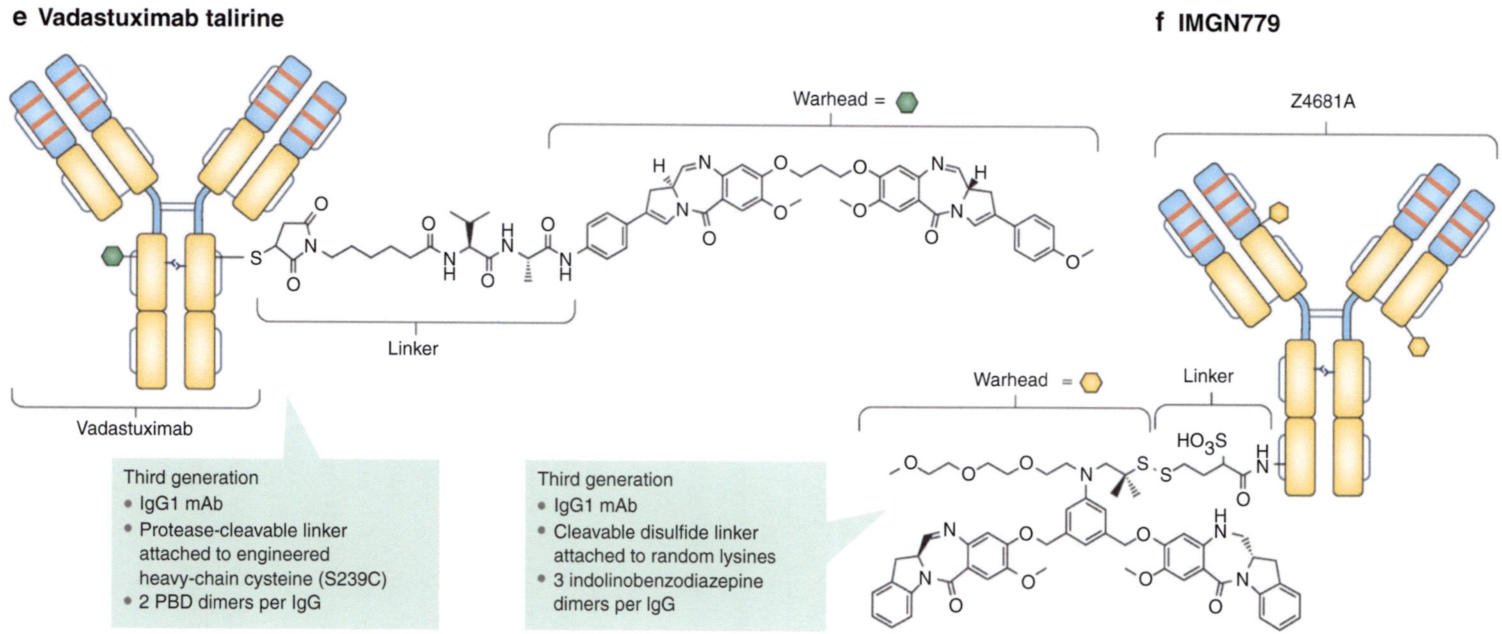

Fig. 54.3 | Structures of selected first-, second- and third-generation ADCs. Schematic overview of selected antibody-drug conjugates (ADCs). Cytotoxic warheads are conjugated to human, humanized, or chimeric monoclonal antibodies (mAbs) via a range of linker types. (**a**) | The first-generation ADC gemtuzumab ozogamicin is a conjugate of a humanized immunoglobulin G4 (IgG4) mAb specific for CD33 and two to three calicheamicin moieties per IgG, which are attached via cleavable hydrazone linkers to random lysine residues. (**b**) | The second-generation ADC trastuzumab emtansine consists of conjugate of a humanized IgG1 mAb specific for human epidermal growth factor receptor 2 (HER2) and three to four DM1 moieties per IgG, which are attached via non-cleavable thioether linkers to random lysine residues. (**c**) | The second-generation ADC brentuximab vedotin is a conjugate of a chimeric IgG1 mAb specific for CD30 and four monomethyl auristatin E (MMAE) moieties per IgG, which are attached to the hinge region through a protease-cleavable linker. (**d**) | The third-generation ADC MEDI4276 consists of a biparatopic antibody that targets two non-overlapping epitopes on HER2, conjugated to four tubulysin moieties per antibody through a maleimidocaproyl linker. (**e**) | The third-generation ADC vadastuximab talirine is a conjugate of a humanized IgG1 mAb specific for CD33 and two pyrrolobenzodiazepine (PBD) moieties per IgG, which are attached to engineered cysteines (S239C) in the heavy chain through a protease-cleavable linker. (**f**) | The third-generation ADC IMGN779 is a conjugate of a humanized IgG1 mAb specific for CD33 and three indolinobenzodiazepine moieties per IgG, which are attached to random lysine residues by a cleavable disulfide linker. scFv, single-chain variable fragment. (Reprinted from Beck et al. [13]. With permission from Nature Publishing Group.)

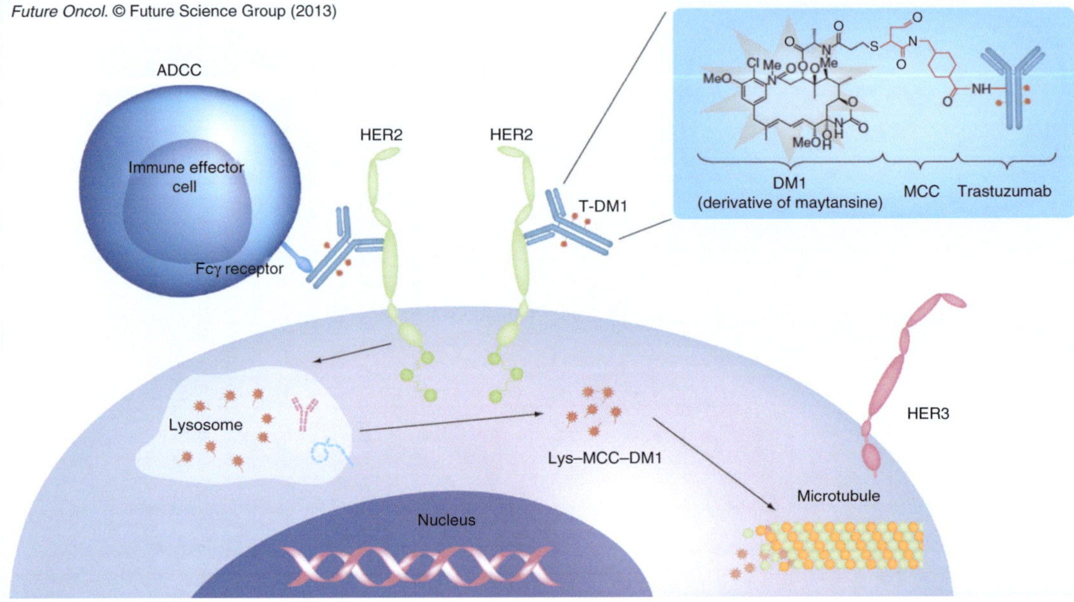

Fig. 54.4 Structure of trastuzumab emtansine and mechanisms of action. After T-DM1 binds HER2, the HER2-T-DM1 complex undergoes internalization, followed by lysosomal degradation. This process results in the intracellular release of DM1 containing catabolites that bind to tubulin and prevent microtubule polymerization, as well as suppress microtubule dynamic instability. T-DM1 has also been shown to retain mechanisms of action of trastuzumab, including disruption of the HER3/PI3K/AKT signaling pathway and FCγ receptor-mediated engagement of immune effector cells that leads to antibody-dependent cellular cytotoxicity. ADCC antibody-dependent cellular cytotoxicity, Lys lysine, T-DM1 trastuzumab emtansine. (Reprinted from Peddi and Hurvitz [14]. With permission from Future Medicine Ltd)

other hand, small molecules of molecular weight less than 900 daltons such as imatinib (mol. wt., 507 daltons) can pass through the cell membrane through active transport (e.g., organic cation transporter 1 (OCT1; SLC22A1)) to target molecules inside a cell.

Although how targeted therapy drugs function is beyond the scope of this chapter (for details refer to Refs. [4–6]); they fall into the following categories:

(a) Agents that target cell signaling pathways or so-called signal transduction inhibitors, e.g., HER2 signaling downregulation by trastuzumab and suppression of the PI3K/Akt pathway.
(b) Agents that target a unique phenotype of the cancer cell through cytotoxic effect, e.g., rituximab directed against B-cell CD20 antigen (Fig. 54.2).
(c) Therapies that stop the growth of hormone-sensitive tumors, e.g., antiestrogen drugs such as tamoxifen for breast cancer and antiandrogen drugs such as enzalutamide for prostate cancer.
(d) Angiogenesis inhibitors that block blood vessel growth to cancers, e.g., bevacizumab for inhibiting VEGF receptor (Fig. 54.5b).
(e) Antibody-drug conjugates (e.g., trastuzumab emtansine) or radioactively attached antibody particles (e.g., ibritumomab tiuxetan

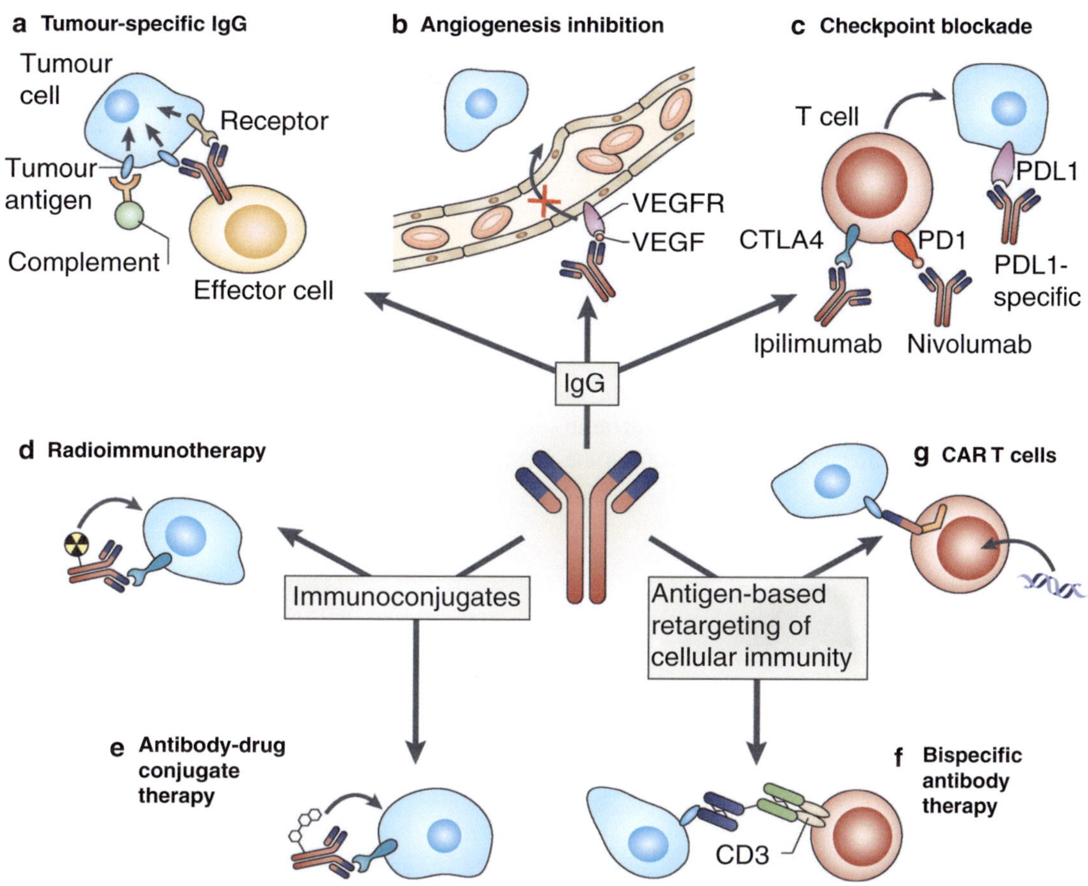

Fig. 54.5 | Monoclonal antibody-based cancer therapeutic strategies. Successful monoclonal antibody (mAb) therapeutics have been based on a number of strategies. Immunoglobulin G (IgG) molecules that bind to target cancer cells (part **a**) can mediate antibody-dependent cellular cytotoxicity (ADCC) by immune effector cells, induce complement-mediated cytotoxicity (CMC) or result in the direct signaling-induced death of cancer cells (e.g., Herceptin and rituximab). IgG mAbs can also be used to inhibit angiogenesis (part **b**) (e.g., bevacizumab) or to block inhibitory signals (part **c**), thereby resulting in a stronger antitumor T-cell response (e.g., ipilimumab and nivolumab). Radioimmunoconjugates (part **d**) (e.g., 131I-tositumomab and ibritumomab tiuxetan) deliver radioisotopes to the cancer cells, whereas antibody-drug conjugates (part **e**) (e.g., brentuximab vedotin and trastuzumab emtansine) deliver highly potent toxic drugs to the cancer cells. mAb variable regions are also used to retarget immune effector cells toward cancer cells through the use of bispecific mAbs that recognize cancer cells with one arm and activating antigens on immune effector cells with the other arm (part **f**) (e.g., blinatumomab) or through a gene therapy approach in which DNA for a mAb variable region fused to signaling peptides is transferred to T cells, thereby rendering them chimeric antigen receptor (CAR) T cells (part **g**) specific for the tumor. CD3 T-cell surface glycoprotein CD3 ε-chain, CTLA4 cytotoxic T lymphocyte-associated antigen 4, PD-1 programmed cell death protein 1, PDL1 PD-1 ligand, VEGF vascular endothelial growth factor, VEGFR VEGF receptor. (Reprinted from Weiner [15]. With permission from Nature Publishing Group)

(Zevalin)) to kill cancer cells through their direct cell-specific toxicity (Fig. 54.3).

(f) Immunotherapies that trigger or reactivate the body's immune system to destroy cancer cells, e.g., ipilimumab or nivolumab through CTLA4 or PD-1 blockade (Fig. 54.5c).

(g) Epigenetic drugs that inhibit enzymes called histone deacetylases through epigenetic regulation of gene expression, e.g., vorinostat for the treatment of patients with cutaneous T-cell lymphoma.

(h) Proteasome inhibitors, e.g., bortezomib. In myeloma cells, the boron atom in bortezomib binds the catalytic site of the 26S proteasome with high affinity and specificity to inhibit protein expression and degradation of ubiquitylated proteins. In normal cells, proteasomes regulate protein expression and function by degradation of ubiquitylated proteins. Proteasomes also clean up the cell of abnormal or misfolded proteins (Fig. 54.6).

In FDA-approved targeted therapies in oncology based on the indication (disease), an anticancer drug, predictive biomarker, and company are summarized in Table 54.1. The table has been constructed using various public resources such as the US National Cancer Institute, My Cancer

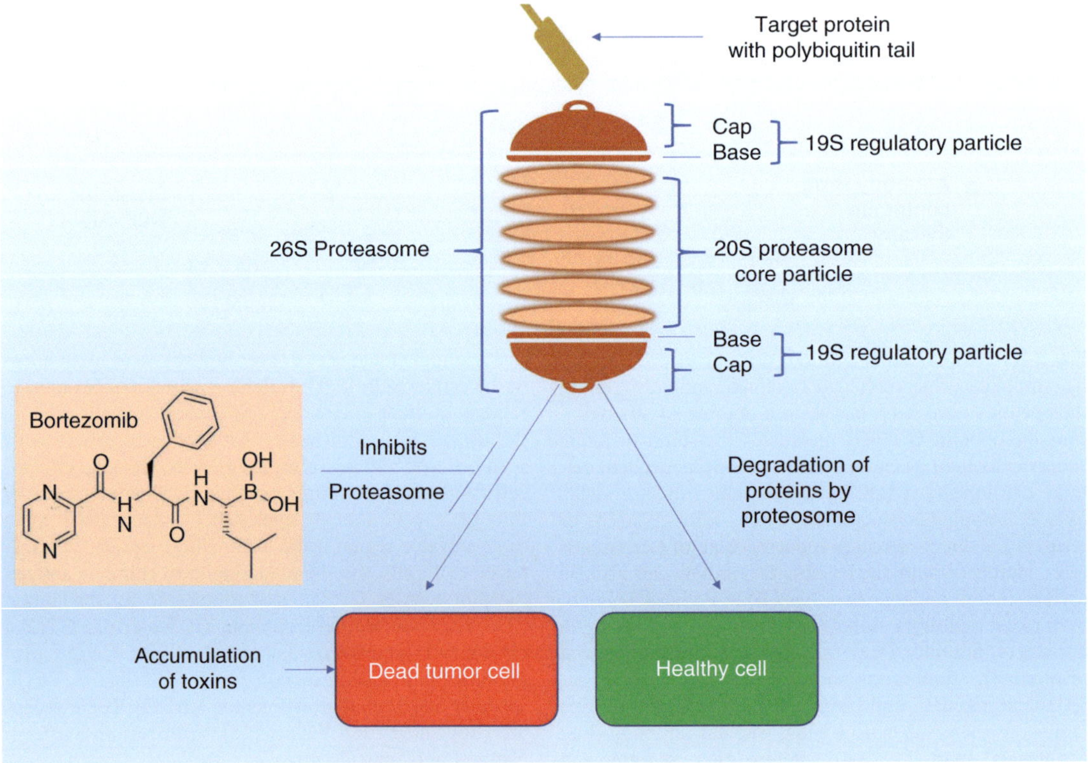

Fig. 54.6 Bortezomib (Velcade®) is a proteasome inhibitor for treating multiple myeloma. The boron atom in bortezomib binds the catalytic site of the 26S proteasome with high affinity and specificity. In normal cells, the proteasome degrades ubiquitylated proteins and cleanses the cell of abnormal or misfolded proteins. In cancer cells, bortezomib blocks the degradation of abnormal or misfolded proteins (toxins) resulting in the accumulation of toxins. This results in cell death. (Adapted from Anchoori et al. [16]. With permission from Elsevier]

Genome, OncoKB Precision Oncology Knowledge Database, pharmaceutical company websites, and targeted therapy books. For details, see [7–9]. Some targeted drugs do not have predictive biomarkers. Examples of such drugs include sipuleucel-T (Provenge®) and denileukin diftitox (Ontak®). Many rare cancers (see Ref. [9]) also do not have targeted therapy [9].

In the table below, therapeutic drugs with international nonproprietary names (INN) that end with the letters "-mab" are monoclonal antibodies. Small-molecule inhibitors usually end with the stem "-ib" indicating protein inhibitory properties (e.g., imatinib for tyrosine kinase inhibitor). Monoclonal antibodies that end with INN sub-stem "-ximab" indicate chimeric human-mouse antibodies (e.g., rituximab, cetuximab), and monoclonal antibodies that end with INN sub-stems"-zumab" (trastuzumab, bevacizumab) and "-mumab" (e.g., panitumumab, ipilimumab) are for humanized mouse antibodies and fully human antibodies, respectively. For upto date information on FDA approved drugs, see references [17, 18]

Table 54.1 FDA-approved targeted therapies in oncology

Indication	Drug		
For details, refer to the company*homepage	Generic (and brand name)	Therapeutic target	Company*
Breast cancer (BC)			
"Off-label use" with chemotherapy for treating HER2-negative locally recurrent metastatic breast cancer	Bevacizumab (Avastin®)	Vascular endothelial growth factor (VEGF)	Genentech/Roche
For the treatment of estrogen receptor-positive metastatic breast cancer in postmenopausal women with disease progression following antiestrogen therapy	Fulvestrant (Faslodex®)	Estrogen receptor (ER)	AstraZeneca
For treating ER+/EGFR+/HER2+ breast cancer patients	Lapatinib (Tykerb®)	Epidermal growth factor receptor (EGFR) and human epidermal growth factor receptor 2 (HER2)	Novartis GlaxoSmithKline plc.
For treating HER2-positive metastatic BC	Trastuzumab (Herceptin®) (see also Fig. 54.3b)	Human epidermal growth factor receptor 2 (HER2 also referred to as ERBB2)	Genentech/Roche
For treating postmenopausal women with advanced ER-/PR-positive, HER2-negative BC in conjunction with exemestane	Everolimus (Afinitor®)	Mammalian target of rapamycin (mTOR)	Novartis
For treating postmenopausal women with metastatic BC who are estrogen receptor-positive or in unknown tumors where hormonal status is not known	Toremifene citrate (Fareston®)	Estrogen receptor (ER)	Kyowa Kirin, Inc.
For treatment of early and advanced estrogen receptor-positive breast cancer in men and women	Tamoxifen citrate (Nolvadex)	Estrogen receptor (ER)	AstraZeneca
For treating HER2-positive metastatic breast cancer in combination with trastuzumab and docetaxel who have not received prior anti-HER2 therapy or chemotherapy	Pertuzumab (Perjeta®)	HER2 (ERBB2/neu)	Genentech/Roche

Table 54.1 (continued)

Indication For details, refer to the company*homepage	Drug Generic (and brand name)	Therapeutic target	Company*
To treat metastatic HER2-positive breast cancer after prior treatment with trastuzumab and taxane	Ado-trastuzumab emtansine (Kadcyla®) (see Fig. 54.3b)	HER2 (ERBB2/neu)	Genentech/Roche
For treating metastatic breast cancer in menopausal women who are ER+ and/or PR+, HER2-, in combination with letrozole or fulvestrant	Palbociclib (Ibrance®)	Cyclin-dependent kinase CDK4 and CDK6	Pfizer
For the treatment of postmenopausal women with hormone receptor-positive 2 (HER2) or HER2-negative advanced or metastatic breast cancer in combination with an aromatase inhibitor as initial endocrine-based therapy	Ribociclib (Kisqali®)	Cyclin-dependent kinase CDK4 and CDK6	Novartis
Adjuvant treatment of adult patients with early-stage HER2 overexpressed/amplified breast cancer	Neratinib (Nerlynx™)	Her2	Puma Biotechnology Inc.
In combination with fulvestrant for women with hormone receptor-positive (HR+) or hormone receptor-negative 2 (HER2−) advanced or metastatic breast cancer (MBC) with disease progression following endocrine therapy	Abemaciclib (Verzenio™)	CDK4/CDK6	Eli Lilly
Brain tumors			
For adult glioblastoma patients whose cancer has progressed after prior treatment (also called recurrent glioblastoma)	Bevacizumab (Avastin®)	Vascular endothelial growth factor (VEGF)	Genentech/Roche
For treating pediatric neuroblastoma in combination with granulocyte-macrophage colony-stimulating factor (GM-CSF), interleukin-2 (IL-2), and 13-cis-retinoic acid (RA)	Dinutuximab (Unituxin™)	Glycolipid antigen GD2	United Therapeutics Corporation
Nonresectable subependymal giant cell astrocytoma associated with tuberous sclerosis	Everolimus (Afinitor®)	Mammalian target of rapamycin (mTOR)	Novartis
Bladder cancer			
For the treatment of patients with locally advanced or metastatic urothelial carcinoma	Atezolizumab (Tecentriq™)	PD-L1	Genentech/Roche
For the treatment of patients with locally advanced or metastatic urothelial carcinoma	Nivolumab (Opdivo®)	PD-1	Bristol-Myers Squibb
For the treatment of patients with locally advanced or metastatic urothelial carcinoma	Durvalumab (IMFINZI™)	PD-L1	AstraZeneca
For the treatment of patients with locally advanced or metastatic urothelial carcinoma	Pembrolizumab (Keytruda®)	PD-1	Merck

(continued)

Table 54.1 (continued)

Indication For details, refer to the company*homepage	Drug Generic (and brand name)	Therapeutic target	Company*
Endocrine system			
Metastatic medullary thyroid cancer	Cabozantinib (Cometriq®)	c-MET, VEGFR2	Exelixis, Ipsen
Advanced medullary thyroid cancer	Vandetanib (Caprelsa®)	VEGFR2, EGFR, RET	Sanofi Genzyme
Differentiated thyroid carcinoma	Sorafenib (Nexavar®)	VEGFR, PDGFR	Bayer
Differentiated thyroid carcinoma	Lenvatinib mesylate (Lenvima®)	VEGFR 1, 2, and 3; FGFR 1, 2, 3, and 4; PDGFR alpha, c-KIT, and RET	Eisai
Gastrointestinal tract cancers			
HER2+ metastatic cancer of the stomach or gastroesophageal junction	Trastuzumab (Herceptin®)	HER2	Genentech/Roche
Locally advanced or metastatic gastric cancer and gastroesophageal junction (GEJ) adenocarcinoma	Ramucirumab (Cyramza®)	VEGFR2	Eli Lilly
Metastatic colorectal cancer (KRAS wild-type)	Cetuximab (Erbitux®)	EGFR	Bristol-Myers Squibb
EGFR-expressing metastatic colorectal cancer with KRAS wild-type gene	Panitumumab (Vectibix®)	EGFR	Amgen
Metastatic colorectal cancer whose disease has spread while on first-line treatment	Bevacizumab (Avastin®)	Vascular endothelial growth factor (VEGF)	Genentech/Roche
Metastatic colorectal cancer in combination with chemotherapy	Ziv-aflibercept (Zaltrap®)	VEGF	Sanofi-Genzyme
Patients with RAS wild-type metastatic colorectal cancer (CRC) who have been previously treated with chemotherapy	Regorafenib (Stivarga®)	VEGFR2	Bayer
Also used for locally advanced, unresectable, or metastatic gastrointestinal stromal tumors (GIST) who have been previously treated with imatinib mesylate and sunitinib malate		KIT, PDGFR, ABL (for GIST tumors)	
Metastatic colorectal cancer in combination with chemotherapy or after prior therapy with bevacizumab	Ramucirumab (Cyramza®)	VEGFR2	Eli Lilly
Gastrointestinal stromal tumors (KIT+)	Imatinib mesylate (Gleevec®)	KIT, PDGFR, ABL	Novartis
GIST not controlled by the Gleevec® (imatinib mesylate) or in patients who cannot take Gleevec	Sunitinib (Sutent®)	PDGFR, VEGFR, c-KIT (CD117)	Pfizer
Locally advanced (regional), unresectable (stage III) hepatocellular carcinoma and/or metastatic (stage IV) hepatocellular carcinoma	Sorafenib (Nexavar®)	VEGFR, PDGFR	Bayer/Onyx
In combination with chemotherapy is indicated for the first-line treatment of patients with locally advanced, unresectable, or metastatic pancreatic cancer	Erlotinib (Tarceva®)	EGFR (HER1/ERBB1)	Genentech/Roche and Astellas Pharma Inc.

Table 54.1 (continued)

Indication For details, refer to the company*homepage	Drug Generic (and brand name)	Therapeutic target	Company*
For treating pancreatic cancer known as pancreatic neuroendocrine tumor (PNET) that has progressed and cannot be treated with surgery	Everolimus (Afinitor®)	Mammalian target of rapamycin (mTOR)	Novartis
For treating pancreatic neuroendocrine tumor (PNET) that has progressed and cannot be treated with surgery	Sunitinib (Sutent®)	PDGFR, VEGFR, c-KIT (CD117)	Pfizer
Progressive GI and lung neuroendocrine tumor	Everolimus (Afinitor®)	Mammalian target of rapamycin (mTOR)	Novartis
Genitourinary system cancers			
To treat adults with advanced renal cell carcinoma (RCC) when other drugs such as Sutent® [sunitinib] and/or Nexavar® [sorafenib] does not work	Bevacizumab (Avastin®)	VEGF	Genentech/Roche
Advanced RCC	Sorafenib (Nexavar®)	VEGFR, PDGFR	Bayer/Onyx
Advanced RCC	Sunitinib (Sutent®)	PDGFR, VEGFR, c-KIT (CD117)	Pfizer
Advanced RCC	Pazopanib (Votrient®)	VEGFR, PDGFR, c-KIT, FGFR	Novartis
Advanced RCC	Temsirolimus (Torisel®)	Mammalian target of rapamycin (mTOR)	Pfizer
Advanced RCC	Everolimus (Afinitor®)	mTOR	Novartis
Advanced RCC	Axitinib (Inlyta®)	VEGFR1, VEGFR2, VEGFR3	Pfizer
Renal cell carcinoma patients who have received prior anti-angiogenic therapy	Nivolumab (Opdivo®)	PD-1	Bristol-Myers Squibb
Advanced RCC	Cabozantinib (Cabometyx™)	MET; VEGFR1, 2, and 3; AXL	Exelixis
Advanced RCC patients who have been treated with everolimus	Lenvatinib mesylate (Lenvima®)	VEGFR 1, 2, and 3 FGFR 1, 2, 3, and 4 PDGFR alpha, c-KIT, and RET	Eisai
For the treatment of patients with metastatic castration-resistant prostate cancer (CRPC)	Enzalutamide (Xtandi®)	Androgen receptor	Astellas Medivation
Previously treated advanced urothelial carcinoma	Atezolizumab (Tecentriq™)	PD-L1	Genentech/Roche
To treat patients with locally advanced or metastatic urothelial carcinoma	Durvalumab (IMFINZI™)	PD-L1	AstraZeneca
Gynecologic cancers			
Platinum-resistant ovarian cancer	Bevacizumab (Avastin®)	Vascular endothelial growth factor (VEGF)	Genentech/Roche
Advanced ovarian cancer patients with abnormal inherited *BRCA1 and BRCA 2* genes	Olaparib (Lynparza™)	DNA repair enzyme poly(ADP-ribose) polymerase (PARP)	AstraZeneca

(continued)

Table 54.1 (continued)

Indication For details, refer to the company*homepage	Drug Generic (and brand name)	Therapeutic target	Company*
Treatment of advanced ovarian cancer patients with deleterious *BRCA*1 and *BRCA* 2 mutations (germline and/or somatic) who have been treated with two or more chemotherapies	Rucaparib (Rubraca™)	DNA repair enzyme poly(ADP-ribose) polymerase (PARP) 1	Clovis Oncology
Advanced cervical cancer	Bevacizumab (Avastin®)	Vascular endothelial growth factor (VEGF)	Genentech/Roche
Recurrent epithelial ovarian, fallopian tube, or primary peritoneal cancer	Niraparib (Zejula™)	DNA repair enzyme poly(ADP-ribose) polymerase (PARP1 and PARP2)	Tesaro
Head and neck cancer			
Locally or regionally advanced squamous cell carcinoma of the head and neck	Cetuximab (Erbitux®)	EGFR (HER1/ERBB1)	Eli Lilly
Advanced squamous cell carcinoma of the head and neck (SCCHN) on or after platinum-based therapy	Pembrolizumab (Keytruda®)	PD-1	Merck
Metastatic or recurrent SCCHN on or after platinum-based therapy	Nivolumab (Opdivo®)	PD-1	Bristol-Myers Squibb
Hematologic malignancies (lymphomas, leukemias, and plasma cell tumors)			
Philadelphia chromosome-positive (Ph+) chronic myeloid leukemia (CML) Ph+ acute lymphoblastic leukemia (Ph+ ALL)	Dasatinib (Sprycel®)	BCR/ABL (the "Philadelphia chromosome") and Src	Bristol-Myers Squibb Otsuka America Pharmaceutical, Inc.
Ph+ chronic myeloid leukemia (CML)	Nilotinib (Tasigna®)	BCR/ABL	Novartis
Ph+ chronic myelogenous leukemia	Bosutinib (Bosulif®)	BCR/ABL and Src	Pfizer
CD20-positive non-Hodgkin's lymphoma	Rituximab (Rituxan®)	CD20	Genentech/Roche and Biogen
CD20-positive diffuse large B-cell non-Hodgkin's lymphoma			
CD20-positive chronic lymphocytic leukemia			
B-cell chronic lymphocytic leukemia	Alemtuzumab (Campath®)	CD52	Genzyme
Previously untreated patients with chronic lymphocytic leukemia	Ofatumumab (Arzerra®)	CD20	Novartis
Previously untreated patients with chronic lymphocytic leukemia	Obinutuzumab (Gazyva®)	CD20	Genentech/Roche
Relapsed or refractory follicular lymphoma			
Mantle cell lymphoma	Ibrutinib (Imbruvica®)	Bruton's tyrosine kinase (BTK)	Pharmacyclics, an AbbVie company, Janssen, Johnson & Johnson
Chronic lymphocytic leukemia			
Small lymphocytic leukemia with 17p deletion			
Chronic lymphocytic leukemia	Idelalisib (Zydelig®)	Phosphatidylinositol-4,5-bisphosphate 3-kinase catalytic subunit delta isoform (PI3Kδ) coded by *PIK3CD* gene	Gilead
Follicular B-cell non-Hodgkin's lymphoma			
Small lymphocytic lymphoma			

Table 54.1 (continued)

Indication For details, refer to the company*homepage	Drug Generic (and brand name)	Therapeutic target	Company*
For the treatment of Philadelphia chromosome-negative relapsed or refractory B-cell precursor acute lymphoblastic leukemia (ALL)	Blinatumomab (Blincyto®)	CD19	Amgen
Chronic lymphocytic leukemia with 17p deletion	Venetoclax (Venclexta™)	BCL2	Genentech/Roche
Classical Hodgkin's lymphoma	Nivolumab (Opdivo®)	PD-1	Bristol-Myers Squibb
Classical Hodgkin's lymphoma	Pembrolizumab (Keytruda®)	PD-1	Merck
Relapsed or refractory, low-grade, or follicular B-cell non-Hodgkin's lymphoma (NHL)	Ibritumomab Tiuxetan (Zevalin®)	CD20-directed radiotherapeutic yttrium-90 antibody	Spectrum Pharmaceuticals, Inc.
Classical Hodgkin's lymphoma Systemic anaplastic large cell lymphoma	Brentuximab Vedotin (Adcetris®)	CD30	Seattle Genetics
Cutaneous T-cell lymphoma	Vorinostat (Zolinza®)	Blocks histone deacetylase (HDAC)	Merck
Cutaneous T-cell lymphoma Peripheral T-cell lymphoma	Romidepsin (Istodax®)	Blocks histone deacetylase (HDAC)	Celgene
Multiple myeloma Mantle cell lymphoma	Bortezomib (Velcade®)	Proteasome	Millennium Pharmaceuticals
Relapsed or refractory peripheral T-cell lymphoma	Belinostat (Beleodaq®)	HDAC	Spectrum Pharmaceuticals
Previously untreated chronic lymphocytic leukemia Relapsed or refractory follicular lymphoma	Obinutuzumab (Gazyva®)	CD20	Genentech/Roche
Multiple myeloma	Panobinostat (Farydak®)	HDAC	Novartis
Relapsed or refractory multiple myeloma	Carfilzomib (Kyprolis®)	20S proteasome	Amgen
Multiple myeloma	Daratumumab (Darzalex™)	CD38	Janssen Oncology
Multiple myeloma	Ixazomib citrate (Ninlaro®)	Proteasome	Millennium Pharmaceuticals
Multiple myeloma	Elotuzumab (Empliciti™)	SLAMF7 (CD319)	Bristol-Myers Squibb
Polycythemia vera and intermediate or high-risk myelofibrosis	Ruxolitinib phosphate (Jakafi®)	JAK1/JAK2	Incyte

(continued)

Table 54.1 (continued)

Indication For details, refer to the company*homepage	Drug Generic (and brand name)	Therapeutic target	Company*
For treating Ph+ CML in the chronic phase Relapsed or refractory Ph+ ALL Myelodysplastic or myeloproliferative diseases associated with PDGFR gene rearrangements Aggressive systemic mastocytosis without the D816V c-KIT mutation or with c-KIT mutational status unknown Hypereosinophilic syndrome (HES) and/or chronic eosinophilic leukemia (CEL) who have the FIP1L1-PDGFRα fusion kinase and for patients with HES and/or CEL who are FIP1L1-PDGFRα fusion kinase negative or unknown Unresectable, recurrent, and/or metastatic dermatofibrosarcoma protuberans	Imatinib mesylate (Gleevec®)	KIT, PDGFR, ABL	Novartis
For treating Ph+ CML Ph+ ALL	Ponatinib (Iclusig®)	ABL, FGFR1–FGFR3, FLT3, VEGFR2	Takeda Ariad
For treating multicentric Castleman disease; also known as angiofollicular lymph node hyperplasia	Siltuximab (Sylvant®)	IL-6 signaling pathway	Janssen Biotech Inc.
Used for treating cytokine release syndrome, a side effect of CAR-T-cell therapies In Japan, used in treating Castleman disease, also known as angiofollicular lymph node hyperplasia	Tocilizumab (Actemra®)	IL-6 signaling pathway	Genentech
Relapsed or refractory B-cell precursor acute lymphoblastic leukemia (ALL)	Inotuzumab ozogamicin (Besponsa®)	CD22	(Pfizer/Wyeth)
Adult patients with relapsed or refractory acute myeloid leukemia (AML) with an isocitrate dehydrogenase 2 (IDH2) mutation	Enasidenib (Idhifa®)	Isocitrate dehydrogenase 2 (IDH2) mutation	Celgene Corp.
For treating relapsing or refractory ALL	Tisagenlecleucel (Kymriah™)	CD19	Novartis
For treating adult patients with relapsed follicular lymphoma (FL) who have received at least two prior systemic therapies	Copanlisib (Aliqopa™)	PI3K	Bayer
For treating AML in adults and children over the age of 2	Gemtuzumab ozogamicin (Mylotarg™) (see Fig. 54.3a)	CD33	Pfizer Oncology

Table 54.1 (continued)

Indication For details, refer to the company*homepage	Drug Generic (and brand name)	Therapeutic target	Company*
Acute myeloid leukemia (AML)	Midostaurin (RYDAPT®)	*FLT3* mutation	Novartis
Aggressive systemic mastocytosis (ASM), systemic mastocytosis with associated hematological neoplasm (SM-AHN)			
Mast cell leukemia (MCL)			
For treating adult patients with relapsed or refractory large B-cell lymphoma, including diffuse large B-cell lymphoma (DLBCL), primary mediastinal large B-cell lymphoma, high-grade B-cell lymphoma, and DLBCL arising from follicular lymphoma	Axicabtagene ciloleucel (YESCARTA™)	CD19-directed genetically modified autologous T-cell immunotherapy	Gilead Sciences (Kite Pharma)
Lung cancer			
Advanced non-squamous non-small cell lung cancer (NSCLC) in people who have not received chemotherapy	Bevacizumab (Avastin®)	Vascular endothelial growth factor (VEGF)	Genentech/Roche
Metastatic ALK+ or ROS1+ NSCLC	Crizotinib (Xalkori®)	Anaplastic lymphoma kinase (ALK), proto-oncogene receptor tyrosine kinase ROS1	Pfizer Oncology & EMD Serono
Metastatic NSCLC with EGFR mutations	Erlotinib hydrochloride (Tarceva®)	EGFR	Genentech/Roche
Metastatic NSCLC with EGFR mutations other than exon 19 deletions or exon 21 (L858R) substitution mutations	Gefitinib (Iressa®)	EGFR	AstraZeneca
First-line treatment of metastatic EGFR mutation-positive NSCLC	Afatinib dimaleate (Gilotrif®)	HER1/ERBB1, HER2/ERBB2	Boehringer Ingelheim
Metastatic ALK+ NSCLC	Ceritinib (Zykadia™)	ALK	Novartis
Metastatic NSCLC with EGFR or ALK mutations	Ramucirumab (Cyramza®)	VEGFR2	Eli Lilly
First-line treatment of metastatic NSCLC patients with high PD-L1 expression and with no EGFR or ALK mutations	Nivolumab (Opdivo®)	PD-1	Bristol-Myers Squibb
First-line treatment of metastatic NSCLC patients with high PD-L1 expression and with no EGFR or ALK mutations	Pembrolizumab (Keytruda®)	PD-1	Merck
Metastatic NSCLC with EGFR mutation	Osimertinib (Tagrisso™)	EGFR	AstraZeneca
Metastatic NSCLC	Necitumumab (Portrazza™)	EGFR (HER1/ERBB1)	Eli Lilly
Metastatic ALK+ NSCLC	Alectinib (Alecensa®)	ALK	Genentech/Roche
Metastatic NSCLC	Atezolizumab (Tecentriq™)	PD-L1	Genentech/Roche

(continued)

Table 54.1 (continued)

Indication For details, refer to the company*homepage	Drug Generic (and brand name)	Therapeutic target	Company*
ALK+ metastatic NSCLC who have progressed on or are intolerant to crizotinib	Brigatinib (Alunbrig™)	ALK	Takeda
Sarcomas of soft tissue and bone			
Giant cell tumor of the bone that is unresectable	Denosumab (Xgeva®)	RANKL (receptor activator of nuclear factor kappa B)	Amgen
Advanced soft tissue sarcoma	Pazopanib (Votrient®)	VEGFR, PDGFR, c-KIT, FGFR	Novartis
Adult patients with soft tissue sarcoma	Olaratumab (Lartruvo™)	PDGFR-α	Eli Lilly
Dermatofibrosarcoma protuberans	Imatinib mesylate (Gleevec®)	KIT, PDGFR, ABL	Novartis
Skin			
Melanoma	Ipilimumab (Yervoy®)	CTLA-4	Bristol-Myers Squibb
Melanoma (with BRAF V600 mutation)	Vemurafenib (Zelboraf®)	BRAF	Roche
Melanoma (with BRAF V600 mutation)	Trametinib (Mekinist®)	MEK	Novartis
Melanoma (with BRAF V600 mutation)	Dabrafenib (Tafinlar®)	BRAF	Novartis
Melanoma	Pembrolizumab (Keytruda®)	PD-1	Merck
Melanoma	Nivolumab (Opdivo®)	PD-1	Bristol-Myers Squibb
Melanoma (with BRAF V600E or V600 K mutation)	Cobimetinib (Cotellic™)	MEK	Genentech/Roche
Basal cell carcinoma	Vismodegib (Erivedge®)	PTCH, smoothened	Genentech/Roche
Basal cell carcinoma	Sonidegib (Odomzo®)	Smoothened	Novartis
Merkel cell carcinoma	Avelumab (Bavencio®)	PD-1	Pfizer/Merck KGaA

References

1. U.S. Food and Drug Administration. Countries and Regions Covered by OIP Offices. http://www.fda.gov/AboutFDA/CentersOffices/OfficeofGlobalRegulatoryOperationsandPolicy/OfficeofInternationalPrograms/ucm342377.htm. Accessed 16 Dec 2016.
2. Jordan VC. Tamoxifen: catalyst for the change to targeted therapy. Eur J Cancer. 2008;44(1):30–8.
3. American Society of Clinical Oncology. Cancer Progress Timeline. http://cancerprogress.net/timeline/major-milestones-against-cancer. Accessed 16 Dec 2016.
4. Rahma OE, Kunk PR, Khleif SN. In: Khleif S, Rixe O, Skeel RT, editors. Skeel's handbook of cancer therapy. 9th ed. Philadelphia: Wolters Kluver; 2016. p. 17–62.
5. List of therapeutic monoclonal antibodies. Wikipedia. https://en.wikipedia.org/wiki/List_of_therapeutic_monoclonal_antibodies. Accessed 21 Dec 2016.
6. Carter PJ. Antibody drug nomenclature: what is INN a name? WHO has been changing them? Antibody Engineering & Therapeutics, Dec. 9th 2015. San Diego, CA. http://www.antibodysociety.org/wordpress/wp-content/uploads/2015/12/Carter-IBC-INN-talk-Dec-2015-FINAL.pdf. Accessed 21 Dec 2016.
7. Jones TD, Carter PJ, Plückthun A, et al. The INNs and outs of antibody nonproprietary names. MAbs. 2016;8(1):1–9.
8. OncoKB. OncoKB Team. http://oncokb.org/#/team. Accessed 19 Mar 2017.

9. Cancer Research UK. Your cancer type. http://www.cancerresearchuk.org/about-cancer/type/rare-cancers/rare-cancers-name/. Accessed 19 Mar 2017.
10. Weisber E, et al. Second generation inhibitors of BCR-ABL for the treatment of imatinib resistant chronic myeloid leukaemia. Nat Rev Cancer. 2007;7:345–56.
11. Alberts B, Johnson A, Lewis J, et al. Cancer. In: Alberts B, Johnson A, Lewis J, Raff M, Roberts K, Walter P, editors. Molecular biology of the cell. 5th ed. New York: Garland Science; 2008. p. 1205–67.
12. Seyfizadeh N, et al. A molecular perspective on rituximab: a monoclonal antibody for B-cell non-Hodgkin lymphoma and other affections. Crit Rev Oncol/Hematol. 2016;97:275–90.
13. Beck A, et al. Strategies and challenges for the next generation of antibody-drug conjugates. Nat Rev Drug Discov. 2017;16(5):315–37.
14. Peddi PF, Hurvitz SA. Trastuzumab emtansine: the first targeted chemotherapy for treatment of breast cancer. Future Oncol. 2013;9(3) https://doi.org/10.2217/fon.13.7.
15. Weiner GJ. Building better monoclonal antibody-based therapeutics. Nat Rev Cancer. 2015; 15(6):361–70.
16. Anchoori RK, Karanam B, Peng S, et al. A bis-Benzylidine piperidone targeting proteasome ubiquitin receptor RPN13/ADRM1 as a therapy for cancer. Cancer Cell. 2013;24(6):791–805.
17. https://www.centerwatch.com/drug-information/fda-approved-drugs/. Accessed July 24, 2018
18. https://www.fda.gov/Drugs/DevelopmentApprovalProcess/DrugInnovation/ucm592464.htm. Accessed July 24, 2018

Index

S. Badve, G. L. Kumar (eds.), *Predictive Biomarkers in Oncology*,
https://doi.org/10.1007/978-3-319-95228-4

I

J

K

Q

R

S

T